Essentials of Biology

Sylvia S. Mader

FOR MY CHILDREN...

McGraw-Hill **Higher Education**

Boston   Burr Ridge, Il   Dubuque, IA   Madison, WI   New York   San Francisco   St. Louis
Bangkok   Bogotá   Caracas   Kuala Lumpur   Lisbon   London   Madrid   Mexico City
Milan   Montreal   New Delhi   Santiago   Seoul   Singapore   Sydney   Taipei   Toronto

The McGraw-Hill Companies

Mc Graw Hill Higher Education

ESSENTIALS OF BIOLOGY

Published by McGraw-Hill, a business unit of The McGraw-Hill Companies, Inc., 1221 Avenue of the Americas, New York, NY 10020. Copyright © 2007 by The McGraw-Hill Companies, Inc. All rights reserved. No part of this publication may be reproduced or distributed in any form or by any means, or stored in a database or retrieval system, without the prior written consent of The McGraw-Hill Companies, Inc., including, but not limited to, in any network or other electronic storage or transmission, or broadcast for distance learning.

Some ancillaries, including electronic and print components, may not be available to customers outside the United States.

This book is printed on recycled, acid-free paper containing 10% postconsumer waste.

1 2 3 4 5 6 7 8 9 0 VNH/VNH 0 9 8 7 6 5

ISBN-13 978–0–07–110696–2
ISBN-10 0–07–110696–0

The credits section for this book begins on page 623 and is considered an extension of the copyright page.

www.mhhe.com

Brief Contents

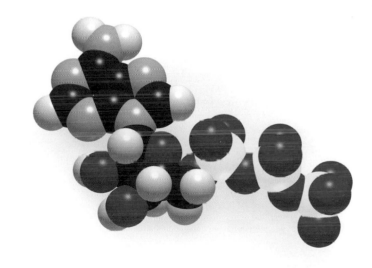

Contents

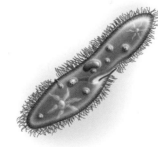

CHAPTER 5

The Dynamic Cell 69

CHAPTER 6

Energy for Life 83

CHAPTER 7

Energy for Cells 97

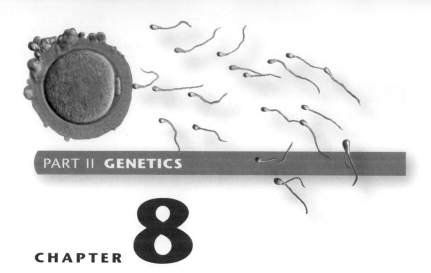

PART II GENETICS

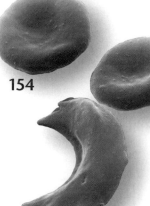

CHAPTER **12**

Gene Regulation and Cancer 179

CHAPTER **13**

Genetic Counseling 195

PART III EVOLUTION

CHAPTER **14**

Darwin and Evolution 215

CHAPTER **15**

Evolution on a Small Scale 231

PART V PLANT STRUCTURE AND FUNCTION

CHAPTER 20

Plant Anatomy and Growth   337

CHAPTER 21

Plant Responses and Reproduction   355

PART VI ANIMAL STRUCTURE AND FUNCTION

CHAPTER 22

Being Organized and Steady   377

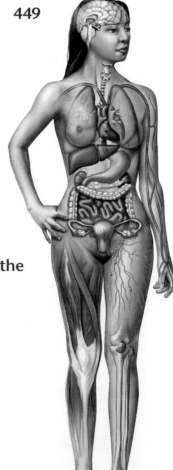

CHAPTER **27**

The Control Systems   471

CHAPTER **28**

Sensory Input and Motor Output   493

CHAPTER **29**

Reproduction and Development   511

PART VII ECOLOGY

CHAPTER **30**

Ecology of Populations   533

CHAPTER **31**

Communities and Ecosystems   551

CHAPTER **32**

Human Impact on the Biosphere   575

Appendix   A

Appendix   B

Preface

Is it possible to engage students while introducing the principles of biology within the time span of one semester? *Essentials of Biology* is the text that meets the challenge!

My goal in writing *Essentials of Biology* was twofold: to explain the principles of biology clearly and to illustrate them in a captivating, easy-to-understand manner. *Essentials of Biology* is traditional in its approach, organized around the major concepts of biology—the theory of evolution, the cell theory, the gene theory, the theory of homeostasis, and the theory of ecosystems. However, adopters will notice from the outset a decided effort to reach out to today's student. The design and illustration program of the book is appropriate for those who are accustomed to being visually stimulated. The writing style is conversational and inviting in its tone. In short, this book offers the essence of biology without the amount of detail found in other introductory biology texts. Even though this book is succinct, it retains the hallmark features of a Mader book: clear writing, well-developed visuals, a great pedagogical system, and logical organization of chapters.

Because biology is now characterized by new revelations almost daily, it is sometimes difficult to decide what to include in a text. While it is tempting to concentrate on the new, it remains clear that today's students need a good foundation in the basics, just as previous students did. *Essentials of Biology* stresses the principles of biology but uses pertinent applications to increase appreciation and to show that biology is a science relevant to everyday life. Genetics comes alive in Chapter 13, which shows how both Mendelian and molecular genetics can be used to counsel clients about how genetic disorders can be detected, controlled, and/or treated. Chapter 25, a nutrition chapter, shows students how the knowledge of chemistry is beneficial when making dietary decisions to achieve the proper weight and remain healthy. Stem cell research, human diseases, and reproductive choices, all topics of interest to students, are also included. Environmental concerns are addressed, and Chapter 32 concentrates on how human activities impact the biosphere.

An Overview of the Text

The introductory chapter provides students with a preview of biological principles before they take up individual topics. Included are the characteristics of life, an overview of life's diversity, and a discussion of the scientific process.

Part I The Cell

In this part students are introduced to a bit of chemistry before considering the anatomy and physiology of the cell. Cells, like organisms, must acquire and use matter and energy in order to maintain their existence.

Part II Genetics

DNA, the composition of genes, is passed on during cellular and organismal reproduction. Patterns of inheritance are pertinent, but today, the treatment of genetic disorders and cancer is dependent upon molecular genetics. With the sequencing of the human genome, new ways are being found to expand the field of medicine.

Part III Evolution

Biology cannot be understood without a knowledge of evolution. This part begins with a chapter that explains the main points of Darwin's theory and examines the variety of evidence that supports evolution. The mechanisms necessary to microevolution and macroevolution are then considered before the history of life on Earth is reviewed.

Part IV Diversity of Life

The major categories of life are presented and their possible relationships are explored. Students need to be aware of the variety of life with which we share this planet. An evolutionary tree for plants and another for animals become icons for appropriate chapters.

Part V Plant Structure and Function

These chapters instill in the students an appreciation of botany. The flowering plant is used as the representative organism to study the basics of plant anatomy.

Part VI Animal Structure and Function

A comparative theme, which uses the human animal as the representative organism, runs through this part. The emphasis is on homeostasis, which is introduced and explored in Chapter 22.

Part VII Ecology

This part moves from population dynamics through the interactions of populations in communities and ecosystems. The last chapter shows how human activities stress the biosphere and gives reasons why biodiversity should be preserved.

Acknowledgments

Many dedicated and talented individuals assisted in the development of *Essentials of Biology*. I am very grateful for the help of so many professionals at McGraw-Hill who were involved in bringing this book to fruition. In particular, let me thank Margaret Horn, the developmental editor who lent her talents and advice to all those who worked on this text. The biology editor was Thomas Lyon, who steadfastly encouraged and supported this project. The project manager, Jayne Klein, faithfully and carefully steered the book through the publication process. Tamara Maury, the marketing manager, tirelessly promoted the text and educated the sales reps on its message

The design of the book is the result of the creative talents of Rick Noel and many others who assisted in deciding the appearance of each element in the text. Precision Graphics followed their guidelines as they created and reworked each illustration, emphasizing pedagogy and beauty to arrive at the best presentation on the page. Lori Hancock and Connie Mueller did a superb job of finding just the right photographs and micrographs.

My staff, consisting of Evelyn Jo Hebert and Beth Butler, worked faithfully as they helped proof the chapters and made sure all was well before the book went to press. As always, my family was extremely patient with me as I remained determined to meet every deadline on the road to publication. My husband, Arthur Cohen, is also a teacher of biology. The many discussions we have about the minutest detail to the gravest concept are invaluable to me.

I am very much indebted to the following reviewers whose suggestions and expertise were so valuable as I developed *Essentials of Biology*.

Sylvester Allred
Northern Arizona University

Paul E. Arriola
Elmhurst College

Tammy Atchison
Pitt Community College

James S. Backer
Daytona Beach Community College

Gail F. Baker
LaGuardia Community College

Sirakaya Beatrice
Pennsylvania State University

Carla Bundrick Benejam
California State University—Monterey Bay

Charles L. Biles
East Central University

Donna H. Bivans
Pitt Community College

Steven G. Brumbaugh
Greenriver Community College

Neil Buckley
SUNY—Plattsburgh

Nancy Butler
Kutztown University

Michelle Cawthorn
Georgia Southern University

Van D. Christman
Brigham Young University

Genevieve C. Chung
Broward Community College

Kimberly Cline-Brown
University of Northern Iowa

Mary C. Colavito
Santa Monica College

Mark A. Coykendall
College of Lake County

Don C. Dailey
Austin Peay State University

Cathy A. Davison
Empire State College

Bonnie L. Dean
West Virginia State University

William R. DeMott
Indiana-Purdue University—Fort Wayne

Amy Stinnett Dewald
Eureka College

Lee C. Drickamer
Northern Arizona University

Marie D. Dugan
Broward Community College

James W. DuMond, Jr.
Texas State University

Kathryn A. Durham
Lorain County Community College

Andrew R. Dyer
University of South Carolina—Aiken

Steven E. Fields
Winthrop University

Lynn Firestone
Brigham Young University—Idaho

Susan Fisher
Ohio State University

Edison R. Fowlks
Hampton University

Dennis W. Fulbright
Michigan State University

Ron Gaines
Cameron University

John R. Geiser
Western Michigan University

Beatriz Gonzalez
Santa Fe Community College

Andrew Goyke
Northland College

Richard Gringer
Augusta State University

Lonnie J. Guralnick
Western Oregon University

William F. Hanna
Massasoit Community College

Lisa K. Johansen
University of Colorado at Denver

Ragupathy Kannan
University of Arkansas—Fort Smith

Arnold Karpoff
University of Louisville

Darla E. Kelly
Orange Coast College

Elaine B. Kent
California State University—Sacramento

Scott L. Kight
Montclair State University

Kristin Lenertz
Black Hawk College

Melanie Loo
California State University—Sacramento

Michelle Malott
Minnesota State University— Moorhead

Paul Mangum
Midland College

Mara Manis
Hillsborough Community College

Karen Benn Marshall
Montgomery College— Takoma Park

Cynthia Conaway Mauroidis
Northwest State Community College

Elizabeth McPartlan
De Anza College

Dwight Meyer
Queensborough Community College

Rod Nelson
University of Arkansas—Fort Smith

Donald J. Padgett
Bridgewater State College

Tricia L. Paramore
Hutchinson Community College

Brian K. Paulson
California University of Pennsylvania

Debra K. Pearce
Northern Kentucky University

Lisa Rapp
Springfield Technical Community College

Jill Raymond
Rock Valley Community College

Cara Shillington
Eastern Michigan University

Lee Sola
Glendale Community College

John D. Sollinger
Southern Oregon University

Andrew Storfer
Washington State University

Janis G. Thompson
Lorain County Community College

Briana Timmerman
University of South Carolina

James R. Triplett
Pittsburg State University

Paul Twigg
University of Nebraska—Kearney

Garland Rudolph Upchurch, Jr.
Texas State University, San Marcos

James A. Wallis II
St. Petersburg College, Tarpon Springs Campus

Cosima B. Wiese
College Misericordia

Melissa Zwick
Longwood University

I am also grateful to the following who made significant contributions to *Essentials of Biology.*

Nancy Butler
Kutztown University of Pennsylvania

Stephen D. Ebbs
Southern Illinois University

Lynn Firestone
Brigham Young University—Idaho

Patrick Galliart
Northern Illinois University

Shelley Jansky
University of Wisconsin—Madison

Kimberly Lyle-Ippolito
Anderson University

Cherie McKeever
Montana State University College of Technology—Great Falls

Donna H. Mueller
Drexel University

Kathleen Pelkki
Saginaw Valley State University

Wendy Schiff
St. Louis Community College—Meramec

Kent Thomas
Wichita State University

Wendy Vermillion
Columbus State Community College

Jennifer Warner
University of North Carolina—Charlotte

Nicole Welch
Middle Tennessee State University

For the Instructor

McGraw-Hill offers a variety of tools and technology products to support *Essentials of Biology.* Instructors can obtain teaching aids by calling the Customer Service Department at (800) 338-3987 or by contacting their local McGraw-Hill sales representative.

Essentials of Biology Laboratory Manual

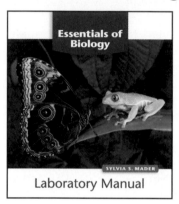

The *Essentials of Biology Laboratory Manual* is written by Dr. Sylvia Mader. With few exceptions, each chapter in the text has an accompanying laboratory exercise in the manual. Every laboratory has been written to help students learn the fundamental concepts of biology and the specific content of the chapter to which the lab relates, as well as gain a better understanding of the scientific method.

ISBN-13: 978-0-07-340341-0 (ISBN-10: 0-07-340341-5)

Digital Content Manager

This collection of multimedia resources provides tools for rich visual support of your lectures. You can utilize artwork from the text in multiple formats to create customized classroom presentations, visually based tests and quizzes, dynamic course website content, or attractive printed support materials. The following digital assets are available either on a cross-platform CD-ROM or on a DVD and are grouped by chapters:

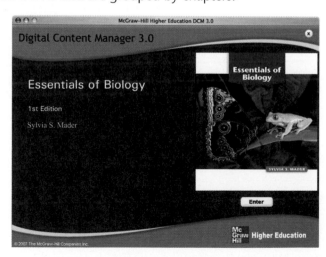

Art Libraries. Full-color digital files of all illustrations in the book, plus the same art saved in unlabeled and gray scale version, can be readily incorporated into lecture presentations, exams, or custom-made classroom materials.

Photos Library. All photos from the text are available in digital format.

Active Art Library. Illustrations depicting key processes have been converted to a format that allows the artwork to be edited inside of PowerPoint. Each piece can be broken down to its core elements, grouped or ungrouped, and edited to create customized illustrations.

Animations Library. The next generation of biology animations is now available! New animations bring key processes to life and offer total flexibility. Designed to be used in lectures, you can pause, rewind, fast-forward, and turn the audio on or off to create dynamic lecture presentations. Many of the animations are also available with Spanish narration and audio.

Tables Library. Every table that appears in the text is provided in electronic format.

Additional Photos Library. Over 700 photos not found in *Essentials of Biology* are available for use in creating lecture presentations.

PowerPoint Lecture Outlines. A ready-made presentation that combines lecture notes and illustrations is written for each chapter. They can be used as they are, or the instructor can customize them to preferred lecture topics and organization.

PowerPoint Art Slides. Art, photos, and tables from each chapter have been pre-inserted into blank PowerPoint slides to which you can add your own notes.

CD-ROM ISBN-13: 978-0-07-297442-3 (ISBN-10: 0-07-297442-7)
DVD ISBN-13: 978-0-07-326194-2 (ISBN-10: 0-07-326194-7)

Instructor's Testing and Resource CD-ROM

This cross-platform CD-ROM provides these resources for the instructor:

Instructor's Manual contains learning objectives, extended lecture outlines, lecture enrichment and student activities suggestions, and critical thinking questions. In addition, there is an explanation of text changes and reorganization as well as information on new and revised illustrations and tables.

Test Bank offers questions that can be used for homework assignments or the preparation of exams.

Computerized Test Bank utilizes testing software to quickly create customized exams. This user-friendly program allows instructors to sort questions by format or level of difficulty; edit existing questions or add new ones; and scramble questions and answer keys for multiple versions of the same test.

CPS Question Bank for use with the eInstruction Classroom Performance System is included on this CD-ROM, ISBN-13: 978-0-07-297444-7 (ISBN-10: 0-07-297444-3)

eInstruction Classroom Performance System (CPS)

Wireless technology brings interactivity into the classroom or lecture hall. Instructors and students receive immediate feedback through wireless response pads that are easy to use and engage students. eInstruction can be used by instructors to:

- Take attendance
- Administer quizzes and tests
- Create a lecture with intermittent questions
- Manage lectures and student comprehension through use of the CPS grade book
- Integrate interactivity into their PowerPoint presentations

Transparencies

This set of overhead transparencies includes every piece of line art in the textbook plus every table. The images are printed with better visibility and contrast than ever before, and labels are large and bold for clear projection.
ISBN-13: 978-0-07-297441-6 (ISBN-10: 0-07-297441-9)

ARIS

McGraw-Hill's ARIS Assessment, Review, and Instruction System—for *Essentials of Biology* is a complete online tutorial, electronic homework, and course management system designed for greater ease of use than any other system available. Free with adoption of McGraw-Hill's *Essentials of Biology* text, instructors can create and share course materials and assignments with colleagues with a few clicks of the mouse. All PowerPoint lectures, assignments, quizzes, tutorials, and interactives are directly tied to text-specific materials in *Essentials of Biology*, but instructors can also edit questions, import their own content, and create announcements and due dates for assignments. ARIS has automatic grading and reporting of easy-to-assign homework, quizzing, and testing. All student activity within McGraw-Hill's ARIS is automatically recorded and available to the instructor through a fully integrated grade book that can be downloaded to Excel.

The *Essentials of Biology* ARIS site at www.mhhe.com/maderessentials offers access to a vast array of premium online content to fortify the learning and teaching experience for students and instructors.

Instructor Edition

In addition to all of the resources for students, the Instructor Edition of the Online Learning Center has these assets:

- **eInstruction Classroom Performance System (CPS) Question Bank** A set of questions for use with the CPS is provided for every textbook chapter to assist instructors in quickly assessing student comprehension of the concepts
- **Animations** The next generation of biology animations is available with *Essentials of Biology*. Full-color presentations of key biological processes have been brought to life via animation. These animations offer flexibility for instructors. Designed to be used in lectures, you can pause, rewind, fast-forward, and turn the audio on or off. Many of the animations are also available with Spanish narration and audio.
- **Laboratory Resource Guide** This preparation guide provides set-up instructions, sources for materials and supplies, time estimates, special requirements, and suggested answers to all questions in the *Essentials of Biology Laboratory Manual*.
- **PageOut** McGraw-Hill's exclusive tool for creating your own website for your general biology course. It requires no knowledge of coding and is hosted by McGraw-Hill.
- **Active Art Demo** Teaches you how to use the Active Art that is on the Digital Content Manager CD-ROM.
- **Case Studies** Offers suggestions on how to use Case Studies in your classroom.

McGraw-Hill: Biology Digitized Video Clips

McGraw-Hill is pleased to offer adopting instructors a new presentation tool—digitized biology video clips on DVD! Licensed from some of the highest-quality science video producers in the world, these brief segments range from about five seconds to just under three minutes in length and cover all areas of general biology from cells to ecosystems. Engaging and informative, McGraw-Hill's digitized biology video clips will help capture students' interest while illustrating key biological concepts and processes such as mitosis, how cilia and flagella work, and how some plants have evolved into carnivores.
ISBN-13: 978-0-07-312155-0 (ISBN-10: 0-07-312155-X)

Mader Micrograph Slides

This set contains one hundred 35mm slides of many of the photomicrographs and electron micrographs in the text.
ISBN-13: 978-0-07-239977-6 (ISBN-10: 0-07-239977-5)

Learning Supplements

For the Student

Student Study Guide

Dr. Sylvia Mader has written the *Student Study Guide* that accompanies *Essentials of Biology*, thereby ensuring close coordination with the text. Each text chapter has a corresponding study guide chapter that includes a chapter review, a review of the key terms in the chapter, study exercises and questions for each section of the chapter, and a chapter test. Answers to all questions are provided to give students immediate feedback. Students who make use of the *Student Study Guide* should find that performance increases dramatically.
ISBN-13: 978-0-07-321774-1 (ISBN-10: 0-07-321774-3)

ARIS

McGraw-Hill's ARIS—Assessment, Review, and Instruction System—for *Essentials of Biology* at www.mhhe.com/maderessentials offers access to a vast array of premium online content to fortify the learning experience.

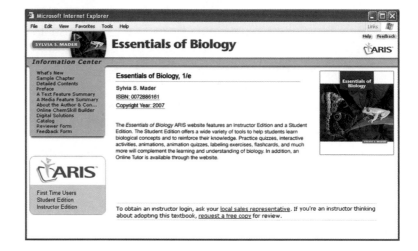

Student Edition

The Student Edition of ARIS features a wide variety of tools to help students learn biological concepts and to reinforce their knowledge:

Online study aids are organized according to the major sections of each chapter. **Practice quizzes, interactive activities, labeling exercises, flashcards,** and much more will complement the learning and understanding of biology

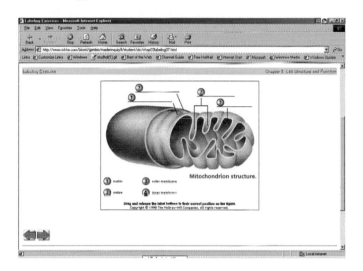

Essential Study Partner This collection of interactive study modules contains hundreds of animations, learning activities, and quizzes designed to help students grasp complex concepts.

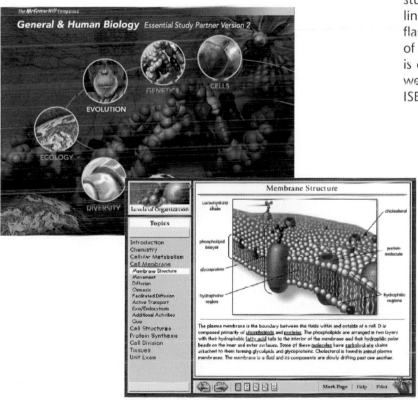

Animations Full-color presentations of key biological processes have been brought to life via animation. You can pause, rewind, fast-forward, and turn the audio on or off. Many of the animations are also available with Spanish narration and audio.

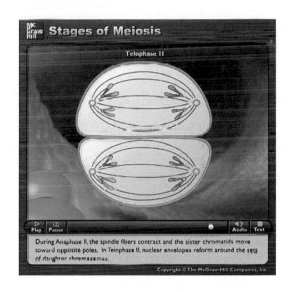

Animation Quizzes Quizzes based on the new animations help you assess your understanding of the concepts.

Online Tutoring The tutorial service is moderated by qualified instructors. Help with difficult concepts is only an email away!

Student Interactive CD-ROM

This interactive CD-ROM is an indispensable resource for studying topics covered in the text. It includes chapter outlines, chapter-based quizzes, animations of complex processes, flashcards, PowerPoint® lecture outlines, and PowerPoint® slides of all art and photos found in the textbook. All of the material is organized chapter-by-chapter. Direct links to the text's ARIS website and to the Essential Study Partner are also provided. ISBN-13: 978-0-07-321775-8 (ISBN-10: 0-07-321775-1)

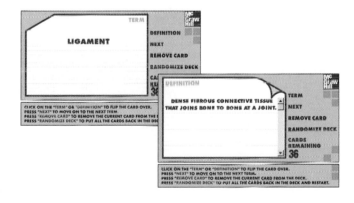

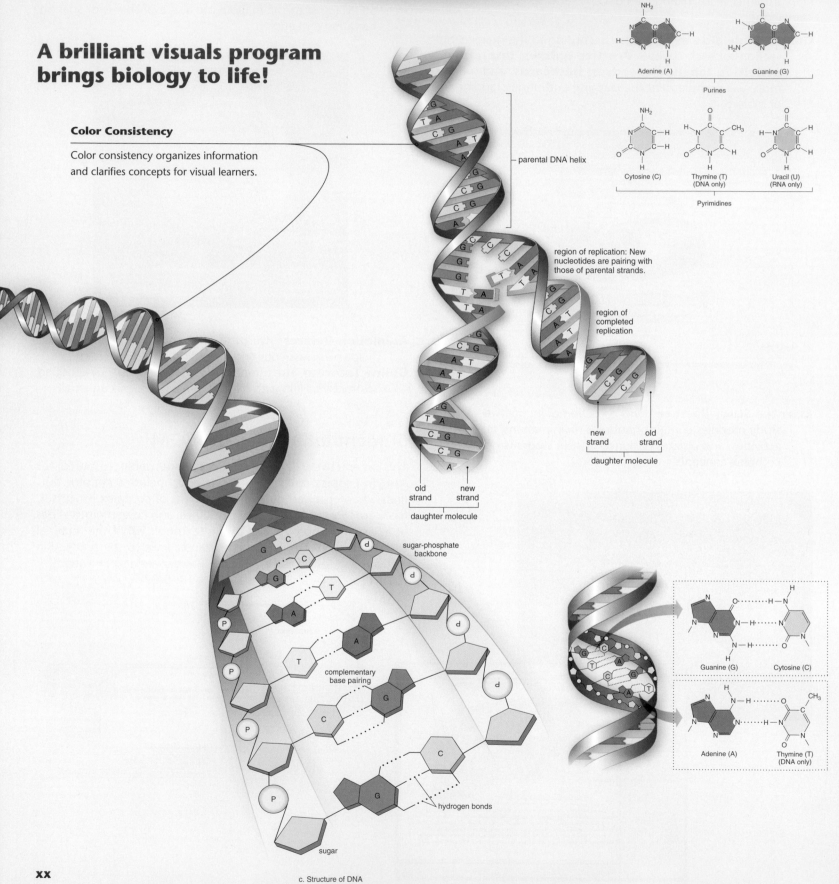

A brilliant visuals program brings biology to life!

Color Consistency

Color consistency organizes information and clarifies concepts for visual learners.

Nitrogen-containing bases

Adenine (A)

Guanine (G)

Purines

Cytosine (C)

Thymine (T)
(DNA only)

Uracil (U)
(RNA only)

Pyrimidines

parental DNA helix

region of replication: New nucleotides are pairing with those of parental strands.

region of completed replication

new strand

old strand

daughter molecule

old strand

new strand

daughter molecule

sugar-phosphate backbone

complementary base pairing

hydrogen bonds

sugar

Guanine (G)   Cytosine (C)

Adenine (A)   Thymine (T)
(DNA only)

c. Structure of DNA

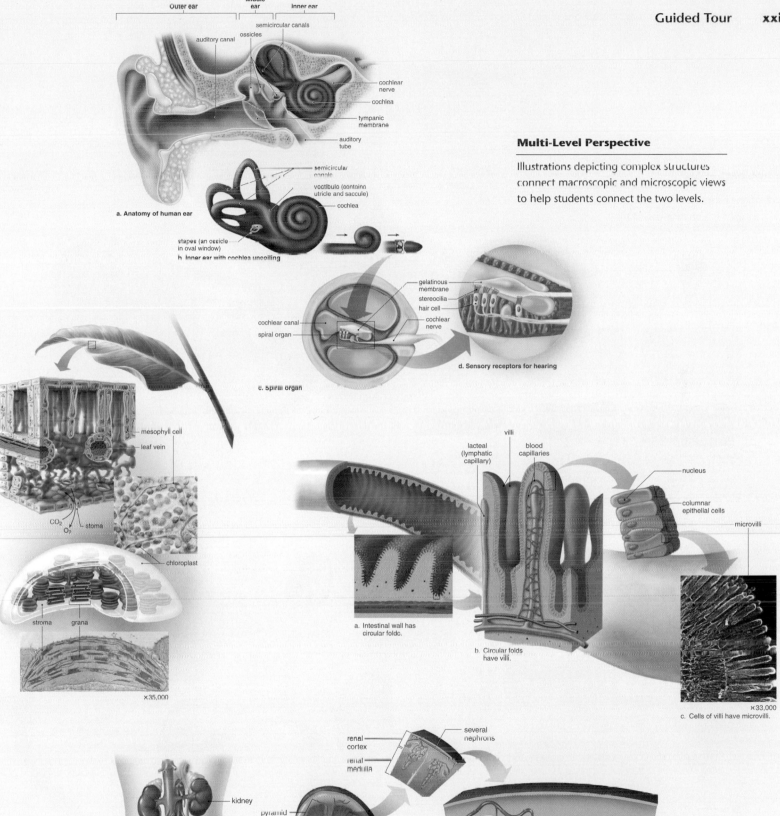

Outer ear | Middle ear | Inner ear

semicircular canals
ossicles
auditory canal
cochlear nerve
cochlea
tympanic membrane
auditory tube

a. Anatomy of human ear

semicircular canals
vestibule (contains utricle and saccule)
cochlea
stapes (an ossicle in oval window)

b. Inner ear with cochlea uncoiling

Multi-Level Perspective

Illustrations depicting complex structures connect macroscopic and microscopic views to help students connect the two levels.

gelatinous membrane
stereocilia
hair cell
cochlear nerve
cochlear canal
spiral organ

d. Sensory receptors for hearing

c. Spiral organ

mesophyll cell
leaf vein
CO_2
O_2
stoma
chloroplast
stroma
grana

×35,000

lacteal (lymphatic capillary)
villi
blood capillaries
nucleus
columnar epithelial cells
microvilli

a. Intestinal wall has circular folds.

b. Circular folds have villi.

×33,000
c. Cells of villi have microvilli.

renal cortex
renal medulla
several nephrons

kidney
ureter
urinary bladder
urethra

a. Urinary system

pyramid
renal artery
renal vein
renal pelvis
ureter

b. Kidney

nephron capsule
proximal tubule
peritubular capillary
distal tubule
loop of the nephron
collecting duct

c. One nephron

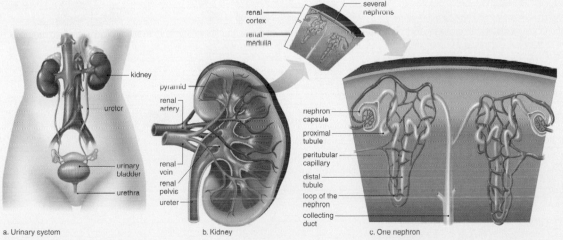

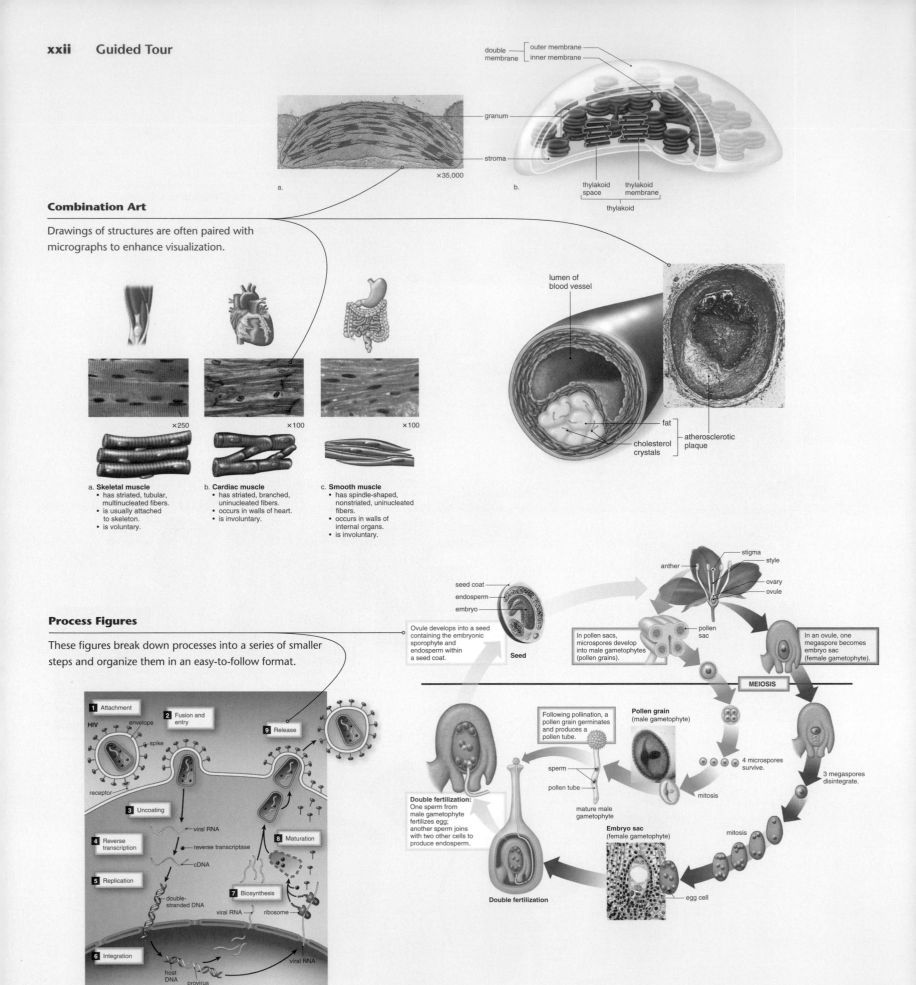

double
membrane ⎡ outer membrane
⎣ inner membrane

granum

stroma

×35,000

a.

b.

thylakoid
space

thylakoid
membrane

thylakoid

Combination Art

Drawings of structures are often paired with
micrographs to enhance visualization.

lumen of
blood vessel

fat ⎤
⎥ atherosclerotic
cholesterol ⎦ plaque
crystals

×250

×100

×100

a. **Skeletal muscle**
 • has striated, tubular,
  multinucleated fibers.
 • is usually attached
  to skeleton.
 • is voluntary.

b. **Cardiac muscle**
 • has striated, branched,
  uninucleated fibers.
 • occurs in walls of heart.
 • is involuntary.

c. **Smooth muscle**
 • has spindle-shaped,
  nonstriated, uninucleated
  fibers.
 • occurs in walls of
  internal organs.
 • is involuntary.

Process Figures

These figures break down processes into a series of smaller
steps and organize them in an easy-to-follow format.

stigma
style
anther
ovary
ovule

seed coat
endosperm
embryo

Ovule develops into a seed
containing the embryonic
sporophyte and
endosperm within
a seed coat.

Seed

In pollen sacs,
microspores develop
into male gametophytes
(pollen grains).

pollen
sac

In an ovule, one
megaspore becomes
embryo sac
(female gametophyte).

MEIOSIS

1 Attachment
2 Fusion and
entry
9 Release

HIV
envelope
spike

receptor

3 Uncoating

4 Reverse
transcription

viral RNA

reverse transcriptase

cDNA

8 Maturation

5 Replication

double-
stranded DNA

7 Biosynthesis

viral RNA
ribosome

6 Integration

host
DNA

provirus
DNA

viral RNA

Nucleus

Following pollination, a
pollen grain germinates
and produces a
pollen tube.

Pollen grain
(male gametophyte)

sperm

pollen tube

4 microspores
survive.

3 megaspores
disintegrate.

mitosis

Double fertilization:
One sperm from
male gametophyte
fertilizes egg;
another sperm joins
with two other cells to
produce endosperm.

mature male
gametophyte

Embryo sac
(female gametophyte)

mitosis

Double fertilization

egg cell

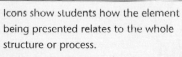

Icons

Icons show students how the element being presented relates to the whole structure or process.

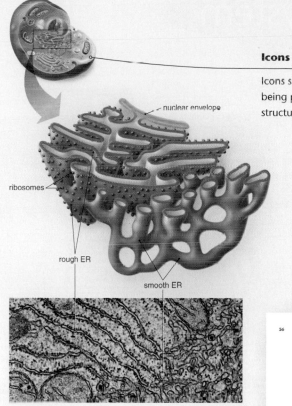

nuclear envelope

ribosomes

rough ER

smooth ER

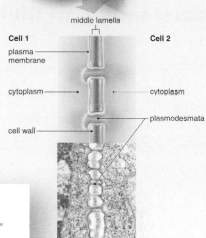

cell wall

plasmodesmata

cell wall

middle lamella

Cell 1

plasma membrane

cytoplasm

cell wall

Cell 2

cytoplasm

plasmodesmata

×53,000

Integrated Page Layouts

The innovative page layouts integrate text, art, and photos, enhancing visual appeal and pedagogical value and thereby making it easier for students to understand the material being presented.

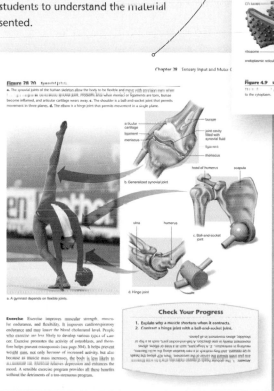

The Learning System

Features That Will Facilitate Your Understanding of Biology

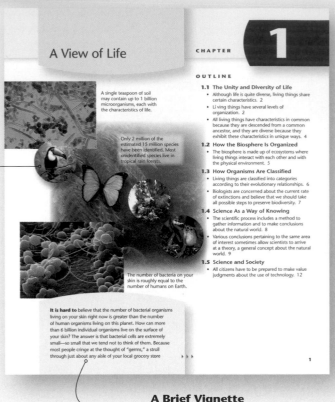

A View of Life

CHAPTER 1

OUTLINE

1.1 The Unity and Diversity of Life
- Although life is quite diverse, living things share certain characteristics. 2
- Living things have several levels of organization. 2
- All living things have characteristics in common because they are descended from a common ancestor, and they are diverse because they exhibit these characteristics in unique ways. 4

1.2 How the Biosphere Is Organized
- The biosphere is made up of ecosystems where living things interact with each other and with the physical environment. 5

1.3 How Organisms Are Classified
- Living things are classified into categories according to their evolutionary relationships. 6
- Biologists are concerned about the current rate of extinctions and believe that we should take all possible steps to preserve biodiversity. 7

1.4 Science As a Way of Knowing
- The scientific process includes a method to gather information and to make conclusions about the natural world. 8
- Various conclusions pertaining to the same area of interest sometimes allow scientists to arrive at a theory, a general concept about the natural world. 9

1.5 Science and Society
- All citizens have to be prepared to make value judgments about the use of technology. 12

A single teaspoon of soil may contain up to 1 billion microorganisms, each with the characteristics of life.

Only 2 million of the estimated 15 million species have been identified. Most unidentified species live in tropical rain forests.

The number of bacteria on your skin is roughly equal to the number of humans on Earth.

It is hard to believe that the number of bacterial organisms living on your skin right now is greater than the number of human organisms living on this planet. How can more than 6 billion individual organisms live on the surface of your skin? The answer is that bacterial cells are extremely small—so small that we tend not to think of them. Because most people cringe at the thought of "germs," a stroll through just about any aisle of your local grocery store

1

Captivating Illustrations

open the chapter.

An Outline

lists the major topics for the chapter.

A Brief Vignette

relates the illustrations to students' lives.

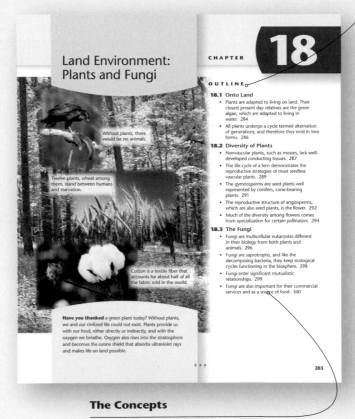

Land Environment: Plants and Fungi

CHAPTER 18

OUTLINE

18.1 Onto Land
- Plants are adapted to living on land. Their closest present day relatives are the green algae, which are adapted to living in water. 284
- All plants undergo a cycle termed alternation of generations, and therefore they exist in two forms. 286

18.2 Diversity of Plants
- Nonvascular plants, such as mosses, lack well-developed conducting tissues. 287
- The life cycle of a fern demonstrates the reproductive strategies of most seedless vascular plants. 289
- The gymnosperms are seed plants well represented by conifers, cone-bearing plants. 291
- The reproductive structure of angiosperms, which are also seed plants, is the flower. 292
- Much of the diversity among flowers comes from specialization for certain pollinators. 294

18.3 The Fungi
- Fungi are multicellular eukaryotes different in their biology from both plants and animals. 296
- Fungi are saprotrophs, and like the decomposing bacteria, they keep ecological cycles functioning in the biosphere. 298
- Fungi enter significant mutualistic relationships. 299
- Fungi are also important for their commercial services and as a source of food. 300

Without plants, there would be no animals.

Twelve plants, wheat among them, stand between humans and starvation.

Cotton is a textile fiber that accounts for about half of all the fabric sold in the world.

Have you thanked a green plant today? Without plants, we and our civilized life could not exist. Plants provide us with our food, either directly or indirectly, and with the oxygen we breathe. Oxygen also rises into the stratosphere and becomes the ozone shield that absorbs ultraviolet rays and makes life on land possible.

283

The Concepts

related to each topic are page referenced.

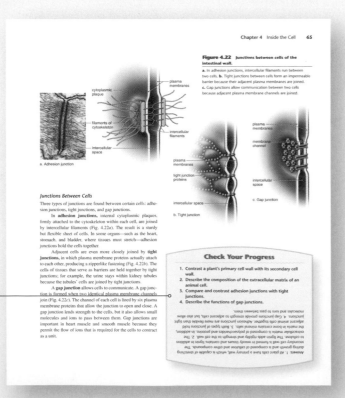

Chapter 4 Inside the Cell 65

Figure 4.22 Junctions between cells of the intestinal wall.

a. In adhesion junctions, intercellular filaments run between two cells. **b.** Tight junctions between cells form an impermeable barrier because their adjacent plasma membranes are joined. **c.** Gap junctions allow communication between two cells because adjacent plasma membrane channels are joined.

cytoplasmic plaque

plasma membranes

filaments of cytoskeleton

intercellular filaments

plasma membranes

membrane channel

plasma membranes

a. Adhesion junction

tight junction proteins

intercellular space

intercellular space

b. Tight junction

c. Gap junction

Junctions Between Cells

Three types of junctions are found between certain cells: adhesion junctions, tight junctions, and gap junctions.

In **adhesion junctions**, internal cytoplasmic plaques, firmly attached to the cytoskeleton within each cell, are joined by intercellular filaments (Fig. 4.22a). The result is a sturdy but flexible sheet of cells. In some organs—such as the heart, stomach, and bladder, where tissues must stretch—adhesion junctions hold the cells together.

Adjacent cells are even more closely joined by **tight junctions**, in which plasma membrane proteins actually attach to each other, producing a zipperlike fastening (Fig. 4.22b). The cells of tissues that serve as barriers are held together by tight junctions; for example, the urine stays within kidney tubules because the tubules' cells are joined by tight junctions.

A **gap junction** allows cells to communicate. A gap junction is formed when two identical plasma membrane channels join (Fig. 4.22c). The channel of each cell is lined by six plasma membrane proteins that allow the junction to open and close. A gap junction lends strength to the cells, but it also allows small molecules and ions to pass between them. Gap junctions are important in heart muscle and smooth muscle because they permit the flow of ions that is required for the cells to contract as a unit.

Check Your Progress

1. Contrast a plant's primary cell wall with its secondary cell wall.
2. Describe the composition of the extracellular matrix of an animal cell.
3. Compare and contrast adhesion junctions with tight junctions.
4. Describe the functions of gap junctions.

Check Your Progress

Questions follow main sections of the text and help students assess their understanding of the material presented.

Chapter Summary

An extensive chapter summary is organized according to the major sections in the chapter. Brief statements, lists, tables, and artwork help students review the important topics and concepts.

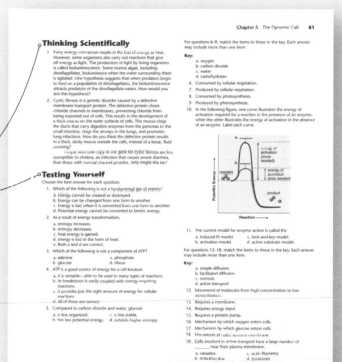

End-of-Chapter Study Tools

Thinking Scientifically

Critical thinking questions give students an opportunity to reason as a scientist. Detailed answers to these questions are found in the Answer Appendix in the textbook.

Testing Yourself

Objective and art-based questions allow students to review material and prepare for tests. Answers to these questions are given in the Answer Appendix in the textbook.

Website Reminder

This reminder directs you to the book's website for additional quiz questions and other study aids.

Bioethical Issue

A bioethical issue is presented at the end of the chapter. These short readings discuss a variety of controversial topics that confront our society. Appropriate questions in the reading help students fully consider the issue and arrive at an opinion.

Understanding the Terms

The boldface terms in the chapter are page referenced, and a matching exercise allows students to test their knowledge of the terms.

A View of Life

A single teaspoon of soil may contain up to 1 billion microorganisms, each with the characteristics of life.

Only 2 million of the estimated 15 million species have been identified. Most unidentified species live in tropical rain forests.

The number of bacteria on your skin is roughly equal to the number of humans on Earth.

It is hard to believe that the number of bacterial organisms living on your skin right now is greater than the number of human organisms living on this planet. How can more than 6 billion individual organisms live on the surface of your skin? The answer is that bacterial cells are extremely small— so small that we tend not to think of them. Because most people cringe at the thought of "germs," a stroll through just about any aisle of your local grocery store will reveal

▶ ▶ ▶

a slew of products that claim to be antibacterial. Advertisers tend to make us think that all bacteria are harmful. But only a handful of bacterial species are actually dangerous to humans, while the overwhelming majority are beneficial. The billions of bacteria that live on your skin right now are termed normal flora (or resident bacteria), and they essentially occupy all the available space on your skin surface. Should a detrimental bacterium come along, it will literally have no room to settle and multiply. Because the normal flora protect you in this way, we certainly don't want to do away with them.

The bacterial cells occupying your skin have a lot in common with the cells that make up your body. This chapter introduces the common characteristics, the diversity, and the organization of living things, as well as explaining how scientists perform their studies.

1.1 The Unity and Diversity of Life

From huge menacing sharks to miniscule exotic orchids, life is very diverse. Despite life's diversity, all living things, also called **organisms,** share certain characteristics, and these characteristics give us insight into the nature of life and help distinguish organisms from nonliving things.

Living Things Are Organized

The complex organization of living things begins with small molecules that join to form larger molecules within a **cell,** the smallest, most basic unit of life. Although a cell is alive it is made from nonliving molecules.

Some organisms are **unicellular,** being only a single cell. Plants and animals are **multicellular** and are composed of many types of cells. In multicellular organisms, similar cells combine to form **tissues.** Tissues make up **organs,** as when various tissues combine to form a heart or a leaf. Organs work together in **organ systems;** for example, the heart and blood vessels form the cardiovascular system. Various organ systems work together within complex organisms (Fig. 1.1).

Later in this chapter, we will see that levels of biological organization extend beyond the individual organism. For example, all the members of one species in a particular area belong to a population. Zebras form one of the many populations on an African plain. All the populations make up a community. The community of populations interacts with the physical environment to form an ecosystem. Finally, all the Earth's ecosystems make up the biosphere.

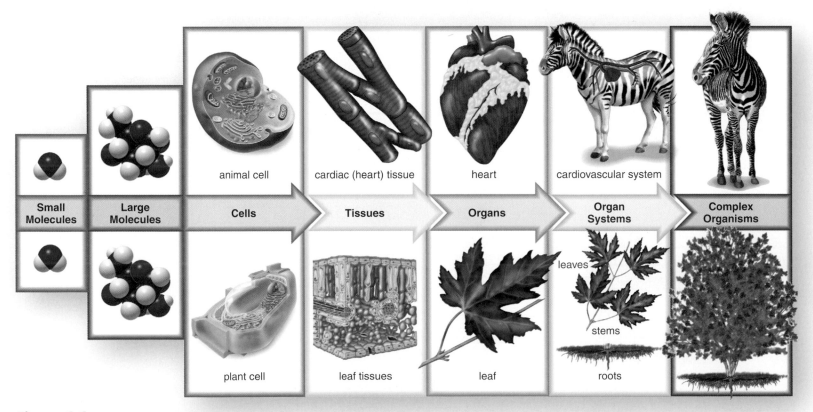

Figure 1.1 **Levels of biological organization.**

Living Things Acquire Materials and Energy

Living things cannot maintain their organization or carry on life's activities without an outside source of materials and energy (Fig. 1.2). Food provides nutrient molecules, which are used as building blocks or energy sources. **Energy** is the capacity to do work, and it takes work to maintain the organization of the cell and the organism. When cells use nutrient molecules to make their parts and products, they carry out a sequence of chemical reactions. The term **metabolism** encompasses all the chemical reactions that occur in a cell.

The ultimate source of energy for nearly all life on Earth is the sun. Plants and certain other organisms are able to capture solar energy and carry on **photosynthesis,** a process that transforms solar energy into the chemical energy of nutrient molecules. Animals and plants get energy by metabolizing nutrient molecules made by photosynthesizers.

Remaining Homeostatic

For metabolic processes to continue, living things need to keep themselves stable in temperature, moisture level, acidity, and other physiological factors. This is **homeostasis**—the maintenance of internal conditions within certain boundaries.

Many organisms depend on behavior to regulate their internal environment. A chilly lizard may raise its internal temperature by basking in the sun on a hot rock. When it starts to overheat, it scurries for cool shade. Other organisms have control mechanisms that do not require any conscious activity. When a student is so engrossed in her textbook that she forgets to eat lunch, her liver releases stored sugar to keep the blood sugar level within normal limits. Hormones regulate sugar storage and release, but in other instances the nervous system is involved in maintaining homeostasis.

Living Things Respond

Living things find energy and/or nutrients by interacting with their surroundings. Even unicellular organisms can respond to their environment. The beating of microscopic hairs or the snapping of whiplike tails moves them toward or away from light or chemicals. Multicellular organisms can manage more complex responses. A monarch butterfly can sense the approach of fall and begin its flight south where resources are still abundant. A vulture can smell meat a mile away and soar toward dinner.

The ability to respond often results in movement. The leaves of a plant turn toward the sun, and animals dart toward safety. Appropriate responses help ensure survival of the organism and allow it to carry on its daily activities. Altogether, we call these activities the *behavior* of the organism.

Living Things Reproduce and Develop

Life comes only from life. Every type of living thing can **reproduce,** or make another organism like itself. Bacteria and other types of unicellular organisms simply split in two. In multicellular organisms,

a.

b.

Figure 1.2   Acquiring nutrient materials and energy.

a. Osprey, a type of hawk, eating a fish. **b.** Humans harvesting crops.

the reproductive process usually begins with the pairing of a sperm from one partner and an egg from the other partner. The union of sperm and egg, followed by many cell divisions, results in an immature individual, which grows and develops through various stages to become an adult.

An embryo develops into a whale or a yellow daffodil or a human being because of the specific set of genes inherited from its parents (Fig. 1.3). In all organisms, the **genes** are made of long DNA (deoxyribonucleic acid) molecules, but even so the genes are different between species and individuals. DNA provides the blueprint or instructions for the organization and metabolism of the particular organism. All cells in a multicellular organism contain the same set of genes, but only certain ones are turned on in each type of specialized cell.

Living Things Have Adaptations

Adaptations are modifications that make organisms suited to their way of life. Some hawks have the ability to catch fish (Fig. 1.2), and others are best at catching rabbits (Fig. 1.4). Hawks can fly in part because they have hollow bones to reduce their weight and flight muscles to depress and elevate their wings. When a hawk dives, its strong feet take the first shock of the landing, and its long, sharp claws reach out and hold onto the prey.

Organisms become modified over time by a process called **natural selection.** Certain members of a **species,** defined as a group of interbreeding individuals, may inherit a genetic change that causes them to be better suited to a particular environment. These members can be expected to produce more surviving offspring that also have the favorable characteristic. In this way, the attributes of the species' members change over time.

Descent with Modification

The unity of living things extends beyond those characteristics that are observable by the human eye. As mentioned, every organism's genes are composed of DNA, and organisms carry out the same metabolic reactions to acquire energy and maintain their organization. This unity suggests that all living things are descended from a common ancestor—the first cell or cells.

Evolution is descent with modification. One species can be a common ancestor to several species, each adapted to a particular set of environmental conditions. Specific adaptations allow species to play particular roles in an ecosystem. The diversity of life-forms is best understood in terms of the many different ways in which organisms carry on their life functions within the ecosystem where they live, acquire energy, and reproduce.

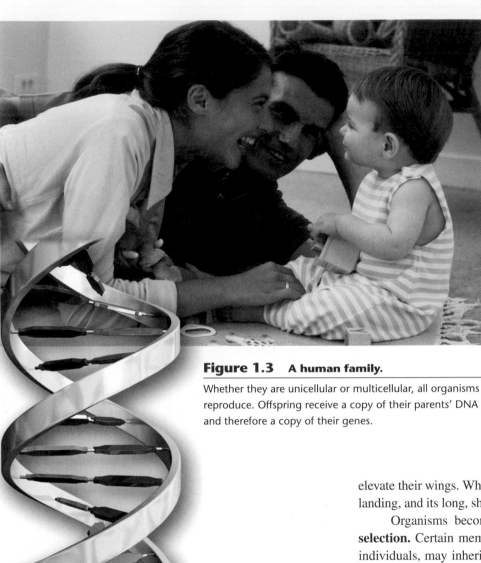

Figure 1.3   A human family.

Whether they are unicellular or multicellular, all organisms reproduce. Offspring receive a copy of their parents' DNA and therefore a copy of their genes.

DNA

Check Your Progress

1. **List the levels of organization of a multicellular animal, beginning with molecules.**
2. **List five features common to organisms.**
3. **Define evolution.**

Answers: 1. molecules-cells-tissues-organs-organ systems-organism 2. Organisms are organized, acquire materials and energy, respond to external stimuli, reproduce and develop, and have adaptations. 3. Evolution is descent of species from common ancestors, with genetic modifications that make each species more suited to its environment.

1.2 How the Biosphere Is Organized

The organization of life extends beyond the individual to the **biosphere,** the zone of air, land, and water at the surface of the Earth where living organisms are found. Individual organisms belong to a **population,** all the members of a species within a particular area. The populations within a **community** interact among themselves and with the physical environment (soil, atmosphere, etc.), thereby forming an **ecosystem.**

Ecosystem

One example of an ecosystem is a North American grassland, which is inhabited by populations of rabbits, hawks, and various types of grasses, among many others. These populations interact with each other by forming food chains in which one population feeds on another. For example, rabbits feed on grasses, while hawks feed on rabbits and other organisms.

As Figure 1.4 shows, ecosystems are characterized by chemical cycling and energy flow, both of which begin when *producers,* such as grasses, take in solar energy and inorganic nutrients to produce food (organic nutrients) by photosynthesis. Chemical cycling (aqua arrows) occurs as chemicals move from one *consumer* population to another in a food chain, until with death, *decomposers* return inorganic nutrients to the producers once again. Energy, on the other hand, flows from the sun through plants and the other members of the food chain as they feed on one another. The energy gradually dissipates and returns to the atmosphere as heat (red arrows). Because energy does not cycle, ecosystems could not stay in existence without solar energy and the ability of photosynthesizers to absorb it.

Biosphere

Climate largely determines where different ecosystems are found in the biosphere. For example, deserts exist in areas of minimal rain, while forests require much rain. The two most biologically diverse ecosystems—tropical rain forests and coral reefs—occur where solar energy is most abundant. The human population tends to modify these and all ecosystems for its own purposes. Humans clear forests or grasslands in order to grow crops; later, they build houses on what was once farmland; and finally, they convert small towns into cities. As coasts are developed, humans send sediments, sewage, and other pollutants into the sea and onto coral reefs.

Tropical rain forests, coral reefs, and most ecosystems are severely threatened as the human population increases in size. Some coral reefs are 50 million years old, and yet in just a few decades, human activities have destroyed 10% of all reefs and seriously degraded another 30%. At this rate, nearly three-quarters could be destroyed within 50 years. Similar statistics are available for tropical rain forests.

It has long been clear that human beings depend on healthy ecosystems for food, medicines, and various raw materials. We are only now beginning to realize that we depend on them even more for the services they provide. The workings of ecosystems ensure that environmental conditions are suitable for the continued existence of humans.

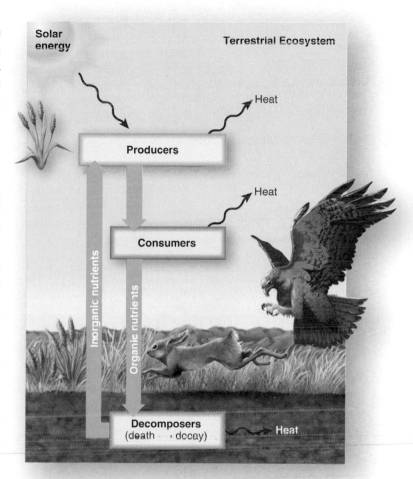

Figure 1.4   A grassland, a terrestrial ecosystem.

In an ecosystem, chemical cycling (blue arrows) and energy flow begin when plants use solar energy and inorganic nutrients to produce their own food (organic nutrients). Chemicals and energy are passed from one population to another in a food chain. Eventually, energy dissipates as heat (red arrows). With the death and decomposition of organisms, inorganic nutrients are returned to living plants once more.

Check Your Progress

Contrast a community with an ecosystem.

Answer: A community consists of different populations that interact in a particular place. An ecosystem is a community together with the physical environment with which it interacts.

×20,000

×64,000

Figure 1.5 **Domain Archaea.**

Archaea are capable of living in extreme environments. These are exterior and interior views of *Methanosarcina mazei.*

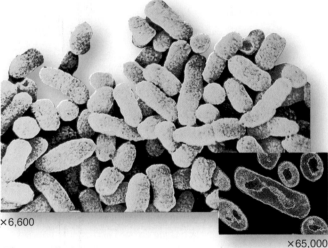

×6,600

×65,000

Figure 1.6 **Domain Bacteria.**

Bacteria are structurally simple but metabolically diverse. These are exterior and interior views of *Escherichia coli.*

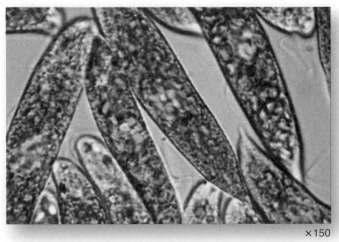

×150

Figure 1.7 **Domain Eukarya, Kingdom Protista.**

Many protists are unicellular. This is *Euglena.*

1.3 How Organisms Are Classified

Because life is so diverse, it is helpful to have a system that groups organisms into categories. **Taxonomy** is the discipline of identifying and classifying organisms according to certain rules. Taxonomy makes sense out of the bewildering variety of life on Earth by classifying organisms according to their presumed evolutionary relationship. As more is learned about evolutionary relationships between species, taxonomy changes. Taxonomists are even now making observations and performing experiments that will one day bring about changes in the classification system adopted by this text.

Categories of Classification

The classification categories, going from least inclusive to most inclusive, are **species, genus, family, order, class, phylum, kingdom,** and **domain.** Each successive category above species contains more types of organisms than the preceding one. Species placed within one genus share many specific characteristics and are the most closely related, while species placed in the same domain share only general characteristics. For example, all species in the genus *Pisum* look pretty much the same—that is, like pea plants—but species in the plant kingdom can be quite varied, as is evident when we compare grasses to trees. By the same token, only modern humans are in the genus *Homo,* but many types of species, from tiny hydras to huge whales, are members of the animal kingdom. Species placed in different domains are the most distantly related.

Domains

Biochemical evidence suggests that there are only three domains: **domain Bacteria, domain Archaea,** and **domain Eukarya.** Both domain Bacteria and domain Archaea contain prokaryotes. Prokaryotes are unicellular, and they lack the membrane-bounded nucleus found in the eukaryotes of domain Eukarya.

Prokaryotes are structurally simple (Figs. 1.5 and 1.6) but metabolically complex. Archaea live in aquatic environments that lack oxygen or are too salty, too hot, or too acidic for most other organisms. Perhaps these environments are similar to those of the primitive Earth, and archaea are representative of the first cells that evolved. Bacteria are found almost anywhere—in the water, soil, and atmosphere, as well as on our skin and in our mouths and large intestines. Although some bacteria cause diseases, others perform many services, both environmental and commercial. Bacteria are used to conduct genetic research in our laboratories, produce innumerable products in our factories, and help purify water in our sewage treatment plants.

Kingdoms

Taxonomists haven't yet decided how to categorize archaea and bacteria into kingdoms. Domain Eukarya, on the other hand, has four kingdoms with which you may be familiar. **Protists** (kingdom Protista) range from unicellular to a few multicellular organisms (Fig. 1.7). Some are photosynthesizers, and others must ingest their food. Among the **fungi** (kingdom Fungi) are the familiar molds and mushrooms that, along with many types of bacteria, help decompose dead organisms (Fig. 1.8). **Plants** (kingdom Plantae) are well known as multicellular photosynthesizers (Fig. 1.9). **Animals** (kingdom Animalia) are multicellular organisms that ingest their food (Fig. 1.10).

Scientific Naming

Biologists give each living thing a two-part scientific name called a **binomial name.** For example, the scientific name for the garden pea is *Pisum sativum.* The first word is the genus, and the second word is the specific epithet of a species within a genus. The genus may be abbreviated (e.g., *P. sativum*). Scientific names are universally used by biologists to avoid confusion. Common names tend to overlap, and often they are in the language of the people who use that particular name. But scientific names are based on Latin, a universal language that not too long ago was well known by most scholars.

Biodiversity

Classifying organisms helps keep track of biodiversity because biologists classify all known species. Even so, **biodiversity** not only includes the total number of species, but also the variability of organisms' genes and the variability of ecosystems in the biosphere.

The present number of species has been estimated to be as high as 15 million, but so far, under 2 million have been identified, named, and classified. **Extinction** is the death of a species or a larger taxonomic group. It is estimated that presently as many as 400 species per day are lost because of human activities. For example, several species of fishes have all but disappeared from the coral reefs of Indonesia and along the African coast because of overfishing. Many biologists are alarmed about the present rate of extinction and believe it may eventually rival the rates of the five mass extinctions that have occurred during our planet's history. The dinosaurs became extinct during the last mass extinction, 65 million years ago.

It has been suggested that the primary bioethical issue of our time is preservation of ecosystems, because in that way we preserve biodiversity. Just as a native fisherman who assists in overfishing a reef is doing away with his own food source, we as a society are contributing to the destruction of our home, the biosphere, when we destroy ecosystems. If instead we adopt a conservation ethic that preserves ecosystems, we are helping to ensure the continued existence of all species, including our own.

Figure 1.8   **Domain Eukarya, Kingdom Fungi.**

Fungi are multicellular and break down organic debris. This is a shaggy mane mushroom.

Figure 1.9   **Domain Eukarya, Kingdom Plantae.**

Plants are multicellular photosynthesizers. This is a rose.

Figure 1.10   **Domain Eukarya, Kingdom Animalia.**

Animals are multicellular and ingest their food. This is a European lynx.

Check Your Progress

1. List the eight classification categories, from least to most inclusive.
2. Explain why scientists prefer to refer to organisms by their scientific names rather than their common name.
3. List the four kingdoms in domain Eukarya.

Answers: 1. species, genus, family, order, class, phylum, kingdom, domain. 2. For various reasons, the same organism can have different common names, or the same common name may be given to several different organisms. Each organism has only one scientific name, and that name is never used for any other organism. 3. Protista, Fungi, Plantae, Animalia.

a. Botanist

b. Biochemist

Science As a Way of Knowing

Biology, with its numerous branches, is the scientific study of life (Fig. 1.11). Religion, aesthetics, ethics, and science are all ways that human beings have of finding order in the natural world. Science differs from the other fields by its process, which often involves the use of the scientific method (Fig. 1.12).

Observation

The scientific method begins with **observations.** We can observe with our noses that dinner is almost ready, observe with our fingertips that a surface is smooth and cold, and observe with our ears that a piano needs tuning. Scientists also extend the ability of their senses by using instruments; for example, the microscope enables them to see objects that could never be seen by the naked eye. Finally, scientists may expand their understanding even further by taking advantage of the knowledge and experiences of other scientists. For instance, they may look up past studies on the Internet or at the library, or they may write or speak to others who are researching similar topics.

Hypothesis

After making observations and gathering knowledge about a phenomenon, a scientist uses inductive reasoning. **Inductive reasoning** occurs whenever a person uses creative thinking to combine isolated facts into a cohesive whole. Chance alone can help a scientist arrive at an idea. The most famous case pertains to the antibiotic penicillin. While examining a petri dish containing the mold *Penicillium,* Alexander Fleming observed an area around the mold that was free of bacteria. Fleming thought the mold might be producing an antibacterial substance. We call such a possible explanation for a natural event a **hypothesis.** Fleming's hypothesis was supported by further study.

All of a scientist's past experiences, no matter what they might be, may influence the formation of a hypothesis. But a scientist only considers hypotheses that can be tested by experiments or further observations. Moral and religious beliefs, while very important to our lives, differ between cultures and through time, and are not always testable.

Experiments/Further Observations

The manner in which a scientist intends to conduct an experiment is called the **experimental design.** A good experimental design ensures that scientists are testing what they want to test and that their results will be meaningful. When an experiment is done in a laboratory, all conditions can be kept constant except for an **experimental variable,** which is deliberately changed. One or more test groups are exposed to the experimental variable, but one other group, called the **control group,** is not. If, by chance, the control group shows the same results as the test group, the experimenter knows the results are invalid.

Scientists often use a **model,** a representation of an actual object. Modeling occurs when scientists use software to decide how human activities will

c. Ecologists

Figure 1.11   Biologists.

Biologists work in a variety of settings. For example, (**a**) some botanists work in greenhouses; (**b**) some biologists, such as this biochemist, work in laboratories; and (**c**) many ecologists and environmentalists collect data in the field.

affect climate, or when they use mice instead of humans for, say, testing a new drug. Ideally, a medicine that is effective in mice should still be tested in humans. Whenever a model is used to study a phenomenon, any conclusion is in need of further testing. Someday, a scientist might devise a way to test the conclusion further.

Data

The results of an experiment are referred to as the **data.** Mathematical data are often displayed in the form of a graph or table. Sometimes studies rely on statistical data. Let's say an investigator wants to know if eating onions can prevent women from getting osteoporosis (weak bones). The scientist conducts a survey asking women about their onion-eating habits and then correlates these data with the condition of their bones. Other scientists critiquing this study would want to know: How many women were surveyed? How old were the women? What were their exercise habits? What criteria were used to determine the condition of their bones? And what is the probability that the data are in error? The greater the variance in the data, the greater the probability of error. In any case, even if the data do suggest a correlation, scientists would want to know the ingredient in onions that has a direct biochemical or physiological effect on bones. In this way scientists are skeptics who always pressure one another to keep investigating.

Conclusion

Scientists must analyze the data in order to reach a **conclusion** about whether a hypothesis is supported or not. Because science progresses, the conclusion of one experiment can lead to the hypothesis for another experiment (Fig. 1.12). In other words, results that do not support one hypothesis can often help a scientist formulate another hypothesis to be tested. Scientists report their findings in scientific journals so that their methodology and data are available to other scientists. Experiments and observations must be *repeatable*—that is, the reporting scientist and any scientist who repeats the experiment must get the same results, or else the data are suspect.

Scientific Theory

The ultimate goal of science is to understand the natural world in terms of **scientific theories,** which are accepted explanations for how the world works. Some of the basic theories of biology are the cell theory, which says that all organisms are composed of cells; the gene theory, which says that inherited information dictates the form, function, and behavior of organisms; and the theory of evolution, which says that all organisms have a common ancestor, and each one is adapted to a particular way of life.

The theory of evolution is considered the unifying concept of biology because it pertains to many different aspects of organisms. For example, the theory of evolution enables scientists to understand the history of life, the variety of organisms, and the anatomy, physiology, and development of organisms. The theory of evolution has been a very fruitful scientific theory, meaning that it has helped scientists generate new testable hypotheses.

Because this theory has been supported by so many observations and experiments for over 100 years, some biologists refer to the **principle** of evolution, a term sometimes used for theories that are generally accepted by an overwhelming number of scientists. Others prefer the term law instead of principle.

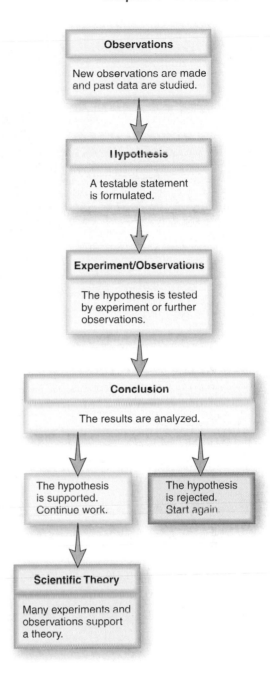

Figure 1.12   Flow diagram for the scientific method.

On the basis of new and/or previous observations, a scientist formulates a hypothesis. The hypothesis is tested by further experiments and/or observations, and new data either support or do not support the hypothesis. Following an experiment, a scientist often chooses to retest the same hypothesis or to test a related hypothesis. Conclusions from many different but related experiments may lead to the development of a scientific theory. For example, studies pertaining to development, anatomy, and fossil remains all support the theory of evolution.

a.

b.

Figure 1.13 A controlled study.

a. Pea plants growing in clay pots at the end of the summer. **b.** After the pea plants were mixed into the soil, winter wheat was planted in the same pots.

Example of a Controlled Study

Some investigators were concerned about the excessive use of nitrogen fertilizer to grow crops. Water pollution occurs when rain runoff washes fertilizer into nearby bodies of water. Nitrogen from a fertilizer can make well water on farms toxic to drink, especially for infants.

The investigators hypothesized that if pea plants were grown and turned over in the soil *before* winter wheat was planted there, they would act as a "natural fertilizer," eliminating the need to add nitrogen fertilizer. If so, wheat could be grown successfully in an environmentally friendly way.

The Study

In this study, the investigators decided on the following experimental design:

Control Group

Winter wheat was planted in pots of soil that received no prior treatment.

Test Groups

1. Winter wheat was grown in clay pots in soil treated with an inorganic nitrogen fertilizer (45 kg/ha).
2. Winter wheat was grown in clay pots in soil treated with twice as much nitrogen fertilizer (90 kg/ha).
3. Pea plants were grown in clay pots (Fig. 1.13*a*). The pea plants were then turned over into the soil, and winter wheat was planted in the same pots (Fig. 1.13*b*).

To ensure a controlled experiment, the conditions for the control group and the test groups were identical; the plants were exposed to the same environmental conditions and watered equally (Fig. 1.13*c*). During the following spring, the wheat plants were dried and weighed to determine the wheat yield in each of the pots.

The Results After the first year, the wheat yield for test groups 1 and 2 was greater than for the control group (Fig. 1.13*d*). To the surprise of investigators,

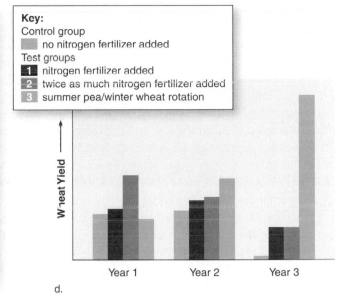

Key:
Control group
☐ no nitrogen fertilizer added
Test groups
1 nitrogen fertilizer added
2 twice as much nitrogen fertilizer added
3 summer pea/winter wheat rotation

c.

d.

Figure 1.13 A controlled study, continued.

c. Winter wheat was grown in both control and test pots. **d.** Three years of data are given in this bar graph.

wheat production following summer planting of peas did not demonstrate as high a yield as even the control group (aqua bar compared to gray bar.)

CONCLUSION: The hypothesis is not supported. Wheat yield following the growth of peas is not as high as the yield following the use of nitrogen fertilizer at the end of year 1.

Continuing the Experiment

The researchers decided to continue the experiment using the same design and the same pots as before, to see if the buildup of residual soil nitrogen from pea plants would eventually increase the yield of wheat. Figure 1.13*d* shows the results after year 2: The wheat yield following the growth of peas (aqua bar) is better than that for the other groups.

CONCLUSION: The hypothesis is supported. At the end of two years, the summer planting of peas prior to the planting of wheat does result in a higher wheat yield.

The researchers continued their experiment for still another year. After year 3, the wheat yield had decreased in both the control group and the inorganic nitrogen fertilizer groups, while the wheat yield following the planting of peas was dramatically better (aqua bar compared to the other bars in year 3, Fig. 1.13*d*).

The researchers suggested that their results showed that the soil had gradually improved after the planting of the peas. The researchers published their results in a scientific journal[1] because any alternative to the use of nitrogen fertilizer would be beneficial to our society, which depends on wheat as a staple food. Also, a regimen that improves rather than depletes the soil would enable agriculture to continue unabated in the future.

[1] Bidlack, J. E., Rao, S. C., and Demezas, D. H. 2001. Nodulation, nitrogenase activity, and dry weight of chickpea and pigeon pea cultivars using different *Bradyrhizobium* strains. *Journal of Plant Nutrition* 24:549–60.

Check Your Progress

1. **Explain how a scientist uses the scientific method to create a hypothesis.**
2. **Compare and contrast the control group with the test groups in a scientific study.**

Answers: 1. First, a scientist makes observations about some phenomenon in the natural world. Then, s/he uses inductive reasoning to formulate a hypothesis, a possible explanation for the phenomenon. Finally, s/he uses the information gained to create an explanation for the phenomenon. 2. In an experiment, the control group is not exposed to the experimental variable, but the test groups are exposed to the experimental variable. Otherwise, the control groups and the test groups were exposed to identical conditions; if there would be no basis for comparing the results.

1.5   Science and Society

Many scientists work in the field or the laboratory collecting data and coming to conclusions that sometimes seem remote from our everyday lives. Other scientists are interested in using the findings of past and present scientists to produce a product or develop a technique that does affect our lives. The application of scientific knowledge for a practical purpose is called *technology*. For example, virology, the study of viruses and their molecular chemistry, led to the discovery of new drugs that extend the life-spans of people who have AIDS. And cell biology research discovers the causes of cancer, allowing physicians to develop various cancer treatments.

Most technologies have benefits but also drawbacks. Research has led to modern agricultural practices that are helping to feed the burgeoning world population. However, the use of nitrogen fertilizers leads to water pollution, and the use of pesticides, as you may know, kills not only pests but also other types of organisms. The scientist Rachel Carson wrote the book *Silent Spring* to make the public aware of the harmful environmental effects of pesticide use.

Who should decide how, and even whether, a technology is put to use? Making value judgments is not a part of science. Ethical and moral decisions must be made by all people. Therefore, the responsibility for how to use the fruits of science must reside with people from all walks of life, and not with scientists alone. Scientists should provide the public with as much information as possible, but all citizens, including scientists, should make decisions about the use of technologies.

In the future we may need to decide whether we want to stop producing bioengineered organisms if they prove to be harmful to the environment. Also, through gene therapy, we are developing the ability to cure diseases and to alter the genes of our offspring. Perhaps one day we might even be able to clone ourselves. Should we do these things? So far, as a society, we continue to believe in the sacredness of human life, and therefore we have passed laws against doing research with fetal tissues or using fetal tissues to cure human ills. Even if the procedure for human cloning is perfected, we may also continue to rule against its use. Each of us must wrestle with this and other bioethical issues, and make decisions that we hope are beneficial to society.

THE CHAPTER IN REVIEW

Summary

1.1   The Unity and Diversity of Life

Living things, often called organisms, share several common characteristics. Organisms

- are organized (have levels of organization).
- acquire materials and energy.
- respond to external stimuli.
- reproduce and develop.
- have adaptations.

Evolution explains the unity and diversity of life. Descent from a common ancestor explains why organisms share some characteristics, and adaptation to various ways of life explains the diversity of life-forms.

1.2   How the Biosphere Is Organized

Populations within a community interact with one another and with the physical environment, forming an ecosystem. With an ecosystem, nutrients cycle, but energy flows unidirectionally and eventually becomes heat.

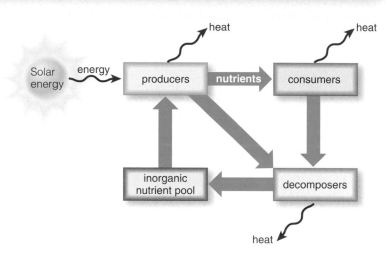

1.3   How Organisms Are Classified

According to the rules of taxonomy, species belong to the following categories (from the most inclusive to the least inclusive):

Domains
- Archaea: prokaryotes
- Bacteria: prokaryotes
- Eukarya: eukaryotes

Kingdoms in domain Eukarya:
- Protista: unicellular to multicellular organisms with various modes of nutrition
- Fungi: live on debris; molds and mushrooms
- Plantae: multicellular photosynthesizers
- Animalia: multicellular organisms that ingest food

To classify a particular organism, scientists use a consistent sequence of categories. A binomial name consists of the genus and the specific epithet. For example, here is how the names *Homo sapiens* and *Zea mays* are derived:

LEVELS OF CLASSIFICATION		
Category	**Human**	**Corn**
Domain	Eukarya	Eukarya
Kingdom	Animalia	Plantae
Phylum	Chordata	Anthophyta
Class	Mammalia	Liliopsida
Order	Primates	Commelinales
Family	Hominidae	Poaceae
Genus	Homo	Zea
*Species**	H. sapiens	Z. mays

* To specify an organism, you must use the full binomial name, such as *Homo sapiens*.

1.4   Science As a Way of Knowing

The scientific process includes a series of systemic steps known as the scientific method:

- Observations, which use the senses and may also include studies done by others
- A hypothesis (a statement to be tested)
- New observations and experiments
- A conclusion reached by analyzing data to determine whether the results support or do not support the hypothesis.

A hypothesis confirmed by many different studies becomes known as a theory. Examples are the cell theory, the gene theory, and the theory of evolution, which is the unifying concept of biology.

1.5   Science and Society

Scientific findings often lead to the development of a technology that can be of service to human beings. Technologies have both benefits and drawbacks. Every member of society needs to be prepared to participate in deciding whether, and how, a technology should be used.

Thinking Scientifically

1. While shopping for fertilizer for the tomato plants in your garden, you have a choice between a name-brand and a generic brand. The labels indicate that they are identical in composition, but the advertising for the name-brand fertilizer claims that it is superior to all other fertilizers. How would you set up an experiment to test that claim?

2. Two peach varieties (Biscoe and Encore) were treated with a plant growth regulator that influences fruit softening. The three treatment application dates were 21, 14, and 7 days before harvest. Which variety showed effects that increased consistently from 21 to 7 days before harvest? Did the growth regulator work equally effectively in both cultivars? Which treatment was most effective?

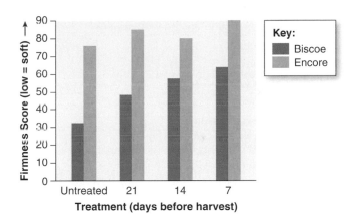

Testing Yourself

Choose the best answer for each question.

1. The smallest living unit is a(n)

   a. atom.   c. cell.
   b. molecule.   d. tissue.

2. In the following diagram, label the levels of biological organization:

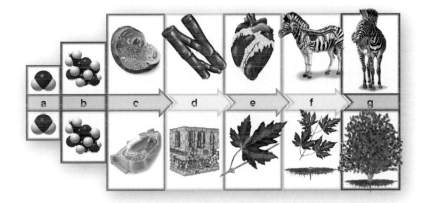

3. All of the chemical reactions that occur in a cell are called
   - a. homeostasis.
   - b. metabolism.
   - c. heterostasis.
   - d. cytoplasm.

4. The process of turning solar energy into chemical energy is called
   - a. work.
   - b. metabolism.
   - c. photosynthesis.
   - d. respiration.

For questions 5–9, choose one answer from the following key.

Key:
   - a. unicellular organism
   - b. multicellular organism
   - c. neither
   - d. both

5. Capable of responding to the environment.

6. Reproduces by uniting an egg with a sperm.

7. Contains tissues.

8. Contains genes.

9. Creates energy.

10. In an ecosystem,
    - a. energy flows and nutrients cycle.
    - b. energy cycles and nutrients flow.
    - c. energy and nutrients flow.
    - d. energy and nutrients cycle.

11. Taxonomy is a biological discipline in which organisms are grouped according to their
    - a. geographic location.
    - b. size.
    - c. common ancestries.
    - d. gene number.

12. The first cells to evolve in the primitive Earth were most likely members of the domain
    - a. Archaea.
    - b. Bacteria.
    - c. Eukarya.

13. Choose the correct way to represent the scientific name of corn.
    - a. Zea Mays
    - b. zea mays
    - c. Zea mays
    - d. zea Mays
    - e. *Zea Mays*
    - f. *zea mays*
    - g. *Zea mays*
    - h. *zea Mays*

Go to www.mhhe.com/maderessentials for more quiz questions.

Bioethical Issue

Clinical trials on drug efficacy are typically carried out on two groups of patients. One group receives the drug, while the other (the control group) receives a placebo. The control group determines the extent to which psychology affects people who believe they are taking the new drug.

Suppose a new cancer drug has been developed. Many people sign up to participate in a clinical trial to test the drug's ability to slow the progression of the disease. These people know that they may receive a placebo in place of the real drug, but they participate because they have run out of options. Is it right to give the placebo to some of the participants, knowing that their disease is going to progress during the course of the trial? What if the drug was one that alleviated pain in chronic pain sufferers? In this case, no disease progression would occur during the course of the trial, but participants taking the placebo would have to contend with serious pain.

Understanding the Terms

adaptation 4	homeostasis 3
animal 6	hypothesis 8
binomial name 7	inductive reasoning 8
biodiversity 7	kingdom 6
biology 8	metabolism 3
biosphere 5	model 8
cell 2	multicellular 2
class 6	natural selection 4
community 5	observation 8
conclusion 9	order 6
control group 8	organ 2
data 9	organism 2
domain 6	organ system 2
domain Archaea 6	photosynthesis 3
domain Bacteria 6	phylum 6
domain Eukarya 6	plant 6
ecosystem 5	population 5
energy 3	principle 9
evolution 4	protist 6
experimental design 8	reproduce 3
experimental variable 8	scientific theory 9
extinction 7	species 4, 6
family 6	taxonomy 6
fungus 6	tissue 2
gene 4	unicellular 2
genus 6	

Match the terms to these definitions:

a. _____ A scientist's suggested explanation for a natural event.

b. _____ A representation of a natural phenomenon studied by scientists when the real object is impossible to study.

c. _____ A modification that helps equip organisms for their way of life.

d. _____ A group of interbreeding individuals.

e. _____ The discipline of classifying organisms according to certain rules.

f. _____ The region of the Earth's surface where all organisms are found.

g. _____ Information collected from a scientific experiment.

The Chemical Basis of Life

CHAPTER

2

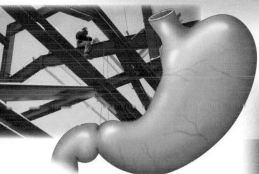

The pH of your stomach is acidic enough to dissolve steel.

The chemical composition of human blood closely resembles that of seawater.

UV radiation can cause skin cancer.

OUTLINE

2.1 The Nature of Matter

- Matter is composed of 92 naturally occurring elements, each composed of atoms. 16
- Atoms have subatomic particles: neutrons, protons, and electrons. 17
- Atoms of the same type that differ by the number of neutrons are called isotopes. 18
- Atoms react with one another by giving up, gaining, or sharing electrons. 19
- Bonding between atoms results in molecules with distinctive chemical properties and shapes. 21

2.2 Water's Importance to Life

- The existence of organisms is dependent on the chemical and physical properties of water. 22
- Organisms are sensitive to the hydrogen ion concentration [H^+] of solutions, which can be indicated using the pH scale. 25

Radiation causes cancer, but it can also be used to cure cancer.

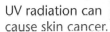

You are probably aware that the effects of exposure to extreme levels of radiation in humans can range from sunburns, to hair loss, to cancer, and maybe even to death. What is it about exposure to radioactivity that causes such damage to living cells? The atoms that make up all of the molecules in living cells are composed of subatomic particles. When we refer to a "radioactive atom," what we really mean is that the atom is not stable and has a tendency to release subatomic particles. If subatomic particles are released from an atom, they can cause major damage. When organisms are exposed to radioactivity, loose subatomic particles penetrate their cells and cause cellular molecules to form *free radicals*. These free radicals damage other molecules such as DNA within the cell. When DNA is damaged, a mutation results. Cancer can be the result of mutations; hence, exposure to enough radioactivity often causes cancer in humans. A high enough level of radiation causes so much damage to cells that they (and the individual) die. The cells most susceptible to free radical damage are those that naturally divide the fastest, such as skin, hair, and digestive cells.

Chemistry is the very essence of all organisms. You may have never thought of your body as simply a pile of chemicals, or considered how the chemistry of your body changes when you eat or drink certain foods, or realized the changes that occur in cells when they are exposed to radioactivity. Understanding these basic principles of chemistry will greatly enhance your ability to understand biology and provide a foundation on which to build further knowledge.

2.1 The Nature of Matter

Kiss your sweetheart, pat your dog, catch a bus, mow your lawn—everything you touch is matter. **Matter** refers to anything that takes up space and has mass. It is helpful to know that matter can exist as a solid, a liquid, or a gas. Then we can realize that not only are we matter, but so are the water we drink and the air we breathe.

All matter, both nonliving and living, is composed of elements. Formally speaking, an **element** is a substance that cannot be broken down into another substance by ordinary chemical means. There are only 92 naturally occurring elements, and each of these differs from the others in its properties. (A *property* is a physical or chemical characteristic, such as density, solubility, melting point, and reactivity.)

Both the Earth's crust and all organisms are made up of elements, but these elements occur in different proportions (Fig. 2.1). Six elements—carbon, hydrogen, nitrogen, oxygen, phosphorus, and sulfur—are of special significance to us. They make up about 98% of the body weight of most organisms. The acronym CHNOPS helps us remember these six elements, whose properties are essential to the uniqueness of living things, from cells to organisms.

a.

carbon (C) 18%
oxygen (O) 65%
hydrogen (H) 10%
nitrogen (N) 5%
calcium (Ca) 2%
phosphorus (P) 1.1%
other elements, including sulfur 0.9%

b.

aluminum (Al) 8.1%
iron (Fe) 5%
calcium (Ca) 3.6%
sodium (Na) 2.8%
potassium (K) 2.6%
magnesium (Mg) 2.1%
other elements 1.5%
silicon (Si) 27.7%
oxygen (O) 46.6%

Figure 2.1 **Elements in the Earth's crust and living organisms.**

If analyzed at the level of atoms, the Earth's crust has a very different composition from human beings. **a.** By weight, human beings are mostly composed of oxygen, carbon, hydrogen, nitrogen, calcium, and phosphorus. **b.** The Earth's crust, containing rock, soil, sand, and other materials, is mostly oxygen, silicon, and aluminum atoms.

Atomic Structure

The *atomic theory* states that elements consist of tiny particles called **atoms.** Because each element consists of only one kind of atom, the same name is given to an element and its atoms. This name is represented by one or two letters, called the **atomic symbol.** For example, the symbol H stands for a hydrogen atom, and the symbol Na (for *natrium* in Latin) stands for a sodium atom.

From our discussion of elements, you might expect each atom to have a certain mass. The **mass number** of an atom is dependent upon the presence of three types of subatomic particles: **neutrons,** which have no electrical charge; **protons,** which have a positive charge; and **electrons,** which have a negative charge. Protons and neutrons are located within the center of an atom, which is called its **nucleus,** while electrons move about the nucleus.

Figure 2.2 shows the arrangement of the subatomic particles in a helium atom, which has only two electrons. In Figure 2.2*a,* the stippling shows the probable location of electrons, and in Figure 2.2*b,* the circle represents the approximate location of electrons. If we could draw an atom the size of a baseball stadium, the nucleus would be like a gumball in the center of the stadium, and the electrons would be tiny specks whirling about in the upper stands. Most of an atom is empty space. Usually, we can only indicate where the electrons are expected to be. In our analogy, the electrons might very well stray outside the stadium at times.

In effect, the mass number of an atom is just about equal to the sum of its protons and neutrons. Protons and neutrons are assigned one atomic mass unit each. Electrons are so small that their mass is assumed to be zero in most calculations. The term mass number is used, rather than atomic weight because mass is constant while weight changes according to the gravitational force of a planet. The gravitational force of the Earth is greater than that of the moon; therefore, substances weigh less on the moon even though their mass has not changed.

All atoms of an element have the same number of protons. This is called the atom's **atomic number.** The number of protons makes an atom unique. In Figure 2.3, the atomic number is written above the atomic symbol. The mass number is written below the atomic symbol. For example, the carbon atom is shown in this way:

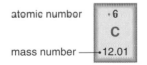

atomic number ⎯ ·6
C
mass number ⎯ ·12.01

The Periodic Table

Once chemists discovered a number of the elements, they began to realize that the elements' chemical and physical characteristics recur in a predictable manner. The periodic table (Fig. 2.3) was developed as a way to display the elements, and therefore the atoms, according to these characteristics. Every atom is in a particular period (the horizontal rows) and in a particular group (the vertical columns). The atoms in group 8 are called the *noble gases* because they are gases that rarely react with another atom. Notice that helium is a noble gas.

The atomic number written above the atomic symbol tells you the number of positively charged protons, and also the number of negatively charged electrons if the atom is electrically neutral. To determine the usual number of neutrons, subtract the number of protons from the mass number, written below the atomic symbol, and take the closest whole number.

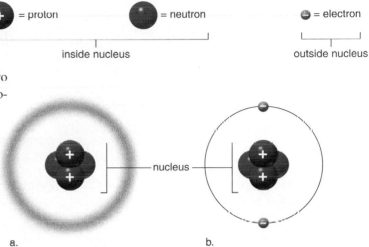

= proton   = neutron   = electron

inside nucleus   outside nucleus

nucleus

a.   b.

Figure 2.2 **Two models of helium (He).**

Atoms contain subatomic particles, which are located as shown in these two simplified models of helium. Protons are positively charged, neutrons have no charge, and electrons are negatively charged. Protons and neutrons are found within the nucleus, and electrons are outside the nucleus. **a.** This model shows electrons as a negatively charged cloud around the nucleus. **b.** In this model, the average location of electrons is sometimes represented by a circle.

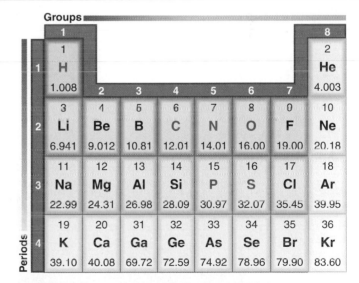

Figure 2.3 **A portion of the periodic table.**

In the periodic table, the elements, and therefore the atoms that compose them, are in the order of their atomic numbers but arranged in periods (horizontal rows) and groups (vertical columns). All the atoms in a particular group have certain chemical characteristics in common. The six elements previously mentioned (CHNOPS) are highlighted in red.

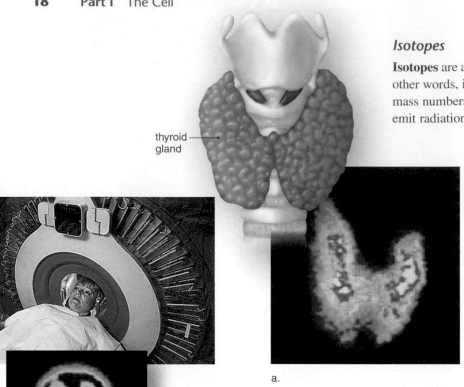

thyroid gland

a.

Figure 2.4   PET scans.

In a PET scan, red indicates areas of greatest metabolic activity, and blue means areas of least activity. Computers analyze the data from different sections of an organ. **a.** A longitudinal scan of the thyroid resembles its overall shape. **b.** A cross section of the brain is a powerful diagnostic tool for the physician.

b.

a.   b.

Figure 2.5   High levels of radiation.

a. Radiation kills microbes. After irradiation, peaches (bottom) no longer spoil and can be kept for a longer length of time.
b. Physicians treat patients with high levels of radioactive isotopes to kill cancer cells.

Isotopes

Isotopes are atoms of the same element that differ in the number of neutrons. In other words, isotopes have the same number of protons, but they have different mass numbers. A nucleus with excess neutrons is unstable and may decay and emit radiation. The radiation given off by radioactive isotopes can be detected in various ways. Most people are familiar with the use of a Geiger counter to detect radiation.

Uses of Radioactive Isotopes   The importance of chemistry to biology and medicine is nowhere more evident than in the many uses of radioactive isotopes. The chemical behavior of a radioactive isotope is essentially the same as that of the stable isotopes of an element. This means that you can put a small amount of radioactive isotope in a sample and it becomes a **tracer** by which to detect molecular changes.

Specific tracers are used in imaging the body's organs and tissues. For example, positron-emission tomography (PET) is a way to determine the comparative activity of tissues. Radioactively labeled glucose that emits a subatomic particle known as a positron is injected into the body. The radiation given off is detected by sensors and analyzed by a computer. The result is a color image that shows which tissues took up glucose and are metabolically active (Fig. 2.4a,b). Although not shown in Figure 2.4, a PET scan can help diagnose a malfunctioning thyroid, a brain tumor, Alzheimer disease, epilepsy, or whether a stroke has occurred.

Radioactive substances in the environment can harm cells, damage DNA, and cause cancer. The release of radioactive particles following a nuclear power plant accident can have far-reaching and long-lasting effects on human health. However, the effects of radiation can also be put to good use (Fig. 2.5). The ability of radiation to kill cells is often used to increase shelf life of fruits (Fig. 2.5a) and to destroy cancer cells (Fig. 2.5b). Packets of radioactive isotopes can be placed in the body so that the subatomic particles emitted destroy only cancer cells, with little risk to the rest of the body.

Radiation from radioactive isotopes has been used for many years to sterilize medical and dental equipment. Now the possibility exists that it can also be used to sterilize the U.S. mail to free it of possible pathogens, such as anthrax spores introduced by terrorists.

Arrangement of Electrons in an Atom

Although it is not possible to determine the precise location of an individual electron at any given moment, it is useful to construct models of atoms that show electrons at discrete energy levels about the nucleus (Fig. 2.6). It seems reasonable to suggest that negatively charged electrons are attracted to the positively charged nucleus, and therefore it takes an increasing amount of energy to push them farther away from the nucleus. Electrons in outer shells, therefore, contain more energy than those in inner shells.

Each energy level contains a certain number of electrons. In the models shown in Figure 2.6, the energy levels (**electron shells**) are drawn as concentric rings about the nucleus. For atoms up through number 20 (i.e., calcium), the first shell closest to the nucleus can contain two electrons; thereafter, each additional

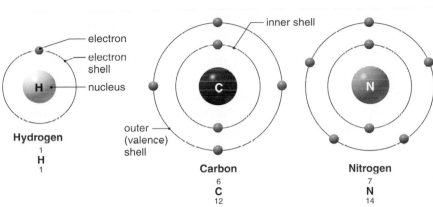

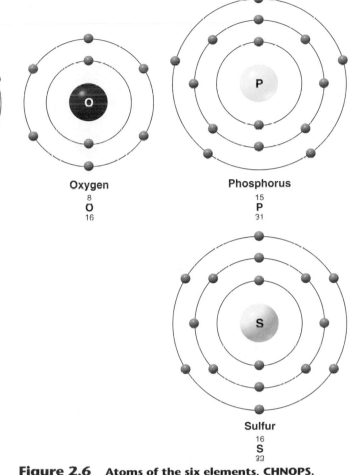

Figure 2.6 Atoms of the six elements, CHNOPS.

Electrons orbit the nucleus at particular energy levels (electron shells): The first shell contains up to two electrons, and each shell thereafter can contain up to eight electrons (until we consider atoms with an atomic number above 20). Each shell is to be filled before electrons are placed in the next shell. Why does carbon have only two shells while phosphorus and sulfur have three shells?

shell can contain eight electrons. For these atoms, each lower level is filled with electrons before the next higher level contains any electrons.

The sulfur atom, with an atomic number of 16, has two electrons in the first shell, eight electrons in the second shell, and six electrons in the third, or outer, shell. Revisit the periodic table (see Fig. 2.3), and note that sulfur is in the third period. In other words, the period tells you how many shells an atom has. Also note that sulfur is in group 6. The group tells you how many electrons an atom has in its outer shell.

If an atom has only one shell, the outer shell is complete when it has two electrons. If an atom has two or more shells, the outer shell is most stable when it has eight electrons; this is called the **octet rule.** As mentioned previously, atoms in group 8 of the periodic table are called the noble gases because they do not ordinarily undergo reactions. Atoms with fewer than eight electrons in the outer shell react with other atoms in such a way that each has a completed outer shell after the reaction. Atoms can give up, accept, or share electrons in order to have eight electrons in the outer shell. In other words, the number of electrons in an atom's outer shell, called the **valence shell,** determines its chemical reactivity. The size of an atom is also important. Both carbon (C) and silicon (Si) atoms are in group 4, and therefore can bond with four other atoms in order to achieve eight electrons in their outer shells. But carbon is in period 2 and silicon is in period 3. The smaller atom, carbon, can bond to other carbon atoms and form long-chained molecules, while the larger silicon atom is unable to do so.

Types of Chemical Bonds

A group of atoms bonded together is called a **molecule.** When a molecule contains atoms of more than one element, it can be called a **compound.** For example, sodium (Na) can combine with chlorine (Cl) to form sodium chloride (NaCl), the compound we know as table salt.

Check Your Progress

1. Contrast mass number with atomic number.
2. List some uses of radioactive isotopes in biology and medicine.
3. How do group 3 elements differ in the periodic table? How do period 3 elements differ?

Answers: 1. Mass number is approximately the sum of the protons and neutrons in an atom. Atomic number is the number of protons in an atom. 2. Uses of radioactive isotopes include imaging of body parts (e.g., PET scans, thyroid imaging), sterilization of medical equipment, cancer therapy, and increased storage life of produce. 3. The third horizontal row in the periodic table contains atoms that sequentially increase by one in the number of electrons in the outer shell. Group 3 elements (the third vertical column in the table) contain atoms that all have three electrons in the outer shell.

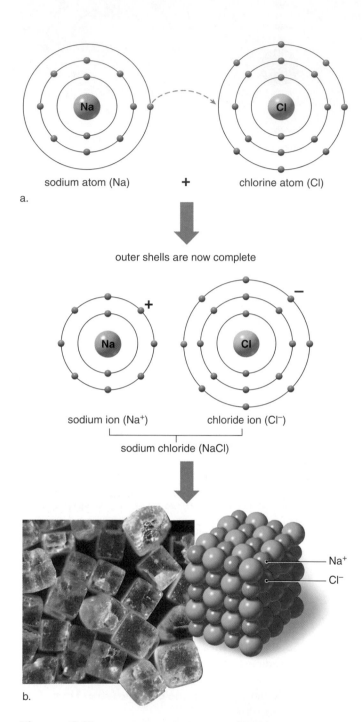

sodium atom (Na)   **+**   chlorine atom (Cl)

a.

outer shells are now complete

sodium ion (Na⁺)   chloride ion (Cl⁻)

sodium chloride (NaCl)

Na⁺

Cl⁻

b.

Figure 2.7   **Formation of sodium chloride.**

a. During the formation of sodium chloride, an electron is transferred from the sodium atom to the chlorine atom. At the completion of the reaction, each atom has eight electrons in the outer shell, but each also carries a charge as shown. **b.** In a sodium chloride crystal, ionic bonding between Na⁺ and Cl⁻ causes the atoms to assume a three-dimensional lattice shape.

Ionic Bonding

The reaction between sodium and chlorine atoms is an example of how charged atoms called **ions** form. Consider that sodium (Na), with only one electron in its third shell, usually gives up an electron (Fig. 2.7a). Once it does so, the second shell, with eight electrons, becomes its outer shell. Chlorine (Cl), on the other hand, tends to take on an electron, because its outer shell only has seven electrons. If chlorine gets one more electron it has a completed outer shell. So, when a sodium atom and a chlorine atom react, an electron is transferred from sodium to chlorine. Now both atoms have eight electrons in their outer shells.

This electron transfer causes these atoms to become ions. The sodium ion has one more proton than it has electrons; therefore, it has a net charge of $+1$ (symbolized by Na^+). The chloride ion has one more electron than it has protons; therefore, it has a net charge of -1 (symbolized by Cl^-). Negatively charged ions often have names that end in "ide," and thus Cl^- is called a chloride ion. In the periodic table, atoms in groups 1 and 2 and groups 6 and 7 become ions when they react with other atoms. Atoms in groups 2 and 6 always transfer two electrons. For example, calcium becomes Ca^{2+}, while oxygen becomes O^{2-}.

Ionic compounds are held together by an attraction between negatively and positively charged ions, called an **ionic bond.** A sodium chloride crystal illustrates the solid form of a salt (Fig. 2.7b). **Salts** can exist as a dry solid, but when placed in water, they release ions as they dissolve. NaCl separates into Na^+ and Cl^-. Ionic compounds are most commonly found in this dissociated (ionized) form in biological systems because these systems are 70–90% water.

Covalent Bonding

A **covalent bond** results when two atoms share electrons in order to have a completed outer shell. In a hydrogen atom, the outer shell is complete when it contains two electrons. If hydrogen is in the presence of a strong electron acceptor, it gives up its electron to become a hydrogen ion (H^+). But if this is not possible, hydrogen can share with another atom, and thereby have a completed outer shell. For example, one hydrogen atom can share with another hydrogen atom. In this case, the two orbitals overlap and the electrons are shared between them—that is, you count the electrons as belonging to both atoms:

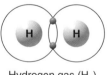

Hydrogen gas (H₂)

Rather than drawing an orbital model like the above, scientists often use simpler ways to indicate molecules. A *structural formula* uses straight lines, as in H—H. A *molecular formula* omits the lines that indicate bonds and simply shows the number of atoms involved, as in H_2.

Sometimes, atoms share more than two electrons to complete their octets. A double covalent bond occurs when two atoms share two pairs of electrons, as in this molecule of oxygen gas:

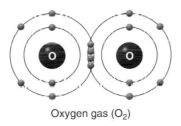

Oxygen gas (O₂)

In order to show that oxygen gas (O₂) contains a double bond, the structural formula would be written as O=O.

It is also possible for atoms to form triple covalent bonds, as in nitrogen gas (N₂), which can be written as N≡N. Single covalent bonds between atoms are quite strong, but double and triple bonds are even stronger.

The molecule methane results when carbon binds to four other atoms (Fig. 2.8). In methane, each bond actually points to one corner of a tetrahedron. The best model to show this arrangement is a ball-and-stick model (Fig. 2.8c). Space-filling models (Fig. 2.8d) come closest to showing the actual shape of a molecule. The shapes of molecules help dictate the functional roles they play in organisms.

Chemical Reactions

Chemical reactions are very important to organisms. For example, we have already noted that the process of photosynthesis enables plants to make molecular energy available to themselves and other organisms. An overall equation for the photosynthetic reaction indicates that some bonds are broken and others are formed:

$6\ CO_2$	+	$6\ H_2O$	$\longrightarrow$	$C_6H_{12}O_6$	+	$6\ O_2$
carbon dioxide		water		glucose		oxygen

This equation says that six molecules of carbon dioxide react with six molecules of water to form one glucose molecule and six molecules of oxygen. The **reactants** (molecules that participate in the reaction) are shown on the left of the arrow, and the **products** (molecules formed by the reaction) are shown on the right. Notice that the equation is "balanced"—that is, the same number of each type of atom occurs on both sides of the arrow.

Note the glucose molecule in the equation above. Glucose is more complex than any molecule discussed so far. The arrangement of its atoms is shown below:

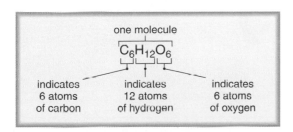

Methane (CH₄)

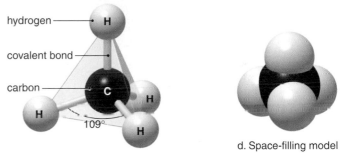

b. Structural model

a. Electron model showing covalent bonds

c. Ball-and-stick model

d. Space-filling model

Figure 2.8 Shapes of covalently bonded molecules.

An electron model (**a**) and a structural model (**b**) show that methane (CH₄) contains one carbon atom bonded to four hydrogen atoms. **c.** The ball-and-stick model shows that when carbon bonds to four other atoms, as in methane, each bond actually points to one corner of a tetrahedron. **d.** The space-filling model is a three-dimensional representation of the molecule.

Check Your Progress

1. Contrast an ionic bond with a covalent bond.
2. Methane is formed by covalent bonding between carbon and hydrogen. How many hydrogen atoms would be bonded to the carbon atom?

Answers: 1. An ionic bond is created when one atom gives up electron(s) and another gains electron(s) so that both atoms have their outer shells filled. A covalent bond is formed when two atoms share electrons to fill their outer shell. 2. Carbon requires four electrons to fill its outer shell, so it bonds with four hydrogen atoms.

2.2 Water's Importance to Life

Life began in water, and it is the single most important molecule on Earth. All organisms are 70–90% water; their cells consist of membranous compartments enclosing aqueous solutions. Water has unique properties that make it a life-supporting substance. These properties of water stem from the structure of the molecule.

The Structure of Water

Atoms differ in their **electronegativity**—that is, their affinity for electrons in a covalent bond. For example, in water, oxygen shares electrons with two hydrogen atoms. The covalent bonds are angled, and the molecule is bent roughly into a Λ shape because the oxygen is more electronegative than the hydrogens. The shared electrons spend more time orbiting the oxygen nucleus than the hydrogen nuclei, and this unequal sharing of electrons makes water a **polar** molecule. The point of the Λ (oxygen) is the negative (−) end, and the two hydrogens are the positive (+) end (Fig. 2.9a).

The polarity of water molecules causes them to be attracted to one another. The positive hydrogen atoms in one molecule are attracted to the negative oxygen atoms in other water molecules. This attraction is called a **hydrogen (H) bond,** and each water molecule can engage in as many as four H bonds (Fig. 2.9b). The covalent bond is much stronger than an H bond, but the sheer number of H bonds in water makes the H bond strong overall. The properties of water are due to its polarity and its ability to form hydrogen bonds.

Properties of Water

Water is so familiar that we take it for granted, but without water, life as we know it would not exist. The properties of water that support life are solvency, cohesion and adhesion, high surface tension, high heat capacity, high heat of vaporization, and varying density.

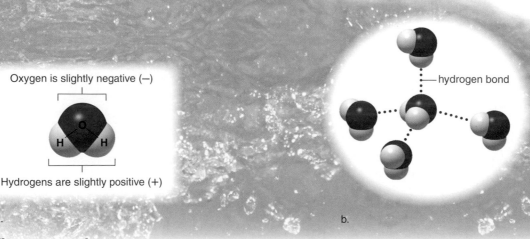

Oxygen is slightly negative (−)

Hydrogens are slightly positive (+)

a.

hydrogen bond

b.

Figure 2.9   The structure of water.

a. The space-filling model shows the Λ shape of a water molecule. Oxygen attracts the shared electrons more than hydrogen atoms do, and this causes the molecule to be polar: The oxygen carries a slight negative charge and the hydrogens carry a slight positive charge. **b.** The positive hydrogens form hydrogen bonds with the negative oxygen in nearby molecules. Each water molecule can be joined to other water molecules by as many as four hydrogen bonds.

Water is a solvent. Because of its polarity and H-bonding ability, water dissolves a great number of substances. Molecules that are attracted to water are said to be **hydrophilic** (*hydro*, water; *phil*, love). Nonionized and nonpolar molecules that are not attracted to water are said to be **hydrophobic** (*hydro*, water; *phob*, fear).

When a salt such as sodium chloride (NaCl) is put into water, the negative ends of the water molecules are attracted to the sodium ions, and the positive ends of the water molecules are attracted to the chloride ions. This attraction causes the sodium ions and the chloride ions to break up, or dissociate, in water:

The salt NaCl dissociates in water.

Water dissolves many polar nonionic substances, such as glucose, by forming H bonds with them. When ions and molecules disperse in water, they move about and collide, allowing reactions to occur. As mentioned, cellular fluids are aqueous solutions of various substances, and so are the oceans. Without the dissolving power of water, aquatic organisms could not take up the substances they need from the water.

Water is cohesive and adhesive. Because of hydrogen bonding, water molecules cling together, and water exists as a liquid under ordinary conditions of temperature and pressure. The strong *cohesion* of water molecules is apparent because water flows freely, and yet the molecules do not separate from each other. The positive and negative poles of water molecules cause water to adhere to polar surfaces; therefore, water exhibits *adhesion*.

Liquid water is an excellent transport system, both outside and within living organisms. Unicellular organisms rely on external water to transport nutrient and waste molecules, but multicellular organisms often contain internal vessels in which water transports nutrients and wastes. For example, the liquid portion of our blood, which transports dissolved and suspended substances about the body, is 90% water. Cohesion and adhesion allow water to fill a tubular vessel. In plants, because of cohesion and adhesion, water is able to rise to the top of even very tall trees (Fig. 2.10). Water evaporating from the leaves creates a tension that pulls a continuous water column up from the roots to the leaves of the plant.

Water has a high surface tension. The stronger the force between molecules in a liquid, the greater the surface tension. As with cohesion, hydrogen bonding causes water to have a high surface tension. This property makes it possible for humans to skip rocks on water. An insect called a water strider can even walk on the surface of a pond without breaking through (Fig. 2.11). The high surface tension of water is important to other forms of surface aquatic life as well. For example, it keeps plant debris resting on the surface, providing shelter and food for those organisms that live near the surface.

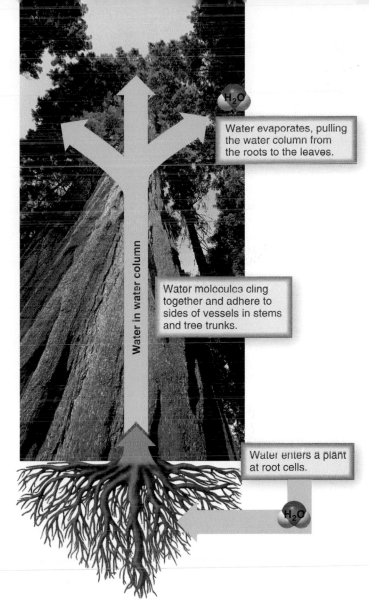

Water evaporates, pulling the water column from the roots to the leaves.

Water molecules cling together and adhere to sides of vessels in stems and tree trunks.

Water enters a plant at root cells.

Figure 2.10   Cohesion and adhesion of water molecules.

How does water rise to the top of tall trees? Water-filled vessels extend from the roots to the leaves. When water evaporates from the leaves, this water column is pulled upward due to the cohesion of water molecules with one another and the adhesion of water molecules to the sides of the vessel.

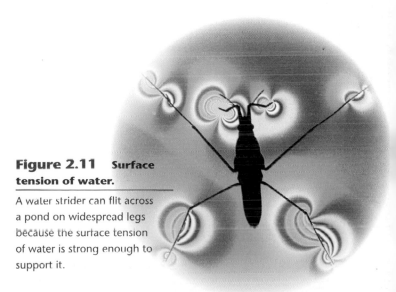

Figure 2.11   Surface tension of water.

A water strider can flit across a pond on widespread legs because the surface tension of water is strong enough to support it.

Figure 2.12 **Heat of vaporization.**

At room temperature, water is a liquid. **a.** Water boils at 100°C, vaporizes, and takes up a large amount of heat. **b.** It takes much body heat to vaporize sweat, which is mostly liquid water, and this helps keep our bodies cool when the temperature rises.

a.

b.

Water has a high heat capacity.   The many hydrogen bonds that link water molecules allow water to absorb heat without greatly changing in temperature. Water's high heat capacity is important not only for aquatic organisms but for all organisms. Because the temperature of water rises and falls slowly, terrestrial organisms are better able to maintain their normal internal temperatures and are also protected from rapid temperature changes.

Water also has a high heat of vaporization: It takes a great deal of heat to break the hydrogen bonds in water so that it becomes gaseous and evaporates into the environment. If the heat given off by our metabolic activities were to go directly into raising our body temperature, death would follow. Instead, the heat is dispelled as sweat evaporates (Fig. 2.12).

Because of water's high heat capacity and high heat of vaporization, temperatures along the Earth's coasts are moderate. During the summer, the ocean absorbs and stores solar heat, and during the winter, the ocean releases it slowly. In contrast, the interior regions of the continents experience abrupt changes in temperature.

Water is less dense as ice.   Unlike other substances, water expands as it freezes, which explains why cans of soda burst when placed in a freezer and how roads in northern climates become bumpy because of "frost heaves" in the winter. It also means that ice is less dense than liquid water, and therefore ice floats on liquid water (Fig. 2.13).

If ice did not float on water, it would sink, and ponds, lakes, and perhaps even the ocean would freeze solid, making life impossible in the water and also on land. Instead, bodies of water always freeze from the top down. When a body of water freezes on the surface, the ice acts as an insulator to prevent the water

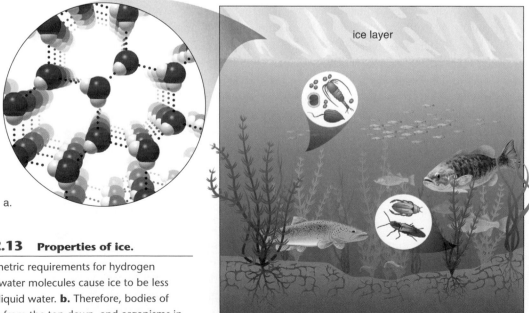

a.

ice layer

b.

Figure 2.13   **Properties of ice.**

a. The geometric requirements for hydrogen bonding of water molecules cause ice to be less dense than liquid water. **b.** Therefore, bodies of water freeze from the top down, and organisms in ponds and lakes are protected during the winter.

below it from freezing. This protects aquatic organisms so that they can survive the winter. As ice melts in the spring, it draws heat from the environment, helping to prevent a sudden change in temperature that might be harmful to life.

Acids and Bases

Figure 2.14 shows that when water dissociates, it releases an equal number of **hydrogen ions (H^+)** and **hydroxide ions (OH^-)** as in the following reaction:

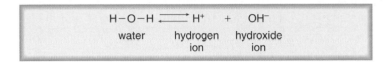

Acidic Solutions (High H⁺ Concentration)

Lemon juice, vinegar, tomatoes, and coffee are all acidic solutions. What do they have in common? Acidic solutions have a sharp or sour taste, and therefore we sometimes associate them with indigestion. To a chemist, **acids** are substances that dissociate in water, releasing hydrogen ions (H^+). For example, an important acid is hydrochloric acid (HCl), which dissociates in this manner:

$$HCl \longrightarrow H^+ + Cl^-$$

Dissociation is almost complete; therefore, HCl is called a strong acid. If hydrochloric acid is added to a beaker of water, the number of hydrogen ions (H^+) increases greatly (Fig. 2.15).

Basic Solutions (Low H⁺ Concentration)

Milk of magnesia and ammonia are common basic solutions that most people have heard of. Basic solutions have a bitter taste and feel slippery when in water. To a chemist, **bases** are substances that either take up hydrogen ions (H^+) or release hydroxide ions (OH^-). For example, an important base is sodium hydroxide (NaOH), which dissociates in this manner:

$$NaOH \longrightarrow Na^+ + OH^-$$

Dissociation is almost complete; therefore, sodium hydroxide is called a strong base. If sodium hydroxide is added to a beaker of water, the number of hydroxide ions increases (Fig. 2.16).

It is not recommended that you taste any strong acid or base, because they are quite destructive to cells. Any container of household cleanser, such as ammonia, has a poison symbol and carries a strong warning not to ingest the product.

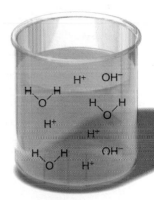

Figure 2.14   Dissociation of water molecules.

Dissociation produces an equal number of hydrogen ions (H^+) and hydroxide ions (OH^-). (These illustrations are not meant to be mathematically accurate.)

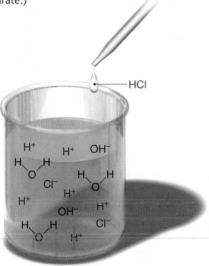

Figure 2.15   Addition of hydrochloric acid (HCl).

HCl releases hydrogen ions (H^+) as it dissociates. The addition of HCl to water results in a solution with more H^+ than OH^-.

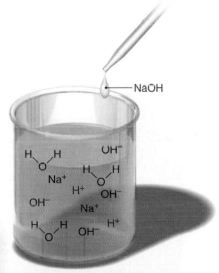

Figure 2.16   Addition of sodium hydroxide (NaOH), a base.

NaOH releases OH^- as it dissociates. The addition of NaOH to water results in a solution with more OH^- than H^+.

Check Your Progress

Contrast an acid with a base.

Answer: An acid dissociates in water to release hydrogen ions. A base either takes up hydrogen ions or releases hydroxide ions.

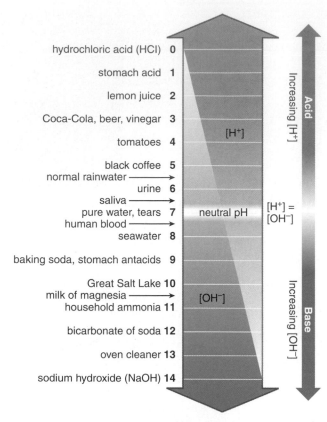

Figure 2.17   **The pH scale.**

The proportionate amount of hydrogen ions to hydroxide ions is indicated by the diagonal line. Any solution with a pH above 7 is basic, while any solution with a pH below 7 is acidic.

a.

Figure 2.18   **Effects of acid deposition.**

The burning of gasoline derived from oil, a fossil fuel, leads to acid deposition, which causes (**a**) trees to die and (**b**) statues to deteriorate.

b.

pH and the pH Scale

pH[1] is a mathematical way of indicating the number of hydrogen ions in a solution. The **pH scale** is used to indicate the acidity or basicity (alkalinity) of a solution. The pH scale ranges from 0 to 14 (Fig. 2.17). A pH of 7 represents a neutral state in which the hydrogen ion and hydroxide ion concentrations are equal, as in pure water. A pH below 7 is acidic because the hydrogen ion concentration, commonly expressed in brackets as $[H^+]$, is greater than the hydroxide concentration, $[OH^-]$. A pH above 7 is basic because $[OH^-]$ is greater than $[H^+]$. Further, as we move down the pH scale from pH 7 to pH 0, each unit has 10 times the acidity $[H^+]$ of the previous unit. As we move up the scale from 7 to 14, each unit has 10 times the basicity $[OH^-]$ of the previous unit.

The pH scale was devised to eliminate the use of cumbersome numbers. For example, the hydrogen ion concentrations of these solutions are on the left, and the pH is on the right:

$[H^+]$ (moles per liter)		pH
0.000001	$= 1 \times 10^{-6}$	6 (acid)
0.0000001	$= 1 \times 10^{-7}$	7 (neutral)
0.00000001	$= 1 \times 10^{-8}$	8 (base)

The effect of pH on organisms is dramatically illustrated by the phenomenon known as acid deposition. When fossil fuels are burned, sulfur dioxide and nitrogen oxides are produced, and they combine with water in the atmosphere to form acids. These acids then come in contact with organisms and objects, leading to damage or even death (Fig. 2.18).

Buffers and pH

In the human body, pH needs to be kept within a narrow range in order to maintain health. A **buffer** is a chemical or a combination of chemicals that keeps pH within normal limits. Buffers resist pH changes because they can take up excess hydrogen ions (H^+) or hydroxide ions (OH^-). Many commercial products, such as Bufferin, shampoos, and deodorants, are buffered as an added incentive for us to buy them.

The pH of our blood is usually about 7.4, and this normal level is maintained in part by a buffer consisting of a combination of carbonic acid and bicarbonate ions. Carbonic acid (H_2CO_3) is a weak acid that minimally dissociates. The following reaction shows how carbonic acid dissociates and can re-form:

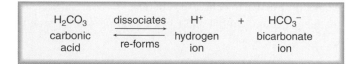

$$H_2CO_3 \underset{\text{re-forms}}{\overset{\text{dissociates}}{\rightleftharpoons}} H^+ + HCO_3^-$$

carbonic acid   hydrogen ion   bicarbonate ion

When hydrogen ions (H^+) are added to blood, this reaction occurs:

$$H^+ + HCO_3^- \rightarrow H_2CO_3$$

When hydroxide ions (OH^-) are added to blood, this reaction occurs:

$$OH^- + H_2CO_3 \rightarrow HCO_3^- + H_2O$$

These reactions prevent any significant change in blood pH.

---

[1] pH is defined as the negative log of the hydrogen ion concentration $[H^+]$. A log is the power to which 10 must be raised to produce a given number.

THE CHAPTER IN REVIEW

Summary

2.1 The Nature of Matter

Definitions to remember:

- **Matter:** Takes up space, has mass, and is composed of elements
- **Element:** A fundamental constituent of matter; contains atoms
- **Nucleus:** The center of an atom, contains protons (+) and neutrons
- **Mass number:** Protons plus neutrons
- **Isotopes:** Atoms of same type that differ in the number of neutrons
- **Atomic number:** Number of protons equals number of electrons when atom is neutral
- **Electrons (−)**
  First shell: Complete with two electrons
  Other shells: Complete with eight electrons assuming an atomic number of 20 or below.
  Valence shell: Number of electrons in outer shell determines reactivity of atom.
- **Octet rule:** Atoms react with one another in order to have a completed outer shell with eight electrons

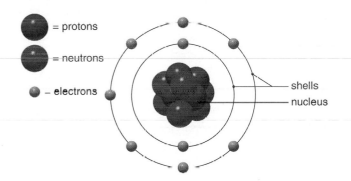

= protons
= neutrons
= electrons
shells
nucleus

- **Ionic bond:** Attraction between oppositely charged ions. Ions form when atoms lose or gain one or more electrons to achieve a completed outer shell.
- **Covalent bond:** Sharing of electrons between two atoms. There are single covalent bonds (sharing one pair of electrons), double (sharing two pairs of electrons), and triple (sharing three pairs of electrons).
- **Polar covalent bond:** Sharing of electrons is not even and as a result, a molecule has a negative pole and a positive pole.
- **Hydrogen bond:** Weak attraction between slightly positive hydrogen atoms of one molecule and slightly negative oxygen or nitrogen atoms within same or different molecules.

2.2 Water's Importance to Life

Water is a polar molecule. Its polarity and hydrogen bonding account for its unique properties, which can be summarized as follows:

Properties	Chemical Reasons	Effect
Water is a solvent.	Polarity	Water facilitates chemical reactions.
Water is cohesive and adhesive.	Hydrogen bonding; polarity	Water serves as a transport medium.
Water has a high surface tension.	Hydrogen bonding	The surface tension of water is hard to break.
Water has a high heat capacity.	Hydrogen bonding	Water protects organisms from rapid changes in temperature.
Water has a high heat of vaporization.	Hydrogen bonding	Water helps organisms resist overheating.
Water is less dense as ice.	Hydrogen bonding changes	Ice floats on liquid water.

Acids and Bases

Water dissociates to produce an equal number of hydrogen ions and hydroxide ions. This is neutral pH. This illustration shows the range of acidic and basic solutions:

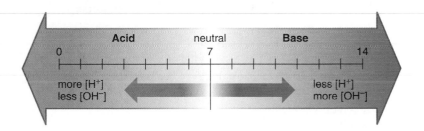

Cells are sensitive to pH changes. Biological systems contain buffers that help keep the pH within a normal range.

Thinking Scientifically

If you watch a bird dive into the water, you will see that the surface of the water is smooth and continuous as the bird enters. However, when the bird flies back into the air, many drops of water fly into the air as well (see Fig. 2.9). Explain this observation based on properties of water.

Testing Yourself

Choose the best answer for each question.

1. Which of the following is not a component of an atom?

   a. proton   c. neutron
   b. positron   d. electron

2. The most abundant element by weight in the human body is

   a. carbon.   c. oxygen.
   b. hydrogen.   d. nitrogen.

3. This rule states that the outer electron shell is most stable when it contains eight electrons.

   a. stability rule   c. octet rule
   b. atomic rule   d. shell rule

4. How many electrons does nitrogen require to fill its outer shell?

   a. 0   c. 2
   b. 1   d. 3

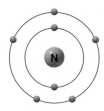

Nitrogen

5. Molecules held together by _____ bonds tend to dissociate in biological systems due to the water content in those systems.

   a. covalent   c. hydrogen
   b. ionic   d. nitrogen

6. Water flows freely, but does not separate into individual molecules because water is

   a. cohesive.   c. hydrophobic.
   b. hydrophilic.   d. adhesive.

7. Water can absorb a large amount of heat without much change in temperature because it has

   a. a high surface tension.
   b. a high heat capacity.
   c. ten times as many hydrogen ions.
   d. ten times as many hydroxide ions.

Go to www.mhhe.com/maderessentials for more quiz questions.

Bioethical Issue

Acid precipitation is produced when atmospheric water is polluted by sulfur dioxide and nitrous oxide emissions. These emissions are mostly produced by the burning of fossil fuels. Because the emissions are airborne, the acid precipitation may occur far from where the fossil fuels are being burned. For example, most of the acid precipitation in Scandinavian countries comes from emissions produced in Great Britain, central Europe, and Russia. Similarly, much of the acid precipitation in southeastern Canada results from industrial activity in the midwestern and eastern United States.

Do you think the countries producing pollution should compensate their neighbors for the damage being done? Are countries at least obligated to minimize the effects of their air pollutants on other countries?

Under the Kyoto Protocol of 1997, some industrialized nations made legally binding commitments to reduce fossil fuel emissions. The countries that have signed realize that their industrial activities affect other countries. While they are not compensating other countries for damage done by their pollutants, they have agreed to take steps to limit the emission of pollutants. Although the United States produces more fossil fuel emissions per capita than any other country in the world, it opted out of the Kyoto Protocol in 2001.

Understanding the Terms

acid 25	ionic bond 20
atom 17	isotope 18
atomic number 17	mass number 17
atomic symbol 17	matter 16
base 25	molecule 19
buffer 26	neutron 17
compound 19	nucleus 17
covalent bond 20	octet rule 19
electron 17	pH 26
electronegativity 22	pH scale 26
electron shell 18	polar 22
element 16	product 21
hydrogen (H) bond 22	proton 17
hydrogen ion (H$^+$) 25	reactant 21
hydrophilic 23	salt 20
hydrophobic 23	tracer 18
hydroxide ion (OH$^-$) 25	valence shell 19
ion 20	

Match the terms to these definitions:

a. _____ Subatomic particle that moves in shells around the nucleus of an atom.

b. _____ Anything that takes up space and has mass.

c. _____ A molecule that has positive and negative ends.

d. _____ Having an affinity for water.

e. _____ Atoms of the same element that differ in the number of neutrons.

f. _____ Substance having an attached radioactive isotope that allows a researcher to track its whereabouts in a biological system.

g. _____ An atom that has become charged due to electron transfer.

The Organic Molecules of Life

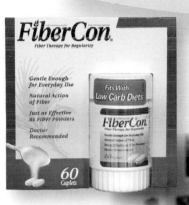

Some human cells can produce up to 10,000 different types of proteins.

Dietary fiber found in supplements, green vegetables, and grains has no nutritional value.

Cholesterol found in foods such as egg yolks has many important functions, including the production of sex hormones.

Without cholesterol your body would not function well. What is it about cholesterol that can cause problems in our bodies? Although blood tests now distinguish between "good" cholesterol and "bad" cholesterol, in reality there is only one molecule called cholesterol. Cholesterol is absolutely essential to the normal functioning of our cells. It is a part of the cell's outer membrane, and it also serves as a precursor to hormones

▶ ▶ ▶

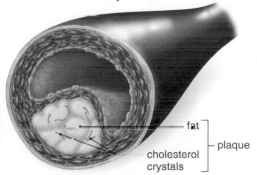

fat

plaque

cholesterol crystals

such as the sex hormones testosterone and estrogen. Without cholesterol, your cells would be in major trouble.

So when does cholesterol become "bad"? Your body packages cholesterol in high-density lipoproteins (HDL) or low-density lipoproteins (LDL), depending on such factors as genetics, your diet, your activity level, and whether you smoke. Cholesterol packaged as HDL is primarily on its way from the tissues to the liver for recycling. This reduces the likelihood of deposits called plaques forming in the arteries; thus, HDL cholesterol is "good." LDL is carrying cholesterol to the tissues; high levels of LDL-cholesterol contribute to the development of plaque that can result in heart disease, making LDL the "bad" form. When cholesterol levels are measured, the ratio of LDL to HDL is determined. Exercising, improving the diet, and not smoking can lower LDL. In addition, a variety of prescription medications lower LDL levels.

The molecules that make up our bodies include carbohydrates, lipids, proteins, and nucleic acids. The building blocks for nearly all of these molecules come from our diet. These molecules give cells the ability to perform many of their complex duties. In this chapter you will learn about the structure and function of the major molecules of cells.

3.1 Organic Molecules

If you decide to take a chemistry course, you will most likely have a choice of either inorganic chemistry or organic chemistry. **Inorganic chemistry** is often thought of as the chemistry of the nonliving world, and **organic chemistry** is the chemistry of the living world (Fig. 3.1). But water, the molecule that makes up 70–90% of a cell, is an inorganic molecule. Why should that be? For a molecule to be **organic,** it must contain carbon and hydrogen; otherwise, it is an inorganic molecule. The chemistry of carbon accounts for the molecules that make up the structure of cells and organisms.

The Carbon Atom

A bacterial cell contains some 5,000 different organic molecules, and a plant or animal cell has twice that number. What is there about carbon that makes organic molecules so diverse and also so complex? Carbon, with a total of six electrons, has four electrons in the outer shell. In order to acquire four electrons to complete its outer shell, a carbon atom almost always shares electrons with—you guessed it—CHNOPS, the elements that make up most of the weight of living things (see p. 16).

Figure 3.1 Organic molecules as structural materials.

Some organic molecules are purely structural in nature. Related organic molecules (**a**) help hold a plant stem erect, (**b**) make the shell of a crab hard, and (**c**) support the wall of a bacterium.

b.

c.   ×38,000   a.

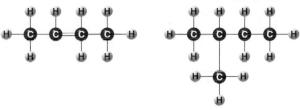

Carbon chains can vary in length, and/or have double bonds, and/or be branched.

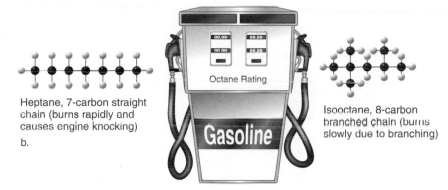

Heptane, 7-carbon straight chain (burns rapidly and causes engine knocking)
b.

Octane Rating

Gasoline

Isooctane, 8-carbon branched chain (burns slowly due to branching)

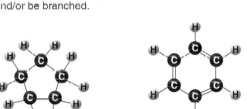

Carbon chains can form rings of different sizes and have double bonds.
a.

Figure 3.2 **Hydrocarbons are highly versatile.**

Hydrocarbons contain only hydrogen and carbon. **a.** Even so, they can be quite varied according to the number of carbons, the placement of any double bonds, possible branching, and possible ring formation. **b.** Carbon-to-carbon bonds are a source of energy. Gasoline contains heptane and isooctane—the higher the octane rating (percentage), the smoother your car engine will run.

Because carbon is small and needs to acquire four electrons, carbon can bond with as many as four other elements, and this spells diversity. But even more significant to the shape of biomolecules, and therefore their function, is the fact that carbon often shares electrons with another carbon atom. The C—C bond is stable, and the result is that carbon chains can be quite long. Hydrocarbons are chains of carbon atoms that are bonded only to hydrogen atoms. Any carbon atom of a hydrocarbon molecule can start a branch chain, and a hydrocarbon can also turn back on itself to form a ring compound (Fig. 3.2). Carbon can form double bonds with itself and other atoms. Some carbon molecules, called **isomers,** have the same number and kinds of atoms but a different arrangement of atoms. Isomers are another example of how the chemistry of carbon leads to variations in organic molecules.

The Carbon Skeleton and Functional Groups

The carbon chain of an organic molecule is called its skeleton, or backbone. This terminology is appropriate because just as your skeleton accounts for your shape, so does the carbon skeleton of an organic molecule account for its shape. The reactivity of an organic molecule is largely dependent on the attached functional groups (Fig. 3.3). A **functional group** is a specific combination of bonded atoms that always reacts in the same way, regardless of the particular carbon skeleton. As in Figure 3.3, it is even acceptable to use an *R* to stand for the rest of the molecule because only the functional group is involved in the reaction.

Notice that functional groups occur in certain types of compounds. For example, sugars contain many polar —OH groups. Thus, although all hydrocarbons are nonpolar and **hydrophobic** (not soluble in water), glucose with several —OH groups is **hydrophilic** (soluble in water; see Chapter 2). Because cells are mainly composed of water, the ability to interact with and be soluble in water profoundly affects the function of organic molecules in cells.

Organic molecules containing carboxyl groups (—COOH) are both polar (hydrophilic) and acidic. They tend to ionize and release hydrogen ions in solution:

$$-COOH \rightarrow -COO^- + H^+$$

Functional Groups		
Group	**Structure**	**Found in**
Hydroxyl	R—O—H	Alcohols, sugars
Carboxyl	R—C with =O and O—H	Amino acids, fatty acids
Amino	R—N with H and H	Amino acids, proteins
Sulfhydryl	R—S—H	Amino acid cysteine, proteins
Phosphate	—P ; R—O—P—O—H with O and O	ATP nucleic acids
R = remainder of molecule		

Figure 3.3 **Functional groups.**

Molecules with the same carbon skeleton can still differ according to the type of functional group attached. Many of these functional groups are polar, helping to make the molecule soluble in water. In this illustration, the remainder of the molecule, the hydrocarbon chain, is represented by an *R*.

Figure 3.4 **Carbohydrates.**

The familiar carbohydrates, sugar and starch, are present in the foods shown. A diet loaded with these carbohydrates may make you prone to diabetes type 2 and perhaps to other illnesses as well. In contrast, moderate amounts of multigrain foods are better for you and provide fiber to keep you regular.

Figure 3.5 **Lipid foods.**

A diet rich in fat can indeed make you fat. However, you do need some fat in your diet. Choose vegetable oils over the animal fat in lard, butter, and other dairy products. The fats in vegetable oils even lower your risk of cardiovascular disease if they haven't been hydrogenated, as they are in processed foods.

Figure 3.6 **Protein foods.**

It's surprising how little protein is needed in the diet. Just 6 ounces a day will provide you with all the amino acids you need. Beef, but not chicken, can be loaded with animal fat. Fish, however, contains beneficial oils that lower the incidence of cardiovascular disease. Also, the diet can include an egg a day, usually with no ill effects.

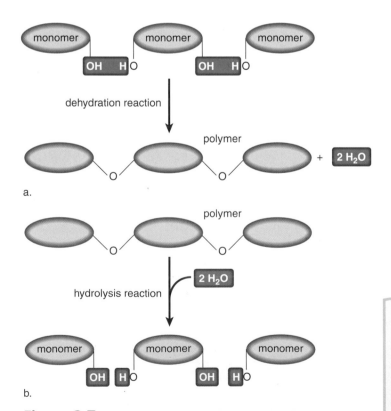

Figure 3.7 **Synthesis and breakdown of polymers.**

a. In cells, synthesis often occurs when monomers bond during a dehydration reaction (removal of H_2O). **b.** Breakdown occurs when the monomers in a polymer separate because of a hydrolysis reaction (the addition of H_2O).

3.2 The Organic Molecules of Cells

Despite their great diversity, organic molecules in living things are grouped into only four categories: carbohydrates, lipids, proteins, and nucleic acids. You are very familiar with these molecules because certain foods are known to be rich in carbohydrates, lipids, and proteins, as illustrated in Figures 3.4, 3.5, and 3.6. When you digest these foods, they break down into subunit molecules. Your body then takes these subunits and builds from them the large macromolecules that make up your cells. Many different foods also contain nucleic acids, the type of molecule that forms your genes.

Macromolecules are constructed by linking together a large number of the same type of subunits, called **monomers.** Linking many monomers results in a **polymer.** A protein can contain hundreds of amino acid monomers, and a nucleic acid can contain hundreds of nucleotide monomers. How can poly-

Check Your Progress

1. Describe the properties of a carbon atom that make it an ideal foundation for life.
2. List the four classes of organic molecules.

Answers: 1. The carbon atom can bond with up to four different elements. Carbon-to-carbon bonds are stable, so long chains can be built. These chains can be variable in length and branching patterns. Some carbon molecules can create isomers. 2. Carbohydrates, lipids, proteins, and nucleic acids.

mers get so large? Just as a train increases in length when boxcars are hitched together one by one, so a polymer gets longer as monomers bond to one another.

A cell uses the same type of reaction to synthesize any type of macromolecule. It is called a **dehydration reaction** because the equivalent of a water molecule, that is, an —OH (hydroxyl group) and an —H (hydrogen atom), is removed as the reaction occurs (Fig. 3.7a). To break down a macromolecule, a cell and also the digestive tract uses an opposite type of reaction: During a **hydrolysis reaction,** an —OH group from water attaches to one subunit, and an —H from water attaches to the other subunit (Fig. 3.7b). In other words, water is used to break the bond holding monomers together.

Carbohydrates

Carbohydrates are almost universally used as an immediate energy source in living things, but they also play structural roles in a variety of organisms. The term carbohydrate refers to either a single sugar molecule or two sugar molecules bonded together. Typically, the sugar glucose is a monomer for carbohydrate polymers.

Monosaccharides: Ready Energy

Monosaccharides (*mono,* one; *saccharide,* sugar) have only a single sugar molecule; therefore, they are simple sugars. A simple sugar can have a carbon backbone of three to seven carbons. The word *carbohydrate* might make you think that every carbon atom is bonded to an H and an —OH. This is not strictly correct, as you can see by examining the structural formula for glucose (Fig. 3.8). Still, sugars do have many polar —OH groups, which make them soluble in water.

Glucose, with six carbon atoms, has a molecular formula of $C_6H_{12}O_6$. Glucose has two important isomers, called *fructose* and *galactose,* but even so, we usually think of glucose when we see the formula $C_6H_{12}O_6$. That's because glucose has a special place in the chemistry of organisms. Glucose is transported in the blood of animals, and it is also the molecule that is broken down in nearly all types of organisms as an immediate source of energy. In other words, cells use glucose as the energy source of choice.

Ribose and **deoxyribose,** with five carbon atoms, are significant because they are found in the nucleic acids RNA and DNA, respectively. RNA and DNA are discussed later in this chapter.

Disaccharides: Varied Uses

A **disaccharide** (*di,* two; *saccharide,* sugar) contains two monosaccharides bonded together. The brewing of beer relies on maltose, a disaccharide usually derived from barley. During the production of beer, yeast breaks down maltose to two units of glucose and then uses glucose as an energy source. Because yeasts are using a process, called *fermentation,* ethyl alcohol is produced (Fig. 3.9).

Sucrose, a disaccharide acquired from sugar beets and sugarcane, is of special interest because we use it at the table and in baking as a sweetener. Our body digests sucrose to its components, glucose and fructose. (Later, the fructose is changed to glucose, our usual energy source.) If the body doesn't need more energy at the moment, the glucose is metabolized to fat! That's why eating sugary desserts can make you fat.

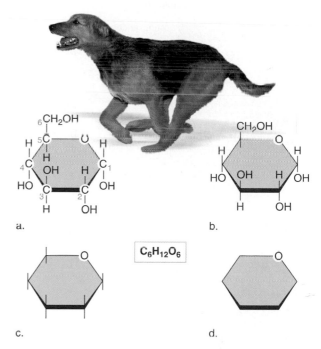

a.   b.

$C_6H_{12}O_6$

c.   d.

Figure 3.8   Glucose.

Each of these structural formulas represents glucose. **a.** The carbon skeleton and all attached groups are shown. **b.** The carbon skeleton is omitted. **c.** The carbon skeleton and attached groups are omitted. **d.** Only the ring shape of the molecule remains.

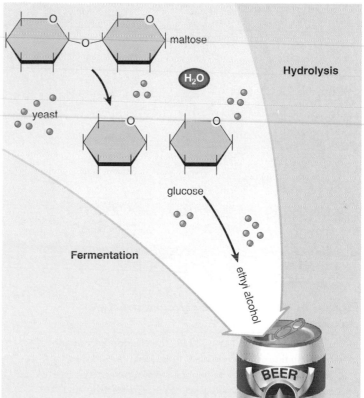

maltose

Hydrolysis

H_2O

yeast

glucose

Fermentation

ethyl alcohol

BEER

Figure 3.9   Breakdown of maltose, a disaccharide.

Maltose is the energy source for yeast during the production of beer. Yeasts differ as to the amount of maltose they convert to alcohol, so selection of the type of yeast is important.

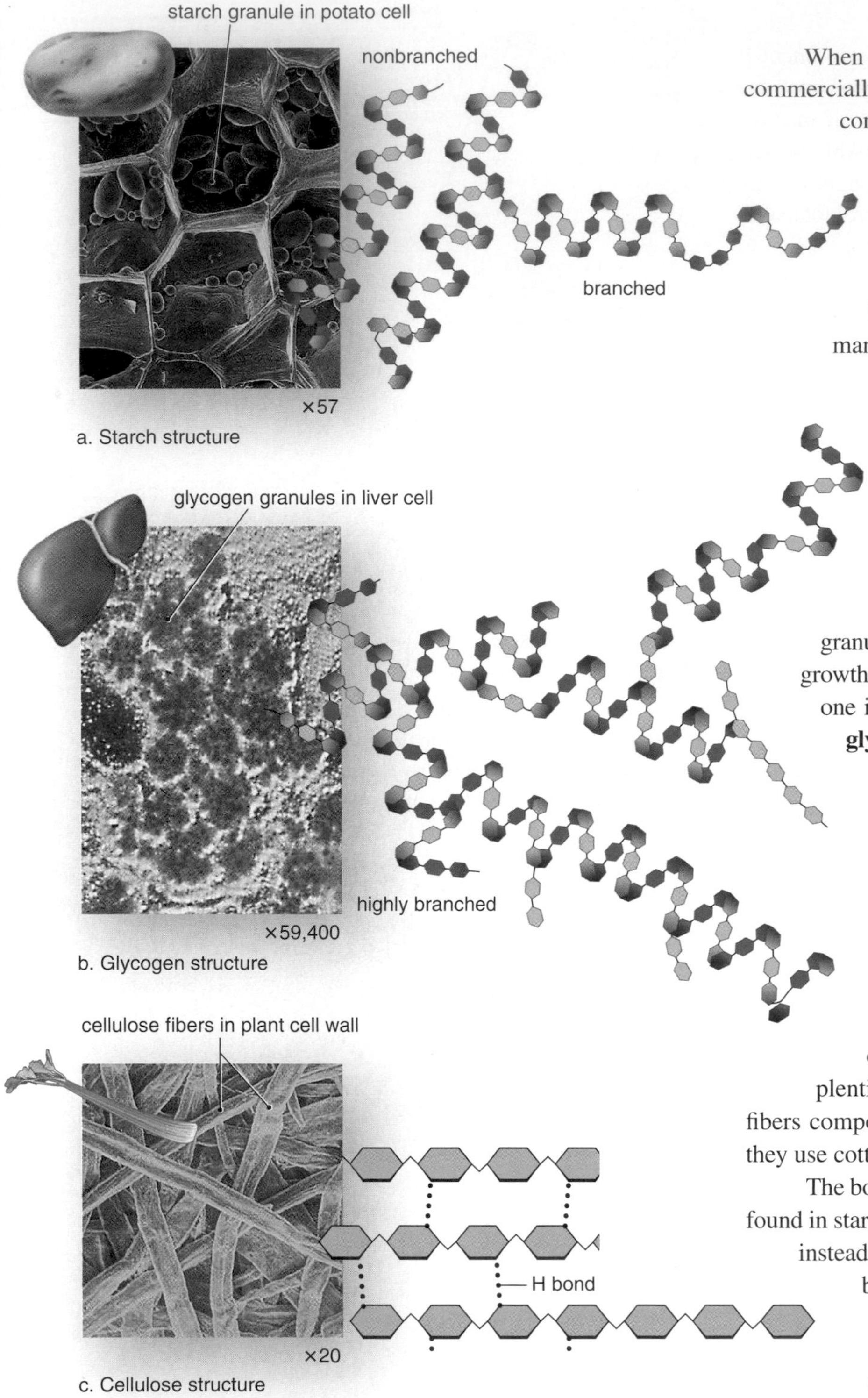

starch granule in potato cell

nonbranched

branched

×57

a. Starch structure

glycogen granules in liver cell

highly branched

×59,400

b. Glycogen structure

cellulose fibers in plant cell wall

H bond

×20

c. Cellulose structure

Figure 3.10   Starch and glycogen structure and function.

a. Glucose is stored in plants as starch. The electron micrograph shows the location of starch in plant cells. Starch is a chain of glucose molecules that can be nonbranched or branched.
b. Glucose is stored in animals as glycogen. The electron micrograph shows glycogen deposits in a portion of a liver cell. Glycogen is a highly branched polymer of glucose molecules.
c. In plant cell walls, each cellulose fiber contains several microfibrils. Each microfibril contains many polymers of glucose hydrogen-bonded together. Three such polymers are shown.

When you make lemonade at home, you add sucrose. But drinks made commercially often contain high-fructose corn syrup (HFCS). In the 1980s, a commercial method was developed for converting the glucose in corn syrup to the much sweeter-tasting fructose. Nutritionists are not in favor of eating highly *processed* foods that are rich in sucrose, HFCS, and white starches. They say these foods provide "empty" calories, meaning that although they supply energy, they don't supply any of the vitamins, minerals, and fiber needed in the diet. In contrast, minimally processed foods provide glucose, starch, and many other types of nutritious molecules.

Polysaccharides as Energy Storage Molecules

Polysaccharides are polymers of monosaccharides. Some types of polysaccharides function as short-term energy storage molecules because they are much larger than a sugar and are relatively insoluble. Polysaccharides cannot easily pass through the plasma membrane and are kept (stored) within the cell.

Plants store glucose as **starch.** The cells of a potato contain granules where starch resides during winter until energy is needed for growth in the spring. Notice in Figure 3.10*a* that starch exists in two forms—one is nonbranched and the other is branched. Animals store glucose as **glycogen,** which is even more branched (Fig. 3.10*b*). Branching makes a polysaccharide subject to hydrolytic attack by enzymes.

The storage and release of glucose from liver cells is controlled by hormones. After we eat, the pancreas releases the hormone insulin, which promotes the storage of glucose as glycogen.

Polysaccharides as Structural Molecules

Some types of polysaccharides function as structural components of cells. **Cellulose** is the most abundant of all the carbohydrates, which in turn are the most abundant of all the organic molecules on Earth. Plant cell walls contain cellulose and therefore cellulose is plentiful in the wood of tree trunks. The seeds of cotton plants have long fibers composed mostly of cellulose. Humans use wood for construction, and they use cotton fibers to make a cloth of the same name.

The bonds joining the glucose subunits in cellulose are different from those found in starch and glycogen (Fig. 3.10*c*). As a result, the molecule is not helical; instead, the long glucose chains are held parallel to each other by hydrogen bonding to form strong microfibrils and then fibers. The fibers crisscross within plant cell walls for even more strength.

The digestive juices of animals can't hydrolyze cellulose, but some microorganisms can digest it. Cows and other ruminants (cud-chewing animals) have a special internal pouch where microorganisms break down cellulose to glucose. In humans, cellulose has the benefit of serving as dietary fiber, which maintains regular elimination.

Chitin, which is found in the exoskeleton of crabs and related animals such as lobsters and insects, is also a polymer of glucose. However, each glucose subunit has an amino group (—NH$_2$) attached to it. The linkage between the glucose molecules is like that found in cellulose; therefore, chitin is not digestible by humans but still has many good uses. Seeds are coated with chitin, and this protects them from attack by soil fungi. Because chitin also has antibacterial and antiviral properties, it is processed and used in medicine as a wound dressing and suture material. Chitin is even useful during the production of cosmetics and various foods.

Lipids

Although molecules classified as **lipids** are quite varied, they have one characteristic in common: They are all insoluble in water due to their nonpolar hydrocarbon chains. Salad dressings are rich in vegetable oils. Even after shaking, the vegetable oil will separate out from the water.

Fats (such as bacon fat, lard, and butter) and also oils (such as corn oil, olive oil, and coconut oil) are well-known lipids. In animals, fat is used for both insulation and long-term energy storage. Fat below the skin of marine mammals is called blubber; around the waist of humans, it is often referred to as a "spare tire." Instead of fat, plants use oils for long-term energy storage. In animals, the secretions of oil glands help waterproof skin, hair, and feathers (Fig. 3.11).

Several other types of molecules are lipids. Phospholipids, which are constructed similarly to fats and oils, are important components of the plasma membrane that surrounds cells. Steroids, with a structure entirely different from that of fats and oils, are an important type of lipid in the bodies of animals. Cholesterol, the molecule mentioned at the beginning of this chapter, and some hormones, such as the sex hormones, are examples of steroids.

Fats and Oils: Long-term Energy Storage

Fats and **oils** contain two types of subunit molecules: glycerol and fatty acids (Fig. 3.12). **Glycerol** is a compound with three —OH groups. The —OH groups are polar; therefore, glycerol is soluble in water. A **fatty acid** has a long chain of carbon atoms bonded only to hydrogen, with a carboxyl group at one end. A fat or oil forms when the carboxyl portions of three fatty acids react with the — OH groups of glycerol. This is a dehydration reaction because in addition to a fat molecule,

Check Your Progress

1. Describe the significance of glucose in living systems.
2. Explain why humans cannot utilize the glucose in cellulose.
3. Compare and contrast cellulose with chitin.

Answers: 1. Most organisms use glucose as the energy source of choice. 2. Humans do not produce digestive juices capable of breaking the bonds attaching glucose subunits together in cellulose. 3. Both cellulose and chitin are composed of glucose subunits linked together in the same way. In contrast to cellulose, chitin has an amino group attached to each glucose molecule. Cellulose is found in plant cell walls, while chitin is in the exoskeleton of some animals.

Figure 3.11   Preening in birds.

As a bird preens, it transfers oil to its feathers from a gland at the base of its tail. Because the oil repels water, the feathers become waterproof.

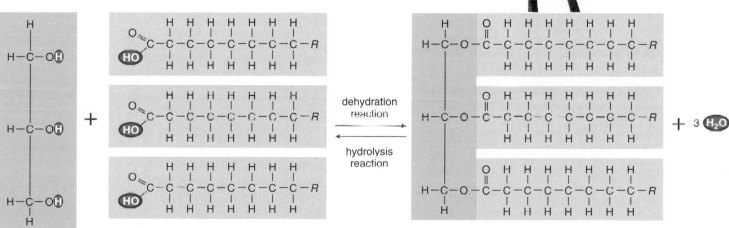

Glycerol   3 Fatty acids   Fat   3 Waters

Figure 3.12   Synthesis and breakdown of fat.

Following a dehydration reaction, glycerol is bonded to three fatty acid molecules, and water is given off. Following a hydrolysis reaction, the bonds are broken due to the addition of water. *R* represents the remainder of the molecule, which in this case is a continuation of the hydrocarbon chain, composed of 16 or 18 carbons.

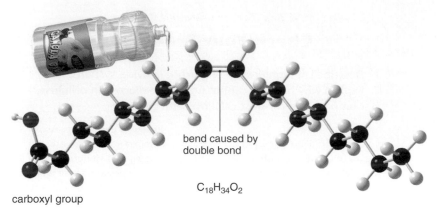

carboxyl group $C_{18}H_{34}O_2$

bend caused by double bond

a. Oleic acid, a monounsaturated fatty acid (one double bond) found in canola oil.

butter

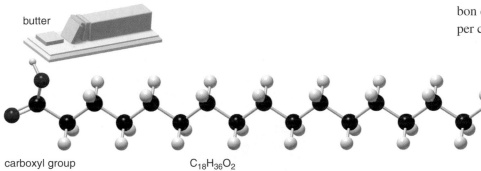

carboxyl group $C_{18}H_{36}O_2$

b. Stearic acid, a saturated fatty acid (no double bonds) found in butter.

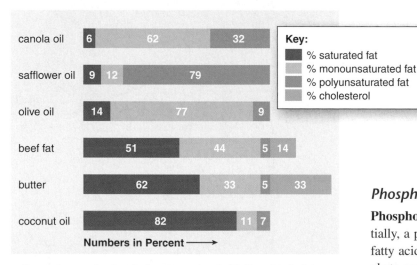

c. Percentages of saturated and unsaturated fatty acids in fats and oils.

Figure 3.13   Fatty acids.

A fatty acid has a carboxyl group attached to a long hydrocarbon chain. **a.** If there are double bonds between some of the carbons in the chain, the fatty acid is monounsaturated. **b.** If there are no double bonds, the fatty acid is saturated. A diet high in saturated fats appears to contribute to diseases of the heart and blood vessels. **c.** Which fat or oil do you judge to be the most healthy to use?

three molecules of water result. Fats and oils are degraded during a hydrolysis reaction, in which water is added to the molecule (see Fig. 3.12). Because three long fatty acids are attached to each glycerol molecule, fats and oils are called *triglycerides* and pack a lot of energy in one molecule.

Fatty Acids

Most of the fatty acids in cells contain 16 or 18 carbon atoms per molecule, although smaller ones are also found. Fatty acids are either saturated or unsaturated. **Saturated fatty acids** have no double bonds between the carbon atoms. The carbon chain is saturated, so to speak, with all the hydrogens it can hold. **Unsaturated fatty acids** have double bonds in the carbon chain wherever the number of hydrogens is less than two per carbon atom.

In general, oils are liquids at room temperature because they contain unsaturated fatty acids. Notice in Figure 3.13a that the double bond creates a bend in the fatty acid chain. Such kinks prevent close packing between the hydrocarbon chains and account for the fluidity of oils. On the other hand, butter, which contains saturated fatty acids, is a solid at room temperature (Fig. 3.13b).

Saturated fats in particular are associated with a cardiovascular disease called atherosclerosis, in which lipid material, called plaque, accumulates inside blood vessels. Plaque contributes to high blood pressure and heart attacks. Even more harmful than naturally occurring saturated fats are the so-called *trans fats,* which are vegetable oils hydrogenated commercially to make them solid. Trans fats are often found in processed foods—margarine, baked goods, and fried foods in particular. Unsaturated oils, particularly *monounsaturated* (one double bond) oils but also *polyunsaturated* (many double bonds) oils, have been found to be protective against atherosclerosis (Fig. 3.13c).

Phospholipids: Membrane Components

Phospholipids, as implied by their name, contain a phosphate group. Essentially, a phospholipid is constructed like a fat, except that in place of the third fatty acid attached to glycerol, there is a charged phosphate group. The phosphate group is usually bonded to another polar group, indicated by *R* in Figure 3.14a. This portion of the molecule is the polar head, while the hydrocarbon chains of the fatty acids are the nonpolar tails.

Because phospholipids have hydrophilic (polar) heads and hydrophobic (nonpolar) tails, they tend to arrange themselves so that only the polar heads are adjacent to a watery medium. Between two compartments of water, such as the outside and inside of a cell, phospholipids become a bilayer in which the hydrophilic heads project outward and the hydrophobic tails project inward. The bulk of the plasma membrane that surrounds cells consists of a fairly fluid phospholipid bilayer, as do all the other membranes in the cell (Fig. 3.14b). A plasma membrane is absolutely essential to the structure and function of a cell, and thus phospholipids are vitally important to humans and other organisms.

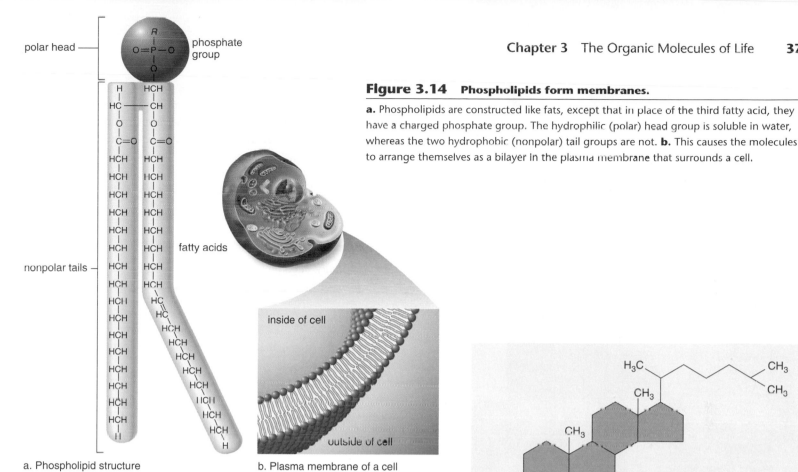

a. Phospholipid structure

b. Plasma membrane of a cell

Figure 3.14 **Phospholipids form membranes.**

a. Phospholipids are constructed like fats, except that in place of the third fatty acid, they have a charged phosphate group. The hydrophilic (polar) head group is soluble in water, whereas the two hydrophobic (nonpolar) tail groups are not. **b.** This causes the molecules to arrange themselves as a bilayer in the plasma membrane that surrounds a cell.

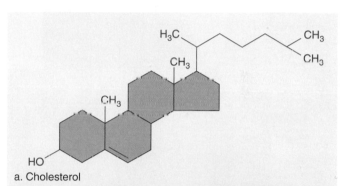

a. Cholesterol

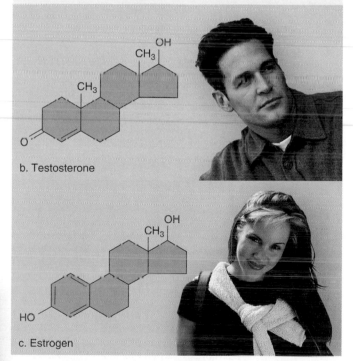

b. Testosterone

c. Estrogen

Steroids: Four Fused Rings

Steroids are lipids that have entirely different structures from those of fats. Steroid molecules have skeletons of four fused hydrocarbon rings, shown in red in Figure 3.15. Each type of steroid differs primarily in the types of functional groups attached to the carbon skeleton.

Cholesterol (Fig. 3.15a) is a component of an animal cell's plasma membrane, and it is the precursor of other steroids, such as the sex hormones testosterone and estrogen. The male sex hormone, testosterone, is formed primarily in the testes, and the female sex hormone, estrogen, is formed primarily in the ovaries (Fig. 3.15b,c). Testosterone and estrogen differ only by the functional groups attached to the same carbon skeleton, and yet they have a profound effect on the body and the sexuality of humans and other animals.

Figure 3.15 **Steroid diversity.**

a. All the steroids are derived from cholesterol, an important component of the plasma membrane. Cholesterol and all steroid molecules have four adjacent rings, but their effects on the body largely depend on the attached groups indicated in red. The different effects of **(b)** testosterone and **(c)** estrogen on the body are due to different functional groups attached to the same carbon skeleton.

Check Your Progress

1. Compare and contrast a saturated fatty acid with an unsaturated fatty acid.
2. Explain why phospholipids form a bilayer in a watery medium.

Answers: 1. A saturated fatty acid contains no double bonds between carbon atoms, while an unsaturated fatty acid contains one or more double bonds. 2. Phospholipids arrange themselves so that their hydrophilic heads are adjacent to water, while the hydrophobic tails point inward toward each other.

Most likely you are aware that some people take anabolic steroids, synthetic testosterone, to increase muscle mass. The result is usually unfortunate. The presence of the steroid in the body upsets the normal hormonal balance: The testes atrophy, and males may develop breasts; females tend to grow facial hair and lose hair on their heads. Because the taking of steroids gives athletes an unfair advantage and destroys their health—heart, kidney, liver, and psychological disorders are common—they are banned by professional athletic associations.

Proteins

Proteins are of primary importance in the structure and function of cells. Here are some of their many functions:

Support Some proteins are structural proteins. Examples include the protein in spider webs; keratin, the protein that makes up hair and fingernails; and collagen, the protein that lends support to skin, ligaments, and tendons (Fig. 3.16a).

Metabolism Some proteins are **enzymes.** They bring reactants together and thereby speed chemical reactions in cells. They are specific for one particular type of reaction and can function at body temperature.

Transport Channel and carrier proteins in the plasma membrane allow substances to enter and exit cells. Other proteins transport molecules in the blood of animals—for example, **hemoglobin** is a complex protein that transports oxygen (Fig. 3.16b).

Defense Some proteins, called antibodies, combine with disease-causing agents to prevent them from destroying cells and upsetting homeostasis, the relative constancy of the internal environment.

a. Structural proteins

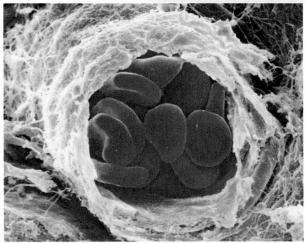

b. Transport proteins

c. Contractile proteins

Figure 3.16 Types of proteins.

a. The protein in hair, fingernails, and spider webs (keratin) is a structural protein, as is collagen. Collagen injections can reduce wrinkles and scarring by providing extra support. **b.** Hemoglobin, a major protein in red blood cells, is involved in transporting oxygen. **c.** Contractile proteins, actin and myosin, cause muscles to move.

Regulation Hormones are regulatory proteins. They serve as intercellular messengers that influence the metabolism of cells. For example, the hormone insulin regulates the content of glucose in the blood and in cells, while growth hormone determines the height of an individual.

Motion The contractile proteins actin and myosin allow parts of cells to move and cause muscles to contract (Fig. 3.16c). Muscle contraction enables animals to move from place to place.

The structures and functions of cells differ according to the type of protein they contain. Muscle cells contain actin and myosin; red blood cells contain hemoglobin; support cells contain collagen; and so forth.

Amino Acids: Subunits of Proteins

A significant carbon atom in an **amino acid** bonds to a hydrogen atom and also to three other groups of atoms (Fig. 3.17a). The name *amino acid* is appropriate because one of these groups is an —NH₂ (amino group) and another is a —COOH (an acid group as discussed on page 31). The third group is the *R* group for an amino acid.

Amino acids differ according to their particular *R* group. The *R* groups range in complexity from a single hydrogen atom to a complicated ring compound. The unique chemical properties of an amino acid depend on those of the *R* group. For example, some *R* groups are polar and some are not. Also, the amino acid cysteine has an *R* group that ends with a sulfhydryl (—SH) group, which often serves to connect one chain of amino acids to another by a disulfide bond, —S—S—. Four amino acids commonly found in cells are shown in Figure 3.17b.

Peptides

Figure 3.18 shows how two amino acids join by a dehydration reaction between the carboxyl group of one and the amino group of another. The resulting covalent bond between two amino acids is called a **peptide bond.** The atoms associated with the peptide bond share the electrons unevenly because oxygen is more electronegative than nitrogen. Therefore, the hydrogen attached to the nitrogen has a slightly positive charge (δ^+), while the oxygen has a slightly negative charge (δ^-):

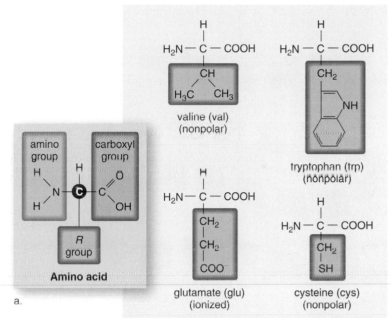

a.

b.

Figure 3.17   Amino acids.

a. Structure of an amino acid. **b.** Proteins contain 20 different kinds of amino acids, four of which are shown here. Amino acids differ by the particular *R* group (blue) attached to the central carbon. Some *R* groups are nonpolar and hydrophobic, some are polar and hydrophilic, and some are ionized and hydrophilic.

Figure 3.18   Synthesis and degradation of a peptide.

Following a dehydration reaction, a peptide bond joins two amino acids, and water is given off. Following a hydrolysis reaction, the bond is broken due to the addition of water.

A **peptide** is two or more amino acids covalently bonded together, and a **polypeptide** is a chain of many amino acids joined by peptide bonds. A protein may contain more than one polypeptide chain; therefore, you can see why a protein could be composed of a very large number of amino acids. Each polypeptide has its own normal sequence. The amino acid sequence determines the final three-dimensional shape of the protein. Proteins that have an abnormal sequence of amino acids often have the wrong shape and cannot function properly.

Shape of Proteins

All proteins have levels of structure. These levels are called the primary, secondary, tertiary, and quaternary structures (Fig. 3.19). A protein's sequence of amino acids is called its primary structure. Consider that an almost infinite number of words can be constructed by changing the number and sequence of the 26 letters in our alphabet. In the same way, many different proteins can result by varying the number and sequence of just 20 amino acids. After a chain of amino acids forms, it coils or folds in a particular way. This shape is called the protein's secondary structure. The polypeptide can have the spiral shape of a *helix*, or it can turn back on itself, forming a *pleated sheet*. Regardless, hydrogen bonding between peptide bonds maintains the secondary structure of a protein. **Fibrous proteins** are structural proteins that have only a secondary structure. *Keratin*, the fibrous protein in hair, fingernails, horns, and feathers, has many helical regions. The silk threads spun by spiders are largely pleated sheets, as is collagen, the protein that gives shape to the skin, tendons, ligaments, cartilage, and bones of animals.

Globular proteins have a tertiary structure. Their rounded, three-dimensional shape results from the folding and twisting of the secondary structure. Various types of bonding between *R* groups maintain the tertiary structure. Enzymes are globular proteins. An enzyme has an optimum temperature and pH at which it works best and maintains its normal shape. A rise in temperature or a change in pH can disrupt the interactions between the *R* groups and change the shape of the enzyme. When an enzyme loses its shape, it is **denatured** and can't function any more.

Some proteins, such as hemoglobin and certain enzymes, have a quaternary structure, which means they consist of more than one polypeptide, each with its own primary, secondary, and tertiary structure.

Figure 3.19 **Levels of protein organization.**

All proteins have a primary structure. Fibrous proteins have a secondary structure; they are either helices (e.g., keratin, collagen) or pleated sheets (e.g., silk). Globular proteins always have a tertiary structure, and most have a quaternary structure (e.g., hemoglobin, enzymes).

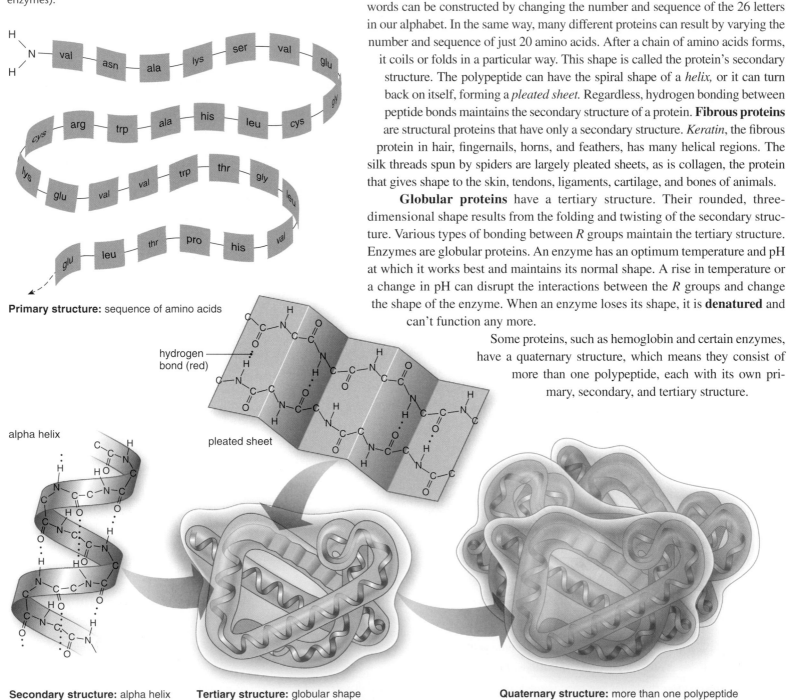

Primary structure: sequence of amino acids

Secondary structure: alpha helix and pleated sheet

Tertiary structure: globular shape

Quaternary structure: more than one polypeptide

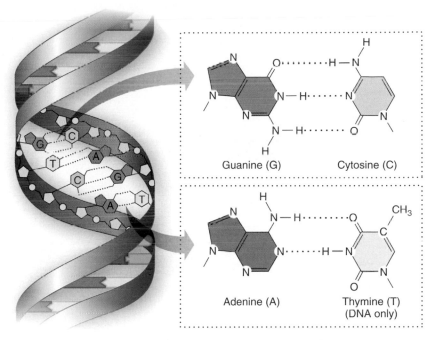

a. Nucleotide

b. DNA structure with base pairs: G with C and A with T

Nucleic Acids

DNA (*deoxyribonucleic acid***)** and **RNA (***ribonucleic acid***)** are the **nucleic acids** in cells. Early investigators called them nucleic acids because they were first detected in the nucleus. As already stated, DNA stores genetic information. Each DNA molecule contains many genes, and genes specify the sequence of the amino acids in proteins. RNA is the helper that takes this information to the site of protein synthesis.

Nucleic acids are polymers in which the monomer is called a **nucleotide.** All nucleotides are composed of three parts: a phosphate, a 5-carbon sugar, and a nitrogen-containing base (Fig. 3.20*a*). The phosphate is simply —PO_4 . The sugar is deoxyribose in DNA and ribose in RNA, which accounts for their names. (Deoxyribose has one less oxygen than does ribose.)

As you can see in Figure 3.20*b*, DNA is a double helix, meaning that two strands spiral around one another. In each strand, the backbone of the molecule is composed of phosphates bonded to sugars, and the bases project to the side. Interestingly, the base guanine (G) is always paired with cytosine (C), and the base adenine (A) is always paired with thymine (T). This is called *complementary base pairing.* Complementary base pairing is very important when DNA makes a copy of itself, a process called replication.

RNA differs from DNA not only by its sugar but also because it uses the base uracil instead of thymine. Whereas DNA is double-stranded, RNA is single-stranded (Fig. 3.20*c*).Complementary base pairing also allows DNA to pass genetic information to RNA. The information is in the sequence of bases. DNA has a triplet code, and every three bases stands for one of the 20 amino acids in cells. Once you know the sequence of bases in a gene, you know the sequence of amino acids in a polypeptide. Many scientists around the world are involved in the Human Genome Project. Not long ago, members of this project determined the sequence of all the base pairs in all of our 25,000 genes. They hope this information will lead to cures for many human illnesses that are caused by proteins that don't function as they should.

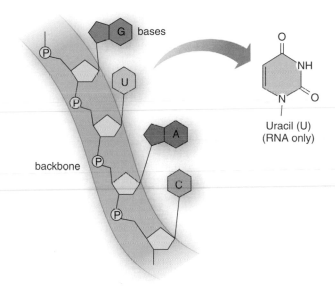

c. RNA structure with bases G, U, A, C

Figure 3.20   DNA and RNA structure.

a. Structure of a nucleotide. **b.** Structure of DNA and its bases.
c. Structure of RNA in which uracil replaces thymine.

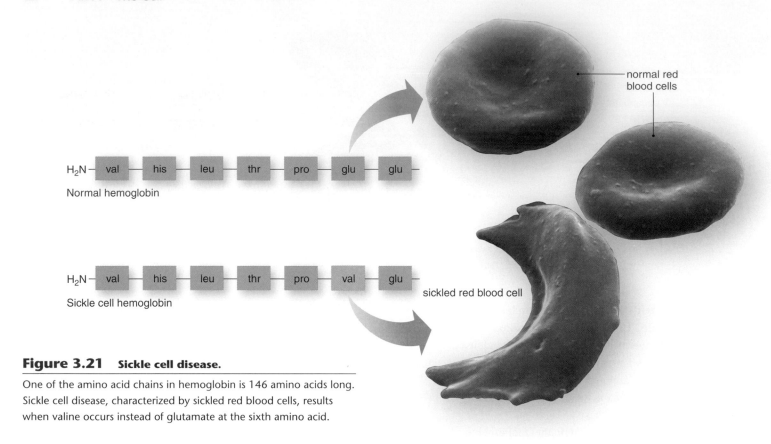

Figure 3.21   Sickle cell disease.

One of the amino acid chains in hemoglobin is 146 amino acids long. Sickle cell disease, characterized by sickled red blood cells, results when valine occurs instead of glutamate at the sixth amino acid.

Relationship Between Proteins and Nucleic Acids

We have learned that the functional group of an amino acid determines its behavior, and that the order of the amino acids within a polypeptide determines its shape. The shape of a protein determines its function. The structure and function of cells are determined by the types of proteins they contain. The same is true for organisms—that is, they differ with regard to their proteins. Next, we learned that DNA, the genetic material, bears instructions for the sequence of amino acids in polypeptides. The proteins of organisms differ because their genes differ.

The relationship between a gene, a protein, and illness is illustrated by sickle cell disease. In sickle cell disease, the individual's red blood cells are sickle-shaped because in one particular spot, the amino acid valine (val) appears where the amino acid glutamate (glu) should be (Fig. 3.21). This substitution makes red blood cells lose their normally round, flexible shape and become hard and rigid. When these abnormal red blood cells go through the small blood vessels, they clog the flow of blood and break apart. This condition can cause pain, organ damage, and a low red blood cell count (called anemia), and is often fatal.

What's the root of the problem? The individual inherited a faulty code for an amino acid in one of the hemoglobin's polypeptides.

Check Your Progress

1. What is the primary structure of a protein?
2. a. What does the peptide bond have to do with the secondary structure of a protein? b. What type of bonding maintains the tertiary structure of a protein?
3. List the three components of a nucleotide.
4. Explain why complementary base pairing is important for nucleic acids.

Answers: 1. The protein's sequence of amino acids is its primary structure. 2. a. The peptide bond includes a partially negative oxygen and a partially negative hydrogen. Therefore, hydrogen bonding between peptide bonds accounts for a protein's secondary structure. b. Bonding between R groups, such as disulfide linkages. 3. Phosphate, 5-carbon sugar, and nitrogen-containing base. 4. Base pairing is important when DNA replicates and also for producing RNA molecules that are copies of genes.

Summary

3.1   Organic Molecules

In order to be organic, a molecule must contain carbon and hydrogen. Organic molecules have different chemical properties depending on the carbon skeleton and attached functional groups. Some functional groups are nonpolar (hydrophobic), and others are polar (hydrophilic).

3.2   The Organic Molecules of Cells

Polymers (many monomers bonded together) are synthesized by a dehydration reaction and degraded by a hydrolysis reaction.

Living things are composed of four types of molecules: carbohydrates, lipids, proteins, and nucleotides. The characteristics of these molecules are summarized in the following chart:

Organic molecules	Examples	Monomers	Functions
Carbohydrates	Monosaccharides, disaccharides, polysaccharides	Glucose	Immediate energy and stored energy; structural molecules
Lipids	Fats, oils, phospholipids, steroids	Glycerol, Fatty acid	Long-term energy storage; membrane components
Proteins	Structural, enzymatic, carrier, hormonal, contractile	Amino acid	Support, metabolic, transport, regulation, motion
Nucleic acids	DNA, RNA	Nucleotide	Storage of genetic information

Carbohydrates

- **Glucose** is a monosaccharide that serves as blood sugar and as a monomer of starch, glycogen, and cellulose. Its isomers are fructose and galactose.
- **Sucrose** is a disaccharide (composed of glucose and fructose) that we know as table sugar.
- **Starch,** a polymer of glucose, stores energy in plants.
- **Glycogen,** a polymer of glucose, stores energy in animals.
- **Maltose** is a disaccharide that results from the digestion of starch and glycogen.
- **Cellulose** is a polymer of glucose that makes up the structure of plant cell walls.

Lipids

Fats and oils, which are composed of glycerol and fatty acids, are called triglycerides. Triglycerides made of saturated fatty acids (having no double bonds) are solids and are called fats. Triglycerides composed of unsaturated fatty acids (having double bonds) are liquids and are called oils.

Phospholipids have the same structure as triglycerides, except that a group containing phosphate takes the place of one fatty acid. Phospholipids make up the plasma membrane as well as other cellular membranes.

Steroids are composed of four fused hydrocarbon rings. Cholesterol, a steroid, is a component of the plasma membrane. The sex hormones testosterone and estrogen are steroids.

Proteins

Proteins are polymers of amino acids. A peptide is composed of two amino acids; a polypeptide contains many amino acids. A peptide bond is a polar covalent bond.

A protein has levels of structure:

- Primary structure is the primary sequence of amino acids.
- Secondary structure is a helix or pleated sheet.
- Tertiary structure forms due to bending and twisting of the secondary structure, as seen in globular proteins.
- Quaternary structure occurs when a protein has more than one polypeptide.

Nucleic Acids

Cells and organisms differ because of their proteins, which are coded for by genes composed of nucleic acids. Nucleic acids are polymers of nucleotides.

DNA (deoxyribonucleic acid) contains deoxyribose sugar and is the chemical that makes up our genes. DNA is double-stranded and shaped like a double helix. The two strands of DNA are joined by complementary base pairing: Adenine (A) pairs with thymine (T), and guanine (G) pairs with cytosine (C).

RNA (ribonucleic acid) serves as a helper to DNA during protein synthesis. Its sugar is ribose, and it contains the base uracil in place of thymine. RNA is single-stranded and thus does not form a double helix. Table 3.1 compares the structure of DNA and RNA.

Table 3.1

DNA STRUCTURE COMPARED TO RNA STRUCTURE

	DNA	RNA
Sugar	Deoxyribose	Ribose
Bases	Adenine, guanine, thymine, cytosine	Adenine, guanine, uracil, cytosine
Strands	Double-stranded with base pairing	Single-stranded
Helix	Yes	No

Thinking Scientifically

1. Nutritionists advise us to avoid consuming oils from tropical plants because, compared to oils from temperate plants, they contain high levels of saturated fatty acids. Let's try to determine a physiological reason for this difference in fatty acid composition.

   Look at the following graph and answer questions a–f.

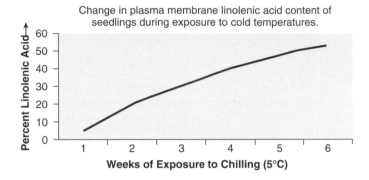

Change in plasma membrane linolenic acid content of seedlings during exposure to cold temperatures.

a. What plant cell structure are we considering here?

b. Linolenic acid is an unsaturated fatty acid. What happens to its level following exposure of plants to cold temperatures?

c. If you put a product with a high level of unsaturated fats (such as canola oil) and a product with a high level of saturated fats (such as butter) in your refrigerator (chilling temperature), how would they differ in consistency?

d. Does chilling alter protein structure and function? In other words, does chilling cause denaturation? (Think of whether meat or eggs are damaged by chilling.)

e. Therefore, which of the two membrane components—fatty acids or proteins—will be most dramatically affected by chilling temperatures?

f. Finally, how does fatty acid composition relate to chilling tolerance?

2. In order to understand the relationship between enzyme structure and function, researchers often study mutations that result in altered enzyme structure. In one bacterial enzyme, function is retained if a substituted amino acid has a nonpolar *R* group, but function is lost if the substituted amino acid has a polar *R* group. Why might that be?

Testing Yourself

Choose the best answer for each question.

1. Which of the following is an organic molecule?
   a. CO_2
   b. H_2O
   c. $C_6H_{12}O_6$
   d. O_2
   e. More than one of these is correct.

2. Carbon requires how many electrons to complete its outer shell?
   a. 2
   b. 3
   c. 4
   d. Any one of these is correct.

3. A hydrocarbon chain
   a. contains carbon and hydrogen atoms only.
   b. is hydrophobic.
   c. is hydrophilic.
   d. Both a and b are correct.

4. Carbon chains can vary in
   a. length.
   b. number of double bonds.
   c. branching pattern.
   d. All of these are correct.

5. Organic molecules containing carboxyl groups are
   a. nonpolar.
   b. acidic.
   c. basic.
   d. More than one of these is correct.

6. Monomers are attached together to create polymers when a hydroxyl group and a hydrogen atom are _____ in a _____ reaction.
   a. added, dehydration
   b. removed, dehydration
   c. added, hydrolysis
   d. removed, hydrolysis

7. Which of the following is a disaccharide?
   a. glucose
   b. ribose
   c. fructose
   d. sucrose

For questions 8–15, match the items to those in the key. Answers can be used more than once.

Key:
   a. carbohydrate
   b. lipid
   c. protein
   d. nucleic acid

8. Cellulose, the major component of plant cell walls.

9. Keratin, found in hair, fingernails, horns, and feathers.

10. Steroids such as cholesterol and sex hormones.

11. Composed of nucleotides.

12. Insoluble in water due to hydrocarbon chains.

13. Sometimes undergoes complementary base pairing.

14. May contain pleated sheets and helices.

15. May be a ring of six carbon atoms attached to hydroxyl groups.

16. A triglyceride contains
   a. glycerol and three fatty acids.
   b. glycerol and three sugars.
   c. protein and three fatty acids.
   d. protein and three sugars.

17. Variations in three-dimensional shapes among proteins are due to bonding between the
   a. amino groups.
   b. ion groups.
   c. *R* groups.
   d. H atoms.

18. The following picture illustrates a(n)
   a. saturated fatty acid.
   b. monounsaturated fatty acid.
   c. polyunsaturated fatty acid.
   d. All of these are correct.

19. The three-dimensional structure of a protein that contains two or more polypeptides is the
    a. primary structure.   c. tertiary structure.
    b. secondary structure.   d. quaternary structure.

For questions 20–24, match the structure in the following diagram with the name of the functional group.

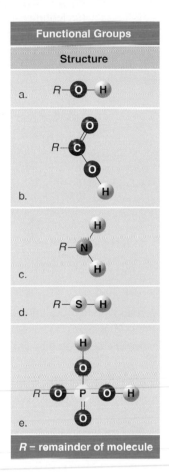

20. Phosphate
21. Carboxyl
22. Hydroxyl
23. Sulfhydral
24. Amino
25. Carbon and oxygen can be found in
    a. amino acids.
    b. glucose.
    c. starch.
    d. All of these are correct.
26. Plants store glucose as
    a. maltose.
    b. glycogen.
    c. starch.
    d. None of these are correct.

27. The polysaccharide found in plant cell walls is
    a. glucose.   c. maltose.
    b. starch.   d. cellulose.
28. Phosphates can be found in
    a. DNA.   c. glucose.
    b. RNA.   d. Both a and b are correct.
29. All _____ are _____.
    a. proteins, enzymes
    b. sugars, monosaccharides
    c. enzymes, proteins
    d. sugars, polysaccharides
30. _____ is the precursor of _____.
    a. Estrogen, cholesterol
    b. Cholesterol, glucose
    c. Testosterone, cholesterol
    d. Cholesterol, testosterone and estrogen
31. An example of a hydrocarbon would be
    a. heptane.
    b. glucose.
    c. maltose.
    d. None of these are correct.
32. A 5-carbon sugar is associated with
    a. glucose.
    b. a protein.
    c. DNA.
    d. lipids.
33. Nucleotides
    a. contain sugar, a nitrogen-containing base, and a phosphate molecule.
    b. are the monomers for fats and polysaccharides.
    c. join together by covalent bonding between the bases.
    d. are found in DNA, RNA, and proteins.
34. The joining of two adjacent amino acids is called a
    a. peptide bond.
    b. dehydration reaction.
    c. covalent bond.
    d. All of these are correct.
35. Which of the following pertains to an RNA nucleotide and not to a DNA nucleotide?
    a. contains the sugar ribose
    b. contains a nitrogen-containing base
    c. contains a phosphate molecule
    d. becomes bonded to other nucleotides by condensation
36. Saturated fatty acids and unsaturated fatty acids differ in
    a. the number of carbon-to-carbon bonds.
    b. their consistency at room temperature.
    c. the number of hydrogen atoms present.
    d. All of these are correct.
37. Which of the following is an example of a polysaccharide used for energy storage?
    a. cellulose
    b. glycogen
    c. cholesterol
    d. glucose

Go to www.mhhe.com/maderessentials for more quiz questions.

Bioethical Issue

Almost everything we buy is wrapped in plastic. Plastic is an ideal packaging material because it protects the product and is inexpensive. However, as our uses for plastic grow, problems associated with its manufacture and disposal also grow. Plastics are made from fossil fuels, a nonrenewable resource. In addition, the polymers used to produce plastic cannot be broken down by microorganisms, and therefore plastics are not biodegradable. Thus, researchers have been trying to create biodegradable plastics from a renewable resource—plants.

Starch can be made into a bioplastic (biodegradable plastic), but it absorbs water, so it becomes deformed when wet. Alternatively, starch can be converted to lactic acid, and then the lactic acid molecules can be linked together to form a more stable bioplastic called polylactide. As you might expect, this extra processing adds to the cost of the plastic. Biodegradable plastics are 2–10 times more expensive than conventional plastic. In addition, in order to get them to decompose, they must be composted instead of sealed in a landfill.

Would you pay more for products wrapped in bioplastic to benefit the environment? How much more? Would you make the effort to separate bioplastics from the remainder of your garbage so they could be sent to a composting site? Who should pay for more bioplastics research—the federal government or companies that manufacture plastic?

Understanding the Terms

amino acid   39
carbohydrate   33
cellulose   34
chitin   34
cholesterol   37
dehydration reaction   33
denatured   40
deoxyribose   33
disaccharide   33
DNA
   (deoxyribonucleic acid)   41
enzyme   38
fat   35
fatty acid   35
fibrous protein   40
functional group   31
globular protein   40
glucose   33
glycerol   35
glycogen   34
hemoglobin   38
hydrolysis reaction   33
hydrophilic   31
hydrophobic   31
inorganic chemistry   30
isomer   31
lipid   35
monomer   32
monosaccharide   33
nucleic acid   41
nucleotide   41
oil   35
organic   30
organic chemistry   30
peptide   40
peptide bond   39
phospholipid   36
polymer   32
polypeptide   40
polysaccharide   34
protein   38
ribose   33
RNA (ribonucleic acid)   41
saturated fatty acid   36
starch   34
steroid   37
unsaturated fatty acid   36

Match the terms to these definitions:

a. _____ Molecules with identical molecular formulas but different arrangements of atoms.

b. _____ Macromolecule created by linking together many monomers.

c. _____ Four fused hydrocarbon rings.

d. _____ A molecule that speeds a reaction by bringing reactants together.

e. _____ A glycerol molecule attached to two fatty acids and a phosphate group.

f. _____ The condition of an enzyme when it has lost its shape and can no longer function.

Inside the Cell

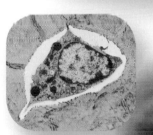

A bone cell encased within bone.

Cone cell in eye accounts for color vision.

The human body is composed of up to 100 trillion cells.

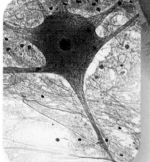

A nerve cell within the brain.

An egg cell protected within an ovarian follicle.

A cell caught in the act of committing suicide, a necessary part of development.

Red blood cells coursing through a blood vessel.

Red blood cells are the most abundant cell type in your body, and they perform a critical job—delivering oxygen throughout your body and picking up carbon dioxide waste from its cells. Along with other types of blood cells, red blood cells are produced from precursor cells located in the marrow of your bones. Red blood cells do not resemble a typical animal cell in structure because they are chock full of hemoglobin, the pigment that makes them red and enables

▶ ▶ ▶

OUTLINE

them to carry oxygen. Because red blood cells do not have the usual contents of an animal cell, they only live about 120 days before being destroyed by the liver and the spleen. Thus, the body continually faces the daunting task of replacing the three million red blood cells it loses every day. If this process of replacing the red blood cells is impaired, as occurs in a condition called aplastic anemia, a person can quickly die.

Cells are the fundamental unit of all living things. Although the diversity of organisms is incredible, the cells of all organisms share many similarities. In fact, there are only a few structural differences between most types of cells. In this chapter you will learn about the major types of cells and how they are similar as well as different. Learning about the structure of cells will later help you understand their complex functioning.

4.1   Cells Under the Microscope

Cells are extremely diverse, but nearly all require a microscope to see them (Fig. 4.1). When we speak of "a cell" or "the cell," we mean a cell that can be discussed in general terms. Our own bodies are composed of several hundred cell types, and each type is present billions of times over. There are nerve cells, muscle cells, gland cells, and bone cells, to name a few. Each type of cell is specialized to perform a particular function—a nerve cell conducts, a muscle cell contracts, a gland cell secretes, and a bone cell supports.

Cells are complex and yet tiny. The light microscope, invented in the seventeenth century, allows us to see cells but not much of their complexity. That's because the properties

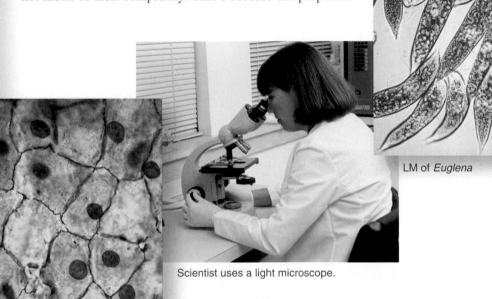

LM of *Euglena*

TEM of human lymphocyte

LM of human epithelial cells

Scientist uses a light microscope.

Scientist uses an electron microscope.

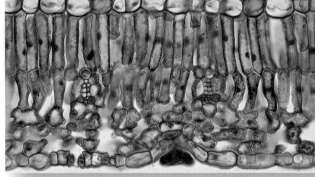

LM of leaf cells

Figure 4.1   Diversity of cells.

All organisms are composed of cells, but it takes a microscope to see most of them. Today, three basic types of microscopes are in use: the light microscope, the transmission electron microscope, and the scanning electron microscope. The light microscope and the transmission electron microscope reveal the insides of cells, while the scanning electron microscope shows three-dimensional surface features. Pictures resulting from the use of the light microscope are called light micrographs (LM), and those that result from the use of the transmission electron microscope are called transmission electron micrographs (TEM). Use of the scanning electron microscope results in scanning electron micrographs (SEM).

Figure 4.2   Relative sizes of some living things and their components.

This diagram not only gives you an idea of the relative sizes of organisms, cells, and their components, but it also shows which of them can be seen with an electron microscope, a light microscope, and the unaided eye. These sizes are given in metric units, with each higher unit ten times greater than the lower unit. See Appendix B for the complete metric system.

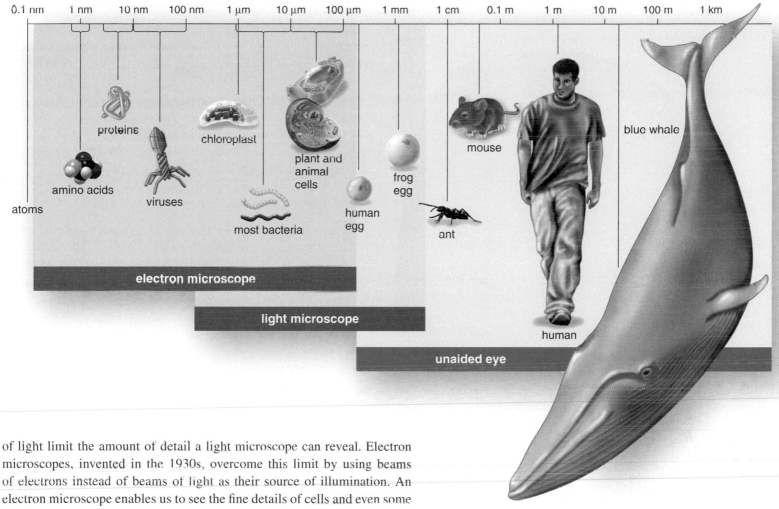

of light limit the amount of detail a light microscope can reveal. Electron microscopes, invented in the 1930s, overcome this limit by using beams of electrons instead of beams of light as their source of illumination. An electron microscope enables us to see the fine details of cells and even some of the larger molecules within them. Figure 4.2 compares the visual ranges of the electron microscope, the light microscope, and the unaided eye.

Why are cells so small? To answer this question, consider that a cell needs a surface area large enough to allow adequate nutrients to enter it and to rid itself of wastes. If you were to cut a large cube into smaller cubes, the smaller cube would have a lot more surface area per volume than the large cube. For example, a 4-cm cube has a **surface-area-to-volume ratio** of only 1.5:1, but a 1 cm cube has a surface-area-to-volume ratio of 6:1. Therefore, small cells, not large cells, are most likely to have an adequate surface area for exchanging wastes and nutrients.

A chicken's egg is a cell several centimeters in diameter, but it is not actively metabolizing. If development begins, cell division occurs and provides the surface area needed for exchange of materials. Animal cells that specialize in absorption have ways to increase their surface-area-to-volume ratio. Often they have projections called microvilli (sing., microvillus), which increase the surface area.

Check Your Progress

Explain why a large surface-to-volume ratio is needed for the proper functioning of cells.

Answer: A cell needs a relatively large surface area for absorption of nutrients and secretion of wastes.

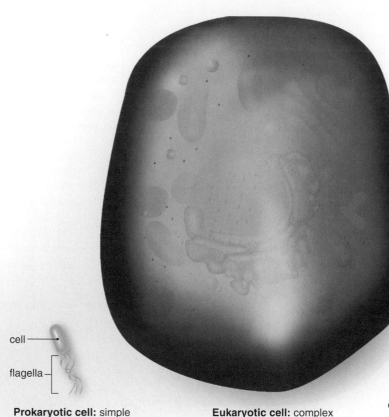

cell

flagella

Prokaryotic cell: simple
internal structure

Eukaryotic cell: complex
internal structure

Figure 4.3   Comparison of prokaryotic cells and eukaryotic cells.

Prokaryotic cells are much smaller than eukaryotic cells. They are also less complex in structure, as we shall see.

4.2   The Two Main Types of Cells

All cells have an outer membrane called a plasma membrane. The plasma membrane encloses a semifluid substance called the **cytoplasm** and the cell's genetic material. The plasma membrane regulates what enters and exits cells; the cytoplasm carries on chemical reactions; and the genetic material provides the information needed for growth and reproduction. When a cell reproduces, it produces more cells. The concepts that all organisms are composed of cells, and that cells come only from preexisting cells, are the two central tenets of the **cell theory.**

Cells are divided into two main types according to the way their genetic material is organized. **Prokaryotic cells** are so named because they lack a membrane-bounded nucleus. Their DNA is located in a region of the cytoplasm called the nucleoid. The other type of cell, called a **eukaryotic cell,** has a nucleus that houses its DNA.

Prokaryotic Cells

Prokaryotic cells are much smaller in size and simpler in structure than eukaryotic cells (Fig. 4.3). Their small size allows them to exist in great numbers in the air, in bodies of water, in the soil, and even on you. They are an extremely successful group of organisms, whose evolutionary history dates back to the first cells on Earth.

Bacteria, a type of prokaryotic cell, are well known because they cause some serious diseases, including tuberculosis, throat infections, and gonorrhea. Even so, the biosphere would not long continue without bacteria. Many bacteria decompose dead remains and contribute to ecological cycles. Bacteria also assist humans in another way—we use them to manufacture all sorts of products, from industrial chemicals to foodstuffs and drugs. The active cultures found in a container of yogurt are bacteria that can be beneficial to us. Also, much of our knowledge about how DNA specifies the sequence of amino acids in proteins was learned by doing experiments utilizing *E. coli,* a bacterium that lives in the human intestine. The fact that this information applies to all prokaryotes and eukaryotes, even ourselves, reveals the remarkable unity of living things and gives evidence that all organisms share a common descent.

Bacterial Structure

In bacteria, the cytoplasm is surrounded by a plasma membrane, a cell wall, and possibly a capsule (Fig. 4.4). The cytoplasm contains a variety of different enzymes. These organic catalysts speed many types of chemical reactions that are required to maintain an organism. Both prokaryotes and eukaryotes have the same type of plasma membrane; its structure is discussed on page 52. The **cell wall** maintains the shape of the cell, even if the cytoplasm should happen to take up an abundance of water. The **capsule** is a protective layer of polysaccharides lying outside the cell wall.

The DNA of a bacterium is located in a single coiled chromosome that resides in a region called the **nucleoid.** The many proteins specified by bacterial DNA are synthesized on tiny particles called **ribosomes.** A bacterial cell contains thousands of ribosomes.

The appendages of a bacterium, namely flagella, fimbriae, and sex pili, are made of protein. Motile bacteria can propel themselves in water because they have **flagella** (sing., flagellum), which are usually about 20 nm in diameter and 1–70 nm long. The bacterial flagellum has a filament, a hook, and a basal body, which is a series of rings anchored in the cell wall and plasma membrane. The hook rotates 360° within the basal body, and this rotary motion propels bacteria—the bacterial flagellum does not move back and forth like a whip. Sometimes flagella occur only at the two ends of a cell, and other times they are dispersed randomly over the surface.

Fimbriae are small, bristlelike fibers that sprout from the cell surface. They don't have anything to do with motility, but they do help bacteria attach to a surface. **Sex pili** are rigid tubular structures used by bacteria to pass DNA from cell to cell. Bacteria can also take up DNA directly from the external medium or by way of viruses.

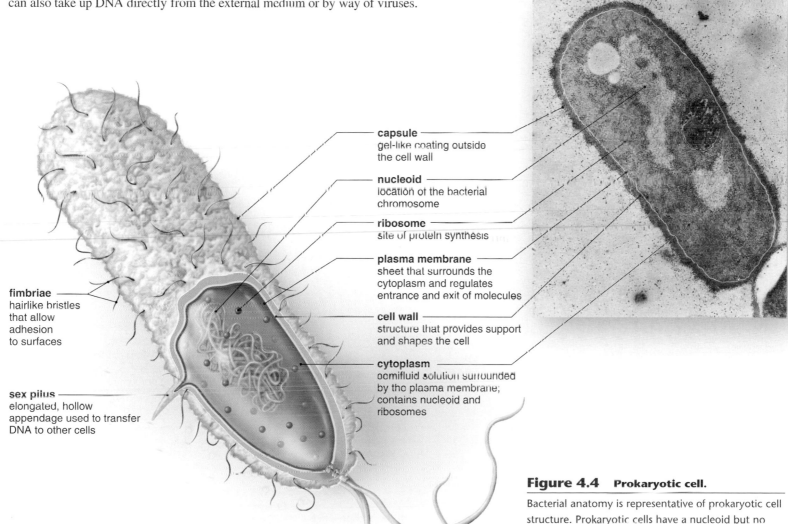

capsule
gel-like coating outside the cell wall

nucleoid
location of the bacterial chromosome

ribosome
site of protein synthesis

plasma membrane
sheet that surrounds the cytoplasm and regulates entrance and exit of molecules

cell wall
structure that provides support and shapes the cell

cytoplasm
semifluid solution surrounded by the plasma membrane; contains nucleoid and ribosomes

fimbriae
hairlike bristles that allow adhesion to surfaces

sex pilus
elongated, hollow appendage used to transfer DNA to other cells

flagellum
rotating filament that propels the cell

Figure 4.4   Prokaryotic cell.

Bacterial anatomy is representative of prokaryotic cell structure. Prokaryotic cells have a nucleoid but no membrane-bounded nucleus.

Check Your Progress

1. Briefly describe the structure of the plasma membrane.
2. List six types of proteins found in the plasma membrane.

Answers: 1. The plasma membrane is composed of a phospholipid bilayer, with the fatty acid tails pointing inward and many embedded proteins. 2. Channel, transport, cell recognition, receptor, enzymatic, and junction proteins.

The **plasma membrane** marks the boundary between the outside and inside of a cell. Its integrity and function are necessary to the life of the cell because it regulates the passage of molecules and ions into and out of the cell.

In both prokaryotes and eukaryotes, the plasma membrane is a phospholipid bilayer (see Fig. 3.14). The polar heads of the phospholipids face toward the outside of the cell and toward the inside of the cell where there is a watery medium. The nonpolar tails face inward toward each other, where there is no water. Cholesterol molecules, if present, lend support to the membrane, which has the consistency of olive oil. Short chains of sugars are attached to the outer surface of some proteins, forming glycoproteins. The sugar chain helps a protein perform its particular function.

The fluid-mosaic model states that the protein molecules embedded in the membrane have a pattern (form a mosaic) within the phospholipid bilayer (Fig. 4.5). The pattern varies according to the particular membrane and also within the same membrane at different times. When you consider that the plasma membrane of a red blood cell contains over 50 different types of proteins, you can see why the membrane is said to be a mosaic.

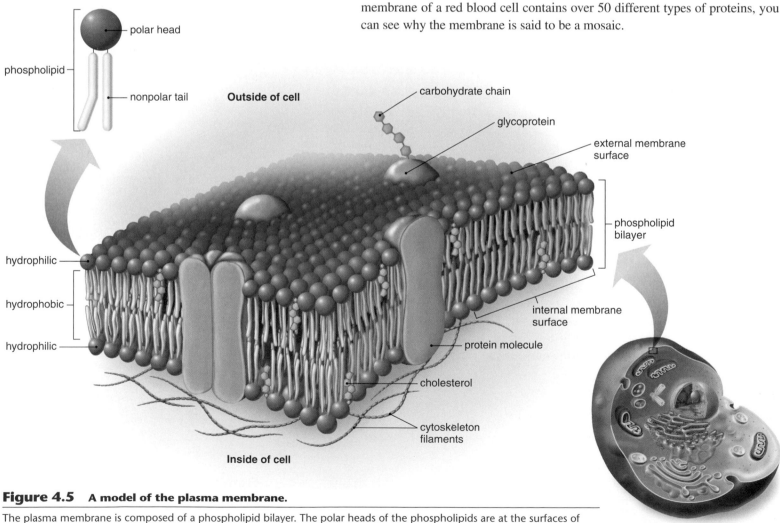

Figure 4.5   A model of the plasma membrane.

The plasma membrane is composed of a phospholipid bilayer. The polar heads of the phospholipids are at the surfaces of the membrane; the nonpolar tails make up the interior of the membrane. Proteins embedded in the membrane have various functions (see Fig. 4.6).

Functions of Membrane Proteins

The proteins on the internal surface of the plasma membrane often help stabilize and shape the membrane. The embedded proteins also perform specific functions. There are several types of embedded proteins.

Channel Proteins

Channel proteins permit the passage of molecules through the membrane. Via a channel, a specific substance can simply move across the membrane (Fig. 4.6a). For example, a channel protein allows hydrogen ions to flow across the inner mitochondrial membrane. Without this movement of hydrogen ions, ATP would never be produced.

Transport Proteins

Transport proteins are also involved in the passage of molecules through the membrane. They combine with a substance and help it move across the membrane (Fig. 4.6b). For example, a transport protein conveys sodium and potassium ions across a nerve cell membrane. Without this transport protein, nerve conduction would be impossible.

Cell Recognition Proteins

Cell recognition proteins are glycoproteins (Fig. 4.6c). Among other functions, these proteins enable our bodies to recognize a pathogen invasion so that an immune reaction can occur. Without this recognition, pathogens would be able to freely invade the body.

Receptor Proteins

A receptor protein has a shape that allows a specific molecule to bind to it (Fig. 4.6d). The binding of this molecule causes the protein to change its shape, and thereby bring about a cellular response. The coordination of the body's organs is totally dependent on such signaling molecules. For example, the liver stores glucose after it is signaled to do so by insulin.

Enzymatic Proteins

Some plasma membrane proteins are enzymatic proteins that carry out metabolic reactions directly (Fig. 4.6e). Without enzymes, some of which are attached to the various membranes of the cell, a cell would never be able to perform the degradative and synthetic reactions that are important to its function.

Junction Proteins

As discussed on page 65, proteins are also involved in forming various types of junctions between cells (Fig. 4.6f). The junctions assist cell-to-cell communication.

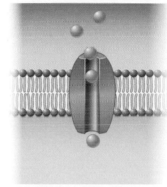
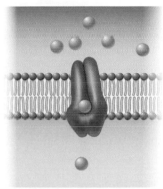

a. Channel protein   b. Transport protein

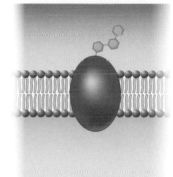
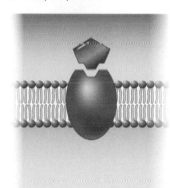

c. Cell recognition protein   d. Receptor protein

e. Enzymatic protein   f. Junction proteins

Figure 4.6   Membrane protein diversity.

These are some of the types of proteins embedded in the plasma membrane.

4.4 Eukaryotic Cells

As discussed in Chapter 1, protists, fungi, plants, and animals are all composed of eukaryotic cells. Unlike prokaryotic cells, eukaryotic cells have a membrane-bounded **nucleus** that houses their DNA. The DNA is located in several chromosomes.

Eukaryotic cells are much larger than prokaryotic cells, and therefore, as you might expect from the previous discussion, they have less surface area per volume than prokaryotic cells. Wouldn't this circumstance be a hindrance to their very existence? No, because their interior is compartmentalized into small structures called **organelles** that differ in structure and function.

In this chapter, the organelles are divided into four categories:

1. The nucleus and the ribosomes. The nucleus communicates with ribosomes in the cytoplasm.
2. Organelles of the endomembrane system (ES). Each of these organelles has its own particular set of enzymes and produces its own products. The

×10,000

a.

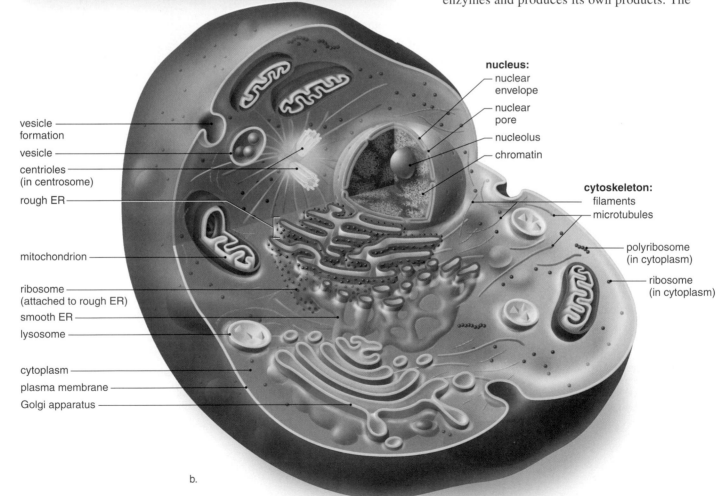

b.

Figure 4.7 **Structure of a typical animal cell.**

a. False-colored TEM of an animal cell. **b.** Generalized drawing.

products are carried between the ES organelles by transport vesicles, small membranous sacs that keep the products from entering the cytoplasm.

3. The energy-related organelles. The energy-related organelles—chloroplasts in plant cells and mitochondria in plant and animal cells—are self-contained. They even have their own genetic material, and their ribosomes resemble those of prokaryotic cells.

4. The cytoskeleton. The **cytoskeleton** is a lattice of protein filaments and tubules that maintains the shape of the cell and assists in the movement of organelles. Without an efficient means of moving organelles and their products, eukaryotic cells could not exist. The manner in which vesicles and other organelles move along the tracks provided by the cytoskeleton will be discussed in more detail later.

Each structure in the typical animal and plant cells shown in Figures 4.7 and 4.8 has been given a particular color, and this color will be used to represent the same structure throughout this text.

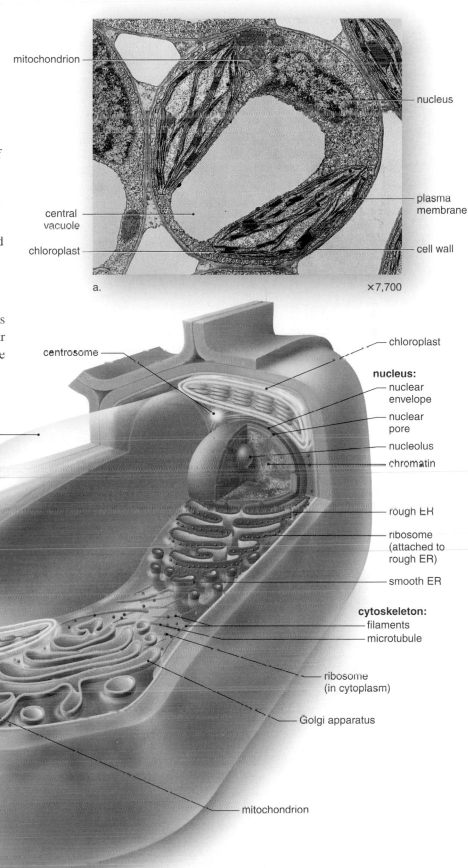

Figure 4.8 Structure of a typical plant cell.

a. False-colored TEM of a plant cell. **b.** Generalized drawing.

Nucleus and Ribosomes

The nucleus stores genetic information, and the ribosomes in the cytoplasm use this information to carry out protein synthesis.

The Nucleus

Because of its large size, the nucleus is a noticeable structure in the eukaryotic cell (Fig. 4.9). The nucleus contains **chromatin** within a semifluid nucleoplasm. Chromatin looks grainy, but actually it is a network of strands. Just before the cell divides, the chromatin condenses and undergoes coiling into rodlike structures called **chromosomes.** All of the cells of an organism contain the same number of chromosomes except for the egg and sperm, which have half this number.

Chromatin (and therefore chromosomes) is composed of DNA, protein, and some RNA. The genes are composed of DNA, and are located in the chromosomes. The nucleus therefore contains the genetic information that is passed on from cell to cell and from generation to generation.

RNA, which exists in several forms, is produced in the nucleus and has a sequence of bases that mirrors the sequence in DNA. A **nucleolus** is a dark region of chromatin where a type of RNA called ribosomal RNA (rRNA) is produced, and where rRNA joins with proteins to form the subunits of ribosomes. **Ribosomes** are small bodies in the cytoplasm where protein synthesis occurs. Another type of RNA, called messenger RNA (mRNA), acts as an intermediary for DNA. It helps specify the sequence of amino acids during protein synthesis. The proteins of a

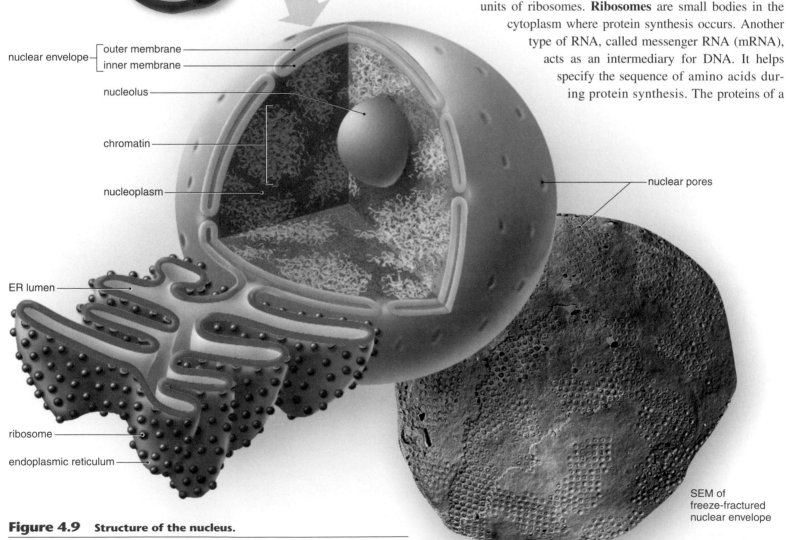

nuclear envelope
outer membrane
inner membrane
nucleolus
chromatin
nucleoplasm
nuclear pores
ER lumen
ribosome
endoplasmic reticulum
SEM of freeze-fractured nuclear envelope

Figure 4.9   **Structure of the nucleus.**

The nuclear envelope contains pores that allow substances to pass from the nucleus to the cytoplasm.

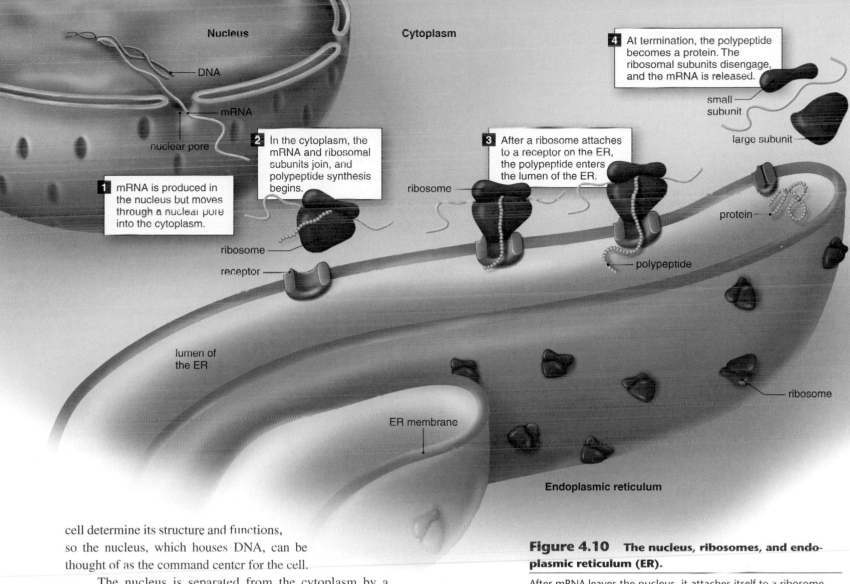

Nucleus

Cytoplasm

DNA

mRNA

nuclear pore

1 mRNA is produced in the nucleus but moves through a nuclear pore into the cytoplasm.

2 In the cytoplasm, the mRNA and ribosomal subunits join, and polypeptide synthesis begins.

3 After a ribosome attaches to a receptor on the ER, the polypeptide enters the lumen of the ER.

4 At termination, the polypeptide becomes a protein. The ribosomal subunits disengage, and the mRNA is released.

small subunit

large subunit

ribosome

ribosome

receptor

protein

polypeptide

lumen of the ER

ER membrane

ribosome

Endoplasmic reticulum

cell determine its structure and functions, so the nucleus, which houses DNA, can be thought of as the command center for the cell.

The nucleus is separated from the cytoplasm by a double membrane known as the **nuclear envelope.** Even so, the nucleus communicates with the cytoplasm. The nuclear envelope has **nuclear pores** of sufficient size (100 nm) to permit the passage of ribosomal subunits and mRNA out of the nucleus into the cytoplasm, and the passage of proteins from the cytoplasm into the nucleus.

Ribosomes

Ribosomes are found in both prokaryotes and eukaryotes. In both types of cells, ribosomes are composed of two subunits, one large and one small. Each subunit has its own mix of proteins and rRNA. As mentioned, ribosomes are sites of protein synthesis. They receive mRNA from the nucleus, which carries a coded message from DNA. The mRNA encodes the correct sequence of amino acids in a polypeptide.

In eukaryotic cells, some ribosomes occur freely within the cytoplasm, either singly or in groups called polyribosomes. Polypeptides synthesized by these ribosomes are used in the cytoplasm. Many ribosomes are attached to the **endoplasmic reticulum (ER),** an organelle composed of many saccules and channels. After the ribosome binds to a receptor at the ER, the polypeptide being synthesized enters the lumen of the ER (Fig. 4.10).

Figure 4.10   The nucleus, ribosomes, and endoplasmic reticulum (ER).

After mRNA leaves the nucleus, it attaches itself to a ribosome, and polypeptide synthesis begins. When a ribosome combines with a receptor at the endoplasmic reticulum (ER), the polypeptide enters the lumen of the ER through a channel in the receptor. Exterior to the ER, the ribosome splits, releasing the mRNA while a protein takes shape inside the ER lumen.

Check Your Progress

1. List the components of the nucleus and give a function for each.
2. Where are ribosomes found in the cell, and what do they do?

Answers: 1. nuclear envelope—defines the nucleus; nuclear pore—allows substances to move into and out of nucleus; nucleolus—formation of ribosomal RNA (rRNA); chromatin—becomes chromosomes and contains DNA. 2. Ribosomes are found attached to the ER and in the cytoplasm. In the cytoplasm, they occur either singly or as polyribosomes. Ribosomes carry out protein synthesis.

Endomembrane System

The **endomembrane system** consists of the nuclear envelope, the membranes of the endoplasmic reticulum, the Golgi apparatus, and many small membranous sacs called **vesicles.** This system helps compartmentalize the cell, so that particular enzymatic reactions are restricted to specific regions. Transport vesicles carry molecules from one part of the system to another.

Endoplasmic Reticulum

The endoplasmic reticulum (ER) consists of a complicated system of membranous channels and saccules (flattened vesicles). It is physically continuous with the outer membrane of the nuclear envelope (Fig. 4.11).

Rough ER is studded with ribosomes on the side of the membrane that faces the cytoplasm; therefore, rough ER is able to synthesize proteins. It also modifies proteins after they have entered the central enclosed region of the ER, called the lumen. The rough ER forms vesicles in which large molecules are transported to other parts of the cell. Often these vesicles are on their way to the plasma membrane or the Golgi apparatus (described below).

Smooth ER, which is continuous with rough ER, does not have attached ribosomes. Smooth ER synthesizes lipids, such as phospholipids and steroids. The functions of smooth ER are dependent on the particular cell. In the testes, it produces testosterone, and in the liver, it helps detoxify drugs. Regardless of any specialized function, rough and smooth ER also form transport vesicles that carry molecules to other parts of the cell, notably the Golgi apparatus.

Golgi Apparatus

The **Golgi apparatus,** named for its discoverer, Camillo Golgi, consists of a stack of three to twenty slightly curved, flattened saccules resembling pancakes (Fig. 4.12). The Golgi apparatus receives vesicles and alters their contents as molecules move through saccules in the Golgi. For example, the Golgi can substitute one sugar for another when a sugar chain is attached to a protein. This action can alter recognition of the protein and change its destination in the cell.

The Golgi apparatus packages its products in new vesicles. In animal cells, some of these vesicles are lysosomes, which are discussed next. Some transport vesicles proceed to the plasma membrane, where they discharge their contents during **secretion.**

Lysosomes

Lysosomes are vesicles produced by the Golgi apparatus that digest molecules and even portions of the cell itself. Sometimes, after engulfing molecules outside the cell, a vesicle formed at the plasma membrane fuses with a lysosome. The contents of the vesicle are digested by lysosomal enzymes. In Tay-Sachs disease, a genetic disorder, lysosomes are missing an enzyme for a particular lipid molecule. The cells become so full of storage bodies that the individual dies, usually in childhood. Someday it may be possible to provide the missing enzyme for these children.

Lysosomes also participate in **apoptosis,** or programmed cell death, which is a normal part of development. When a tadpole becomes a frog, lysosomes digest away the cells of the tail. In humans, the fingers of an embryo are at first webbed, but they are freed from one another as a result of lysosomal action.

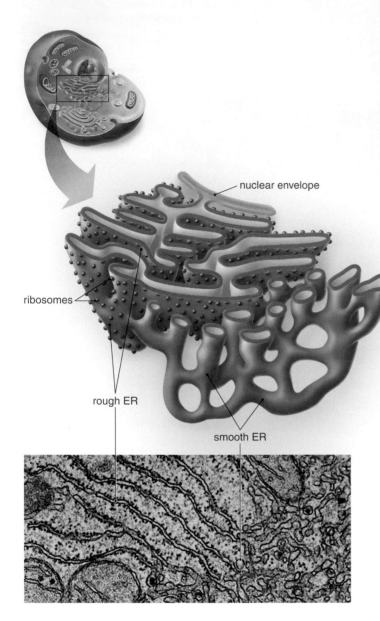

nuclear envelope

ribosomes

rough ER

smooth ER

Figure 4.11   Endoplasmic reticulum (ER).

Ribosomes are present on rough ER, which consists of flattened saccules, but not on smooth ER, which is more tubular. Proteins are synthesized and modified by rough ER; lipids are synthesized by smooth ER, which can have several other functions as well.

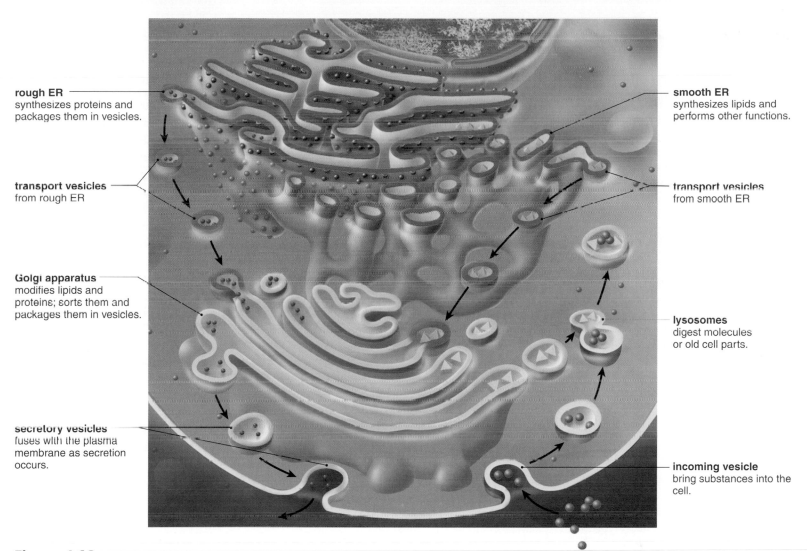

rough ER
synthesizes proteins and
packages them in vesicles.

transport vesicles
from rough ER

Golgi apparatus
modifies lipids and
proteins; sorts them and
packages them in vesicles.

secretory vesicles
fuses with the plasma
membrane as secretion
occurs.

smooth ER
synthesizes lipids and
performs other functions.

transport vesicles
from smooth ER

lysosomes
digest molecules
or old cell parts.

incoming vesicle
bring substances into the
cell.

Figure 4.12 **Endomembrane system.**

The organelles in the endomembrane system work together to carry out the functions noted. Plant cells do not have lysosomes.

Vacuoles

Vacuoles, like vesicles, are membranous sacs, but vacuoles are larger than
vesicles. The vacuoles of some protists are quite specialized; they include
contractile vacuoles for ridding the cell of excess water, and digestive
vacuoles for breaking down nutrients (Fig. 4.13a). Vacuoles usually store
substances, such as nutrients or ions. Plant vacuoles contain not only water,
sugars, and salts, but also pigments and toxic molecules (Fig. 4.13b). The
pigments are responsible for many of the red, blue, or purple colors of
flowers and some leaves. The toxic substances help protect a plant from
herbivorous animals.

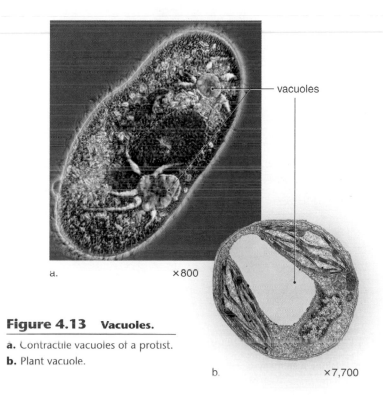

vacuoles

a.   ×800

Figure 4.13 **Vacuoles.**

a. Contractile vacuoles of a protist.
b. Plant vacuole.

b.   ×7,700

Check Your Progress

1. Contrast rough endoplasmic reticulum with smooth endoplasmic reticulum.
2. Describe the relationship between the components of the endomembrane system.

Answers: 1. Rough ER contains ribosomes, while smooth ER does not. Rough ER synthesizes proteins and modifies them, while smooth ER synthesizes lipids, among other activities. 2. Transport vesicles from the ER proceed to the Golgi apparatus. The Golgi apparatus modifies their contents and repackages them in new vesicles, some of which carry out secretion, and some of which are lysosomes.

Energy-Related Organelles

Chloroplasts and mitochondria are the two eukaryotic membranous organelles that specialize in energy conversion. **Chloroplasts** use solar energy to synthesize carbohydrates.

Mitochondria (sing., mitochondrion) break down carbohydrate-derived products to produce **adenosine triphosphate (ATP)** molecules. The production of ATP is of great importance because ATP serves as a carrier of energy in cells. The energy of ATP is used whenever a cell synthesizes molecules, transports molecules, or carries out a special function such as muscle contraction or nerve conduction. Without a constant supply of ATP, no cell could long exist.

Chloroplasts

The chloroplast, a plant cell organelle, is quite large, being twice as wide and as much as five times the length of a mitochondrion. Chloroplasts have a three-membrane system. They are bounded by a double membrane, which includes an outer membrane and an inner membrane. The large inner space, called the **stroma,** contains a concentrated mixture of enzymes and thylakoids. The **thylakoids** are disk-like sacs formed from the third membrane. A stack of thylakoids is called a **granum.** The lumens of thylakoid sacs are believed to form a large internal compartment called the thylakoid space (Fig. 4.14). The pigments that capture solar energy are located in the thylakoid membrane, and the enzymes that synthesize carbohydrates are in the stroma. The carbohydrates produced by chloroplasts serve as organic nutrient molecules for plants and indeed for all living things on Earth.

The discovery that chloroplasts have their own DNA and ribosomes supports a hypothesis that chloroplasts are derived from single-celled algae that entered a eukaryotic cell in the distant past. As shown in Figure 4.8, plant cells contain *both* mitochondria and chloroplasts, as do algal cells.

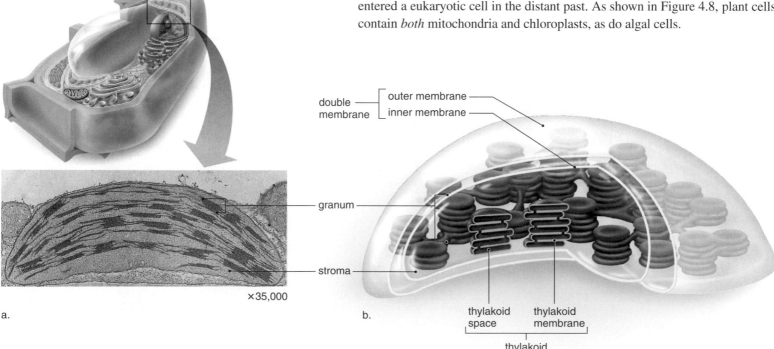

×35,000

a.

b.

double membrane — outer membrane — inner membrane

granum

stroma

thylakoid space   thylakoid membrane

thylakoid

Figure 4.14   Chloroplast structure.

a. Electron micrograph of a chloroplast, which carries out photosynthesis. **b.** Generalized drawing in which the outer and inner membranes have been cut away to reveal the grana, each of which is composed of a stack of membranous sacs called thylakoids. In some grana, but not all, it is obvious that the thylakoid spaces are interconnected.

Mitochondria

Even though mitochondria are smaller than chloroplasts, they can be seen with the light microscope in both plant and animal cells. We think of mitochondria as having a shape like that shown in Figure 4.15, but actually they often change shape, becoming longer and thinner or shorter and broader. Mitochondria can form long, moving chains, or they can remain fixed in one location—often where energy is most needed. For example, they are packed between the contractile elements of cardiac cells and wrapped around the interior of a sperm's flagellum.

Like chloroplasts, mitochondria are bounded by a double membrane. The inner membrane is highly convoluted into **cristae** that project into the matrix. These cristae increase the surface area of the inner membrane so much that in a liver cell they account for about one-third of the total membrane in the cell. The inner membrane encloses the **matrix,** which contains mitochondrial DNA and ribosomes. It has been suggested that mitochondria are derived from bacteria that took up residence in an early eukaryotic cell.

Mitochondria are often called the powerhouses of the cell because they produce most of the ATP the cell utilizes. (ATP is discussed on page 60.) Their matrix is a highly concentrated mixture of enzymes that break down carbohydrates and other nutrient molecules. These reactions supply the chemical energy that permits ATP synthesis to take place. The entire process, which also involves the cytoplasm, is called **cellular respiration** because oxygen is needed and carbon dioxide is given off.

Check Your Progress

Compare and contrast the structure and function of chloroplasts with those of mitochondria.

Answer: Structure: The two main parts of a chloroplast are the thylakoids and the stroma; the two main parts of a mitochondrion are the cristae and the matrix. Function: Chloroplasts are larger than mitochondria and capture energy from the sun to build carbohydrates. Mitochondria break down carbohydrates to release energy for ATP production.

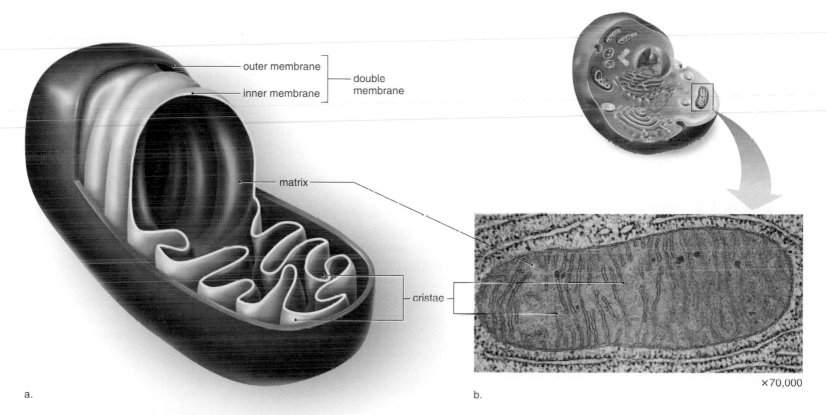

Figure 4.15 labels: outer membrane · inner membrane } double membrane · matrix · cristae

a.

b.

×70,000

Figure 4.15   Mitochondrion structure.

a. Generalized drawing in which the outer membrane and portions of the inner membrane have been cut away to reveal the cristae. **b.** Electron micrograph of a mitochondrion, which is involved in cellular respiration.

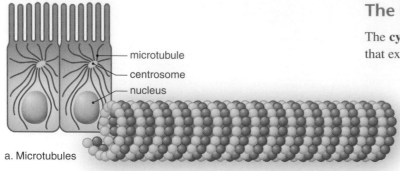

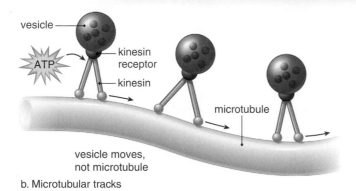

a. Microtubules

b. Microtubular tracks

Figure 4.16   Microtubules.

a. Microtubules (tubulin dimer cylinders) radiate out from the centrosome. **b.** Motor molecules, such as kinesin shown here, change the location of an organelle by moving it along a microtubule. One end of kinesin binds to an organelle, and the other end attaches, detaches, and reattaches to the microtubule. ATP supplies the energy for movement.

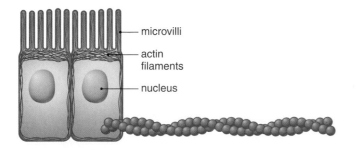

Figure 4.17   Actin filaments.

Actin filaments (helical actin polymers) are organized into bundles or networks just under the plasma membrane, where they lend support to the shape of a cell.

The Cytoskeleton

The **cytoskeleton** is a network of interconnected protein filaments and tubules that extends from the nucleus to the plasma membrane in eukaryotic cells. The cytoskeleton can be compared to the bones and muscles of an animal. Bones and muscles give an animal structure and produce movement. Similarly, the elements of the cytoskeleton maintain cell shape and allow the cell and its organelles to move. The cytoskeleton is dynamic in that the elements can assemble and disassemble as appropriate. The cytoskeleton includes microtubules, intermediate filaments, and actin filaments.

Microtubules

Each **microtubule** is a small, hollow cylinder (Fig. 4.16a). When microtubules assemble, tubulin molecules come together by twos that are arranged in rows. Microtubules have 13 rows of tubulin dimers surrounding what appears in electron micrographs to be an empty central core.

Microtubule assembly is controlled by a microtubule organizing center called the **centrosome,** which lies near the nucleus. Microtubules radiate from the centrosome, helping to maintain the shape of the cell and acting as tracks along which organelles can move. Whereas the motor molecule myosin is associated with actin filaments, the motor molecules kinesin and dynein are associated with microtubules. First, an organelle, such as a vesicle, combines with a motor molecule, and then the motor molecule attaches, detaches, and reattaches further along the cytoskeletal element, such as a microtubule. In this way, the organelle moves along the microtubular track (Fig. 4.16b).

Intermediate Filaments

Intermediate filaments are intermediate in size between actin filaments and microtubules. They are ropelike assemblies of fibrous polypeptides that typically run between the nuclear envelope and the plasma membrane. The network they form supports both the nucleus and the plasma membrane.

The protein making up intermediate filaments is different from one cell type to another. In the skin, intermediate filaments made of the protein keratin give great mechanical strength to skin cells.

Actin Filaments

Each **actin filament** consists of two chains of globular actin monomers twisted about one another in a helical manner. Actin filaments support the cell and any projections, such as microvilli. They form a dense, complex web just under the plasma membrane (Fig. 4.17).

When a muscle cell contracts, myosin acts as a motor molecule that pulls actin filaments toward the middle of the cell. Even in nonmuscle cells, myosin interacting with actin filaments produces movement as when cells move in an amoeboid fashion and/or engulf large particles. During animal cell division, the two new cells form when actin, in conjunction with myosin, pinches off the cells from one another. In plant cells, actin filaments apparently form the tracks along which chloroplasts circulate in the cytoplasm.

Centrioles

Centrioles are short cylinders with a 9 + 0 pattern of microtubule triplets—that is, nine sets of triplets occur in a ring, and none are in the middle of the cylinder (Fig. 4.18). In animal cells and most protists, two centrioles lie at right angles to one another in the middle of a centrosome. A centrosome, as mentioned previ-

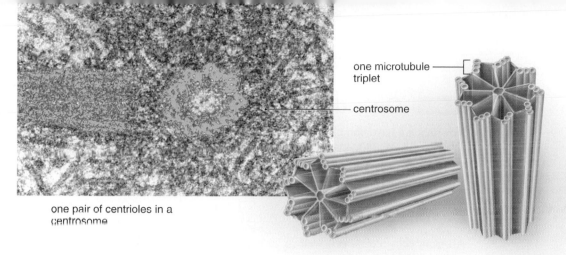

one pair of centrioles in a centrosome

one microtubule triplet

centrosome

Figure 4.18   Centrioles.

A pair of centrioles lies to one side of the nucleus. Their exact function is unknown, however possibilities do exist.

ously, is the major microtubule organizing center for the cell. Therefore, it is possible that centrioles are also involved in the process by which microtubules assemble and disassemble.

Before an animal cell divides, the centrioles replicate, and the members of each pair are at right angles to one another. Then each pair becomes part of a separate centrosome. During cell division, the centrosomes move apart and most likely play a role in organizing a microtubular apparatus (the spindle), which assists the movement of chromosomes. In any case, each new cell eventually has its own centrosome and pair of centrioles.

Plant and fungal cells have the equivalent of a centrosome, but this structure does not contain centrioles, suggesting that centrioles are not necessary for the assembly of cytoplasmic microtubules.

Cilia and Flagella

In eukaryotes, **cilia** and **flagella** are hairlike projections that can move either in an undulating fashion, like a whip, or stiffly, like an oar (Fig. 4.19a). Cells that have these organelles are capable of movement. Cilia are much shorter than flagella, but they have a similar construction. Both are membrane-bounded cylinders enclosing a matrix area. In the matrix are nine microtubule doublets arranged in a circle around two central microtubules, called the 9 + 2 pattern of microtubules (Fig. 4.19b). Cilia and flagella move when the microtubule doublets slide past one another.

Each cilium and flagellum has a basal body lying in the cytoplasm at its base. Basal bodies have the same circular arrangement of microtubule triplets as centrioles and are believed to be derived from them. Whether centrioles are involved in organizing the microtubules in cilia and flagella is still a matter of study.

cilia in bronchial wall

flagella of sperm

a.

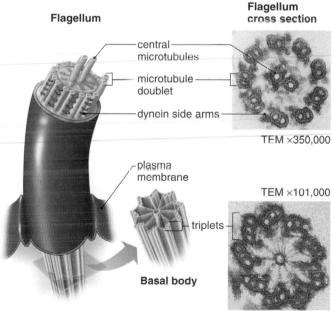

Flagellum

central microtubules

microtubule doublet

dynein side arms

Flagellum cross section

TEM ×350,000

plasma membrane

triplets

Basal body

TEM ×101,000

Basal body cross section

b.

Figure 4.19   Cilia and flagella.

a. Cilia in the bronchial wall and the flagella of sperm are organelles capable of movement. The cilia in our bronchi sweep mucus and debris back up into the throat where it can be swallowed or ejected. The flagella of sperm allow them to swim to the egg. **b.** Cilia and flagella have a distinct pattern of microtubules bounded by a plasma membrane. Both have a basal body with the same pattern of microtubules as a centriole. Therefore, it is believed that centrioles give rise to basal bodies.

Check Your Progress

1. List the components of the cytoskeleton.
2. Explain the structure of cilia and flagella.
3. Give an example of a cell that has cilia and one that has flagella. Describe the functions of these cells.

Answers: 1. Microtubules, intermediate filaments, and actin filaments. 2. Cilia and flagella are both composed of microtubules arranged in a particular pattern and enclosed by the plasma membrane. 3. Cells lining the respiratory tract have cilia that sweep mucus and debris back up into the throat where it can be swallowed or ejected; sperm have flagella that allow them to swim to the egg.

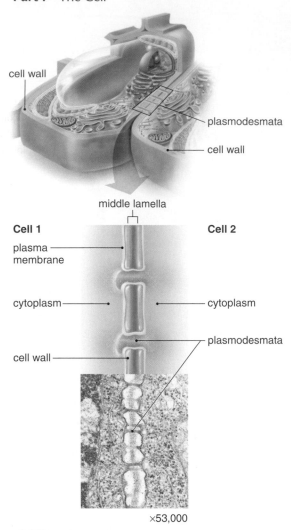

Figure 4.20   Plasmodesmata.

Plant cells are joined by membrane-lined channels called plasmodesmata that contain cytoplasm. Through these channels, water and small molecules can pass from cell to cell.

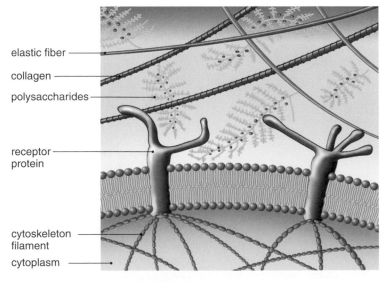

Figure 4.21   Animal cell extracellular matrix.

The extracellular matrix supports an animal cell and also affects its behavior.

4.5   Outside the Eukaryotic Cell

Now that we have completed our discussion of the plasma membrane and the cell contents, you might think we have finished our tour of the cell. However, most cells also have extracellular structures formed from materials the cell produces and transports across its plasma membrane.

Plant Cell Walls

All plant cells have a **cell wall.** A *primary cell wall* contains cellulose fibrils and noncellulose substances, and these allow the wall to stretch when the cell is growing. Adhesive substances are abundant outside the cell wall in the middle lamella, a layer that holds two plant cells together.

Some cells in woody plants have a *secondary cell wall* that forms inside the primary cell wall. The secondary wall has a greater quantity of cellulose fibrils, which are laid down at right angles to one another. Lignin, a substance that adds strength, is a common ingredient of secondary cell walls in woody plants.

In a plant, living cells are connected by **plasmodesmata** (sing., plasmodesma), numerous narrow, membrane-lined channels that pass through the cell wall (Fig. 4.20). Cytoplasmic strands within these channels allow direct exchange of some materials between adjacent plant cells and eventually among all the cells of a plant. The plasmodesmata allow only water and small solutes to pass freely from cell to cell.

Cell Surfaces in Animals

Animal cells do not have a cell wall, but they do have two other exterior surface features of interest: (1) An extracellular matrix exists outside the cell, and (2) various junctions occur between some cell types.

Extracellular Matrix

An **extracellular matrix** is a meshwork of polysaccharides and proteins in close association with the cell that produced them (Fig. 4.21). Collagen and elastic fibers are two well-known structural proteins in the extracellular matrix.

Collagen gives the matrix strength, and elastic fibers give it resilience. Other proteins play a dynamic role by forming "highways" that direct the migration of cells during development. Sometimes the proteins bind to receptors in a cell's plasma membrane, permitting communication between the extracellular matrix and the cytoplasm of the cell.

Polysaccharides in the extracellular matrix provide a rigid packing gel for the various matrix proteins. At present, we know that this gel permits rapid diffusion of nutrients, metabolites, and hormones between blood and tissue cells. Most likely, the polysaccharides regulate the activity of molecules that bind to receptors in the plasma membrane.

The extracellular matrix of various tissues may be quite flexible, as in cartilage, or rock solid, as in bone. The extracellular matrix of bone is hard because in addition to the components mentioned, mineral salts, notably calcium salts, are deposited outside the cell.

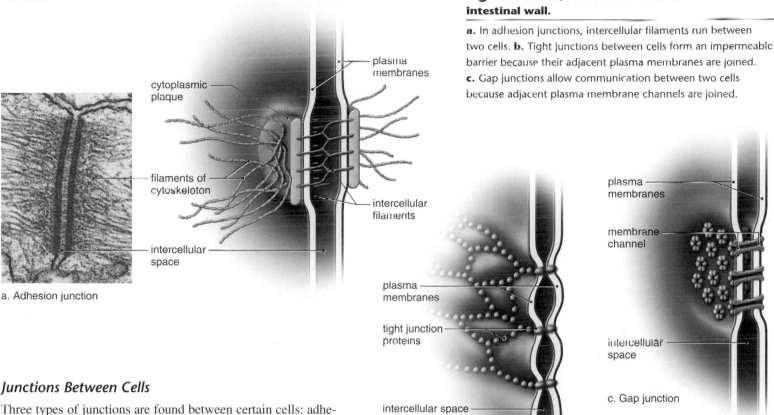

Figure 4.22 Junctions between cells of the intestinal wall.

a. In adhesion junctions, intercellular filaments run between two cells. **b.** Tight junctions between cells form an impermeable barrier because their adjacent plasma membranes are joined. **c.** Gap junctions allow communication between two cells because adjacent plasma membrane channels are joined.

a. Adhesion junction

b. Tight junction

c. Gap junction

Junctions Between Cells

Three types of junctions are found between certain cells: adhesion junctions, tight junctions, and gap junctions.

In **adhesion junctions,** internal cytoplasmic plaques, firmly attached to the cytoskeleton within each cell, are joined by intercellular filaments (Fig. 4.22a). The result is a sturdy but flexible sheet of cells. In some organs—such as the heart, stomach, and bladder, where tissues must stretch—adhesion junctions hold the cells together.

Adjacent cells are even more closely joined by **tight junctions,** in which plasma membrane proteins actually attach to each other, producing a zipperlike fastening (Fig. 4.22b). The cells of tissues that serve as barriers are held together by tight junctions; for example, the urine stays within kidney tubules because the tubules' cells are joined by tight junctions.

A **gap junction** allows cells to communicate. A gap junction is formed when two identical plasma membrane channels join (Fig. 4.22c). The channel of each cell is lined by six plasma membrane proteins that allow the junction to open and close. A gap junction lends strength to the cells, but it also allows small molecules and ions to pass between them. Gap junctions are important in heart muscle and smooth muscle because they permit the flow of ions that is required for the cells to contract as a unit.

Check Your Progress

1. Contrast a plant's primary cell wall with its secondary cell wall.
2. Describe the composition of the extracellular matrix of an animal cell.
3. Compare and contrast adhesion junctions with tight junctions.
4. Describe the functions of gap junctions.

Answers: 1. All plant cells have a primary wall, which is capable of stretching during growth and is composed of cellulose and other compounds. The secondary cell wall is formed in woody tissues and contains lignin in addition to cellulose. The lignin adds rigidity and strength to the cell wall. 2. The extracellular matrix is composed of polysaccharides and proteins. In addition, the matrix in bone contains mineral salts. 3. Both types of junctions hold adjacent animal cells together. Adhesion junctions are more flexible than tight junctions. 4. Gap junctions provide strength to adjacent cells, but also allow molecules and ions to pass between them.

Summary

4.1 Cells Under the Microscope

- Cells are microscopic in size. Although a light microscope allows you to see cells, it cannot make out the detail that an electron microscope can.
- Cells must remain small in order to have an adequate amount of surface area per cell volume.

4.2 The Two Main Types of Cells

- All cells have a plasma membrane, cytoplasm, and genetic material.
- Prokaryotic cells do not have a membrane-bounded nucleus. Eukaryotic cells have a membrane-bounded nucleus as well as various membranous organelles.
- Bacteria are representative of the prokaryotes. They have a cell wall and capsule, in addition to a plasma membrane. Their DNA is in the nucleoid. They have many ribosomes and three possible appendages.

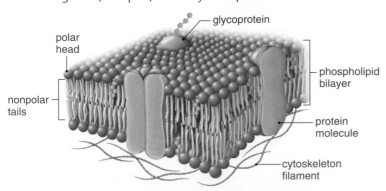

4.3 The Plasma Membrane

The plasma membrane of both prokaryotes and eukaryotes is a phospholipid bilayer.

- The phospholipid bilayer regulates the passage of molecules and ions into and out of the cell.
- The fluid-mosaic model of membrane structure shows that the embedded proteins form a mosaic (varying) pattern.
- The types of embedded proteins include channel, transport, cell recognition, receptor, and enzymatic proteins.

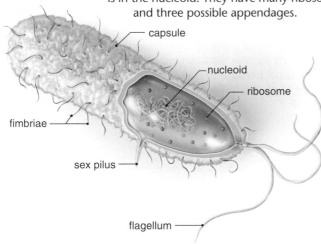

4.4 Eukaryotic Cells

Eukaryotic cells, which are much larger than prokaryotic cells, have the following organelles:

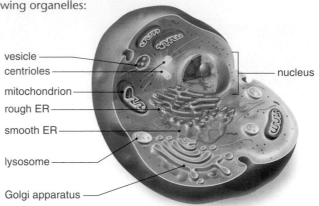

Nucleus and Ribosomes

- The nucleus houses chromatin, which contains DNA, the genetic material. During division, chromatin becomes condensed into chromosomes. In the nucleolus, ribosomal RNA is produced, and it joins with proteins to form the subunits of ribosomes. These subunits exit the nucleus at nuclear pores.
- Ribosomes in the cytoplasm synthesize proteins with the help of mRNA, a DNA intermediary.

Endomembrane System

- Rough ER produces proteins, which are modified in the ER before they are packaged in transport vesicles.
- Smooth ER has various metabolic functions depending on the cell type, but it also forms vesicles.
- The Golgi apparatus receives transport vesicles and modifies, sorts, and repackages proteins into vesicles that fuse with the plasma membrane as secretion occurs.
- Lysosomes are produced by the Golgi apparatus. They contain enzymes that carry out intracellular digestion.

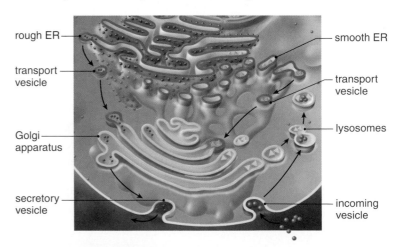

Vacuoles and Vesicles

- Vacuoles are large membranous sacs specialized for storage, contraction, digestion, and other functions, depending on the type of organism and cells in which they occur.
- Vesicles are small membranous sacs.

Energy-Related Organelles
- Chloroplasts capture the energy of the sun and carry on photosynthesis, which produces carbohydrates.
- Mitochondria break down carbohydrate-derived products as ATP is produced during cellular respiration.

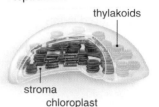

thylakoids — matrix

— cristae

stroma
chloroplast   mitochondrion

Cytoskeleton
- The cytoskeleton maintains cell shape and allows the cell and the organelles to move.
- Microtubules radiate from the centrosome and are present in centrioles, cilia, and flagella.
- Intermediate filaments support the nuclear envelope and the plasma membrane and probably participate in cell-to-cell junctions.
- Actin filaments, the thinnest filaments, interact with the motor molecule myosin to bring about contraction.

Centrioles
- Centrioles are short cylinders with a 9 + 0 pattern of microtubule triplets.
- In animal cells and those of most protists, two centrioles lie at right angles just outside the nucleus.
- Centrioles are not present in plant cells.

Cilia and Flagella
- Cilia and flagella are hairlike projections that allow some cells to move.
- Both have a cylindrical construction with an internal 9 + 2 pattern of microtubules.
- Their basal bodies resemble centrioles.

The following table summarizes the contents of bacterial, animal, and plant cells:

COMPARISON OF BACTERIAL, ANIMAL, AND PLANT CELLS	Bacteria	Animal	Plant
Cell wall	Yes	No	Yes
Plasma membrane	Yes	Yes	Yes
Nucleus	No	Yes	Yes
Nucleolus	No	Yes	Yes
Ribosomes	Yes	Yes	Yes
Endoplasmic reticulum	No	Yes	Yes
Golgi apparatus	No	Yes	Yes
Lysosomes	No	Yes	No
Mitochondria	No	Yes	Yes
Chloroplasts	No	No	Yes
Cytoskeleton	No	Yes	Yes
Centrioles	No	Yes	No
9 + 2 cilia or flagella	No	Often	Some

4.5   Outside the Eukaryotic Cell

Plant cells have a freely permeable cell wall, with cellulose as its main component. Small membrane-lined channels called plasmodesmata span the cell wall and contain strands of cytoplasm that allow materials to pass from one cell to another.

Animal cells typically have an extracellular matrix that contains proteins and polysaccharides produced by the cell. The matrix helps support cells and aids in communication between cells. Some animal cells have junctions. Adhesion junctions and tight junctions help hold cells together; gap junctions allow passage of small molecules between cells.

Thinking Scientifically

1. The 1992 movie *Lorenzo's Oil* recounts the story of six year-old Lorenzo Odone and his battle with adrenoleukodystrophy (ALD), which is caused by a malfunctioning organelle called a peroxisome. A peroxisome is an organelle that ordinarily contains enzymes capable of breaking down long-chain fatty acids. In ALD, the enzyme is missing and when long-chain fatty acids accumulate muscle weakness and loss of control result.
a. How might a dietitian be able to help a person with ALD?
b. What organelle discussed in this chapter is most similar in structure and function to a peroxisome?

2. Protists of the phylum Apicomplexa cause malaria and contribute to infections associated with AIDS. These parasites are unusual because they contain plastids. (Chloroplasts are a type of plastid but most plastids store materials.) Scientists have discovered that an effective antibiotic that kills the parasite inhibits the functioning of the plastid. Can you conclude that the plastid is necessary to the life of the parasite? Why or why not?

Testing Yourself
Choose the best answer for each question.

1. Which of the following is not found in a prokaryotic cell?
   a. cytoplasm   d. ribosome
   b. plasma membrane   e. mitochondrion
   c. nucleoid

2. Which of the following can only be viewed with an electron microscope?
   a. virus   c. bacteria
   b. chloroplast   d. human egg

3. The plasma membrane has the consistency of
   a. butter.   c. olive oil.
   b. peanut butter.   d. water.

4. The plasma membrane is a mosaic of
   a. phospholipids and waxes.
   b. waxes and polysaccharides.
   c. polysaccharides and proteins.
   d. proteins and phospholipids.

For questions 5–8, match the items to those in the key.

Key:

     a. channel proteins   d. receptor proteins
     b. transport proteins   e. enzymatic proteins
     c. cell recognition proteins   f. junction proteins

5. Important component of the immune system.

6. Assist with cell-to-cell communication.

7. Combine with molecules to carry them across plasma membranes.

8. Act as signaling molecules.

9. Eukaryotic cells compensate for a low surface-to-volume ratio by

    a. taking up materials from the environment more efficiently.
    b. lowering their rate of metabolism.
    c. compartmentalizing their activities into organelles.
    d. reducing the number of activities in each cell.

10. What is synthesized by the nucleolus?

    a. mitochondria   c. transfer RNA
    b. ribosomal subunits   d. DNA

For questions 11–14, match the items to those in the key.

Key:

     a. single membrane
     b. double membrane
     c. no membrane

11. Ribosome

12. Nucleus

13. Vacuole

14. Mitochondrion

15. The organelle that can modify the sugars on a protein, determining the protein's destination in the cell is the

    a. ribosome.   c. Golgi apparatus.
    b. vacuole.   d. lysosome.

16. Plant vacuoles may contain

    a. flower color pigments.
    b. toxins that protect plants against herbivorous animals.
    c. sugars.
    d. All of these are correct.

17. The nonmembrane component of a mitochondrion is called the

    a. cristae.   c. matrix.
    b. thylakoid.   d. granum.

18. Label the parts of the following cell.

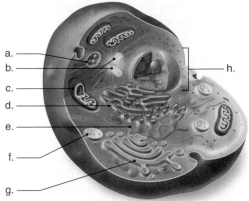

a.
b.
c.
d.
e.
f.
g.
h.

Go to **www.mhhe.com/maderessentials** for more quiz questions.

Bioethical Issue

It is common to grow large cultures of bacterial cells for the purpose of extracting commercially important compounds produced naturally by the bacteria. However, we can now introduce human genes into bacteria, allowing them to produce human proteins. These proteins can then be extracted, purified, and used to treat human disorders. For example, human growth hormone is produced in this way and used to treat dwarfism.

    Do you believe it is right to transfer human genes into bacterial cells in order to produce large quantities of human proteins for medical purposes? Would your opinion change if the genes were put into the cells of a mammal such as a goat or a pig?

Understanding the Terms

actin filament   62
adenosine triphosphate
  (ATP)   60
adhesion junction   65
apoptosis   58
capsule   50
cell   48
cell theory   50
cellular respiration   61
cell wall   50, 64
centriole   62
centrosome   62
chloroplast   60
chromatin   56
chromosome   56
cilium (pl., cilia)   63
cristae   61
cytoplasm   50
cytoskeleton   55, 62
endomembrane system   58
endoplasmic reticulum (ER)   57
eukaryotic cell   50
extracellular matrix   64
fimbriae   51
flagellum (pl., flagella)   51, 63
gap junction   65
Golgi apparatus   58

granum   60
lysosome   58
matrix   61
microtubule   62
mitochondrion   60
nuclear envelope   57
nuclear pore   57
nucleoid   51
nucleolus   56
nucleus   54
organelle   54
plasma membrane   52
plasmodesmata   64
polyribosome   57
prokaryotic cell   50
ribosome   51, 56
rough ER   58
secretion   58
sex pili   51
smooth ER   58
stroma   60
surface-area-to-volume
  ratio   49
thylakoid   60
tight junction   65
vacuole   59
vesicle   58

Match the terms to these definitions:

a. _____ The concepts that all organisms are composed of cells, and cells only come from preexisting cells.

b. _____ The collection of membranes that compartmentalize the cell so that enzymatic reactions are confined to specific regions.

c. _____ Programmed cell death.

d. _____ A stack of thylakoids.

e. _____ Composed of a stack of flattened saccules resembling pancakes.

f. _____ A lattice of protein fiber that maintains the shape of the cell and assists in the movement of organelles.

g. _____ A protective layer of polysaccharides outside the cell wall of bacteria.

The Dynamic Cell

Body temperature is maintained by energy conversions in cells.

Penicillin, a life-saving antibiotic, works by inhibiting the activity of certain enzymes.

Drinking excessive amounts of water can have a toxic effect on cells.

Although we normally think of all the wonderful things water does for our bodies, drinking an excessive amount of it during a short period of time can be toxic. In 2001, a 4-year-old girl in Utah died after her adoptive parents forced her to drink an extreme amount of fluids. Drinking too much water has also been reported in people who are under the influence of drugs such as Ecstasy.

OUTLINE

▶ ▶ ▶

In humans, water intake typically equals water output, primarily by the kidneys. When huge amounts of water are consumed very quickly, the kidneys sometimes can't keep up, and thus cells swell. This swelling, especially in the brain, causes symptoms similar to those associated with intoxication, and the condition is called *water intoxication.*

What makes cells swell? As you will learn in this chapter, the process of osmosis causes water to move from the side of a cell containing more water to the side containing less water. In the case of water intoxication, first there is much more water outside the cells than inside the cells. Then, when this water moves into the cells to achieve equilibrium, the cells swell and may literally explode and die.

This chapter also explores how other molecules move in and out of cells, basic properties of energy, and the functions of enzymes. Only after understanding these principles will you appreciate the processes of photosynthesis and cellular respiration. These processes are vital for maintaining life and will be described in Chapters 6 and 7.

5.1 What Is Energy?

Energy is defined as the capacity to do work—to make things happen. Without a source of energy, we humans would not be here on Earth—nor would any other living thing. The biosphere of which we are a part gets its energy from the sun, and thereafter, one form of energy is changed to another form as life processes take place.

The two basic forms of energy are potential energy and kinetic energy. **Potential energy** is stored energy, and **kinetic energy** is the energy of motion. Potential energy is constantly being converted to kinetic energy, and vice versa. Let's look at the example in Figure 5.1. The food a cross-country skier has for breakfast contains chemical energy, which is a form of potential energy. When our skier hits the slopes, she may have to ascend a hill. During her climb, the potential energy of food is converted to the kinetic energy of motion, a type of mechanical energy. Once she reaches the hilltop, kinetic energy has been converted to the potential energy of location. As she skis down the hill, this potential energy is converted to the kinetic energy of motion again. But with each conversion, some energy is lost as heat.

Measuring Energy

Chemists use a unit of measurement called the joule to measure energy, but it is common to measure food energy in terms of calories. A **calorie** is the amount of heat required to raise the temperature of 1 gram of water by one degree Celsius. This isn't much energy, so the caloric value of food is listed in nutrition labels and in diet charts in terms of **kilocalories** (1,000 calories). In this text, we will use *Calorie* to mean 1,000 calories.

Two Energy Laws

Two **energy laws** govern energy flow and help us understand the principles of energy conversion. The first law, called the law of conservation of energy, tells us:

Energy cannot be created or destroyed, but it can be changed from one form to another.

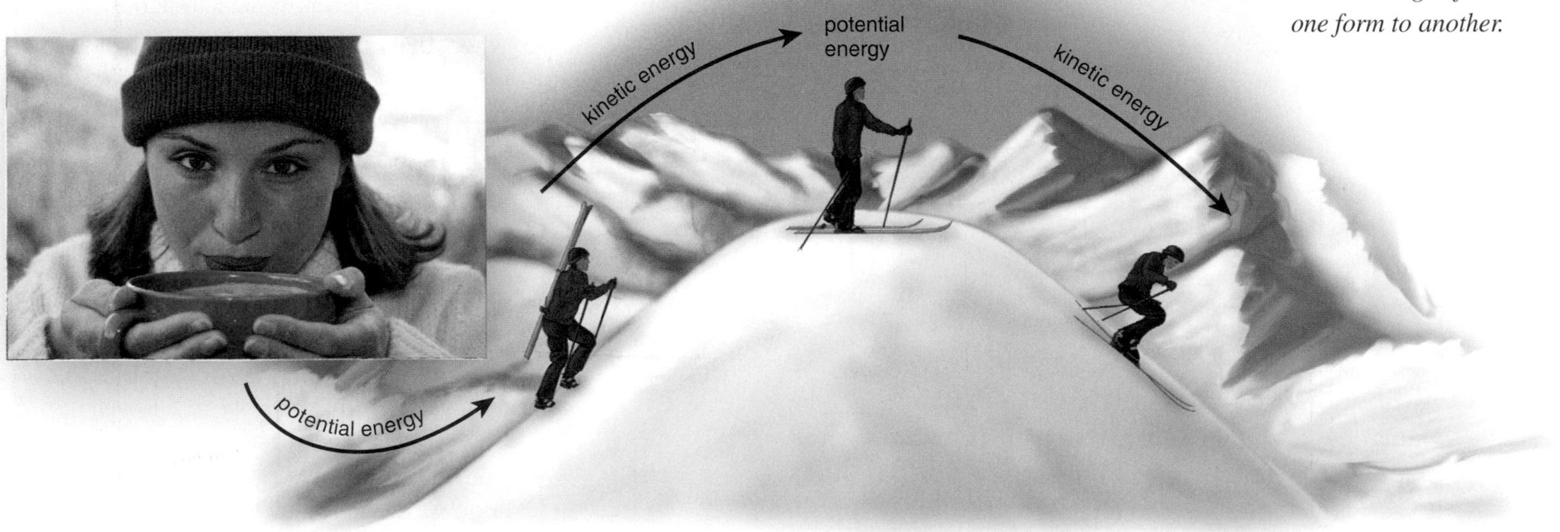

Figure 5.1 **Potential energy versus kinetic energy.**

Food contains potential energy that a skier can convert to kinetic energy in order to climb a hill. Height is potential energy due to location, which the skier converts to kinetic energy as she descends the hill. With every conversion to kinetic energy, some potential energy is lost as heat.

Relating the law to our previous example, we know that the skier had to acquire energy by eating food before she could climb the hill, and that energy conversions occurred before she reached the bottom of the hill again.

The second energy law tells us: *Energy cannot be changed from one form to another without a loss of usable energy.* Many forms of energy are usable, such as the energy of the sun, food, and ATP. **Heat** is diffuse energy and the least usable form. Every energy conversion results in a loss of usable energy in the form of heat. For example, no doubt the skier worked up a sweat and gave off heat as she climbed the hill. This heat represents a loss of usable energy.

Entropy

The second energy law can be stated another way. Every energy transformation leads to an increase in the amount of disorganization or disorder. The term **entropy** refers to the relative amount of disorganization. The only way to maintain or bring about order is to add more energy to a system. To take an example from your own experience, you know that a tidy room is more organized and less stable than a messy room, which is disorganized and more stable (Fig. 5.2a). In other words, your room is much more likely to stay messy than it is to stay tidy. Why? Unless you continually add energy to keep your room organized and neat, it will inevitably become less organized and messy.

Because our universe is a *closed system* in which energy cannot be created, all energy transformations, including those in cells, increase the total entropy of the universe. Figure 5.2b shows a process that occurs in cells because it proceeds from a more ordered state to a more disordered state. Just as a tidy room tends to become messy, hydrogen ions (H^+) that have accumulated on one side of a membrane tend to move to the other side unless they are prevented from doing so by the addition of energy. Why? Because when H^+ ions are distributed randomly on either side of the membrane, no additional energy is needed to keep them that way, and the entropy, or disorder, of their arrangement has increased. The result is a more stable arrangement of H^+ ions.

What about reactions in cells that apparently proceed from disorder to order? For example, we know that plant cells can make glucose out of carbon dioxide and water. How do they do it? Energy provided by the sun allows plants to make glucose, which is a highly organized molecule. Even this process, however, involves a loss of some potential energy. When light energy is converted to chemical energy in plant cells, some of the sun's energy is always lost as heat. In other words, the organization of a cell has a constant energy cost that results in an increase in the entropy of the universe.

Entropy: Second Law of Thermodynamics

| Tidy bedroom | Messy bedroom |

a.

- more organized
- more potential energy
- less stable
- less entropy

- less organized
- less potential energy
- more stable
- more entropy

| Unequal distribution of hydrogen ions | Equal distribution of hydrogen ions |

b.

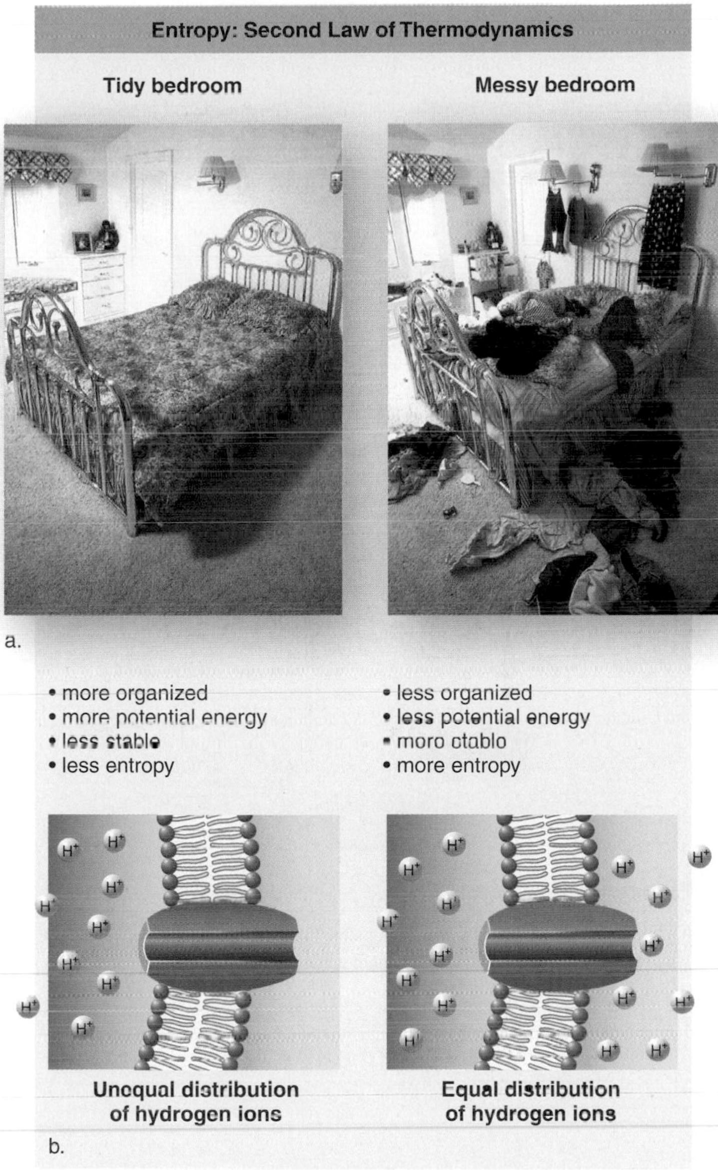

Figure 5.2 Cells and entropy.

The second energy law tells us that entropy (disorder) always increases. Therefore (**a**) a tidy room tends to become messy and disorganized, and (**b**) hydrogen ions (H^+) on one side of a membrane tend to move to the other side so that the ions are randomly distributed. Both processes result in a loss of potential energy and an increase in entropy.

Check Your Progress

1. Contrast potential energy with kinetic energy.
2. Explain how the second energy law is related to entropy.

Answers: 1. Potential energy is stored energy, while kinetic energy is energy of motion. *2.* One way to express the second energy law is to say that every energy transformation increases disorder. Entropy is the tendency toward disorder.

5.2 ATP: Energy for Cells

ATP (adenosine triphosphate) is the energy currency of cells. Just as you use coins to purchase all sorts of products, a cell uses ATP to carry out nearly all of its activities, including synthesizing proteins, transporting ions across the plasma membranes, and causing organelles and cilia to move.

Structure of ATP

ATP is a nucleotide, the type of molecule that serves as a monomer for the construction of DNA and RNA. ATP's name, adenosine triphosphate, means that it contains the sugar ribose, the nitrogen-containing base adenine, and three phosphate groups (Fig. 5.3). The three phosphate groups are negatively charged and repel one another. It takes energy to overcome their repulsion, and thus these phosphate groups make the molecule unstable. ATP easily loses the last phosphate group because the breakdown products, ADP (adenosine diphosphate) and a separate phosphate group symbolized as (P), are more stable than ATP. This reaction is written as: ATP → ADP + (P). ADP can also lose a phosphate group to become AMP (adenosine monophosphate).

Use and Production of ATP

The continual breakdown and regeneration of ATP is known as the ATP cycle (Fig. 5.4). ATP stores energy for only a short period of time before it is used in a reaction that requires energy. Then ATP is rebuilt from ADP + (P). Each ATP molecule undergoes about 10,000 cycles of synthesis and breakdown every day. Our bodies use some 40 kg (about 88 lb) of ATP daily, and the amount on hand at any one moment is sufficiently high to meet only current metabolic needs.

ATP's instability, the very feature that makes it an effective energy donor, keeps it from being an energy storage molecule. Instead, the many H—C bonds of carbohydrates and fats make them the energy storage molecules of choice. Their energy is extracted during cellular respiration and used to rebuild ATP, mostly within mitochondria. You will learn in Chapter 7 that the breakdown of one molecule of glucose permits the buildup of some 38 molecules of ATP. During cellular respiration, only 39% of the potential energy of glucose is converted to the potential energy of ATP; the rest is lost as heat.

The production of ATP is still worthwhile for the cell for the following reasons:

1. ATP is suitable for use in many different types of cellular reactions.
2. When ATP becomes ADP + (P), the amount of energy released is more than the amount needed for a biological purpose, but not overly wasteful.
3. The structure of ATP allows its breakdown to be *coupled* to an energy-requiring reaction, as described next.

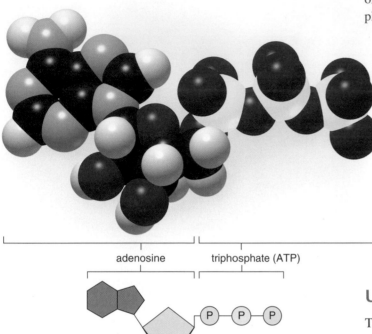

Figure 5.3   ATP.

ATP, the universal energy currency of cells, is composed of adenosine and three phosphate groups (called a triphosphate).

adenosine   triphosphate (ATP)

Figure 5.4   The ATP cycle.

When ATP is used as an energy source, a phosphate group is removed by hydrolysis. ATP is regenerated in mitochondria as cellular respiration occurs.

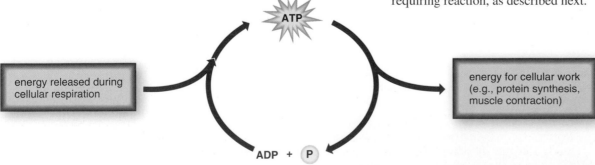

ATP

energy released during cellular respiration

energy for cellular work (e.g., protein synthesis, muscle contraction)

ADP + P

Coupled Reactions

Coupled reactions are reactions that occur in the same place, at the same time, and in such a way that an energy-releasing reaction can drive an energy-requiring reaction. Usually the energy-releasing reaction is ATP breakdown. Because the cleavage of ATP's phosphate group releases more energy than the amount consumed by the energy-requiring reaction, entropy has increased and both reactions will proceed. The simplest way to represent a coupled reaction is like this:

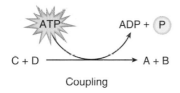

Coupling

This reaction tells you that coupling occurs, but it does not show how coupling is achieved. A cell has two main ways to couple ATP breakdown to an energy-requiring reaction: ATP is used to energize a reactant or to change its shape. Both are often achieved by transferring a phosphate group to the reactant. For example, when polypeptide synthesis occurs at a ribosome, an enzyme transfers a phosphate group from ATP to each amino acid in turn, and this transfer activates the amino acid, causing it to bond with another amino acid.

Figure 5.5 shows how ATP breakdown provides the necessary energy for muscular contraction. During muscle contraction, myosin filaments pull actin filaments to the center of the cell, and the muscle shortens. First, myosin combines with ATP, and only then does ATP break down to ADP + Ⓟ. The release of ADP + Ⓟ from the molecule causes myosin to change shape and pull on the actin filament.

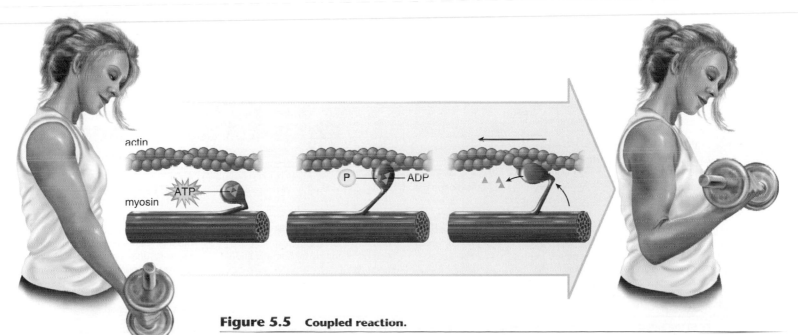

Figure 5.5 Coupled reaction.

Muscle contraction occurs only when it is coupled to ATP breakdown. Myosin combines with ATP prior to its breakdown. Release of ADP + Ⓟ causes myosin to change position and pull on an actin filament.

Energy Conversions

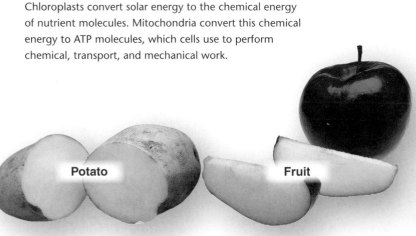

heat   solar energy

chloroplast

CO_2 and H_2O

O_2

Chemical energy
(carbohydrate)

mitochondrion

heat

ATP   Chemical work
Transport work
Mechanical work

The Flow of Energy

In the biosphere, the activities of chloroplasts and mitochondria enable energy to flow from the sun through all living things. During photosynthesis, the chloroplasts in plants capture solar energy and use it to convert water and carbon dioxide into carbohydrates, which serve as food for themselves and for other organisms. During cellular respiration, mitochondria complete the breakdown of carbohydrates and use the released energy to build ATP molecules.

Notice in Figure 5.6 that cellular respiration requires oxygen and produces carbon dioxide and water, the very molecules taken up by chloroplasts. It is actually the cycling of molecules between chloroplasts and mitochondria that allows a flow of energy from the sun through all living things. This flow of energy maintains the levels of biological organization from molecules to organisms to ecosystems. In keeping with the energy laws, useful energy is lost with each chemical transformation, and eventually the solar energy captured by plants is lost in the form of heat. In this way, living things are dependent upon an input of solar energy.

Human beings are also involved in the cycling of molecules between plants and animals and in the flow of energy from the sun. We inhale oxygen and eat plants and their stored carbohydrates, or we eat other animals that have eaten plants. Oxygen and nutrient molecules enter our mitochondria, which produce ATP and release carbon dioxide and water. Without a supply of energy-rich foods, we could not produce the ATP molecules needed to maintain our bodies and carry on activities (Fig. 5.7).

Figure 5.6   Flow of energy.

Chloroplasts convert solar energy to the chemical energy of nutrient molecules. Mitochondria convert this chemical energy to ATP molecules, which cells use to perform chemical, transport, and mechanical work.

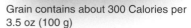

Potato

Potatoes contain about 90 Calories per 3.5 oz (100 g)

Potatoes contain starch, which stores energy for the growing plant. When a potato is eaten, starch molecules are broken down to simple sugars in the body before their energy is released.

Fruit

Apples contain about 60 Calories per 3.5 oz (100 g)

Fruits such as apples contain natural sugars that provide energy. The body converts excess sugar into fat, which is stored energy that can be used later.

Grain

Grain contains about 300 Calories per 3.5 oz (100 g)

Grain is an efficient energy source because it directly stores the energy of photosynthesis. Grains are a good source of energy for humans and animals.

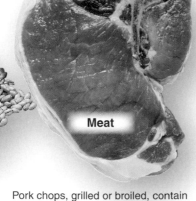

Meat

Pork chops, grilled or broiled, contain about 215 Calories per 3.5 oz (100 g)

Meat is a good energy source; much of its energy is stored in fat. However, meat isn't an efficient energy source because only a small amount of the energy taken in by a living pig, for example, is converted to energy usable by the human or animal that eats the pork chop.

Figure 5.7   Energy for life.

Our food provides all the energy our bodies need. The energy in food is measured in Calories. To drive a car requires about 61 Calories per hour. To run at a speed of 1 mile in 7 minutes takes about 865 Calories per hour. If you take in more Calories than you burn, you will gain weight.

5.3   Metabolic Pathways and Enzymes

Reactions do not occur haphazardly in cells; they are usually part of a **metabolic pathway,** a series of linked reactions. Metabolic pathways begin with a particular reactant and terminate with an end product. In the pathway, one reaction leads to the next reaction, which leads to the next reaction, and so forth in an organized, highly structured manner. This arrangement makes it possible for one pathway to lead to others, and for metabolic energy to be captured and utilized in small increments rather than all at once.

A metabolic pathway can be represented by the following diagram:

$$\begin{array}{cccccc} E_1 & E_2 & E_3 & E_4 & E_5 & E_6 \\ A \rightarrow & B \rightarrow & C \rightarrow & D \rightarrow & E \rightarrow & F \rightarrow G \end{array}$$

In this diagram, the letters A–F are reactants and the letters B–G are products. In other words, the product from the previous reaction becomes the reactant of the next reaction. The letters E_1–E_6 represent enzymes.

An **enzyme** is a protein molecule that functions as an organic catalyst to speed a chemical reaction. Enzymes can only speed reactions that are possible to begin with. In the cell, an enzyme is analogous to a mutual friend who causes two people to meet and interact, because an enzyme brings together particular molecules and causes them to react with one another

The reactants in an enzymatic reaction are called the **substrates** for that enzyme. In the first reaction, A is the substrate for E_1, and B is the product. Now B becomes the substrate for E_2, and C is the product. This process continues until the final product (G) forms.

Energy of Activation

Molecules frequently do not react with one another unless they are activated in some way. In the lab, for example, in the absence of an enzyme, activation is very often achieved by heating a mixture to increase the number of effective collisions between molecules. The energy needed to cause molecules to react with one another is called the **energy of activation (E_a).** Figure 5.8 compares the E_a without an enzyme to the E_a with an enzyme, illustrating that enzymes lower the amount of energy required for a reaction to occur. Enzymes lower the energy of activation by bringing the substrates into contact and even by participating in the reaction at times.

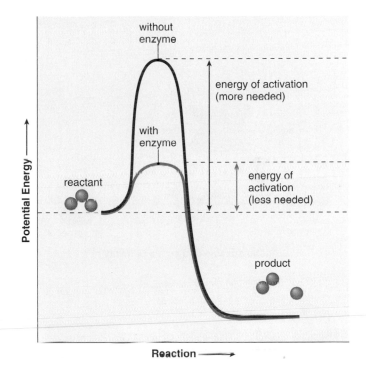

Figure 5.8   Energy of activation (E_a).

Enzymes speed the rate of reactions because they lower the amount of energy required for the reactants to react. Even reactions like this one, in which the energy of the product is less than the energy of the reactant, speed up when an enzyme is present.

Check Your Progress

Explain why reactions in a cell are usually part of a metabolic pathway.

Answer: Each reaction produces a product that can be used for another reaction. Therefore, each reaction can lead to others, and energy is used in small increments.

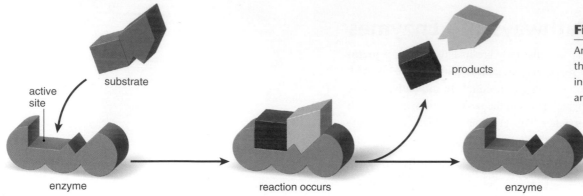

Figure 5.9   Enzymatic action.

An enzyme has an active site where the substrates and enzyme fit together in such a way that the substrates are oriented to react. Following the reaction, the products are released, and the enzyme is free to act again.

An Enzyme's Active Site

In most instances, only one small part of the enzyme, called the **active site,** accommodates the substrate(s). At the active site, the substrate fits into the enzyme seemingly like a key fits a lock. However, the active site undergoes a slight change in shape in order to accommodate the substrate(s). This is called the **induced fit model** because the enzyme is induced to undergo a slight alteration to achieve optimum fit (Fig. 5.9).

The change in the shape of the active site facilitates the reaction that next occurs. After the reaction has been completed, the product(s) is released, and the active site returns to its original state, ready to bind to another substrate molecule. Only a small amount of each enzyme is actually needed in a cell because enzymes are not used up by the reaction.

Enzyme Inhibition

Enzyme inhibition occurs when an active enzyme is prevented from combining with its substrate. Enzyme inhibitors are often poisonous to certain organisms. Cyanide, for example, is an inhibitor of the enzyme cytochrome *c* oxidase, which performs a vital function in all cells because it is involved in making ATP. Penicillin, in contrast, blocks the active site of an enzyme unique to bacteria. Therefore, penicillin is a poison for bacteria but not for humans.

The activity of almost every enzyme in a cell is regulated by **feedback inhibition.** In the simplest case, when a product is in abundance, it competes with the substrate for the enzyme's active site. As the product is used up, inhibition is reduced, and more product can be produced. In this way, the concentration of the product always stays within a certain range.

Most metabolic pathways are regulated by more complex types of feedback inhibition (Fig. 5.10). In these instances, when the end product is plentiful, it binds to a site other than the active site of the first enzyme in the pathway. This binding changes the shape of the active site, preventing the enzyme from binding to its substrate. Without the activity of the first enzyme, the entire pathway shuts down.

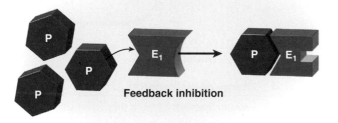

Active enzyme and active pathway

Feedback inhibition

Inactive enzyme and inactive pathway

Figure 5.10   Feedback inhibition.

This type of feedback inhibition occurs when the end product (P) of an active enzyme pathway is plentiful and binds to the first enzyme (E₁) of the pathway at a site other than the active site. This changes the shape of the active site so that the substrate (S) can no longer bind to the enzyme. Now the entire pathway becomes inactive.

Check Your Progress

Explain how an enzyme facilitates a reaction.

Answer: The active site of the enzyme undergoes a slight change in shape, allowing it to fit together with the substrate. That change in shape allows the reaction to occur with a lower energy of activation.

5.4   Cell Transport

The plasma membrane regulates the passage of molecules into and out of the cell. This function is crucial because the life of the cell depends on maintenance of its normal composition. The plasma membrane can carry out this function because it is differentially permeable, meaning that certain substances can freely pass through the membrane while others are transported across. Basically, substances enter a cell in one of three ways: passive transport, active transport, and bulk transport. Although there are different types of passive transport, in all cases substances move from a higher to a lower concentration, and no energy is required. Active transport moves substances against a concentration gradient and requires energy. Bulk transport requires energy, but movement of the large substances involved is independent of concentration gradients.

Passive Transport: No Energy Required

Actually, no membrane is required for simple diffusion. During **simple diffusion,** molecules move down their concentration gradient until equilibrium is achieved and they are distributed equally. Simple diffusion occurs because molecules are in motion, but it is a *passive* form of transport because a cell does not need to expend energy for it to happen. Small, noncharged molecules, such as oxygen, carbon dioxide, glycerol, alcohol, and water, are able to slip between the phospholipid molecules making up the plasma membrane. Therefore, these molecules can diffuse across the membrane.

Figure 5.11 demonstrates simple diffusion. Water is present on two sides of a membrane, and dye is added to one side. The dye particles move in various directions, but the net movement is toward the opposite side of the membrane. Eventually, the dye is dispersed, with no net movement of dye in either direction. A **solution** contains both a solute and a solvent. In this case, the dye is called the **solute,** and the water is called the **solvent.** Solutes are usually solids, and solvents are usually liquids.

Dissolved gases can diffuse through the phospholipid bilayer, and indeed this is the mechanism by which oxygen enters cells and carbon dioxide exits them. Also, oxygen enters blood from the air sacs of the lungs, and carbon dioxide moves in the opposite direction by diffusion.

Ions and polar molecules, such as glucose and amino acids, are often assisted across the plasma membrane by transport proteins. This process is called **facilitated diffusion.** Even those proteins that simply provide a channel for passage are usually specific to the solute. In these cases, the transport proteins most likely undergo a change in shape as the solute enters the cell.

Osmosis

Diffusion of water across a differentially permeable membrane is called **osmosis.** To illustrate osmosis, a tube containing a 5% salt solution and covered at one end by a membrane is placed in a beaker that contains water only (Fig. 5.12a). The beaker has a higher concentration of water molecules than the tube does because the tube also contains a solute. Water can cross the membrane, but the solute cannot. Therefore, there will be a net movement of water across the membrane from the beaker to the inside of the tube. Theoretically, the solution inside the tube will rise until there is an equal concentration of water on both sides of the membrane (Fig. 5.12b,c). What would happen if the beaker contained a 2% salt solution? Water would still diffuse into the tube, because the tube at 5% salt would still contain a lower concentration of water molecules than the beaker at 2%.

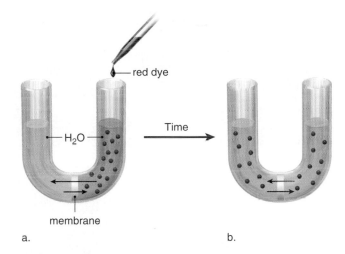

Figure 5.11   Simple diffusion demonstration.

Simple diffusion is spontaneous, and no energy is required to bring it about. **a.** Red dye is added to water separated by a membrane. The dye molecules can pass through the membrane. The dye molecules move randomly about, but over time the net movement of dye is toward the region of lower concentration. **b.** Eventually, the dye molecules are equally distributed throughout the container and there is no net movement of dye in either direction.

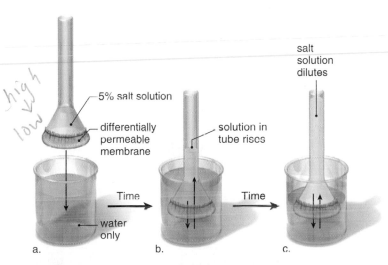

Figure 5.12   Osmosis demonstration.

a. A tube covered at the broad end by a differentially permeable membrane contains a 5% salt solution, and the beaker contains only water. The salt ions are unable to pass through freely, but the water molecules can pass through the membrane. **b.** Therefore, there is a net movement of water toward the inside of the tube, where the percentage of water molecules is lower than the outside. **c.** The level of the solution rises in the tube because of the incoming water.

Red blood cells

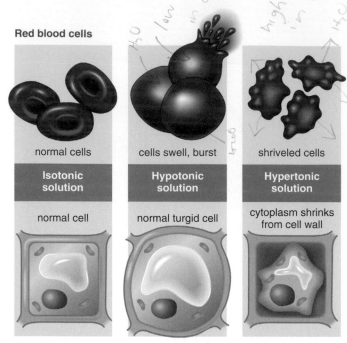

normal cells | cells swell, burst | shriveled cells

Isotonic solution | **Hypotonic solution** | **Hypertonic solution**

normal cell | normal turgid cell | cytoplasm shrinks from cell wall

Plant cells

Figure 5.13   Osmosis in animal and plant cells.

In an isotonic solution, cells neither gain nor lose water. In a hypotonic solution, cells gain water. Red blood cells swell to bursting, and plant cells become turgid. In a hypertonic solution, cells lose water. Red blood cells shrivel, and plant cell cytoplasm shrinks away from the cell wall.

The Effect of Osmosis on Cells

Osmosis can affect the size and shape of cells, as shown in Figure 5.13. In the laboratory, cells are normally placed in **isotonic solutions** (*iso,* same as) in which the cell neither gains nor loses water—that is, the concentration of water is the same on both sides of the membrane. In medical settings, a 0.9% solution of the salt sodium chloride (NaCl) is known to be isotonic to red blood cells; therefore, intravenous solutions usually have this concentration.

Cells placed in a **hypotonic solution** (*hypo,* less than) gain water. Outside the cell, the concentration of solute is less, and the concentration of water is greater, than inside the cell. Animal cells placed in a hypotonic solution expand and sometimes burst. The term *lysis* refers to disrupted cells; *hemolysis,* then, is disrupted red blood cells.

When a plant cell is placed in a hypotonic solution, the large central vacuole gains water, and the plasma membrane pushes against the rigid cell wall as the plant cell becomes *turgid.* The plant cell does not burst because the cell wall does not give way. Turgor pressure in plant cells is extremely important in maintaining their erect position.

Cells placed in a **hypertonic solution** (*hyper,* more than) lose water. Outside the cell, the concentration of solute is more, and the concentration of water is less, than inside the cell. Animal cells placed in a hypertonic solution shrink. For example, meats are sometimes preserved by being salted. Bacteria are killed not by the salt, but by the lack of water in the meat.

When a plant cell is placed in a hypertonic solution, the plasma membrane pulls away from the cell wall as the large central vacuole loses water. This is an example of **plasmolysis,** shrinking of the cytoplasm due to osmosis. The dead plants you may see along a roadside could have died due to exposure to a hypertonic solution during the winter, when salt was used on the road.

Check Your Progress

1. **Compare and contrast simple diffusion with facilitated diffusion.**
2. **Describe the relationship between a solute, a solvent, and a solution.**

Active Transport: Energy Required

During **active transport,** molecules or ions move through the plasma membrane, accumulating on one side of the cell (Fig. 5.14). For example, iodine collects in the cells of the thyroid gland; glucose is completely absorbed from the digestive tract by the cells lining the digestive tract; and sodium can be almost completely withdrawn from urine by cells lining the kidney tubules. In these instances, molecules have moved against their concentration gradients, a situation that requires both a transport protein and ATP. Therefore, cells involved in active transport, such as kidney cells, have a large number of mitochondria near their plasma membranes to generate the ATP.

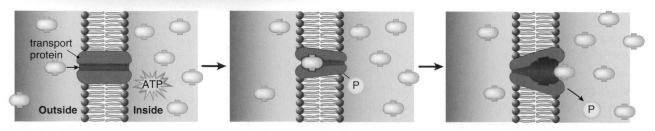

Figure 5.14   Active transport.

During active transport, a transport protein uses energy to move a solute across the plasma membrane toward a higher concentration. Note that the transport protein changes shape during the process.

The passage of salt (NaCl) across a plasma membrane is of primary importance in cells because the salt causes water to move to that side of the plasma membrane. First, sodium ions are actively transported across a membrane, and then chloride ions simply diffuse through channels that allow their passage. Chloride ion channels malfunction in persons with cystic fibrosis, leading to the symptoms of this inherited disorder. Proteins engaged in active transport are often called *pumps*. The **sodium-potassium pump**, vitally important to nerve conduction, undergoes a change in shape that allows it to combine alternately with sodium ions and potassium ions.

Bulk Transport

Macromolecules, such as polypeptides, polysaccharides, or polynucleotides, are too large to be moved by transport proteins. Instead, vesicle formation takes them into or out of a cell. For example, digestive enzymes and hormones use molecules transported out of the cell by **exocytosis** (Fig. 5.15*a*). In cells that synthesize these products, secretory vesicles accumulate near the plasma membrane. These vesicles release their contents only when the cell is stimulated by a signal received at the plasma membrane, a process called regulated secretion.

When cells take in substances by vesicle formation, the process is known as **endocytosis** (Fig. 5.15*b*). If the material taken in is large, such as a food particle or another cell, the process is called **phagocytosis**. Phagocytosis is common in unicellular organisms, such as amoebas. It also occurs in humans. Certain types of human white blood cells are amoeboid—that is, they are mobile like an amoeba, and are able to engulf debris such as worn-out red blood cells or bacteria. When an endocytic vesicle fuses with a lysosome, digestion occurs. In Chapter 26, we will see that this process is a necessary and preliminary step toward the development of immunity to bacterial diseases.

Pinocytosis occurs when vesicles form around a liquid or around very small particles. White blood cells, cells that line the kidney tubules and the intestinal wall, and plant root cells all use pinocytosis to ingest substances.

During **receptor-mediated endocytosis,** receptors for particular substances are found at one location in the plasma membrane. This location is called a coated pit because there is a layer of protein on its intracellular side (Fig. 5.16). Receptor-mediated endocytosis is selective and much more efficient than ordinary pinocytosis. It is involved when substances move from maternal blood into fetal blood at the placenta, for example.

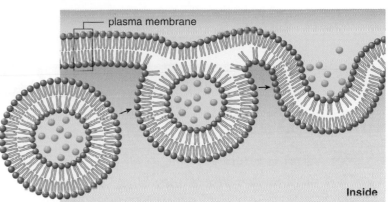

a. Exocytosis

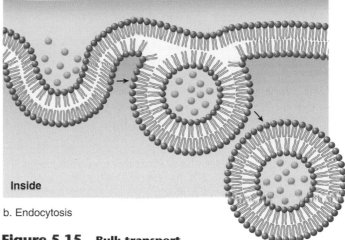

b. Endocytosis

Figure 5.15 **Bulk transport.**

During exocytosis (**a**) and endocytosis (**b**), vesicle formation transports substances out of or into a cell, respectively.

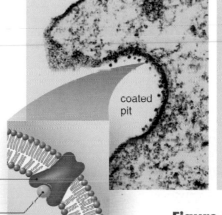
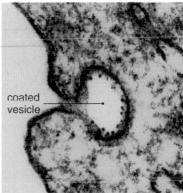

coated pit

coated vesicle

receptor protein

molecule

Figure 5.16 **Receptor-mediated endocytosis.**

During receptor-mediated endocytosis, molecules first bind to specific receptor proteins that are in a coated pit. The vesicle that forms contains the molecules and their receptors.

Check Your Progress

Compare and contrast exocytosis and endocytosis.

Answer: Both use vesicles to transport materials across the plasma membrane. Molecules are transported out by exocytosis and in by endocytosis.

Summary

5.1 What Is Energy?

Potential energy can be converted to kinetic energy, and vice versa. Solar and chemical energy are forms of potential energy; mechanical energy is a form of kinetic energy. Two energy laws hold true universally:

- Energy cannot be created or destroyed, but can be transferred or transformed.
- One form of energy cannot be completely converted into another form without a loss of usable energy. Therefore, the entropy of the universe is increasing, and only a constant input of energy maintains the organization of living things.

5.2 ATP: Energy for Cells

Energy flows from the sun through chloroplasts and mitochondria, which produce ATP. Because ATP has three adjoining negative phosphate groups, it is a high-energy molecule that tends to break down to ADP + P. ATP breakdown is coupled to various energy-requiring cellular reactions, including protein synthesis, active transport, and muscle contraction. Cellular respiration provides the energy for the production of ATP. The following diagram summarizes the ATP cycle:

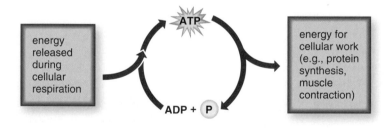

5.3 Metabolic Pathways and Enzymes

A metabolic pathway is a series of reactions that proceed in an orderly, step-by-step manner. Each reaction requires an enzyme that is specific to its substrate. Enzymes bring substrates together at an enzyme's active site, and speed reactions by lowering the energy of activation. The activity of most enzymes and metabolic pathways is regulated by feedback inhibition.

5.4 Cell Transport

The plasma membrane is differentially permeable; some substances can freely cross the membrane, and some must be assisted across if they are to enter the cell.

Passive transport requires no metabolic energy and moves substances from a higher to a lower concentration.

- In simple diffusion, molecules move from an area of higher concentration to the area of lower concentration. Some molecules cross plasma membranes by simple diffusion.

- In facilitated diffusion, molecules diffuse across a plasma membrane with the assistance of transport proteins.
- Osmosis is the simple diffusion of water across a membrane toward the area of lower concentration of water. Cells in an isotonic solution neither gain nor lose water; cells in a hypotonic solution gain water; and cells in a hypertonic solution lose water.
- Some molecules diffuse with the assistance of membrane proteins. This process is termed facilitated diffusion.

Active transport requires metabolic energy (ATP) and moves substances from a lower to a higher concentration across a membrane.

- A transport protein acts as a pump that causes a substance to move against its concentration gradient. For example, the sodium-potassium pump carries Na^+ to the outside of the cell and K^+ to the inside of the cell.

Bulk transport requires vesicle formation and metabolic energy. It occurs independent of concentration gradients.

- Exocytosis transports macromolecules out of a cell via vesicle formation and often results in secretion.
- Endocytosis transports macromolecules into a cell via vesicle formation.
- Phagocytosis is a type of endocytosis that transports large particles; pinocytosis transports liquids or small particles; and receptor-mediated endocytosis is a form of pinocytosis, which makes use of receptor proteins in the plasma membrane.

The following diagram illustrates the different types of passive and active transport:

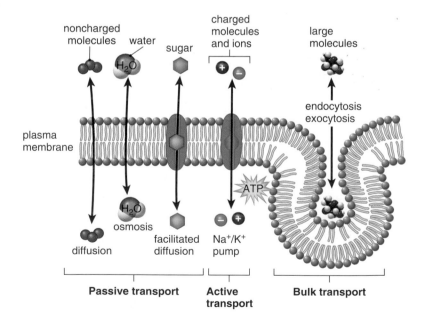

Thinking Scientifically

1. Every energy conversion results in the loss of energy as heat. However, some organisms also carry out reactions that give off energy as light. The production of light by living organisms is called bioluminescence. Some marine algae, including dinoflagellates, bioluminesce when the water surrounding them is agitated. One hypothesis suggests that when predators begin to feed on a population of dinoflagellates, the bioluminescence attracts predators of the dinoflagellate eaters. How would you test this hypothesis?

2. Cystic fibrosis is a genetic disorder caused by a defective membrane transport protein. The defective protein closes chloride channels in membranes, preventing chloride from being exported out of cells. This results in the development of a thick mucus on the outer surfaces of cells. This mucus clogs the ducts that carry digestive enzymes from the pancreas to the small intestine, clogs the airways in the lungs, and promotes lung infections. How do you think the defective protein results in a thick, sticky mucus outside the cells, instead of a loose, fluid covering?

    People with one copy of the gene for cystic fibrosis are less susceptible to cholera, an infection that causes severe diarrhea, than those with normal channel proteins. Why might this be?

Testing Yourself

Choose the best answer for each question.

1. Which of the following is not a fundamental law of energy?

    a. Energy cannot be created or destroyed.
    b. Energy can be changed from one form to another.
    c. Energy is lost when it is converted from one form to another.
    d. Potential energy cannot be converted to kinetic energy.

2. As a result of energy transformation,

    a. entropy increases.
    b. entropy decreases.
    c. heat energy is gained.
    d. energy is lost in the form of heat.
    e. Both a and d are correct.

3. Which of the following is not a component of ATP?

    a. adenine   c. phosphate
    b. glucose   d. ribose

4. ATP is a good source of energy for a cell because

    a. it is versatile—able to be used in many types of reactions.
    b. its breakdown is easily coupled with energy-requiring reactions.
    c. it provides just the right amount of energy for cellular reactions.
    d. All of these are correct.

5. Compared to carbon dioxide and water, glucose

    a. is less organized.   c. is less stable.
    b. has less potential energy. d. exhibits higher entropy.

For questions 6–9, match the items to those in the key. Each answer may include more than one item.

Key:

    a. oxygen
    b. carbon dioxide
    c. water
    d. carbohydrates

6. Consumed by cellular respiration.   c
7. Produced by cellular respiration.   b
8. Consumed by photosynthesis.   a
9. Produced by photosynthesis.   d

10. In the following figure, one curve illustrates the energy of activation required for a reaction in the presence of an enzyme, while the other illustrates the energy of activation in the absence of an enzyme. Label each curve.

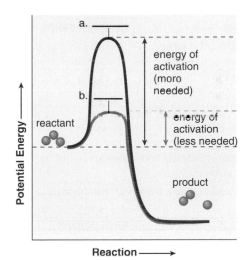

11. The current model for enzyme action is called the

    a. induced fit model.   c. lock-and-key model.
    b. activation model.   d. active substrate model.

For questions 12–18, match the items to those in the key. Each answer may include more than one item.

Key:

    a. simple diffusion
    b. facilitated diffusion
    c. osmosis
    d. active transport

12. Movement of molecules from high concentration to low concentration.   a b
13. Requires a membrane.   b
14. Requires energy input.   d
15. Requires a protein pump.   d b
16. Mechanism by which oxygen enters cells.   a
17. Mechanism by which glucose enters cells.   b d
18. Movement of water across a membrane.   c

19. Cells involved in active transport have a large number of _____ near their plasma membrane.

    a. vacuoles   c. actin filaments
    b. mitochondria   d. lysosomes

20. The movement of substances from maternal blood to fetal blood at the placenta occurs as a result of
    a. receptor-mediated endocytosis.
    b. substrate-mediated endocytosis.
    c. receptor-mediated diffusion.
    d. substrate-mediated diffusion.

21. Energy is used for
    a. simple diffusion.
    b. osmosis.
    c. active transport.
    d. None of these are correct.

22. Which of these are methods of endocytosis?
    a. phagocytosis
    b. pinocytosis
    c. receptor-mediated endocytosis
    d. All of these are correct.

23. Isotonic solutions have _____ concentration(s) of water and _____ concentration(s) of solute.
    a. the same, the same
    b. the same, different
    c. different, the same
    d. different, different

24. A coated pit is associated with
    a. simple diffusion.
    b. osmosis.
    c. receptor-mediated endocytosis.
    d. pinocytosis.

25. Pumps used in active transport are made of
    a. sugars.   c. lipids.
    b. proteins.   d. cholesterol.

26. Which of these is correct?
    a. Energy can be transformed into another type as shown by the production of ATP in mitochondria.
    b. Energy can be created as shown by the production of ATP in mitochondria.
    c. Energy transformations cannot occur; therefore, ecosystems need a constant input of solar energy.
    d. Both a and c are correct.

27. ATP is a modified
    a. protein.   c. nucleotide.
    b. amino acid.   d. fat.

28. The amount of energy needed to get a chemical reaction started is known as the
    a. starter energy.
    b. energy of activation.
    c. reaction energy.
    d. product energy.

29. Entropy is a term used to indicate the relative amount of
    a. organization.
    b. disorganization.
    c. enzyme action.
    d. None of these are correct.

30. Enzymes catalyze reactions by
    a. bringing the reactants together.
    b. lowering the activation energy.
    c. Both a and b are correct.
    d. Neither a nor b is correct.

31. The active site of an enzyme
    a. is identical to that of any other enzyme.
    b. is the part of the enzyme where its substrate can fit.
    c. can be used over and over again.
    d. is not affected by environmental factors such as pH and temperature.
    e. Both b and c are correct.

Go to www.mhhe.com/maderessentials for more quiz questions.

Bioethical Issue

Some organophosphates are agricultural pesticides that inhibit the enzyme acetylcholinesterase. This enzyme breaks down the nerve-excitation chemical acetylcholine. When the enzyme is inhibited by organophosphates, the nervous system becomes overstimulated. Exposure to low levels of organophosphates is believed to cause sleeplessness, anxiety, and depression. Because organophosphates interfere with the nervous system of humans and other animals in agricultural areas, some people believe that their use should be banned. Others argue that they should not be eliminated from use against insects because they would be replaced by pesticides for which modes of action are not as well characterized, making them even more harmful. Currently, large-scale agricultural systems in the United States require the use of pesticides, so the complete elimination of pesticides is not an option.

Do you believe organophosphates should be banned from use in agriculture? Or is the risk of exposure to organophosphates worth the benefits of an abundant, inexpensive food supply?

Understanding the Terms

active site 76	isotonic solution 78
active transport 78	kilocalorie 70
calorie 70	kinetic energy 70
coupled reaction 73	metabolic pathway 75
endocytosis 79	osmosis 77
energy 70	phagocytosis 79
energy laws 70	pinocytosis 79
energy of activation 75	plasmolysis 78
entropy 71	potential energy 70
enzyme 75	receptor-mediated
enzyme inhibition 76	endocytosis 79
exocytosis 79	simple diffusion 77
facilitated diffusion 77	sodium-potassium pump 79
feedback inhibition 76	solute 77
heat 71	solution 77
hypertonic solution 78	solvent 77
hypotonic solution 78	substrate 75
induced fit model 76	

Match the terms to these definitions:

a. _____ The tendency toward disorder.

b. _____ Shrinking of the cytoplasm due to osmosis.

c. _____ A series of linked reactions in a cell.

d. _____ A protein that acts as an organic catalyst.

e. _____ The reactant in an enzymatic pathway.

Energy for Life

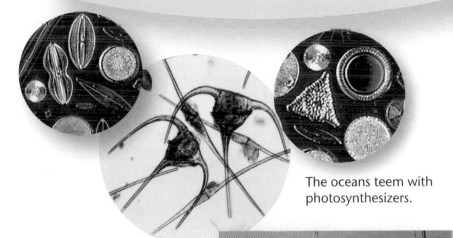

The oceans teem with photosynthesizers.

Plants grow fastest when exposed to red and blue light.

Solar energy

H_2O

H_2O

H_2O

O_2

Plants produce oxygen by splitting water molecules.

Most people are aware that plants produce oxygen gas. Further, they often think plants produce this oxygen as some sort of altruistic act to benefit humans and other animals! But the facts don't bear this out. During the process of photosynthesis, plants are actually making food for themselves. To do this, they reduce the carbon dioxide you exhale to carbohydrates after converting energy from the sun into chemical energy. This conversion process

▶ ▶ ▶

is complex, and depends on the energizing of electrons removed from water. In the process, water splits and releases oxygen. This oxygen is the gas we and other animals depend upon. Thus, plants provide the oxygen you inhale and the glucose you use as a source of energy only because they are in the process of photosynthesizing to keep themselves alive. Plants are solar powered, and indirectly so are all other living things on Earth, including ourselves.

Understanding the process of photosynthesis and its critical importance to nearly all living organism, will help you appreciate just how dependent we humans are on plants. In this chapter, you will learn about basic plant cell structure and how plants perform photosynthesis.

6.1 Overview of Photosynthesis

Photosynthesis transforms solar energy into the chemical energy of a carbohydrate. Photosynthetic organisms, including plants, algae, and cyanobacteria, produce an enormous amount of carbohydrate (Fig. 6.1). If the amount of carbohydrate were instantly converted to coal, and the coal loaded into standard railroad cars (each car holding about 50 tons), the photosynthesizers of the biosphere would fill more than 100 cars *per second* with coal.

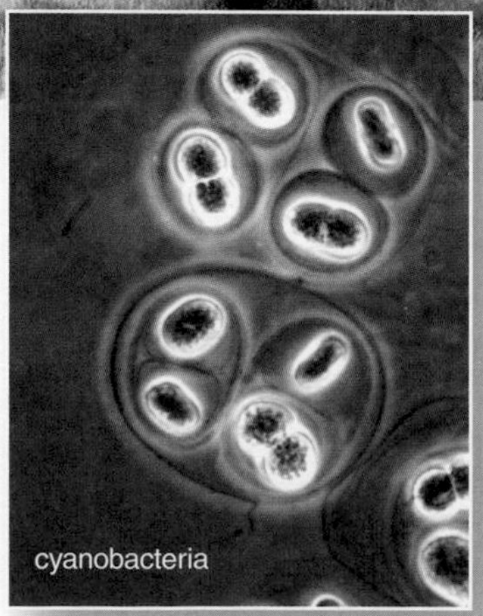

photosynthesis transforms solar energy into the chemical energy of a carbohydrate.

garden plants

trees

mosses

diatoms

kelp

cyanobacteria

Figure 6.1 **Photosynthetic organisms.**

Photosynthetic organisms include plants, such as trees, flowers, and mosses, which typically live on land; photosynthetic protists, such as *Euglena,* diatoms, and kelp, which typically live in water; and cyanobacteria, a type of bacterium that lives everywhere.

It is no wonder, then, that photosynthetic organisms are able to sustain themselves and, with a few exceptions,[1] all of the other living things on Earth. To appreciate this, consider that most food chains lead back to plants. In other words, producers, which have the ability to synthesize carbohydrates, feed not only themselves but also consumers, which must take in preformed organic molecules.

Our analogy about photosynthetic products becoming coal is apt because the bodies of plants formed the coal we burn today. This process occurred hundreds of thousands of years ago, and that is why coal is called a fossil fuel. The wood of trees is also commonly used as fuel. In addition, the fermentation of plant materials produces alcohol, which can be used to fuel automobiles directly or as a gasoline additive.

Flowering Plants as Photosynthesizers

The green portions of plants, particularly the leaves, carry on photosynthesis. The raw materials for photosynthesis are **water** and **carbon dioxide**. The roots of a plant absorb water, which then moves in vascular tissue up the stem to a leaf. Water exits into a leaf by way of leaf veins. The leaf of a flowering plant contains mesophyll tissue in which cells are specialized to carry on photosynthesis. Carbon dioxide in the air enters a leaf through small openings called **stomata** (sing., stoma). Carbon dioxide and water diffuse into mesophyll cells and then into **chloroplasts,** the organelles that carry on photosynthesis (Fig. 6.2).

In a chloroplast, a double membrane surrounds a fluid called the **stroma.** A third membrane system within the stroma forms flattened sacs called **thylakoids,** which in some places are stacked to form **grana** (sing., granum), so called because early microscopists thought they looked like piles of seeds. The space within each thylakoid is believed to be connected to the space within every other thylakoid, thereby forming an inner compartment within chloroplasts called the *thylakoid space.*

Chlorophyll and other pigments reside within the membranes of the thylakoids. These pigments are capable of absorbing solar energy, the energy that drives photosynthesis. The stroma is an enzyme-rich solution in which carbon dioxide is first attached to an organic compound and then reduced to a carbohydrate.

As you well know, human beings, and indeed nearly all organisms, release carbon dioxide into the air. This same carbon dioxide enters leaves and is converted to a carbohydrate. Carbohydrate in the form of glucose is the chief organic source of energy for most organisms.

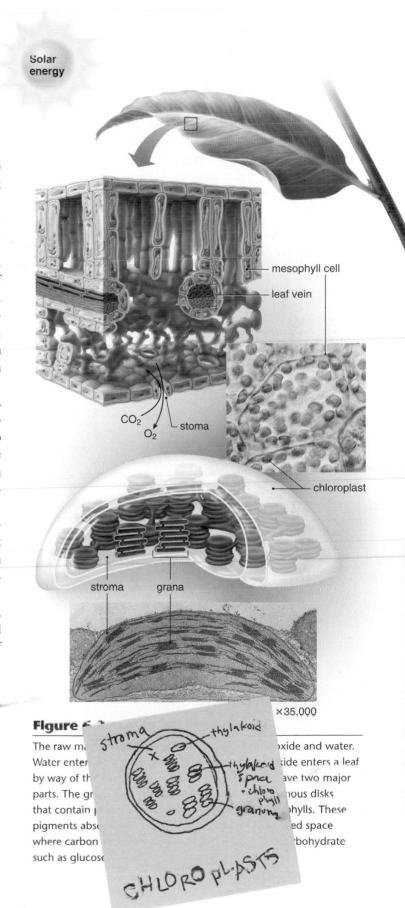

Solar energy

mesophyll cell

leaf vein

CO_2
O_2
stoma

chloroplast

stroma   grana

×35,000

Figure 6.2

The raw m... ...xide and water. Water enter... ...xide enters a leaf by way of th... ...ave two major parts. The gr... ...nous disks that contain p... ...hylls. These pigments abs... ...ed space where carbon... ...rbohydrate such as glucose...

Handwritten note: stomata are small openings, that let in carbon dioxide

Handwritten note: Chloroplast, in mesophyll cells, is where photosynth... is carried out.

Handwritten diagram: stroma — thylakoid — thylakoid space — chlorophyll — granum — CHLOROPLASTS

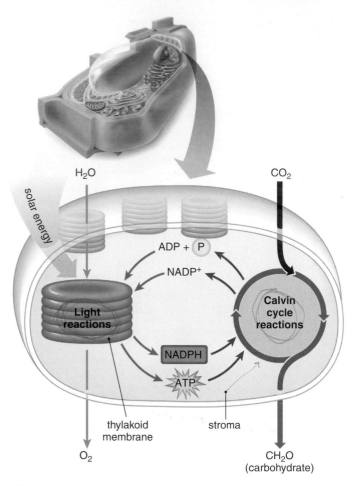

Figure 6.3   The photosynthetic process.

The process of photosynthesis consists of the light reactions and the Calvin cycle reactions. The light reactions, which produce ATP and NADPH, occur in the thylakoid membrane. These molecules are used in the Calvin cycle in the stroma to reduce carbon dioxide to a carbohydrate.

The Photosynthetic Process

The overall equation for photosynthesis is sometimes written like this:

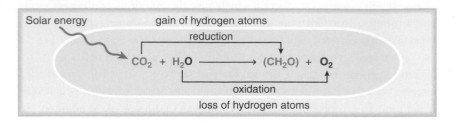

This equation tells us that photosynthesis begins with the end products of cellular respiration, CO_2 and H_2O, which are low-energy molecules. In contrast, a high-energy carbohydrate, symbolized here by CH_2O, is an end product of photosynthesis. To form a carbohydrate, hydrogen atoms have been removed from water and added to CO_2. To reduce CO_2, energy is required, and this energy is provided by the sun. The oxygen released by photosynthesis, so necessary to cellular respiration, is simply a by-product of the oxidation of water.

Notice that if you multiplied (CH_2O) by six you would get $C_6H_{12}O_6$, which is the formula for glucose. Some prefer to emphasize that glucose is an end product of photosynthesis.

Two Sets of Reactions

An overall equation for photosynthesis tells us the beginning reactant and the end products of the pathway. But much goes on in between. The word photosynthesis suggests that the process requires two sets of reactions. *Photo,* which means light, refers to the reactions that capture solar energy, and *synthesis* refers to the reactions that produce a carbohydrate. The two sets of reactions are called the **light reactions** and the **Calvin cycle reactions** (Fig. 6.3).

Light Reactions   The following events occur in the thylakoid membrane during the light reactions:

- Chlorophyll within the thylakoid membranes absorbs solar energy and energizes electrons.
- ATP is produced from ADP + ⓟ with the help of an electron transport chain.
- NADP$^+$,[1] an enzyme helper, accepts electrons and becomes NADPH.

Calvin Cycle Reactions   The following events occur in the stroma during the Calvin cycle reactions:

- CO_2 is taken up by one of the substrates in the cycle.
- ATP and NADPH from the light reactions reduce CO_2 to a carbohydrate.

In the following sections, we discuss details of these two reactions.

Check Your Progress

1. List three major groups of photosynthetic organisms.
2. What molecules are required in order for photosynthesis to begin, and what molecule is the most significant end product?

Answers: 1. Plants, algae, and cyanobacteria. 2. During photosynthesis, carbon dioxide and water become a carbohydrate.

[1]NADP = Nicotinamide adenine dinucleotide phosphate.

6.2 Light Reactions

During the light reactions, the pigments within the thylakoid membranes absorb solar (radiant) energy. Solar energy can be described in terms of its wavelength and its energy content. Figure 6.4 lists the different types of radiant energy, from the shortest wavelength, gamma rays, to the longest, radio waves. *Visible light* is only a small portion of this spectrum.

Visible light contains various wavelengths of light, as can be proven by passing it through a prism; the different wavelengths appear to us as the colors of the rainbow, ranging from violet (the shortest wavelength) to blue, green, yellow, orange, and red (the longest wavelength). The energy content is highest for violet light and lowest for red light.

Only about 42% of the solar radiation that hits the Earth's atmosphere ever reaches the surface, and most of this radiation is within the visible-light range. Higher-energy wavelengths are screened out by the ozone layer in the atmosphere, and lower-energy wavelengths are screened out by water vapor and CO_2 before they reach the Earth's surface. Both the organic molecules within organisms and certain life processes, such as vision and photosynthesis, are adapted to the radiation that is most prevalent in the environment.

Photosynthetic Pigments

The pigments found within most types of photosynthesizing cells are **chlorophylls** and **carotenoids.** These pigments are capable of absorbing portions of visible light. Both chlorophyll *a* and chlorophyll *b* absorb violet, blue, and red wavelengths better than those of other colors. Because green light is reflected and only minimally absorbed, leaves appear green to us. Accessory pigments such as the carotenoids appear yellow or orange because they are able to absorb light in the violet-blue-green range, but not the yellow-orange range. These pigments and others become noticeable in the fall when chlorophyll breaks down and the other pigments are uncovered (Fig. 6.5).

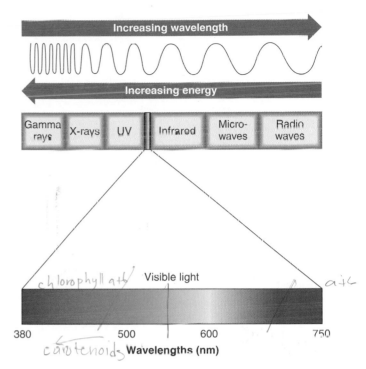

Figure 6.4   Radiant energy.

Radiant energy exists in a range of wavelengths that extends from the very short wavelengths of gamma rays through the very long wavelengths of radio waves. Visible light, which drives photosynthesis, is expanded to show its component colors. The components differ according to wavelength and energy content.

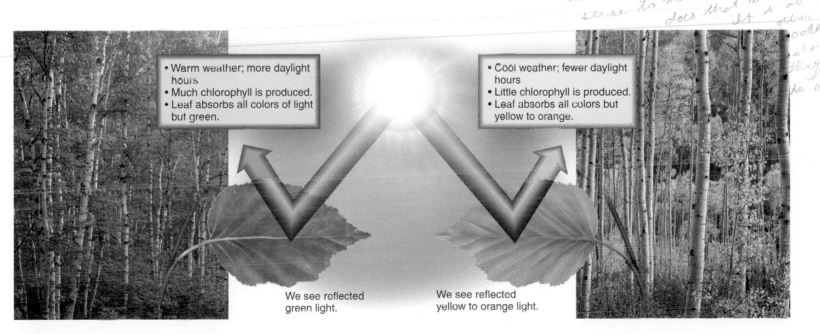

- Warm weather; more daylight hours
- Much chlorophyll is produced.
- Leaf absorbs all colors of light but green.

- Cool weather; fewer daylight hours
- Little chlorophyll is produced.
- Leaf absorbs all colors but yellow to orange.

We see reflected green light.

We see reflected yellow to orange light.

Figure 6.5   Leaf colors.

During the summer, leaves appear green because the chlorophylls absorb other portions of the visible spectrum better than they absorb green. During the fall, chlorophyll breaks down, and the carotenoids that remain cause leaves to appear yellow to orange because they do not absorb these colors.

The Electron Pathway of the Light Reactions

The light reactions consist of an electron pathway that is necessary to the production of ATP and NADPH (Fig. 6.6). This pathway uses two **photosystems,** called photosystem I (PS I) and photosystem II (PS II). The photosystems are named for the order in which they were discovered, not for the order in which they participate in the photosynthetic process. Here is a summary of how the pathway works:

- *The photosystems* PS II and PS I consist of a pigment complex (contains chlorophyll and carotenoid molecules) and an electron acceptor. The pigment complex serves as an "antenna" for gathering solar energy, which is then passed from one pigment to the other until it is concentrated in a particular pair of chlorophyll *a* molecules, called the reaction center.

- *PS II splits water* Due to the absorption of solar energy, electrons (e^-) in the reaction center of PS II become so energized that they escape and move to a nearby electron-acceptor molecule. Replacement electrons are removed from water, which splits, releasing oxygen to the atmosphere. The electron acceptor sends energized electrons, received from the reaction center, down an electron transport chain.

- *The electron transport chain* In an **electron transport chain,** a series of carriers pass electrons from one to the other, releasing energy that is stored in the form of a hydrogen ion (H^+) gradient. Later, ATP is produced (see page 89).

- *PS I produces NADPH* When the PS I pigment complex absorbs solar energy, energized electrons leave its reaction center and are captured by a different electron acceptor. (Low-energy electrons from the electron transport chain adjacent to PS II replace those lost by PS I.) The electron acceptor in PS I passes its electrons to a $NADP^+$ molecule. $NADP^+$ accepts electrons and becomes NADPH.

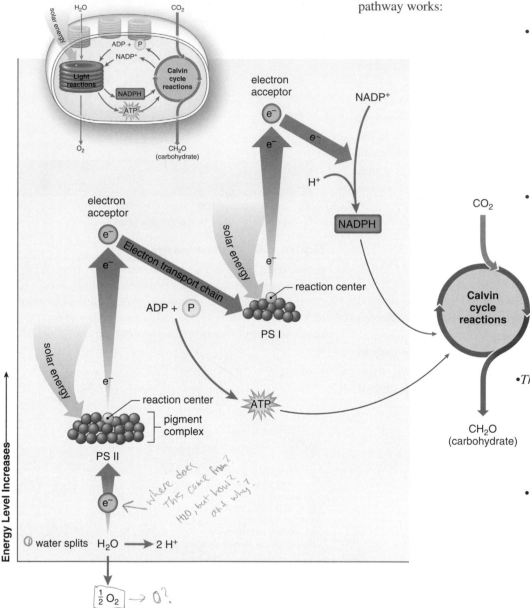

Figure 6.6   **The electron pathway.**

Energized electrons (replaced from water, which splits to release oxygen) leave PS II and pass down an electron transport chain, leading to the formation of ATP. Energized electrons (replaced from PS II) leave PS I and pass to $NADP^+$, which then combines with H^+, becoming NADPH.

Organization of the Thylakoid Membrane

PS II, PS I, and the electron transport chain are located within molecular complexes in the thylakoid membrane (Fig. 6.7). Also present is an ATP synthase complex.

ATP Production

During photosynthesis, the thylakoid space acts as a reservoir for hydrogen ions (H^+). First, each time water is oxidized, two H^+ remain in the thylakoid space. Second, as the electrons move from carrier to carrier down the electron transport chain, the electrons give up energy. This energy is used to pump H^+ from the stroma into the thylakoid space. Therefore, there are many more H^+ ions in the thylakoid space than in the stroma, and an H^+ gradient has been established.

The H^+ ions now flow down their concentration gradient, across the thylakoid membrane at the ATP synthase complex. This causes the enzyme **ATP synthase** to change its shape and produce ATP from ADP + (P).

NADPH Production

Enzymes often have non-protein helpers called **coenzymes.** $NADP^+$ is a coenzyme that accepts electrons from a substrate, becoming NADPH. When it gives up electrons to a substrate, the substrate is reduced. During the light reactions, $NADP^+$ receives electrons at the end of the electron pathway in the thylakoid membrane, and then it picks up a hydrogen ion to become NADPH.

Figure 6.7   Organization of a thylakoid.

Molecular complexes of the electron transport chain within the thylakoid membrane pump hydrogen ions from the stroma into the thylakoid space. When hydrogen ions flow back out of the space into the stroma through the ATP synthase complex, ATP is produced from ADP + (P). $NADP^+$ accepts two electrons and joins with H^+ to become NADPH.

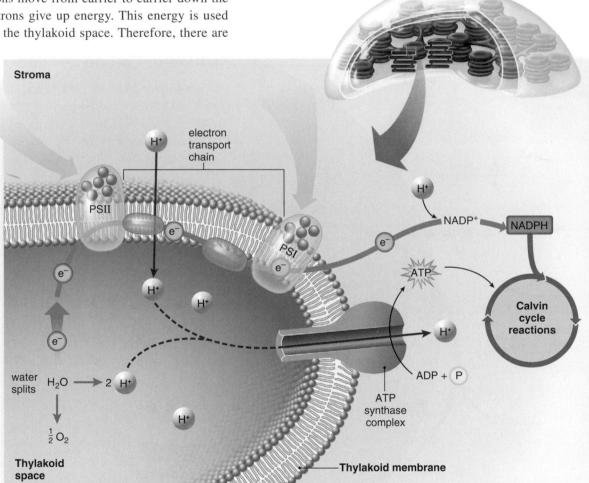

Check Your Progress

1. List the two sets of reactions that are carried out during photosynthesis. *light reactions, calvin cycle*
2. What two molecules are produced as a result of the electron pathway of the light reactions? *ATP & NADPH*

Answers: 1. Light reactions and Calvin cycle reactions. 2. ATP and NADPH.

6.3 Calvin Cycle Reactions

The Calvin cycle is a series of reactions that produce carbohydrate before returning to the starting point (Fig. 6.8). The end product of the Calvin cycle is often considered to be glucose, as mentioned. The cycle is named for Melvin Calvin who, with colleagues, used the radioactive isotope ^{14}C as a tracer to discover the reactions that make up the cycle.

This series of reactions utilizes carbon dioxide from the atmosphere to produce carbohydrate. The Calvin cycle includes (1) carbon dioxide fixation, (2) carbon dioxide reduction, and (3) regeneration of the first substrate, RuBP (ribulose 1,5-bisphosphate).

Fixation of Carbon Dioxide

Carbon dioxide fixation is the first step of the Calvin cycle. During this reaction, CO_2 from the atmosphere is attached to RuBP, a 5-carbon molecule. The enzyme for this reaction is called **RuBP carboxylase (rubisco),** and the result is a 6-carbon molecule that splits into two 3-carbon molecules.

Reduction of Carbon Dioxide

Reduction of CO_2 is the sequence of reactions that uses NADPH from the light reactions, and also uses some ATP from the same source. Carbon dioxide is reduced to a carbohydrate as $R–CO_2$ becomes $R–CH_2O$. Electrons and energy are needed for this reduction reaction, and these are supplied by NADPH and ATP, respectively.

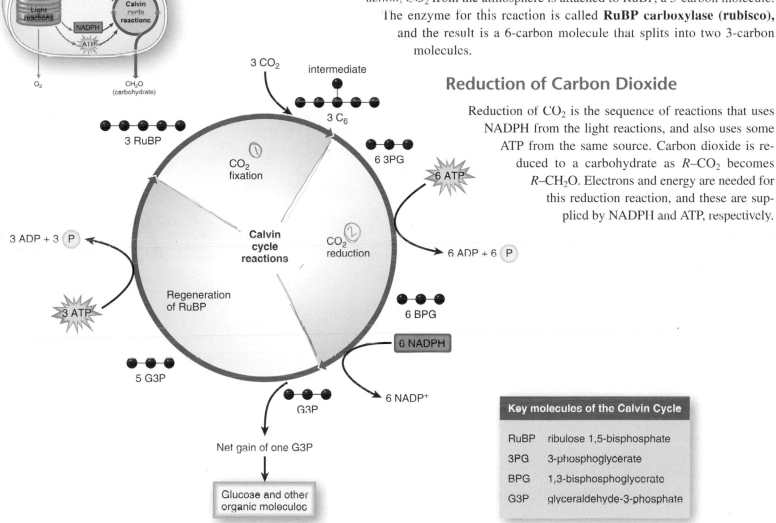

Figure 6.8 The Calvin cycle reactions.

The Calvin cycle is divided into three portions: CO_2 fixation, CO_2 reduction, and regeneration of RuBP. Because five G3P are needed to re-form three RuBP, it takes three turns of the cycle to achieve a net gain of one G3P. Two G3P molecules are needed to form glucose.

Regeneration of RuBP

The product of the Calvin cycle is actually glyceraldehyde-3-phosphate (G3P), which is used to form glucose. Notice that the Calvin cycle reactions in Figure 6.8 are multiplied by three because it takes three turns of the Calvin cycle to allow one G3P to exit. Why? Because, for every three turns of the Calvin cycle, five molecules of G3P are used to re-form three molecules of RuBP, a C_5 molecule. These reactions also utilize some of the ATP produced by the light reactions.

The Importance of the Calvin Cycle

Compared to animal cells, algae and plants have enormous biochemical capabilities. From a G3P molecule, they can make all the molecules they need (Fig. 6.9). A plant can utilize the hydrocarbon skeleton of G3P to form fatty acids and glycerol, which are combined in plant oils. We are all familiar with corn oil, sunflower oil, and olive oil used in cooking. Also, when nitrogen is added to the hydrocarbon skeleton derived from G3P, amino acids are formed.

Notice also that glucose phosphate is among the organic molecules that result from G3P metabolism. Glucose is the molecule that plants and animals most often metabolize to produce ATP molecules to meet their energy needs. Glucose phosphate can be combined with fructose (and the phosphate removed) to form sucrose, the molecule that plants use to transport carbohydrates from one part of the body to another. Glucose phosphate is also the starting point for the synthesis of starch and cellulose. Starch is the storage form of glucose. Some starch is stored in chloroplasts, but most starch is stored in amyloplasts in plant roots. Cellulose is a structural component of plant cell walls; it serves as fiber in our diet because we are unable to digest it.

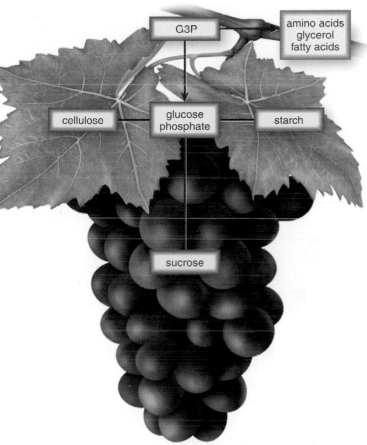

Figure 6.9   The fate of G3P.

G3P is the first reactant in a number of plant cell metabolic pathways. Two G3Ps are needed to form glucose phosphate; glucose is often considered the end product of photosynthesis. Sucrose is the transport sugar in plants; starch is the storage form of glucose; and cellulose is a major constituent of plant cell walls.

Check Your Progress

1. What are three major steps of the Calvin cycle? *Cbs Hvyhm*,
2. Where does the Calvin cycle get the NADPH and the ATP it uses to reduce carbon dioxide to a carbohydrate?
3. List the products that a plant cell can make from G3P, the product of the Calvin cycle.
4. In what part of a chloroplast do the light reactions occur? In what part do the Calvin cycle reactions occur?

Answers: 1. Carbon dioxide fixation, carbon dioxide reduction, and regeneration of RuBP. *2.* From the light reactions. *3.* Glucose, sucrose, starch, cellulose, fatty acids, glycerol, and amino acids. *4.* The light reactions occur in the thylakoid membrane, and the Calvin cycle reactions occur in the stroma.

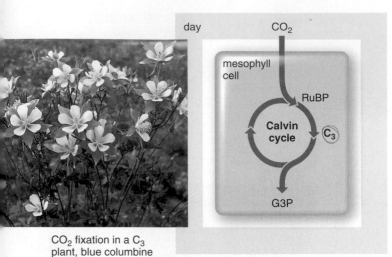

CO₂ fixation in a C₃ plant, blue columbine

Figure 6.10   Carbon dioxide fixation in C₃ plants.

In C₃ plants, such as columbines, CO_2 is taken up by the Calvin cycle directly in mesophyll cells, and the first detectable molecule is a C₃ molecule (red).

6.4   Other Types of Photosynthesis

Plants are physically adapted to their environments. In the cold, windy climates of the north, evergreen trees have small, narrow leaves that look like needles. In the warm, wet climates of the south, some evergreen trees have large, flat leaves to catch the rays of the sun.

In the same way, plants are metabolically adapted to their environments. Where temperature and rainfall tend to be moderate, plants carry on C₃ photosynthesis, and are therefore called C₃ plants. In a **C₃ plant,** the first detectable molecule after CO_2 fixation is a C₃ molecule composed of three carbon atoms (Fig. 6.10). Look again at the Calvin cycle (see Fig. 6.8), and notice that the C₆ molecule formed when RuBP combines with carbon dioxide immediately breaks down to two C₃ molecules.

If the weather is hot and dry, stomata close, preventing the loss of water. (Water loss might cause the plant to wilt and die.) Now CO_2 decreases, and O_2, a by-product of photosynthesis, increases in leaf spaces. In C₃ plants, this O_2 competes with CO_2 for the active site of *rubisco*, the first enzyme of the Calvin cycle, and less C₃ is produced. Such yield decreases are of concern because many food crops are C₃ plants.

In hot, dry climates, successful plants tend to carry on C₄ photosynthesis instead of C₃ photosynthesis. In a **C₄ plant,** the first detectable molecule following CO_2 fixation is a C₄ molecule composed of four carbon atoms. C₄ plants are able to avoid the uptake of O_2 by rubisco. Let's explore how C₄ plants do this.

C₄ Photosynthesis

The anatomy of a C₄ plant is different from that of a C₃ plant. In a C₃ leaf, mesophyll cells arranged in parallel rows contain well-formed chloroplasts. The Calvin cycle reactions occur in the chloroplasts of mesophyll cells (Fig. 6.11). In a C₄ leaf, chloroplasts are located in the mesophyll cells, but they are also located in bundle sheath cells, which surround the leaf vein. Further, the mesophyll cells are arranged concentrically around the bundle sheath cells, shielding the bundle sheath cells from leaf spaces including O_2 from photosynthesis:

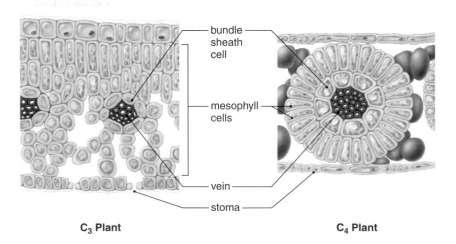

C₃ Plant   **C₄ Plant**

In C_4 plants, the Calvin cycle reactions occur in the bundle sheath cells and not in the mesophyll cells. Therefore, CO_2 from the air is not fixed by the Calvin cycle. Instead, CO_2 is fixed by a C_3 molecule, and the C_4 that results is modified and then pumped into the bundle sheath cells (Fig. 6.11). Now CO_2 enters the Calvin cycle. This represents partitioning of the pathways in space.

It takes energy to pump molecules, and you would think that the C_4 pathway would be disadvantageous. Yet in hot, dry climates, the net photosynthetic rate of C_4 plants such as sugarcane, corn, and Bermuda grass is two to three times that of C_3 plants such as wheat, rice, and oats. Why do C_4 plants enjoy such an advantage? The answer is that when the weather is hot and dry and stomata close, rubisco is not exposed to oxygen, and yield is maintained.

When the weather is moderate, C_3 plants ordinarily have the advantage, but when the weather becomes hot and dry, C_4 plants have their chance to take over, and we can expect them to predominate. In the early summer, C_3 plants such as Kentucky bluegrass and creeping bent grass predominate in lawns in the cooler parts of the United States, but by midsummer, crabgrass, a C_4 plant, begins to take over.

CAM Photosynthesis

Another type of photosynthesis is called **CAM,** which stands for crassulacean-acid metabolism. It gets its name from the Crassulaceae, a family of flowering succulent (water-containing) plants that live in warm, arid regions of the world. CAM was first discovered in these plants, but now it is known to be prevalent among most succulent plants that grow in desert environments, including cactuses.

Whereas a C_4 plant represents partitioning in space — that is, carbon dioxide fixation occurs in mesophyll cells, and the Calvin cycle reactions occur in bundle sheath cells—CAM is partitioning by the use of time. During the night, CAM plants use a C_3 molecule to fix some CO_2, forming C_4 molecules. These molecules are stored in large vacuoles in mesophyll cells. During the day, the C_4 molecules release CO_2 to the Calvin cycle when NADPH and ATP are available from the light reactions (Fig. 6.12).

The primary advantage of this partitioning again relates to the conservation of water. CAM plants open their stomata only at night, and therefore only at that time is atmospheric CO_2 available. During the day, the stomata close. This conserves water, but CO_2 cannot enter the plant.

Photosynthesis in a CAM plant is minimal because of the limited amount of CO_2 fixed at night, but it does allow CAM plants to live under stressful conditions.

Evolutionary Trends

C_4 plants most likely evolved in, and are adapted to high light intensities, high temperatures. However, C_4 plants are more sensit probably do better than C_4 plants b

CAM plants, on the other hand type of plant when the environment ingly, CAM is quite widespread and of flowering plants, including cactuse bromeliads. It is also found among no some ferns and cone-bearing trees.

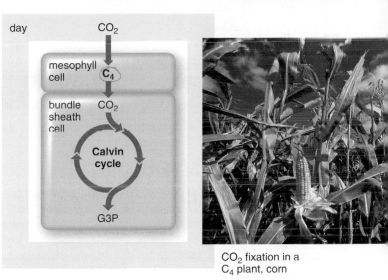

day CO_2

mesophyll cell C_4

bundle sheath cell CO_2

Calvin cycle

G3P

CO_2 fixation in a C_4 plant, corn

Figure 6.11 **Carbon dioxide fixation in C_4 plants.**

C_4 plants, such as corn, form a C_4 molecule (red) in mesophyll cells prior to releasing CO_2 to the Calvin cycle in bundle sheath cells.

night CO_2

C_4

CO_2

Calvin cycle

G3P

day

CO_2 fixation in a CAM plant, pineapple

Figure 6.12 **Carbon dioxide fixation in a CAM plant.**

CAM plants, such as pineapple, fix CO_2 at night, forming a C_4 molecule (red) that is released to the Calvin cycle during the day.

Check Your Progress

1. Name some plants that use a method of photosynthesis other than C_3 photosynthesis.
2. Explain why C_4 photosynthesis is advantageous in hot, dry conditions.

Answers: 1. C_4 plants include many grasses, sugarcane, and corn; CAM plants include cactuses, stonecrops, orchids, and bromeliads. 2. Stomata close under hot, dry conditions, increasing the concentration of oxygen relative to carbon dioxide in the leaf. The spatial separation of carbon fixation and the Calvin cycle reactions in C_4 plants prevents oxygen from competing with carbon dioxide for an active site on the enzyme rubisco.

[handwritten note:]
Pineapple use Cam and corn uses C4
② rubisco is hot exposed to oxygen when stomata close.
This entire stuff is confusing! the book and the teacher don't explain it very well

[handwritten note right margin:] how are ... learning? No one ... this?

Summary

6.1   Overview of Photosynthesis

Cyanobacteria, algae, and plants carry on photosynthesis, a process in which water is oxidized and carbon dioxide is reduced using solar energy. The end products of photosynthesis include carbohydrate and oxygen:

$$solar\ energy + CO_2 + 2\,H_2O \longrightarrow (CH_2O) + H_2O + O_2$$

In plants, photosynthesis takes place in chloroplasts. A chloroplast is bounded by a double membrane and contains two main components: the semiliquid stroma and the membranous grana made up of thylakoids. During photosynthesis, the light reactions take place in the thylakoid membrane, and the Calvin cycle reactions take place in the stroma:

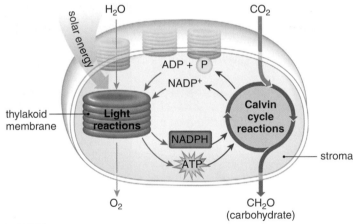

6.2   Light Reactions

The light reactions use solar energy in the visible-light range. Pigment complexes in two photosystems (PS II and PS I) absorb various wavelengths of light, and the energy is concentrated in the reaction centers of the photosystems. The light reactions involve the following electron pathway:

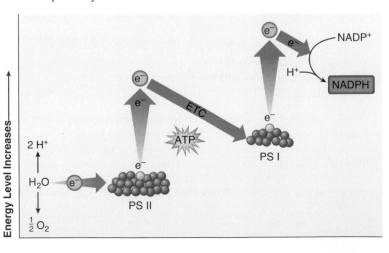

- Solar energy enters PS II, and energized electrons are picked up by an electron acceptor. The oxidation (splitting) of water replaces these electrons in the reaction center. O_2 is released to the atmosphere, and hydrogen ions (H^+) remain in the thylakoid space.
- As electrons pass down an electron transport chain (ETC), the release of energy allows the carriers to pump H^+ into the thylakoid space. The buildup of H^+ establishes an electrochemical gradient.
- When solar energy is absorbed by PS I, energized electrons leave and are ultimately received by $NADP^+$, which also combines with H^+ from the stroma to become NADPH. Electrons from PS II replace those lost by PS I.
- When H^+ flows down its concentration gradient through the channel present in ATP synthase complexes, ATP is synthesized from ADP and Ⓟ by ATP synthase.

6.3   Calvin Cycle Reactions

The Calvin cycle consists of the following stages:

CO_2 fixation   The enzyme rubisco fixes CO_2 to RuBP, producing a 6-carbon molecule that immediately breaks down to two C_3 molecules.

CO_2 reduction CO_2 (incorporated into an organic molecule) is reduced to carbohydrate (CH_2O). This step requires the NADPH and some of the ATP from the light reactions.

Regeneration of RuBP   For every three turns of the Calvin cycle, the net gain is one G3P molecule; the other five G3P molecules are used to re-form three molecules of RuBP. This step also requires the energy of ATP. G3P is then converted to all the organic molecules a plant needs. It takes two G3P molecules to make one glucose molecule.

6.4   Other Types of Photosynthesis

C₄ Photosynthesis

When the weather is hot and dry, C_3 plants are at a disadvantage because stomata close and O_2 from photosynthesis competes with CO_2 for the active site of rubisco. C_4 plants avoid this drawback. In C_4 plants, CO_2 fixation occurs in mesophyll cells, and the Calvin cycle reactions occur in bundle sheath cells. In mesophyll cells, a C_3 molecule fixes CO_2, and the result is a C_4 molecule, for which this type of photosynthesis is named. A modified form of this molecule is pumped into bundle sheath cells, where CO_2 is released to the Calvin cycle. This represents partitioning in space.

CAM Photosynthesis

CAM plants live in hot, dry environments. At night when stomata remain open, a C_3 molecule fixes CO_2 to produce a C_4 molecule. The next day, CO_2 is released and enters the Calvin cycle within the same cells. This represents a partitioning of pathways in time: Carbon dioxide fixation occurs at night, and the Calvin cycle occurs during the day.

Thinking Scientifically

1. Greenhouse growers of horticultural plants have long known that carbon dioxide can be a limiting factor for photosynthesis. That is, if plants are supplied with higher than normal levels of carbon dioxide, they will grow faster. Based on the data graphed below, answer the following questions:

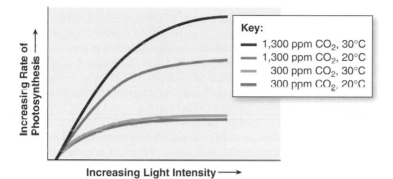

Key:
— 1,300 ppm CO_2, 30°C
— 1,300 ppm CO_2, 20°C
— 300 ppm CO_2, 30°C
— 300 ppm CO_2, 20°C

Increasing Rate of Photosynthesis ↑

Increasing Light Intensity →

   At 300 ppm (parts per million) CO_2 (roughly the level in the atmosphere), did an increase in temperature result in an increase in photosynthetic rate? Did an increase in light intensity at 300 ppm CO_2 result in an increase in photosynthetic rate?
   At 1,300 ppm CO_2 (a CO_2-enriched atmosphere), did an increase in temperature result in an increase in photosynthetic rate? Did an increase in light intensity at 1,300 ppm CO_2 result in an increase in photosynthetic rate? Is CO_2 a major limiting factor for growth in this experiment?

2. Because of greenhouse studies similar to the one outlined in question 1, some scientists have suggested that global warming will increase rates of photosynthesis, causing plants to take up higher levels of CO_2, and therefore decreasing atmospheric CO_2 levels. However, studies at the ecosystem level indicate that higher atmospheric CO_2 levels, combined with higher temperatures, do not result in higher levels of CO_2 fixed by photosynthesis. How might plant responses in an ecosystem differ from those in a greenhouse?

3. In 1882, T. W. Engelmann carried out an ingenious experiment to demonstrate that chlorophyll absorbs light in the blue and red portions of the spectrum. He placed a single filament of a green alga in a drop of water on a microscope slide. Then he passed light through a prism and onto the string of algal cells. The slide also contained aerobic bacterial cells. After some time, he peered into the microscope and saw the bacteria clustered around the regions of the algal filament that were receiving blue light and red light, as shown in the following illustration. Why do you suppose the bacterial cells were clustered in this manner?

Oxygen-seeking bacteria

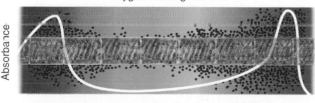

Absorbance

Filament of green alga

Testing Yourself

Choose the best answer for each question.

1. The raw materials for photosynthesis are
   a. oxygen and water.
   b. oxygen and carbon dioxide.
   c. carbon dioxide and water.
   d. carbohydrates and water.
   e. carbohydrates and carbon dioxide.

2. During photosynthesis, carbon dioxide is _____ and water is _____.
   a. reduced, oxidized   c. reduced, reduced
   b. oxidized, reduced   d. oxidized, oxidized

3. When electrons in the reaction center of PS I are passed to an energy acceptor molecule, they are replaced by electrons that have been given up by
   a. oxygen.   c. carbon dioxide.
   b. glucose.   d. water.

4. PS I, PS II, and the electron transport chain are located in the
   a. thylakoid membrane.
   b. stroma.
   c. outer chloroplast membrane.
   d. cell's nucleus.

5. During the light reactions of photosynthesis, ATP is produced when hydrogen ions move
   a. down a concentration gradient from the thylakoid space to the stroma.
   b. against a concentration gradient from the thylakoid space to the stroma.
   c. down a concentration gradient from the stroma to the thylakoid space.
   d. against a concentration gradient from the stroma to the thylakoid space.

6. Oxygen is generated by the
   a. light reactions.
   b. Calvin cycle.
   c. light reactions and the Calvin cycle reactions.
   d. None of these are correct.

7. The Calvin cycle requires _____ from the light reactions.
   a. carbon dioxide and water
   b. ATP and NADPH
   c. carbon dioxide and ATP
   d. ATP and water
   e. NADH and water

8. The product of the Calvin cycle and a reactant in many plant metabolic pathways is
   a. ribulose 1,5-bisphosphate, RuBP.
   b. adenosine diphosphate, ADP.
   c. 3-phosphoglycerate, 3PG.
   d. glyceraldehyde-3-phosphate, G3P.
   e. carbon dioxide, CO_2.

9. Use these terms to label the following diagram of the light reactions.

ATP c
thylakoid membrane f
electron transport chain b
✶ stroma a
thylakoid space g
NADPH d
ATP synthase complex e

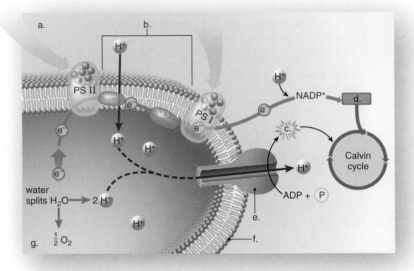

10. In C_4 plants, a 4-carbon molecule

   a. is the product of the Calvin cycle.
   b. is the first electron carrier in the light reactions.
   c. is the storage form of carbohydrates.
   d. precedes the Calvin cycle.

11. A leaf of a C_4 plant differs from that of a C_3 plant because it

   a. has more stomata.
   b. has more layers of mesophyll.
   c. contains mesophyll cells in a concentric ring around the bundle sheath cells.
   d. has needlelike leaves

For questions 12–15, match the items to those in the key. Answers can be used more than once, and each question can have more than one answer.

Key:

   a. C_3 plant
   b. C_4 plant
   c. CAM plant

12. Carbon fixation occurs in mesophyll cells.

13. The Calvin cycle reactions occur in bundle sheath cells.

14. Carbon fixation and the Calvin cycle reactions are separated by space.

15. Carbon fixation and the Calvin cycle reactions are separated by time.

Go to www.mhhe.com/maderessentials for more quiz questions.

Bioethical Issue

In the United States, solar energy to grow food is greatly supplemented by fossil fuel energy. Even before crops are sowed, there is an input of fossil fuel energy for the production of seeds, tools, fertilizers, pesticides, and transportation of these to the farm. Fuel is then needed to plant seeds, apply fertilizers and pesticides, irrigate, and harvest crops. After harvesting, more fuel is used to process and package crops.

At this time, the fossil fuel energy to grow food is several hundred times its caloric content because of the limited amount of land devoted to agriculture, plus we use high-yielding plants that require additional care. It takes about 20 times the amount of energy to keep cattle in feedlots and feed them grain as it does to range-feed them. Because the combustion of fossil fuel energy contributes to environmental problems, such as global warming and air pollution, we should take steps to reduce the amount of fossil fuel energy used to grow food. What can be done? We could devote as much land as possible to farming and animal husbandry, and plant breeders could sacrifice some yield to develop plants that would require less fossil fuel energy. If cattle were range-fed, manure could substitute, in part, for chemical fertilizers. Biological control—the use of natural enemies to control pests—would cut down on pesticide use and possibly improve the health of farm families. The use of solar energy and wind power would also reduce the use of fossil fuel energy.

As consumers, we could eat blemished fruits and vegetables, thus eliminating the need for some pesticides. We could also reduce our consumption of processed foods, eat less meat, and avoid using electrically powered devices when preparing food at home.

Understanding the Terms

ATP synthase   89
C_3 plant   92
C_4 plant   92
Calvin cycle reaction   86
CAM   93
carbon dioxide (CO_2)
   fixation   90
carotenoid   87
chlorophyll   85, 87
chloroplast   85

coenzyme   89
electron transport chain   88
grana (sing., granum)   85
light reaction   86
photosynthesis   84
photosystem   88
RuBP carboxylase (rubisco)   90
stomata (sing., stoma)   85
stroma   85
thylakoid   85

Match the terms to these definitions:

a. ___Stroma___ The fluid component of a chloroplast.

b. ___stomata___ Carbon dioxide enters leaves through these openings.

c. ___ATP synthase___ This enzyme produces adenosine triphosphate.

d. ___thylakoid___ Hydrogen ions accumulate here during the light reactions.

✶ e. ___coenzyme___ These nonprotein molecules help enzymes catalyze reactions.

f. ___CO_2 fixation___ The first step of the Calvin cycle.

✶ g. ___photosystem___ A cluster of pigment molecules that absorbs energy during the light reactions.

h. ___carotenoid___ The photosynthetic pigment that absorbs violet, blue, and green light.

i. ___CAM___ These plants fix carbon at night and carry out Calvin cycle reactions during the day.

Energy for Cells

Lack of oxygen for cellular respiration makes muscles ache.

Carbon monoxide and cyanide kill by stopping cellular respiration.

If a diet sounds too good to be true, it probably is.

In American society, we think a lot about weight loss. Everywhere we look, advertisements claim that some product will rev up our slow metabolism and cure our weight problems for good. In the meantime, our society has more obese individuals than ever. A "miracle cure" for being heavy often comes in the form of a very odd diet or a nutritional supplement. Have you ever considered the science behind how these products claim to work? When

▶ ▶ ▶

you do, you will see that many of these claims simply can't be verified. Yet millions of people spend a fortune hoping for a "miracle," and many of them suffer health problems as a result of fad diets or the use of nutritional supplements.

Numerous nutritional supplements are supposed to allow you to eat everything you want, with no need to exercise. It sounds nice, but the reality is that our cells don't work that way! In order to acquire the energy you need to live, cells break down glucose in a process called cellular respiration. Both the glucose and the oxygen needed for cellular respiration are provided by the process of photosynthesis in plants. In this chapter, you will learn how cells use glucose to produce the ATP they need. Understanding this process will then help you make informed decisions about weight loss, diets, and nutritional supplements.

7.1 Cellular Respiration

Whether you go skiing, take an aerobics class, or just hang out, ATP molecules provide the energy needed for your muscles to contract. ATP molecules are produced during cellular respiration, a process that requires the participation of mitochondria. Cellular respiration is aptly named because just as you take in oxygen (O_2) and give off carbon dioxide (CO_2) during breathing, so do the mitochondria in your cells (Fig. 7.1). In fact, cellular respiration, which occurs in all cells of the body, is the reason you breathe.

Oxidation of substrates is a fundamental part of cellular respiration. In living things, oxidation doesn't occur by the addition of oxygen (O_2). Instead, oxidation is the removal of hydrogen atoms from a molecule. As cellular respiration occurs, hydrogen atoms are removed from glucose (and glucose products) and transferred to oxygen atoms, forming carbon dioxide (CO_2) and water (H_2O):

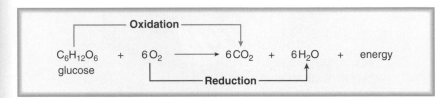

$$C_6H_{12}O_6 \;+\; 6\,O_2 \longrightarrow 6\,CO_2 \;+\; 6\,H_2O \;+\; \text{energy}$$

The breakdown of glucose releases a lot of energy. If you mistakenly burn sugar in a skillet, the energy escapes into the atmosphere as heat. A cell is more sophisticated than that. In a cell, glucose is broken down slowly—not all at once—and the energy given off isn't all lost as heat. Hydrogen atoms are removed bit by bit, and this allows energy to be captured and used to make ATP molecules.

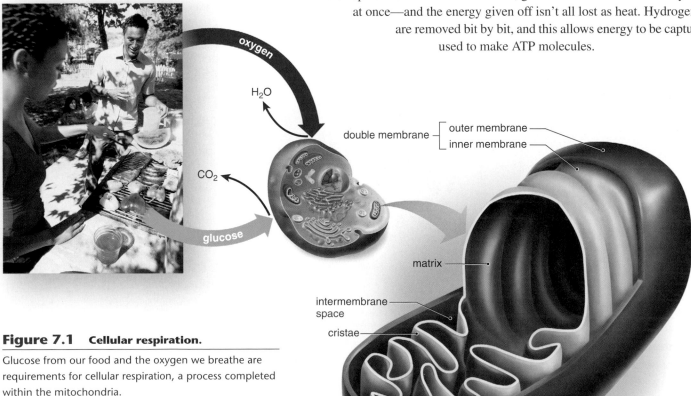

Figure 7.1   Cellular respiration.

Glucose from our food and the oxygen we breathe are requirements for cellular respiration, a process completed within the mitochondria.

Phases of Complete Glucose Breakdown

The enzymes that carry out oxidation during cellular respiration are assisted by non-protein helpers called **coenzymes.** As glucose is oxidized, the coenzymes NAD^+ and FAD^1 receive hydrogen atoms ($H^+ + e^-$) and become NADH and $FADH_2$, respectively (Fig. 7.2).

During these phases, notice where CO_2 and H_2O are produced.

- **Glycolysis,** which occurs in the cytoplasm outside the mitochondria, is the breakdown of glucose to 2 molecules of pyruvate. Oxidation results in NADH, and there is enough energy left over for a net gain of 2 ATP molecules.

- The **preparatory (prep) reaction** takes place in the matrix of mitochondria. Pyruvate is broken down to a 2-carbon acetyl group carried by **coenzyme Λ (CoA).** Oxidation of pyruvate results in not only NADH, but also CO_2.

- The **citric acid cycle** also takes place in the matrix of mitochondria. As oxidation occurs, NADH and $FADH_2$ result and more CO_2 is released. The citric acid cycle is able to produce 2 ATP per glucose molecule.

- The **electron transport chain** is a series of electron carriers in the cristae of mitochondria. NADH and $FADH_2$ give up electrons to the chain. Energy is released and captured as the electrons move from a higher energy to a lower energy state. Later, this energy will be used for the production of ATP. Oxygen (O_2) finally shows up here as the last acceptor of electrons from the chain. Combination with hydrogen ions (H^+) produces water (H_2O).

^{1} NAD = Nicotinamide adenine dinucleotide; FAD = Flavin adenine dinucleotide

Check Your Progress

1. Why is breathing necessary to cellular respiration?
2. Explain why glucose is broken down slowly, rather than quickly, during cellular respiration.
3. List the four phases of complete glucose breakdown.

Answers: 1. Breathing takes in oxygen needed for cellular respiration and rids the body of carbon dioxide, a waste product of cellular respiration. 2. Slow breakdown allows much of the released energy to be captured and utilized by the cell. 3. Glycolysis, the preparatory reaction, the citric acid cycle, and the electron transport chain.

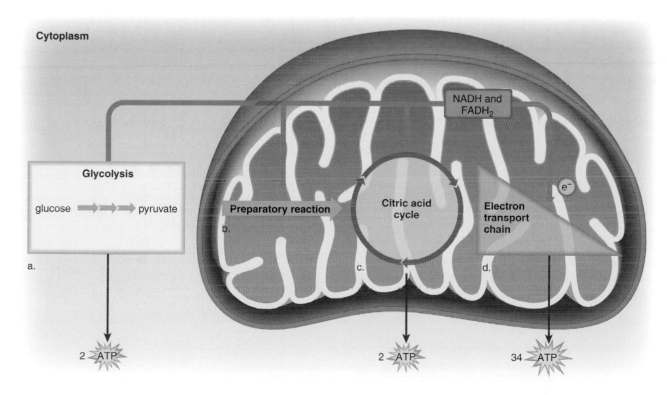

Figure 7.2
The four phases of complete glucose breakdown.

a. The enzymatic reactions of glycolysis take place in the cytoplasm. **b.** The preparatory reaction, (**c**) the citric acid cycle, and (**d**) the electron transport chain occur in mitochondria.

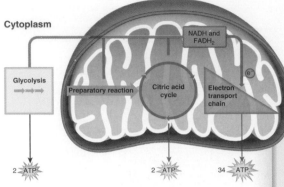

Cytoplasm

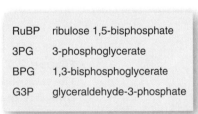

RuBP   ribulose 1,5-bisphosphate

3PG   3-phosphoglycerate

BPG   1,3-bisphosphoglycerate

G3P   glyceraldehyde-3-phosphate

Figure 7.3   Glycolysis.

This metabolic pathway begins with glucose and ends with pyruvate. A net gain of 2 ATP molecules can be calculated by subtracting those expended during the energy-investment steps from those produced during the energy-harvesting steps.

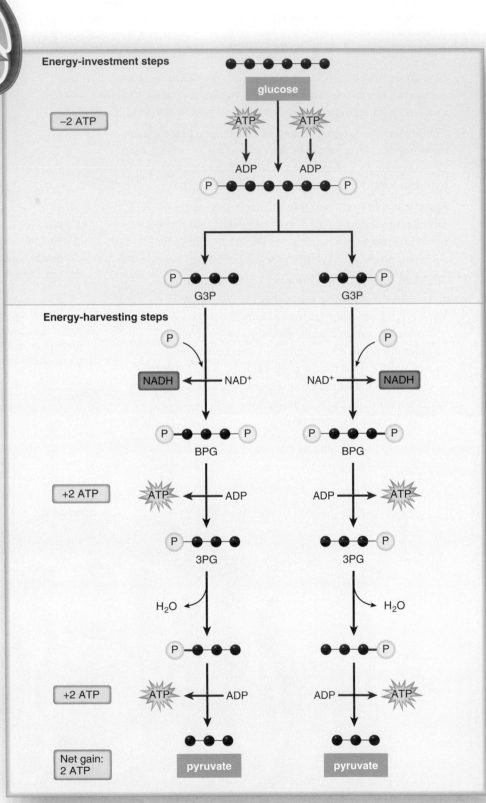

Energy-investment steps

glucose

−2 ATP

ATP   ATP

ADP   ADP

G3P   G3P

Energy-harvesting steps

P   P

NADH ← NAD+   NAD+ → NADH

BPG   BPG

+2 ATP   ATP ← ADP   ADP → ATP

3PG   3PG

H_2O H_2O

+2 ATP   ATP ← ADP   ADP → ATP

Net gain: 2 ATP   pyruvate   pyruvate

7.2   Outside the Mitochondria: Glycolysis

In eukaryotes, such as plants and animals, glycolysis takes place within the cytoplasm outside the mitochondria. During glycolysis, glucose, a C_6 molecule, is broken down to 2 molecules of pyruvate, a C_3 molecule. Glycolysis is divided into (1) the energy-investment steps when ATP is used, and (2) the energy-harvesting steps, when NADH and ATP are produced (Fig. 7.3).

Energy-Investment Steps

During the energy-investment steps, 2 ATP transfer phosphate groups to substrates, and 2 ADP + (P) result. In other words, ATP has been broken down, not built up. However, the phosphate groups activate the substrates so that they can undergo reactions.

Energy-Harvesting Steps

During the energy-harvesting steps, substrates are oxidized by the removal of hydrogen atoms, and 2 NADH result.

Oxidation produces substrates with energized phosphate groups, which are used to synthesize 4 ATP. As a phosphate group is transferred to ADP, ATP results. The process is called substrate-level ATP synthesis (Fig. 7.4).

What is the net gain of ATP from glycolysis? Confirm that 2 ATP are used to get started, and 4 ATP are produced by substrate-level ATP synthesis. Therefore, there is a net gain of 2 ATP from glycolysis.

If oxygen is available, pyruvate, the end product of glycolysis, enters mitochondria, where it undergoes further breakdown. If oxygen is not available, pyruvate undergoes reduction. In humans, if oxygen is not available, pyruvate is reduced to lactate, as discussed on page 107.

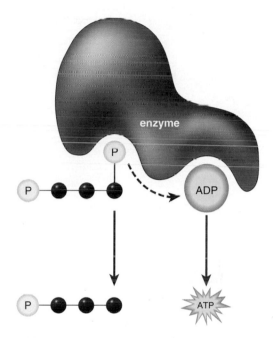

Figure 7.4   Substrate-level ATP synthesis.

The net gain of 2 ATP from glycolysis is the result of substrate-level ATP synthesis. At an enzyme's active site, ADP acquires an energized phosphate group from a substrate, and ATP results.

Glycolysis	
inputs	outputs
glucose	2 pyruvate
2 NAD⁺	2 NADH
2 ATP	
ADP + (P)	4 ATP
	2 ATP — net

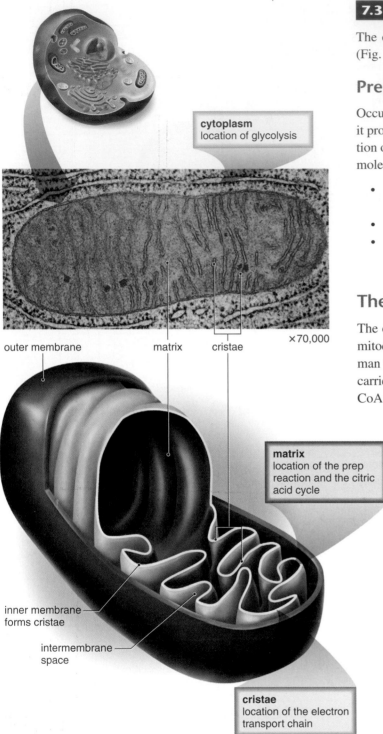

cytoplasm
location of glycolysis

outer membrane   matrix   cristae

×70,000

matrix
location of the prep
reaction and the citric
acid cycle

inner membrane
forms cristae

intermembrane
space

cristae
location of the electron
transport chain

Figure 7.5   Mitochondrion structure and function.

A mitochondrion is bounded by a double membrane. The inner membrane
invaginates to form the shelflike cristae. Glycolysis takes place in the cytoplasm
outside the mitochondria. The preparatory reaction and the citric acid cycle
occur within the mitochondrial matrix. The electron transport chain is located
on the cristae of a mitochondrion.

7.3   Inside the Mitochondria

The other three phases of cellular respiration occur inside the mitochondria
(Fig. 7.5).

Preparatory Reaction

Occurring in the matrix, the **preparatory (prep) reaction** is so called because
it produces a substrate that can enter the citric acid cycle. The preparatory reaction occurs twice per glucose molecule because glycolysis results in 2 pyruvate
molecules. During the prep reaction:

- Pyruvate is oxidized, and a CO_2 molecule is given off. This is part of the
  CO_2 we breathe out!
- NAD^+ accepts a hydrogen atom, and NADH results.
- A C_2 acetyl group is attached to coenzyme A (CoA), forming **acetyl-
  CoA.**

The Citric Acid Cycle

The **citric acid cycle** is a cyclical metabolic pathway located in the matrix of
mitochondria (Fig. 7.6). It was originally called the *Krebs cycle* to honor the
man who first studied it. At the start of the citric acid cycle, the C_2 acetyl group
carried by CoA joins with a C_4 molecule, and a C_6 citrate molecule results. The
CoA returns to the preparatory reaction to be used again.

During the citric acid cycle:

- The acetyl group is oxidized, and the rest of the CO_2 we
  breathe out per glucose molecule is released.
- Both NAD^+ and FAD accept hydrogen atoms, resulting in
  NADH and $FADH_2$.
- Substrate-level ATP synthesis occurs (see Fig. 7.4), and an
  ATP results.

Because the citric acid cycle turns twice for each original glucose
molecule, the inputs and outputs of the citric acid cycle per glucose
molecule are as follows:

Citric acid cycle	
inputs	outputs
2 acetyl-CoA	4 CO_2
6 NAD^+	6 NADH
2 FAD	2 $FADH_2$
2 ADP + 2 P	2 ATP

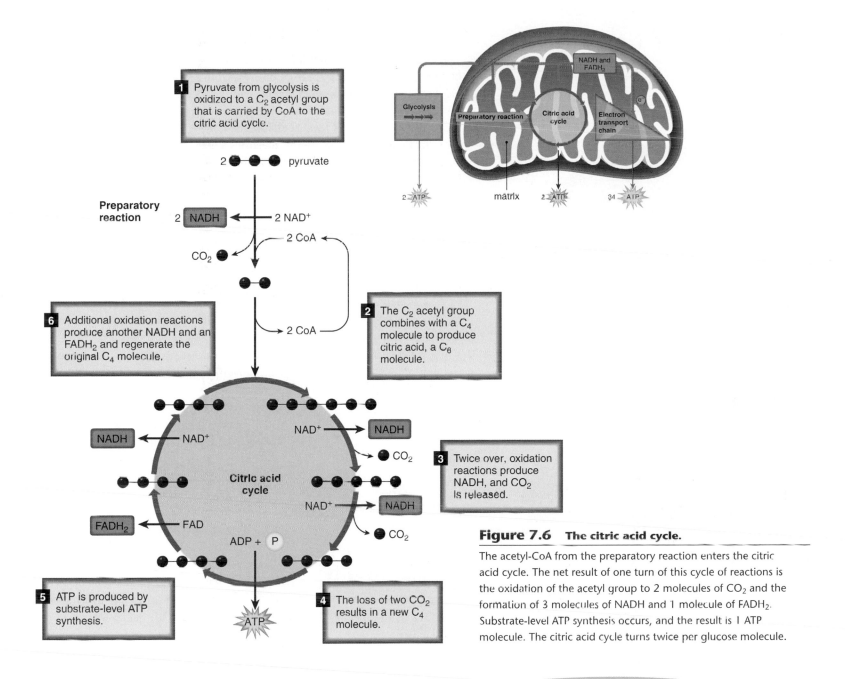

1 Pyruvate from glycolysis is oxidized to a C_2 acetyl group that is carried by CoA to the citric acid cycle.

2 ●●● pyruvate

Preparatory reaction

2 NADH ← 2 NAD$^+$

2 CoA ←

CO_2 ●

2 CoA →

2 The C_2 acetyl group combines with a C_4 molecule to produce citric acid, a C_6 molecule.

6 Additional oxidation reactions produce another NADH and an FADH$_2$ and regenerate the original C_4 molecule.

Citric acid cycle

NADH ← NAD$^+$

NAD$^+$ → NADH

● CO_2

3 Twice over, oxidation reactions produce NADH, and CO_2 is released.

NAD$^+$ → NADH

● CO_2

FADH$_2$ ← FAD

ADP + P

ATP

5 ATP is produced by substrate-level ATP synthesis.

4 The loss of two CO_2 results in a new C_4 molecule.

Figure 7.6 The citric acid cycle.

The acetyl-CoA from the preparatory reaction enters the citric acid cycle. The net result of one turn of this cycle of reactions is the oxidation of the acetyl group to 2 molecules of CO_2 and the formation of 3 molecules of NADH and 1 molecule of FADH$_2$. Substrate-level ATP synthesis occurs, and the result is 1 ATP molecule. The citric acid cycle turns twice per glucose molecule.

Check Your Progress

1. A C_2 acetyl group enters the citric acid cycle. Where does it come from?
2. What are the products of the citric acid cycle as a result of further breakdown of glucose?

Answers: 1. The C_2 acetyl group comes from the prep reaction. *2.* The citric acid cycle turns twice per glucose molecule, producing 2 CO_2, 3 NADH, 1 FADH$_2$, 1 ATP per turn.

The Electron Transport Chain

The electron transport chain located in the cristae of mitochondria is a series of carriers that pass electrons from one to the other. NADH and $FADH_2$ deliver electrons to the chain. Consider that the hydrogen atoms attached to NADH and $FADH_2$ consist of an e^- and an H^+. The members of the electron transport chain accept only electrons (e^-) and not hydrogen ions (H^+).

In Figure 7.7, high-energy electrons enter the chain, and low-energy electrons leave the chain. When NADH gives up its electrons, the next carrier gains the electrons and is reduced. This oxidation-reduction reaction starts the process, and each of the carriers in turn becomes reduced and then oxidized as the electrons move down the system. As the pair of electrons is passed from carrier to carrier, energy is released and captured for ATP production. The final acceptor of electrons is oxygen (O_2), the very O_2 we breathe in. It's remarkable to think that the role of oxygen in cellular respiration is to keep the electrons moving from the first to the last carrier. Why can oxygen play this role? Because oxygen attracts electrons to a greater degree than the carriers of the chain. Once oxygen accepts electrons it combines with H^+, and the other end product of cellular respiration (i.e., water) results (see the equation on page 98).

When each NADH delivers electrons to the first carrier of the electron transport chain, enough energy is captured by the time the electrons are received by O_2 to permit the production of three ATP molecules. When each $FADH_2$ delivers electrons to the electron transport chain, only 2 ATP are produced.

Once NADH has delivered electrons to the electron transport chain, NAD^+ is regenerated and can be used again. In the same manner, FAD is regenerated and can be used again. The recycling of coenzymes, and for that matter ADP, increases cellular efficiency since it does away with the need to synthesize NAD^+, FAD, and ADP anew.

The Cristae of a Mitochondrion

The carriers of the electron transport chain are located in molecular complexes within the inner mitochondrial membrane. ATP synthesis is carried out by ATP synthase complexes also located in this membrane (Fig. 7.8).

The carriers of the electron transport chain accept electrons from NADH or $FADH_2$ and then pass them from one to the other by way of two additional

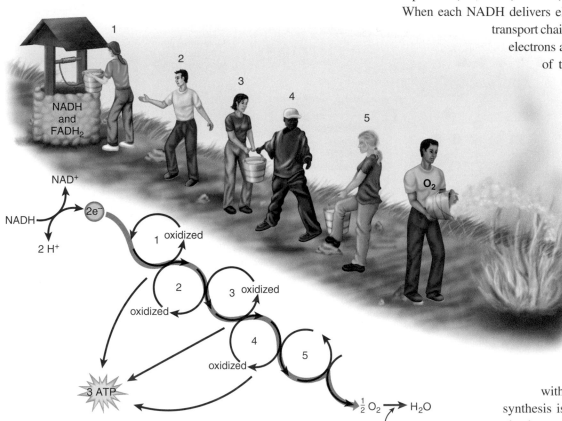

Figure 7.7   The electron transport chain.

An electron transport chain operates like a bucket brigade. Each electron carrier (#1–5) is alternatively reduced (orange) and oxidized as if the electrons were a bucket being passed from person to person. As oxidation-reduction occurs, energy is released that will be used to make ATP.

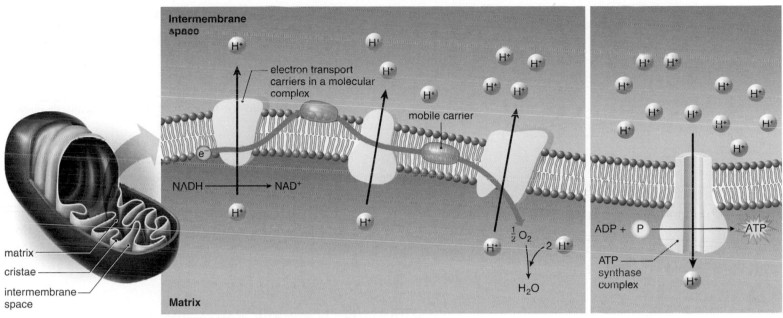

Intermembrane space

electron transport carriers in a molecular complex

mobile carrier

NADH → NAD⁺

$\frac{1}{2}$ O₂

2 H⁺

H₂O

matrix
cristae
intermembrane space

Matrix

ADP + P → ATP

ATP synthase complex

a. Electron transport chain

b. ATP synthesis

Figure 7.8   The organization of cristae.

Molecular complexes that contain the electron transport carriers are located in the cristae as are ATP synthase complexes. **a.** As electrons move from one carrier to the other, hydrogen ions (H⁺) are pumped from the matrix into the intermembrane space. **b.** As hydrogen ions flow back down a concentration gradient through an ATP synthase complex, ATP is synthesized by the enzyme ATP synthase.

mobile carriers (orange arrow). What happens to the hydrogen ions (H⁺) carried by NADH and FADH₂? The complexes use the energy released by oxidation-reduction to pump H⁺ from the mitochondrial matrix into the **intermembrane space** located between the outer and inner membrane of a mitochondrion. The pumping of H⁺ into the intermembrane space establishes an unequal distribution of H⁺ ions; in other words, there are many H⁺ in the intermembrane space and few in the matrix of a mitochondrion.

The energy stored in the H⁺ gradient is now used to drive forward ATP synthesis. The cristae of mitochondria (like the thylakoid membrane of chloroplasts) contain an *ATP synthase complex* that allows H⁺ to return to the matrix. The flow of H⁺ through the ATP synthase complex brings about a conformational change, which causes the enzyme ATP synthase to synthesize ATP from ADP + Ⓟ. ATP leaves the matrix by way of a channel protein. This ATP remains in the cell and is used for cellular work.

Check Your Progress

Explain how the electron transport chain results in the synthesis of ATP.

Answer: As electrons move from one carrier to another in the cristae, energy is released, and this energy is used to pump hydrogen ions from the matrix to the intermembrane space. The flow of hydrogen ions back down the concentration gradient into the matrix drives the synthesis of ATP by ATP synthase.

Phase	NADH	FADH$_2$	ATP Yield
Glycolysis	2	–	2
Prep reaction	2	–	–
Citric acid cycle	6	2	2
Electron transport chain	10 ⟶	2 ⟶	30 4
Total ATP			38

Figure 7.9   Calculating ATP energy yield per glucose molecule.

Substrate-level ATP synthesis during glycolysis and the citric acid cycle accounts for 4 ATP. The electron transport chain produces a maximum of 34 ATP, and the maximum total is 38 ATP. Some cells, however, produce only 36 ATP per glucose molecule or even less.

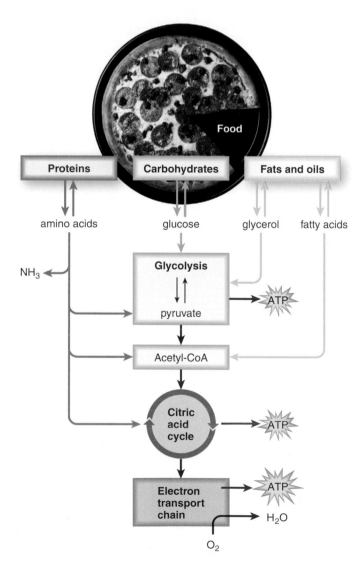

Figure 7.10   Alternative metabolic pathways.

All the types of food in a pizza can be used to generate ATP.

Energy Yield from Glucose Metabolism

Figure 7.9 calculates the ATP yield for the complete breakdown of glucose to CO_2 and H_2O. Per glucose molecule, there is a net gain of 2 ATP from glycolysis, which takes place in the cytoplasm. The citric acid cycle, which occurs in the matrix of mitochondria, accounts for 2 ATP per glucose molecule. This means that a total of 4 ATP form due to substrate-level ATP synthesis outside the electron transport chain.

Most of the ATP produced comes from the electron transport chain and the ATP synthase complex. Per glucose molecule, 10 NADH and 2 FADH$_2$ take electrons to the electron transport chain. The maximum number of ATP produced by the chain is therefore 34 ATP, and the maximum number produced by both the chain and substrate-level ATP synthesis is 38. However, for reasons beyond the scope of this book, the maximum number of ATP produced per glucose molecule in some cells is only 36 ATP or lower. A yield of 36–38 ATP represents about 40% of the available energy in a glucose molecule. The rest of the energy is lost in the form of heat.

Alternative Metabolic Pathways

Let's say you are on a low-carbohydrate diet. Will you then run out of ATP? No, because your cells can also utilize other energy sources—the components of fats and oils, namely glycerol and fatty acids, and amino acids, which are derived from proteins (Fig. 7.10).

Because glycerol is a carbohydrate, it enters the process of cellular respiration during glycolysis. Fatty acids can be metabolized to acetyl groups, which enter the citric acid cycle. A fatty acid with a chain of 18 carbons can make three times the number of acetyl groups as does glucose. For this reason, fats are an efficient form of stored energy—there are three long fatty acid chains per fat molecule. The complete breakdown of glycerol and fatty acids to carbon dioxide and water results in many more ATP molecules per molecule than does the breakdown of glucose.

Only the hydrocarbon backbone of amino acids, not the amino group, can be used by the cellular respiration pathways. The amino group becomes ammonia (NH_3), which becomes part of urea, the primary excretory product of humans. Just where the hydrocarbon backbone from an amino acid begins degradation to produce ATP molecules depends on its length. Figure 7.10 shows that the hydrocarbon backbone from an amino acid can enter cellular respiration pathways at pyruvate, at acetyl-CoA, or during the citric acid cycle.

The smaller molecules in Figure 7.10 can also be used to synthesize larger molecules. In such instances ATP is used instead of generated. You already know that amino acids can be employed to synthesize proteins. Also, some substrates of the citric acid cycle can become amino acids through the addition of an amino group. Of the 20 most common amino acids, humans have the ability to synthesize 11 amino acids in this way, but we cannot synthesize the other 9. These nine are called the *essential amino acids*, meaning that they must be present in the diet or else we suffer a protein deficiency.

Similarly, substrates from glycolysis can become glycerol, and acetyl groups can be used to produce fatty acids. When glycerol and three fatty acids join, a fat results. This explains why you can gain weight from eating carbohydrate-rich foods.

7.4　Fermentation

Fermentation is the anaerobic breakdown of glucose resulting in the buildup of 2 ATP and a toxic end product (Fig. 7.11). During fermentation in animal cells, the pyruvate formed by glycolysis accepts 2 hydrogen atoms and is reduced to lactate. Notice in Figure 7.11 that 2 NADH pass hydrogen atoms to pyruvate, reducing it. Why is it beneficial for pyruvate to be reduced to lactate when oxygen is not available? The answer is that this reaction regenerates NAD^+, which can then pick up more electrons during the earlier reactions of glycolysis. This keeps glycolysis going, during which ATP is produced by substrate-level ATP synthesis.

The 2 ATP produced by fermentation represent only a small fraction of the potential energy stored in a glucose molecule. Following fermentation, most of this potential energy is still waiting to be released. Despite its low yield of only 2 ATP, fermentation is essential. It can provide a rapid burst of ATP, and muscle cells are more apt than other cells to carry on fermentation. When our muscles are working vigorously over a short period of time, as when we run, fermentation is a way to produce ATP even though oxygen is temporarily in limited supply.

However, one of its end products, lactate, is toxic to cells. At first, blood carries away all the lactate formed in muscles. But eventually, lactate begins to build up, changing the pH and causing the muscles to "burn" and then to fatigue so that they no longer contract. When we stop running, our bodies are in **oxygen deficit,** as signified by the fact that we continue to breathe very heavily for a time. Recovery is complete when all the lactate is transported to the liver, where it is reconverted to pyruvate. Some of the pyruvate is oxidized completely, and the rest is converted back to glucose.

Microorganisms and Fermentation

Bacteria utilize fermentation to produce an organic acid, such as lactate, or an alcohol and CO_2, depending on the type of bacterium.

Yeasts are good examples of microorganisms that generate ethyl alcohol and CO_2 when they carry out fermentation. When yeast is used to leaven bread, the CO_2 makes the bread rise. When yeast is used to ferment grapes for wine production or to ferment wort—derived from barley—for beer production, ethyl alcohol is the desired product. However, the yeast are killed by the very product they produce.

The inputs and outputs of fermentation are as follows:

Fermentation	
inputs	outputs
glucose	2 lactate
	or
2 ATP	2 alcohol and 2 CO_2
ADP + P	2 ATP net

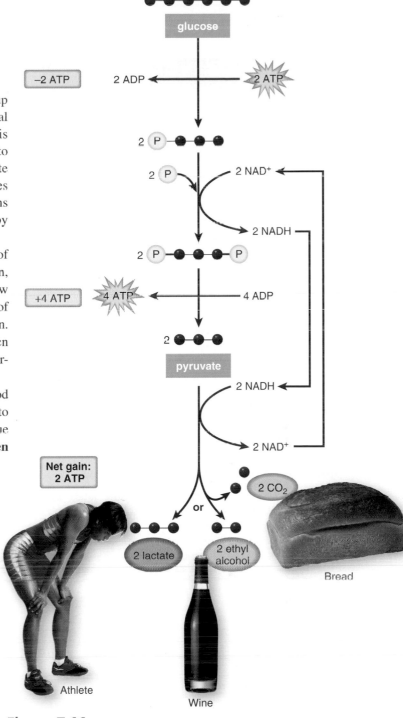

Figure 7.11　Fermentation.

Fermentation consists of glycolysis followed by a reduction of pyruvate by NAD^+, which returns to the glycolytic pathway to pick up more hydrogen atoms.

Check Your Progress

What are the drawbacks and benefits of fermentation?

Answer: Drawbacks: Most of the energy in a glucose molecule is unused and it results in a toxic end product. Benefits: The 2 ATP gained can be used as a burst of energy when oxygen is not available for complete glucose breakdown.

Summary

7.1   Cellular Respiration

During cellular respiration, glucose from food is oxidized to CO_2, which we exhale. Oxygen (O_2), which we breathe in, is reduced to H_2O. When glucose is oxidized, energy is released. Cellular respiration captures the energy of oxidation and uses it to produce ATP molecules. The following equation gives an overview of these events:

$$C_6H_{12}O_6 \ + \ 6\,O_2 \longrightarrow 6\,CO_2 \ + \ 6\,H_2O \ + \ \text{ATP}$$
glucose

7.2   Outside the Mitochondria: Glycolysis

Glycolysis, the breakdown of glucose to 2 molecules of pyruvate, is a series of enzymatic reactions that occur in the cytoplasm. During glycolysis:

- Glucose is oxidized by removal of hydrogen atoms.
- When NAD^+ accepts these electrons, NADH results.

Breakdown releases enough energy to immediately give a net gain of 2 ATP by substrate-level ATP synthesis. The inputs and outputs of glycolysis are summarized here:

Glycolysis	
inputs	outputs
glucose	2 pyruvate
2 NAD^+	2 NADH
2 ATP	
ADP + P	4 ATP
	2 ATP net

When oxygen is available, pyruvate from glycolysis enters a mitochondrion.

7.3   Inside the Mitochondria

Preparatory Reaction

During the preparatory reaction in the matrix:

- Oxidation occurs as CO_2 is removed from pyruvate.
- NAD^+ accepts hydrogen atoms, and NADH results.
- An acetyl group, the end product, combines with CoA.

This reaction takes place twice per glucose molecule.

Citric Acid Cycle

Acetyl groups enter the citric acid cycle, a series of reactions occurring in the mitochondrial matrix. During one turn of the cycle, oxidation results in 2 CO_2 molecules, 3 NADH molecules, and 1 FADH. One turn also produces 1 ATP molecule. The cycle must turn twice per glucose molecule.

Electron Transport Chain

The final stage of cellular respiration involves the electron transport chain located in the cristae of the mitochondria. The chain is a series of electron carriers that accept electrons (e^-) from NADH and $FADH_2$ and pass them along until they are finally received by oxygen, which combines with H^+ to produce water.

The carriers of the electron transport chain are located in molecular complexes on the cristae of mitochondria. These carriers capture energy from the passage of electrons and use it to pump H^+ into the intermembrane space of the mitochondrion. When H^+ flows down its gradient into the matrix through ATP synthase complexes, energy is released and used to form ATP molecules from ADP and P.

Energy Yield

Of the maximum 38 ATP formed by complete glucose breakdown, 4 are the result of substrate-level ATP synthesis, and the rest are produced as a result of the electron transport chain and ATP synthase:

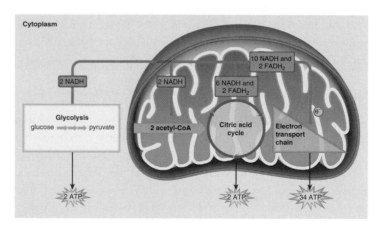

Alternative Metabolic Pathways

Besides carbohydrates, glycerol and fatty acids from fats, and amino acids from proteins can undergo cellular respiration by entering glycolysis and/or the citric acid cycle. These metabolic pathways also provide substrates for the synthesis of fats and proteins.

7.4   Fermentation

Fermentation involves glycolysis followed by the reduction of pyruvate by NADH, either to lactate or to alcohol and CO_2. The reduction of pyruvate regenerates NAD^+, which can accept more hydrogen atoms during glycolysis.

- Although fermentation results in only 2 ATP molecules, it still provides a quick burst of ATP energy for short-term, strenuous muscular activity.
- The accumulation of lactate puts the individual in oxygen deficit, which is the amount of oxygen needed when lactate is completely metabolized to CO_2 and H_2O.

Thinking Scientifically

1. Occasionally, you'll hear a news story about a bin of grain that has undergone spontaneous combustion, resulting in a spectacular fire. It may seem odd that wet grain is more likely to burn than dry grain. However, the grain contains living plant seeds that are physiologically more active when moist than when dry. In addition, the surfaces of the kernels of grain are covered with microorganisms that increase their growth rates when moist. Explain how the consumption of oxygen by these organisms can contribute to a grain-bin fire.

2. One of the major risk factors for diabetes in the elderly is insulin resistance, which is decreased tissue sensitivity to the action of insulin. Tissues then compensate by increasing insulin secretion. Insulin resistance can result from the accumulation of fatty acids in muscle and liver tissue. Researchers have recently found a connection between fatty acid accumulation and mitochondrial function in elderly people. Logically, what might be this connection? Using this knowledge, how might elderly people reduce their risk of diabetes?

Testing Yourself

Choose the best answer for each question.

1. During cellular respiration, _____ is oxidized and _____ is reduced.

   a. glucose, oxygen
   b. glucose, water
   c. oxygen, water
   d. water, oxygen
   e. oxygen, carbon dioxide

2. The products of cellular respiration are energy and

   a. water.
   b. oxygen.
   c. water and carbon dioxide.
   d. oxygen and carbon dioxide.
   e. oxygen and water.

3. During the energy-harvesting steps of glycolysis, which are produced?

   a. ATP and NADH   c. ATP and NAD
   b. ADP and NADH   d. ADP and NAD

4. The end product of glycolysis is

   a. phosphoenol pyruvate.   c. phosphoglyceraldehyde.
   b. glucose.   d. pyruvate.

5. Acetyl-CoA is the end product of

   a. glycolysis.   c. the citric acid cycle.
   b. the preparatory reaction. d. the electron transport chain.

6. The citric acid cycle results in the release of

   a. carbon dioxide.   c. oxygen.
   b. pyruvate.   d. water.

7. The following reactions occur in the matrix of the mitochondria:

   a. glycolysis and the preparatory reaction
   b. the preparatory reaction and the citric acid cycle
   c. the citric acid cycle and the electron transport chain
   d. the electron transport chain and glycolysis

8. Match the descriptions below to the lettered events in the preparatory reaction and citric acid cycle.

   Pyruvate is broken down to an acetyl group.
   Acetyl group is taken up and a C_6 molecule results.
   Oxidation results in NADH and CO_2.
   ATP is produced by substrate-level ATP synthesis.
   Oxidation produces more NADH and $FADH_2$.

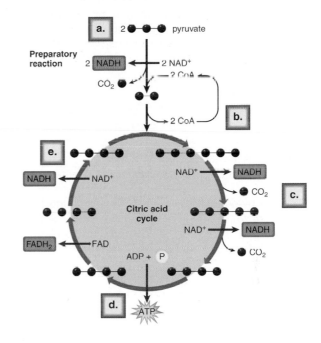

9. The strongest and final electron acceptor in the electron transport chain is

   a. NADH.   c. oxygen.
   b. $FADH_2$.   d. water.

For questions 10–16, match the items to those in the key. Answers can be used more than once, and each question can have more than one answer.

Key:
   a. glycolysis
   b. preparatory reaction
   c. citric acid cycle
   d. electron transport chain

10. Produces ATP.

11. Uses ATP.

12. Produces NADH.

13. Uses NADH.

14. Produces carbon dioxide.

15. Occurs in cytoplasm.

16. Occurs in mitochondria.

17. The carriers in the electron transport chain undergo

    a. oxidation only.
    b. reduction only.
    c. oxidation and reduction.
    d. the loss of hydrogen ions.
    e. the gain of hydrogen ions.

18. The final acceptor for hydrogen atoms during fermentation is
    a. O_2.
    c. FAD.
    b. acetyl CoA.
    d. pyruvic acid.

19. Which of the following do not enter the cellular respiration pathways?
    a. fats
    c. nucleic acids
    b. amino acids
    d. carbohydrates

20. When animals carry out fermentation, they produce _____, while yeasts produce _____.
    a. lactate, malate
    b. lactate, ethyl alcohol
    c. malate, ethyl alcohol
    d. malate, lactate
    e. ethyl alcohol, lactate

21. Fermentation does not yield as much ATP as cellular respiration does because fermentation
    a. generates mostly heat.
    b. makes use of only a small amount of the potential energy in glucose.
    c. creates by-products that require large amounts of ATP to break down.
    d. creates ATP molecules that leak into the cytoplasm and are broken down.

22. Which type of human cells carries on the most fermentation?
    a. fat
    b. muscle
    c. nerve
    d. bone

23. Cellular respiration cannot occur without
    a. sodium.
    b. oxygen.
    c. lactate.
    d. All of these are correct.

24. The metabolic process that produces the most ATP molecules is
    a. glycolysis.
    b. the citric acid cycle.
    c. the electron transport chain.
    d. fermentation.

25. The greatest contributor of electrons to the electron transport chain is
    a. oxygen.
    b. glycolysis.
    c. the citric acid cycle.
    d. the preparatory reaction.
    e. fermentation.

26. Substrate-level ATP synthesis takes place in
    a. glycolysis and the citric acid cycle.
    b. the electron transport chain and the preparatory reaction.
    c. glycolysis and the electron transport chain.
    d. the citric acid cycle and the preparatory reaction.

27. Which of the following is not true of fermentation? Fermentation
    a. has a net gain of only 2 ATP.
    b. occurs in the cytoplasm.
    c. donates electrons to the electron transport chain.
    d. begins with glucose.
    e. is carried on by yeast.

28. Match the terms to their definitions. Only four of these terms are needed:

    anaerobic   oxygen deficit
    citric acid cycle   pyruvate
    fermentation   preparatory reaction

    a. Occurs in mitochondria and produces CO_2, ATP, NADH, and $FADH_2$.
    b. Growing or metabolizing in the absence of oxygen.
    c. End product of glycolysis.
    d. Anaerobic breakdown of glucose that results in a gain of 2 ATP and end products such as alcohol and lactate.

Go to www.mhhe.com/maderessentials for more quiz questions.

Bioethical Issue

For millennia, humans have taken advantage of a product of fermentation in microbes—ethyl alcohol. However, alcohol is toxic to human cells, and alcohol abuse is a serious problem from many perspectives. For example, a woman who consumes large amounts of alcohol during pregnancy may cause her child to have fetal alcohol syndrome. Children with this syndrome suffer from mental retardation and physical problems. Some people believe the unborn child has a right to be protected from harm and argue that intervention is justified if a pregnant woman drinks heavily despite her doctor's orders. In extreme cases, it may be necessary to incarcerate the woman throughout the pregnancy. The alternative point of view is that every person has a right to freedom of choice, and society has no authority to intervene in the life of a pregnant woman. In addition, forcing a pregnant woman to follow medical treatment against her will, for the sake of her fetus, is imposing an obligation that we do not impose on others—that is, other members of our society are not forced to change their lifestyles purely for the sake of others.

Do you think society has an obligation to do whatever is necessary to prevent women from drinking excessively during pregnancy? Will this lead to attempts to control the lives of pregnant women in other ways as well, such as requiring them to exercise more or to abstain from caffeine?

Understanding the Terms

acetyl-CoA 102	glycolysis 99
citric acid cycle 99, 102	intermembrane space 105
coenzyme A (CoA) 99	oxygen deficit 107
electron transport chain 99	preparatory (prep) reaction 99, 102
fermentation 107	

Match the terms to these definitions:

a. _____ First step in cellular respiration.

b. _____ This metabolic pathway breaks down pyruvate in the mitochondrial matrix.

c. _____ Reduction of pyruvate when oxygen is not available.

d. _____ Hydrogen ions are pumped into this region during the electron transport chain.

e. _____ Acetyl-CoA is needed during this phase of cellular respiration.

Cellular Reproduction

CHAPTER

8

Cancer is characterized by the over-division of cells.

Heart attacks are caused by the death of cardiac muscle cells.

The brain has 35 billion nerve cells—but when these cells die (green), they are not replaced.

Many of our body tissues and organs can repair themselves if they become injured or damaged. This is because their cells are capable of undergoing mitosis, a process of nuclear division that enables cells and tissues to grow and be repaired. Exceptions are nerve cells, the cells that make up the nervous system. Injuries to the brain and spinal cord are very serious because nerve cells are designed to last a lifetime and do not ordinarily undergo ▶ ▶ ▶

mitosis. When nerve cells die, they are not likely to be replaced. The brain has approximately 35 billion nerve cells, and the loss of some of these is normal. However, in the case of brain injury or a condition such as Alzheimer disease, nerve cells die at a rapid rate, and brain function declines.

Scientists want to find a way to coerce differentiated nerve cells into undergoing mitosis. They also believe that human **stem cells**, which do divide, can be coaxed into becoming new nerve cells. Although the use of stem cells derived from human embryos is controversial and limited by legislation in the United States, adult stem cells are easily accessible in the bone marrow and already show promise of becoming nerve cells under the proper circumstances. All cells in the body of an organism have the same genetic composition, which explains why stem cells can potentially be used to form all cell types.

In this chapter, you will learn about the process of mitosis, how it is regulated, and what happens if it occurs either not at all or too frequently.

8.1 The Basics of Cellular Reproduction

We humans, like other multicellular organisms, begin life as a single cell. In nine short months, however, we become trillions of cells because cellular reproduction has occurred over and over again. Even after we are born, cellular reproduction doesn't stop—it continues as we grow (Fig. 8.1*a*), and when we are adults, it replaces worn-out or damaged tissues (Fig. 8.1*b*). Right now, your body is producing thousands of new red blood cells, skin cells, and cells that line your respiratory and digestive tracts. If you suffer a cut, cellular reproduction helps repair the injury.

Cellular reproduction is also necessary for the reproduction of certain organisms. When an amoeba splits, two new individual amoebas are produced. This process is called *asexual reproduction* because it doesn't require a sperm and an egg (Fig. 8.1*c*). The next chapter concerns the production of egg and sperm, which are needed for sexual reproduction.

One way to emphasize the importance of cellular reproduction is to say that "all cells come from cells." You can't have a new cell without a pre-existing cell, and you can't have a new organism without a pre-existing organism (Fig. 8.1*d*). Cellular reproduction is necessary for the production of both new cells and new organisms.

×50

c. Amoebas reproduce

b. Tissues repair

a. Childen grow

d. Zygotes develop

Figure 8.1 Cellular reproduction.

Cellular reproduction occurs (**a**) when small children grow to be adults and also (**b**) when mature organisms repair their tissues. Cellular reproduction also occurs when (**c**) unicellular organisms reproduce, and (**d**) when zygotes, the product of sperm and egg fusion, develop into multicellular organisms capable of independent existence.

Chromosomes

Cellular reproduction always involves two important processes: growth and cell division. During growth, a cell duplicates the contents of its cytoplasm and its DNA. Then, during division, the cytoplasm and the DNA of the *parent cell* are distributed to the so-called *daughter cells*. (These terms have nothing to do with gender; they are simply a way to designate the beginning cell and the resulting cells.)

The passage of DNA to the daughter cells is critical because cells cannot continue to live without a copy of the genetic material. Especially in eukaryotic cells, passage of DNA to the daughter cells presents a problem because of the large quantity of DNA in the nucleus. For example, a human cell contains about 2 meters of DNA and a nucleus is only 5 to 8 micrometers (μm) in diameter. During cellular reproduction, DNA is packaged into chromosomes, which allow DNA to be distributed to the daughter cells. A **chromosome** contains DNA, and it also contains proteins that help package the DNA and possibly function in utilizing the DNA as well.

Chromatin to Chromosomes

When a eukaryotic cell is not undergoing cell division, the DNA and associated proteins have the appearance of thin threads called **chromatin.** Closer examination reveals that chromatin is periodically wound around a core of eight protein molecules so that it looks like beads on a string. The protein molecules are **histones,** and each bead is called a **nucleosome** (Fig. 8.2).

Just before cell division occurs, the chromatin coils tightly into a fiber that has several nucleosomes to a turn. Then the fiber coils again before it loops back and forth and condenses to produce highly compacted chromosomes. Each species has a characteristic number of chromosomes; a human cell has 46 chromosomes. We can easily see chromosomes with a light microscope because just before division occurs a chromosome is 10,000 times more compact than is chromatin.

Another important event, that occurs in preparation for partition of chromosomes is **DNA replication,** when DNA makes a copy of itself. By the time we can clearly see the chromosomes, they are duplicated. A duplicated chromosome is composed of two identical halves called **sister chromatids** held together at a constricted region called a **centromere.** Each sister chromatid contains an identical DNA double helix.

Check Your Progress

1. When does cellular reproduction occur in an animal?
2. In what ways is cellular reproduction necessary to the continued existence of a mature organism and the production of a new organism?
3. Contrast the appearance of chromatin with a chromosome that is ready to undergo cellular reproduction.

Answers: 1. When an animal grows, and when tissues are repaired. 2. Mature organisms have to replenish worn out cells and repair injuries. Unicellular organisms divide in order to reproduce, and multicellular organisms develop from a single cell. 3. Chromatin is less condensed than a chromosome, and its DNA is not duplicated. A chromosome ready to undergo cellular reproduction is condensed and is duplicated.

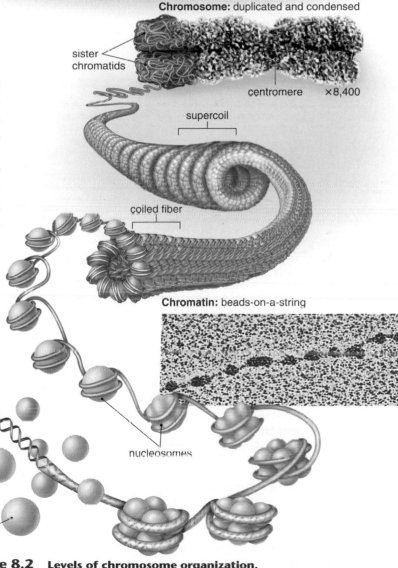

Chromosome: duplicated and condensed

sister chromatids

centromere   ×8,400

supercoil

coiled fiber

Chromatin: beads-on-a-string

nucleosomes

histones

Figure 8.2   Levels of chromosome organization.

Histones are responsible for packaging chromatin so that it fits into the nucleus and so that it becomes highly condensed when cell division occurs.

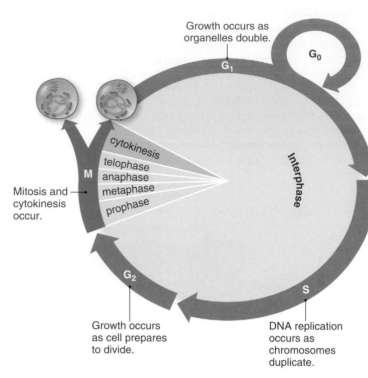

Figure 8.3 shows labels: Growth occurs as organelles double. G_1 • G_0 • cytokinesis, telophase, anaphase, metaphase, prophase • M • Mitosis and cytokinesis occur. • Interphase • G_2 • S • Growth occurs as cell prepares to divide. • DNA replication occurs as chromosomes duplicate.

Figure 8.3   The cell cycle.

Cells go through a cycle that consists of four stages: G_1, S, G_2, and M. The major activity for each stage is given. Some cells can exit G_1 and enter a G_0 stage.

Check Your Progress

1. **During what stage of the cell cycle does DNA replicate? What does a chromosome look like following DNA replication?**
2. **What happens during the M stage of the cell cycle?**

Answers: 1. DNA replication occurs during the S stage of the cell cycle. Following replication, the chromosomes are composed of two sister chromatids held together at a centromere. 2. Mitosis (nuclear division in which the chromosome number stays constant) and cytokinesis (division of cytoplasm) occur.

8.2 ▶ The Cell Cycle

We have already indicated that cellular reproduction involves duplication of cell contents followed by cell division. For cellular reproduction to be orderly, you would expect the first event to occur before the second event, and that's just what happens during the so-called cell cycle. The **cell cycle** is an orderly sequence of stages that takes place between the time a new cell has arisen from division of a parent cell to the point when it has given rise to two daughter cells. Duplication of cell contents occurs during the stage called interphase.

Interphase

As Figure 8.3 shows, most of the cell cycle is spent in **interphase.** This is the time when a cell performs its usual functions, depending on its location in the body. The amount of time the cell takes for interphase varies widely. Some cells, such as nerve and muscle cells, typically remain in interphase and cell division is permanently arrested. These cells are said to have entered a G_0 stage. Embryonic cells complete the entire cell cycle in just a few hours. In contrast, interphase alone in a rapidly dividing mammalian cell, such as an adult stem cell, may last for about 20 hours, which is 90% of the cell cycle.

DNA replication occurs in the middle of interphase and serves as a way to divide interphase into three stages: G_1, S, and G_2. G_1 is the stage before DNA replication, and G_2 is the stage following DNA synthesis. Originally, G stood for "gap," but now that we know how metabolically active the cell is, it is better to think of G as standing for "growth." Protein synthesis is very much a part of these growth stages.

During G_1, a cell doubles its organelles (such as mitochondria and ribosomes) and accumulates materials that will be used for DNA replication. Following G_1, the cell enters the S stage. The *S* stands for synthesis, and certainly DNA synthesis is required for DNA replication. At the beginning of the S stage, each chromosome has one DNA double helix. At the end of this stage, each chromosome is composed of two sister chromatids, each having one double helix. Another way of expressing these events is to say that DNA replication results in duplicated chromosomes.

Following the S stage, G_2 is the stage that extends from the completion of DNA replication to the onset of mitosis. During this stage, the cell synthesizes proteins that will be needed for cell division, such as the protein found in microtubules. The role of microtubules in cell division is described in a later section.

M (Mitotic) Stage

Cell division occurs during the M stage, which encompasses both division of the nucleus and division of the cytoplasm. The type of nuclear division associated with the cell cycle is called **mitosis,** which accounts for why this stage is called the M stage. As a result of mitosis, the daughter nuclei are identical to the parent cell and to each other—they all have the same number and kinds of chromosomes. Division of the cytoplasm, which starts even before mitosis is finished, is called **cytokinesis.**

8.3 Mitosis and Cytokinesis

Each sister chromatid of a duplicated chromosome carries the same genetic information because its DNA double helix has the same sequence of base pairs as did the original chromosome. Thus, it is proper, once the chromatids have separated, to call them *daughter chromosomes* (Fig. 8.4). Because each original chromosome goes through the same process of DNA replication followed by separation of the chromatids to form daughter chromosomes, the daughter nuclei produced by mitosis are genetically identical to each other and to the parent nucleus. In the simplest of terms, if the parent nucleus has 4 chromosomes, each daughter nucleus also has 4 chromosomes of exactly the same type. One way to keep track of the number of chromosomes in drawings is to count the number of centromeres, because every chromosome has a centromere.

Every animal has an even number of chromosomes because each parent contributed half of the chromosomes to the new individual. In drawings of mitosis, some chromosomes are colored red and some are colored blue to represent that half of the chromosomes are derived from those contributed by one parent and the other half are derived from chromosomes from the other parent.

The Spindle

While it may seem easy to separate the chromatids of only 4 duplicated chromosomes, imagine the task when there are 46 chromosomes, as in humans, or 78, as in dogs. Certainly it is helpful that chromosomes be highly condensed before the task begins, but clearly some mechanism is needed to complete separation in an organized manner. Most eukaryotic cells rely on a **spindle,** a cytoskeletal structure, to pull the chromatids apart. A spindle has spindle fibers made of microtubules that are able to assemble and disassemble. First, the microtubules assemble to form the spindle that takes over the center of the cell and separates the chromatids. Later, they disassemble.

A **centrosome** is the primary microtubule organizing center of a cell. Centrosome duplication occurs at the start of the S phase of the cell cycle and is completed by G_2. During the first part of the M stage, the centrosomes separate and move to opposite sides of the nucleus, where they form the poles of the spindle. As the nuclear envelope breaks down, spindle fibers take over the center of the cell. Certain ones overlap at the **spindle equator,** which is midway between the poles. Others attach to duplicated chromosomes in a way that ensures the separation of the sister chromatids and their proper distribution to the daughter cells. Whereas the chromosomes will be inside the newly formed daughter nuclei, a centrosome will be just outside.

Traditionally, mitosis is divided into a sequence of events, even though it is a continuous process. We will describe mitosis as having four phases: **prophase, metaphase, anaphase,** and **telophase.** These phases are labeled in Figure 8.5 for a dividing plant nucleus. Plant cells have centrosomes but they are not clearly visible especially because they lack centrioles. In an animal cell, each centrosome has two centrioles and an array of microtubules called an **aster.** Otherwise the descriptions on the next two pages apply to both animal and plant cells.

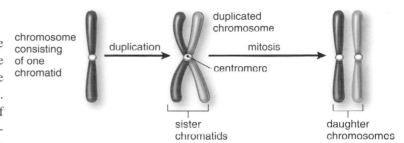

Figure 8.4 Overview of mitosis.

Counting the number of centromeres tells you the number of chromosomes.

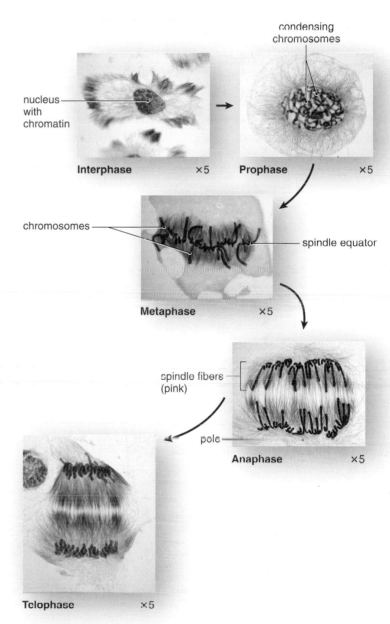

Figure 8.5 Mitosis in a plant cell.

The chromosomes are stained blue, and microtubules of the spindle fibers are stained pink in these photos of a dividing African blood lily. For a description of these phases see Figure 8.6.

Phases of Mitosis in Animal Cells

Separation of chromatids during mitosis requires the phases listed on the previous page and described in Figure 8.6. Although mitosis is divided into these phases, it is a continuous process. Before studying the descriptions of these phases in Figure 8.6, recall that before mitosis begins, DNA has replicated. Each double helix is located in a chromatid; therefore, the chromosomes consist of sister chromatids attached at a centromere. Notice that some chromosomes are colored red and some are colored blue. The red chromosomes were inherited from one parent and the blue chromosomes from the other parent.

During mitosis, a spindle arises that will separate the sister chromatids of each duplicated chromosome. Once separated, the sister chromatids become daughter chromosomes. In this way, the resulting daughter nuclei are identical to

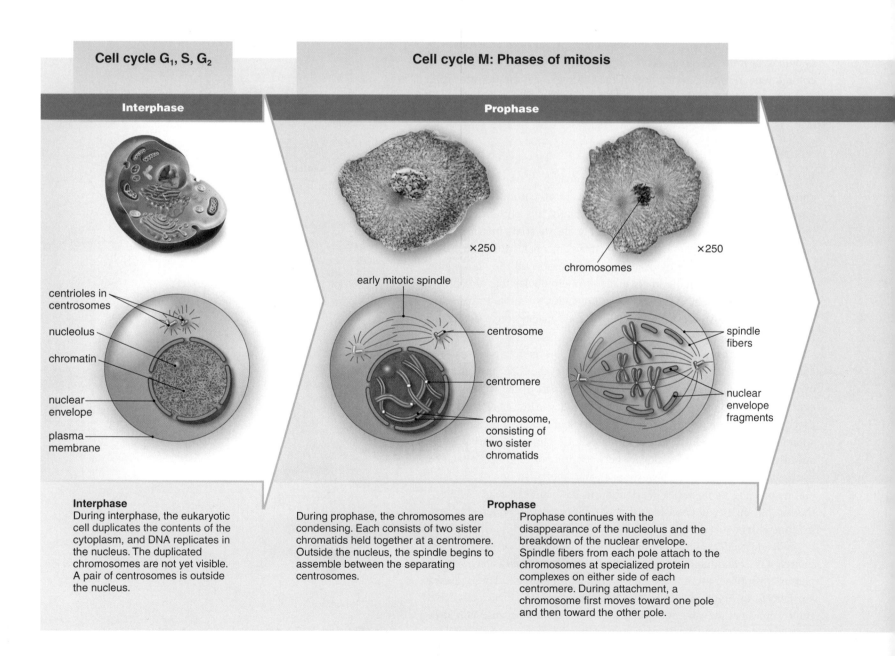

Cell cycle G₁, S, G₂

Cell cycle M: Phases of mitosis

Interphase

Prophase

×250

×250

chromosomes

early mitotic spindle

centrioles in centrosomes

nucleolus

chromatin

nuclear envelope

plasma membrane

centrosome

centromere

chromosome, consisting of two sister chromatids

spindle fibers

nuclear envelope fragments

Interphase
During interphase, the eukaryotic cell duplicates the contents of the cytoplasm, and DNA replicates in the nucleus. The duplicated chromosomes are not yet visible. A pair of centrosomes is outside the nucleus.

Prophase
During prophase, the chromosomes are condensing. Each consists of two sister chromatids held together at a centromere. Outside the nucleus, the spindle begins to assemble between the separating centrosomes.

Prophase continues with the disappearance of the nucleolus and the breakdown of the nuclear envelope. Spindle fibers from each pole attach to the chromosomes at specialized protein complexes on either side of each centromere. During attachment, a chromosome first moves toward one pole and then toward the other pole.

Figure 8.6   Phases of mitosis in animal cells.

The red chromosomes were inherited from one parent and the blue chromosomes from the other parent.

each other and to the parental nucleus. In Figure 8.6, the daughter nuclei not only have four chromosomes; they each have one red short, one blue short, one red long, and one blue long, the same as the mother cell had.

Mitosis is usually followed by division of the cytoplasm, or cytokinesis. Cytokinesis begins during telophase and continues after the nuclei have formed in the daughter cells. The cell cycle is now complete, and the daughter cells enter interphase, during which the cell will grow and DNA will replicate once again. In rapidly dividing mammalian stem cells, the cell cycle lasts for 24 hours. Mitosis and cytokinesis require only one hour of this time period.

Check Your Progress

1. During what phase of mitosis are duplicated chromosomes aligned at the spindle equator?
2. During what phase of mitosis are daughter chromosomes moving toward the spindle poles?

Answers: 1. Metaphase. 2. Anaphase.

Phases of mitosis

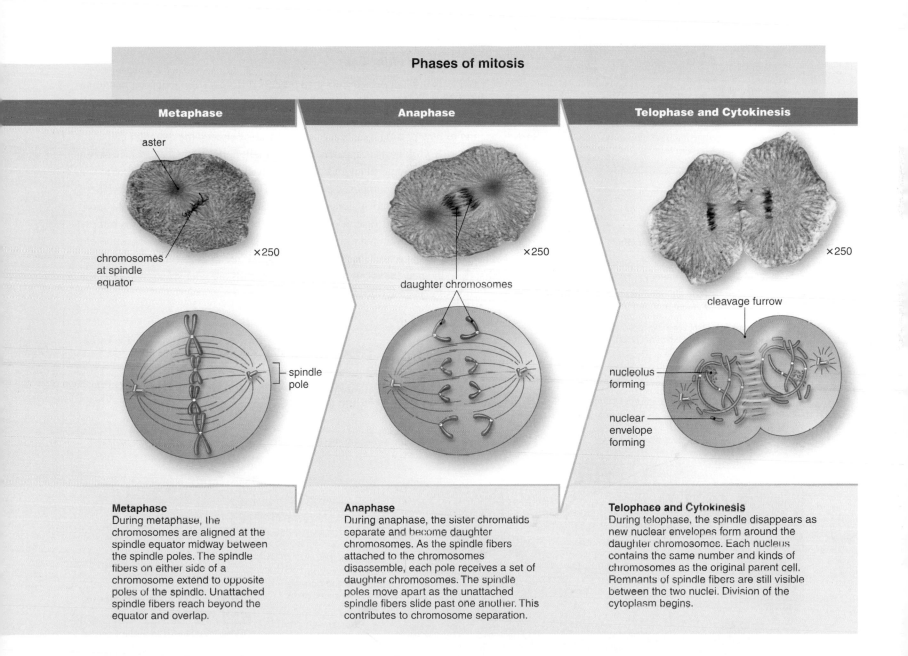

| Metaphase | Anaphase | Telophase and Cytokinesis |

aster

chromosomes at spindle equator

×250

daughter chromosomes

×250

cleavage furrow

×250

spindle pole

nucleolus forming

nuclear envelope forming

Metaphase
During metaphase, the chromosomes are aligned at the spindle equator midway between the spindle poles. The spindle fibers on either side of a chromosome extend to opposite poles of the spindle. Unattached spindle fibers reach beyond the equator and overlap.

Anaphase
During anaphase, the sister chromatids separate and become daughter chromosomes. As the spindle fibers attached to the chromosomes disassemble, each pole receives a set of daughter chromosomes. The spindle poles move apart as the unattached spindle fibers slide past one another. This contributes to chromosome separation.

Telophase and Cytokinesis
During telophase, the spindle disappears as new nuclear envelopes form around the daughter chromosomes. Each nucleus contains the same number and kinds of chromosomes as the original parent cell. Remnants of spindle fibers are still visible between the two nuclei. Division of the cytoplasm begins.

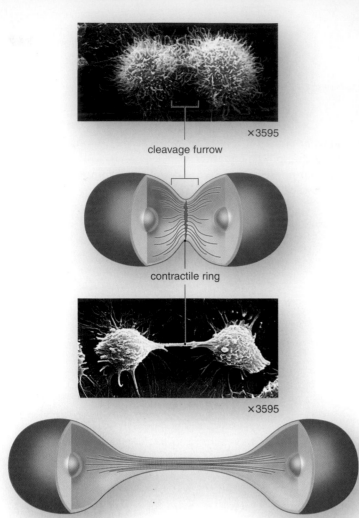

×3595

cleavage furrow

contractile ring

×3595

Figure 8.7 Cytokinesis in animal cells.

A single cell becomes two cells by a furrowing process. A cleavage furrow appears as early as anaphase, and a contractile ring, composed of actin filaments, gradually gets smaller until there are two cells.

Copyright by R. G. Kessel and C. Y. Shih, *Scanning Electron Microscopy in Biology: A Students' Atlas on Biological Organization,* Springer-Verlag, 1974.

Cytokinesis in Animal and Plant Cells

Cytokinesis accompanies mitosis in most cells but not all. When mitosis occurs but cytokinesis doesn't occur, the result is a multinucleated cell. For example, skeletal muscle cells in vertebrate animals and the embryo sac in flowering plants are multinucleated.

Cytokinesis in Animal Cells

In animal cells, a cleavage furrow, which is an indentation of the membrane between the two daughter nuclei, begins as anaphase draws to a close. The cleavage furrow deepens when a band of actin filaments, called the contractile ring, slowly forms a circular constriction between the two daughter cells. The action of the contractile ring can be likened to pulling a drawstring ever tighter about the middle of a balloon. A narrow bridge between the two cells is visible during telophase, and then the contractile ring continues to separate the cytoplasm until there are two independent daughter cells (Fig. 8.7).

Cytokinesis in Plant Cells

Cytokinesis in plant cells occurs by a process different from that seen in animal cells (Fig. 8.8). The rigid cell wall that surrounds plant cells does not permit cytokinesis by furrowing. Instead, cytokinesis in plant cells involves the building of new plasma membrane and cell walls between the daughter cells.

Cytokinesis is apparent when a small, flattened disk appears between the two daughter plant cells. In electron micrographs, it is possible to see that the disk is composed of vesicles. The Golgi apparatus produces these vesicles, which move along microtubules to the region of the disk. As more vesicles arrive and fuse, a cell plate can be seen. The **cell plate** is simply newly formed plasma membrane that expands outward until it reaches the old plasma membrane and fuses with it. The new membrane releases molecules that form the new plant cell walls. These cell walls are later strengthened by the addition of cellulose fibrils.

Check Your Progress

What explanation can you give for the differences in cytokinesis between plant cells and animal cells?

Answer: The rigid cell wall of plant cells does not permit cytokinesis by furrowing as in animal cells.

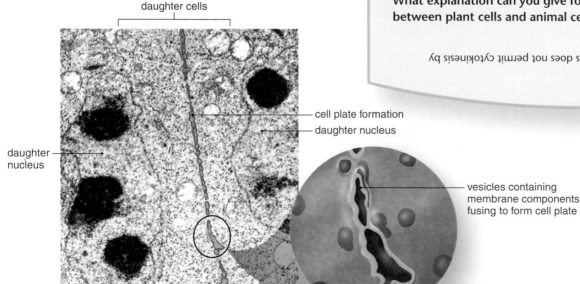

daughter cells

daughter nucleus

cell plate formation

daughter nucleus

vesicles containing membrane components fusing to form cell plate

Figure 8.8 Cytokinesis in plant cells.

During cytokinesis in a plant cell, a cell plate forms midway between two daughter nuclei and extends to the original plasma membrane.

8.4 The Cell Cycle Control System

In order for a cell to reproduce successfully, the cell cycle must be controlled. The importance of cell cycle control can be appreciated by comparing the cell cycle to the events that occur in an automatic washing machine. The washer's control system starts to wash only when the tub is full of water, doesn't spin until the water has been emptied, delays the most vigorous spin until rinsing has occurred, and so forth. Similarly, the cell cycle's control system ensures that the G_1, S, G_2, and M stages occur in order and only when the previous stage has been successfully completed.

Cell Cycle Checkpoints

Just as a washing machine will not begin to agitate the load until the tub is full, the cell cycle has **checkpoints** that can delay the cell cycle until all is well. The cell cycle has many checkpoints, but we will consider only three: G_1, G_2 and the mitotic checkpoint (Fig. 8.9).

The *G_1 checkpoint* is especially significant because if the cell cycle passes this checkpoint, the cell is committed to divide. If the cell does not pass this checkpoint, it can enter G_0, during which it performs specialized functions but does not divide. If the DNA is damaged beyond repair, the internal signaling protein p53 can stop the cycle at this checkpoint. First, p53 attempts to initiate DNA repair, but if that is not possible, it brings about the death of the cell by apoptosis (see Fig. 8.11).

The cell cycle hesitates at the *G_2 checkpoint* ensuring that DNA has replicated. This prevents the initiation of the M stage unless the chromosomes are duplicated. Also, if DNA is damaged, as from exposure to solar radiation or X rays, arresting the cell cycle at this checkpoint allows time for the damage to be repaired so that it is not passed on to daughter cells.

Another cell cycle checkpoint occurs during the mitotic stage. The cycle hesitates to make sure the chromosomes are going to be distributed accurately to the daughter cells. The cell cycle does not continue until every chromosome is ready for the nuclear division process.

Internal and External Signals

The checkpoints of the cell cycle are controlled by internal and external signals. A **signal** is a molecule that stimulates or inhibits an event. Signals outside the cell are called external signals and those inside the cell are called internal signals. Inside the cell, enzymes called **kinases** remove phosphate from ATP and add it to another molecule. Just before the S stage, a protein called S-cyclin combines with a kinase called S-kinase, and synthesis of DNA takes place. Just before the M stage, M-cyclin combines with a kinase called M-kinase, and mitosis occurs. **Cyclins** are so named because their quantity is not constant. They increase in amount until they combine with a kinase, but this is a suicidal act. The kinase not only activates a protein that drives the cell cycle, but it also activates various enzymes, one of which destroys the cyclin (Fig. 8.10).

Some external signals, such as growth factors and hormones, stimulate cells to go through the cell cycle. An animal and its organs grow larger if the cell cycle occurs. Growth factors also stimulate repair of tissues. Even cells that are arrested in G_0 will finish the cell cycle if stimulated to do so by growth factors.

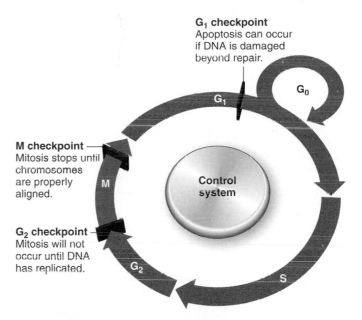

Figure 8.9 **Cell cycle checkpoints.**

Internal and external signals determine whether the cell is ready to proceed past cell cycle checkpoints. Three important checkpoints are designated.

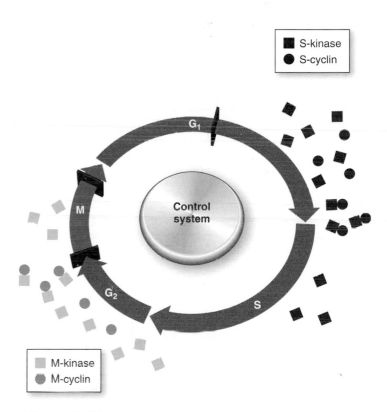

Figure 8.10 **Internal signals of the cell cycle.**

S-cyclin must combine with S-kinase for the cell cycle to begin DNA replication. M-cyclin must combine with M-kinase for the cell cycle to start mitosis.

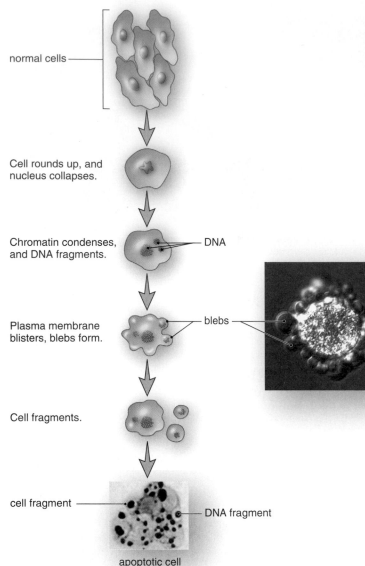

normal cells

Cell rounds up, and nucleus collapses.

Chromatin condenses, and DNA fragments. — DNA

Plasma membrane blisters, blebs form. — blebs

Cell fragments.

cell fragment — — DNA fragment

apoptotic cell

Figure 8.11 **Apoptosis.**

Apoptosis is a sequence of events that results in a fragmented cell. The fragments are phagocytized by white blood cells and neighboring tissue cells.

For example, epidermal growth factor (EGF) stimulates skin in the vicinity of an injury to finish the cell cycle, thereby repairing the damage.

Hormones act on tissues at a distance, and some signal cells to divide. For example, at a certain time in the menstrual cycle of females, the hormone estrogen stimulates cells lining the uterus to divide and prepare the lining for implantation of a fertilized egg.

The cell cycle can be inhibited by cells coming into contact with other cells, and by the shortening of chromosomes. In cell culture, cells will divide until they line a container in a one-cell-thick sheet. Then they stop dividing, a phenomenon termed **contact inhibition.** Researchers are beginning to discover the external signals that result in this inhibition. Some years ago, it was noted that mammalian cells in cell culture divide about 70 times, and then they die. Cells seem to "remember" the number of times they divided before, and stop dividing when the usual number of cell divisions is reached. It's as if *senescence,* the aging of cells, is dependent on an internal battery-operated clock that winds down and then stops. We now know that senescence is due to the shortening of telomeres. A **telomere** is a repeating DNA base sequence (TTAGGG) at the end of the chromosomes that can be as long as 15,000 base pairs. Telomeres have been likened to the protective caps on the ends of shoelaces. Rather than keeping chromosomes from unraveling, however, telomeres stop chromosomes from fusing to each other. Each time a cell divides, some portion of a telomere is lost; when telomeres become too short, the chromosomes fuse and can no longer duplicate. Then the cell is "old" and dies by a process called apoptosis.

Apoptosis

Apoptosis is often defined as programmed cell death because the cell progresses through a typical series of events that bring about its destruction (Fig. 8.11). The cell rounds up and loses contact with its neighbors. The nucleus fragments, and the plasma membrane develops blisters. Finally, the cell breaks into fragments, and its bits and pieces are engulfed by white blood cells and/or neighboring cells. A remarkable finding of the past few years is that cells routinely harbor the enzymes, now called caspases, that bring about apoptosis. These enzymes are ordinarily held in check by inhibitors, but are unleashed by either internal or external signals.

Cell division and apoptosis are two opposing processes that keep the number of cells in the body at an appropriate level. In other words, cell division increases, and apoptosis decreases, the number of **somatic cells** (body cells).

Both the cell cycle and apoptosis are normal parts of growth and development. An organism begins as a single cell that repeatedly undergoes the cell cycle to produce many cells, but eventually some cells must die in order for the organism to take shape. For example, when a tadpole becomes a frog, the tail disappears as apoptosis occurs. In humans, the fingers and toes of an embryo are at first webbed, but later they are freed from one another as a result of apoptosis.

Apoptosis occurs all the time, particularly if an abnormal cell that could become cancerous appears. Death through apoptosis prevents a tumor from developing.

Check Your Progress

1. Explain the significance of checkpoints in the cell cycle.
2. What is a signal, and in general what two kinds of signals control the cell cycle?

Answers: 1. It is important for each step of the cycle to be completed correctly before the next one begins, and checkpoints allow the cell to make sure that happens. 2. A signal is a molecule that promotes or inhibits an event. Internal and external signals control the cell cycle.

8.5   The Cell Cycle and Cancer

Cancer is a disease of the cell cycle, in that the cell cycle is out of control, and cellular reproduction occurs repeatedly without end. Cancers are classified according to their location. *Carcinomas* are cancers of the epithelial tissue that lines organs; *sarcomas* are cancers arising in muscle or connective tissue (especially bone or cartilage); and *leukemias* are cancers of the blood. A high rate of cell division makes a tissue susceptible to cancer because it increases the chances of a mutation (change in DNA base sequence) that causes a cell to divide uncontrollably or to ignore apoptosis signals. In this chapter, we consider some general characteristics of cancer cells. Chapter 12 explores specific changes that lead to cancerous growth.

Characteristics of Cancer Cells

Carcinogenesis, the development of cancer, is gradual; it may be decades before a tumor is visible. Cancer cells then have these characteristics:

Cancer cells lack differentiation. Cancer cells are nonspecialized and do not contribute to the functioning of a body part. A cancer cell does not look like a differentiated epithelial, muscle, nervous, or connective tissue cell; instead, it looks distinctly abnormal. As mentioned, normal cells can enter the cell cycle about 70 times, and then they die. Cancer cells can enter the cell cycle repeatedly, and in this way they are immortal.

Cancer cells have abnormal nuclei. The nuclei of cancer cells are enlarged and may contain an abnormal number of chromosomes. The chromosomes are also abnormal; some parts may be duplicated, or some may be deleted. In addition, gene amplification (extra copies of specific genes) is seen much more frequently than in normal cells. Ordinarily, cells with damaged DNA undergo apoptosis, but cancer cells fail to undergo apoptosis even though they are abnormal.

Cancer cells form tumors. Normal cells anchor themselves to a substratum and/or adhere to their neighbors. Then, they exhibit contact inhibition and stop dividing. Cancer cells, on the other hand, have lost all restraint; they pile on top of one another and grow in multiple layers, forming a **tumor.** They have a reduced need for growth factors, and they no longer respond to inhibitory signals. As cancer develops, the most aggressive cell becomes the dominant cell of the tumor.

Cancer cells promote angiogenesis and undergo metastasis. A *benign tumor* is usually encapsulated and, therefore, will never invade adjacent tissue. *Cancer in situ* is a tumor in its place of origin, but it is not encapsulated and will eventually invade surrounding tissues. Cancer cells produce enzymes that allow tumors to invade underlying tissues. *Invasive tumors* produce cancer cells that travel through the blood and lymph to start tumors elsewhere in the body. Malignancy is present once **metastasis** has established new *metastatic tumors* distant from the primary tumor (Fig. 8.12).

Angiogenesis, the formation of new blood vessels, is required to bring nutrients and oxygen to a cancerous tumor. Some modes of cancer treatment are aimed at preventing angiogenesis from occurring.

The patient's prognosis (probable outcome) is dependent on (1) whether the tumor has invaded surrounding tissues, and (2) whether there are metastatic tumors in distant parts of the body.

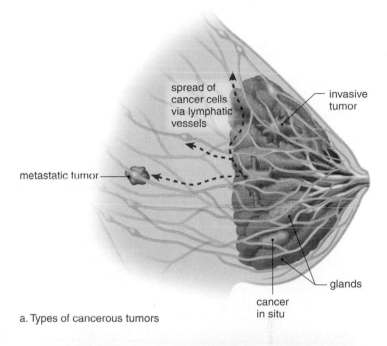

spread of cancer cells via lymphatic vessels

invasive tumor

metastatic tumor

cancer in situ

glands

a. Types of cancerous tumors

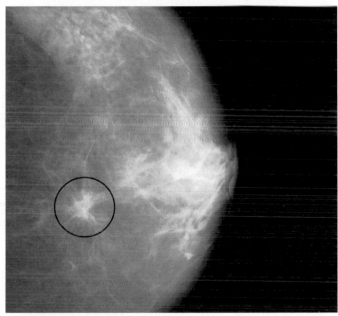

b. Mammogram showing tumor.

Figure 8.12   Development of breast cancer.

a. Breast cancer begins as a cancer in situ. The tumor may become invasive and invade lymphatic or blood vessels. Metastasis occurs when new metastatic tumors occur some distance from the original tumor. **b.** A mammogram (X-ray image of the breast) can find a tumor (circled) too small to be felt.

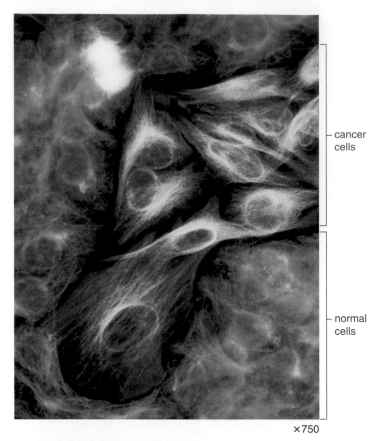

×750

Figure 8.13 **Cancer cells.**

This micrograph contrasts cancer cells with normal cells.

cancer cells

normal cells

Cancer Treatment

Cancer treatments either remove the tumor or interfere with the ability of cancer cells to reproduce. For many solid tumors, removal by surgery is often the first line of treatment. When the cancer is detected at an early stage, surgery may be sufficient to cure the patient by removing all cancerous cells.

Because cancer cells are rapidly dividing, they become susceptible to radiation therapy and chemotherapy (Fig. 8.13). The goal of radiation is to kill cancer cells within a specific tumor by directing high-energy beams at the tumor. DNA is damaged to the point that replication can no longer occur, and the cell undergoes apoptosis. Chemotherapy is a way to kill cancer cells that have spread throughout the body. Like radiation, most chemotherapeutic drugs lead to the death of cells by damaging their DNA or interfering with DNA synthesis. Others interfere with the functioning of the mitotic spindle. The drug vinblastine, first obtained from a flowering plant called the periwinkle, prevents the spindle from forming. Taxol, extracted from the bark of the Pacific yew tree, prevents the spindle from functioning as it should. Unfortunately, radiation and chemotherapy often damage cells other than cancer cells, leading to side effects such as nausea and hair loss.

Hormonal therapy also prevents cell division but in a different way. These drugs are designed to prevent cancer cell growth by preventing the cells from receiving signals necessary for their continued growth and division. For example, the drug tamoxifen modifies the receptor for estrogen so that this hormone cannot bind to it and promote the cell cycle. Other drug therapies are also being investigated. One proposed therapy that is now well under way uses antiangiogenic drugs designed to inhibit angiogenesis by breaking up the network of capillaries in the vicinity of the tumor. Having its blood supply reduced confines and reduces a tumor.

Prevention of Cancer

Evidence is clear that the risk of certain types of cancer can be reduced by adopting protective behaviors and following recommended dietary guidelines.

Protective Behaviors

To prevent the development of cancer, people are advised to avoid smoking, sunbathing, and excessive alcohol consumption.

Cigarette smoking accounts for about 30% of all cancer deaths. Smoking is responsible for 90% of lung cancer cases among men and 79% among women—about 87% altogether. People who smoke two or more packs of cigarettes a day have lung cancer mortality rates 15 to 25 times greater than those of nonsmokers. Smokeless tobacco (chewing tobacco or snuff) increases the risk of cancers of the mouth, larynx, throat, and esophagus.

Skin cancers are considered sun-related. Sun exposure is a major factor in the development of the most dangerous type of skin cancer, melanoma, and the incidence of this cancer increases in people living near the equator. Similarly, excessive radon gas[1] exposure in homes increases the risk of lung cancer, especially in cigarette smokers. It is best to test your home and, if necessary, take the proper remedial actions.

[1]Radon gas results from natural radioactive breakdown of uranium in soil, rocks, and water.

Cancers of the mouth, throat, esophagus, larynx, and liver occur more frequently among heavy drinkers, especially when accompanied by tobacco use (cigarettes, cigars, or chewing tobacco).

Protective Diet

The risk of cancer (especially colon, breast, and uterine cancers) is 55% greater among obese women and 33% greater among obese men, compared to people of normal weight. Weight loss in these groups can therefore reduce cancer risk. In addition, the following dietary guidelines are recommended (Fig. 8.14):

- Increase consumption of foods that are rich in vitamins A and C. Beta-carotene, a precursor of vitamin A, is found in dark green leafy vegetables, carrots, and various fruits. Vitamin C is present in citrus fruits. These vitamins are called antioxidants because in cells they prevent the formation of free radicals (organic ions having an unpaired electron) that can possibly damage DNA. Vitamin C also prevents the conversion of nitrates and nitrites into carcinogenic nitrosamines in the digestive tract.
- Avoid salt-cured or pickled foods because they may increase the risk of stomach and esophageal cancers. Smoked foods, such as ham and sausage, contain chemical carcinogens similar to those in tobacco smoke. Nitrites are sometimes added to processed meats (e.g., hot dogs and cold cuts) and other foods to protect them from spoilage; as mentioned, nitrites are sometimes converted to cancer-causing nitrosamines in the digestive tract.
- Include in the diet vegetables from the cabbage family, which includes cabbage, broccoli, brussels sprouts, kohlrabi, and cauliflower. Consuming these vegetables may reduce the risk of gastrointestinal and respiratory tract cancers.

Check Your Progress

List the major protective strategies you can employ to reduce your risk of cancer.

Answer: Avoid tobacco use, reduce sun exposure, reduce radon exposure, avoid heavy drinking, avoid obesity, consume foods rich in vitamins A and C, avoid chemically preserved foods, and include vegetables from the cabbage family in your diet.

Figure 8.14   The right diet helps prevent cancer.

Foods that help protect against cancer include fruits and dark green leafy vegetables, including swiss chard (*left*), which are rich in vitamins A and C, and vegetables from the cabbage family, including broccoli.

Summary

8.1   The Basics of Cellular Reproduction

Cellular reproduction occurs when growth and repair of tissues take place. Cellular reproduction is also necessary to both asexual and sexual reproduction. Following fertilization, a single cell becomes a multicellular organism by cellular reproduction.

When a cell is not dividing, chromatin looks like beads on a string because DNA (the string) is wound around histones. The bead is called a nucleosome. Just before cell division, duplicated chromatin condenses to form duplicated chromosomes consisting of two chromatids held together at a centromere. During cell division, one-half of the cytoplasm and a copy of the cell's DNA are passed on to each of two daughter cells.

8.2   The Cell Cycle

The cell cycle consists of interphase and the M (mitotic) stage:

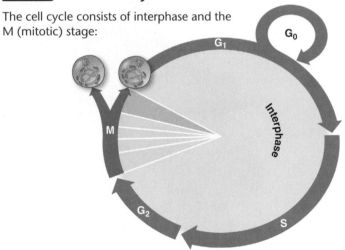

Interphase has the following stages:

- **G_1**   The cell doubles its organelles and accumulates materials that will be used for DNA replication.
- **S**   DNA replicates.
- **G_2**   The cell synthesizes proteins that will be needed for cell division.

The M (mitotic) stage consists of mitosis and cytokinesis. As a result of mitosis, the daughter nuclei are genetically identical to the parent nucleus and to each other. If the parent nucleus has 4 chromosomes, the daughter nuclei each have 4 chromosomes.

8.3   Mitosis and Cytokinesis

Cell division consists of mitosis and cytokinesis.

Mitosis
Mitosis has four phases:

- During **prophase,** the chromosomes are condensing. Outside the nucleus, the spindle begins to assemble between the separating centrosomes. Prophase continues with the disappearance of the nucleolus and the breakdown of the nuclear envelope. Spindle microtubules from each pole attach to the chromosomes in the region of a centromere.

- During **metaphase,** the chromosomes are aligned at the spindle equator midway between the spindle poles.
- During **anaphase,** the sister chromatids separate and become daughter chromosomes. As the microtubules attached to the chromosomes disassemble, each pole receives a set of daughter chromosomes.
- During **telophase,** the spindle disappears as new nuclear envelopes form around the daughter chromosomes. Each nucleus contains the same number and kinds of chromosomes as the original parent nucleus. Division of the cytoplasm begins.

Cytokinesis
Cytokinesis differs for animal and plant cells.

- In animal cells, a furrowing process involving actin filaments divides the cytoplasm.
- In plant cells, a cell plate forms from which the plasma membrane and cell wall develop.

8.4   The Cell Cycle Control System

Checkpoints are interacting signals that promote or inhibit the continuance of the cell cycle. Important checkpoints include these three:

- At G_1, the cell can enter G_0 or undergo apoptosis if DNA is damaged beyond repair. If the cell cycle passes this checkpoint, the cell is committed to complete the cell cycle.
- At G_2, the cell checks to make sure DNA has replicated properly.
- At the mitotic checkpoint, the cell makes sure the chromosomes are ready to be partitioned to the daughter cells.

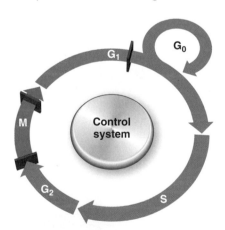

Both internal and external signals converge on checkpoints to control the cell cycle. Cyclin-kinase complexes are internal signals that promote either DNA replication or mitosis. Growth factors and hormones are external signals that promote the cell cycle in connection with growth. Inhibitory external signals are responsible for contact inhibition. Internally, when telomeres are too short, the cell cycle stops because the chromosomes fuse together.

When DNA cannot be repaired, apoptosis occurs as enzymes called caspases bring about destruction of the nucleus and the rest of the cell.

8.5   The Cell Cycle and Cancer

Cancer cells are undifferentiated, divide repeatedly, have abnormal nuclei, do not require growth factors, and are not constrained by their neighbors. After forming a tumor, cancer cells metastasize and start new tumors elsewhere in the body. Certain behaviors, such as avoiding smoking and sunbathing and adopting a diet rich in fruits and vegetables, are protective against cancer.

Thinking Scientifically

1. The survivors of the atomic bombs that were dropped on Hiroshima and Nagasaki have been the subjects of long-term studies of the effects of ionizing radiation on cancer incidence. The frequencies of different types of cancer in these individuals varied across the decades. In the 1950s, high levels of leukemia and cancers of the lung and thyroid gland were observed. The 1960s and 1970s brought high levels of breast and salivary gland cancers. In the 1980s, rates of colon cancer were especially high. Why do you suppose the rates of different types of cancer varied across time?

2. During prophase of mitosis, the nuclear membrane breaks down. This is undoubtedly a complex, energy-consuming process that has to be carried out again, in reverse, at the end of mitosis. Why do you suppose the nuclear membrane must break down at the beginning of mitosis? What mechanism can you envision for the dismantling of the nuclear membrane?

Testing Yourself

Choose the best answer for each question.

1. A chromosome contains
   a. DNA and RNA.
   b. DNA and protein.
   c. DNA only.
   d. DNA, RNA, and protein.

2. The two identical halves of a duplicated chromosome are called
   a. chromosome arms.
   b. nucleosomes.
   c. chromatids.
   d. homologues.

For questions 3–6, match the items to those in the key. Answers can be used more than once.

Key:
   a. G_1 stage of interphase
   b. G_2 stage of interphase
   c. S stage of interphase

3. This stage follows mitosis.
4. DNA is replicated during this stage.
5. Organelles are doubled in number during this stage.

6. During this stage, the cell produces proteins that will be needed for cell division.

7. Label the stages and phases of the cell cycle on the following diagram. Include anaphase, cytokinesis, G_1, G_2, metaphase, prophase, S, and telophase.

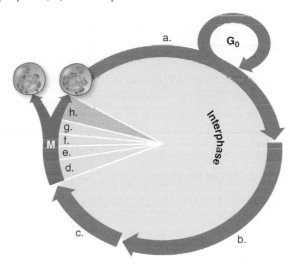

For questions 8–13, match the items to those in the key. Answers can be used more than once, and each question can have more than one answer.

Key:
   a. prophase
   b. metaphase
   c. anaphase
   d. telophase

8. The nucleolus disappears, and the nuclear membrane breaks down.
9. The spindle disappears, and the nuclear envelope forms.
10. Sister chromatids separate.
11. Spindle poles move apart as spindle fibers slide past each other.
12. Chromosomes are aligned on the spindle equator.
13. Each chromosome is composed of two sister chromatids.
14. Plants cannot use a cleavage furrow to undergo cytokinesis because they
    a. lack a plasma membrane.
    b. have a cell wall.
    c. have too many chromosomes.
    d. are too small.
15. Which DNA checkpoint allows damaged DNA to be repaired before it is passed on to daughter cells?
    a. G_1
    b. G_2
    c. S
    d. mitotic

16. Which of the following is not an event of apoptosis?

    a. loss of contact with neighboring cells
    b. blistering of the plasma membrane
    c. increase in the number of mitochondria
    d. fragmentation of the nucleus

17. Which of the following is not a feature of cancer cells?

    a. exhibit contact inhibition
    b. have enlarged nuclei
    c. stimulate the formation of new blood vessels
    d. are capable of traveling through blood and lymph

18. The spindle begins to assemble during

    a. prophase.
    b. metaphase.
    c. anaphase.
    d. interphase.

19. Which is not one of the mitotic stages?

    a. anaphase
    b. telophase
    c. interphase
    d. metaphase

20. What is the checkpoint for the completion of DNA synthesis?

    a. G_1
    b. S
    c. G_2
    d. M

21. Programmed cell death is called

    a. mitosis.
    b. meiosis.
    c. cytokinesis.
    d. apoptosis.

22. In human beings, mitosis is necessary for

    a. growth and repair of tissues.
    b. formation of the gametes.
    c. maintaining the chromosome number in all body cells.
    d. the death of unnecessary cells.
    e. Both a and c are correct.

23. In which phase of mitosis is evidence of cytokinesis present?

    a. prophase
    b. metaphase
    c. anaphase
    d. telophase
    e. Both c and d are correct.

24. Which of these is not a behavior that could help prevent cancer?

    a. maintaining a healthy weight
    b. eating more dark green leafy vegetables, carrots, and various fruits
    c. not smoking
    d. maintaining estrogen levels through hormone replacement therapy
    e. consuming alcohol only in moderation

Go to www.mhhe.com/maderessentials for more quiz questions.

Bioethical Issue

Embryonic stem cells are cells from young embryos which divide indefinitely. They are of interest to researchers because they are undifferentiated and have the ability to develop into a variety of cell types, including brain, heart, bone, muscle, and skin cells. They can help scientists to understand how cells change as they mature. In addition, they offer the hope of curing cell-based diseases such as diabetes, Parkinson's disease, and heart disease. Embryonic stem cell research involves the destruction of human embryos. Opponents of this research argue that embryos should not be destroyed because they have the same rights as all human beings. Proponents counter that the embryos used for research are excess embryos from in vitro fertilization procedures, and they would be destroyed anyway. Therefore, they should be used to alleviate human suffering. Do you think that human embryos should be used for stem cell research?

Understanding the Terms

anaphase 115	histone 113
angiogenesis 121	interphase 114
apoptosis 120	kinase 119
aster 115	metaphase 115
cancer 121	metastasis 121
carcinogenesis 121	mitosis 114
cell cycle 114	nucleosome 113
cell plate 118	prophase 115
centromere 113	signal 119
centrosome 115	sister chromatid 113
checkpoint 119	somatic cell 120
chromatin 113	spindle 115
chromosome 113	spindle equator 115
contact inhibition 120	telomere 120
cyclin 119	telophase 115
cytokinesis 114	tumor 121
DNA replication 113	

Match the terms to these definitions:

a. _____ The protein molecules in nucleosomes.

b. _____ Chromatids are held together at this constricted region of a chromosome.

c. _____ Component of the cytoskeleton that pulls chromatids apart.

d. _____ Microtubules are organized in this region of a cell.

e. _____ A molecule that stimulates or inhibits an event.

Sexual Reproduction

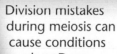
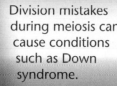

CHAPTER 9

Division mistakes during meiosis can cause conditions such as Down syndrome.

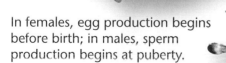

A couple can potentially create 70,368,744,000,000 genetically different zygotes.

In females, egg production begins before birth; in males, sperm production begins at puberty.

Because of a process called meiosis, two individuals can create offspring that are genetically different from themselves and from each other. In humans, more than 70 trillion different genetic combinations are possible from the mating of two individuals.

Meiosis is the type of cell division that occurs during sexual reproduction. The process plus maturation result in four cells, called gametes. In humans, the male gametes are sperm, and the female gametes are eggs. Meiosis that helps produce sperm is called spermatogenesis, and meiosis that helps produce eggs is called oogenesis.

Although spermatogenesis and oogenesis follow the same steps overall, they differ in some ways. One major difference pertains to the age at which the process begins and ends. In males, sperm production does not begin until puberty, but then continues throughout a male's lifetime. In females, the process of producing eggs has started before the female is even born and ends around the age of 50, a time called menopause.

Another difference concerns the number of gametes that can be produced. In males, sperm production is unlimited, whereas in females, the number of potential egg cells is usually set at birth. Furthermore, meiosis in women need not go to completion, and very few cells ever finish the entire process. Many of the birth control methods available to females, such as the pill or "the patch," work by manipulating hormones so that cells called oocytes do not complete meiosis and an egg is never released for fertilization.

In this chapter, you will see how meiosis is involved in the production of gametes. You will also learn how meiosis compares with mitosis, the cell division process discussed in Chapter 8.

9.1 The Basics of Meiosis

Human beings, like most other animals, practice sexual reproduction in which two parents pass chromosomes to their offspring. Because each child receives a unique combination of the parents' chromosomes, children are not exactly like either parent. How does sexual reproduction bring about the distribution of chromosomes to offspring in a way that ensures not only the correct number of chromosomes but also a unique combination?

Let's begin by examining the chromosomes of one of the parents—for instance, the father. To view the chromosomes, a cell can be photographed just prior to division so that a picture of the chromosomes is obtained. The picture can be entered into a computer and the chromosomes electronically arranged by numbered pairs. The members of a pair are called **homologous chromosomes,** or **homologues,** because they have the same size, shape, and constriction (location of the centromere) (Fig. 9.1). Homologous chromosomes also have the same characteristic banding pattern upon staining because they contain genes for the same traits. Alternate forms of a gene are called **alleles.** For example, if a gene codes for the trait of finger length, the allele on one homologue may be for short fingers, and the allele at the same location on the other homologue may be for long fingers. Both sister chromatids of the same homologue carry the same alleles because sister chromatids are identical. One member of a homologous pair was derived from a chromosome inherited from the male parent, and the other member was derived from a chromosome inherited from the female parent.

Figure 9.1 Homologous chromosomes.

In body cells, the chromosomes occur in pairs called homologous chromosomes. In this micrograph of stained chromosomes from a human cell, the pairs have been numbered. These chromosomes are duplicated, and each one is composed of two sister chromatids.

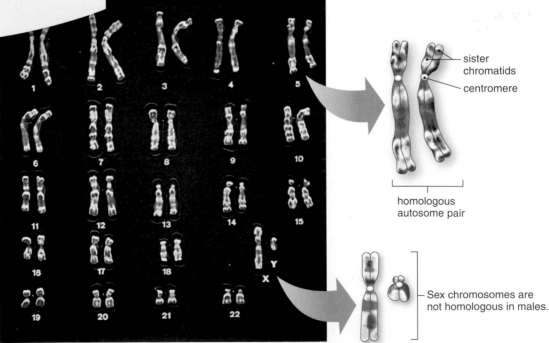

The 46 chromosomes of a male

Both males and females normally have 23 pairs of chromosomes, but one of these pairs is of unequal length in males. The larger chromosome of this pair is the X chromosome, and the smaller is the Y chromosome. In contrast, females have two X chromosomes. The X and Y chromosomes are called the **sex chromosomes** because they contain the genes that determine gender. The other chromosomes, known as **autosomes,** include all the pairs of chromosomes except the X and Y chromosomes. Twenty-three pairs of chromosomes, or 46 altogether, is called the **diploid (2n) number** of chromosomes in humans. Half this number is the **haploid (n) number** of chromosomes.

Check Your Progress

1. Describe what is meant by homologous chromosomes.
2. Explain why homologous chromosomes occur in pairs.

Answers: 1. Homologous chromosomes have the same size, shape, centromere position, and genes. *2.* One member of the pair came from the maternal parent and the other from the paternal parent.

The Human Life Cycle

The term **life cycle** in sexually reproducing organisms refers to all the reproductive events that occur from one generation to the next. The human life cycle involves two types of nuclear division: mitosis and meiosis (Fig. 9.2).

During development and after birth, mitosis is involved in the continued growth of the child and the repair of tissues at any time. As a result of mitosis, each somatic (body) cell has the diploid number of chromosomes.

During sexual reproduction, a special type of nuclear division called **meiosis** reduces the chromosome number from the diploid to the haploid number in such a way that the gametes (sperm and egg) have one chromosome derived from each homologous pair of chromosomes. In males, meiosis is a part of **spermatogenesis,** which occurs in the testes and produces sperm. In females, meiosis is a part of **oogenesis,** which occurs in the ovaries and produces eggs. After the sperm and egg join during fertilization, the resulting cell, called the **zygote,** again has homologous pairs of chromosomes. The zygote then undergoes mitosis and differentiation of cells to become a fetus, and eventually a new human being.

Meiosis is important because, if it did not halve the chromosome number, the gametes would contain the same number of chromosomes as the body cells, and the number of chromosomes would double with each new generation. Within a few generations, the cells of sexually reproducing organisms would be nothing but chromosomes! Because of meiosis, however, the chromosome number stays constant in each generation.

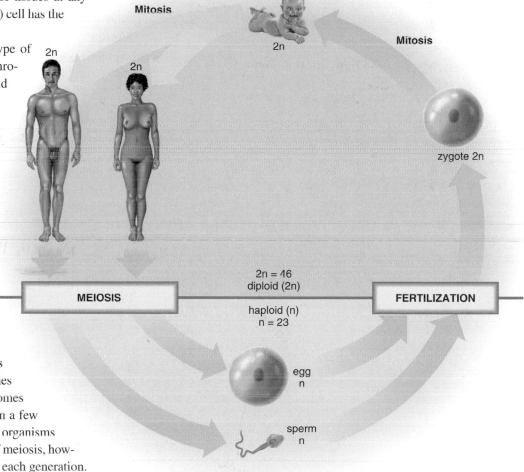

Figure 9.2   Life cycle of humans.

Meiosis in males is a part of sperm production, and meiosis in females is a part of egg production. When a haploid sperm fertilizes a haploid egg, the zygote is diploid. The zygote undergoes mitosis as it eventually develops into a newborn child. Mitosis continues throughout life during growth and repair.

Check Your Progress

Explain how mitosis and meiosis contribute to the human life cycle.

Answer: Mitosis is responsible for growth before and after birth and for repair of tissues. Meiosis results in gametes that unite at fertilization, producing the first cell of a new human being.

Overview of Meiosis

Meiosis results in four daughter cells because it consists of two divisions, called **meiosis I** and **meiosis II** (Fig. 9.3). Before meiosis I begins, each chromosome has duplicated and is composed of two sister chromatids. During meiosis I, the homologous chromosomes of each pair come together and line up side-by-side. What draws them together is still unknown. This so-called **synapsis** results in a **tetrad,** an association of four chromatids that stay in close proximity until they separate. Following synapsis and during meiosis I, the homologous chromosomes of each pair separate. This separation means that one chromosome from each homologous pair goes to each daughter nucleus. No rules restrict which chromosome goes to which daughter nucleus, and therefore, all possible combinations of chromosomes may occur within the gametes.

Following meiosis I, the daughter nuclei have half the number of chromosomes, and the chromosomes are still duplicated. Therefore, no duplication of chromosomes is needed between meiosis I and meiosis II. The chromosomes are **dyads** because each one is composed of two sister chromatids. During meiosis II, the sister chromatids of each dyad in the two daughter ~~nuclei~~ separate, and the resulting four new daughter cells have the haploid number of chromosomes. If the parent cell has 4 chromosomes, then following meiosis, each daughter cell has 2 chromosomes. (Remember that counting the centromeres tells you the number of chromosomes in a nucleus.)

Because of meiosis, the gametes have all possible combinations of chromosomes. Notice in Figure 9.3 that the final pair of daughter cells above does not have the same chromosomes as the pair below. Why not? Because the homologous chromosomes of each pair separated during meiosis I. Other chromosome combinations are possible in addition to those depicted. It's possible that the daughter cells could have only red or only blue chromosomes in Figure 9.3. (The red and blue signify that the chromosomes have come from different parents.)

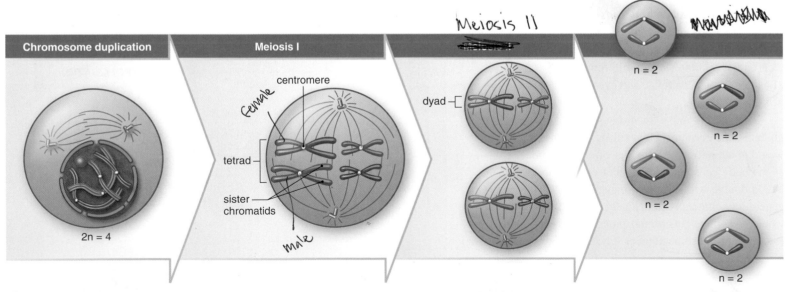

Figure 9.3   Overview of meiosis.

Following duplication of chromosomes, the parent cell undergoes two divisions, meiosis I and meiosis II. During meiosis I, homologous chromosomes separate, and during meiosis II, chromatids separate. The final daughter cells are haploid. (The blue chromosomes were originally inherited from one parent, and the red chromosomes were originally inherited from the other parent.)

Crossing-Over

Aside from allowing gametes to have different chromosomes, sexual reproduction can also lead to a difference in alleles on the chromosomes. This change is not due to mutations, but results from "shuffling" of the genetic material. A tetrad, as you just learned, is made up of the sister chromatids of two homologous chromosomes. When a tetrad forms during synapsis, the nonsister chromatids may exchange genetic material, an event called **crossing-over** (Fig. 9.4).

Recall that the homologous chromosomes carry genes for traits, such as finger length, but that one allele might call for short fingers, for example, while the other allele calls for long fingers. Therefore, when the nonsister chromatids exchange genetic material, the sister chromatids then have a different combination of alleles, and the resulting gametes will be genetically different. So in Figure 9.4, even though two of the gametes have the same chromosomes, these chromosomes may not have the same combination of alleles as before because of crossing-over. Crossing-over increases the variability of the gametes and, therefore, of the offspring.

The Importance of Meiosis

Meiosis is important first of all because the chromosome number stays constant in each new generation of individuals. When a haploid sperm fertilizes a haploid egg, the new individual has the diploid number of chromosomes.

Second, the new individual is ensured a different combination of alleles than either parent because:

1. Crossing-over can result in different alleles on the sister chromatids of a homologous pair of chromosomes.
2. Meiosis produces gametes that have all possible combinations of the haploid number of chromosomes.
3. At fertilization, a new combination of chromosomes can occur, and a zygote can have any one of a vast number of combinations of chromosomes. In humans, $(2^{23})^2$ or 70,368,744,000,000 chromosomally different zygotes are possible, even assuming no crossing-over.

Check Your Progress

1. Describe the chromosome number of the next generation if meiosis did not occur.
2. Describe the major differences between meiosis I and meiosis II.
3. What are two ways in which meiosis results in gametes that are genetically different?

Answers: 1. The chromosome number would double in the next generation. 2. During meiosis I, the parent cell is diploid, the homologous chromosomes separate, and the daughter cells are haploid. During meiosis II, the parent cell is haploid, the chromatids separate, and the daughter cells have the same number of chromosomes as the parent cell. 3. The daughter cells have different combinations of the haploid number of chromosomes, and crossing-over alters the types of alleles on the daughter chromosomes.

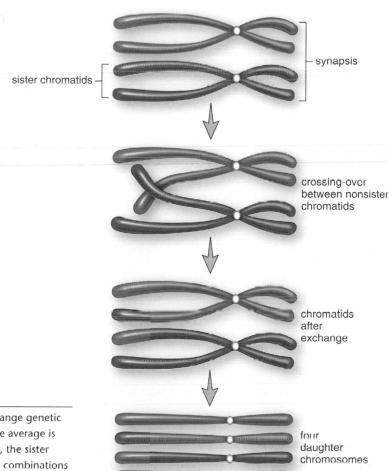

synapsis

sister chromatids

crossing-over between nonsister chromatids

chromatids after exchange

four daughter chromosomes

Figure 9.4 Crossing-over.

When homologous chromosomes are in synapsis, the nonsister chromatids exchange genetic material. This illustration shows only one crossover per chromosome pair, but the average is slightly more than two per homologous pair in humans. Following crossing-over, the sister chromatids of a dyad may no longer be identical and instead may have different combinations of alleles.

9.2 The Phases of Meiosis

The same four stages of mitosis—prophase, metaphase, anaphase, and telophase—occur during both meiosis I (Fig. 9.5) and meiosis II (Fig. 9.6).

The First Division—Meiosis I

To help you recall the events of meiosis, keep in mind what meiosis accomplishes—namely, the production of gametes that have a reduced chromosome number and are genetically different from each other and from the parent cell. During Prophase I, the nuclear envelope fragments and the nucleolus disappears as the spindle appears but more important the condensing homologous chromosomes undergo synapsis to produce tetrads. The formation of tetrads helps prepare the homologous chromosomes for separation; it also allows crossing-over to occur between nonsister chromatids. Crossing-over "shuffles" the alleles on chromosomes.

During metaphase I, the tetrads are attached to the spindle and aligned at the spindle equator. It does not matter which homologous chromosome faces which pole; therefore, all possible combinations of chromosomes can occur in the gametes. Following separation of the homologous chromosomes during anaphase I and re-formation of the nuclear envelopes during telophase, the daughter nuclei are haploid: Each daughter cell contains only one chromo-

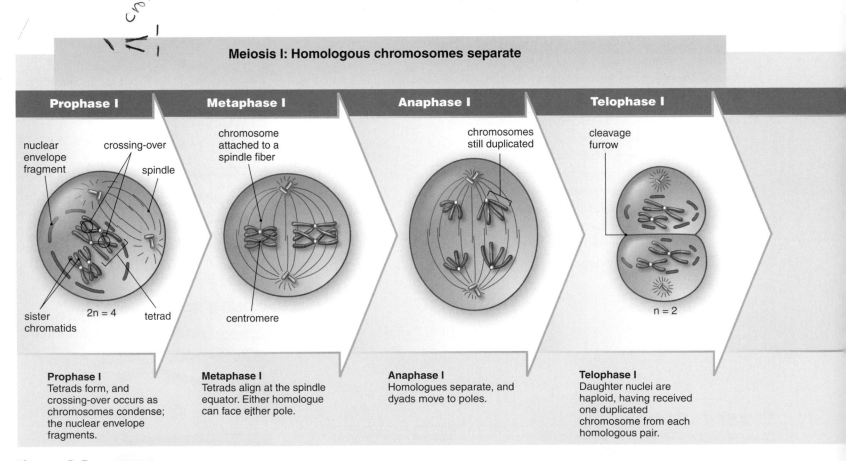

Figure 9.5 **Meiosis I.**

As a result of meiosis I, the daughter nuclei are haploid and are genetically dissimilar. They may contain a different combination of chromosomes and a different combination of alleles on the sister chromatids.

some from each homologous pair. The chromosomes are dyads, and each still has two sister chromatids. No replication of DNA occurs during a period of time called **interkinesis.**

The Second Division—Meiosis II

When you think about it, the events of meiosis II are the same as those for mitosis, except the cells are haploid. At the beginning of prophase II, a spindle appears while the nuclear envelope fragments and the nucleolus disappears. Dyads (one dyad from each pair of homologous chromosomes) are present, and each attaches to the spindle. During metaphase II, the dyads are lined up at the spindle equator. During anaphase II, the sister chromatids of each dyad separate and move toward the poles. Each pole receives the same number and kinds of chromosomes. In telophase II, the spindle disappears as nuclear envelopes form.

During cytokinesis, the plasma membrane pinches off to form two complete cells, each of which has the haploid, or n, number of chromosomes. The gametes are genetically dissimilar because they can contain different combinations of chromosomes and because crossing-over changed which alleles are together on a chromosome. Because each cell from meiosis I undergoes meiosis II, four daughter cells altogether are produced from the original diploid parent cell.

Check Your Progress

1. Describe the significance of tetrad formation during meiosis I.
2. How does the chromosome number differ between a cell that is entering meiosis and a cell that has completed meiosis I?

Answers: 1. Tetrad formation helps prepare homologous chromosomes for separation and allows crossing-over to occur. 2. A cell that has completed meiosis I has half as many chromosomes as it did at the beginning of meiosis.

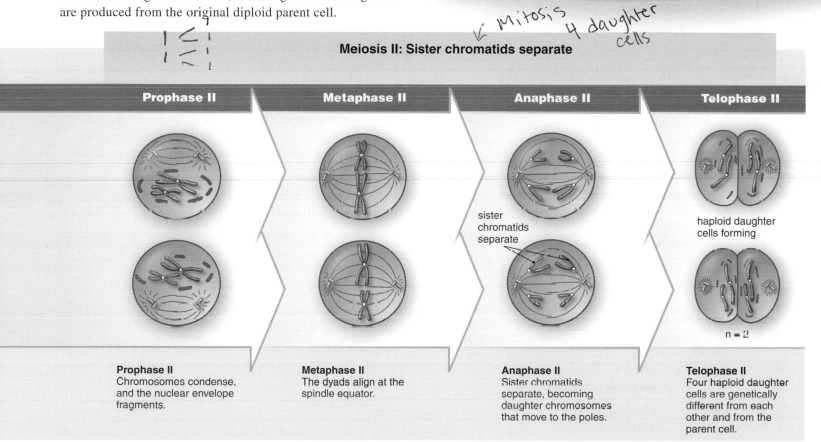

Meiosis II: Sister chromatids separate

| Prophase II | Metaphase II | Anaphase II | Telophase II |

sister chromatids separate

haploid daughter cells forming

n = 2

Prophase II
Chromosomes condense, and the nuclear envelope fragments.

Metaphase II
The dyads align at the spindle equator.

Anaphase II
Sister chromatids separate, becoming daughter chromosomes that move to the poles.

Telophase II
Four haploid daughter cells are genetically different from each other and from the parent cell.

Figure 9.6 Meiosis II.

During meiosis II, daughter chromosomes consisting of one chromatid each move to the poles. Following meiosis II, there are four haploid daughter cells. Comparing the number of centromeres in the daughter cells with the number in the parent cell at the start of meiosis I verifies that the daughter cells are haploid. Notice that the daughter cells are genetically dissimilar from each other and from the parent cell.

Table 9.1	Meiosis I Compared to Mitosis
Meiosis I	**Mitosis**
Prophase I	*Prophase*
Pairing of homologous chromosomes; crossing-over	No pairing of chromosomes; no crossing-over
Metaphase I	*Metaphase*
Tetrads at spindle equator	Dyads at spindle equator
Anaphase I	*Anaphase*
Homologues of each tetrad separate, and dyads move to poles	Sister chromatids separate, becoming daughter chromosomes that move to the poles
Telophase I	*Telophase*
Two haploid daughter cells not identical to parent cell	Two diploid daughter cells, identical to the parent cell

Table 9.2	Meiosis II Compared to Mitosis
Meiosis II	**Mitosis**
Prophase II	*Prophase*
No pairing of chromosomes	No pairing of chromosomes
Metaphase II	*Metaphase*
Haploid number of dyads at spindle equator	Diploid number of dyads at spindle equator
Anaphase II	*Anaphase*
Sister chromatids separate, becoming daughter chromosomes that move to the poles	Sister chromatids separate, becoming daughter chromosomes that move to the poles
Telophase II	*Telophase*
Four haploid daughter cells, not genetically identical to each other or the parent cell	Two daughter cells, genetically identical to the parent cell

9.3 Meiosis Compared to Mitosis

Figure 9.7 compares meiosis to mitosis. Notice that:

- Meiosis requires two nuclear divisions, but mitosis requires only one nuclear division.
- Meiosis produces four daughter nuclei, and there are four daughter cells following cytokinesis. Mitosis followed by cytokinesis results in two daughter cells.
- Following meiosis, the four daughter cells are haploid and have half the chromosome number as the parent cell. Following mitosis, the daughter cells have the same chromosome number as the parent cell.
- Following meiosis, the daughter cells are genetically dissimilar to each other and to the parent cell. Following mitosis, the daughter cells are genetically identical to each other and to the parent cell.

The specific differences between these nuclear divisions can be categorized according to process and occurrence.

Process

To summarize the processes, Table 9.1 compares meiosis I to mitosis and Table 9.2 compares meiosis II to mitosis.

Meiosis I Compared to Mitosis

The following events distinguish meiosis I from mitosis:

- During meiosis I, tetrads form, and crossing-over occurs during prophase I. These events do not occur during mitosis.
- During metaphase I of meiosis, tetrads align at the spindle equator. The paired chromosomes have a total of four chromatids each. During metaphase in mitosis, dyads align at the spindle equator.
- During anaphase I of meiosis, the homologous chromosomes of each tetrad separate, and dyads (with centromeres intact) move to opposite poles. During anaphase of mitosis, sister chromatids separate, becoming daughter chromosomes that move to opposite poles.

Meiosis II Compared to Mitosis

The events of meiosis II are just like those of mitosis except that in meiosis II, the cells have the haploid number of chromosomes.

Occurrence

Meiosis occurs only at certain times in the life cycle of sexually reproducing organisms. In humans, meiosis occurs only in the testes and ovaries, where it is involved in the production of gametes. The function of meiosis is to provide gamete variation and to keep the chromosome number constant generation after generation. With fertilization, the full chromosome number is restored. Because unlike gametes fuse, fertilization also leads to variation among the offspring.

Mitosis is more common because it occurs in all tissues during embryonic development and also during growth and repair. The function of mitosis is to keep the chromosome number constant in all the cells of the body so that every cell has the same genetic material. Consider that reproductive cloning results in an individual with the same genes as the parent because a single diploid nucleus gives genes to the new individual (see Fig. 12.2).

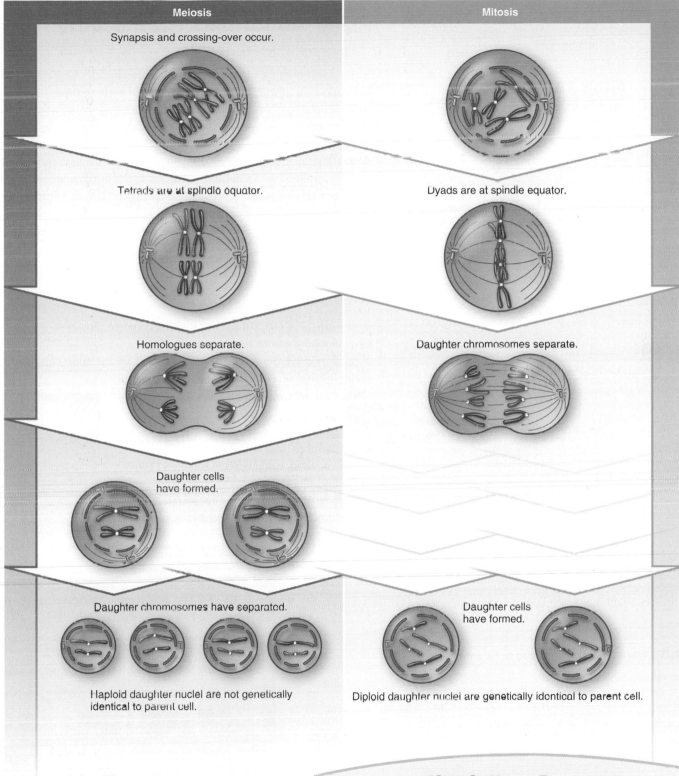

Meiosis	Mitosis
Synapsis and crossing-over occur.	
Tetrads are at spindle equator.	Dyads are at spindle equator.
Homologues separate.	Daughter chromosomes separate.
Daughter cells have formed.	
Daughter chromosomes have separated.	Daughter cells have formed.
Haploid daughter nuclei are not genetically identical to parent cell.	Diploid daughter nuclei are genetically identical to parent cell.

Figure 9.7   Meiosis compared to mitosis.

Check Your Progress

Why does meiosis, but not mitosis, produce daughter cells with half the number of chromosomes?

Answer: The homologous chromosomes pair and separate only during metaphase I of meiosis and not during metaphase of mitosis.

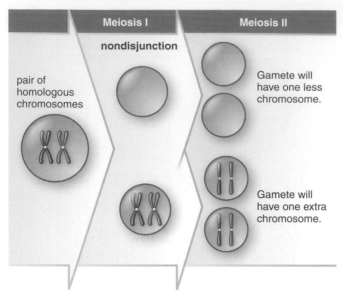

a. Nondisjunction during meiosis I

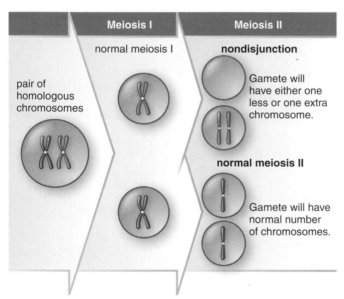

b. Nondisjunction during meiosis II

Figure 9.8 **Nondisjunction during meiosis.**

Because of nondisjunction, gametes either lack a chromosome or have an extra chromosome. Nondisjunction can occur (**a**) during meiosis I if homologous chromosomes fail to separate, and (**b**) during meiosis II if the sister chromatids fail to separate completely.

9.4 Abnormal Chromosome Inheritance

The normal number of chromosomes in human cells is 46, but occasionally humans are born with an abnormal number of chromosomes because improper meiosis occurred. **Nondisjunction** occurs during meiosis I when both members of a homologous pair go into the same daughter cell, or during meiosis II when the sister chromatids fail to separate and both daughter chromosomes go into the same gamete (Fig. 9.8). If an egg that ends up with 24 chromosomes instead of 23 is fertilized with a normal sperm, the result is a *trisomy,* so called because one type of chromosome is present in three copies. If an egg that has 22 chromosomes instead of 23 is fertilized by a normal sperm, the result is a *monosomy,* so called because one type of chromosome is present in a single copy.

Down Syndrome

Down syndrome is *trisomy 21,* in which an individual has three copies of chromosome 21. In most instances, the egg contained two copies of this chromosome instead of one. (In 23% of the cases studied, however, the sperm had the extra chromosome 21.)

Down syndrome is easily recognized by the following characteristics: short stature; an eyelid fold; stubby fingers; a wide gap between the first and second toes; large, fissured tongue; round head; palm crease; and unfortunately, mental retardation, which can sometimes be severe (Fig. 9.9).

The chance of a woman having a Down syndrome child increases rapidly with age, starting at about age 40. The frequency of Down syndrome is 1 in 800 births for mothers

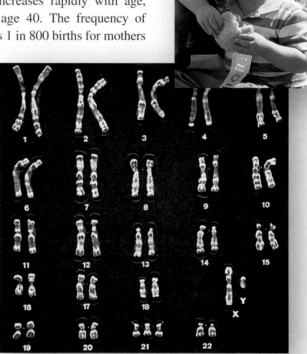

Figure 9.9 **Down syndrome.**

Down syndrome occurs when the egg or the sperm has an extra chromosome 21 due to nondisjunction in either meiosis I or meiosis II. Characteristics include a wide, rounded face and narrow, slanting eyelids. Mental retardation to varying degrees is usually present.

under 40 years of age and 1 in 80 for mothers over 40 years of age. However, most Down syndrome babies are born to women younger than age 40, because this is the age group having the most babies.

Abnormal Sex Chromosome Number

Nondisjunction during oogenesis or spermatogenesis can result in gametes that have too few or too many X or Y chromosomes. Figure 9.8 can be used to illustrate nondisjunction of the sex chromosomes during oogenesis if we assume that the chromosomes shown represent X chromosomes.

Just as having an abnormal number of autosomes can result in a condition such as Down syndrome, having additional X or Y chromosomes, or lacking them, also causes certain syndromes. Newborns with an abnormal sex chromosome number are more likely to survive than those with an abnormal autosome number, and the explanation is quite surprising. Normal females, like normal males, have only one functioning X chromosome. The other X chromosome (or any extra X chromosomes) become an inactive mass called a **Barr body** (after the person who discovered it).

In humans, the presence of a Y chromosome, not the number of X chromosomes, almost always determines maleness. The *SRY* (a gene on the Y chromosome) on the short arm of the Y chromosome produces a hormone called testis-determining factor, which plays a critical role in the development of male genitals. No matter how many X chromosomes are involved, an individual with a Y chromosome is a male, assuming a functional *SRY* is on the Y chromosome.

A person with **Turner syndrome** (XO) is a female. The O signifies the absence of a second sex chromosome. Turner syndrome females are short, with a broad chest and webbed neck. The ovaries, oviducts, and uterus are very small and underdeveloped. Turner females do not undergo puberty or menstruate, and their breasts do not develop (Fig. 9.10*a*). However, some have given birth following in vitro fertilization using donor eggs. These women usually have normal intelligence and can lead fairly normal lives if they receive hormone supplements.

A person with **Klinefelter syndrome** (XXY) is a male. A Klinefelter male has two or more X chromosomes in addition to a Y chromosome. The extra X chromosomes become Barr bodies.

In Klinefelter males, the testes and prostate gland are underdeveloped. There is no facial hair, but some breast development may occur (Fig. 9.10*b*). Affected individuals have large hands and feet and very long arms and legs. They are usually slow to learn but not mentally retarded unless they inherit more than two X chromosomes.

As with Turner syndrome, it is best for parents to know as soon as possible that their child has Klinefelter syndrome because much can be done to help the child lead a normal life.

Check Your Progress

1. Describe how a zygote could receive an abnormal chromosome number.
2. Contrast monosomy with trisomy.

Answers: 1. If nondisjunction occurs and improper separation has occurred in meiosis I or II, the gametes and then the zygote have an abnormal chromosome number. 2. Monosomy results when an individual is missing one chromosome. Trisomy results when an individual has one extra chromosome.

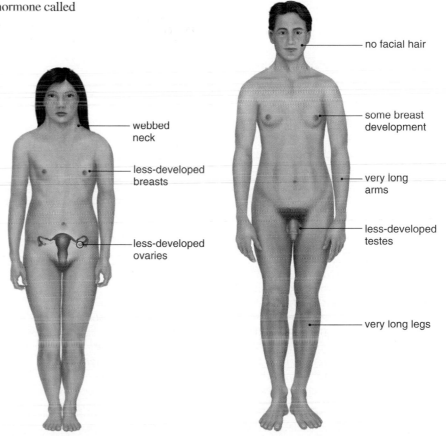

a. A female with Turner (XO) syndrome   b. A male with Klinefelter (XXY) syndrome

Figure 9.10   Abnormal sex chromosome number.

a. A female with Turner syndrome (XO) has a short thick neck, short stature, and lack of breast development. **b.** A male with Klinefelter syndrome (XXY) has immature sex organs and some development of the breasts.

Summary

9.1 The Basics of Meiosis

Parents pass chromosomes to their children, who resemble them but not exactly because they receive a unique combination of their parents' chromosomes and alleles.

Diploid Versus Haploid Diploid cells in humans have 22 homologous pairs of autosomes and 1 pair of sex chromosomes for a total of 46 chromosomes. Males are XY and females are XX. Haploid cells have 22 autosomes and one sex chromosome, either an X or a Y for a total of 23 chromosomes.

The Human Life Cycle
The human life cycle has two types of nuclear division:

- **Mitosis** ensures that every body cell has 23 pairs of chromosomes. It occurs during growth and repair.
- **Meiosis** occurs during formation of gametes. It ensures that the gametes are haploid and have 23 chromosomes, one from each of the pairs of chromosomes.

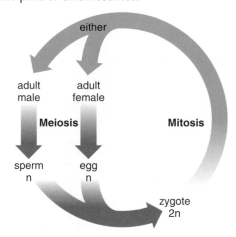

Overview of Meiosis
Meiosis has two nuclear divisions called meiosis I and meiosis II. Therefore, meiosis results in four daughter cells.

- **Meiosis I** Synapsis and formation of tetrads leads to separation of homologous chromosomes. Because no restrictions govern which member goes to which pole, the gametes will contain all possible combinations of chromosomes.
- **Meiosis II** The chromosomes are still duplicated. The chromatids separate, forming daughter chromosomes. The daughter cells are haploid. If the parent cell has 4 chromosomes, each of the daughter cells has 2 chromosomes. All possible combinations of chromosomes occur among the daughter cells.

Crossing-Over Crossing-over between nonsister chromatids during meiosis I "shuffles" the alleles on the chromosomes so that each sister chromatid of a homologue has a different mix of alleles.

Importance of Meiosis Meiosis is important because the chromosome number stays constant between the generations of individuals and because the daughter cells, and therefore the gametes, are genetically different.

9.2 The Phases of Meiosis

Meiosis I
- **Prophase I** Tetrads form, and crossing-over occurs as chromosomes condense; the nuclear envelope fragments.
- **Metaphase I** Tetrads align at the spindle equator. Either homologue can face either pole.
- **Anaphase I** Homologues of each tetrad separate, and dyads move to the poles.
- **Telophase I** Daughter nuclei are haploid, having received one duplicated chromosome from each homologous pair.

Meiosis II
- **Prophase II** Chromosomes condense, and the nuclear envelope fragments.
- **Metaphase II** The haploid number of dyads align at the spindle equator.
- **Anaphase II** Sister chromatids separate, becoming daughter chromosomes that move to the poles.
- **Telophase II** Four haploid daughter cells are genetically different from each other and from the parent cell.

9.3 Meiosis Compared to Mitosis

The following diagrams show the differences between meiosis and mitosis. Duplication occurs before each begins.

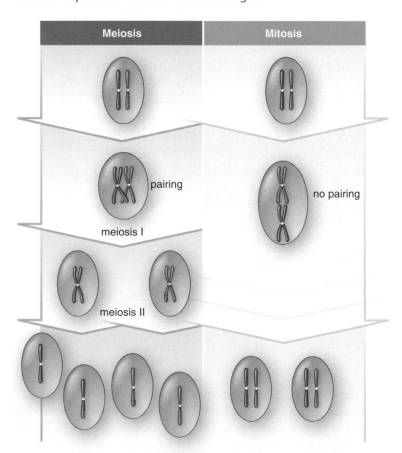

Meiosis

- In humans, meiosis occurs in the testes and ovaries where it produces the gametes.
- Because meiosis has two nuclear divisions, it produces four daughter cells. The daughter cells are haploid and genetically different from each other and from the parent cell.

Mitosis

- Mitosis occurs in the body cells and accounts for growth and repair. Because mitosis has one nuclear division, it produces two daughter cells. The daughter cells are diploid and are genetically identical to each other and to the parent cell.

Meiosis I Compared with Mitosis

Meiosis I differs from mitosis in the following ways:

- During prophase I, tetrads form and crossing-over occurs. No such events occur in mitosis.
- During metaphase I, tetrads are at the spindle equator. During mitosis, dyads are at the spindle equator.
- During anaphase I, homologous chromosomes separate. During mitosis, sister chromatids separate.

Meiosis II Compared with Mitosis

- The events of meiosis II are the same as those of mitosis except that the cells undergoing meiosis II are haploid, whereas those undergoing mitosis are diploid.

9.4   Abnormal Chromosome Inheritance

Nondisjunction accounts for the inheritance of an abnormal chromosome number. Nondisjunction can occur during meiosis I if homologous chromosomes fail to separate, and both go into one daughter cell and the other daughter cell receives neither. It can also occur during meiosis II if chromatids fail to separate, and both go into one daughter cell and the other daughter cell receives neither.

Down Syndrome   In Down syndrome, an example of an autosomal syndrome, the individual inherits three copies of chromosome 21.

Abnormal Sex Chromosome Number

Examples of syndromes caused by inheritance of abnormal sex chromosomes are Turner syndrome (XO) and Klinefelter syndrome (XXY).

Turner Syndrome   An XO individual inherits only one X chromosome. This individual survives because even in an XX individual, one X becomes a Barr body.

Klinefelter Syndrome   An XXY individual inherits two X (or more) chromosomes and also a Y chromosome. This individual survives because one (or more) X chromosome becomes a Barr body.

Thinking Scientifically

1. Some plants contain mutations in the genes that control the movement of chromosomes at meiosis. These mutant plants produce gametes that are not reduced in chromosome number. So, egg and sperm cells have the same number of chromosomes as the plants that produced them. Suppose a mutant plant containing 10 pairs of chromosomes is crossed with a normal plant (also containing 10 pairs of chromosomes). How many chromosomes would their offspring have? If two mutant plants were crossed, how many chromosomes would their offspring have?

2. In the nineteenth century, physicians noticed that people with Down syndrome were often the youngest children in large families. Some physicians suggested that the disorder is due to "maternal reproductive exhaustion." How would you disprove this hypothesis? What might be a reasonable explanation for the relationship between maternal age and incidence of Down syndrome?

Testing Yourself

Choose the best answer for each question.

1. A human cell contains _____ pair(s) of sex chromosomes.
   - a. 1
   - b. 2
   - c. 22
   - d. 23

2. Mitosis _____ chromosome number, while meiosis _____ the chromosome number of the daughter cells.
   - a. maintains, increases
   - b. increases, maintains
   - c. increases, decreases
   - d. maintains, decreases

For questions 3–6, match the items to those in the key. Answers can be used more than once.

Key:
   - a. haploid number
   - b. 2 times the haploid number
   - c. diploid number
   - d. 2 times the diploid number

3. How many chromatids are in a cell as it enters meiosis?
4. How many chromosomes are in each daughter cell following meiosis I?
5. How many chromatids are in each daughter cell following meiosis I?
6. How many chromosomes are in each daughter cell following meiosis II?

7. The two major functions of meiosis are to
   - a. maintain chromosome number and create genetically uniform products.
   - b. reduce chromosome number and create genetically uniform products.
   - c. reduce chromosome number and create genetically variable products.
   - d. maintain chromosome number and create genetically variable products.

For questions 8–15, match the items to those in the key.

Key:
   - a. prophase I
   - b. metaphase I
   - c. anaphase I
   - d. telophase I
   - e. prophase II
   - f. metaphase II
   - g. anaphase II
   - h. telophase II

8. A cleavage furrow forms, resulting in haploid nuclei. Each chromosome contains two chromatids.
9. Tetrads form, and crossing-over occurs.

10. Dyads align at the spindle equator.

11. Four haploid daughter cells are created.

12. Homologous chromosomes move to opposite poles.

13. Sister chromatids separate.

14. Tetrads align on the spindle equator.

15. Chromosomes in haploid nuclei condense.

16. At each letter in the following diagram, indicate how mitosis differs from meiosis.

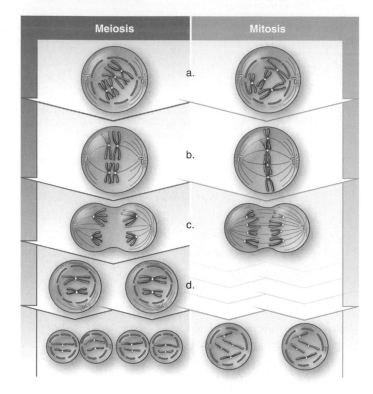

Meiosis	Mitosis
	a.
	b.
	c.
	d.

17. The condition that results when an egg carrying 23 chromosomes unites with a sperm carrying 22 chromosomes is called

   a. monosomy.   c. trisomy.
   b. isomy.   d. tetrasomy.

18. An individual with Turner syndrome is conceived when a normal gamete unites with

   a. an egg that was produced by nondisjunction during oogenesis.
   b. a sperm that was produced by nondisjunction during spermatogenesis.
   c. Either a or b is correct.
   d. Both a and b are correct.

Go to www.mhhe.com/maderessentials for more quiz questions.

Bioethical Issue

The dizzying array of assisted reproductive technologies has progressed from simple in vitro fertilization to the ability to freeze eggs, sperm, or even embryos for future use. Older women who never had the opportunity to freeze their eggs can still have children if they use donated eggs—perhaps today harvested from a fetus.

Legal complications abound, ranging from which mother has first claim to the child—the surrogate mother, the woman who donated the egg, or the primary caregiver—to which partner has first claim to frozen embryos following a divorce. Legal issues involving who has the right to use what techniques have rarely been discussed, much less decided upon. Some clinics will help anyone, male or female, no questions asked, as long as they have the ability to pay. Most clinics are heading toward being able to do any type of procedure, including guaranteeing the sex of the child and making sure the child will be free from a particular genetic disorder. It would not be surprising if, in the future, zygotes could be engineered to have any particular trait desired by the parents.

Even today, eugenic (good gene) goals are evidenced by the fact that reproductive clinics advertise for egg and sperm donors, primarily in elite college newspapers. The questions become, is it too late for us as a society to make ethical decisions about reproductive issues? Should we come to a consensus about what techniques should be allowed and who should be able to use them? Should a woman investigate a couple before she has a child for them?

Understanding the Terms

allele   128
autosome   129
Barr body   137
crossing-over   131
diploid (2n) number   129
Down syndrome   136
dyad   130
haploid (n) number   129
homologous
   chromosome   128
homologue   128
interkinesis   133
Klinefelter syndrome   137

life cycle   129
meiosis   129
meiosis I   130
meiosis II   130
nondisjunction   136
oogenesis   129
sex chromosome   129
spermatogenesis   129
synapsis   130
tetrad   130
Turner syndrome   137
zygote   129

Match the terms to these definitions:

a. _____ The set of events that produces egg cells.

b. _____ A chromosome consisting of two sister chromatids.

c. _____ The side-by-side pairing of chromosomes in early meiosis.

d. _____ The exchange of genetic material between nonsister chromatids.

e. _____ Improper separation of chromosomes at meiosis.

f. _____ The syndrome expressed by an XO individual.

g. _____ An inactive X chromosome.

Patterns of Inheritance

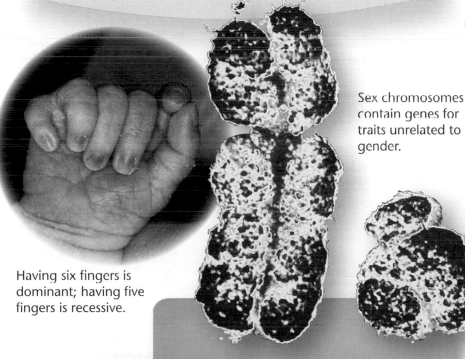

Sex chromosomes contain genes for traits unrelated to gender.

Having six fingers is dominant; having five fingers is recessive.

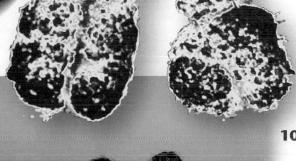

Environmental factors can influence how your genes are expressed.

Just how much influence do genes have over the development of particular traits including genetic diseases? In cases such as sickle cell disease and cystic fibrosis, the genes have complete influence over their development, and the probable occurrence of the trait is easy to predict. In other cases, the influence of the genes alone do not forecast a particular trait. Rather, environmental factors also influence whether certain specific traits develop. Determining

▶ ▶ ▶

OUTLINE

the pattern of inheritance for traits that are not completely genetic is a difficult task.

Tests are available for some alleles associated with genetic diseases. For example, a woman can be tested to see if she has the *BRCA-1* and *BRCA-2* alleles, both of which are associated with breast and ovarian cancer. However, the allele alone is not sufficient to cause the development of cancer—environmental factors are also involved. Therefore, even if a woman tests positive for these alleles, she may not develop cancer. Unfortunately, it is impossible to predict for sure whether or not cancer will occur. Still, some women have resorted to extraordinary methods, such as radical mastectomy to remove all breast tissue, to prevent the possibility of cancer.

In this chapter, you will learn about the basic principles of inheritance. Stress will be placed on the rules of genetics that allow us to predict the chances of a trait being passed on from one generation to the next. However, we have to bear in mind that sometimes environmental influences affect the passage of traits.

10.1 Mendel's Laws

Today, most people know that DNA is the genetic material, and they may have heard that scientists have determined the DNA base sequence of human chromosomes. In contrast, they may not know about Gregor Mendel, an Austrian monk who developed certain laws of heredity after doing crosses between garden pea plants in 1860 (Fig. 10.1). Gregor Mendel investigated genetics at the organismal level, and this is still the level that intrigues most of us on a daily basis. We observe, for example, that facial and other features run in families, and we would like some way of explaining this observation. And so, it is appropriate to begin our study of genetics at the organismal level and to learn Mendel's laws of heredity.

When Mendel began his work, most plant and animal breeders acknowledged that both sexes contributed equally to a new individual. However, they were unable to account for the presence of definite variations (differences) among the members of a family, generation after generation. Mendel's model of heredity does account for such variations. Therefore, Mendel's model is compatible with the theory of evolution, which states that various combinations of traits are tested by the environment, and those combinations that lead to reproductive success are the ones that are passed on.

a.

Figure 10.1   Mendel working in his garden.

a. Mendel grew and tended the garden peas, *Pisum sativum,* he used for his experiments. **b.** Mendel selected these seven traits for study. Before he did any crosses, he made sure the parents bred true. All the offspring either had the dominant trait or the recessive trait.

	Trait		
	Stem length	Pod shape	
Dominant	Tall	Inflated	
Recessive	Short	Constricted	

b.

Mendel's Experimental Procedure

Mendel's parents were farmers, so he no doubt acquired the practical experience he needed to grow pea plants during childhood. Mendel was also a mathematician; most likely, his background in mathematics prompted him to use a statistical basis for his breeding experiments. He prepared for his experiments carefully and conducted preliminary studies with various animals and plants. He then chose to work with the garden pea, *Pisum sativum.*

The garden pea was a good choice. The plants were easy to cultivate and had a short generation time. And although peas normally self-pollinate (pollen goes only to the same flower), they could be cross-pollinated by hand. Many varieties of peas were available, and Mendel chose 22 of them for his experiments. When these varieties self-pollinated, they were *true-breeding*—meaning that the offspring were like the parent plants and like each other. In contrast to his predecessors, Mendel studied the inheritance of relatively simple and easily detected traits, such as seed shape, seed color, and flower color, and he observed no intermediate characteristics among the offspring. Figure 10.2 shows Mendel's procedure.

As Mendel followed the inheritance of individual traits, he kept careful records. He used his understanding of the mathematical laws of probability to interpret his results and to arrive at a theory that has since been supported by innumerable experiments. This theory is called a *particulate theory of inheritance* because it is based on the existence of minute particles we now call genes. Inheritance involves the reshuffling of the same genes from generation to generation.

Figure 10.2 Garden pea anatomy and traits.

a. In the garden pea, pollen grains produced in the anther contain sperm, and ovules in the ovary contain eggs. **b.** When Mendel performed crosses, he brushed pollen from one type pea plant onto the stigma of another type pea plant. After sperm fertilized the eggs, the ovules developed into seeds (peas). For a cross involving a parent with round yellow seeds and a parent with wrinkled yellow seeds, he observed and counted a total of 7,324 peas.

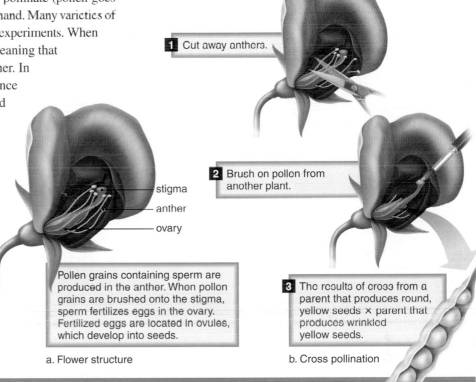

1 Cut away anthers.

2 Brush on pollen from another plant.

stigma
anther
ovary

Pollen grains containing sperm are produced in the anther. When pollen grains are brushed onto the stigma, sperm fertilizes eggs in the ovary. Fertilized eggs are located in ovules, which develop into seeds.

3 The results of cross from a parent that produces round, yellow seeds × parent that produces wrinkled yellow seeds.

a. Flower structure

b. Cross pollination

Seed shape	Seed color	Flower position	Flower color	Pod color
Round	Yellow	Axial	Purple	Green
Wrinkled	Green	Terminal	White	Yellow

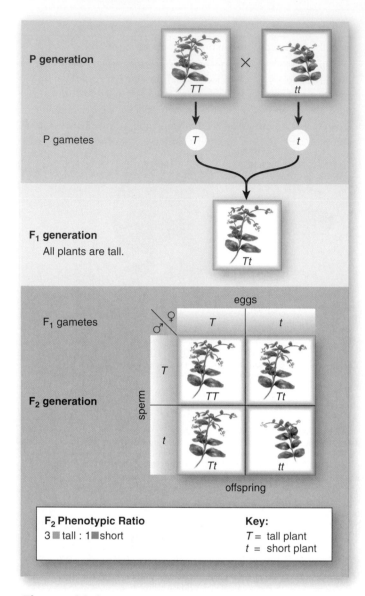

Figure 10.3 **One-trait cross.**

The P generation plants differ in one regard—length of the stem. The F₁ generation plants are all tall, but the factor for short has not disappeared because ¹/₄ of the F₂ generation are short. The 3:1 ratio allowed Mendel to deduce that individuals have two discrete and separate genetic factors for each trait.

One-Trait Inheritance

After ensuring that his pea plants were true-breeding—for example, that his tall plants always had tall offspring and his short plants always had short offspring—Mendel was ready to perform a cross between these two strains. Mendel called the original parents the *P generation,* the first-generation offspring the *F₁* (for filial) *generation,* and the second-generation offspring the *F₂ generation.* The diagram in Figure 10.3 representing Mendel's F₁ cross is called a Punnett square. In a **Punnett square,** all possible types of sperm are lined up vertically, and all possible types of eggs are lined up horizontally, or vice versa, so that every possible combination of gametes occurs within the square.

As Figure 10.3 shows, when Mendel crossed tall pea plants with short pea plants, all the F₁ offspring resembled the tall parent. Did this mean that the other characteristic, shortness, had disappeared permanently? No, because when Mendel allowed the F₁ plants to self-pollinate, ¾ of the F₂ generation were tall and ¼ were short, a 3:1 ratio. Therefore, the F₁ plants had been able to pass on a factor for shortness—it didn't just disappear. Perhaps the F₁ plants were tall because tallness was dominant to shortness?

Mendel's mathematical approach led him to interpret his results differently than previous breeders had done. He reasoned that a 3:1 ratio among the F₂ offspring was possible only if (1) the F₁ parents contained two separate copies of each hereditary factor, one dominant and the other recessive; (2) the factors separated when the gametes were formed, and each gamete carried only one copy of each factor; and (3) random fusion of all possible gametes occurred upon fertilization. Only in this way would shortness reoccur in the F₂ generation.

One-Trait Testcross

Mendel's experimental use of simple dominant and recessive characteristics allowed him to test his interpretation of his crosses. To see if the F₁ carries a recessive factor, Mendel crossed his F₁ generation tall plants with true-breeding, short plants. He reasoned that half the offspring would be tall and half would be short (Fig. 10.4*a*). And, indeed, those were the results he obtained; therefore, his hypothesis that factors segregate when gametes are formed was supported.

Today, a one-trait **testcross** is used to determine whether or not an individual with the dominant trait has two dominant factors for a particular trait.

Figure 10.4 **One-trait testcross.**

Crossing an individual with a dominant appearance (phenotype) with a recessive individual indicates the genotype. **a.** If a parent with the dominant phenotype has only one dominant factor, the results among the offspring are 1:1. **b.** If a parent with the dominant phenotype has two dominant factors, all offspring have the dominant phenotype.

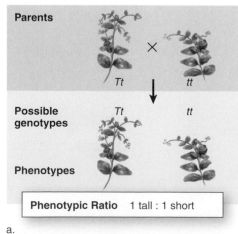

a.

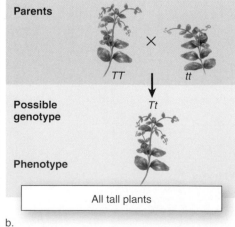

b.

This is not possible to tell by observation because an individual can exhibit the dominant appearance while having only one dominant factor. Figure 10.4*b* shows that if the individual has two dominant factors, all the offspring will be tall.

After doing one-trait crosses, Mendel arrived at his first law of inheritance—the **law of segregation,** which is a cornerstone of his particulate theory of inheritance. The law of segregation states the following:

- Each individual has two factors for each trait.
- The factors segregate (separate) during the formation of the gametes.
- Each gamete contains only one factor from each pair of factors.
- Fertilization gives each new individual two factors for each trait.

The Modern Genetics View

Mendel's law of segregation holds for all organisms, including humans. In modern terms, we say that a trait is controlled by two **alleles,** alternate forms of a gene. The **dominant allele** is so named because of its ability to mask the expression of the other allele, called the **recessive allele.** The dominant allele is identified by an uppercase (capital) letter, and the recessive allele by the same but lowercase (small) letter. In humans, for example, alleles for finger length might be *S* for short fingers and *s* for long fingers. Alleles occur on a homologous pair of chromosomes at a particular location called the **gene locus** (Fig. 10.5).

As studied in the previous chapter, meiosis is the type of cell division that reduces the chromosome number. During meiosis I, the homologous chromosomes of a tetrad separate. Therefore, the process of meiosis explains Mendel's law of segregation and why only one allele for each trait is in a gamete, as we will see shortly.

When an organism has two identical alleles, it is termed **homozygous.** All the gametes from a homologous dominant parent, such as *SS,* contain an allele for short fingers (*S*), and all gametes produced by the homozygous recessive parent contain an allele for long fingers (*s*). Therefore, all the offspring from this couple will have one allele for short fingers and another for long fingers, as in *Ss.* We say that this individual is **heterozygous.**

Genotype Versus Phenotype   It is obvious from our discussion that two organisms with different allelic combinations for a trait can have the same outward appearance. For instance, *SS* and *Ss* individuals both have short fingers. For this reason, it is necessary to distinguish between the alleles present in an organism and the appearance of that organism.

The word **genotype** refers to the alleles an individual receives at fertilization. Genotype may be indicated by letters or by short, descriptive phrases. Genotype *SS* is called homozygous dominant; genotype *ss* is homozygous recessive; and genotype *Ss* is heterozygous.

The word **phenotype** refers to the physical appearance of the individual. The homozygous dominant (*SS*) individual and the heterozygous (*Ss*) individual both show the dominant phenotype and have short fingers, while the homozygous recessive individual shows the recessive phenotype and has long fingers (Table 10.1).

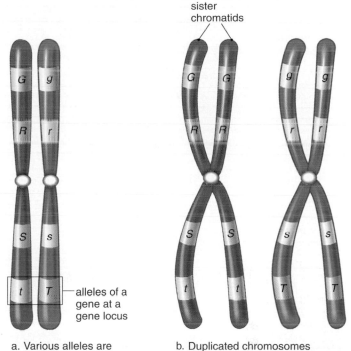

a. Various alleles are located at specific loci.

b. Duplicated chromosomes show that sister chromatids have identical alleles.

Figure 10.5   Homologous chromosomes.

a. The letters represent alleles—that is, alternate forms of a gene. Each allelic pair, such as *Gg* or *Tt,* is located on homologous chromosomes at a particular gene locus. **b.** Sister chromatids carry the same alleles in the same order.

Table 10.1	Genotype Versus Phenotype	
Genotype	Genotype	Phenotype
SS	Homozygous dominant	Short fingers
Ss	Heterozygous	Short fingers
ss	Homozygous recessive	Long fingers

Check Your Progress

1. **Contrast homozygous with heterozygous.**
2. **Contrast genotype with phenotype.**

Answers: 1. A homozygous individual has two identical alleles of a gene. A heterozygous individual has two different alleles of a gene. 2. Genotype is the genetic makeup of an individual, while phenotype is the physical appearance of the individual.

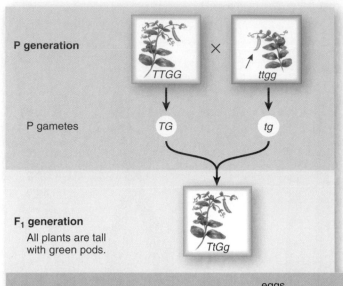

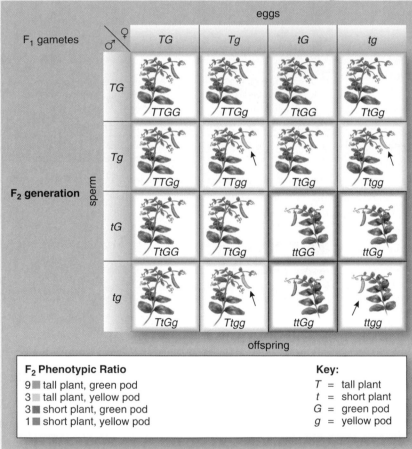

F₂ Phenotypic Ratio

9 ■ tall plant, green pod
3 ■ tall plant, yellow pod
3 ■ short plant, green pod
1 ■ short plant, yellow pod

Key:

T = tall plant
t = short plant
G = green pod
g = yellow pod

Figure 10.6 **Two-trait cross done by Mendel.**

P generation plants differ in two regards—length of the stem and color of the pod. The F₁ generation shows only the dominant phenotypes, but all possible phenotypes appear among the F₂ generation. The 9:3:3:1 ratio allowed Mendel to deduce that factors segregate into gametes independently of other factors. (Arrow indicates yellow pods.)

Two-Trait Inheritance

Mendel performed a second series of crosses in which he crossed true-breeding plants that differed in two traits. For example, he crossed tall plants having green pods with short plants having yellow pods (Fig. 10.6). The F₁ plants showed both dominant characteristics (tall with green pods). As before, Mendel then allowed the F₁ plants to self-pollinate. Two possible results could occur in the F₂ generation:

1. If the dominant factors (*TG*) always go into the F₁ gametes together, and the recessive factors (*tg*) always stay together, then two phenotypes would result among the F₂ plants—tall plants with green pods and short plants with yellow pods.
2. If the four factors segregate into the F₁ gametes independently, then four phenotypes would result among the F₂ plants—tall plants with green pods, tall plants with yellow pods, short plants with green pods, and short plants with yellow pods.

Figure 10.6 shows that Mendel observed four phenotypes among the F₂ plants, supporting the second hypothesis. Therefore, Mendel formulated his second law of heredity—the **law of independent assortment**—which states:

- Each pair of factors segregates (assorts) independently of the other pairs.
- All possible combinations of factors can occur in the gametes.

Two-Trait Testcross

The fruit fly *Drosophila melanogaster,* less than one-fifth the size of a housefly, is a favorite subject for genetic research because it has several mutant characteristics that are easily determined. The **wild-type** fly—the type you are most likely to find in nature—has long wings and a gray body, while some mutant flies have short (vestigial) wings and black (ebony) bodies. The key for a cross involving these traits is L = long wings, l = short wings, G = gray body, and g = black body.

A two-trait testcross can be used to determine whether an individual is homozygous dominant or heterozygous for either of the two traits. Because it is not possible to determine the genotype of a long-winged, gray-bodied fly by inspection, the genotype may be represented as $L\_\_ G\_\_$.

In a two-trait testcross, an individual with the dominant phenotype for both traits is crossed with an individual with the recessive phenotype for both traits because this fly has a *known* genotype. For example, a long-winged, gray-bodied fly is crossed with a short-winged, black-bodied fly. The heterozygous parent fly (*LlGg*) forms four different types of gametes. The homozygous parent fly (*llgg*) forms only one type of gamete:

P:	*LlGg*	×	*llgg*
Gametes:	*LG*		*lg*
	Lg		
	lG		
	lg		

As Figure 10.7 shows, all possible phenotypes occur among the offspring. This 1:1:1:1 phenotypic ratio shows that the *L__G__* fly is heterozygous for both traits and has the genotype *LlGg*. Such an individual is called a **dihybrid.** A Punnett square can also be used to predict the chances of an offspring having a particular phenotype. What are the chances of an offspring with long wings and a gray body? The chances are ¼, or 25%. What are the chances of an offspring with short wings and gray body? The chances are ¼, or 25%, and so forth.

If the *L__G__* fly had been homozygous for both traits, then no offspring would have short wings or a black body when it was crossed with one having the recessive phenotype. If the *L__G__* fly is heterozygous for one trait but not the other, can you predict the expected phenotypic ratio assuming the other fly is doubly recessive?

Mendel's Laws and Probability

When we use a Punnett square to calculate the results of genetic crosses, we assume that each gamete contains one allele for each trait (law of segregation) and that collectively the gametes have all possible combinations of alleles (law of independent assortment). Further, we assume that the male and female gametes combine at random—that is, all possible sperm have an equal chance to fertilize all possible eggs. Under these circumstances, it is possible to use the rules of probability to calculate the expected phenotypic ratio. The **rule of multiplication** says that the chance of two (or more) independent events occurring together is the product of their chances of occurring separately. For example, the chance of getting tails when you toss a coin is ½. The chance of getting two tails when you toss two coins at once is ½ × ½ = ¼. The more two-coin tosses you do, the more likely you'll see two tails in 25% of your total tosses.

Let's use the rule of multiplication to calculate the expected results in Figure 10.7. Because each allele pair separates independently, we can treat the cross as two separate one-trait crosses:

Ll × *ll*: Probability of *ll* offspring = ½
Gg × *gg*: Probability of *gg* offspring = ½

The probability of the offspring's genotype being *llgg*:

½ *ll* × ½ *gg* = ¼

And the same results are obtained for the other possible genotypes among the offspring in Figure 10.7.

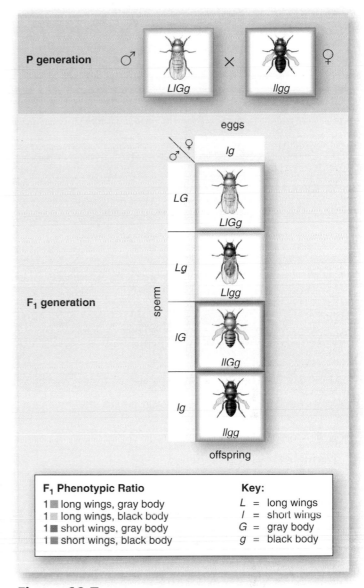

Figure 10.7 **Two-trait testcross.**

If a fly heterozygous for both traits is crossed with a fly that is recessive for both traits, the expected ratio of phenotypes is 1:1:1:1.

F₁ Phenotypic Ratio
1 ▪ long wings, gray body
1 ▪ long wings, black body
1 ▪ short wings, gray body
1 ▪ short wings, black body

Key:
L = long wings
l = short wings
G = gray body
g = black body

Check Your Progress

1. A black mouse with a bent tail reproduces with a yellow mouse with a straight tail. Of 42 offspring, all are black with straight tails. Which traits are dominant?
2. A black mouse with a straight tail reproduces with a yellow mouse with a bent tail. The phenotypic ratio among the offspring is 1:1:1:1. What is the genotype of the first mouse?
3. What would the result be if the test individual in Figure 10.7 were a. homozygous dominant for both traits? b. Homozygous dominant for one trait but heterozygous for the other?

Answers: 1. Black and straight tail. 2. Heterozygous for both traits. 3. a. All long winged, gray bodied flies. b. either 1 long wings, gray body : 1 long wings, black body or 1 long wings, black body : 1 short wings, gray body : 1 short wings, gray body.

Dominant   **Recessive**

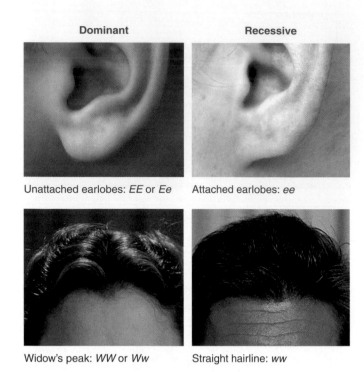

Unattached earlobes: *EE* or *Ee*   Attached earlobes: *ee*

Widow's peak: *WW* or *Ww*   Straight hairline: *ww*

Mendel's Laws and Meiosis

Today, we are aware that Mendel's laws relate to the process of meiosis. In Figure 10.8, a human cell has two pairs of homologues, recognized by length—one pair of homologues is short and the other is long. (The color signifies that we inherit chromosomes from our parents; one homologue of each pair is the "paternal" chromosome, and the other is the "maternal" chromosome.) When the homologues separate (segregate), each gamete receives one member from each pair. The homologues separate (assort) independently; it does not matter which member of each pair goes into which gamete. In the simplest of terms, a gamete in Figure 10.8 can receive one short and one long chromosome of either color. Therefore, all possible combinations of chromosomes are in the gametes.

The alleles *E* for unattached earlobes and *e* for attached earlobes are on one pair of homologues, and the alleles *W* for widow's peak and *w* for straight hairline are on the other pair of homologues. Because there are no restrictions as to which homologue goes into which gamete, a gamete can receive either an *E* or an *e* and either a *W* or a *w* in any combination. In the end, collectively, the gametes will have all possible combinations of alleles.

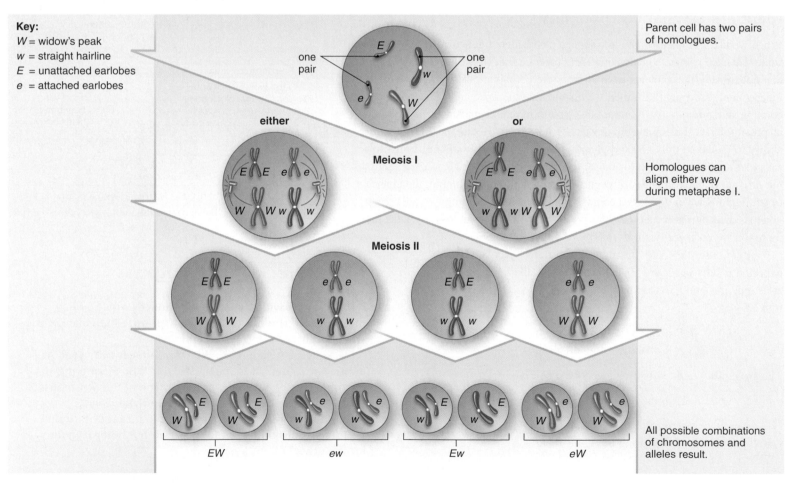

Key:
W = widow's peak
w = straight hairline
E = unattached earlobes
e = attached earlobes

Parent cell has two pairs of homologues.

either   **or**

Meiosis I

Homologues can align either way during metaphase I.

Meiosis II

All possible combinations of chromosomes and alleles result.

EW   *ew*   *Ew*   *eW*

Figure 10.8   Mendel's laws and meiosis.

A human cell has 23 pairs of homologous chromosomes (homologues), of which two pairs are represented here. The homologues, and the alleles they carry, segregate independently during gamete formation. Therefore, all possible combinations of chromosomes and alleles occur in the gametes.

10.2 Beyond Mendel's Laws

Since Mendel's day, variations in the dominant/recessive relationship he described have been discovered. These variations make it clear that the concept of the genotype should be expanded to include all the genes of an individual. The idea of just two alleles per trait, one dominant and one recessive, is too restrictive because various genes work together to bring about a phenotype. In addition, the phenotype is sometimes influenced by the environment.

Incomplete Dominance

Incomplete dominance is exhibited when the heterozygote has an intermediate phenotype between that of either homozygote. In the flowering plant known as the four-o'clock, a cross between red-flowered four-o'clocks and white-flowered four-o'clocks produces offspring with pink flowers (Fig. 10.9). But this is not an example of blending inheritance, because when the pink-flowered plants self-pollinate, ¼ of the offspring have red flowers, ¼ have white flowers, and the rest have pink flowers. The reappearance of the original phenotypes makes it clear that we are still dealing with particulate inheritance of the type described by Mendel.

In humans, people with curly hair have the homozygous recessive genotype, while those with straight hair have the homozygous dominant condition. The heterozygote has wavy hair. We can explain incomplete dominance by assuming that only the dominant allele codes for a gene product and that the single dose of the product gives the intermediate result.

Multiple-Allele Traits

In ABO blood group inheritance, there are three alleles that determine the presence or absence of antigens on red blood cells and therefore blood type:

$$I^A = \text{A antigen on red blood cells}$$
$$I^B = \text{B antigen on red blood cells}$$
$$i = \text{Neither A nor B antigen on red blood cells}$$

Each person has only two of the three possible alleles, and both I^A and I^B are dominant over i. Therefore, there are two possible genotypes for type A blood ($I^A I^A$, $I^A i$) and two possible genotypes for type B blood ($I^B I^B$, $I^B i$). But, I^A and I^B are fully expressed in the presence of each other. Therefore, if a person inherits one of each of these alleles, that person will have type AB blood. Type O blood can only result from the inheritance of two i alleles. Figure 10.10 shows that matings between certain genotypes can have surprising results in terms of blood type.

Notice that human blood type inheritance is an example of **codominance,** another type of inheritance that differs from Mendel's findings because more than one allele is fully expressed. When an individual has blood type AB, both A and B antigens appear on the red blood cells. The two different capital letters signify that both alleles are coding for an antigen.

$C^R C^R$ $C^R C^W$ $C^W C^W$

Figure 10.9 Incomplete dominance in four-o'clocks.

In incomplete dominance, the heterozygote is intermediate between the homozygotes. In this case, the heterozygote is pink, whereas the homozygotes are red or white.

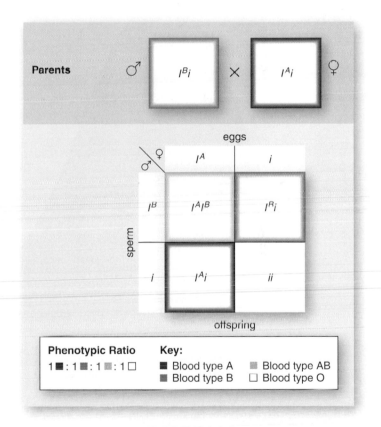

Figure 10.10 Inheritance of ABO blood type.

A mating between individuals with type A blood and type B blood can result in any one of the four blood types. Why? Because the parents are $I^B i$ and $I^A i$. The i allele is recessive; both I^A and I^B are dominant.

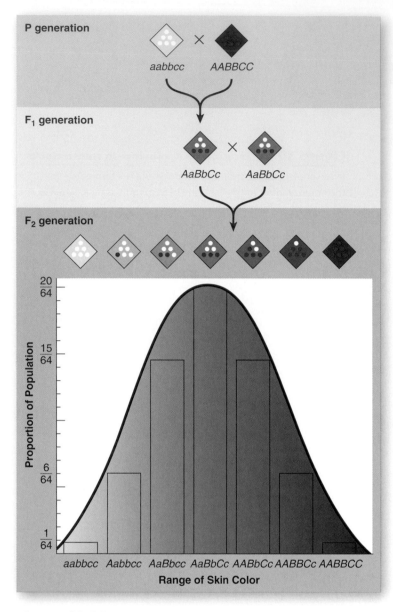

Figure 10.11   **Skin color.**

In this model, skin color is controlled by three pairs of alleles. When two F_1 individuals with genotypes *AaBbCc* are crossed, seven phenotypes are seen among the F_2 generation. If environmental effects are considered, a bell-shaped curve results because of many intervening phenotypes.

Figure 10.12   **Coat color in Himalayan rabbits.**

Hair growing under an ice pack in these rabbits is black. The dark color on the ears, nose, and feet of these rabbits is believed to be due to a lower body temperature in these areas.

Polygenic Inheritance

Polygenic inheritance occurs when a trait is governed by two or more sets of alleles. The individual has a copy of all allelic pairs, possibly located on many different pairs of chromosomes. Each dominant allele has a quantitative effect on the phenotype, and these effects are additive. The result is a continuous variation of phenotypes, resulting in a distribution that resembles a bell-shaped curve. The more genes involved, the more continuous are the variations and distribution of the phenotypes. Also, environmental effects are involved; in the case of skin color, differences in sun exposure bring about a bell-shaped curve (Fig. 10.11).

Multifactorial traits are ones that are controlled by polygenes subject to environmental influences. Disorders, such as cleft lip and/or palate, clubfoot, congenital dislocations of the hip, hypertension, diabetes, schizophrenia, and even allergies and cancers are likely due to the combined action of many genes plus environmental influences. In recent years, reports have surfaced that all sorts of behaviors, including alcoholism, phobias, and even suicide, can be associated with particular genes. No doubt, behavioral traits are somewhat controlled by genes, but again, it is impossible at this time to determine to what degree. And very few scientists would support the idea that these behavioral traits are predetermined by our genes.

Environment and the Phenotype

The relative importance of genetic and environmental influences on the phenotype can vary, but in some instances the environment seems to have an extreme effect. In the water buttercup, the submerged part of the plant has a different appearance from the part above water. This difference is thought to be related to a difference in water intake by the cells.

Temperature can also have a dramatic effect on the phenotypes of plants and animals. Primroses have white flowers when grown above 32°C and red flowers when grown at 24°C. The coats of Siamese cats and Himalayan rabbits are darker in color at the ears, nose, paws, and tail. Himalayan rabbits are known to be homozygous for the allele *ch*, which is involved in the production of melanin. Experimental evidence suggests that the enzyme encoded by this gene is active only at a low temperature and that, therefore, black fur only occurs at the extremities where body heat is lost to the environment (Fig. 10.12). When the animal is placed in a warmer environment, new fur on these body parts is light in color.

These examples lend support to the belief that human traits controlled by polygenes are also subject to environmental influences. Therefore, many investigators are trying to determine what percentage of various traits is due to nature (inheritance) and what percentage is due to nurture (the environment). Some studies use twins separated from birth, because if identical twins in different environments share the same trait, that trait is most likely inherited. Identical twins are more similar in their intellectual talents, personality traits, and levels of lifelong happiness than are fraternal twins separated from birth. Biologists conclude that all behavioral traits are partly heritable and that genes exert their effects by acting together in complex combinations susceptible to environmental influences.

Pleiotropy

Pleiotropy occurs when a single gene has more than one effect. Often, this leads to a *syndrome*, a group of symptoms that appear together and indicate the presence of a particular genetic mutation. For example, persons with Marfan syndrome have disproportionately long arms, legs, hands, and feet; a weakened aorta; and poor eyesight (Fig. 10.13). All of these characteristics are due to the production of abnormal connective tissue. Marfan syndrome has been linked to a mutated gene (*FBN₁*) on chromosome 15 that ordinarily specifies a functional protein called fibrillin. Fibrillin is essential for the formation of elastic fibers in connective tissue. Without the structural support of normal connective tissue, the aorta can burst, particularly if the person is engaged in a strenuous sport, such as volleyball or basketball. Flo Hyman may have been the best American woman volleyball player ever, but she fell to the floor and died when only 31 years old because her aorta gave way during a game. Now that coaches are aware of Marfan syndrome, they are on the lookout for it among very tall basketball players. Chris Weisheit, whose career was cut short after he was diagnosed with Marfan syndrome, said, "I don't want to die playing basketball."

Many other disorders, including sickle cell disease and porphyria, are examples of pleiotropic traits. Porphyria is caused by a chemical insufficiency in the production of hemoglobin, the pigment that makes red blood cells red. The symptoms of porphyria are photosensitivity, strong abdominal pain, port-wine-colored urine, and paralysis in the arms and legs. Many members of the British royal family in the late 1700s and early 1800s suffered from this disorder, which can lead to epileptic convulsions, bizarre behavior, and coma.

Check Your Progress

1. Polygenic inheritance is controlled by three pairs of alleles. What are the two extreme genotypes?
2. What's the difference in the pattern of inheritance for polygenic inheritance and pleiotropy?
3. If both parents have type AB blood, no child would have what blood type?

Answers: 1. AABBCC and aabbcc. 2. Several genes affect one trait in polygenic inheritance, while one gene has several phenotypic effects in pleiotropy. 3. Type O blood.

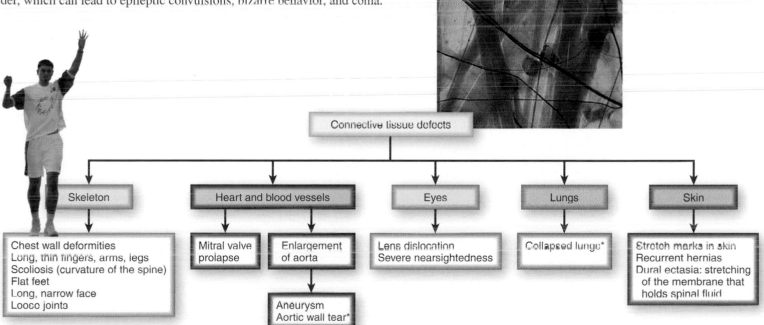

Figure 10.13 **Marfan syndrome, multiple effects of a single human gene.**

Individuals with Marfan syndrome exhibit defects in the connective tissue throughout the body. Important changes occur in the skeleton, heart, blood vessels, eyes, lungs, and skin. All these conditions are due to mutation of the gene *FBN₁*, which codes for a constituent of connective tissue. *Most life-threatening.

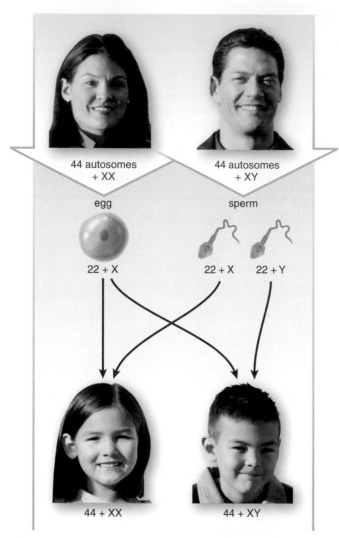

Figure 10.14 **Inheritance of gender in human beings.**

Sperm determine the gender of an offspring because sperm can carry an X or a Y chromosome. Males inherit a Y chromosome.

10.3 Sex-Linked Inheritance

Geneticists of the early twentieth century were convinced that the genes are on the chromosomes because the genes and chromosomes behave similarly during gamete formation (see Fig. 10.8). They also knew that the chromosomes differ between the sexes. As you learned in Chapter 9, the sex chromosomes in females are XX, and those in males are XY (Fig. 10.14). Notice that males produce two different types of gametes—those that contain an X and those that contain a Y. Therefore, the contribution of the male determines the sex of the new individual.

The much shorter Y chromosome contains only about 26 genes, and most of these genes are concerned with sex differences between men and women. One of the genes on the Y chromosome, *SRY* does not have a copy on the X chromosome. If the functional *SRY* gene is present, the individual becomes a male, and if it is absent, the individual becomes a female. No genes determine femaleness; it is the "default setting."

In addition to genes that determine sex, the X chromosome carries genes for traits that have nothing to do with the gender of the individual. By tradition, the term **X-linked** refers to such genes carried on the X chromosome. The Y chromosome does not carry these genes, and that makes for an interesting inheritance pattern.

X-Linked Alleles

We have already mentioned that the fruit fly is a favorite subject for genetic studies. Flies can be easily and inexpensively raised in simple laboratory glassware; females mate only once and then lay hundreds of eggs during their lifetimes; and the generation time is short, taking only about ten days when conditions are favorable.

Drosophila flies have the same sex chromosome pattern as humans, which facilitates our understanding of a cross performed by early geneticists. When a mutant male with white eyes was crossed with a red-eyed female, the F_1 all had red eyes:

	♀		♂
P	red-eyed	×	white-eyed
F_1	red-eyed		red-eyed

From these results, researchers knew that red eyes are the dominant characteristic and white eyes are the recessive characteristic. In the F_2 generation, the expected 3:1 ratio resulted, but all of the white-eyed flies were males:

	♀		♂
$F_1 \times F_1$	red-eyed	×	red-eyed
F_2	red-eyed		1 red-eyed : 1 white-eyed

Obviously, a major difference between the male flies and the female flies was their sex chromosomes. Could it be possible that an allele for eye color was on the Y chromosome but not on the X? This idea was quickly discarded because usually females have red eyes, and they have no Y chromosome. Next investigators hypothesized that perhaps an allele for eye color was on the X, but not on

Chapter 10 Patterns of Inheritance **153**

the Y chromosome, and this explanation turned out to match the results obtained in the experiment (Fig. 10.15). These results support the chromosome theory of inheritance by showing that the behavior of a specific allele corresponds exactly with that of a specific chromosome—the X chromosome in *Drosophila*.

Notice that X-linked alleles have a different pattern of inheritance than alleles on the autosomes. The Y chromosome is blank for these alleles, and so the inheritance of a Y chromosome cannot offset the inheritance of an X-linked recessive allele. For the same reason, *males always receive an X-linked recessive mutant allele from their female parent;* they receive the Y chromosome, which is blank for the allele, from the male parent.

An X-Linked Problem

When solving autosomal genetics problems involving fruit flies, the alleles and genotypes are represented as follows:

Key: L = long wings   Genotypes: *LL, Ll, ll*
l = short wings

As noted in Figure 10.15, however, an X-linked gene is represented by attaching the allele to an X:

Key: X^R – red eyes
X^r = white eyes

The possible genotypes in both males and females are as follows:

Genotypes: $X^R X^R$ = red-eyed female
$X^R X^r$ = red-eyed female
$X^r X^r$ – white-eyed female
$X^R Y$ = red-eyed male
$X^r Y$ = white-eyed male

Notice that three genotypes are possible for females, but only two are possible for males. Females can be heterozygous $X^R X^r$, in which case they are carriers. **Carriers** usually do not exhibit a recessive trait, but they are capable of passing on a recessive allele for a trait. Males cannot be carriers; if the dominant allele is on the single X chromosome, they show the dominant phenotype, and if the recessive allele is on the single X chromosome, they show the recessive phenotype.

Males have white eyes when they receive the mutant recessive allele from the female parent. Females can only have white eyes when they receive a recessive allele from both parents.

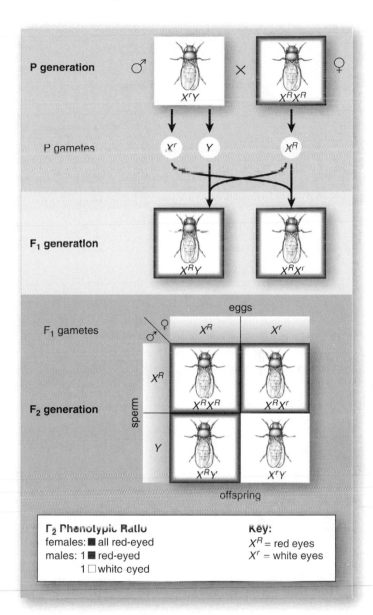

Figure 10.15 X-linked inheritance.

Once researchers deduced that the allele for red/white eye color is only on the X chromosome in *Drosophila*, they were able to explain their experimental results. Males with white eyes in the F_2 inherit the recessive allele only from the female parent; they receive a blank Y chromosome from the male parent.

Check Your Progress

1. Why do males produce two different gametes with respect to sex chromosomes?
2. Why are sex-linked recessive conditions more likely in males than females?

Answers: 1. Being XY, ½ of the gametes carry an X and ½ carry a Y. 2. Only the X carries the alleles (Y is blank), and therefore recessive alleles are always expressed in males.

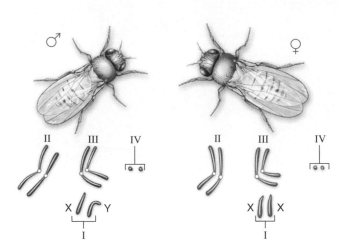

Figure 10.16 **Chromosomes of *Drosophila*.**

The DNA of the *Drosophila* chromosomes contains about 165 million base pairs and an estimated 14,000 genes. However, *Drosophila* has only four pairs of chromosomes: Three pairs are autosomes (II-IV) and one pair of sex chromosomes (I). Therefore, many alleles must be on each chromosome.

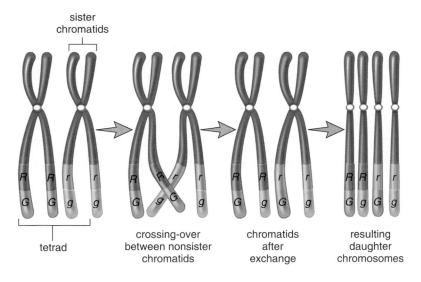

Figure 10.17 **Crossing-over.**

When homologous chromosomes are in synapsis, the nonsister chromatids exchange genetic material. Following crossing-over, recombinant chromosomes occur. Recombinant chromosomes are found in recombinant gametes.

10.4 Inheritance of Linked Genes

After completing a great number of *Drosophila* crosses, investigators discovered many more mutants and were able to perform various two-trait crosses. However, they did not always get the expected ratios among the offspring due to **gene linkage,** the existence of several alleles on the same chromosome. The alleles on the same chromosome form a **linkage group** because these alleles tend to be inherited together.

Drosophila has thousands of different genes controlling all aspects of its structure, biochemistry, and behavior. Yet it has only four pairs of chromosomes (Fig. 10.16). This paradox alone allows us to reason that each chromosome must carry a large number of genes. For example, it is now known that genes controlling eye color, wing type, body color, leg length, and antennae type are all located on chromosome II, in that sequence. When investigators construct a **chromosome map,** it shows the relative distance between the gene loci on a chromosome.

Constructing a Chromosome Map

Crossing-over, you'll recall, occurs between nonsister chromatids when homologous pairs of chromosomes pair up prior to separation during meiosis. During crossing-over, the nonsister chromatids exchange genetic material, and therefore genes. If crossing-over occurs between two linked alleles of a dihybrid, four types of gametes instead of two are produced (Fig. 10.17). **Recombinant gametes** have a new combination of alleles. The recombinant gametes occur in reduced numbers because crossing-over is infrequent.

To take an example, suppose you are crossing a gray-bodied, red-eyed heterozygous fly with a black-bodied, purple-eyed fly. Because the alleles governing these traits are both on chromosome II, you predict that the results will be 1:1 instead of 1:1:1:1, as is the case for unlinked genes. The reason, of course, is that linked alleles tend to stay together and do not separate independently, as predicted by Mendel's laws. Under these circumstances, the heterozygote forms only two types of gametes and produces offspring with only two phenotypes (Fig. 10.18*a*).

When you perform the cross, however, you find that a very small number of offspring show recombinant phenotypes (i.e., those that are different from the original parents). Specifically, you find that 47% of the offspring have black bodies and purple eyes, 47% have gray bodies and red eyes, 3% have black bodies and red eyes, and 3% have gray bodies and purple eyes (Fig. 10.18*b*). What happened? In this instance, crossing-over led to a very small number of recombinant gametes, and when these were fertilized, recombinant phenotypes were observed in the offspring.

Linkage Data

It stands to reason that the closer together two genes are, the less likely they are to cross over. Numerous experiments have repeatedly shown that this is the case. Therefore, you can use the percentage of recombinant phenotypes to map the chromosomes, because there is a direct relationship between the frequency of crossing-over (the percentage of recombinant phenotypes) and the distance between alleles. In our example (Fig. 10.18), a total of 6% of the offspring are recombinants, and this means the rate of crossing-over is 6%. For

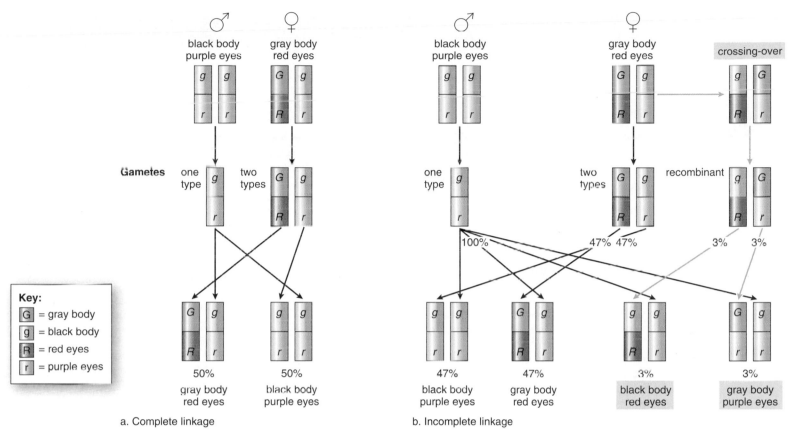

a. Complete linkage

b. Incomplete linkage

Figure 10.18 Complete linkage versus incomplete linkage.

a. Hypothetical cross in which genes for body and eye color on chromosome II of *Drosophila* are completely linked. Instead of the expected 1:1:1:1 ratio among the offspring from this cross, there would be a 1:1 ratio. **b.** Crossing-over occurs, and 6% of the offspring show recombinant phenotypes. The percentage of recombinant phenotypes is used to map the chromosomes (1% = 1 map unit).

the sake of mapping the chromosomes, it is assumed that 1% of crossing-over equals one map unit. Therefore, the allele for black body and the allele for purple eyes are six map units apart.

Suppose you want to determine the order of any three genes on the chromosomes. To do so, you can perform crosses that tell you the map distance between all three pairs of alleles. Assume, for instance, that:

1. the distance between the black-body and purple-eye alleles = 6 map units;
2. the distance between the purple-eye and vestigial-wing alleles = 12.5 map units; and
3. the distance between the black-body and vestigial-wing alleles = 18.5 map units.

Therefore, the order of the alleles must be as shown in Figure 10.9. Black body must be 6 map units away from purple eyes, and purple eyes must be 12.5 map units away from vestigial wings.

Linkage data have been used to map the chromosomes of *Drosophila*, but the possibility of using linkage data to map human chromosomes is limited because we can work only with matings that occur by chance. This, coupled with the fact that humans tend not to have numerous offspring, means that additional methods are needed to sequence the genes on human chromosomes. Today, it is customary also to rely on biochemical methods to map the human chromosomes, as will be discussed in Chapter 13.

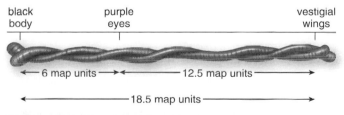

Figure 10.19 Mapping chromosomes.

Genes are arranged linearly on the chromosome at specific gene loci. Using the crossing over data supplied by the text, these genes are sequenced as shown.

THE CHAPTER IN REVIEW

Summary

10.1 Mendel's Laws

In 1860, Gregor Mendel, an Austrian monk, developed two laws of heredity based on crosses utilizing the garden pea.

Law of Segregation
Mendel's law of segregation states the following:

- Each individual has two factors for each trait.
- The factors segregate (separate) during the formation of the gametes.
- Each gamete contains only one factor from each pair of factors.
- Fertilization gives each new individual two factors for each trait.

In the context of genetics today,

- Genes are on the chromosomes; each gene has two alternative forms, called alleles. Letters are used to represent the genotype of an individual. *AA* = homozygous dominant, *Aa* = heterozygous, and *aa* = homozygous recessive.
- Homologues separate during meiosis and the gametes have only one allele for each trait; either an *A* or an *a*.
- Fertilization gives each new individual two alleles for each trait.

Law of Independent Assortment
Mendel's law of independent assortment states the following:

- Each pair of factors segregates (assorts) independently of the other pairs.
- All possible combinations of factors can occur in the gametes.

In the context of genetics today,

- Each pair of homologues separate independently of the other pairs.
- All possible combinations of chromosomes and their alleles occur in the gametes.
- Mendel's laws are consistent with the manner in which homologues and their alleles separate during meiosis.

Common Autosomal Genetic Crosses
A Punnett square allows all types of sperm to fertilize all types of eggs and gives these results

Aa × *Aa*	3:1 phenotypic ratio
Aa × *aa*	1:1 phenotypic ratio
AaBb × *AaBb*	9:3:3:1 phenotypic ratio
AaBb × *aabb*	1:1:1:1 phenotypic ratio

10.2 Beyond Mendel's Laws

In some patterns of inheritance, the alleles are not just dominant or recessive.

Incomplete Dominance
In incomplete dominance, the heterozygote is intermediate between the two homozygotes. For example, the offspring of red × white four-o'clocks produce pink flowers. The red and white phenotypes reappear when pink four o'clocks are crossed.

Multiple Alleles   The multiple-allele inheritance pattern is exemplified in humans by blood type inheritance. Every individual has two out of three possible alleles: I^A, I^B, i. Both I^A and I^B are expressed; therefore, this is also a case of codominance.

Polygenic Inheritance   In polygenic inheritance a trait is controlled by more than one set of alleles. The dominant alleles have an additive effect on the phenotype.

Pleiotropy   In pleiotropy, one gene (consisting of two alleles) has multiple effects on the body. For example, all the disorders common to Marfan syndrome are due to a mutation that leads to a defect in the composition of connective tissue.

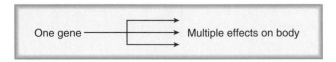

10.3 Sex-Linked Inheritance

Males produce two different types of gametes—those that contain an X and those that contain a Y. The contribution of the male determines the gender of the new individual. XX = female   XY = male

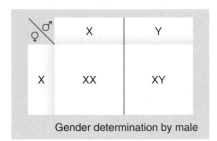

Gender determination by male

Certain alleles are carried on the X chromosome, but the Y is blank. Therefore, a male only needs to inherit one recessive allele on the X chromosome to have a recessive genetic disorder.

Common X-Linked Genetic Crosses
$X^B X^b$ × $X^B Y$ All daughters will be normal, even though they have 50% chance of being carriers, but sons have a 50% chance of being color blind.
$X^B X^B$ × $X^b Y$ All children are normal (daughters will be carriers).

10.4 Inheritance of Linked Genes

All the alleles on a particular chromosome form a linkage group. Crossing-over data can be used to construct a chromosome map, which shows the sequence of alleles along the chromosome.

Thinking Scientifically

1. In peas, genes C and P are required for pigment production in flowers. Gene C codes for an enzyme that converts a compound into a colorless intermediate product. Gene P codes for an enzyme that converts the colorless intermediate product into anthocyanin, a purple pigment. A flower, therefore, will be purple only if it contains at least one dominant allele for each of the two genes (C_ P_). Flowers are white if they do not produce anthocyanin (ccpp). **a.** What phenotypic ratio would you expect in the F_2 generation following a cross between two double heterozygotes (CcPp)? **b.** What phenotypic ratio would you expect if a double heterozygote was crossed with a plant that is homozygous recessive (ccpp) for both genes?

2. Geneticists often look for unusual events to provide insight into genetic mechanisms. In one such instance, researchers studied XX men and XY women. They found that the XX men contained a chromosomal segment normally found in men but not in women, while XY women were missing that region. What gene do you suppose is on that chromosome piece? How did XX men gain that piece and XY women lose it?

Testing Yourself

Choose the best answer for each question.

1. In Mendel's particulate theory of inheritance, the "particles" are
   a. chromosomes.
   b. genes.
   c. plants.
   d. pollen grains.

2. The offspring ratio from a testcross (F_1 × homozygous recessive) should be
   a. all dominant.
   b. ¾ dominant: ¼ recessive.
   c. ½ dominant: ½ recessive.
   d. all recessive.

3. Which of the following is not a component of the law of segregation?
   a. Each gamete contains one factor from each pair of factors in the parent.
   b. Factors segregate during gamete formation.
   c. Following fertilization, the new individual carries two factors for each trait.
   d. Each individual has one factor for each trait.

4. When using a Punnett square to predict offspring ratios, we assume that
   a. each gamete contains one allele of each gene.
   b. the gametes have all possible combinations of alleles.
   c. male and female gametes combine at random
   d. All of these are correct.

5. If you cross a black spaniel with a red spaniel and get a litter of 8 black and 1 red, what is the genotype of the black parent?
   a. BB
   b. Bb
   c. bb
   d. impossible to determine

6. Short hair is dominant over long hair in dogs. If a short-haired dog whose mother was long-haired is crossed with a long-haired dog, what proportion of the offspring will be short-haired?
   a. 25%
   b. 50%
   c. 75%
   d. 100%
   e. impossible to determine

7. When one physical trait is affected by two or more pairs of alleles, the condition is called
   a. incomplete dominance.
   b. codominance.
   c. homozygous dominant.
   d. multiple allele.
   e. polygenic inheritance.

8. Fill in the blank spaces in the following Punnett square.

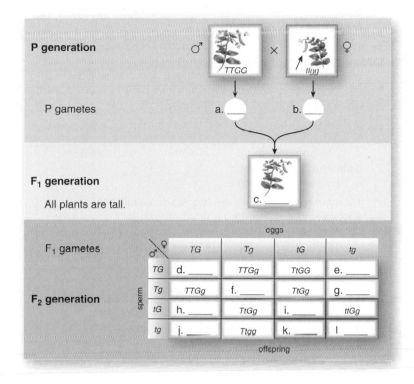

For questions 9–11, consider that coat color and spotting pattern in cocker spaniels depend on two genes. Black (B) is dominant to red, and solid color (S) is dominant to spotted.

9. What phenotypic ratio do you expect in the F_2 generation?
   a. 1 black solid : 1 red solid : 1 black spotted : 1 red spotted
   b. 9 black solid : 3 red solid : 3 black spotted : 1 red spotted
   c. 1 black solid : 3 red solid : 3 black spotted : 9 red spotted
   d. all black solid

10. If you cross a black spotted dog with a black solid dog and get a ratio of 3 black solid : 3 black spotted : 1 red solid : 1 red spotted, what is the genotype of black spotted parent?
    a. BBss
    b. BbSs
    c. Bbss
    d. bbSs
    e. bbss

11. What is the genotype of black solid parent in question 10?
    a. BBSS
    b. BbSS
    c. BbSs
    d. Bbss
    e. bbSs

For questions 12–15, match the cross with the results in the key.

Key:

  a. 3:1 c. 1:1
  b. 9:3:3:1 d. 1:1:1:1

12. $TtYy \times TtYy$

13. $Tt \times Tt$

14. $Tt \times tt$

15. $TtYy \times ttyy$

16. When two monohybrid round squashes are crossed, the offspring ratio is ¼ flat : ½ oblong : ¼ round. Squash shape, therefore, is controlled by incomplete dominance. What offspring ratio would you expect from a cross between a plant with oblong fruit and one with round fruit?

   a. all oblong
   b. all round
   c. ¾ oblong : ¼ round
   d. ¾ round : ¼ oblong
   e. ½ oblong : ½ round

17. If a man of blood group AB marries a woman of blood group A whose father was type O, what phenotypes could their children be?

   a. A only d. A, AB, and B
   b. A, AB, B, and O e. O only
   c. AB only

18. Anemia sometimes results from a mutation in a single gene, causing the blood's oxygen supply to be inadequate. A homozygous recessive person has a number of problems, including lack of energy, fatigue, rapid pulse, pounding heart, and swollen ankles. This is an example of

   a. pleiotropy.
   b. sex-linked inheritance.
   c. incomplete dominance.
   d. polygenic inheritance.
   e. codominance.

For questions 19–22, list the progeny phenotypes from the following key that would result from each of the crosses in *Drosophila.* Red eye color is dominant over white. The gene for eye color is on the X chromosome. Answers can be used more than once.

Key:

   a. red-eyed female
   b. red-eyed male
   c. white-eyed female
   d. white-eyed male

19. Homozygous red-eyed female × white-eyed male.

20. Heterozygous female × white-eyed male.

21. White-eyed female × white-eyed male.

22. Heterozygous female × red-eyed male.

23. Alice and Henry are at the opposite extremes for a polygenic trait. Their children will

   a. be bell-shaped.
   b. be a phenotype typical of a 9:3:3:1 ratio.
   c. have the middle phenotype between their two parents.
   d. look like one parent or the other.

Go to www.mhhe.com/maderessentials for more quiz questions.

Bioethical Issue

You may feel it is ethically wrong to choose which particular embryo can continue development following in vitro fertilization (see page 521). But, what about choosing whether an X-bearing or Y-bearing sperm should fertilize the egg? As you know, the sex of a child depends upon whether an X-bearing sperm or a Y-bearing sperm enters the egg. A new technique has been developed that can separate X-bearing sperm from Y-bearing sperm. First, the sperm are dosed with a DNA-staining chemical. Because the X chromosome has slightly more DNA than the Y chromosome, it takes up more dye. When a laser beam shines on the sperm, the X-bearing sperm shine more brightly than the Y-bearing sperm. A machine sorts the sperm into two groups on this basis. The results are not perfect. Following artificial insemination, there's about an 85% success rate for a girl and about a 65% success rate for a boy.

Some might believe that this is the simplest way to make sure they have a healthy child if the mother is a carrier of an X-linked genetic disorder, such as hemophilia or Duchenne muscular dystrophy. Previously, a pregnant woman with these concerns had to wait for the results of an amniocentesis test and then decide whether to abort the pregnancy if it were a boy. Is it better to increase the chances of a girl to begin with?

Or, do you believe that gender selection is not acceptable for any reason. Even if it doesn't lead to a society with far more members of one sex than another, there could be problems. Once you separate reproduction from the sex act, it might open the door to genetically designing children. On the other hand, is it acceptable to bring a child into the world with a genetic disorder that may cause an early death or lifelong disability? Would it be better to select sperm for a girl, who at worst would be a carrier like her mother?

Understanding the Terms

Match the terms to these definitions:

a. _____ Alternate forms of a gene.

b. _____ Organism that contains two identical alleles.

c. _____ Physical appearance of an individual.

d. _____ Phenotype that is most common in nature.

e. _____ Probability that two or more independent events will happen together is the product of the probabilities that they will occur separately.

f. _____ Genetic system in which the heterozygote is intermediate in phenotype compared to the homozygotes.

g. _____ Genes on the same chromosome.

DNA Biology and Technology

Every type of tissue cell in your body contains a complete copy of your body's DNA.

DNA can be transferred from one kind of species to another. These pigs glow from having received jellyfish bioluminescent genes as embryos.

One recent advance in DNA technology is the production of transgenic animals—animals that contain genes from another species. For example, researchers have created a variety of miniature pigs for the purpose of providing organs for human transplants. A unique feature of these pigs is a yellow snout and hooves due to a gene introduced from jellyfish! This makes the pigs easy to identify.

▶ ▶ ▶

While keeping track of its progress (screen), this machine can copy just a segment of your DNA over and over again until there are millions of copies.

An alternative source of transplant organs is needed because human organs are in limited supply, and many people die each year waiting for transplants. If organs from a normal pig are transplanted into a human, two major problems occur: The organs from pigs are too large for humans and also the organs are quickly rejected. However, transplants from miniature genetically altered pigs avoid both of these complications. First, the miniature size of the pigs makes their organs more appropriate for humans. Second, rejection is avoided because the genes coding for plasma membrane proteins that normally trigger rejection have been "knocked out"—that is, they are not expressed. Pig organs may eventually become a life-saving alternative for people who are unable to acquire human organ transplants.

In this chapter, you will learn about the structure and function of DNA. This knowledge will allow you to appreciate the incredible advances in the field of DNA technology.

11.1 DNA and RNA Structure and Function

Mendel knew nothing about DNA. It took many years for investigators to come to the conclusion that Mendel's factors, now called genes, are on the chromosomes. Then, researchers wanted to show the genes consisted of DNA and not proteins. One experiment involved the use of a virus that attacks bacteria such as E. coli (Fig. 11.1). A virus is composed of an outer capsid made of protein and an inner core of DNA. The use of radioactive tracers showed that DNA and not protein enters bacteria and directs the formation of new viruses. By the early 1950s, investigators not only knew that genes are composed of DNA, they knew that mutated genes result in errors of metabolism. Therefore, DNA in some way must control the functioning of proteins in the cell.

Even though DNA took its name—deoxyribonucleic acid—from the chemical components of its nucleotides, its detailed structure was still to be determined. Finding the structure of DNA was the first step toward understanding how DNA is able to do the following:

- Be variable in order to account for species differences.
- Replicate so that every cell gets a copy during cell division.
- Store information needed to control the cell.
- Undergo mutations, accounting for evolution of new species.

Structure of DNA

Once researchers knew that the genes are composed of DNA, they were racing against time and each other to determine the structure of DNA. They believed that whoever discovered it first would get a Nobel Prize. How James Watson and Francis Crick determined the structure of DNA (and got the Nobel Prize) resembles a mystery, in which each clue was added to the total picture until the breathtaking design of DNA—a double helix—was finally revealed. To achieve this success, Watson and Crick particularly relied on studies done by Erwin Chargaff and Rosalind Franklin.

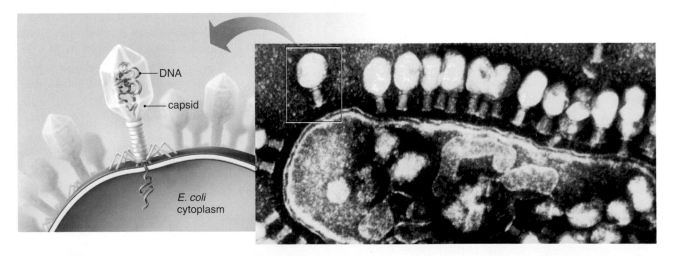

Figure 11.1 **The genes are composed of DNA.**

An early experiment by Alfred Hershey and Martha Chase helped determine that DNA is the genetic material. Their experiment involved a virus which infects bacteria such as E. coli. They wanted to know which part of the virus—the capsid made of protein or the DNA inside the capsid—entered the bacterium. Radioactive tracers showed that DNA, not protein, enters the bacterium and guides the formation of new viruses. Therefore, DNA must be the genetic material.

Chargaff's Rules

Before Erwin Chargaff began his work, it was known that DNA contains four different types of nucleotides based on their nitrogen-containing bases (Fig. 11.2). The bases **adenine (A)** and **guanine (G)** are purines with a double ring, and the bases **thymine (T)** and **cytosine (C)** are pyrimidines with a single ring. With the development of new chemical techniques in the 1940s, Chargaff decided to analyze in detail the base content of DNA.

In contrast to accepted beliefs, Chargaff found that each species has its own percentages of each type of nucleotide. For example, in a human cell, 31% of bases are adenine; 31% are thymine; 19% are guanine; and 19% are cytosine. In all the species Chargaff studied, the amount of A always equaled the amount of T, and the amount of G always equaled the amount of C. These relationships are called *Chargaff's rules*:

1. The amount of A, T, G, and C in DNA varies from species to species.
2. In each species, the amount of A = T and the amount of G = C.

Chargaff's data suggest that DNA has a means to be stable in that A can only pair with T and G can only pair with C. His data also show that DNA can be variable as required for the genetic material. The paired bases may occur in any order, and the amount of variability in their sequences is overwhelming. For example, suppose a chromosome contains 140 million base pairs. Since any of the four possible nucleotide pairs can be present at each pair location, the total number of possible nucleotide pair sequences is $4^{140 \times 10^6}$ or $4^{140,000,000}$.

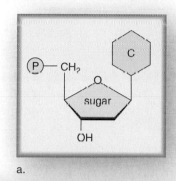

a.

Phosphate

b.

Sugars

deoxyribose
(DNA only)

ribose
(RNA only)

c.

Nitrogen-containing bases

Adenine (A)

Guanine (G)

Purines

Cytosine (C)

Thymine (T)
(DNA only)

Uracil (U)
(RNA only)

Pyrimidines

d.

Figure 11.2 Nucleotide composition of DNA and RNA.

a. All nucleotides contain phosphate, a 5-carbon sugar, and a nitrogen-containing base, such as cytosine (C). **b.** Structure of phosphate. **c.** In DNA, the sugar is deoxyribose; in RNA, the sugar is ribose. **d.** In DNA the nitrogen-containing bases are adenine, guanine, cytosine, and thymine; in RNA the bases are adenine, guanine, cytosine, and uracil.

Check Your Progress

1. Of what benefit was it to early investigators to know that species differ in the amount of A, T, G, and C in their DNA?
2. Of what benefit was it to know that the amount of A = T and the amount of G = C?

Answers: 1. The fact that DNA has a way to be variable made it the most likely candidate to be the genetic material. 2. This pattern suggested that DNA has the stability to be the genetic material.

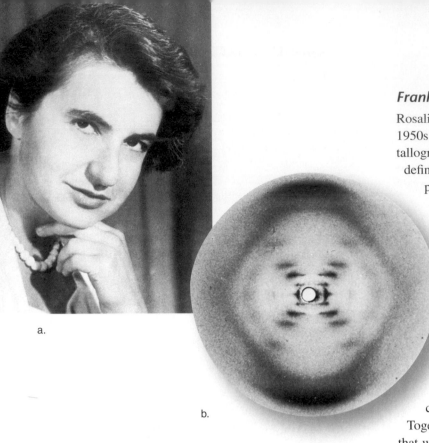

a.

b.

Figure 11.3 X-ray diffraction pattern of DNA.

a. Rosalind Franklin X-rayed DNA using crystallography techniques.
b. The pattern that resulted indicated that DNA is a double helix
(see X pattern in the center) and that some part of the molecule
is repeated over and over again (see the dark portions at top and
bottom). Watson and Crick determined that this repeating feature
was the paired bases.

Franklin's X-Ray Diffraction Data

Rosalind Franklin was a researcher at King's College in London in the early
1950s (Fig. 11.3*a*). She was studying the structure of DNA using X-ray crystallography. When a crystal (a solid substance whose atoms are arranged in a
definite manner) is X-rayed, the X-ray beam is diffracted (deflected), and the
pattern that results shows how the atoms are arranged in the crystal.

First Franklin made a concentrated, viscous solution of DNA and
then saw that it could be separated into fibers. Under the right conditions, the fibers were enough like a crystal that when X-rayed, a
diffraction pattern resulted. The X-ray diffraction pattern of DNA
shows that DNA is a double helix. The helical shape is indicated
by the crossed (X) pattern in the center of the photograph in Figure
11.3*b*. The dark portions at the top and bottom of the photograph
indicate that some portion of the helix is repeated over and over.

The Watson and Crick Model

In 1951, James Watson, a newly graduated biology major, began an
internship at the University of Cambridge, England. There he met Francis Crick, a British physicist, who was interested in molecular structures.
Together, they set out to determine the structure of DNA and to build a model
that would explain how DNA varies from species to species, replicates, stores
information, and undergoes mutations.

Based on available data, they knew that:

1. DNA is a polymer of four types of nucleotides with the bases adenine
   (A) and guanine (G), cytosine (C) and thymine (T).
2. Based on Chargaff's rules, the amount A = T, and the amount of G = C.
3. Based on Rosalind Franklin's X-ray diffraction photograph, DNA is a
   double helix with a repeating pattern.

Using these data, Watson and Crick built an actual model of DNA
out of wire and tin (Fig. 11.4). The model showed that the deoxyribose
sugar–phosphate molecules are bonded to one another and make up the
sides of a twisted ladder. The bases are joined by hydrogen bonds and
make up the rungs of the ladder. Indeed, the pairing of A with T and G
with C—now called **complementary base pairing**—results in rungs of
a consistent width as required by the X-ray diffraction data. Figure 11.5
shows two ways to represent the structure of DNA.

The double-helix model of DNA permits the base pairs to be in any order,
a necessity for genetic variability between species. Also, the model suggests
that complementary base pairing may play a role in the replication of DNA.
As Watson and Crick pointed out in their original paper, "It has not escaped
our notice that the specific pairing we have postulated immediately suggests a
possible copying mechanism for the genetic material."

Figure 11.4 Watson and Crick model of DNA.

James Watson (left) and Francis Crick with their model of DNA.

Figure 11.5   DNA structure.

a. A space-filling model of DNA shows the close stacking of the paired bases, as determined by DNA's X-ray diffraction pattern. **b.** Complementary-base pairing dictates that A is hydrogen-bonded to T (as shown), and G is hydrogen-bonded to C in the same manner. The strands of DNA run counter to one another, with the 3' end of one strand opposite the 5' end of the other strand. **c.** Diagram of the DNA double helix shows that the molecule resembles a twisted ladder. The paired bases can be in any order.

a. Space-filling model

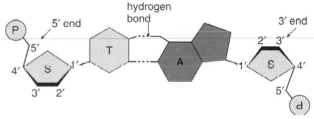

b. Nucleotide pair

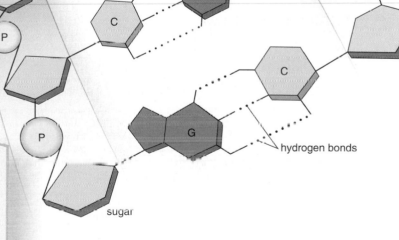

c. Structure of DNA

Check Your Progress

1. What is the significance of the central X in the X-ray diffraction pattern of the molecule?
2. What did Watson and Crick decide about the bases that made their model consistent with both the X-ray diffraction pattern and Chargaff's rules?

Answers: 1. DNA is a double helix. 2. Because A is hydrogen-bonded to T, and G is hydrogen-bonded to C, the amount of A = T and the amount of G = C, as dictated by Chargaff's rules and the rungs are of a consistent width as required by the X-ray diffraction data.

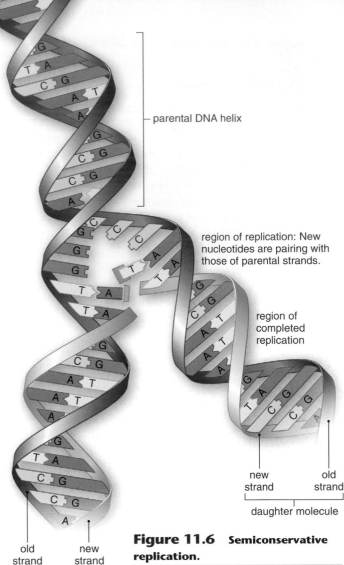

parental DNA helix

region of replication: New nucleotides are pairing with those of parental strands.

region of completed replication

new strand   old strand

daughter molecule

old strand   new strand

daughter molecule

Figure 11.6 **Semiconservative replication.**

After the DNA molecule unwinds, each old strand serves as a template for the formation of a new strand. After replication is complete, there are two daughter DNA molecules identical to each other and to the original double helix.

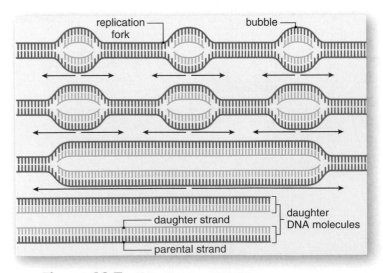

replication fork   bubble

daughter strand

parental strand

daughter DNA molecules

Figure 11.7 **Eukaryotic replication.**

In eukaryotes, replication occurs at numerous replication forks. The bubbles thereby created spread out until they meet.

Replication of DNA

When organisms grow or heal any injuries, cells divide. Each new cell requires an exact copy of DNA or it will not be able to function. **DNA replication** refers to the process of copying a DNA molecule.

During DNA replication, each old DNA strand of the parent molecule serves as a template for a new strand in a daughter molecule (Fig. 11.6). A **template** is most often a mold used to produce a shape complementary to itself. DNA replication is also **semiconservative** because one of the two old strands is conserved, or present, in each daughter molecule.

Before replication begins, the two strands that make up parental DNA are hydrogen-bonded to one another. Replication requires the following steps:

1. **Unwinding.** The old strands that make up the parent DNA molecule are unwound and "unzipped" (i.e., the weak hydrogen bonds between the paired bases are broken). A special enzyme called *helicase* unwinds the molecule.
2. **Complementary base pairing.** New complementary nucleotides, always present in the nucleus, are positioned by the process of complementary base pairing.
3. **Joining.** The complementary nucleotides join to form new strands. Each daughter DNA molecule contains an old strand and a new strand.

Step 3 is carried out by an enzyme complex called **DNA polymerase.** Also, we should note that the enzyme *DNA ligase* seals any breaks in the sugar-phosphate backbone.

In Figure 11.6, the backbones of the parent molecule (original double strand) are purplish. Following replication, the daughter molecules each have an aqua backbone (new strand) and a purplish backbone (old strand). A daughter DNA double helix has the same sequence of base pairs as the parent DNA double helix had originally. Complementary base pairing has allowed this sequence to be maintained.

In eukaryotes, DNA replication begins at numerous origins of replication along the length of the chromosome, and the so-called replication bubbles spread bidirectionally until they meet (Fig. 11.7). Although eukaryotes replicate their DNA at a fairly slow rate—500 to 5,000 base pairs per minute—there are many individual origins of replication. Therefore, eukaryotic cells complete the replication of the diploid amount of DNA (in humans, over 3 billion base pairs) in a matter of hours!

RNA Structure and Function

RNA (ribonucleic acid) is made up of nucleotides containing the sugar ribose, thus accounting for its name. The four nucleotides that make up an RNA molecule have the following bases: adenine (A), **uracil (U),** cytosine (C), and guanine (G). Notice that in RNA, the base uracil replaces the base thymine (Fig. 11.8).

RNA, unlike DNA, is single-stranded, but the single RNA strand sometimes doubles back on itself, allowing complementary base pairing to occur. Similarities and differences between these two nucleic acid molecules are listed in Table 11.1.

In general, RNA is a helper to DNA, allowing protein synthesis to occur according to the genetic information that DNA provides. There are three types of RNA, each with a specific function in protein synthesis.

Messenger RNA

Messenger RNA (mRNA) is produced in the nucleus. DNA serves as a template for its formation during a process called transcription. Which genes are transcribed is tightly controlled in each type of cell and accounts for why some cells are nerve cells and others are muscle cells, for example. Once formed, mRNA carries genetic information from DNA to the ribosomes in the cytoplasm, where protein synthesis occurs.

Transfer RNA

Transfer RNA (tRNA) is also produced in the nucleus, and a portion of DNA also serves as a template for its production. Because DNA serves as a template for both rRNA and tRNA, it is obvious that not all DNA directs protein synthesis.

Appropriate to its name, tRNA transfers amino acids to the ribosomes, where the amino acids are joined, forming a protein. There are 20 different types of amino acids in proteins; therefore, at least 20 tRNAs must be functioning in the cell. Each type of tRNA carries only one type of amino acid.

Ribosomal RNA

In eukaryotic cells, **ribosomal RNA (rRNA)** is produced in the nucleolus of a nucleus, where a portion of DNA also serves as a template for its formation. Ribosomal RNA joins with proteins made in the cytoplasm to form the subunits of **ribosomes,** one large and one small. Each subunit has its own mix of proteins and rRNA. The subunits leave the nucleus and come together in the cytoplasm when protein synthesis is about to begin.

Proteins are synthesized at the ribosomes, which in low-power electron micrographs look like granules often arranged along the *endoplasmic reticulum (ER)*, a system of tubules and saccules within the cytoplasm. Some ribosomes appear free in the cytoplasm or in clusters called polyribosomes. Proteins synthesized by polyribosomes are used in the cytoplasm, and those synthesized by attached ribosomes end up in the ER. From there, the protein is carried in a transport vesicle to the Golgi apparatus for modification and transport to the plasma membrane, where it may leave the cell.

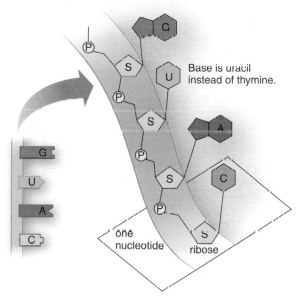

Base is uracil instead of thymine.

one nucleotide   ribose

Figure 11.8   Structure of RNA.

Like DNA, RNA is a polymer of nucleotides. In an RNA nucleotide, the sugar ribose is attached to a phosphate molecule and to a base, either G, U, A, or C. Notice that in RNA, the base uracil (U) replaces thymine as one of the pyrimidine bases. RNA is single-stranded, whereas DNA is double-stranded.

Table 11.1	Comparison of DNA and RNA
DNA-RNA SIMILARITIES	
Both are nucleic acids.	
Both are composed of nucleotides.	
Both have a sugar-phosphate backbone.	
Both have four different types of bases.	
DNA-RNA DIFFERENCES	
DNA	**RNA**
Found in nucleus	Found in nucleus and cytoplasm
Genetic material	Helper to DNA
Sugar is deoxyribose	Sugar is ribose
Bases are A, T, C, G	Bases are A, U, C, G
Double-stranded	Single-stranded
Is transcribed (to give mRNA)	mRNA is translated (to give proteins)

Check Your Progress

1. Why is DNA replication necessary for all organisms?
2. Contrast RNA with DNA.
3. Compare and contrast transfer RNA with ribosomal RNA.

Answers: 1. DNA replication provides a copy of the genetic material for each cell in an organism and for each offspring of the organism. 2. RNA contains ribose, while DNA contains deoxyribose. RNA contains uracil instead of thymine. RNA is usually single-stranded, while DNA is double-stranded. 3. Both are produced in the nucleus and are involved in protein synthesis. Transfer RNA carries amino acids to ribosomes. Ribosomal RNA is a structural component of ribosomes.

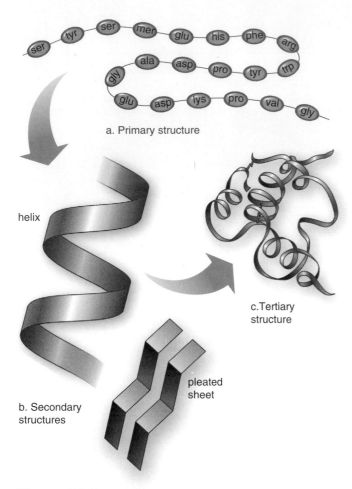

Figure 11.9 **Structure of proteins.**

a. The primary structure of a protein. Each amino acid is designated by three letters that stand for its name. **b.** Portions of the secondary structure can be either a helix or a pleated sheet. **c.** The tertiary structure is the final three-dimensional shape.

11.2 Gene Expression

In the early 1900s, the English physician Sir Archibald Garrod suggested a relationship between inheritance and metabolic diseases. He introduced the phrase *inborn error of metabolism* to dramatize this relationship. Garrod observed that family members often had the same disorder, and he said this inherited defect could be caused by the lack of a functioning enzyme in a metabolic pathway. Since it was already known that enzymes are proteins, Garrod was among the first to hypothesize a link between genes and proteins.

As we shall see, DNA provides the cell with a blueprint for synthesizing proteins. In eukaryotic cells, DNA resides in the nucleus, and protein synthesis occurs in the cytoplasm. Messenger RNA carries a copy of DNA's directions into the cytoplasm, and the other RNA molecules we just discussed are also involved in bringing about protein synthesis. Before describing the mechanics of protein synthesis, let's review the structure of proteins.

Structure and Function of Proteins

Proteins are found in all parts of the body; some are structural proteins, and some are enzymes. Proteins differ because the number and order of their amino acids differ. The unique sequence of amino acids in a protein leads to its particular shape, and the shape of a protein helps determine its function.

Figure 11.9 shows the sequence of amino acids in a portion of adrenocorticotropic hormone, which functions in the human body. Another protein, hemoglobin, is responsible for the red color of red blood cells. Albumins and globulins (antibodies) are well-known plasma proteins. Muscle cells contain the proteins actin and myosin, which give muscles substance and the ability to contract.

Enzymes are organic catalysts that speed reactions in cells. The reactions in cells form metabolic pathways. A pathway can be represented as follows:

$$E_1 \quad E_2 \quad E_3 \quad E_4$$
$$A \rightarrow B \rightarrow C \rightarrow D \rightarrow E$$

In this pathway, the letters are molecules, and the notations over the arrows are enzymes: Molecule A becomes molecule B, and enzyme E_1 speeds the reaction; molecule B becomes molecule C, and enzyme E_2 speeds the reaction; and so forth. Enzymes are *specific:* Enzyme E_1 can only convert A to B, enzyme E_2 can only convert B to C, and so forth. In 1940, George Beadle and Edward Tatum, after performing a series of experiments utilizing red bread mold, concluded that one gene specifies the synthesis of one enzyme. Today, we know that genes also code for structural proteins.

From DNA to RNA to Protein

How does a gene actually specify an enzyme or a protein? What is the mechanism of gene expression? Modern-day molecular biology tells us that during **transcription,** DNA serves as a template for RNA formation. The sequence of bases in DNA determines the sequence of bases in mRNA molecules. Then mRNA moves into the cytoplasm where translation occurs. During **translation,** an mRNA molecule directs the sequence of amino acids in a polypeptide. Every three bases in mRNA is a **codon** that codes for a particular amino acid. mRNA works with rRNA and tRNA to bring about the formation of a protein.

Check Your Progress

1. Explain what is meant by "inborn error of metabolism."
2. Explain how a protein's function is determined by amino acid sequence.

Answers: 1. Family members often have the same disorder, and this inherited defect could be caused by a nonfunctioning enzyme in a metabolic pathway. 2. The amino acid sequence determines the shape of a protein, which in turn determines its function.

The Genetic Code

The sequence of bases in mRNA codes for a particular sequence of amino acids in a polypeptide. Can four bases (A, C, G, U) provide enough combinations to code for 20 amino acids? If the code were a singlet code (one base stands for an amino acid), only four amino acids could be encoded. If the code were a doublet (any two bases stand for one amino acid), it would still not be possible to code for 20 amino acids. But if the code were a triplet, the four bases could supply 64 different triplets, far more than needed to code for 20 different amino acids. It should come as no surprise, then, to learn that the code is a **triplet code.**

Second base

First base		U	C	A	G		Third base
U		UUU ⎤ phenylalanine (phe) UUC ⎦ UUA ⎤ leucine (leu) UUG ⎦	UCU ⎤ UCC ⎥ serine (ser) UCA ⎥ UCG ⎦	UAU ⎤ tyrosine (tyr) UAC ⎦ **UAA** stop **UAG** stop	UGU ⎤ cysteine (cys) UGC ⎦ **UGA** stop UGG tryptophan (trp)		U C A G
C		CUU ⎤ CUC ⎥ leucine (leu) CUA ⎥ CUG ⎦	CCU ⎤ CCC ⎥ proline (pro) CCA ⎥ CCG ⎦	CAU ⎤ histidine (his) CAC ⎦ CAA ⎤ glutamine (gln) CAG ⎦	CGU ⎤ CGC ⎥ arginine (arg) CGA ⎥ CGG ⎦		U C A G
A		AUU ⎤ AUC ⎥ isoleucine (ile) AUA ⎦ **AUG** methionine (met) **(start)**	ACU ⎤ ACC ⎥ threonine (thr) ACA ⎥ ACG ⎦	AAU ⎤ asparagine (asn) AAC ⎦ AAA ⎤ lysine (lys) AAG ⎦	AGU ⎤ serine (ser) AGC ⎦ AGA ⎤ arginine (arg) AGG ⎦		U C A G
G		GUU ⎤ GUC ⎥ valine (val) GUA ⎥ GUG ⎦	GCU ⎤ GCC ⎥ alanine (ala) GCA ⎥ GCG ⎦	GAU ⎤ aspartic acid (asp) GAC ⎦ GAA ⎤ glutamic acid (glu) GAG ⎦	GGU ⎤ GGC ⎥ glycine (gly) GGA ⎥ GGG ⎦		U C A G

Figure 11.10   Messenger RNA codons.

Notice that in this chart, each of the codons are composed of three letters. As an example, find the rectangle where C is the first base and A is the second base. U, C, A, or G can be the third base. CAU and CAC are codons for histidine; CAA and CAG are codons for glutamine.

Each three-letter (nucleotide) unit of an mRNA molecule is called a codon (Fig. 11.10). Sixty-one triplets correspond to a particular amino acid; the remaining three are stop codons, which signal polypeptide termination. The one codon that stands for the amino acid methionine is also a start codon signaling initiation of polypeptide synthesis. Most amino acids have more than one codon, which offers some protection against possibly harmful mutations that could otherwise change the sequence of the bases.

To crack the code, a cell-free experiment was done: Researchers added artificial RNA to a medium containing bacterial ribosomes and a mixture of amino acids. By comparing the bases in the mRNA with the resulting polypeptide, they were able to decipher the code. For example, an mRNA with a sequence of repeating guanines (GGG'GGG'…) would encode a string of glycine amino acids.

The genetic code is just about universal in living things. This suggests that the code dates back to the very first organisms on Earth and that all living things are related.

Transcription

During transcription of DNA, a strand of RNA forms that is complementary to a portion of DNA. While all three classes of RNA are formed by transcription, we will focus on transcription to create mRNA.

mRNA Is Formed   Transcription begins when the enzyme **RNA polymerase** binds tightly to a **promoter,** a region of DNA that contains a special sequence of nucleotides. This enzyme opens up the DNA helix just in front of it so that complementary base pairing can occur. Then RNA polymerase joins the RNA nucleotides, and an mRNA molecule results. When mRNA forms, it has a sequence of bases complementary to DNA; wherever A, T, G, or C is present in the DNA template, U, A, C, or G is incorporated into the mRNA molecule. The resulting **mRNA transcript** is a faithful copy of the sequence of bases in DNA. In Figure 11.11, mRNA is now ready to be processed before it leaves the nucleus for the cytoplasm.

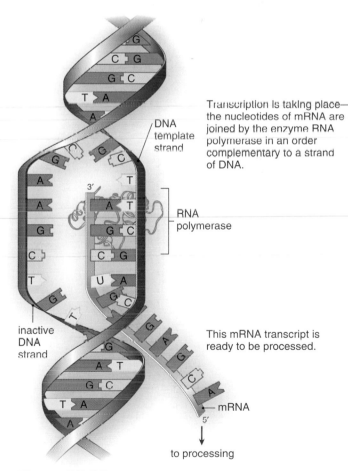

Transcription is taking place—the nucleotides of mRNA are joined by the enzyme RNA polymerase in an order complementary to a strand of DNA.

DNA template strand

RNA polymerase

This mRNA transcript is ready to be processed.

inactive DNA strand

mRNA

to processing

Figure 11.11   Transcription to form mRNA.

During transcription, complementary RNA is made from a DNA template. A portion of DNA unwinds and unzips at the point of attachment of RNA polymerase. A strand of RNA, such as mRNA, is produced when complementary bases join in the order dictated by the sequence of bases in the template DNA strand.

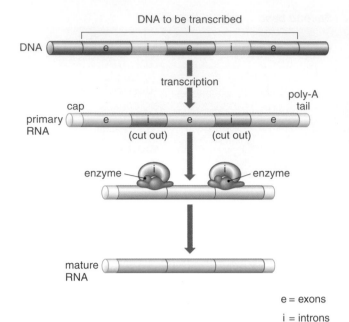

e = exons

i = introns

Figure 11.12   mRNA processing.

Primary mRNA results when both exons and introns are transcribed from DNA. A "cap" and a poly-A "tail" are attached to the ends of the primary RNA transcript, and the introns are removed so that only the exons remain. This mature mRNA molecule moves into the cytoplasm of the cell where translation occurs.

mRNA Is Processed   The newly synthesized *primary mRNA* molecule becomes a *mature mRNA* molecule after processing. Most genes in humans are interrupted by segments of DNA that are not part of the gene. These portions are called *introns* because they are intragene segments. The other portions of the gene are called *exons* because they are ultimately expressed. Only exons result in a protein product.

Primary mRNA contains bases that are complementary to both exons and introns, but during processing two events occur: (1) One end of the mRNA is modified by the addition of a cap, composed of an altered guanine nucleotide, and the other end is modified by the addition of a poly-A tail, a series of adenosine nucleotides. Only mRNAs that have a cap and tail remain active in the cell. (2) The introns are removed, and the exons are joined to form a mature mRNA molecule consisting of continuous exons (Fig. 11.12).

Ordinarily, processing brings together all the exons of a gene. In some instances, however, cells use only certain exons rather than all of them to form the mature RNA transcript. The result is a different protein product in each cell. In other words, alternate mRNA splicing can potentially increase the possible number of protein products from a particular sequence of DNA nucleotides.

Processing occurs in the nucleus of eukaryotic cells. After the mRNA strand is processed, it passes from the cell nucleus into the cytoplasm. There it becomes associated with the ribosomes.

Translation: An Overview

Translation is the second step by which gene expression leads to protein (polypeptide) synthesis. Translation requires several enzymes, mRNA, and the other two types of RNA: transfer RNA and ribosomal RNA.

Transfer RNA Brings Amino Acids to the Ribosomes   Each tRNA is a single-stranded nucleic acid that doubles back on itself such that complementary base pairing results in the shape shown in Figure 11.13. There is at least one tRNA molecule for each of the 20 amino acids found in proteins. The amino acid binds to one end of the molecule. The opposite end of the molecule contains an **anticodon,** a group of three bases that is complementary to a specific codon of mRNA.

After an mRNA transcript is processed, it leaves the nucleus and moves to a ribosome. Now translation begins—the order of codons in mRNA determines the order in which tRNAs bond at the ribosomes. When a tRNA–amino acid complex comes to the ribosome, its anticodon pairs with an mRNA codon. For example, if the codon is ACC, what is the anticodon, and what amino acid will be attached to the tRNA molecule? From Figure 11.10, we can determine this:

Codon (mRNA)	Anticodon (tRNA)	Amino Acid (protein)
ACC	UGG	Threonine

In this way, the order of the codons of the mRNA determines the order that tRNA–amino acid complexes come to a ribosome, and therefore the final sequence of amino acids in a protein.

After translation is complete, a protein contains the sequence of amino acids originally specified by DNA. This is the genetic information that DNA stores and passes on to each cell during the cell cycle, and passes on to the next generation of individuals. DNA's sequence of bases determines the proteins in a cell, and the proteins in turn determine whether an organism is a human being or a giraffe, as an example.

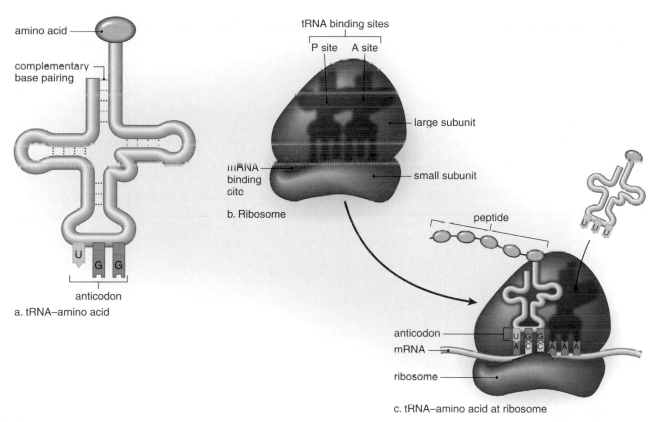

Figure 11.13 **tRNA structure and function.**

a. A tRNA, which is single-stranded but folded, has an amino acid attached to one end and an anticodon at the other end. The anticodon is complementary to a codon. **b.** A ribosome has two binding sites for tRNA, called the P site and the A site. A peptide attached to tRNA at the P site will be passed to an amino acid attached to a tRNA as soon as it arrives at the A site. **c.** The pairing between codon and anticodon at a ribosome ensures that the sequence of amino acids in a polypeptide is the same sequence directed originally by DNA.

Ribosomal RNA Is in Ribosomes Ribosomes are the small structural bodies where translation occurs. Ribosomes are composed of many proteins and several ribosomal RNAs (rRNAs). As stated, in eukaryotic cells, rRNA is produced in a nucleolus within the nucleus. There it joins with proteins manufactured in the cytoplasm to form two ribosomal subunits, one large and one small. The subunits leave the nucleus and join together in the cytoplasm to form a ribosome just as protein synthesis begins.

A ribosome has a binding site for mRNA as well as binding sites for two tRNA molecules at a time (Fig. 11.13). These binding sites facilitate complementary base pairing between tRNA anticodons and mRNA codons. The P binding site is for a tRNA attached to a *peptide*, and the A binding site is for a newly arrived tRNA attached to an *amino* acid.

As soon as the initial portion of mRNA has been translated by one ribosome, and the ribosome has begun to move down the mRNA, another ribosome attaches to the same mRNA. Therefore, several ribosomes are often attached to and translating the same mRNA, allowing the cell to produce thousands of copies of the same protein at a time. The entire complex is called a **polyribosome** (Fig. 11.14.)

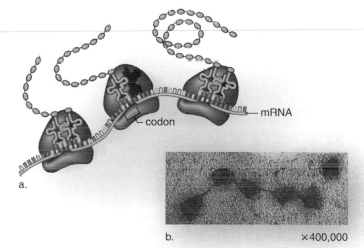

Figure 11.14 **Polyribosome structure and function.**

a. Several ribosomes, collectively called a polyribosome, move along an mRNA at one time. Therefore, several proteins can be made at the same time. **b.** Electron micrograph of a polyribosome.

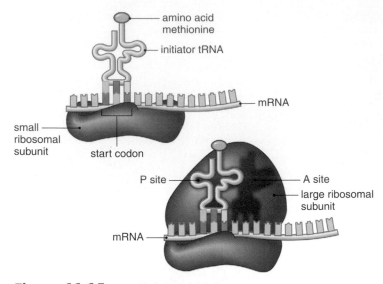

Figure 11.15 **Initiation.**

A small ribosomal subunit binds to mRNA; an initiator tRNA's anticodon binds to its codon, and the large ribosomal subunit completes the ribosome.

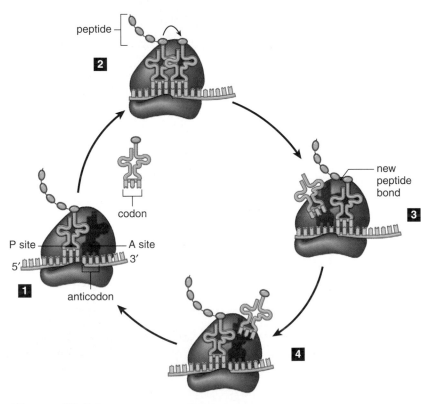

Figure 11.16 **Elongation cycle.**

1 Two tRNAs can be at a ribosome at one time. **2** The tRNA at the P site passes its peptide to the tRNA at the A site. **3** The tRNA at the P sites leaves. **4** The ribosome moves forward (translocation), and the tRNA-peptide complex is now at the P site. A new tRNA–amino acid complex comes to the A site.

Translation Has Three Phases

Polypeptide synthesis has three phases: initiation, an elongation cycle, and termination. Enzymes are required for each of the steps to occur, and energy is needed for the first two steps.

1. During *initiation*, mRNA binds to the smaller of the two ribosomal subunits; then the larger subunit associates with the smaller one.
2. During an *elongation cycle*, a peptide lengthens one amino acid at a time. The incoming tRNA–amino acid complex receives the peptide from the outgoing tRNA. The ribosome then moves forward so that the next mRNA codon is available to receive an incoming tRNA–amino acid complex.
3. *Termination* occurs at a codon that means stop and does not code for an amino acid. The ribosomal subunits and mRNA dissociate. The completed polypeptide is released.

Initiation   During initiation, a small ribosomal subunit, the mRNA, an *initiator tRNA* bound to the amino acid methionine, and a large ribosomal subunit all come together (Fig. 11.15):

- The small ribosomal subunit attaches to the mRNA in the vicinity of the start codon (AUG).
- The anticodon of the initiator tRNA–methionine complex pairs with this codon.
- The large ribosomal subunit joins to the small subunit.

Elongation Cycle   As discussed, a ribosome has two binding sites for tRNA where the tRNA's anticodon binds to a codon of mRNA. During the elongation cycle (Fig. 11.16):

- A tRNA at the P site usually bears a peptide built up previously. (See **1**.)
- This tRNA passes its peptide to tRNA–amino acid at the A site. The tRNA at the P site leaves. (See **2** and **3**.)
- The ribosome moves forward one codon (called **translocation**). The tRNA-peptide is now at the P site, and the codon at the A site is ready for the next tRNA–amino acid. (See **4**.)

The complete cycle—complementary base pairing of new tRNA, transfer of peptide chain, and translocation—is repeated at a rapid rate (about 15 times each second in the bacterium *Escherichia coli*). The outgoing tRNA picks up another amino acid and is ready to return to the ribosome.

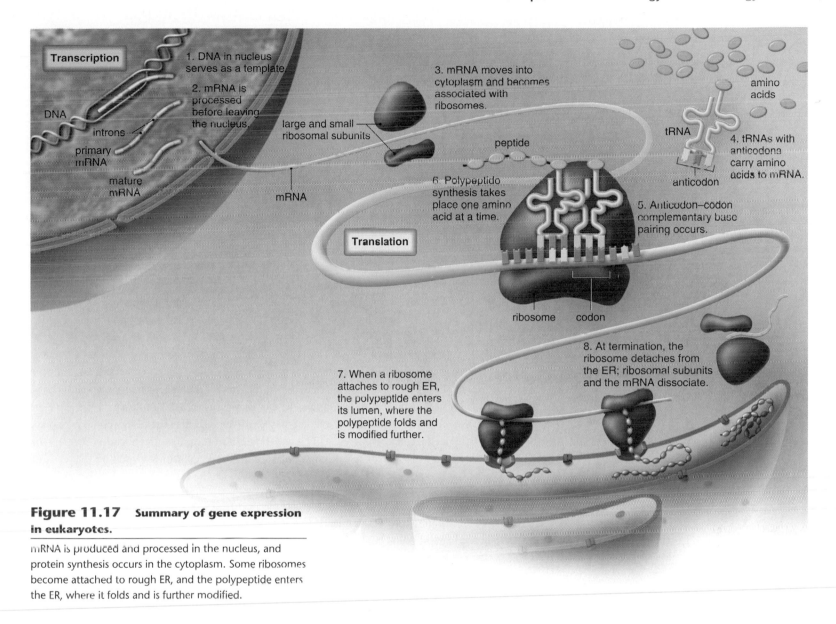

Figure 11.17   Summary of gene expression in eukaryotes.

mRNA is produced and processed in the nucleus, and protein synthesis occurs in the cytoplasm. Some ribosomes become attached to rough ER, and the polypeptide enters the ER, where it folds and is further modified.

Review of Gene Expression

Genes consist of DNA, which codes for all the proteins in a cell. A gene is expressed when its protein product has been made. Figure 11.17 reviews transcription and mRNA processing in the nucleus, as well as translation during protein synthesis in the cytoplasm.

Some ribosomes remain free in the cytoplasm, and others become attached to rough ER. In the latter case, the polypeptide enters the lumen of the ER by way of a channel, where it is further processed by the addition of sugars or lipids. Transport vesicles carry the protein to other locations in the cell, including the Golgi apparatus, which may modify it further and package it in a vesicle for transport out of the cell. Proteins have innumerable functions in cells, from enzymatic to structural. They account for the structure and function of cells, tissues, organs, and the organism.

Check Your Progress

1. Describe the events of transcription.
2. Describe the events of translation.

Answers: 1. During transcription, DNA serves as a template for RNA synthesis. At a promoter region, RNA polymerase opens DNA to make it single-stranded. Then it adds RNA nucleotides to build an RNA strand. 2. During translation, mRNA directs the synthesis of a polypeptide. Messenger RNA has a sequence of codons that dictate the order in which tRNA–amino acid complexes come to a ribosome. During elongation, the peptide is passed from a departing tRNA to an arriving tRNA. In the end, the polypeptide has a sequence of amino acids as dictated by DNA.

a.

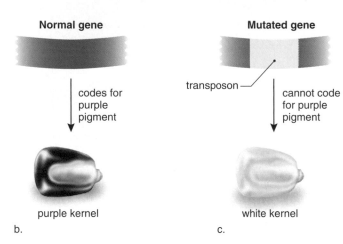

Normal gene   **Mutated gene**

codes for
purple
pigment

transposon

cannot code
for purple
pigment

purple kernel   white kernel

b.   c.

Figure 11.18   Transposon.

a. Barbara McClintock, shown here instructing a student, won the 1983 Nobel Prize in Physiology or Medicine for being the first to discover transposons. She worked with maize (corn). **b.** In corn, a purple-coding gene ordinarily codes for a purple pigment. **c.** A transposon "jumps" into the purple-coding gene. This mutated gene is unable to code for purple pigment, and a white kernel results.

Check Your Progress

Define a gene mutation, and describe three major types of mutations.

Answer: A gene mutation is a change in the sequence of bases in a gene. Frameshift mutations change the reading frame; point mutations change one base; and transposons interrupt DNA.

Genes and Gene Mutations

Whereas early geneticists thought of genes as sections of a chromosome, in molecular genetics, a gene is a sequence of DNA bases that code for a product, most often a protein. Therefore, it follows that a **gene mutation** is a change in the sequence of bases within a gene.

Causes of Gene Mutations

A gene mutation can be caused by an error in replication, a transposon, or an environmental mutagen. Mutations due to DNA replication errors are rare; a frequency of one in 100 million per cell division is often quoted. DNA polymerase, the enzyme that carries out replication, proofreads the new strand against the old strand and detects any mismatched pairs, which are then replaced with the correct nucleotides. **Transposons** are specific DNA sequences that have the remarkable ability to move within and between chromosomes. So-called *jumping genes* have now been discovered in bacteria, fruit flies, and humans, and it is likely all organisms have such elements.

Mutagens are environmental influences that cause mutations. Different forms of radiation, such as radioactivity, X rays, and ultraviolet (UV) light, cause breaks in DNA molecules, which can be incorrectly repaired. Chemical mutagens, such as certain pesticides and chemicals in cigarette smoke, alter the bases in DNA, causing them to pair incorrectly. Mutagens may affect all types of organisms, including humans. If mutagens bring about a mutation in the gametes, the offspring of the individual may be affected. If the mutation occurs in the body cells, cancer may result. The usual rate of mutation is low because *DNA repair enzymes* constantly monitor and repair any irregularities.

Effects of Mutations

In general, we know a mutation has occurred when the organism has a malfunctioning protein that leads to a genetic disorder or to the development of cancer. Most often, genetic disorders, such as cystic fibrosis, are inherited conditions and represent *germ-line mutations* that occurred in the sex cells. *Somatic mutations* that occur in the body cells are most likely the cause of cancer.

Point mutations involve a change in a single DNA nucleotide, and the results can vary depending on the particular base change that occurs. For example, if valine instead of glutamate occurs in the β chain of hemoglobin at one particular location, sickle cell disease results. The abnormal hemoglobin stacks up inside cells, and their sickle shape makes them clog small vessels. Hemorrhaging leads to damage and pain in internal organs and joints.

Codons are read from a specific starting point, as in this sentence: THE CAT ATE THE RAT. If the letter C is deleted from this sentence and the "reading frame" is shifted, we read THE ATA TET HER AT—something that doesn't make sense. If a **frameshift mutation** occurs early in the DNA base sequence of a gene, a completely nonfunctional protein results.

The presence of white kernels in a type of purple corn commonly used in genetic studies is due to a transposon located within a gene coding for a pigment-producing enzyme (Fig. 11.18). In humans, a neurological disorder called Charcot-Marie-Tooth disease is named for the three people who discovered it. This disorder, which causes weakness in the legs, feet, and hands occurs when the transposon, called *mariner,* is present in the human genome. Exactly why its presence causes the disease is being investigated.

11.3 DNA Technology

The knowledge that a gene is a sequence of bases in a DNA strand led to our ability to manipulate the genes of organisms. We can clone genes (make identical copies) and then use the genes for various purposes. For example, cloned genes can be used to determine the difference in base sequence between different forms of a gene in the same or different species. During so called **genetic engineering,** a foreign gene can be inserted into the genome of an organism, which is then called a **transgenic organism.** It's possible to create transgenic bacteria, plants, and animals. Recombinant DNA technology is often used to create transgenic bacteria.

Recombinant DNA Technology

Recombinant DNA (rDNA) contains DNA from two or more different sources (Fig. 11.19). To make rDNA, a researcher needs a **vector,** a piece of DNA that can have foreign DNA added to it. One common vector is a *plasmid,* which is a small accessory ring of DNA found in bacteria.

Two enzymes are needed to introduce foreign DNA into plasmid DNA: (1) **Restriction enzymes** cleave DNA, and (2) **DNA ligase** seals DNA into an opening created by the restriction enzyme. Hundreds of restriction enzymes occur naturally in bacteria, where they cut up (cleave) any viral DNA that enters the cell. They are called restriction enzymes because they *restrict* the growth of viruses, but they can also be used as molecular scissors to cut double-stranded DNA at a specific site. For example, the restriction enzyme *Eco*RI always recognizes and cuts in this manner when DNA has the sequence of bases GAATTC:

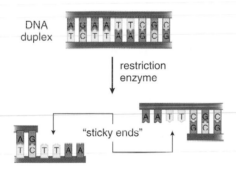

Notice that a gap now exists into which a piece of foreign DNA can be placed if it ends in bases complementary to those exposed by the restriction enzyme. To ensure this, it is only necessary to cleave the foreign DNA with the same restriction enzyme. The single-stranded, but complementary, ends of the two DNA molecules are called "sticky ends" because they can bind a piece of foreign DNA by complementary base pairing. Sticky ends facilitate the insertion of foreign DNA into vector DNA.

DNA ligase, the enzyme that functions in DNA replication to seal breaks in a double-stranded helix, seals the foreign piece of DNA into the plasmid. Bacterial cells take up recombinant plasmids, especially if the cells are treated to make them more permeable. Thereafter, as the plasmid replicates, the gene is cloned.

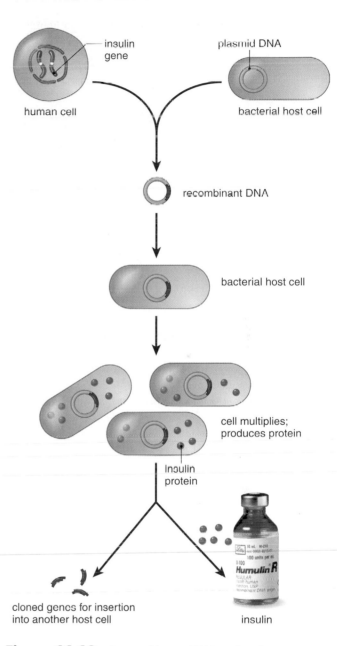

Figure 11.19   Recombinant DNA technology.

Production of insulin is one example of how recombinant DNA technology can benefit humans. Human DNA and plasmid DNA are spliced together. Gene cloning is achieved when a host cell takes up the recombinant plasmid, and as the cell reproduces, the plasmid is replicated. Multiple copies of the gene are now available. If the insulin gene functions normally as expected, insulin may also be retrieved.

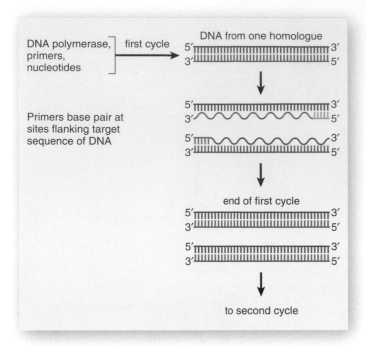

Figure 11.20 **Polymerase chain reaction.**

At the start of PCR, the original double-stranded DNA from all the chromosomes is mixed with DNA polymerase, two primers (gold and pink), that flank the target sequence on both homologues and a supply of the four nucleotides in DNA. At the end of the first cycle, there are two identical copies of double-stranded DNA. Usually 25–30 cycles are run in an automated cycler, resulting in over 30 million copies of the original DNA double strand.

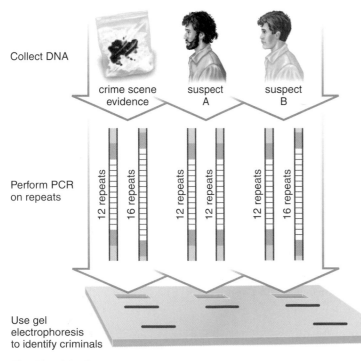

Figure 11.21 **DNA fingerprinting.**

Following PCR, gel electrophoresis reveals the comparative pattern of repeats in the DNA. Here, DNA fingerprinting is being used to identify a criminal.

Polymerase Chain Reaction

The **polymerase chain reaction (PCR)** can create millions of copies of a segment of DNA very quickly in a test tube without the use of a vector or a host cell. The original sample of PCR is usually just a portion of the entire genome (all the genes of an individual). PCR is very specific—it *amplifies* (makes copies of) a targeted DNA sequence that can be less than one part in a million of the total DNA sample!

PCR requires the following: DNA polymerase, the enzyme that carries out DNA replication; a set of primers; and a supply of nucleotides for the new DNA strands. Primers are single-stranded DNA sequences that start the replication process on each DNA strand. PCR is a chain reaction because the targeted DNA is repeatedly replicated as long as the process continues. The process is shown in Figure 11.20. The amount of DNA doubles with each replication cycle; after one cycle, there are two copies of the targeted DNA sequence; after two cycles, there are four copies, and so forth.

PCR has been in use for several years, and now almost every laboratory has automated PCR machines to carry out the procedure. Automation became possible after a temperature-insensitive DNA polymerase was extracted from the bacterium *Thermus aquaticus,* which lives in hot springs. During the PCR cycle, the mixture of DNA and primers is heated to 95°C to separate the two strands of the double helix so that the primers can bind to the single strands of DNA. The polymerase can withstand the high temperature used to separate double-stranded DNA; therefore, replication does not have to be interrupted by the need to add more enzyme.

DNA amplified by PCR is often analyzed for various purposes. Mitochondrial DNA base sequences in modern living populations have been used to decipher the evolutionary history of human populations. Because so little DNA is required for PCR to be effective, it has even been possible to sequence DNA taken from a 76,000-year-old mummified human brain.

Applications

DNA fingerprinting makes use of noncoding sections of DNA that consist of two to five bases repeated over and over again, as in ATCATCATCATC. People differ by how many times such a sequence is repeated, but how can this difference be detected? The greater the number of repeats, the greater the relative amount of DNA that will result after PCR amplication is done. People can be heterozygous for the number of repeats as in Figure 11.21, so the DNA from both homologues has to be amplified separately. Following PCR, the DNA is subjected to gel electrophoresis, a process whereby the DNA samples migrate through a jellylike material according to size. The smaller the amount of DNA, the further it migrates.

DNA fingerprinting has many uses. Medically, it can identify the presence of a viral infection or a mutated gene that could predispose someone to cancer. In forensics, DNA fingerprinting from a single sperm is enough to identify a suspected rapist because the DNA is amplified by PCR. It can also be used to identify the parents of a child or identify the remains of someone who died, such as a victim of the September 11, 2001 terrorist attacks. In the latter instance, skin cells left on a personal item such as a toothbrush or cigarette butt sometimes provided enough DNA for matching tests. The National Football League even uses DNA copied by PCR to mark each of the Super Bowl footballs in order to authenticate them.

Transgenic Organisms

The term **biotechnology** refers to the use of natural biological systems to create a product or achieve some other end desired by human beings. Today, bacteria, plants, and animals can be genetically engineered to make biotechnology products. Various uses of transgenic organisms are shown in Figure 11.22.

Transgenic Bacteria Recombinant DNA technology is used to produce transgenic bacteria, which are grown in huge vats called bioreactors. The gene product is collected from the medium. Products now on the market that are produced by bacteria include insulin, human growth hormone, t-PA (tissue plasminogen activator), and hepatitis B vaccine. Bacteria can be selected for their ability to degrade a particular substance, such as oil, and this ability can then be enhanced by genetic engineering. Similarly, some mining companies are testing genetically engineered organisms that have improved bioleaching capabilities. Bacteria can also be engineered to produce organic chemicals used in industry.

Transgenic Plants Foreign genes have been transferred to cotton, corn, and potato strains to make these plants resistant to pests by causing their cells to produce an insect toxin. Similarly, soybeans have been made resistant to a common herbicide. Some corn and cotton plants are both pest- and herbicide-resistant. These and other genetically engineered crops have increased yields. Plant seeds are also being engineered to produce human proteins, such as hormones, clotting factors, and antibodies.

Transgenic Animals Foreign genes have been inserted into the eggs of animals, often to give them the gene for bovine growth hormone (bGH). The procedure has resulted in larger fishes, cows, pigs, rabbits, and sheep. Gene "pharming," the use of transgenic farm animals to produce pharmaceuticals, is being pursued by a number of firms. Genes that code for therapeutic and diagnostic proteins are incorporated into an animal's DNA, and the proteins appear in the animal's milk.

To achieve a sufficient number of animals that can produce the product, transgenic animals are sometimes cloned. For many years, it was believed that adult vertebrate animals could not be cloned. Although each cell contains a copy of all the genes, certain genes are turned off in mature, specialized cells, and cloning an adult vertebrate would require that all the genes of an adult cell be turned on again, which had long been thought impossible. In 1997, however, Scottish scientists produced a cloned sheep, which they named Dolly. Since then, calves, goats, and many other animals, including some endangered species, have also been cloned. After enucleated eggs from a donor are microinjected with 2n nuclei from a transgenic animal, they are coaxed to begin development in vitro. Development continues in host females until the clones are born. The offspring are clones because all have the identical genotype of the adult that donated the 2n nuclei. Now that scientists have a way to clone animals, this procedure will undoubtedly be used routinely to procure biotechnology products.

Figure 11.22 **Uses of transgenic organisms.**

a. Transgenic bacteria, grown industrially, are used to produce medicines. Transgenic animals and plants increase the food supply. **b.** These salmon (*right*) are much heavier and reproduce much sooner because of the growth hormone genes they received as embryos. **c.** Pests didn't consume unblemished peas because the plants that produced them received genes for pest inhibitors. **d.** Gene guns are often used to insert genes in plant embryos while they are in tissue culture. **e.** Transgenic bacteria can also be used for environmental cleanup. These bacteria are able to break down oil.

Check Your Progress

1. What two techniques allow researchers to clone a segment of DNA?
2. What happens to a cloned gene in order to produce transgenic organisms?

Answers: 1. Recombinant DNA technology is used to produce transgenic bacteria. When the bacteria reproduce, the gene is cloned. PCR can clone a segment of DNA. 2. The gene is placed in a different species from that of the gene donor.

THE CHAPTER IN REVIEW

Summary

11.1   DNA and RNA Structure and Function

Structure of DNA

DNA, the genetic material, has the following structure:

- DNA is a polymer in which four nucleotides differ by their bases. The bases are symbolized by the letters A, G, C, and T.
- Purine A is always paired with T, and G is always paired with C, a pattern called complementary base pairing:

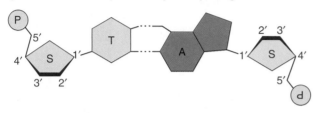

- DNA is a double helix. The deoxyribose sugar–phosphate molecules make up the sides of a twisted ladder, and the paired bases are the rungs of the ladder. Base pairs can be in any order in each type of species.

Replication of DNA

DNA replicates and every new cell gets a copy. During DNA replication, each old strand gives rise to a new strand. Therefore, the process is called semiconservative. Because of complementary base pairing, all these double helix molecules are identical:

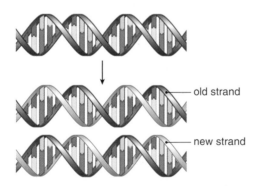

old strand

new strand

RNA Structure and Function

RNA has the following characteristics:

- It is found in the nucleus and cytoplasm of eukaryotic cells.
- It contains the sugar ribose and the bases A, U, C, G.
- It is single-stranded.
- The three forms are mRNA, which carries the DNA message to the ribosomes; tRNA, which transfers amino acids to the ribosomes where protein synthesis occurs; and rRNA, which is found in the ribosomes.

11.2   Gene Expression

DNA specifies the sequence of amino acids in a protein; therefore, gene expression occurs once the protein product is present. Gene expression requires two steps: transcription and translation.

From DNA to RNA to Protein

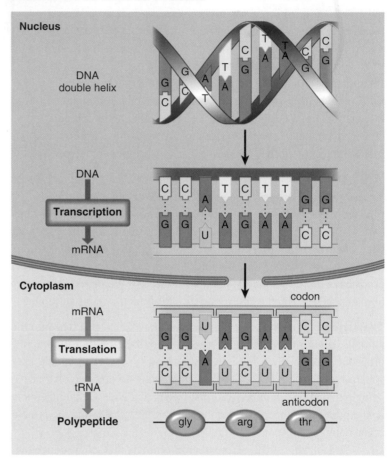

Transcription   During transcription, mRNA forms and is then processed. Processing involves (1) the addition of a cap to one end and the addition of a poly-A tail to the other end, and (2) removal of introns so that only exons remain.

Translation   Translation requires all three types of RNA:

- **mRNA** contains codons in which every three bases code for a particular amino acid except when the codon means start or stop.
- **tRNA** brings amino acids to the ribosomes. One end of the molecule binds to the amino acid, and the other end is the anticodon, three bases that pair with a codon. Each tRNA binds with only one of the 20 types of amino acids in a protein.
- **rRNA** is located in the ribosomes. The P site of a ribosome contains a tRNA attached to a peptide, and the A site contains a tRNA attached to an amino acid.

Three Phases of Translation

The three phases of translation are initiation, the elongation cycle, and termination.

Initiation During chain initiation at the start codon, the ribosomal subunits, the mRNA, and the tRNA-methionine complex come together.

Elongation cycle The chain elongation cycle consists of these events:

- A tRNA at the P site passes its peptide to tRNA–amino acid at the A site. The tRNA at the P site leaves.
- The ribosome moves forward one codon (called translocation), and the codon at the A site is ready for the next tRNA–amino acid.

Termination During chain termination at a stop codon, the ribosome dissociates, the last tRNA departs, and the polypeptide is released.

Genes and Gene Mutations

A gene mutation is a change in the sequence of bases. Mutations can be due to errors in replication, transposons, or environmental mutagens. The results of mutations can vary from no effect to a nonfunctional protein.

11.3 DNA Technology

Recombinant DNA Technology

Recombinant DNA technology uses restriction enzymes to cleave DNA so that a foreign gene can be inserted into a vector, such as a plasmid:

plasmid DNA

human gene

Polymerase Chain Reaction

The polymerase chain reaction occurs in a test tube. Heat separates double-stranded DNA. Primers flank the target DNA, and heat-insensitive DNA polymerase copies the target DNA. The cycle is repeated until hundreds of thousands of copies are produced.

DNA Fingerprinting

DNA fingerprinting helps identify relatives, remains, and criminals. Following PCR if needed, restriction enzymes are used to fragment samples of DNA. When separated by gel electrophoresis, a pattern results that is unique to the individual.

Biotechnology

Biotechnology uses natural biological systems to create a product or to achieve an end desired by human beings. Foreign genes have been transferred to bacteria, to crops, and to farm animals to improve their characteristics and produce commercial products.

Thinking Scientifically

1. Skin cancer is more common than brain cancer. Why might the frequency of cancer be related to the rate of cell division in skin as opposed to the rate of cell division in the brain?

2. There has been much popular interest in re-creating extinct animals from DNA obtained from various types of fossils. However, such DNA is always badly degraded, consisting of extremely short pieces. Explain why it would be impossible to create a dinosaur even if you had 30 intact genes from a dinosaur.

Testing Yourself

1. In the following diagram, label all parts of the DNA molecule.

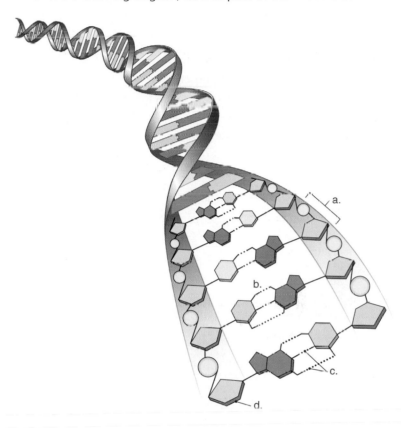

a.

b.

c.

d.

Choose the best answer for each question.

2. Chargaff's rules state that the amount of A, T, G, and C in DNA

   a. varies from species to species, and the amount of A=T and G=C.
   b. varies from species to species, and the amount of A=G and T=C.
   c. is the same from species to species, and the amount of A=T and G=C.
   d. is the same from species to species, and the amount of A=G and T=C.

3. Because each daughter molecule contains one old strand of DNA, DNA replication is said to be

   a. conservative.
   b. preservative.
   c. semidiscontinuous.
   d. semiconservative.

4. Which of the following is not a feature of eukaryotic DNA replication?

   a. Replication bubbles spread bidirectionally.
   b. A new strand is synthesized using an old one as a template.
   c. Complementary base pairing determines which nucleotides should be added to the parental strand.
   d. Each chromosome has one origin of replication.

For questions 5–8, match the items to those in the key. Answers can be used more than once.

Key:

    a. messenger RNA

    b. transfer RNA

    c. ribosomal RNA

5. Produced in the nucleus.

6. Carries amino acids to the ribosome.

7. Carries genetic information from DNA to ribosomes.

8. Produced in the nucleolus.

9. Which is not a feature of an enzyme?

    a. specific in its activity

    b. acts as an organic catalyst

    c. provides structural support to a cell

    d. speeds biochemical reactions

10. Transcription produces _____, while translation produces _____.

    a. DNA, RNA

    b. RNA, polypeptides

    c. polypeptides, RNA

    d. RNA, DNA

11. Label the components of transcription in the following diagram.

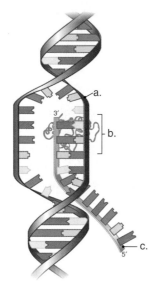

For questions 12–15, match the items to those in the key.

Key:

    a. germ-line mutation

    b. frameshift mutation

    c. point mutation

    d. somatic mutation

12. Occurs in body cells.

13. Base change that changes all the codons from here on.

14. Base change in one codon only.

15. Occurs in the sex cells.

16. Which of the following is not required for the polymerase chain reaction?

    a. DNA polymerase

    b. RNA polymerase

    c. DNA primers

    d. nucleotides

Go to www.mhhe.com/maderessentials for more quiz questions.

Bioethical Issue

Arthur Lee Whitfield had served 22 years of a 63-year sentence for rape when he was notified that he had gained the right to use DNA fingerprinting to establish his innocence. Luckily, biological evidence had been saved from his jury trial, and DNA fingerprinting showed that his DNA did not match that of the rapist. However, this DNA did match that of another inmate who was serving a life sentence for an unrelated rape conviction.

    Considering that DNA fingerprinting can prove the guilt or innocence of suspects, should everyone be required to contribute blood to create a national DNA fingerprint databank? Or does this constitute search without cause, which is illegal in the United States? DNA fingerprinting evidence was ruled inadmissible in the O.J. Simpson trial because it could not be proven that the police had not "planted" Simpson's blood at the crime scene. Do similar suspicions make you think that DNA fingerprinting should not be allowed even to prove the innocence of a person like Whitfield?

Understanding the Terms

adenine (A) 161	polyribosome 169
anticodon 168	promoter 167
biotechnology 175	recombinant DNA (rDNA) 173
codon 166	restriction enzyme 173
complementary base pairing 162	ribosomal RNA (rRNA) 165
cytosine (C) 161	ribosome 165
DNA fingerprinting 174	RNA (ribonucleic acid) 164
DNA ligase 173	RNA polymerase 167
DNA polymerase 164	semiconservative 164
DNA replication 164	template 164
frameshift mutation 172	thymine (T) 161
gene mutation 172	transfer RNA (tRNA) 165
genetic engineering 173	transcription 166
guanine (G) 161	transgenic organism 173
messenger RNA (mRNA) 165	translation 166
mRNA transcript 167	translocation 170
mutagen 172	transposon 172
point mutation 172	triplet code 167
polymerase chain reaction	uracil (U) 164
(PCR) 174	vector 173

Match the terms to these definitions:

a. _____ Double-ring bases in nucleotides.

b. _____ Pairing of adenine with thymine and cytosine with guanine.

c. _____ A mold used to produce a shape complementary to itself.

d. _____ The enzyme that adds new nucleotides during DNA replication.

e. _____ A region of DNA that contains a sequence of nucleotides signaling RNA polymerase to begin transcription.

Gene Regulation and Cancer

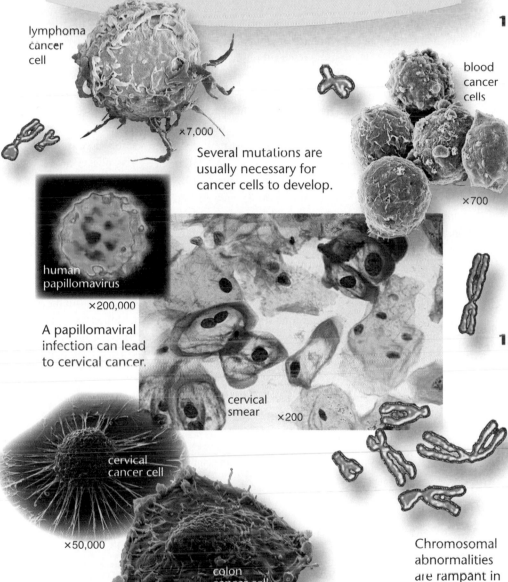

lymphoma cancer cell

×7,000

Several mutations are usually necessary for cancer cells to develop.

blood cancer cells

×700

human papillomavirus

×200,000

A papillomaviral infection can lead to cervical cancer.

cervical smear ×200

cervical cancer cell

×50,000

colon cancer cell ×5,000

Chromosomal abnormalities are rampant in cancer cells.

OUTLINE

The term cancer describes a family of diseases having the common characteristic of uncontrolled cell division. You are probably aware that lifestyle choices such as smoking or tanning might increase your odds of developing certain types of cancer. However, have you ever considered that some viral infections can also increase your chances of cancer? A group of viruses termed oncogenic viruses can contribute to the possibility of certain cancers. These viruses ▶ ▶ ▶

are characterized by their ability to insert viral DNA into human chromosomes. Humans have several genes, called proto-oncogenes, that help regulate cell division. When oncogenic viral DNA inserts a mutated proto-oncogene, called an oncogene, into a human chromosome, the cell cycle may occur repeatedly, and a tumor may result.

Human papillomavirus (HPV) is a well-known oncogenic virus. This virus comes in many forms and often causes benign warts. However, several strains of this virus are known to be oncogenic, and are the primary cause of cervical cancer. Women infected with oncogenic strains of this HPV typically have no symptoms, but an annual Pap smear can detect changes in cervical cells that are consistent with HPV infection. When detected early, infected cells can be removed.

Although the possibility of cancer strikes fear in many individuals, numerous genetic changes are usually required in order for cancer to develop. In this chapter, you will learn about some of the genetic changes that contribute to the development of cancer.

12.1 Control of Gene Expression

The events of the cell cycle and DNA replication ensure that every cell of the body receives a copy of all the genes. This means that every nucleus in your body has the potential to become a complete organism. To demonstrate this potential, let's take a look at the process of reproductive and therapeutic cloning.

Reproductive and Therapeutic Cloning

In **reproductive cloning,** the desired end is an individual that is exactly like the original individual. The cloning of plants has been routine for some time (Fig. 12.1). However, at one time it was thought that the cloning of adult animals would be impossible because investigators found it difficult to have the nucleus of an adult cell "start over," even when it was placed in an enucleated egg cell. (An enucleated egg cell has had its own nucleus removed.)

In March 1997, Scottish investigators announced they had cloned a Dorset sheep, which they named Dolly. How was their procedure different from all the others that had been attempted? Again, an adult nucleus was placed in an enucleated egg cell; however, the donor cells had been starved. Starving the donor cells caused them to stop dividing and go into a resting stage (the G_0 stage of the cell cycle). The G_0 nuclei are amenable to cytoplasmic signals for initiation of development (Fig. 12.2). Now it is common practice to clone farm animals that have desirable traits (Fig. 12.3), and even to clone rare animals that might otherwise become extinct.

In the United States, no federal funds can be used for experiments to clone human beings. And at this point, even cloning farm animals is wasteful—in the case of Dolly, out of 29 clones only one was successful. Also, cloned animals may not be healthy. Dolly was put down by lethal injection in 2003 because she was suffering from lung cancer and crippling arthritis. She had lived only half the normal life span for a Dorset sheep.

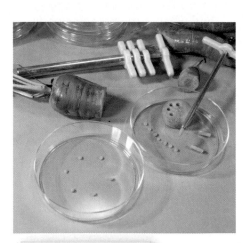

1 Tiny disks are obtained from carrot root.

2 Each disk produces an undifferentiated tissue mass.

3 Many like carrot plantlets are cloned from each tissue mass.

Figure 12.1 **Cloning carrots.**

Carrot cells from a root will produce cloned copies of carrot plants in cell culture. This demonstrates that all the nuclei in a carrot root contain all the genes of a carrot plant.

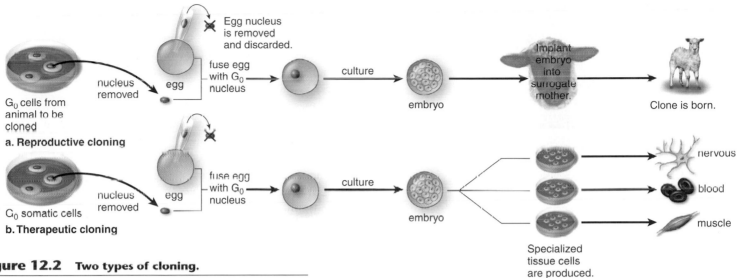

Figure 12.2   **Two types of cloning.**

a. The purpose of reproductive cloning is to produce an individual that is genetically identical to the one that donated a 2n nucleus. The 2n nucleus is placed in an enucleated egg, and after several divisions, the embryo comes to term in a surrogate mother.
b. The purpose of therapeutic cloning is to produce specialized tissue cells. A 2n nucleus is placed in an enucleated egg, and after several divisions, the embryonic cells (called embryonic stem cells) are separated and treated to become specialized cells.

In **therapeutic cloning,** the desired end is not an individual organism, but mature cells of various cell types. The purposes of therapeutic cloning are (1) to learn more about how specialization of cells occurs, and (2) to provide cells and tissues that could be used to treat human illnesses such as diabetes, spinal cord injuries, and Parkinson disease.

Therapeutic cloning can be carried out in several ways. One way follows the same procedure as reproductive cloning, except that embryonic cells, called *embryonic stem cells,* are isolated, and each is subjected to a treatment that causes it to develop into a particular type of cell, such as red blood cells, muscle cells, or nerve cells (see Fig. 12.2*b*). Ethical concerns exist about this type of therapeutic cloning because if the embryo had been allowed to continue development, it would have become an individual. People are especially concerned when researchers short-cut this procedure and use embryos that began life through in vitro fertilization in fertility clinics. Another way to carry out therapeutic cloning is to use *adult stem cells.* Stem cells are found in many organs of an adult's body. For example, the skin has stem cells that constantly divide and produce new skin cells, while the bone marrow has stem cells that produce new blood cells. The goal is to develop techniques that would allow any adult stem cell to become any type of specialized cell. In order to bring this about, researchers need to know how to control gene expression.

Figure 12.3   **Cloned farm animals.**

These pigs were cloned using 2n nuclei from an adult swine. All nuclei contain all the genes of an organism or else cloning wouldn't be possible.

Check Your Progress

Contrast reproductive cloning with therapeutic cloning.

Answer: The goal of reproductive cloning is to create an individual that is genetically identical to the 2n nucleus donor. The goal of therapeutic cloning is to study how cells become specialized and to provide specialized cells for medical procedures.

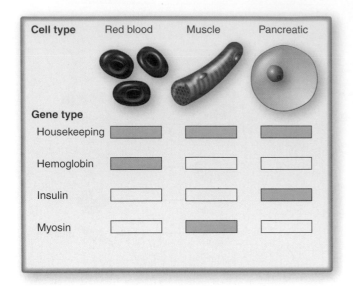

Figure 12.4   Gene expression in specialized cells.

All the cells in your body contain all the genes, but gene expression is controlled, and only certain genes are expressed in each type of cell. Housekeeping genes are expressed in all cells; in addition, cells express those genes that account for their specialization.

Levels of Gene Expression Control

The human body contains many types of cells that differ in structure and function. Each cell type must contain its own mix of proteins that make it different from all other cell types. Therefore, only certain genes are active in cells that perform specialized functions, such as nerve, muscle, gland, and blood cells.

Some of these active genes are called housekeeping genes because they govern functions that are common to many types of cells, such as glucose metabolism. But otherwise, the activity of selected genes accounts for the specialization of cells. In other words, gene expression is controlled in a cell, and this control accounts for its specialization (Fig. 12.4).

Let's begin by taking a look at gene regulation in prokaryotes because prokaryotes utilize only one level of control. They control the transcription of their genes.

Gene Expression in Prokaryotes

The bacterium *Escherichia coli* lives in the intestine and can quickly adjust its production of enzymes according to what we eat. If someone drinks a glass of milk, *E. coli* immediately begins to make three enzymes needed to metabolize lactose. All three enzymes are under the control of one promoter, a short DNA sequence where RNA polymerase first attaches when transcription occurs. The French microbiologists François Jacob and Jacques Monad called such a cluster of bacterial genes, along with the DNA sequences that control their transcription, an **operon.** They received a Nobel prize in 1961 for their investigations because they were the first to show how gene expression was controlled—specifically, how the *lac* operon was controlled (Fig. 12.5).

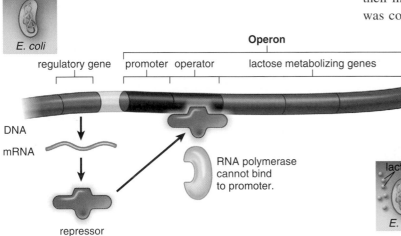

a. **Lactose is absent—operon is turned off.**
Enzymes needed to metabolize lactose are not produced.

Figure 12.5   The *lac* operon.

a. The regulatory gene codes for a repressor. The repressor binds to the operator and prevents RNA polymerase from attaching to the promoter. Therefore, transcription of the structural genes does not occur. **b.** When lactose is present, it binds to the repressor, changing its shape so that it becomes inactive and can no longer bind to the operator. Now RNA polymerase binds to the promoter, and the structural genes are expressed.

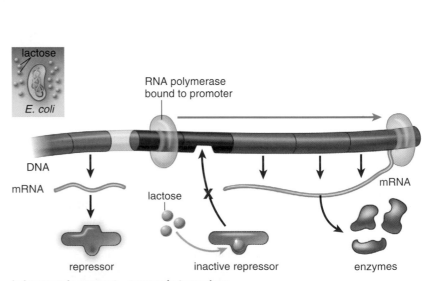

b. **Lactose is present—operon is turned on.**
Enzymes needed to metabolize lactose are produced.

In their model, a **regulatory gene** located outside the operon codes for a **repressor,** a protein with the potential to bind to the operator (a short sequence of DNA near the promoter). The activity of the *lac* operon is not needed much of the time, and the regulatory gene codes for a repressor that ordinarily binds to the operator. Now, RNA polymerase cannot attach to the promoter, and transcription cannot occur.

What turns the operon on when lactose is present? Lactose binds to the repressor, inactivating it; specifically, a change in shape prevents the repressor from attaching to the operator. Now RNA polymerase is able to bind to the promoter, and transcription of the genes needed for lactose metabolism can occur. Binding of RNA polymerase is further ensured by another regulatory protein called CAP, which is not shown in Figure 12.5. In this way, these enzymes are produced only when they are needed to metabolize lactose.

Other bacterial operons, such as those that control amino acid synthesis, are usually turned on. For example, in the *trp* operon, the regulatory gene codes for a repressor that ordinarily is unable to attach to the operator. Therefore, the genes needed to make the amino acid tryptophan are ordinarily expressed. When tryptophan is present, it binds to the repressor. A change in shape activates the repressor and allows it to bind to the operator. Now the operon is turned off.

Check Your Progress

Contrast when the *lac* operon is turned on with when the *trp* operon is turned off.

Answer: The *lac* operon (which breaks down lactose) is turned on only when lactose is present. The *trp* operon (which synthesizes tryptophan) is turned off only when tryptophan is present.

Gene Expression in Eukaryotes

In bacteria, a single promoter serves several genes that make up a transcription unit, while in eukaryotes, each gene has its own promoter where RNA polymerase binds. Bacteria rely purely on transcriptional control, but in eukaryotes, a variety of mechanisms regulate gene expression. These eukaryotic mechanisms affect whether the gene is expressed, the speed with which it is expressed, and how long it is expressed.

Some mechanisms of gene expression occur in the nucleus, and others occur in the cytoplasm (Fig. 12.6). In the nucleus, DNA unpacking, transcription, and mRNA processing all play a role in determining which proteins are made in the cytoplasm.

In the cytoplasm, translation of mRNA into a polypeptide at the ribosomes can occur right away or be delayed. The mRNA can last a long time or be destroyed immediately, and the same holds true for a protein. These mechanisms control the quantity of gene product and/or how long it is active.

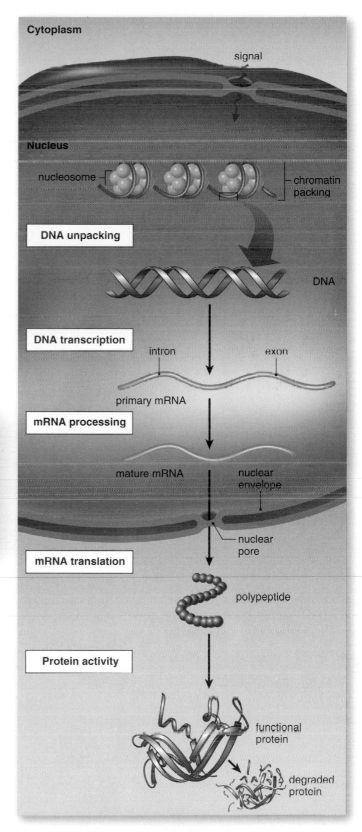

Figure 12.6 **Control of gene expression in eukaryotic cells.**

Gene expression which can be turned on by an external signal (red) is controlled at various levels in eukaryotic cells.

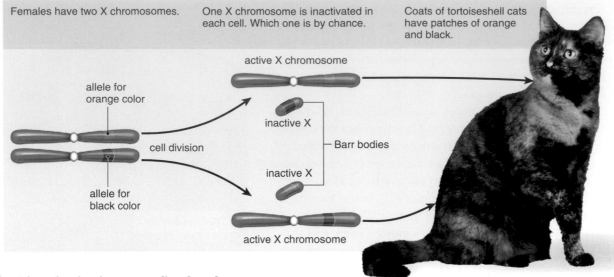

Figure 12.7 **X-inactivation in mammalian females.**

In cats, the alleles for black or orange coat color are carried on the X chromosome. Random X-inactivation occurs in females. Therefore, in heterozygous females, 50% of the cells have an allele for black coat color, and 50% of cells have an allele for orange coat color. The result is calico or tortoiseshell cats that have coats with patches of both black and orange.

Check Your Progress

List the levels of gene expression that can be controlled in eukaryotes.

Answer: DNA unpacking, transcription of DNA, mRNA processing, translation of mRNA, protein activity.

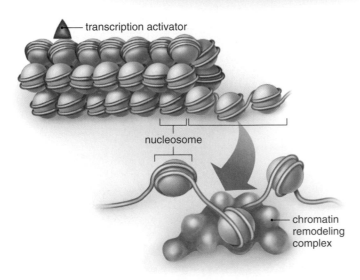

Figure 12.8 **DNA unpacking.**

Compact chromatin is made accessible for transcription when a transcription activator works with a chromatin remodeling complex to push aside the histone portions of nucleosomes.

DNA Unpacking Eukaryotes utilize chromatin packing as a way to keep genes turned off. Some inactive genes are located within darkly staining portions of chromatin, called **heterochromatin.** A dramatic example of inactive heterochromatin is the Barr body in mammalian females. Females have a small, darkly staining mass of condensed chromatin adhering to the inner edge of the nuclear envelope. This structure, called a *Barr body* after its discoverer, is an inactive X chromosome.

How do we know that Barr bodies are inactive X chromosomes that are not producing gene product? Suppose 50% of the cells in a female have one X chromosome active, and 50% have the other X chromosome active. Wouldn't the body of a heterozygous female be a mosaic, with "patches" of genetically different cells? This is exactly what happens. For example, human females who are heterozygous for an X-linked recessive form of ocular albinism have patches of pigmented and nonpigmented cells at the back of the eye. And women who are heterozygous for hereditary absence of sweat glands have patches of skin lacking sweat glands. The female calico cat also provides dramatic support for a difference in X-inactivation in its cells (Fig. 12.7).

Active genes in eukaryotic cells are associated with more loosely packed chromatin, called **euchromatin.** You learned in Chapter 8 that in eukaryotes, a *nucleosome* is a portion of DNA wrapped around a group of histone molecules. When DNA is transcribed, a transcription activator pushes aside the histone portions of nucleosomes so that transcription can begin (Fig. 12.8). In other words, even euchromatin needs further modification before transcription can begin. The presence of nucleosomes limits access to DNA, and euchromatin becomes genetically active when histones no longer bar access to DNA. Only then is it possible for a gene to be turned on and expressed in a eukaryotic cell.

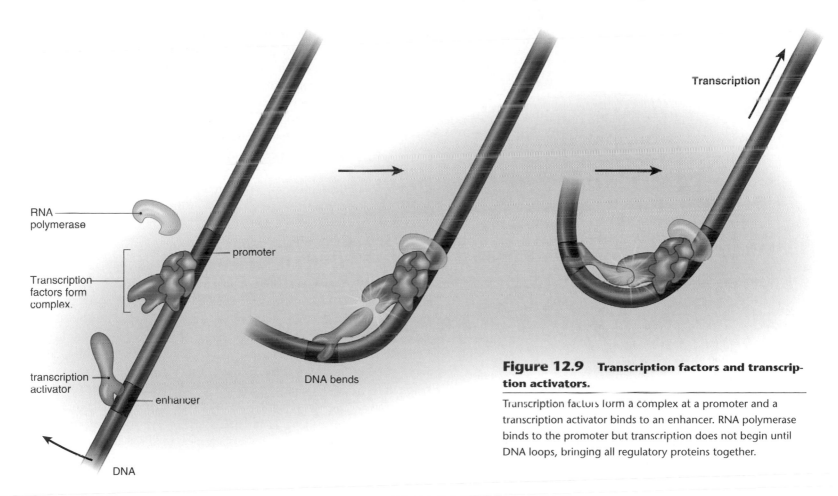

Figure 12.9   Transcription factors and transcription activators.

Transcription factors form a complex at a promoter and a transcription activator binds to an enhancer. RNA polymerase binds to the promoter but transcription does not begin until DNA loops, bringing all regulatory proteins together.

DNA Transcription   Transcription in eukaryotes follows the same principles as in bacteria, except that many more regulatory proteins per gene are involved. The occurrence of so many regulatory proteins allows for greater control, but it also allows for a greater chance of malfunction.

In eukaryotes, **transcription factors** are DNA-binding proteins that help RNA polymerase bind to a promoter. Several translation factors are needed in each case, and if one is missing, transcription cannot take place. All the transcription factors form a *complex* that also helps pull double-stranded DNA apart and even acts to position RNA polymerase so that transcription can begin. The same transcription factors, but not the same mix, are used over again at other promoters, so it is easy to imagine that if one malfunctions, the result could be disastrous to the cell. The genetic disorder Huntington disease is a devastating psychomotor ailment caused by a defect in a transcription factor.

In eukaryotes, **transcription activators** are DNA-binding proteins that speed transcription dramatically. They bind to a DNA region called an **enhancer** that can be quite a distance from the promoter. A hairpin loop in the DNA can bring the transcription activators attached to enhancers into contact with the transcription factor complex (Fig. 12.9).

It's possible for a single regulatory protein to have a decisive effect if it is the last protein needed to turn on a gene. As an example, investigators have found one DNA-binding protein, *MyoD*, that alone causes fibroblasts to become muscle cells in various vertebrates. Another DNA-binding protein, called *Ey*, can bring about the formation of not just a single cell type but a complete eye in flies (Fig. 12.10).

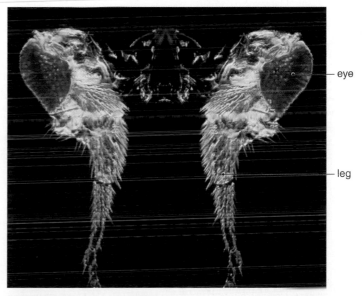

Figure 12.10   Ey gene.

In fruit flies, the expression of the *Ey* gene, in the precursor cells of the leg, triggers the development of an eye on the leg.

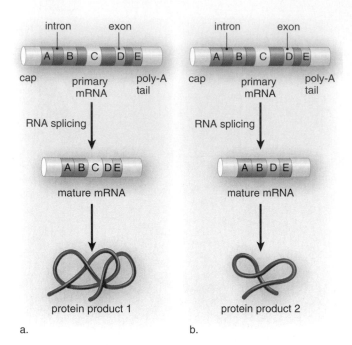

Figure 12.11   Processing of mRNA transcripts.

Because the primary mRNAs are processed differently in these two cells (**a**) and (**b**), distinct proteins result.

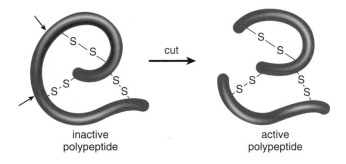

Figure 12.12   Protein activity.

The protein insulin is not active until an enzyme removes a portion of the initial polypeptide. The resulting two polypeptide chains are held together by two disulfide (S—S) bonds between like amino acids.

mRNA Processing   After transcription, messenger RNA (mRNA) is processed before it leaves the nucleus and passes into the cytoplasm. Removal of introns and splicing of exons can vary the type of mRNA that leaves the nucleus. For example, both the hypothalamus and the thyroid gland produce a hormone called calcitonin, but the mRNA that leaves the nucleus is not the same in both types of cells. Radioactive labeling studies show that they vary because of a difference in the exons they contain (Fig. 12.11). Evidence of different patterns of mRNA splicing is found in other cells, such as those that produce neurotransmitters and antibodies.

The speed of transport of mRNA from the nucleus into the cytoplasm can ultimately affect the amount of gene product realized following transcription. Evidence indicates there is a difference in the length of time it takes various mRNA molecules to pass through a nuclear pore.

mRNA Translation   The cytoplasm contains proteins that can control whether translation of mRNA takes place. For example, an initiation factor, known as IF-2, inhibits the start of protein synthesis when it is phosphorylated by a specific protein kinase. Environmental conditions can delay translation. Red blood cells do not produce hemoglobin unless heme, an iron-containing group, is available.

The longer an mRNA remains in the cytoplasm before it is broken down, the more gene product is produced. During maturation, mammalian red blood cells eject their nuclei, and yet they continue to synthesize hemoglobin for several months thereafter. The necessary mRNAs must be able to persist all this time. Differences in the guanine cap or poly-A tail determine how long a particular transcript remains active before it is destroyed by a ribonuclease associated with ribosomes. Hormones cause the stabilization of certain mRNA transcripts. An mRNA called vitellin persists for 3 weeks instead of 15 hours if it is exposed to estrogen.

Protein Activity   Some proteins are not active immediately after synthesis. For example, insulin is a single long polypeptide that folds into a three-dimensional structure. Only then is a sequence of about 30 amino acids enzymatically removed from the middle of the molecule. This leaves two polypeptide chains bonded together by disulfide (S—S) bonds, and an active insulin molecule results (Fig. 12.12).

Feedback control usually regulates the activity of proteins. Many proteins are short-lived in cells because they are degraded or destroyed. The cell has giant enzyme complexes, called proteosomes, that carry out this task.

Check Your Progress

1. Contrast heterochromatin with euchromatin.
2. Explain how mRNA processing can influence the types of gene products made by a cell.
3. Explain how a mammalian red blood cell can synthesize hemoglobin after it ejects its nucleus.

Answers: 1. Heterochromatin is tightly packed and darkly stained, and generally contains inactive genes. Euchromatin is loosely packed and lightly stained, and generally contains active genes. 2. The exons that are spliced out may differ in different cell types, resulting in varying types of products after translation. 3. Its mRNA survives for many months, allowing translation to make hemoglobin.

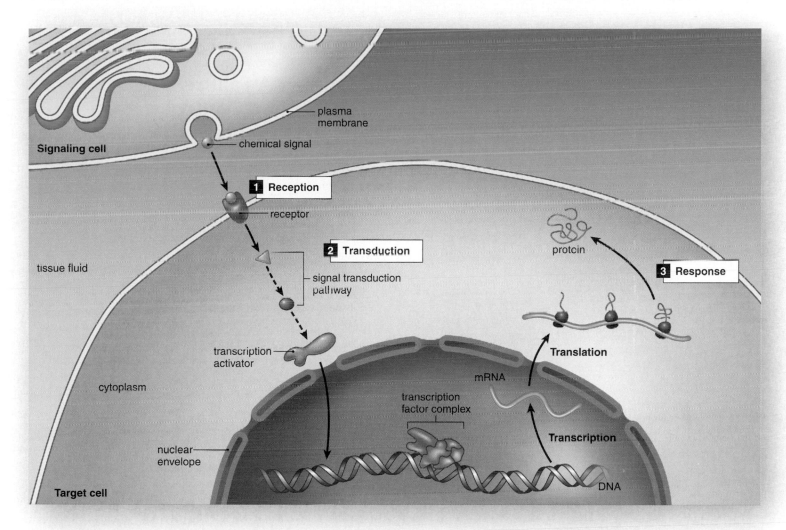

Signaling cell

plasma membrane

chemical signal

1 Reception

receptor

2 Transduction

tissue fluid

signal transduction pathway

protein

3 Response

transcription activator

Translation

cytoplasm

mRNA

transcription factor complex

Transcription

nuclear envelope

DNA

Target cell

Signaling Between Cells in Eukaryotes

In multicellular organisms, cells are constantly sending out chemical signals that influence the behavior of other cells. During animal development, these signals determine the specialized role a cell will play in the organism. Later, the signals help coordinate growth and day-to-day functions. Plant cells also signal each other so that their responses to environmental stimuli, such as direct sunlight, are coordinated.

Typically, a **cell-signaling pathway** begins when a chemical signal binds to a receptor protein in a target cell's plasma membrane. The signal causes the receptor protein to initiate a series of reactions within a **signal transduction pathway.** The end product of the pathway (not the signal) directly affects the metabolism of the cell. For example, growth is only possible if certain genes have been turned on by regulatory proteins.

In Figure 12.13, a signaling cell secretes a chemical signal that binds to a specific receptor located in the receiving cell's plasma membrane. The binding activates a series of reactions within a signal transduction pathway. The last reaction activates a transcription activator that enhances the transcription of a specific gene. Transcription leads to the translation of mRNA and a protein product that, in this case, stimulates the cell cycle so that growth occurs.

A protein called Ras functions in signaling pathways that lead to the transcription of many genes, several of which promote the cell cycle. Ras is normally inactive, but the reception of a growth factor leads to its activation. Ras becomes active when it is phosphorylated.

Figure 12.13 Cell-signaling pathway.

1 A chemical signal is received at a specific receptor protein.

2 A signal transduction pathway terminates when a transcription activator is stimulated. It enhances transcription of a specific gene.

3 Translation of the mRNA transcript follows and a protein product results. The protein product is the response to the signal.

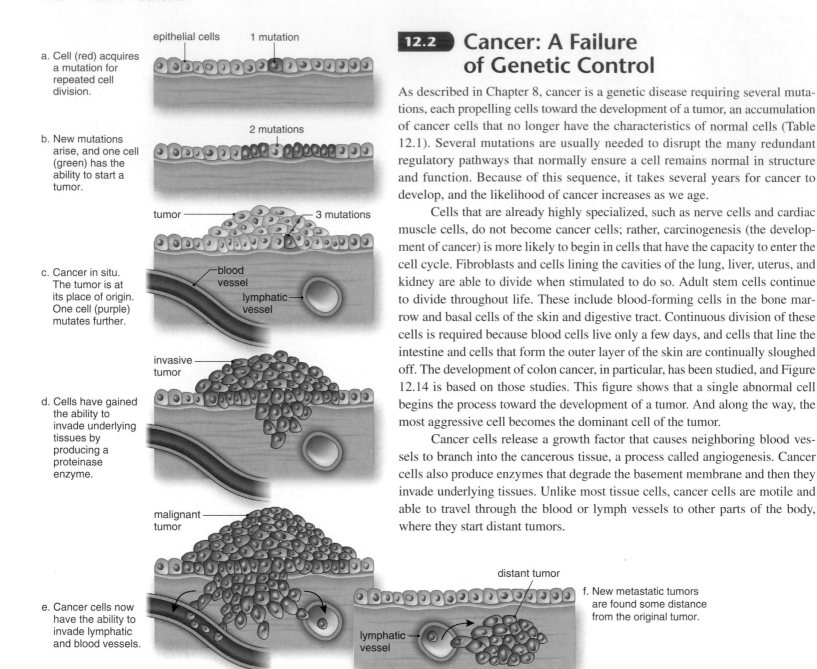

a. Cell (red) acquires a mutation for repeated cell division.

epithelial cells   1 mutation

b. New mutations arise, and one cell (green) has the ability to start a tumor.

2 mutations

c. Cancer in situ. The tumor is at its place of origin. One cell (purple) mutates further.

tumor   3 mutations

blood vessel

lymphatic vessel

d. Cells have gained the ability to invade underlying tissues by producing a proteinase enzyme.

invasive tumor

e. Cancer cells now have the ability to invade lymphatic and blood vessels.

malignant tumor

distant tumor

f. New metastatic tumors are found some distance from the original tumor.

lymphatic vessel

12.2 Cancer: A Failure of Genetic Control

As described in Chapter 8, cancer is a genetic disease requiring several mutations, each propelling cells toward the development of a tumor, an accumulation of cancer cells that no longer have the characteristics of normal cells (Table 12.1). Several mutations are usually needed to disrupt the many redundant regulatory pathways that normally ensure a cell remains normal in structure and function. Because of this sequence, it takes several years for cancer to develop, and the likelihood of cancer increases as we age.

Cells that are already highly specialized, such as nerve cells and cardiac muscle cells, do not become cancer cells; rather, carcinogenesis (the development of cancer) is more likely to begin in cells that have the capacity to enter the cell cycle. Fibroblasts and cells lining the cavities of the lung, liver, uterus, and kidney are able to divide when stimulated to do so. Adult stem cells continue to divide throughout life. These include blood-forming cells in the bone marrow and basal cells of the skin and digestive tract. Continuous division of these cells is required because blood cells live only a few days, and cells that line the intestine and cells that form the outer layer of the skin are continually sloughed off. The development of colon cancer, in particular, has been studied, and Figure 12.14 is based on those studies. This figure shows that a single abnormal cell begins the process toward the development of a tumor. And along the way, the most aggressive cell becomes the dominant cell of the tumor.

Cancer cells release a growth factor that causes neighboring blood vessels to branch into the cancerous tissue, a process called angiogenesis. Cancer cells also produce enzymes that degrade the basement membrane and then they invade underlying tissues. Unlike most tissue cells, cancer cells are motile and able to travel through the blood or lymph vessels to other parts of the body, where they start distant tumors.

Figure 12.14   Development of cancer.

Development of cancer requires several mutations, each contributing to the development of a tumor. Normally cells exhibit contact inhibition and controlled growth. They stop dividing when they are bordered by other cells. Tumor cells lack contact inhibition and pile up forming a tumor. Cancer cells do not fulfill the functions of a specialized tissue and they have abnormal chromosomes. Even so, they do not undergo apoptosis (cell death) and keep on dividing forever. Eventually they form tumors in other parts of the body.

Table 12.1	Normal Cells Compared to Cancer Cells
Normal Cells	**Cancer Cells**
Contact inhibition	No contact inhibition
Controlled growth	Uncontrolled growth (tumor)
Specialized cells	Nonspecialized cells
Normal chromosomes	Abnormal chromosomes
Undergo apoptosis	No apoptosis

Proto-Oncogenes and Tumor Suppressor Genes

Recall that the cell cycle, as discussed on page 114, consists of interphase, followed by mitosis. *Cyclin* is a molecule that has to be present for a cell to proceed from interphase to mitosis. When cancer develops, the cell cycle occurs repeatedly, in large part, due to mutations in two types of genes. Figure 12.15 shows that:

1. **Proto-oncogenes:** code for proteins that promote the cell cycle and inhibit apoptosis. They are often likened to the gas pedal of a car because they cause acceleration of the cell cycle.
2. **Tumor suppressor genes:** code for proteins that inhibit the cell cycle and promote apoptosis. They are often likened to the brakes of a car because they inhibit acceleration of the cell cycle.

Proto-Oncogenes Become Oncogenes

When proto-oncogenes mutate, they become cancer-causing genes called **oncogenes.** These mutations can be called "gain-of-function" mutations because overexpression is the result (Fig. 12.16). Whatever a proto-oncogene does, an oncogene does it better.

A **growth factor** is a signal that activates a cell-signaling pathway, resulting in cell division. Some proto-oncogenes code for a growth factor or for a receptor protein that receives a growth factor. When these proto-oncogenes become oncogenes, receptor proteins are easy to activate and even may be stimulated by a growth factor produced by the receiving cell. Several proto-oncogenes code for Ras proteins that promote mitosis by activating **cyclin.** *Ras* oncogenes are typically found in many different types of cancers. *Cyclin D* is a proto-oncogene that codes for a cyclin directly. When this gene becomes an oncogene, cyclin is readily available all the time.

p53 is a transcription factor instrumental in stopping the cell cycle and activating repair enzymes. If repair is impossible, the p53 protein promotes *apoptosis*, programmed cell death. Apoptosis is an important way carcinogenesis is prevented. A certain proto-oncogene codes for a protein that functions to make p53 unavailable. When this proto-oncogene becomes an oncogene, no matter how much p53 is made, none will be available. Many tumors are lacking in p53 activity.

Tumor Suppressor Genes Become Inactive

When tumor suppressor genes mutate, their products no longer inhibit the cell cycle nor promote apoptosis. Therefore, these mutations can be called "loss of function" mutations (Fig. 12.16).

The retinoblastoma protein (RB) controls the activity of a transcription activator for *cyclin D* and other genes whose products promote entry into the S phase of the cell cycle. When the tumor suppressor gene *p16* mutates, the RB protein is always functional and the result is, again, too much active *cyclin D* in the cell. The cell experiences repeated rounds of DNA synthesis without the occurrence of mitosis.

The protein Bax promotes apoptosis. When a tumor suppressor gene *Bax* mutates, the protein Bax is not present, and apoptosis is less likely to occur. The gene contains a run of eight consecutive G bases, making it subject to mutations.

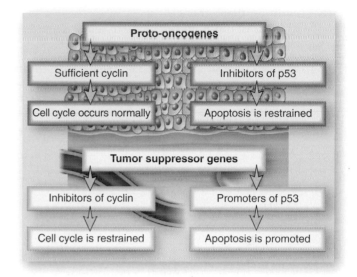

Figure 12.15   Normal cells.

In normal cells, proto-oncogenes code for sufficient cyclin to keep the cell cycle going normally, and they code for proteins that inhibit p53 and apoptosis. Tumor suppressor genes code for proteins that inhibit cyclin, and promote p53 and apoptosis.

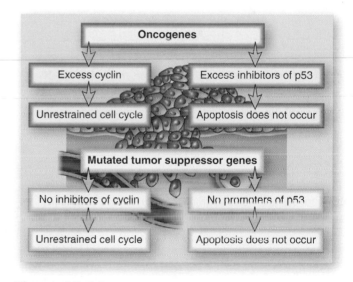

Figure 12.16   Cancer cells.

Notice that oncogenes and mutated tumor suppressor genes have the same end effects: unrestrained cell cycle and apoptosis does not occur.

Other Genetic Changes

Cancer cells also undergo other genetic changes that result in certain other specific characteristics.

Absence of Telomere Shortening

A **telomere** is a repeating DNA sequence (TTAGGG) at the end of the chromosomes that can be as long as 15,000 base pairs. Telomeres have been likened to the protective caps on the ends of shoelaces. Rather than keeping chromosomes from unraveling, however, telomeres stop chromosomes from fusing to each other. Each time a cell divides, some portion of a telomere is lost; when telomeres become too short, the chromosomes fuse and can no longer duplicate. Then the cell is "old" and dies by apoptosis.

Embryonic cells and also certain adult cells, such as stem cells and germ cells, have an enzyme called *telomerase* that can rebuild the telomeres. The gene that codes for telomerase is turned on in cancer cells, which allows them to keep dividing over and over again. Telomerase is believed to become active after a cell has already started proliferating wildly.

Angiogenesis

Angiogenesis, as noted earlier, is the formation of new blood vessels. To grow larger than, say, a pea, a tumor must have a well-developed capillary network to bring it nutrients and oxygen. Some of the growth factors released by cancer cells, such as vascular endothelial growth factor (VEGF), bind to receptors on the epithelial cells of blood vessels. Thereafter, the epithelial cells grow and divide, producing new blood vessels that branch and send capillaries into the center of a tumor.

The new blood vessels supply tumor cells with the nutrients and oxygen they require for rapid growth, but they also rob normal tissues of nutrients and oxygen. Two highly effective drugs, called angiostatin and endostatin, have been shown to inhibit angiogenesis and are now being studied as possible anti-cancer drugs.

Metastasis

A *benign tumor* is a disorganized, usually encapsulated mass that does not invade adjacent tissue. *Cancer in situ* is a tumor located in its place of origin, before any invasion of normal tissue has occurred (Fig. 12.17).

Malignancy is present when **metastasis** establishes new tumors distant from the primary tumor. To metastasize, cancer cells must make their way across a basement membrane and into a blood vessel or lymphatic vessel. Cancer cells produce proteinase enzymes that degrade the basement membrane and allow them to invade underlying tissues.

Due to mutations of particular genes, cancer cells tend to be motile because they have a disorganized internal cytoskeleton and lack intact actin filament bundles. After traveling through the blood or lymph, cancer cells start tumors elsewhere in the body. Not many cancer cells achieve this feat, but when they do, the prognosis for recovery becomes doubtful.

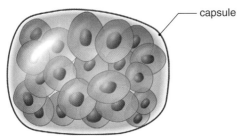

capsule

benign tumor

cancer in situ

a. b.

Figure 12.17   Benign tumor versus cancer in situ.

a. A benign tumor is encapsulated. **b.** Cancer in situ is not encapsulated and is more likely to become a metastatic tumor.

Hereditary Forms of Cancer

The environmental influences discussed in Chapter 8 (radiation, organic chemicals, and viruses) are major risk factors for the development of cancer. These agents can cause mutations in tumor suppressor genes and proto-oncogenes so that cancer is more likely. Although cancer is usually a somatic disease, meaning that it develops only in body cells, some instances of an inherited predisposition to cancer are known.

Inheritance Patterns for Cancer

BRCA1* and *BRCA2   In 1990, DNA linkage studies on large families in which females tended to develop breast cancer identified the first gene associated with breast cancer. Scientists named this gene *b*reast *c*ancer 1 or *BRCA1* (pronounced brak-uh). Later, they found that breast cancer in other families was due to another breast cancer gene they called *BRCA2*. These genes are tumor suppressor genes that are known to behave as if they are autosomal recessive alleles. If one mutated allele is inherited from either parent, a mutation in the other allele is required before the predisposition to cancer is increased. Because the first mutated gene is inherited, it is present in all cells of the body, and then cancer is more likely wherever the second mutation occurs. If the second mutation occurs in the breast, breast cancer may develop (Fig. 12.18). If the second mutation is in the ovary, ovarian cancer may develop if additional cancer-causing mutations occur.

***RB* Gene**   The *RB* gene is also a tumor suppressor gene. It takes its name from its association with an eye tumor called a *retino*blastoma, which first appears as a white mass in the retina. A tumor in one eye is most common because it takes mutations in both alleles before cancer can develop (Fig. 12.19). Children who inherit a mutated allele are more likely to have tumors in both eyes.

***RET* Gene**   An abnormal *RET* gene, which predisposes a person to thyroid cancer, can be passed from parent to child. *RET* is a proto-oncogene known to be inherited in an autosomal dominant manner— only one mutated allele is needed to increase a predisposition to cancer. The remainder of the mutations necessary for thyroid cancer to develop are acquired (not inherited).

Testing for These and Other Genes

Genetic tests are available for the presence of the *BRCA* genes, the *RET* gene, and the *RB* gene. Persons who have inherited these genes may decide to have surgery by choice rather than necessity or be examined frequently for signs of cancer.

Genetic tests are also available for other types of mutated genes that help a physician diagnose cancer. For example, a *ras* oncogene can be detected in stool and urine samples. If the test for *ras* oncogene in the stool is positive, the physician suspects colon cancer, and if the test for *ras* oncogene in the urine is positive, the physician suspects bladder cancer.

Telomerase, you may recall, is the enzyme that keeps telomeres a constant length in cells. The gene that codes for telomerase is turned off in normal cells but is active in cancer cells. Therefore, if the test for the presence of telomerase is positive, the cell is likely cancerous.

SEM of breast cancer cell

Figure 12.18   Breast cancer can run in families.

These three sisters have all had breast cancer. Genetic tests can identify women at risk for breast cancer so that they can choose to have frequent examinations to allow for early detection.

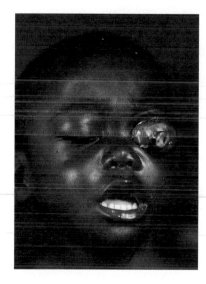

Figure 12.19   Inherited retinoblastoma.

A child is at risk for an eye tumor when a mutated *RB* allele is inherited, even though a second mutation in the normal allele is required before the tumor develops.

Check Your Progress

1. Explain how telomeres influence a cell's life span.
2. Explain why angiogenesis is important for cancer growth.
3. Compare and contrast the *BRCA* gene with the *RB* gene.

Answers: 1. Telomeres become shorter after each cell division. When the telomere becomes too short, chromosomes fuse together and the cell dies. 2. The formation of new blood cells supplies a tumor with the nutrients it needs for growth. 3. Both are tumor suppressor genes. *BRCA* is associated with breast cancer, while *RB* causes retinoblastoma.

THE CHAPTER IN REVIEW

Summary

12.1 Control of Gene Expression

Every cell in the body has a complete set of genes as is proven by the process of cloning, when a 2n nucleus is able to bring about the development of an organism exactly like the 2n nucleus donor or any specialized tissue that matches those of the donor.

Reproductive and Therapeutic Cloning

In reproductive cloning, the desired end is an individual genetically identical to the original individual. To achieve this type of cloning, the early embryo is placed in a surrogate mother until the individual comes to term.

In therapeutic cloning, the cells of the embryo are separated and treated so that they develop into specialized tissues that can be used to treat human disorders. Alternately, adult stem cells can be treated to become specialized tissues.

Levels of Gene Expression Control

The specialization of cells is not due to the presence/absence of genes; it is due to the activity/inactivity of genes.

Gene Expression in Prokaryotes   In prokaryotes, gene expression is controlled at the level of transcription.

- In the *lac* operon, the repressor usually binds to the operator. Then RNA polymerase cannot bind, and the operon is turned off. When lactose is present, it binds to the repressor, and now RNA polymerase binds to the promoter, and the genes for lactose metabolism are transcribed.
- In the *trp* operon, the genes are usually transcribed because the repressor is inactive. When tryptophan is present, the repressor becomes active, and RNA polymerase cannot bind to the promoter.

Gene Expression in Eukaryotes   Gene expression in eukaryotes occurs at various levels:

- DNA unpacking—heterochromatin becomes active.
- DNA transcription—transcription factors present at the promoter, and transcription activators (bound to enhancers) are brought in contact and form a transcription initiation complex.
- mRNA processing—alternate patterns of processing; how quickly mRNA leaves the nucleus.
- Translation of mRNA—how long the mRNA stays active.
- Protein activity—activation of a protein may require further steps and degradation of the protein can occur at once or after a while.

Signaling Between Cells in Eukaryotes   Eukaryotic cells constantly communicate with their neighbors by secreting signals.

- Signals are received at receptor proteins located in the plasma membrane.
- A signal transduction pathway (enzymatic reactions) leads to a change in the behavior of the cell. For example, when stimulated via a transduction pathway, a transcription activator stimulates transcription of a particular gene.

12.2 Cancer: A Failure of Genetic Control

Cancer Is a Genetic Disease

Evidence is strong that cancer is a genetic disease. A series of mutations, particularly in cells that can divide, lead to cancer cells with these characteristics:

- No contact inhibition
- Uncontrolled growth
- Are nonspecialized
- Have abnormal chromosomes
- Do not undergo apoptosis

Proto-Oncogenes and Tumor Suppressor Genes

- Proto-oncogenes ordinarily promote the cell cycle and inhibit apoptosis. Mutation results in a gain of function for these activities.
- Tumor suppressor genes ordinarily suppress the cell cycle and promote apoptosis. Mutation results in loss of function for these activities.

Mutations in both types of cells result in
- Excess cyclin to stimulate the cell cycle
- No p53 to bring about apoptosis

Other Genetic Changes

- Telomeres at ends of chromosomes don't shorten because telomerase is present.
- Angiogenesis, formation of new blood vessels, occurs because cancer cells release VEGF.
- Metastasis, spread of cancer to remote areas, occurs because cells are mobile.

Hereditary Forms of Cancer

Mutagens such as radiation, organic chemicals, and viruses, are also major risk factors for the development of cancer.

Inheritance Patterns for Cancer   Certain genes are known to cause cancer, including:

- *BRCA1* and *BRCA2*, which are tumor suppressor genes. If one mutated gene is inherited and a second mutates, cancer is more likely.
- The *RB* gene is also a tumor suppressor gene. When a mutated allele is inherited, tumors in both eyes are more likely.
- An abnormal *RET* gene, which predisposes a person to thyroid cancer, can also be passed from parent to child.

Testing for Cancer Genes   Genetic tests are available to help a physician diagnose cancer. These genes are turned off in normal cells, but are active in cancer cells. For example, if the test for the presence of telomerase is positive, the cell is cancerous.

Thinking Scientifically

1. A person with Turner syndrome is XO and phenotypically female. Typical symptoms of Turner syndrome include rudimentary sexual development, extra folds of skin on the neck, and short stature. If normal (XX) women have only one active X chromosome, why would someone with a single X chromosome exhibit this syndrome?

2. Most cat breeders know that male tortoiseshell cats are very rare. In addition, these cats are inevitably sterile. Explain these two observations.

Testing Yourself

Choose the best answer for each question.

1. During reproductive cloning, a(n) _____ is placed into a(n) _____.

   a. enucleated egg cell, adult cell nucleus
   b. adult cell nucleus, enucleated egg cell
   c. egg cell nucleus, enucleated adult cell
   d. enucleated adult cell, egg cell nucleus

2. The major challenge to therapeutic cloning using adult stem cells is

   a. finding appropriate cell types.
   b. obtaining enough tissue.
   c. controlling gene expression.
   d. keeping cells alive in culture.

3. Cancer cells are immortal because

   a. they produce telomerase.
   b. their tumor suppressor genes stimulate apoptosis.
   c. they have an unlimited food supply.
   d. they can complete mitosis without cytokinesis.

4. Label this operon with these terms: active repressor, lactose metabolizing genes, operator, RNA polymerase promoter, regulatory gene, mRNA

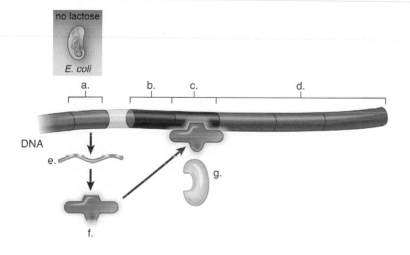

5. Which of the following is not a gene expression control mechanism in eukaryotes?

   a. mRNA processing
   b. rate of ribosome synthesis
   c. longevity of mRNA
   d. speed of passage of mRNA through nuclear pores

6. What types of offspring would you expect from a female tortoiseshell cat and a black male? Remember that this coat color gene is due to heterozygous alleles on the X chromosomes.

   a. tortoiseshell females and black males
   b. all black cats
   c. tortoiseshell females, black females, orange males, black males
   d. orange females and black males

For questions 7–11, match the examples to the gene expression control mechanisms in the key. Each answer can be used more than once.

Key:

   a. DNA unpacking
   b. mRNA processing
   c. translation of mRNA
   d. protein activity

7. Insulin does not become active until 30 amino acids are cleaved from the middle of the molecule.

8. The mRNA for vitellin is longer-lived if it is exposed to estrogen.

9. Genes in Barr bodies are inactivated.

10. Calcitonin is produced in both the hypothalamus and the thyroid gland, but in different forms due to exon splicing.

11. A translation initiation factor inhibits translation when it is phosphorylated by kinase.

12. Suppose a woman is heterozygous for the *RB* gene that causes retinoblastoma. Her husband's mother was homozygous for the mutant *RB* allele, but his father's side of the family has no history of retinoblastoma. What is the probability that this couple will have a child who does not carry the mutant *RB* allele?

   a. ¼   c. ¾
   b. ½   d. 0

13. A cell is likely to be cancerous if it tests positive for a (an)

   a. elongation factor.   c. enhancer.
   b. transcription factor.   d. telomerase.

14. Which of the following correctly describes reproductive cloning?

   a. The desired end is an individual that is a genetic improvement over the original individual.
   b. The desired end is mature cells of various cell types.
   c. An embryonic nucleus is placed in an enucleated and starved embryonic donor cell.
   d. Products of this process show signs of being unhealthy and may age more rapidly.
   e. Ethical concerns surround the embryo, which could become an individual if allowed.

15. The product of therapeutic cloning differs from the product of reproductive cloning in that it is

   a. various mature cells, not an individual.
   b. genetically identical to the somatic cells from the donor.
   c. various mature cells with the ability to become an individual.
   d. an individual that is genetically identical to the donor of the nucleus.
   e. a result of starving donor cells, causing them to stop dividing and go into a resting state.

16. What is a Barr body?

    a. an example of gene regulation using chromatin packing
    b. an inactive X chromosome
    c. an X chromosome not producing gene products
    d. explains tortoiseshell cats with patches of black and orange fur
    e. All of these are correct.

17. How is transcription controlled in eukaryotic cells?

    a. through the use of signal transduction pathways
    b. by means of cell signaling
    c. using transcription factors and activators
    d. when chromatin is packed to keep genes turned on
    e. None of these are correct.

18. An operon is a short sequence of DNA

    a. that prevents RNA polymerase from binding to the promoter.
    b. that prevents transcription from occurring.
    c. and the sequences that control its transcription.
    d. that codes for the repressor protein.
    e. that functions to prevent the repressor from binding to the operator.

19. Which of these correctly describes the function of a regulatory gene for the *lac* operon?

    a. prevents transcription from occurring
    b. a sequence of DNA that codes for the repressor
    c. prevents the repressor from binding to the operator
    d. keeps the operon off until lactose is present
    e. Both b and d are correct.

20. What is the purpose of cell signaling in eukaryotes?

    a. to change the behavior of the receiving cell
    b. to initiate a set of reactions through the receptor protein
    c. to turn on a gene
    d. to transmit a nerve impulse, for example
    e. All of these are correct.

21. Eukaryotic cells signal one another through the use of

    a. repressor proteins acting upon operons.
    b. molecules that bind to a receptor protein in the membrane of a target cell.
    c. transcription factors and activators acting upon RNA polymerase.
    d. RNA polymerase acting upon transcription promoters and enhancers.
    e. nerve impulses that cause a receiving cell to change its behavior.

22. Which of these is not a possible cause of cancer?

    a. radiation   d. genes
    b. metastasis   e. oncogenes
    c. viruses

23. Which of the following is not an example of why cancer cells are a failure of genetic control?

    a. They exhibit contact inhibition and do not grow in multiple layers.
    b. They lack specialization and do not contribute to the function of a body part.
    c. They have enlarged nuclei and may have an abnormal number of chromosomes.
    d. They fail to undergo apoptosis even though they are abnormal cells.
    e. They release a growth factor that causes neighboring blood vessels to branch into the tumor and therefore feed the cancerous tissue.

Go to www.mhhe.com/maderessentials for more quiz questions.

Bioethical Issue

Over the past decade, genetic tests have become available for certain cancer genes. If women test positive for defective *BRCA1* and *BRCA2* genes, they have an increased risk for early-onset breast and ovarian cancer. If individuals test positive for the *APC* gene, they are at greater risk for the development of colon cancer. Other genetic tests exist for rare cancers, including retinoblastoma and Wilms tumor.

   Advocates of genetic testing say that it can alert those who test positive for these mutated genes to undergo frequent mammograms or colonoscopies. Early detection of cancer clearly offers the best chance for successful treatment. Others feel that genetic testing is unnecessary because nothing can presently be done to prevent the disease. Perhaps it is enough for those who have a family history of cancer to schedule frequent checkups, beginning at a younger age.

   People opposed to genetic testing worry that being predisposed to cancer might threaten their job or health insurance. They suggest that genetic testing be confined to a research setting, especially since it is not known which particular mutations in the genes predispose a person to cancer. They are afraid, for example, that a woman with a defective *BRCA1* or *BRCA2* gene might make the unnecessary decision to have a radical mastectomy. The lack of proper counseling also concerns many. In a study of 177 patients who underwent *APC* gene testing for susceptibility to colon cancer, less than 20% received counseling before the test. Moreover, physicians misinterpreted the test results in nearly one-third of the cases.

   Should everyone be aware that genetic testing for cancer is a possibility, or should genetic testing be confined to a research setting? If genetic testing was offered to you, would you take advantage of it?

Understanding the Terms

angiogenesis 190	proto-oncogene 189
cell-signaling pathway 187	regulatory gene 183
cyclin 189	repressor 183
enhancer 186	reproductive cloning 180
euchromatin 184	signal transduction pathway 187
growth factor 189	telomere 190
heterochromatin 184	therapeutic cloning 181
metastasis 190	transcription activator 186
oncogene 189	transcription factor 186
operon 182	tumor suppressor gene 189

Match the terms to these definitions:

a. _____ Set of bacterial genes under the control of a single promoter.

b. _____ Protein that binds to the operator of an operon.

c. _____ Darkly stained and tightly packed chromatin.

d. _____ Protein in eukaryotes that helps RNA polymerase bind to a promoter.

e. _____ DNA region to which transcription activators bind to speed transcription in eukaryotes.

f. _____ Signal that activates a cell-signaling pathway.

g. _____ Repeating DNA sequence at the end of a chromosome.

Genetic Counseling

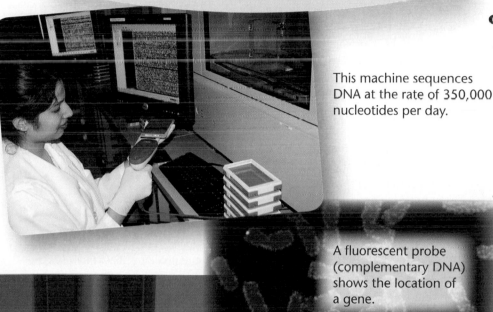

This machine sequences DNA at the rate of 350,000 nucleotides per day.

A fluorescent probe (complementary DNA) shows the location of a gene.

×150

One cell of an embryo can be removed and tested for genetic mutations.

Today it is possible to screen embryos for genetic abnormalities before they are implanted in the uterus. This process is called pre-implantation genetic diagnosis (PGD). Couples who are at high risk of having a child with a genetic disorder can use PGD to increase their chances of having a normal child. The couple donates a number of egg cells and sperm cells, and the embryos are created in a lab by a process called in vitro fertilization. Once the embryos have ▶▶▶

enough cells, one cell from each embryo can be removed and tested for genetic mutations. If an embryo is found to have a mutation for a genetic disorder of concern, it can be discarded. Embryos that appear free of genetic flaws can be implanted into the mother. If all goes well, a healthy baby will be born.

Some people have ethical concerns about using PGD to select genetically healthy embryos. Even more controversial, however, is the use of this technology purely to choose the sex of an embryo. Some couples so desperately want a boy or a girl that they use PGD to guarantee the gender of their child, and some physicians justify this use of the technology for the achievement of "gender-balance" within a family. This use of PGD is legal in the United States, but has been banned in some countries.

In this chapter, you will learn about some of the techniques currently being used to counsel couples about their chances of passing on a genetic disorder, as well as technologies that may be used to test and treat parents and children for genetic disorders in the future.

13.1 Counseling for Chromosomal Disorders

Potential parents are becoming aware that many illnesses are caused by abnormal chromosomal inheritance or by gene mutations. Therefore, more couples are seeking **genetic counseling,** which is available at many major hospitals as a means to determine the risk of inherited disorders in a family. For example, a couple might be prompted to seek counseling after several miscarriages, when several relatives have a particular medical condition, or if they already have a child with a genetic defect. The counselor helps the couple understand the mode of inheritance, the medical consequences of a particular genetic disorder, and the decisions they might wish to make (Fig. 13.1).

Various human disorders result from abnormal chromosome number or structure. When a pregnant woman is concerned that her unborn child might have a chromosomal defect, the counselor may recommend karyotyping the fetus's chromosomes.

Karyotyping

A **karyotype** is a visual display of the chromosomes arranged by size, shape, and banding pattern. Any cell in the body except red blood cells, which lack a nucleus, can be a source of chromosomes for karyotyping. In adults, it is easiest to use white blood cells separated from a blood sample for this purpose. In fetuses, whose chromosomes are often examined to detect a syndrome, cells can be obtained by either amniocentesis or chorionic villi sampling.

Figure 13.1 Genetic counseling.
A genetic counselor uses genetic information about the family to predict the chances a couple will have a child affected by a genetic disorder. If the woman is pregnant, tests can determine whether the child will be born free of genetic disorders.

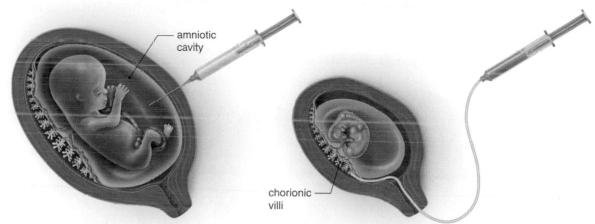

Figure 13.2 **Testing for chromosomal mutations.**

To test the fetus for an alteration in the chromosome number or structure, fetal cells can be acquired by (**a**) amniocentesis or (**b**) chorionic villi sampling. **c.** Karyotyping then will reveal chromosomal mutations. In this case, the karyotype shows that the newborn will have Down syndrome.

a. During amniocentesis, a long needle is used to withdraw amniotic fluid containing fetal cells.

b. During chorionic villi sampling, a suction tube is used to remove cells from the chorion, where the placenta will develop.

Amniocentesis is a procedure for obtaining a sample of amniotic fluid from the uterus of a pregnant woman. Blood tests and the age of the mother are considered when determining whether the procedure should be done. The risk of spontaneous abortion increases by about 0.3% due to amniocentesis, and doctors use the procedure only if it is medically warranted.

Amniocentesis is not usually performed until about the fourteenth to the seventeenth week of pregnancy. A long needle is passed through the abdominal and uterine walls to withdraw a small amount of fluid, which also contains fetal cells (Fig. 13.2a). Tests are done on the amniotic fluid. Karyotyping the chromosomes may be delayed as long as four weeks so that the cells can be cultured to increase their number.

Chorionic villi sampling (CVS) is a procedure for obtaining chorionic cells in the region where the placenta will develop. This procedure can be done as early as the fifth week of pregnancy. A long, thin suction tube is inserted through the vagina into the uterus (Fig. 13.2b). Ultrasound, which gives a picture of the uterine contents, is used to place the tube between the uterine lining and the chorionic villi. Then, a sampling of chorionic cells is obtained by suction. The cells do not have to be cultured, and karyotyping can be done immediately. But testing amniotic fluid is not possible because no amniotic fluid is collected. Also, CVS carries a greater risk of spontaneous abortion than amniocentesis—0.8% compared to 0.3%. The advantage of CVS is getting the results of karyotyping at an earlier date.

After a cell sample has been obtained, the cells are stimulated to divide in a culture medium. A chemical is used to stop mitosis during metaphase when chromosomes are the most highly compacted and condensed. The cells are then killed, spread on a microscope slide, and dried. Stains are applied to the slides, and the cells are photographed. Staining causes the chromosome to have dark and light crossbands of varying widths, and these can be used, in addition to size and shape, to help pair up the chromosomes. Today, technicians use a computer to arrange the chromosomes in pairs. As described in Chapter 9, the karyotype of a person who has Down syndrome usually has three number 21 chromosomes instead of two (Fig. 13.2c).

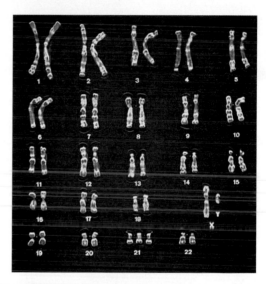

c. Karyotype of a person with Down syndrome. Note the three number 21 chromosomes.

Check Your Progress

1. Describe the relationship between amniocentesis and karyotyping.
2. List the two reasons that chorionic villi sampling allows a karyotype to be determined earlier than amniocentesis.
3. Explain why karyotypes are created from chromosomes at metaphase of mitosis.

Answers: 1. During amniocentesis, a sample of amniotic fluid is collected. Cells from this fluid are cultured and used for karyotyping. 2. Chorionic villi sampling can be performed earlier in pregnancy and does not require a cell culture period. 3. The chromosomes are the most condensed during metaphase.

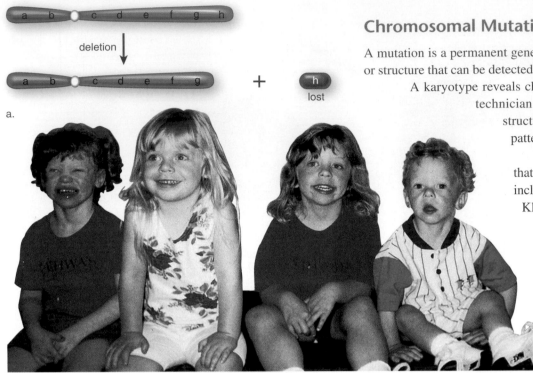

deletion

+ h lost

a.

b.

Figure 13.3 **Deletion.**

a. When chromosome 7 loses an end piece, the result is Williams syndrome. **b.** These children, although unrelated, have the same appearance, health, and behavioral problems that are characteristic of Williams syndrome.

duplication   inversion

a.

b.

Figure 13.4 **Duplication.**

a. When a piece of chromosome 15 is duplicated and inverted, inv dup 15 syndrome results. **b.** Children with this syndrome have poor muscle tone and autistic characteristics.

Chromosomal Mutations

A mutation is a permanent genetic change. A change in chromosome number or structure that can be detected microscopically is a **chromosomal mutation.**

A karyotype reveals changes in chromosome number, and a skilled technician is able to detect differences in chromosome structure based on a change in the normal banding patterns of the chromosomes (see Fig. 13.2c).

In Chapter 9, you learned about syndromes that result from changes in chromosome number, including Down syndrome, Turner syndrome, and Klinefelter syndrome.

Syndromes that result from changes in chromosome structure are due to the breakage of chromosomes and their failure to reunite properly. Various environmental agents—radiation, certain organic chemicals, or even viruses—can cause chromosomes to break apart. Ordinarily, when breaks occur in chromosomes, the segments reunite to give the same sequence of genes. But their failure to do so results in one of several types of mutations: deletion, duplication, translocation, or inversion. Chromosomal mutations can occur during meiosis, and if the offspring inherits the abnormal chromosome, a syndrome may develop.

Deletions and Duplications

A **deletion** occurs when a single break causes a chromosome to lose an end piece, or when two simultaneous breaks lead to the loss of an internal chromosome segment. An individual who inherits a normal chromosome from one parent and a chromosome with a deletion from the other parent no longer has a pair of alleles for each trait, and a syndrome can result.

Williams syndrome occurs when chromosome 7 loses a tiny end piece (Fig. 13.3). Children who have this syndrome look like pixies, with turned-up noses, wide mouths, a small chin, and large ears. Although their academic skills are poor, they exhibit excellent verbal and musical abilities. The gene that governs the production of the protein elastin is missing, and this affects the health of the cardiovascular system and causes their skin to age prematurely. Such individuals are very friendly but need an ordered life, perhaps because of the loss of a gene for a protein that is normally active in the brain.

Cri du chat (cat's cry) syndrome occurs when chromosome 5 is missing an end piece. The affected individual has a small head, is mentally retarded, and has facial abnormalities. Abnormal development of the glottis and larynx results in the most characteristic symptom—the infant's cry resembles that of a cat.

In a **duplication,** a chromosome segment is repeated so that the individual has more than two alleles for certain traits. An inverted duplication is known to occur in chromosome 15. (**Inversion** means that a segment joins in the direction opposite from normal.) Children with this syndrome, called *inv dup 15 syndrome,* have poor muscle tone, mental retardation, seizures, a curved spine, and autistic characteristics, including poor speech, hand flapping, and lack of eye contact (Fig. 13.4).

Translocation

A **translocation** is the exchange of chromosome segments between two nonhomologous chromosomes. A person who has both of the involved chromosomes has the normal amount of genetic material and is healthy, unless the chromosome exchange breaks an allele into two pieces. The person who inherits only one of the translocated chromosomes will no doubt have only one copy of certain alleles and three copies of other alleles. A genetic counselor begins to suspect a translocation has occurred when spontaneous abortions are commonplace and family members suffer from various syndromes. A special microscopic technique allows a technician to determine that a translocation has occurred.

In 5% of Down syndrome cases, a translocation that occurred in a previous generation between chromosomes 21 and 14 is the cause. As long as the two chromosomes are inherited together, the individual is normal. But in future generations a person may inherit two normal chromosomes 21 and the abnormal chromosome 14 that contains a segment of chromosome 21. In these cases, Down syndrome is not related to parental age but instead tends to run in the family of either the father or the mother.

Figure 13.5 shows a father and son who have a translocation between chromosomes 2 and 20. Although they have the normal amount of genetic material, they have the distinctive face, abnormalities of the eyes and internal organs, and severe itching characteristic of Alagille syndrome. People with this syndrome ordinarily have a deletion on chromosome 20; therefore, it can be deduced that the translocation disrupted an allele on chromosome 20 in the father. The symptoms of Alagille syndrome range from mild to severe, so some people may not be aware they have the syndrome. This father did not realize it until he had a child with the syndrome.

Inversion

An inversion occurs when a segment of a chromosome is turned 180°. You might think this is not a problem because the same genes are present, but the reverse sequence of alleles can lead to altered gene activity.

Crossing-over between an inverted chromosome and the noninverted homologue can lead to recombinant chromosomes that have both duplicated and deleted segments. This happens because alignment between the two homologues is only possible when the inverted chromosome forms a loop (Fig. 13.6).

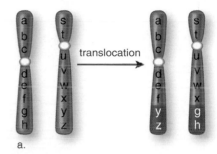
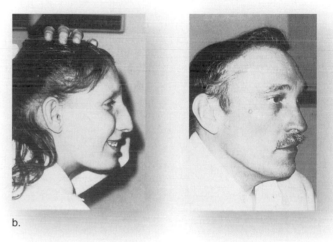

Figure 13.5   Translocation.

a. When chromosomes 2 and 20 exchange segments, Alagille syndrome results because the translocation disrupts an allele on chromosome 20. **b.** Distinctive facial features are one of the characteristics of Alagille syndrome.

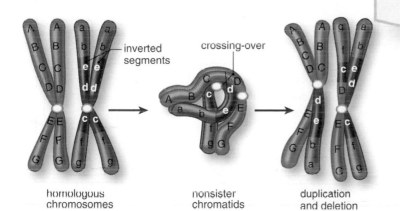

homologous chromosomes   —inverted segments   crossing-over   nonsister chromatids   duplication and deletion in both

Figure 13.6   Inversion.

(Left) A segment is inverted in the sister chromatids of one homologue. Notice that, in the red chromosome, *edc* occurs instead of *cde*. *(Middle)* The nonsister chromatids can align only when the inverted sequence forms an internal loop. After crossing-over, a duplication and a deletion can occur. *(Right)* The inner nonsister chromatid on the left has *AB* and *ab* sequences and neither *fg* nor *FG* genes. The inner nonsister chromatid on the right has *gf* and *GF* sequences and neither *AB* nor *ab* genes.

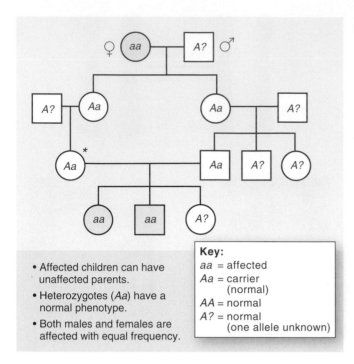

Key:
aa = affected
Aa = carrier
    (normal)
AA = normal
A? = normal
    (one allele unknown)

- Affected children can have unaffected parents.
- Heterozygotes (Aa) have a normal phenotype.
- Both males and females are affected with equal frequency.

Figure 13.7 **Autosomal recessive pedigree.**

The list gives ways to recognize an autosomal recessive disorder. How would you know that the individual at the * is heterozygous?

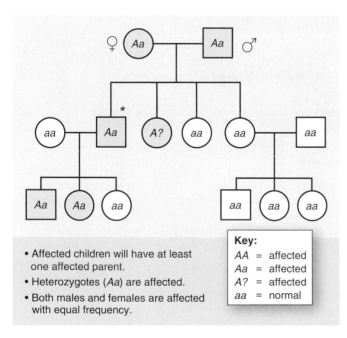

Key:
AA = affected
Aa = affected
A? = affected
aa = normal

- Affected children will have at least one affected parent.
- Heterozygotes (Aa) are affected.
- Both males and females are affected with equal frequency.

Figure 13.8 **Autosomal dominant pedigree.**

The list gives ways to recognize an autosomal dominant disorder. How would you know that the individual at the * is heterozygous?
*See page 213.

13.2 Counseling for Genetic Disorders: The Present

Even if no chromosomal abnormality is likely, amniocentesis still might be done because it is now possible to perform biochemical tests on amniotic fluid to detect over 400 different disorders caused by specific genes in a fetus. The genetic counselor determines ahead of time what tests might be warranted. To do this, the counselor needs to know the medical history of the family in order to construct a pedigree.

Family Pedigrees

A **pedigree** is a chart of a family's history with regard to a particular genetic trait. In the chart, males are designated by squares and females by circles. Shaded circles and squares are affected individuals; they have a genetic disorder. A line between a square and a circle represents a union. A vertical line going downward leads directly to a single child; if there are more children, they are placed off a horizontal line.

From the counselor's knowledge of genetic disorders, he or she might already know the *pattern of inheritance* of a trait—that is, whether it is autosomal dominant, autosomal recessive, or X-linked recessive. The counselor can then determine the chances that any child born to the couple will have the abnormal phenotype.

Pedigrees for Autosomal Disorders

A family pedigree for an *autosomal recessive disorder* is shown in Figure 13.7. In this pattern, a child can be affected when neither parent is affected. Such heterozygous parents are *carriers* because, although they are unaffected, they are capable of having a child with the genetic disorder. If the family pedigree suggests that the parents are carriers for an autosomal recessive disorder, the counselor might suggest confirming this by doing the appropriate genetic test. Then, if the parents so desire, it would be possible to do prenatal testing of the fetus for the genetic disorder.

Figure 13.7 lists other ways that a counselor may recognize an autosomal recessive pattern of inheritance. Notice that in this pedigree, cousins are the parents of three children, two of whom have the disorder. Aside from illustrating that reproduction between cousins is more likely to bring out recessive traits, this pedigree also shows that "chance has no memory"; therefore, each child born to heterozygous parents has a 25% chance of having the disorder. In other words, if a heterozygous couple has four children, each child might have the condition.

Figure 13.8 shows an *autosomal dominant* pattern of inheritance. In this pattern, a child can be unaffected even when the parents are heterozygous and therefore affected. Figure 13.8 lists other ways to recognize an autosomal dominant pattern of inheritance. This pedigree illustrates that when both parents are unaffected, all their children are unaffected. Why? Because neither parent has a dominant gene that causes the condition to be passed on.

Pedigrees for Sex-Linked Disorders

Figure 13.9 gives a pedigree for an *X-linked recessive disorder.* Recall that sons inherit an X-linked recessive allele from their mothers because their fathers give them a Y chromosome. More males than females have the disorder because recessive alleles on the X chromosome are always expressed in males—the Y chromosome lacks an allele. Females who have the condition inherited the allele from both their mother and their father, and all the sons of such a female will have the condition.

If a male has an X-linked recessive condition, his daughters are all carriers even if his partner is normal. Therefore, X-linked recessive conditions often pass from grandfather to grandson. Figure 13.9 lists other ways to recognize an X-linked recessive disorder.

Only a few traits are known to be *X-linked dominant.* If a condition is X-linked dominant, daughters of affected males have a 100% chance of having the condition. Females can pass an X-linked dominant allele to both sons and daughters. If a female is heterozygous and her partner is normal, each child has a 50% chance of escaping an X-linked dominant disorder, depending on which of the mother's X chromosomes is inherited.

A few genetic disorders involve genes carried on the Y chromosome. One gene, called the SRY-determining region, is involved in determining gender during development. Others code for membrane proteins, including an enzyme that regulates the movement of ADP into and ATP out of mitochondria. A counselor would recognize a Y-linked pattern of inheritance because Y-linked disorders are present only in males and are passed directly from father to *all* sons but not to daughters. (Can you explain why?)

Check Your Progress

1. Explain what a genetic counselor can learn from a pedigree.
2. Explain what is meant by "chance has no memory" in pedigree analysis.
3. Explain why more males than females express X-linked recessive disorders.
4. Describe the inheritance pattern of a Y-linked trait.

Answers: 1. The counselor might be able to determine the inheritance pattern of the disorder and the chances that a child born to the couple will have the disorder. 2. The genotype of one child does not influence the genotypes of its siblings. 3. Males always express recessive alleles on the X chromosome. There is no possibility for a second allele to mask the recessive allele. 4. It would be passed from father to all sons, and it is only expressed in males.

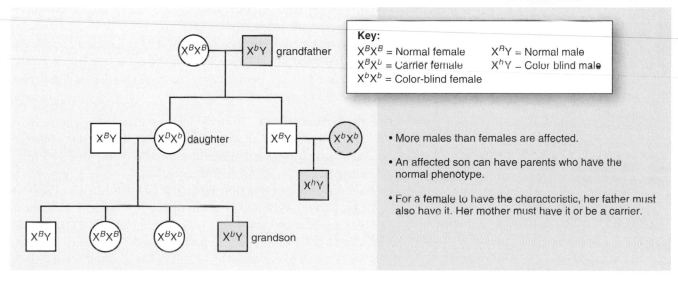

Figure 13.9 X-linked recessive pedigree.

The list gives ways of recognizing an X-linked recessive disorder—in this case, color blindness.

Genetic Disorders of Interest

Medical genetics has traditionally focused on disorders caused by single gene mutations, and we will discuss a few of the better known examples.

Autosomal Disorders

Autosomal disorders are caused by mutated alleles on the autosomal chromosomes (all the chromosomes except the sex chromosomes). Some of these are recessive, and therefore two affected alleles are necessary before an individual has the disorder. Others are dominant, meaning that it takes only one affected allele to cause the disorder. Dominant and recessive inheritance was discussed in Chapter 10.

Tay-Sachs Disease   **Tay-Sachs disease** is a well-known autosomal recessive disorder that occurs usually among Jewish people of central and eastern European descent. Tay-Sachs disease results from a lack of the enzyme hexosaminidase A (Hex A) and the subsequent storage of its substrate, a glycosphingolipid, in lysosomes. Lysosomes build up in many body cells, but the primary sites of storage are the cells of the brain, which accounts for the onset of symptoms and the progressive deterioration of psychomotor functions (Fig. 13.10).

At first, it is not apparent that a baby has Tay-Sachs disease. However, development begins to slow down between four months and eight months of age, and neurological impairment and psychomotor difficulties then become apparent. The child gradually becomes blind and helpless, develops uncontrollable seizures, and eventually becomes paralyzed and dies.

Cystic Fibrosis   **Cystic fibrosis** is an autosomal recessive disorder that occurs among all ethnic groups, but it is the most common lethal genetic disorder among Caucasians in the United States. Research has demonstrated that chloride ions (Cl^-) fail to pass through a plasma membrane channel protein in the cells of these patients. Ordinarily, after chloride ions have passed through the membrane, sodium ions (Na^+) and water follow. It is believed that lack of water then causes abnormally thick mucus in the bronchial tubes and pancreatic ducts, thus interfering with the function of the lungs and pancreas. To ease breathing in affected children, the thick mucus in the lungs must be loosened periodically but still the lungs become infected frequently (Fig. 13.11). Clogged pancreatic ducts prevent digestive enzymes from reaching the small intestine, and to improve digestion, patients take digestive enzymes mixed with applesauce before every meal.

Phenylketonuria   **Phenylketonuria (PKU)** is an autosomal recessive metabolic disorder that affects nervous system development. Affected individuals lack an enzyme that is needed for the normal metabolism of the amino acid phenylalanine, which therefore appears in the urine and the blood. Newborns are routinely tested in the hospital for elevated levels of phenylalanine in the blood. If elevated levels are detected, newborns will develop normally as long as they are limited to a diet low in phenylalanine until the brain is fully developed, usually around the age of seven. Otherwise, severe mental retardation develops. Some doctors recommend that the diet continue for life, but in any case, a pregnant woman with phenylketonuria must also be on the diet in order to protect her unborn child from harm.

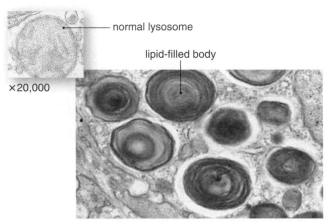

×20,000

— normal lysosome

lipid-filled body

×8,400

Figure 13.10   **Tay-Sachs disease.**

Tay-Sachs disease is a lysosomal storage disease; lipid-filled bodies fill the cell because the enzyme needed to metabolize the lipid is missing from lysosomes. The result is impairment of brain and sensory functions.

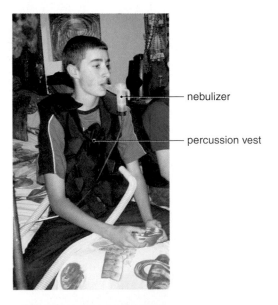

— nebulizer

— percussion vest

Figure 13.11   **Cystic fibrosis.**

This person is undergoing antibiotic and percussion therapy for cystic fibrosis. Antibiotic therapy is used to control lung infections of cystic fibrosis patients. The antibiotic can be aerosolized and administered using a nebulizer. A percussion vest loosens mucus in the lungs.

Sickle Cell Disease Sickle cell disease is an autosomal recessive disorder in which the red blood cells are not biconcave disks like normal red blood cells, but are irregular in shape (Fig. 13.12). In fact, many are sickle-shaped. The defect is caused by an abnormal hemoglobin molecule that differs from normal hemoglobin by one amino acid in the protein globin. The single amino acid change causes hemoglobin molecules to stack up and form insoluble rods, and the red blood cells become sickle-shaped.

Because sickle-shaped cells can't pass along narrow capillary passageways as disk-shaped cells can, they clog the vessels and break down. This is why persons with sickle cell disease suffer from poor circulation, anemia, and low resistance to infection. Internal hemorrhaging leads to further complications, such as jaundice, episodic pain in the abdomen and joints, and damage to internal organs.

Sickle cell heterozygotes have the sickle cell trait, in which the blood cells are normal, unless they experience dehydration or mild oxygen deprivation. Still, at present, most experts believe that persons with the sickle cell trait do not need to restrict their physical activity.

Marfan Syndrome Marfan syndrome (see Chapter 10, page 151), an autosomal dominant disorder, is caused by a defect in an elastic connective tissue protein called fibrillin. This protein is normally abundant in the lens of the eye; the bones of limbs; fingers, and ribs; and the wall of the aorta. Thus, the affected person often has a dislocated lens, long limbs and fingers, and a caved-in chest. The aorta wall is weak and can possibly burst without warning. A tissue graft can strengthen the aorta, but affected individuals still should not overexert themselves.

Huntington Disease Huntington disease is a dominant neurological disorder that leads to progressive degeneration of neurons in the brain (Fig. 13.13). The disease is caused by a single mutated copy of the gene for a protein called huntingtin. Most patients appear normal until they are of middle age and have already had children, who may then also eventually be stricken. Occasionally, the first sign of the disease in the next generation is seen in teenagers or even younger children. There is no effective treatment, and death comes 10 to 15 years after the onset of symptoms.

Several years ago, researchers found that the gene for Huntington disease was located on chromosome 4. A test was developed for the presence of the gene, but few people want to know if they have inherited the gene because there is no cure. But now we know that the disease stems from a mutation that causes the huntingtin protein to have too many copies of the amino acid glutamine. The normal version of huntingtin has stretches of between 10 and 25 glutamines. If huntingtin has more than 36 glutamines, it changes shape and forms large clumps inside neurons. Even worse, it attracts and causes other proteins to clump with it. One of these proteins, called CBP, helps nerve cells survive. Researchers hope to combat the disease by boosting CBP levels.

Check Your Progress

1. Compare and contrast the physiological bases of Tay-Sachs disease and phenylketonuria.
2. Explain why sickle-shaped red blood cells result in a wide array of symptoms in individuals with sickle cell disease.

Answers: 1. Both result from the buildup of a substrate that is not broken down properly. The substance is hexosaminidase A for Tay-Sachs disease and phenylalanine for phenylketonuria. 2. The cells cannot pass through narrow capillaries as well, so they block them and break down.

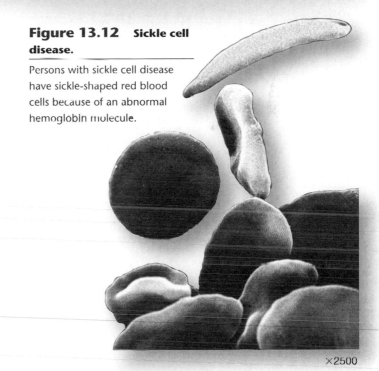

Figure 13.12 Sickle cell disease.

Persons with sickle cell disease have sickle-shaped red blood cells because of an abnormal hemoglobin molecule.

×2500

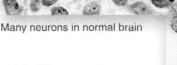

Many neurons in normal brain

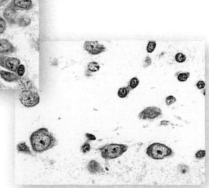

Loss of neurons in Huntington brain

Figure 13.13 Huntington disease.

Huntington disease is characterized by increasingly serious psychomotor and mental disturbances because of a loss of neurons in the brain.

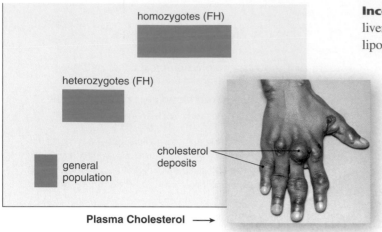

Figure 13.14   Familial hypercholesterolemia (FH).

Familial hypercholesterolemia (FH) is incompletely dominant. Heterozygotes have an abnormally high level of plasma cholesterol, and homozygotes have a higher level still.

Figure 13.15   Muscular dystrophy.

In muscular dystrophy, the calves enlarge because fibrous tissue develops as muscles waste away due to a lack of the protein dystrophin.

Incomplete Dominance   In **familial hypercholesterolemia (FH),** the liver in homozygotes with two mutated alleles completely lacks low-density lipoprotein (LDL) receptors that take up cholesterol from the bloodstream. Heterozygotes have half the normal number of cholesterol receptors. The number of cholesterol receptors inversely parallels the amount of cholesterol in the plasma. Figure 13.14 compares the amount of plasma cholesterol in the general population with that of heterozygotes and homozygotes for FH. Heart disease occurs in both heterozygotes and homozygotes. Homozygotes die of heart disease as children. Heterozygotes may die when they are young or after they have reached middle age.

X-Linked Recessive Disorders

As discussed in Chapter 10, X-linked recessive disorders are caused by mutated alleles on the X chromosome. A son inherits an X-linked recessive condition from his mother.

Color Blindness   **Color blindness** is a common X-linked recessive disorder. About 8% of Caucasian men have red-green color blindness. Most of them see brighter greens as tans, olive greens as browns, and reds as reddish-browns. A few cannot tell reds from greens at all; they see only yellows, blues, blacks, whites, and grays.

Duchenne Muscular Dystrophy   **Duchenne muscular dystrophy** is an X-linked recessive disorder characterized by wasting away of the muscles. The absence of a protein, now called dystrophin, is the cause of the disorder. Much investigative work determined that dystrophin is involved in the release of calcium from the sarcoplasmic reticulum in muscle fibers. The lack of dystrophin causes calcium to leak into the cell, which promotes the action of an enzyme that dissolves muscle fibers. When the body attempts to repair the tissue, fibrous tissue forms (Fig. 13.15), and this cuts off the blood supply so that more and more cells die.

Symptoms such as waddling gait, toe walking, frequent falls, and difficulty in rising may appear as soon as the child starts to walk. Muscle weakness intensifies until the individual is confined to a wheelchair. Death usually occurs by age 20; therefore, affected males are rarely fathers. The recessive allele remains in the population through passage from carrier mother to carrier daughter.

As therapy, immature muscle cells can be injected into muscles, and for every 100,000 cells injected, dystrophin production occurs in 30–40% of muscle fibers.

Hemophilia   There are two common types of **hemophilia,** an X-linked recessive disorder. Hemophilia A is due to the absence or minimal presence of a clotting factor known as factor VIII, and hemophilia B is due to the absence of clotting factor IX. Hemophilia is called the bleeder's disease because the affected person's blood either does not clot or clots very slowly. Although hemophiliacs bleed externally after an injury, they also bleed internally, particularly around joints. Hemorrhages can be stopped with transfusions of fresh blood (or plasma) or concentrates of the clotting protein. Also, factors VIII and IX are now available as biotechnology products.

Testing for Genetic Disorders

Following genetic testing, a genetic counselor can explain to prospective parents the chances a child of theirs will have a disorder that runs in their family. If a woman is already pregnant, the parents may want to know whether the unborn child has the disorder. If the woman is not pregnant, the parents may opt for testing of the embryo or egg before she does become pregnant, as described shortly.

Testing depends on the genetic disorder of interest. In some instances, it is appropriate to test for a particular protein, and in others to test for the mutated gene.

Testing for a Protein

Some genetic mutations lead to disorders caused by a missing enzyme. For example, in the case of Tay-Sachs disease, it is possible to test for the quantity of the enzyme hex A in a sample of cells and from that, determine whether the individual is likely homozygous normal, a carrier, or has Tay-Sachs disease. If the parents are carriers, each child has a 25% chance of having Tay-Sachs disease. This knowledge may lead prospective parents to opt for testing of the embryo or egg, as discussed on the next page.

In the case of PKU, an enzyme is missing, but the test is performed for the substrate of the enzyme, namely phenylalanine. Paper disks containing the newborn's blood are placed on a bacterial culture, and if the bacteria grow around them, the newborn has PKU.

Testing the DNA

Two types of DNA testing are possible: testing for a genetic marker and using a DNA probe.

Genetic Markers   Testing for a genetic marker relies on a difference in the DNA due to the presence of the abnormal allele. As an example, consider that individuals with sickle cell trait or Huntington disease have an abnormality in a gene's base sequence. This abnormality in sequence is a **genetic marker.** As you know, restriction enzymes cleave DNA at particular base sequences. Therefore, the fragments that result from the use of a restriction enzyme will be different for people who are normal than for those who are heterozygous or homozygous for a mutation (Fig. 13.16).

DNA Probes   A **DNA probe** is a single-stranded piece of DNA that will bind to complementary DNA. For the purpose of genetic testing, the DNA probe bears a genetic mutation of interest. A new technology that can test for many genetic disorders at a time uses a **DNA chip,** a very small glass square that contains several rows of DNA probes. Sample DNA is cut into small pieces using restriction enzymes, and the fragments are tagged with a fluorescent dye and converted to single DNA strands before being applied to the chip (Fig. 13.17). Fragments that contain a mutated gene bind to one of the probes, and binding is detected by a laser scanner. Therefore, the results tell whether an individual has particular mutated genes.

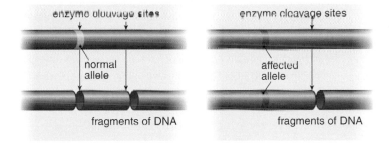

a. Normal fragmentation pattern   b. Genetic disorder fragmentation pattern

Figure 13.16   Use of a genetic marker to test for a genetic mutation.

a. In this example, DNA from a normal individual has certain restriction enzyme cleavage sites. **b.** DNA from another individual lacks one of the cleavage sites, and this loss indicates that the person has a mutated gene. In heterozygotes, half of their DNA would have the cleavage site and half would not have it. (In other instances, the gain in a cleavage site could be an indication of a mutation.)

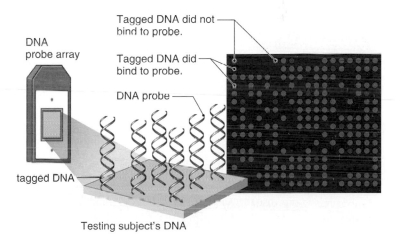

Testing subject's DNA

Figure 13.17   Use of a DNA chip to test for mutated genes.

This DNA chip contains rows of DNA probes for mutations that indicate the presence of a particular genetic disorder. If single-stranded DNA fragments derived from an individual's DNA bind to a probe, the individual has the mutation. Heterozygotes would not have as much binding as homozygotes.

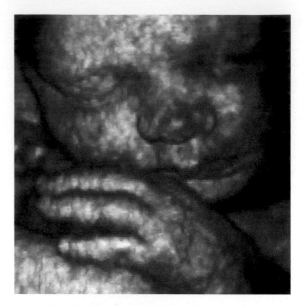

Figure 13.18 **Ultrasound.**

Today, an ultrasound exam can produce three-dimensional images that allow physicians to detect serious abnormalities, such as neural tube abnormalities.

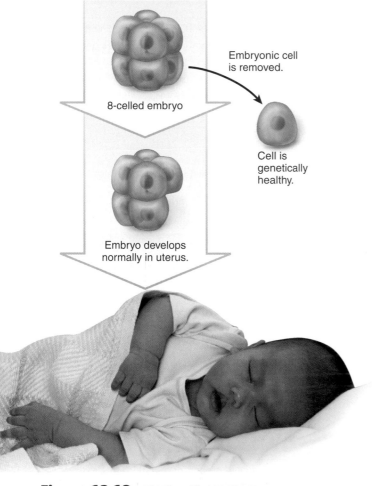

Figure 13.19 **Testing the embryo.**

Genetic analysis is performed on one cell removed from an 8-celled embryo. If this cell is found to be free of the genetic defect of concern, and the 7-celled embryo is implanted in the uterus, it develops into a newborn with a normal phenotype.

(Labels on figure: 8-celled embryo; Embryonic cell is removed.; Cell is genetically healthy.; Embryo develops normally in uterus.)

Testing the Fetus

If a woman is already pregnant, ultrasound can detect serious fetal abnormalities, and it is also possible to obtain and test the DNA of fetal cells for genetic defects.

Ultrasound Ultrasound images help doctors evaluate fetal anatomy. An ultrasound probe scans the mother's abdomen, and a transducer transmits high-frequency sound waves that are transformed into a picture on a video screen. This picture shows the fetus inside the uterus (Fig. 13.18). Ultrasound done after 16 weeks can be used to determine the baby's age and size, and whether there is more than one baby. It's also possible to tell if a baby has a serious condition, such as spina bifida, which results when the spine fails to close properly during the first month of pregnancy. Surgery to close a newborn's spine in such a case is generally performed within 24 hours after birth.

Testing Fetal Cells Fetal cells can be tested for various genetic disorders. If the fetus has an incurable disorder, such as Tay-Sachs disease, the parents may wish to consider an abortion.

For testing purposes, fetal cells may be acquired through amniocentesis or chorionic villi sampling, as described earlier in this chapter. In addition, fetal cells may be collected from the mother's blood. As early as nine weeks into the pregnancy, a small number of fetal cells can be isolated from the mother's blood using a cell sorter. While mature red blood cells lack a nucleus, immature red blood cells do have a nucleus, and they also have a shorter life span than mature red blood cells. Therefore, if nucleated fetal red blood cells are collected from the mother's blood, they are known to be from this pregnancy.

Only about 1/70,000 blood cells in a mother's blood are fetal cells, and therefore PCR has to be used to amplify the DNA from the few cells collected. The procedure poses no risk whatsoever to the fetus.

Testing the Embryo and Egg

As discussed in Chapter 29, in vitro fertilization (IVF) is carried out in laboratory glassware. The physician obtains eggs from the prospective mother and sperm from the prospective father, and places them in the same receptacle, where fertilization occurs. Following IVF, now a routine procedure, it is possible to test the embryo. Prior to IVF, it is possible to test the egg for any genetic defect. In any case, only normal embryos are transferred to the uterus for further development.

Testing the Embryo   If prospective parents are carriers for one of the genetic disorders discussed earlier, they may want assurance that their offspring will be free of the disorder. Testing the embryo will provide this assurance.

Following IVF, the zygote (fertilized egg) divides. When the embryo has six to eight cells, one of these cells can be removed for testing, with no effect on normal development (Fig. 13.19). Only embryos that test negative for the genetic disorders of interest are placed in the uterus to continue developing.

So far, about 1,000 children worldwide have been born free of alleles for genetic disorders that run in their families following embryo testing. In the future, it's possible that embryos who test positive for a disorder could be treated by gene therapy, so that they, too, would be allowed to continue to term.

Testing the Egg   Recall that meiosis in females results in a single egg and at least two polar bodies. Polar bodies, which later disintegrate, receive very little cytoplasm, but they do receive a haploid number of chromosomes, and thus can be useful in genetic testing. When a woman is heterozygous for a recessive genetic disorder, about half the polar bodies receive the mutated allele, and in these instances the egg receives the normal allele. Therefore, if a polar body tests positive for a mutated allele, the egg received the normal allele (Fig. 13.20). Only normal eggs are then used for IVF. Even if the sperm should happen to carry the mutation, the zygote will, at worst, be heterozygous. But the phenotype will appear normal.

If gene therapy becomes routine in the future, it's possible that an egg could be given genes that control traits desired by the parents, such as musical or athletic ability, prior to IVF.

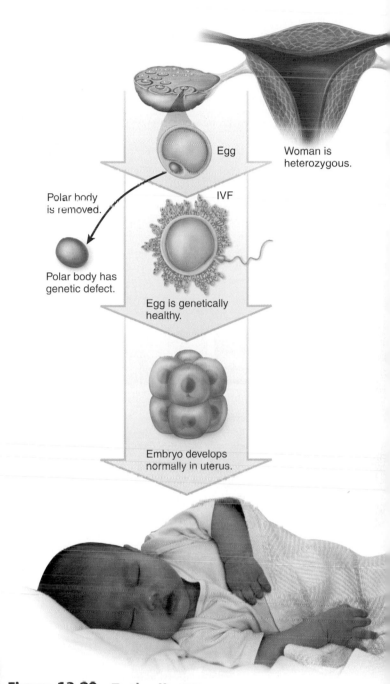

Figure 13.20   Testing the egg.

Genetic analysis is performed on a polar body removed from an egg. If the egg is free of a genetic defect, it is used for IVF, and the embryo is implanted in the uterus for further development.

Check Your Progress

1. List the three ways a doctor can collect fetal cells for genetic testing.
2. Explain how an embryo can be tested for a genetic disorder.
3. Explain how egg cells can be tested for a genetic disorder.

Answers: 1. Amniocentesis, chorionic villi sampling, fetal cells in the mother's blood. 2. When the embryo contains six to eight cells, one cell is removed and used for testing. 3. Meiosis results in the production of an egg cell and at least two polar bodies. The polar bodies can be tested for the genetic disorder, and this information can be used to deduce the genotype of the egg cell.

a.

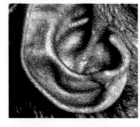
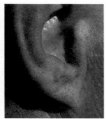

b.

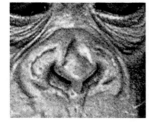

c.

Figure 13.21 **Studying genomic differences between chimpanzees and humans.**

Researchers found that by comparing the human genome to that of a chimpanzee they were able to discover genes that cause human diseases when abnormal. They also concluded that the genes for (**a**) speech, (**b**) hearing, and (**c**) smell may have influenced the evolution of humans.

13.3 Counseling for Genetic Disorders: The Future

In the previous century, researchers discovered the structure of DNA, how DNA replicates, and how protein synthesis occurs. Genetics in the twenty-first century concerns **genomics,** the study of **genomes** (organisms' genes) to better understand how they direct growth and development and otherwise control the structure and function of the cell. The enormity of this task can be appreciated by knowing that, at the very least, humans have 25,000 genes that code for proteins. An abnormality in any one of these proteins can be the cause of a human illness.

Sequencing the Bases of the Human Genome

We now have a working draft of the base pair sequence in the DNA of all our chromosomes. This feat was accomplished by the Human Genome Project, a 13-year effort that involved both university and private laboratories around the world. How did they do it? First, investigators developed a laboratory procedure that would allow them to decipher a short sequence of base pairs, and then they devised an instrument that would carry out this procedure automatically. As the studies proceeded, DNA sequencers were constantly improved, until today we have instruments that can automatically sequence to 350,000 base pairs of DNA per day (see page 159).

Genome Comparisons

Researchers are taking all sorts of avenues to link DNA base sequence differences to illnesses. One study compared the human genome to that of chromosome 22 in chimpanzees. Among the many genes that differed in sequence were three of particular interest: a gene for proper speech development, several for hearing, and several for smell. The gene necessary for proper speech development is thought to have played an important role in human evolution. You can suppose that changes in hearing may have facilitated using language for communication between people. Changes in smell genes are a little more problematic. The researchers speculated that the olfaction genes may have affected dietary changes or sexual selection. Or, they may have been involved in traits, other than just smell (Fig. 13.21). It was a surprise to find that many of the other genes they located and studied are known to cause human diseases if abnormal. Perhaps comparing genomes would be a way of finding genes associated with human diseases.

The genomes of many other organisms, such as a common bacterium, a form of yeast, and a species of mouse, are also in the final-draft stage. There are many similarities between the sequences of DNA bases in humans and in other organisms. From this, we can conclude that we share a large number of genes with much simpler organisms, including bacteria! However, with a genome size of 3 billion base pairs, there are also many differences.

Genetic Profiling

DNA chips (also called DNA microarrays) can rapidly identify an individual's complete genotype, including all the various mutations. This genotype is called the person's **genetic profile.** To get a genetic profile is easy. The patient provides a few cells, often by simply swabbing the inside of a cheek. The DNA is removed from the cells, amplified by PCR if need be, and cut into fragments, which are tagged by a fluorescent dye. A technician then applies the fragments to a DNA chip, and reads the results.

With the help of a genetic counselor, individuals can be educated about their genetic profile. It's possible that a person has or will have a genetic disorder caused by a single pair of alleles. However, polygenic traits are more common, and in these instances, the genetic profile can indicate an increased or decreased risk for a disorder. Risk information can be used to design a program of medical surveillance and to foster a lifestyle aimed at reducing the risk. For example, suppose an individual has mutations common to people with colon cancer. It would be helpful to have an annual colonoscopy so that any abnormal growths can be detected and removed before they become invasive.

Proteomics and Informatics

The genetic profile may also indicate what drug therapy might be most appropriate for an individual. Because drugs tend to be proteins or small molecules that affect the behavior of proteins, the study of protein function is essential to the discovery of better drugs. One day it may be possible to correlate drug treatment to an individual genetic profile to increase efficiency and decrease side effects.

The field of **proteomics** is especially pertinent because it deals with the development of new drugs for the treatment of genetic disorders. Proteomics explores the structure, function, and interaction of cellular proteins. The known sequence of bases in the human genome predicts that at least 25,000 genes are translated into proteins. The translation of all of these genes results in a collection of proteins called the human **proteome.** Computer modeling of the three-dimensional shape of these proteins is an important part of proteomics. Because the primary structure of these proteins is now known, it should be possible to predict their final shape.

Bioinformatics is the application of computer technologies to the study of the genome (Fig. 13.22). Genomics and proteomics produce raw data, and these fields depend upon computer analysis to find significant patterns in the data. As a result of bioinformatics, scientists hope to find cause-and-effect relationships between various genetic profiles and genetic disorders caused by polygenes.

Also, the current genome sequence contains 82 gene "deserts," with no known function. Bioinformatics might find that these regions have functions by correlating any sequence changes with resulting phenotypes. New computational tools will most likely be needed in order to accomplish these goals.

Figure 13.22   Bioinformatics.

New computer programs are being developed to make sense out of the raw data generated by genomics and proteomics. Bioinformatics allows researchers to correlate gene activity and protein function, so that the phenotype can be better understood in molecular terms.

Check Your Progress

1. Describe the goals of genomics research.
2. Describe the benefits of knowing your genetic profile.
3. What do the fields of proteomics and bioinformatics have in common?

Answers: 1. Genomics research attempts to understand how genes direct growth and development in order to control the activities of cells and organisms. 2. A genetic profile will tell you whether or not you are at risk for a genetic disorder. It can also indicate which drugs will be most effective for your particular phenotype. 3. Although proteomics studies proteins and bioinformatics studies genes, both depend on the computer.

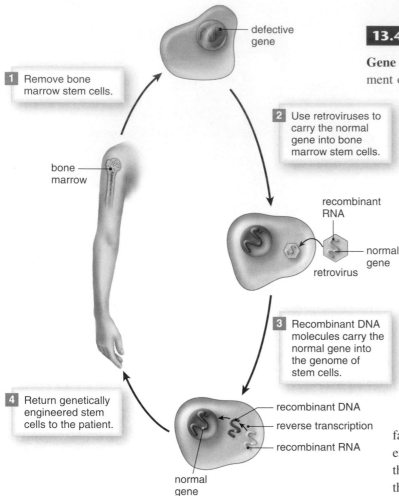

1 Remove bone marrow stem cells.

defective gene

bone marrow

2 Use retroviruses to carry the normal gene into bone marrow stem cells.

recombinant RNA

normal gene

retrovirus

3 Recombinant DNA molecules carry the normal gene into the genome of stem cells.

4 Return genetically engineered stem cells to the patient.

recombinant DNA

reverse transcription

recombinant RNA

normal gene

Figure 13.23   Ex vivo gene therapy in humans.

Bone marrow stem cells are withdrawn from the body, an RNA retrovirus is used to insert a normal gene into the host genome, and then the cells are returned to the body.

a. Cindy Cutshall

b. Donovan Decker

Figure 13.24   Gene therapy patients.

a. This patient was treated with the *ADA* gene to cure severe combined immunodeficiency. **b.** This patient was treated with the *VEGF* gene to alleviate poor coronary circulation.

13.4   Gene Therapy

Gene therapy is the insertion of genetic material into human cells for the treatment of a disorder. It includes procedures that give a patient healthy genes to make up for faulty genes, as well as the use of genes to treat various other human illnesses, such as cardiovascular disease and cancer. Gene therapy includes both ex vivo (outside the body) and in vivo (inside the body) methods.

Ex Vivo Gene Therapy

One example of ex vivo gene therapy is the methodology for treating children who have SCID (severe combined immunodeficiency) (Fig. 13.23). These children lack the enzyme ADA (adenosine deaminase), which is involved in the maturation of T and B cells. In order to carry out gene therapy, bone marrow stem cells are removed from the blood and infected with an RNA retrovirus that carries a normal gene for the enzyme. Then the cells are returned to the patient. Bone marrow stem cells are preferred for this procedure because they divide to produce more cells with the same genes. Patients who have undergone this procedure show significantly improved immune function associated with a sustained rise in the level of ADA enzyme activity in the blood (Fig. 13.24*a*).

Among the many gene therapy trials, one is for the treatment of familial hypercholesterolemia, described earlier in this chapter as an example of incomplete dominance. High levels of plasma cholesterol make the patient subject to fatal heart attacks at a young age. A small portion of the liver is surgically excised and then infected with a retrovirus containing a normal gene for a cholesterol receptor before the tissue is returned to the patient. Several patients have experienced lowered plasma cholesterol levels following this procedure.

Ex vivo gene therapy is being used in the treatment of cancer. In one procedure, immune system cells are removed from a cancer patient and genetically engineered to display tumor antigens. After these cells are returned to the patient, they stimulate the immune system to kill tumor cells.

In Vivo Gene Therapy

As mentioned earlier, cystic fibrosis patients lack a gene that codes for a regulator of a transmembrane carrier for the chloride ion. In gene therapy trials, the gene needed to cure cystic fibrosis is sprayed into the nose or delivered to the lower respiratory tract by adenoviruses or by the use of liposomes (microscopic vesicles that spontaneously form when lipoproteins are put into a solution). Investigators are trying to improve uptake of the genes and are also hypothesizing that a combination of all three vectors might be more successful.

Genes are also being used to treat medical conditions such as poor coronary circulation. It has been known for some time that VEGF (vascular endothelial growth factor) can cause the growth of new blood vessels. The gene that codes for this growth factor can be injected alone, or within a virus, into the heart to stimulate branching of coronary blood vessels. Patients who have received this treatment report that they have less chest pain and can run longer on a treadmill (Fig. 13.24*b*).

In vivo gene therapy is increasingly becoming a part of cancer therapy. Genes are being used to make healthy cells more tolerant of chemotherapy, and to make tumors more vulnerable to chemotherapy. The gene *p53* brings about apoptosis, and introduction of this gene into cancer cells may be a method of killing them off.

Figure 13.25 summarizes how gene therapy is being used to treat various illnesses.

Check Your Progress

1. Explain why bone marrow stem cells are preferred for ex vivo gene therapy.
2. Explain how ex vivo gene therapy is used to treat cancer.
3. Explain how in vivo gene therapy is used to treat cystic fibrosis.

Answers: **1.** Bone marrow stem cells are capable of dividing after genetic alteration to produce cells carrying the alteration. **2.** Immune system cells from the patient are removed and genetically altered so that they display tumor antigens. These altered cells are returned to the patient to stimulate the immune system to kill tumor cells. **3.** The gene needed to cure the disorder is delivered to the respiratory tract in several ways, including via nasal sprays, adenoviruses, or liposomes.

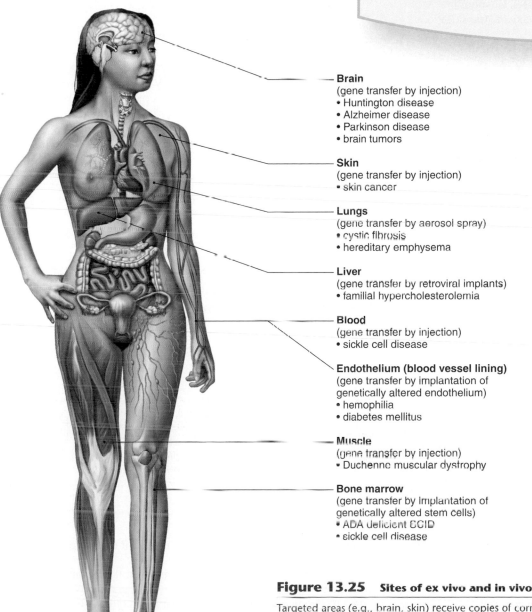

Brain
(gene transfer by injection)
• Huntington disease
• Alzheimer disease
• Parkinson disease
• brain tumors

Skin
(gene transfer by injection)
• skin cancer

Lungs
(gene transfer by aerosol spray)
• cystic fibrosis
• hereditary emphysema

Liver
(gene transfer by retroviral implants)
• familial hypercholesterolemia

Blood
(gene transfer by injection)
• sickle cell disease

Endothelium (blood vessel lining)
(gene transfer by implantation of genetically altered endothelium)
• hemophilia
• diabetes mellitus

Muscle
(gene transfer by injection)
• Duchenne muscular dystrophy

Bone marrow
(gene transfer by implantation of genetically altered stem cells)
• ADA deficient SCID
• sickle cell disease

Figure 13.25 **Sites of ex vivo and in vivo somatic gene therapy.**

Targeted areas (e.g., brain, skin) receive copies of corrected genes by various methods of gene transfer. For many diseases, genetically modified viruses ferry the corrected gene into the body. Many gene therapy treatments are still undergoing trials.

Summary

13.1   Counseling for Chromosomal Disorders

A counselor can detect chromosomal mutations by studying a karyotype of the individual.

Karyotyping

A karyotype is a display of the chromosomes arranged by pairs; the autosomes are numbered from 1 to 22. The sex chromosomes are not numbered.

Chromosomal Mutations

Nondisjunction during meiosis can result in an abnormal number of autosomes or sex chromosomes in the gametes. Changes in chromosome structure include deletions, duplications, translocations, and inversions:

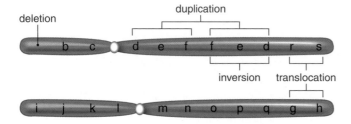

In Williams syndrome, one copy of chromosome 7 has a deletion; in cri du chat syndrome, one copy of chromosome 5 has a deletion; and in inv dup 15 syndrome, chromosome 15 has an inverted duplication. Down syndrome can be due to a translocation between chromosomes 14 and 21 in a previous generation.

An inversion can lead to chromosomes that have a deletion and a duplication when the inverted piece loops back to align with the noninverted homologue and crossing-over follows.

13.2   Counseling for Genetic Disorders: The Present

A counselor can decide the chances of an offspring inheriting a genetic disorder by constructing a pedigree of a disorder that runs in the family.

Family Pedigree

A family pedigree is a visual representation of the history of a genetic disorder in a family. Constructing pedigrees helps a genetic counselor decide whether a genetic disorder that runs in a family is autosomal recessive (see Fig. 13.7) or dominant (see Fig. 13.8); X-linked (see Fig. 13.9); or some other pattern of inheritance.

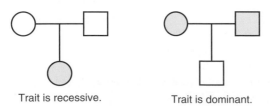

Trait is recessive.   Trait is dominant.

Genetic Disorders of Interest

Autosomal Disorders
- Tay-Sachs disease (a lysosomal storage disease)
- Cystic fibrosis (faulty regulator of chloride channel)
- Phenylketonuria (inability to metabolize phenylalanine)
- Sickle cell disease (sickle-shaped red blood cells)
- Marfan syndrome (defective elastic connective tissue)
- Huntington disease (abnormal huntingtin protein)

Incomplete Dominance Disorders
- Familial hypercholesterolemia (liver cells lack cholesterol receptors)

X-Linked Disorders
- Duchenne muscular dystrophy (absence of dystrophin leads to muscle weakness)
- Hemophilia (inability of blood to clot)

Testing for Genetic Disorders

A counselor can order the appropriate test to detect a disorder.

Testing for Proteins   Persons with Tay-Sachs disease lack the enzyme hex A, and those with PKU have much phenylalanine in their blood.

Testing DNA   Cut DNA with restriction enzymes and then compare the fragment pattern to the normal pattern. See if fragments bind to DNA probes that contain the mutation.

Testing the Fetus   An ultrasound allows for the detection of severe abnormalities, such as spina bifida. Fetal cells can be obtained by amniocentesis or chorionic villi sampling, or by sorting out fetal cells from the mother's blood.

Testing the Embryo   Following in vitro fertilization (IVF), it is possible to test the embryo. A cell is removed from an 8-celled embryo, and if it is found to be genetically healthy, the embryo is implanted in the uterus, where it develops to term.

Testing the Egg   Before IVF, a polar body can be tested. If the woman is heterozygous, and the polar body has the genetic defect, the egg does not have it. Following fertilization, the embryo is implanted in the uterus.

13.3   Counseling for Genetic Disorders: The Future

In the future, a counselor will be able to study the genetic profile of an individual in order to counsel them about the risk of developing a genetic disorder.

Human Genome Project   Thanks to the Human Genome Project, we now know the sequence of the base pairs in human DNA. As a consequence, it will soon be possible to easily determine the genetic profile of individuals.

Genomics and Proteomics   Genomics is the study of the genome, and proteomics is the study of the proteome—all the proteins active in an organism or cell. Proteomics is expected to provide new medicines for genetic disorders.

13.4   Gene Therapy

During gene therapy, a genetic defect is treated by giving the patient a foreign gene.
- Ex vivo therapy   Cells are removed from the patient, treated, and returned to the patient.
- In vivo therapy   A foreign gene is given directly to the patient.

Thinking Scientifically

1. Cystic fibrosis occurs in individuals who have two defective copies of the gene for a protein called cystic fibrosis transmembrane regulator (CFTR). Suppose two people have cystic fibrosis, but one is much more severely affected than the other. In addition, a genetic test for cystic fibrosis is positive for one person, but negative for the other. How do you explain these observations?

2. Recently, gene therapy trials were performed on ten infants with X-linked severe combined immunodeficiency syndrome (XSCID), also known as "bubble boy disease." A normal copy of the gene associated with XSCID was inserted into the virus. The virus was then used to transfer the gene into chromosomes in the patients' cells. The trial was considered a success because the gene was expressed and the children's immune systems were restored. However, researchers were shocked and disappointed when two children developed leukemia. How might you explain the development of leukemia in these gene therapy patients?

Testing Yourself

Choose the best answer for each question.

1. The major advantage of chorionic villi sampling over amniocentesis is that it

   a. allows karyotyping to be done earlier.
   b. produces karyotypes with better images of chromosomes.
   c. carries a lower risk of spontaneous abortion.
   d. produces a sample that is not contaminated by cells from the mother.

2. Human genetic disorders due to changes in chromosome number are caused by

   a. radiation.
   b. chemical mutagens.
   c. nondisjunction at meiosis.
   d. fertilization by two sperm cells.

3. An example of a chromosomal mutation that involves two nonhomologous chromosomes is

   a. an inversion.   c. a deletion.
   b. a duplication.   d. a translocation.

4. Fill in the genotypes a.–i. of the family members in the following pedigree for an autosomal recessive trait. Shaded individuals are affected.

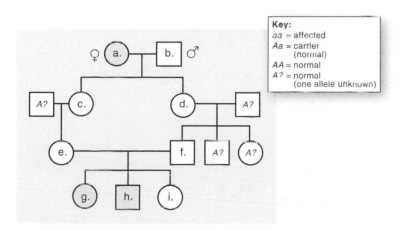

5. Fill in the genotypes a.–h. of the family members in the following pedigree for an autosomal dominant trait. Shaded individuals are affected.

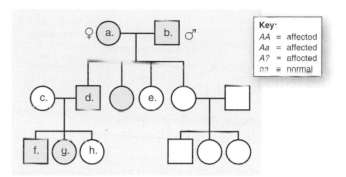

6. Fill in the genotypes a.–e. of the family members in the following pedigree for an X-linked recessive trait. Shaded individuals are affected.

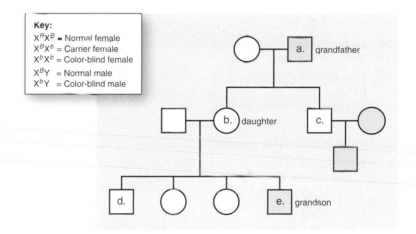

For questions 7–10, match the pedigree characteristics to those in the key. Answers can be used more than once. Some questions may have more than one answer.

Key:

   a. X-linked recessive
   b. X-linked dominant
   c. autosomal recessive
   d. autosomal dominant

7. All daughters and no sons from affected males express the trait.

8. Males express the trait more frequently than females.

9. Affected children always have at least one affected parent.

10. Affected children can have unaffected parents.

11. Disorders that result from the inability to break down a substance are

   a. cystic fibrosis and hemophilia.
   b. Tay-Sachs disease and phenylketonuria.
   c. color blindness and Tay-Sachs disease.
   d. muscular dystrophy and cystic fibrosis.

For questions 12–20, match the descriptions to the conditions in the key. Answers can be used more than once. Some questions may have more than one answer.

Key:

a. Tay-Sachs disease   f. Huntington disease
b. cystic fibrosis   g. familial hypercholesterolemia
c. phenylketonuria   h. color blindness
d. sickle cell disease   i. Duchenne muscular dystrophy
e. Marfan syndrome   j. hemophilia

12. Autosomal dominant disorder.

13. Disorder caused by incomplete dominance.

14. Neurological disorder that leads to progressive degeneration of brain cells.

15. X-linked recessive disorder in which muscle tissue wastes away.

16. The most common lethal genetic disorder among U.S. Caucasians.

17. X-linked recessive disorder.

18. Results from the lack of the enzyme hex A, resulting in the storage of its substrate in lysosomes.

19. Results from the inability to metabolize phenylalanine.

20. Caused by abnormal hemoglobin, resulting in red blood cells that are not round.

21. The current estimate for the number of genes in the human genome is

   a. 1,000.   c. 500,000.
   b. 30,000.   d. 3 million.

22. Adult chromosomes are most easily acquired for study using

   a. amniotic fluid.   d. a blood sample.
   b. placental cells.   e. a hair sample.
   c. a saliva sample.

23. Proteomics is

   a. the application of computer technologies to the study of the genome.
   b. the study of the structure, function, and interaction of cellular proteins.
   c. the study of the human genome to better understand how genes work.
   d. all the genes that occur in a cell.
   e. the study of a person's complete genotype, or genetic profile.

24. A color-blind female and normal male have children. What are the chances that any sons are color-blind? That any daughters are color-blind?

25. Parents who do not have Tay-Sachs disease produce a child who has Tay-Sachs disease. What is the genotype of all persons involved?

26. A 25-year-old man has Huntington disease. Is it possible for him to father a normal son? Explain.

27. A girl has hemophilia. What is her genotype? What are the possible genotypes of the parents?

Go to www.mhhe.com/maderessentials for more quiz questions.

Bioethical Issue

Myriad Genetics, a genomics and genetic testing company, has been awarded a patent on 47 different mutations of the *BRCA1* gene. These DNA sequences are used to detect a major breast cancer gene. The patent gives the company the exclusive right to prevent others from using the sequence to develop genetic tests. Opponents of human DNA patents argue that human genes are being treated like commodities. They suggest that reducing humans to a DNA sequence is an assault on our dignity. Proponents argue that the patent system helps improve human health by stimulating biotechnology companies to invest in genetic screening programs.

Do you agree that human DNA sequences should be protected by patents? If not, do you support the patenting of DNA sequences of other organisms?

Understanding the Terms

amniocentesis   197
bioinformatics   209
chorionic villi sampling   197
chromosomal mutation   198
color blindness   204
cystic fibrosis   202
deletion   198
DNA chip   205
DNA probe   205
Duchenne muscular dystrophy   204
duplication   198
familial hypercholesterolemia (FH)   204
gene therapy   210
genetic counseling   196

genetic marker   205
genetic profile   209
genome   208
genomics   208
hemophilia   204
Huntington disease   203
inversion   198
karyotype   196
Marfan syndrome   203
pedigree   200
phenylketonuria (PKU)   202
proteome   209
proteomics   209
sickle cell disease   203
Tay-Sachs disease   202
translocation   199

Match the terms to these definitions:

a. _____ Visual display of an individual's chromosomes arranged by size.

b. _____ Procedure for sampling cells from the region where the placenta will develop.

c. _____ Exchange of chromosome segments between two nonhomologous chromosomes.

d. _____ Chart of a family's history for a genetic trait.

e. _____ Study of the structure, function, and interaction of cellular proteins.

f. _____ Insertion of genetic material into human cells for the treatment of a disorder.

Darwin and Evolution

14

Only in Hawaii did honeycreepers evolve into over 16 species.

Akiapolaau

Laysan Finch

Due to artificial selection, you can choose from more than 150 breeds of dogs.

What do the many breeds of dogs, the rare buttercups of New Zealand, and a child's resistant ear infection have in common? EVOLUTION! Without **evolution**—change in a line of descent over time—we wouldn't see such a great variety of living things about us. But aside from its many benefits, evolution also sometimes causes problems for humans.

Like dogs and honey-creepers, bacteria evolve. A growing problem in hospitals is the number of bacteria that have become resistant to antibiotics.

▶ ▶ ▶

Some bacteria have evolved to the point that they are resistant to the antibiotics once successfully used to cure the diseases they cause. For example, antibiotics originally cured bacterial ear infections within a few days. Unseen, however, were the one or two bacteria with just the right mutation to resist a particular drug. All the descendants of these bacteria were also resistant, causing the antibiotic to be useless as a cure for this type of ear infection. The antibiotic is considered the *selective agent* because it allowed the resistant bacteria to flourish while killing their relatives.

What was the selective agent for the many breeds of dogs available as pets today? Humans were, of course. Over the years, humans selected which dogs to mate to produce today's varieties. This process is called *artificial selection.*

In nature, the selective agent is the environment, and the process is called *natural selection.* Many microenvironments in New Zealand plus periodic isolation allowed 14 alpine species of buttercups to evolve at different times, but now they are found side by side throughout the same area. In fact, there are more species of buttercups on the two islands of New Zealand than on any of the continents.

This chapter explains how different types of selection lead to evolution and begins by describing the early studies that contributed to our current knowledge of the evolutionary process, particularly the work of Charles Darwin.

14.1 Darwin's Theory of Evolution

Charles Darwin, the father of evolution, was only 22 in 1831 when he accepted the position of naturalist aboard the HMS *Beagle,* a British naval ship about to sail around the world (Fig. 14.1). Darwin had a suitable background for his position. He was an ardent student of nature and had long been a collector of insects. His sensitive nature had prevented him from studying medicine, and he went to divinity school at Cambridge instead. Even so, he attended many lectures in both biology and geology, and he was also tutored in these subjects by a friend and teacher, the Reverend John Henslow. Darwin spent the summer of 1831 doing fieldwork with Adam Sedgwick, a geologist at Cambridge, before Henslow recommended him to the captain of the *Beagle* as a naturalist. The trip was to take five years, and the ship was to traverse the Southern Hemisphere, where we now know that life is most abundant and varied. Along the way, Darwin encountered forms of life very different from those in his native England.

Figure 14.1 **Voyage of the HMS *Beagle.***

The map shows the journey of the HMS *Beagle* around the world. As Darwin traveled along the east coast of South America, he noted that a bird called a rhea looked like the African ostrich. On the Galápagos Islands, marine iguanas, found no other place on Earth, use their large claws to cling to rocks and their blunt snouts for eating seaweed.

Charles Darwin

HMS *Beagle*

marine iguanas

Galápagos Islands

rhea

Darwin's major mission was to expand the navy's knowledge of natural resources (e.g., water and food) in foreign lands. The captain also hoped Darwin would find evidence of the biblical account of creation as it was interpreted at that time. Instead, by the time the trip was over, Darwin had made observations that were inconsistent with current beliefs, and these observations led him to conclude that biological evolution occurs.

Before Darwin

Prior to Darwin, lay people had an entirely different way of looking at the world (Table 14.1). They believed that the Earth was only a few thousand years old and that since the time of creation, species had remained exactly the same. Even so, studying the anatomy of organisms and then classifying them interested many investigators who wished to show that species were created to be suitable to their environment. Explorers and collectors traveled the world and brought back currently existing species to be classified and also **fossils,** the remains of once-living organisms often found in strata. **Strata** are layers of rock or sedimentary material (Fig. 14.2a). Figure 14.3 shows how strata come about in the sea. Even if they form in the sea, they can later be found on land.

A noted zoologist at the time, Georges Cuvier, founded the science of paleontology, the study of fossils. Cuvier was faced with a problem. He believed in the fixity of species; yet, the Earth's strata clearly showed a succession of different life-forms over time. To explain these observations, he hypothesized that a local catastrophe had caused a mass extinction whenever a new stratum of that region showed a new mix of fossils. After each catastrophe, the region was repopulated by species from surrounding areas, and this accounted for the appearance of new fossils in the new stratum. The result of all these catastrophes was change appearing over time. Some of Cuvier's followers, who came to be called *catastrophists,* even suggested that worldwide catastrophes had occurred and that after each of these events, new sets of species had been created.

In contrast to Cuvier, Jean-Baptiste de Lamarck, another biologist, hypothesized that evolution occurs, and he also said adaptation to the environment was the cause of diversity. Therefore, after studying the succession of life-forms in strata, Lamarck concluded that more complex organisms are descended from less complex organisms. To explain the process of adaptation to the environment, Lamarck supported the idea of *inheritance of acquired characteristics,* which proposes that use and disuse of a structure can bring about inherited change. One example Lamarck gave—and the one for which he is most famous—is that the long neck of a giraffe developed over time because animals stretched their necks to reach food high in trees and then passed on a long neck to their offspring (see Fig. 14.8). However, his hypothesis for the inheritance of acquired characteristics has never been

a. Visible strata.

b. Fossil amphibian, dated 280–248 MYA.

Figure 14.2 Fossils and strata.

a. When highways cut through mountains, it is often possible to see a number of strata, layers of rock or sedimentary material that contain fossils (**b**). The oldest fossils are in the lowest stratum.

Check Your Progress

1. Describe the pre-Darwinian view of the world.
2. Describe Cuvier's contribution to evolutionary theory.

Answers: 1. The Earth is only thousands of years old. Species do not change over time. A creator is responsible for adaptation to the environment. Observations should support the current view of the world. 2. Cuvier founded the science of paleontology. Fossils are found in strata, and a change in species can be observed over time.

Table 14.1 Contrast of Worldviews

Pre-Darwinian View	Post-Darwinian View
1. The Earth is relatively young—age is measured in thousands of years.	1. The Earth is relatively old—age is measured in billions of years.
2. Each species is specially created; species don't change, and the number of species remains the same.	2. Species are related by descent—it is possible to piece together a history of life on Earth, showing lines of descent.
3. Adaptation to the environment is the work of a creator who decided the structure and function of each type of organism. Any variations are imperfections.	3. Adaptation to the environment is the interplay of random variations and environmental conditions.
4. Observations are supposed to substantiate the prevailing worldview.	4. Observation and experimentation are used to test hypotheses, including hypotheses about evolution.

substantiated by experimentation. For example, if acquired characteristics were inherited, people who become blind by accident would have blind children, and circumcised males would have boys that lack a foreskin from birth. Modern genetics explains why the idea of acquired characteristics cannot be substantiated. Phenotypic changes acquired during an organism's lifetime do not result in genetic changes that can be passed to subsequent generations.

Darwin's ideas were close to those of Lamarck. For example, Darwin said that living things share common characteristics because they have a common ancestry. One of the most unfortunate interpretations of this statement was that we are descended from apes. For Darwin, however, humans and apes share a common ancestry just as, say, you and your cousins can trace your ancestry back to the same grandparents. In contrast to Lamarck, Darwin's observations led him to conclude that species are suited to the environment through no will of their own but by natural selection. He saw the process of natural selection as the means by which different species come about (see Fig. 14.8).

Figure 14.3 Formation of strata.

a. This diagram shows how water brings sediments into the sea; the sediments then become compacted to form a stratum (sing.). Fossils are often trapped in strata, and as a result of a later geological upheaval, the strata may be located on land. **b.** Fossil remains of freshwater snails, *Turritella,* in a stratum.

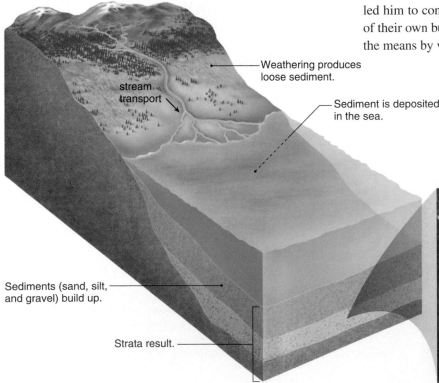

Weathering produces loose sediment.

stream transport

Sediment is deposited in the sea.

Sediments (sand, silt, and gravel) build up.

Strata result.

a.

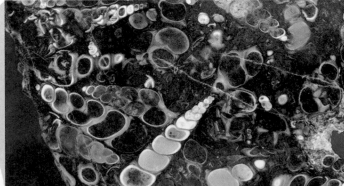

b.

Darwin's Conclusions

Darwin's conclusions that organisms are related through common descent and that adaptation to various environments results in diversity were based on several types of data, including his study of geology and fossils and also his study of biogeography. **Biogeography** is the distribution of life-forms on Earth.

Darwin's Study of Geology and Fossils

Darwin took Charles Lyell's *Principles of Geology* on the *Beagle* voyage. This book presented arguments to support a theory of geological change proposed by James Hutton. In contrast to the catastrophists, Hutton believed the Earth was subject to slow but continuous cycles of erosion and uplift. Weathering causes erosion; thereafter, dirt and rock debris are washed into the rivers and transported to oceans. When these loose sediments are deposited, strata results (see Fig. 14.3a). Then the strata, which often contain fossils, are uplifted from below sea level to form land. Hutton concluded that extreme geological changes can be accounted for by slow, natural processes, given enough time. Lyell went on to propose a theory of **uniformitarianism,** which stated that these slow changes occurred at a uniform rate. Hutton's general ideas about slow and continual geological change are still accepted today, although modern geologists realize that rates of change have not always been uniform. Darwin was not a believer in the idea of uniform change, but he was convinced, as was Lyell, that the Earth's massive geological changes are the result of slow processes, and that the Earth, therefore, must be very old.

On his trip, Darwin observed massive geological changes firsthand. When he explored what is now Argentina, he saw raised beaches for great distances along the coast. When he got to the Andes Mountains, he was impressed by their great height. In Chile, he found fossil marine shells inland, well above sea level, and witnessed the effects of an earthquake that caused the land to rise several feet. While Darwin was making geological observations, he also collected fossil specimens that differed somewhat from modern species (Fig. 14.4). Once Darwin accepted the supposition that the Earth must be very old, he began to think that there would have been enough time for descent with modification—that is, evolution, to occur. Living forms could be descended from extinct forms known only from the fossil record. It would seem that species were not fixed; instead, they change over time.

a.

b.

Figure 14.4   A glyptodont compared to an armadillo.

a. The giant armadillo-like glyptodont is known only by the study of its fossil remains. Darwin found such fossils and came to the conclusion that this extinct animal must be related to living armadillos. The glyptodont weighed 2,000 kilograms. **b.** A modern armadillo weighs about 4.5 kilograms.

Check Your Progress

1. Describe Lamarck's idea of inheritance of acquired characteristics.
2. Why did Lyell think the Earth was very old?

Answers: 1. Use and disuse of structures can produce heritable changes. *2.* Massive geological changes have occurred but at a slow rate.

Figure 14.5   The Patagonian hare.

This animal has the face of a guinea pig and is native to South America, which has no native rabbits. The Patagonian hare has long legs and other adaptations similar to those of rabbits.

Darwin's Study of Biogeography

Darwin could not help but compare the animals of South America to those with which he was familiar in England. For example, instead of rabbits, he found the Patagonian hare in the grasslands of South America. The Patagonian hare has long legs and ears but the face of a guinea pig, a rodent native to South America (Fig. 14.5). Did the Patagonian hare resemble a rabbit because the two types of animals were adapted to the same type of environment—an outcome we call convergent evolution today? Both animals ate grass, hid in bushes, and moved rapidly using long hind legs. Did the Patagonian hare have the face of a guinea pig because of common descent with guinea pigs?

As he sailed southward along the eastern coast of the continent of South America, Darwin saw how similar species replaced each other. For example, the greater rhea (an ostrichlike bird) found in the north was replaced by the lesser rhea in the south. Therefore, Darwin reasoned that related species could be modified according to the environment. When he got to the Galápagos Islands, he found further evidence of this. The Galápagos Islands are a small group of volcanic islands off the western coast of South America. The few types of plants and animals found there were slightly different from species Darwin had observed on the mainland, and even more important, they also varied from island to island according to the particular environments.

Finches   Although some of the the finches on the Galápagos Islands seemed like mainland finches, others were quite different (Fig. 14.6). Today, there are ground-dwelling finches with different-sized beaks, depending on the size of the seeds they feed on, and a cactus-eating finch with a more pointed beak. The beak size of the tree-dwelling finches also varies, according to the size of their insect prey. The most unusual of the finches is a woodpecker-type finch. This bird has a sharp beak to chisel through tree bark but lacks the woodpecker's long tongue that probes for insects. To make up for this, the bird carries a twig or cactus thorn in its beak and uses it to poke into crevices. Once an insect emerges, the finch drops this tool and seizes the insect with its beak.

Later, Darwin speculated as to whether all the different species of finches he had seen could have descended from a mainland finch. In other words, he wondered if a mainland finch was the common ancestor to all the types on the Galápagos Islands. Had **speciation,** formation of a new species, occurred because the islands allowed isolated populations of birds to evolve independently? Could the present-day species have resulted from accumulated changes occurring within each of these isolated populations?

a. Ground-dwelling finch

b. Warbler finch

c. Cactus finch

Figure 14.6   Galápagos finches.

Each of the present-day 13 species of finches has a beak adapted to a particular way of life. **a.** The heavy beak of the large ground-dwelling finch is suited to a diet of seeds. **b.** The beak of the warbler finch is suited to feeding on insects found among ground vegetation or caught in the air. **c.** The long, somewhat decurved beak and split tongue of the cactus finch are suited to probing cactus flowers for nectar.

Tortoises Each of the Galápagos Islands also seemed to have its own type of tortoise, and Darwin began to wonder if this could be correlated with a difference in vegetation among the islands. Long-necked tortoises seemed to inhabit only dry areas, where food was scarce, and most likely the longer neck was helpful in reaching cactuses. In moist regions with relatively abundant ground foliage, short-necked tortoises were found. Had an ancestral tortoise given rise to these different types, each adapted to a different environment? An **adaptation** is any characteristic that makes an organism more suited to its environment.

Natural Selection and Adaptation

Once Darwin decided that adaptations develop over time, he began to think about a mechanism by which adaptations might arise. Eventually, he proposed **natural selection** as the mechanism. Natural selection is a process in which steps 1–3 result in a population adapted to the environment (4):

1. The members of a population have heritable variations (Fig. 14.7).
2. The population produces more offspring than the resources of an environment can support.
3. The more adapted individuals survive and reproduce to a greater extent than those that lack the adaptations.
4. Across generations, a larger proportion of the population becomes adapted to the environment.

Notice that because natural selection utilizes only variations that happen to be provided by genetic changes, it lacks any directedness or anticipation of future needs. Natural selection is an ongoing process because the environment of living things is constantly changing. Extinction (loss of a species) can occur when previous adaptations are no longer suitable to a changed environment.

Organisms Have Variations

With reference to step 1, Darwin emphasized that the members of a population vary in their functional, physical, and behavioral characteristics. Before Darwin, variations were considered imperfections that should be ignored since they were not important to the description of a species (see Table 14.1). Darwin, on the other hand, believed variations were essential to the natural selection process. Darwin suspected—but did not have the evidence we have today—that the occurrence of variations is completely random; they arise by accident and for no particular purpose. Also, new variations are more likely to be harmful than beneficial to an organism.

The variations that make adaptation to the environment possible are those that are passed on from generation to generation. The science of genetics was not yet well established, so Darwin was never able to determine the cause of variations or how they are passed on. Today, we realize that genes, together with the environment, determine the phenotypes of an organism. Mutations, along with chromosomal rearrangements and assortment of chromosomes during meiosis and fertilization, can cause new variations to arise.

Figure 14.7 **Variations in shells of a marine snail,** ***Liguus fascitus.***

For Darwin, variations such as these in a species of snails were highly significant and were required in order for natural selection to result in adaptation to the environment.

Check Your Progress

1. Explain why Darwin's observations about Galápagos finches were significant to him.
2. Outline Darwin's theory of natural selection.

Answers: 1. The finches were all descended from a mainland finch, and yet they were adapted differently. 2. Members of a population have heritable variations. More offspring are produced than resources can support. The most adapted individuals survive and reproduce. Across generations, a larger proportion of a species becomes more adapted to the environment.

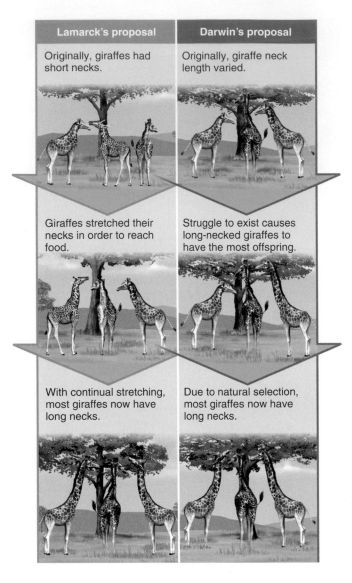

Lamarck's proposal	Darwin's proposal
Originally, giraffes had short necks.	Originally, giraffe neck length varied.
Giraffes stretched their necks in order to reach food.	Struggle to exist causes long-necked giraffes to have the most offspring.
With continual stretching, most giraffes now have long necks.	Due to natural selection, most giraffes now have long necks.

Figure 14.8   Mechanism of evolution.

This diagram contrasts Jean-Baptiste Lamarck's proposal of acquired characteristics with Charles Darwin's proposal of natural selection. Only natural selection is supported by data.

Check Your Progress

What is the main difference between artificial selection and natural selection?

Answer: Artificial selection is carried out by humans, and natural selection is carried out by environmental forces, other than humans.

Organisms Struggle to Exist

With reference to step 2, in Darwin's time a socioeconomist named Thomas Malthus stressed the reproductive potential of human beings. He proposed that death and famine are inevitable because the human population tends to increase faster than the supply of food. Darwin applied this concept to all organisms and saw that the available resources were not sufficient to allow all members of a population to survive. He calculated the reproductive potential of elephants. Assuming a life span of about 100 years and a breeding span of 30 to 90 years, a single female probably bears no fewer than six young. If all these young were to survive and continue to reproduce at the same rate, after only 750 years, the descendants of a single pair of elephants would number about 19 million!

Each generation has the same reproductive potential as the previous generation. Therefore, there is a constant struggle for existence, and only certain members of a population survive and reproduce each generation.

Organisms Differ in Fitness

With reference to step 3, **fitness** is the reproductive success of an individual relative to other members of a population, in a particular environment. The most-fit individuals are the ones that capture a disproportionate amount of resources, and convert these resources into a larger number of viable offspring. Because organisms vary anatomically and physiologically, and because the challenges of local environments vary as well, what determines fitness varies for different populations. For example, among western diamondback rattlesnakes living on lava flows, the most fit are those that are black in color. But among those living on desert soil, the most fit are those with the typical light and dark-brown coloring. Background matching helps an animal both capture prey and avoid being captured; therefore, it is expected to lead to survival and increased reproduction.

Darwin noted that when humans help carry out **artificial selection,** breeders select animals with particular traits to reproduce. For example, prehistoric humans probably noted desirable variations among wolves and selected particular animals for breeding. Therefore, the desired traits increased in frequency in the next generation. This same process was repeated many times over. The result today is the existence of many varieties of dogs, all descended from the wolf. In a similar way, several varieties of vegetables can be traced to a single ancestor. Chinese cabbage, brussels sprouts, and kohlrabi are all derived from a single species, *Brassica oleracea.*

In nature, interactions with the environment determine which members of a population reproduce to a greater degree than other members. In contrast to artificial selection, the result of natural selection is not predesired. Natural selection occurs because certain members of a population happen to have a variation that allows them to survive and reproduce to a greater extent than other members. For example, any variation that increases the speed of a hoofed animal helps it escape predators and live longer; a variation that reduces water loss is beneficial to a desert plant; and one that increases the sense of smell helps a wild dog find its prey. Therefore, we expect the organisms with these characteristics to have increased fitness. Figure 14.8 contrasts Lamarck's ideas with those of Darwin.

Organisms Become Adapted

With reference to step 4 (see page 221), it takes time for adaptations to evolve. We rarely observe the process, but we can observe the end result. We can especially recognize an adaptation when unrelated organisms living in a particular environment display similar characteristics. For example, manatees, penguins, and sea turtles all have flippers, which help them move through the water—also an example of convergent evolution. Adaptations also account for why organisms are able to escape their predators (Fig. 14.9) or why they are suited to their way of life (Fig. 14.10). Natural selection causes adaptive traits to be increasingly represented in each succeeding generation. There are other processes of evolution aside from natural selection (see Chapter 15), but natural selection is the only process that results in adaptation to the environment.

Darwin and Wallace

After the HMS *Beagle* returned to England in 1836, Darwin waited more than 20 years to publish his book *On the Origin of Species*. During the intervening years, he used the scientific process to test his hypothesis that today's diverse life-forms arose by descent from a common ancestor and that natural selection is a mechanism by which species can change and new species can arise. Darwin was prompted to publish his book after reading a similar hypothesis formulated by Alfred Russel Wallace.

Wallace (1823–1913) was an English naturalist who, like Darwin, was a collector at home and abroad. He went on collecting trips, each of which lasted several years, to the Amazon and the Malay Archipelago. After studying the animals on every major island, he divided these islands into a western group, with animals like those of the Orient, and an eastern group, with animals like those of Australia. The dividing line between the islands is a narrow but deep strait now known as the Wallace Line.

While traveling, Wallace also wrote an essay called "On the Law Which Has Regulated the Introduction of New Species." In this essay he said that "every species has come into existence coincident both in time and space with a preexisting closely allied species." A year later, after reading Malthus's treatise on human population increase, Wallace conceived the idea of "survival of the fittest." He quickly completed an essay proposing natural selection as an agent for evolutionary change and sent it to Darwin for comment. Darwin was stunned. Here was the hypothesis he had formulated, but had never dared to publish. He told his friend and colleague, Charles Lyell, that Wallace's ideas were so similar to his own that even Wallace's "terms now stand as heads of my chapters."

Darwin suggested that Wallace's paper be published immediately, even though Darwin himself as yet had nothing in print. Lyell and others who knew of Darwin's detailed work substantiating the process of natural selection suggested that a joint paper be read to the Linnean Society, a renowned scientific society. The title of Wallace's section was "On the Tendency of Varieties to Depart Indefinitely from the Original Type." Darwin presented an abstract of a paper he had written in 1844 and an abstract of *On the Origin of Species*, which was published in 1859.

b.

Figure 14.9 **Adaptations of the alligator bug.**

The alligator bug of the Brazilian rain forest has antipredator adaptations. **a.** The insect blends into its background, but if discovered, the false head, which resembles a miniature alligator, may frighten a predator. **b.** If the predator is not frightened, the insect suddenly reveals huge false eyespots on its hindwings.

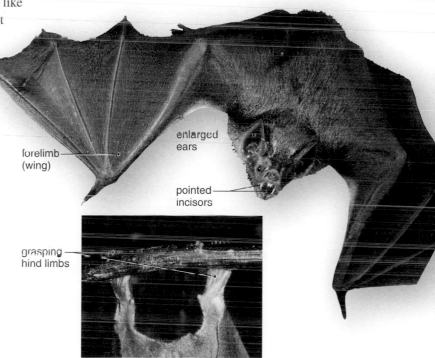

Figure 14.10 **Adaptations of the vampire bat.**

Vampire bats of the rain forests of Central and South America have the adaptations of a nocturnal, winged predator. The bat uses its enlarged ears and echolocation to locate prey in the dark. It bites its prey with two pointed incisors. Saliva containing an anticoagulant called draculin runs into the bite; the bat then licks the flowing blood. The vampire bat's forelimbs are modified to form wings, and it roosts by using its grasping hind limbs.

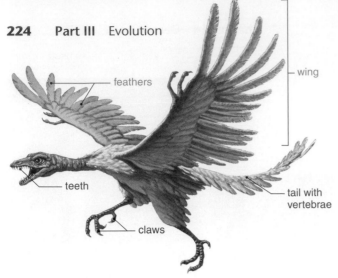

Figure 14.11 Re-creation *Archaeopteryx*.

The fossil record suggests that *Archaeopteryx* had a feather-covered, reptilian-type tail, which shows up well in this artist's representation. (Red labels = reptilian characteristics; green labels = bird characteristics.)

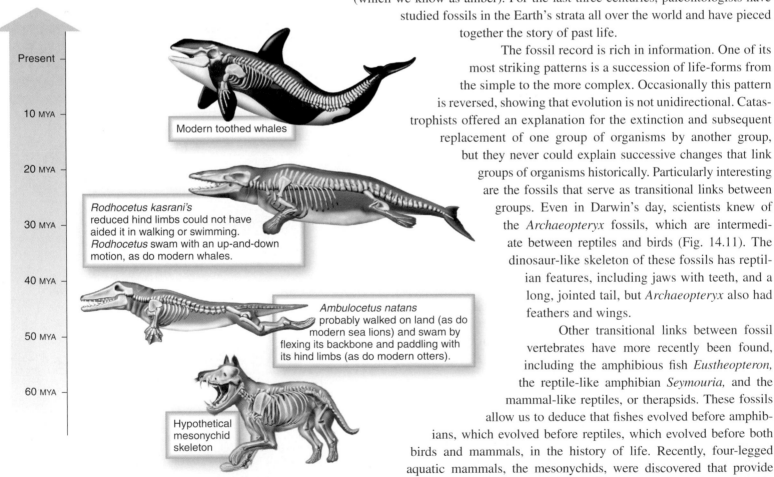

Figure 14.12 Evolution of whales.

The discoveries of *Ambulocetus* and *Rodhocetus* have filled in the gaps in the evolution of whales from extinct hoofed mammals that lived on land to the ocean-dwelling mammal we know today.

14.2 Evidence for Evolution

The Theory of Evolution states that all living things have a common ancestor, but each is adapted to a particular way of life. Many lines of evidence consistently support the hypothesis that organisms are related through common descent. A hypothesis becomes a scientific theory only when a variety of evidence made by independent investigators supports the hypothesis. The theory of evolution is a unifying principle in biology because it can explain so many different observations in various fields of biology. The theory of evolution has the same status in biology that the germ theory of disease has in medicine.

Darwin cited much of the evidence for evolution we will discuss, except that he had no knowledge of the biochemical data that became available after his time.

Fossil Evidence

The **fossil record** is the history of life recorded by remains from the past. Fossils are at least 10,000 years old and include such items as pieces of bone, impressions of plants pressed into shale, and even insects trapped in tree resin (which we know as amber). For the last three centuries, paleontologists have studied fossils in the Earth's strata all over the world and have pieced together the story of past life.

The fossil record is rich in information. One of its most striking patterns is a succession of life-forms from the simple to the more complex. Occasionally this pattern is reversed, showing that evolution is not unidirectional. Catastrophists offered an explanation for the extinction and subsequent replacement of one group of organisms by another group, but they never could explain successive changes that link groups of organisms historically. Particularly interesting are the fossils that serve as transitional links between groups. Even in Darwin's day, scientists knew of the *Archaeopteryx* fossils, which are intermediate between reptiles and birds (Fig. 14.11). The dinosaur-like skeleton of these fossils has reptilian features, including jaws with teeth, and a long, jointed tail, but *Archaeopteryx* also had feathers and wings.

Other transitional links between fossil vertebrates have more recently been found, including the amphibious fish *Eustheopteron*, the reptile-like amphibian *Seymouria*, and the mammal-like reptiles, or therapsids. These fossils allow us to deduce that fishes evolved before amphibians, which evolved before reptiles, which evolved before both birds and mammals, in the history of life. Recently, four-legged aquatic mammals, the mesonychids, were discovered that provide important insights concerning the evolution of whales from land-living, hoofed ancestors (Fig. 14.12). The fossilized whale *Ambulocetus* may have been amphibious, walking on land and still swimming in the sea. *Rodhocetus* swam with an up-and-down tail motion like modern whales do; its reduced hind limbs could not have helped in swimming.

Biogeographical Evidence

Biogeography is the study of the distribution of plants and animals in different places throughout the world. Such distributions are consistent with the hypothesis that, when forms are related, they evolved in one locale and then spread to accessible regions. Therefore, you would expect a different mix of plants and animals whenever geography separates continents, islands, or seas. As previously mentioned, Darwin noted that South America lacked rabbits, even though the environment was quite suitable for them. He concluded that no rabbits lived in South America because rabbits evolved somewhere else and had no means of reaching South America.

To take another example, both cactuses and euphorbia are plants adapted to a hot, dry environment—both are succulent, spiny, flowering plants. Why do cactuses grow in North American deserts and euphorbia grow in African deserts, when each would do well on the other continent? It seems obvious that they just happened to evolve on their respective continents.

The islands of the world are home to many unique species of animals and plants that are found no place else, even when the soil and climate are the same. Why do so many species of finches live on the Galápagos Islands when these same species are not on the mainland? The reasonable explanation is that an ancestral finch migrated to all the different islands. Then geographic isolation allowed the ancestral finch to evolve into a different species on each island.

Also, long ago South America, Antarctica, and Australia were connected (see Fig. 16.13*a*). Marsupials (pouched mammals) and placental mammals arose at this time, but today marsupials are plentiful only in Australia and placental mammals are plentiful in South America. Why are marsupials only plentiful in Australia (Fig. 14.13)? After marsupials arose, Australia separated and drifted away, and marsupials were free to evolve into many different forms because they had no competition from placental mammals. In the Americas, the placentals were able to successfully compete against the marsupials, and the opossum is the only marsupial in the Americas. In some cases, placental and marsupial mammals physically resemble each other—for example, the pouched marsupial mouse and the harvest mouse, the marsupial mole and the common mole, the marsupial wombat and the marmot, and the tasmanian wolf and the wolf. This supports the hypothesis that evolution is influenced by the environment and by the mix of plants and animals in a particular continent—that is, by biogeography, not by design.

Sugar glider, a tree dweller

Kangaroo, a herbivore of plains and forests

Koala, a tree dweller

Coarse-haired wombat, nocturnal and living in burrows

Figure 14.13   Marsupials of Australia.

Each type of marsupial in Australia is adapted to a different way of life. All of them presumably evolved from a common ancestor that entered Australia some 60 million years ago.

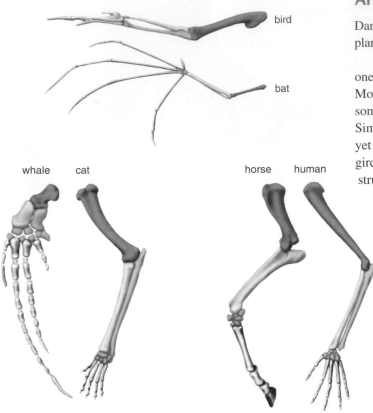

Anatomical Evidence

Darwin was able to show that a common descent hypothesis offers a plausible explanation for vestigial structures and anatomical similarities among organisms.

Vestigial structures are anatomical features that are fully developed in one group of organisms but reduced and nonfunctional in other similar groups. Most birds, for example, have well-developed wings used for flight. However, some bird species (e.g., ostrich) have greatly reduced wings and do not fly. Similarly, whales (see Fig. 14.12) and snakes have no use for hind limbs, and yet extinct whales and even some present-day snakes have remnants of a pelvic girdle and legs. Humans have a tailbone but no tail. The presence of vestigial structures can be explained by the common descent hypothesis. Vestigial structures occur because organisms inherit their anatomy from their ancestors; they are traces of an organism's evolutionary history.

Vertebrate forelimbs are used for flight (birds and bats), orientation during swimming (whales and seals), running (horses), climbing (arboreal lizards), or swinging from tree branches (monkeys). Yet all vertebrate forelimbs contain the same sets of bones organized in similar ways, despite their dissimilar functions (Fig. 14.14). The most plausible explanation for this unity is that the basic forelimb plan belonged to a common ancestor, and then the plan became modified in the succeeding groups as each continued along its own evolutionary pathway. Anatomically similar structures explainable by inheritance from a common ancestor are called **homologous structures**. In contrast, **analogous structures** serve the same function, but are not constructed similarly, and therefore could not have a common ancestry. The wings of birds and insects are analogous structures.

The homology shared by vertebrates extends to their embryological development (Fig. 14.15). At some time during development, all vertebrates have a postanal tail and exhibit paired pharyngeal pouches. In fishes and amphibian larvae, these pouches develop into functioning gills. In humans, the first pair of pouches becomes the cavity of the middle ear and the auditory tube. The second pair becomes the tonsils, while the third and fourth pairs become the thymus and parathyroid glands. Why should pharyngeal pouches, which have lost their original function, develop and then become modified in terrestrial vertebrates?

The most likely explanation is that fishes are ancestral to other vertebrate groups. Anatomical and developmental homologies are independent evidences of a shared common ancestor and an evolutionary relationship between groups of organisms.

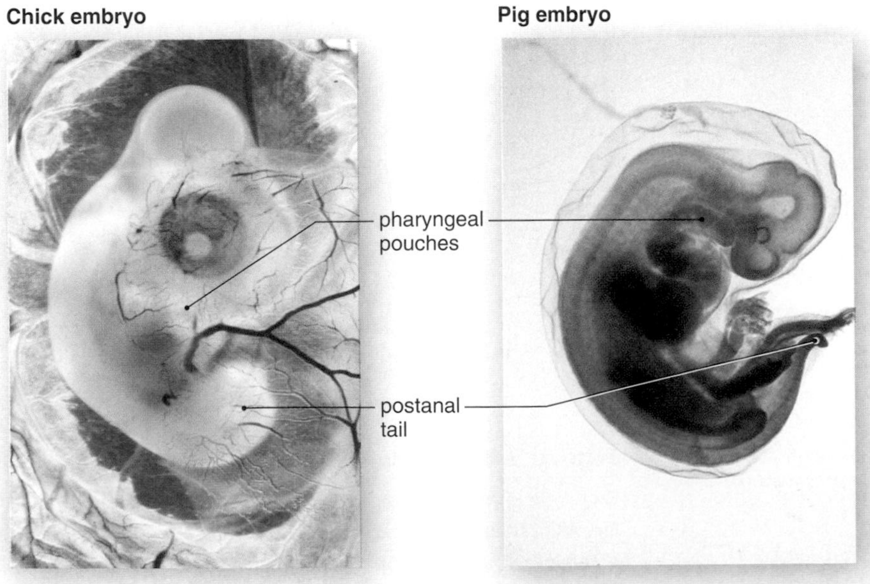

Figure 14.14   Significance of structural similarities.

Although the specific design details of vertebrate forelimbs are different, the same bones are present (color-coded here). This unity of plan is evidence of a common ancestor.

Figure 14.15   Significance of developmental similarities.

At this comparable developmental stage, a chick embryo and a pig embryo have many features in common, which suggests they evolved from a common ancestor.

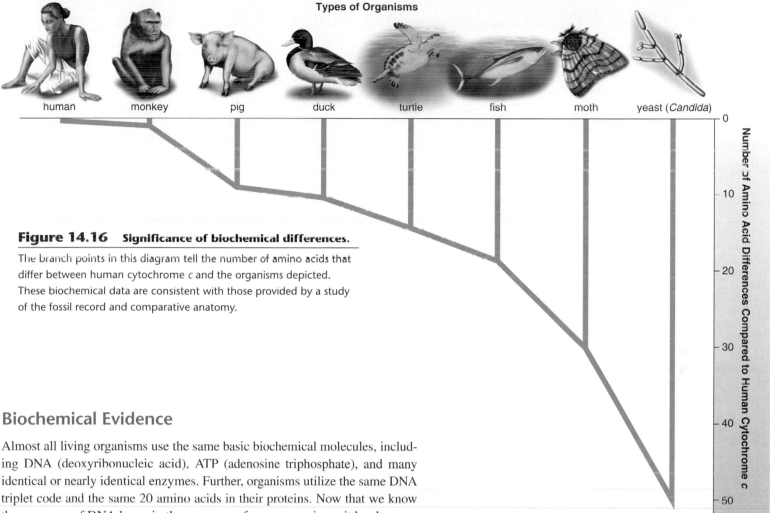

Types of Organisms

human   monkey   pig   duck   turtle   fish   moth   yeast (*Candida*)

Number of Amino Acid Differences Compared to Human Cytochrome *c*

Figure 14.16   Significance of biochemical differences.
The branch points in this diagram tell the number of amino acids that differ between human cytochrome *c* and the organisms depicted. These biochemical data are consistent with those provided by a study of the fossil record and comparative anatomy.

Biochemical Evidence

Almost all living organisms use the same basic biochemical molecules, including DNA (deoxyribonucleic acid), ATP (adenosine triphosphate), and many identical or nearly identical enzymes. Further, organisms utilize the same DNA triplet code and the same 20 amino acids in their proteins. Now that we know the sequence of DNA bases in the genomes of many organisms, it has become clear that humans share a large number of genes with much simpler organisms.

Also of interest, evo-devo researchers (evolutionary developmental biologists) have found that many developmental genes are shared in animals ranging from worms to humans. It appears that life's vast diversity has come about by only slight differences in the same genes. The result has been widely divergent body plans. For example, a similar gene in arthropods and vertebrates determines the back to front axis. But, although the base sequences are similar, the genes have opposite effects. In arthropods, such as fruit flies and crayfish, the nerve cord is toward the front, whereas in vertebrates, such as chicks and humans, the nerve cord is toward the back.

When the degree of similarity in DNA base sequences or the degree of similarity in amino acid sequences of proteins is examined, the data are as expected, assuming common descent. Cytochrome *c* is a molecule used in the electron transport chain of all the organisms shown in Figure 14.16. Data regarding differences in the amino acid sequence of cytochrome *c* show that the sequence in a human differs from that in a monkey by only one amino acid, from that in a duck by 11 amino acids, and from that in *Candida,* a yeast, by 51 amino acids. These data are consistent with other data regarding the anatomical similarities of these organisms, and therefore their relatedness.

Check Your Progress

1. Explain how vestigial structures support the theory of evolution.
2. Contrast homologous structures with analogous structures.
3. With reference to Figure 14.16, why would it make sense that the most distantly related species have greater structural differences and greater DNA differences?

Answers: 1. These structures probably arose from an ancestor that needed them, but they are no longer useful. 2. Homologous structures share an anatomical similarity that reflects a common ancestry. Analogous structures have a common function but do not serve as evidence of a common ancestry. 3. DNA determines the structure and function of organisms.

THE CHAPTER IN REVIEW

Summary

14.1 Darwin's Theory of Evolution

Charles Darwin took a position as a naturalist aboard the HMS *Beagle* and made a trip around the world, largely in the Southern Hemisphere.

Before Darwin
Before Darwin, people believed that the Earth was young, that species didn't change, that any variations were imperfections, and that observations could substantiate this view.

- Cuvier was an early paleontologist who believed that species do not change. He observed species come and go in the fossil record, and he said that these changes were due to catastrophies.
- Lamarck was a zoologist who hypothesized that evolution and adaptation to the environment do occur. He suggested that acquired characteristics could be inherited. For example, he said that giraffes stretched their necks to reach food in trees, and then this change was inherited by the next generation.

Darwin's Conclusions
Darwin's conclusions based on geology and fossils are:

- The Earth is very old, giving time for evolution to occur.
- Living things are descended from extinct forms known only from the fossil record.

Darwin's conclusions based on biogeography are:

- Living things evolve where they are. This explains, for example, why South America has the Patagonian hare, while England has the rabbit.
- Living things are adapted to local environments. This explains why there are many types of finches and tortoises in the Galápagos Islands.

Natural Selection and Adaptation
According to Darwin, the result of natural selection is a population adapted to its local environment:

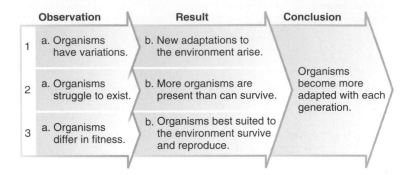

Darwin and Wallace
Alfred Russel Wallace was a naturalist who, like Darwin, traveled to other continents in the Southern Hemisphere. He also collected evidence of common descent, and his reading of Malthus caused him to develop the same mechanism for adaptation (natural selection) as Darwin. Darwin's work was more thorough, as evidenced by his book *On the Origin of Species*.

14.2 Evidence for Evolution

A theory in science is a concept supported by much evidence, and the Theory of Evolution is supported by several types of evidence:

- Fossil record indicates the history of life in general and allows us to trace the descent of a particular group.
- Biogeography shows that the distribution of organisms on Earth is explainable by assuming organisms evolved in one locale.
- Anatomy and development of organisms reveals homologies explainable by common ancestry.
- Biochemical molecules of the same type occur in all organisms. Differences indicate the degree of relatedness.

Thinking Scientifically

1. Recent research indicates that genes involved in male reproduction (sperm production and morphology) evolve at a faster rate than other genes in primates. Explain why this makes sense, based on natural selection.

2. Geneticists often compare DNA sequences of genes among organisms to determine their rate of evolution, called the molecular clock. The molecular clock has been found to run at different rates for different genes. That is, some genes evolve faster than others. Explain why the DNA sequence in one gene might change faster over time than that in another gene.

Testing Yourself

Choose the best answer for each question.

1. _____ developed the idea that acquired characteristics can be inherited.

   a. Darwin
   b. Lamarck
   c. Hutton
   d. Sedgwick
   e. Cuvier

2. The major feature Darwin noticed that differed among species of Galápagos finches was

   a. coloration.
   b. beak size.
   c. leg length.
   d. length of tail feathers.

3. Which of the following is not an example of natural selection?

   a. Insect populations exposed to pesticides become resistant to the chemicals.
   b. Plant species that produce fragrances to attract pollinators produce more offspring.
   c. Rabbits that sprint quickly are more likely to escape predation.
   d. On a tree, leaves that grow in the shade are larger than those that grow in the sun.

4. The variations necessary for natural selection

   a. occur randomly.
   b. are influenced by environment.
   c. can be caused by mutation.
   d. can be caused by recombination during meiosis.
   e. All of these are correct.

5. Which is most likely to be favored during natural selection, but not artificial selection?

   a. fast seed germination rate
   b. short generation time
   c. efficient seed dispersal
   d. lean pork meat production

6. Natural selection is the only process that results in

   a. genetic variation.
   b. adaptation to the environment.
   c. phenotypic change.
   d. competition among individuals in a population.

7. Why was it helpful to Darwin to learn that Lyell concluded the Earth was very old?

   a. An old Earth has more fossils than a new Earth.
   b. It meant there was enough time for evolution to have occurred slowly.
   c. There was enough time for the same species to spread out into all continents.
   d. Darwin said artificial selection occurs slowly.
   e. All of these are correct.

8. In the following diagram, contrast Lamarck's proposal with Darwin's proposal by matching the phrases in the key to the letters in the diagram.

Key:

Originally, giraffe neck length varied.
Giraffes stretched their necks in order to reach food.
Struggle to exist causes long-necked giraffes to have the most offspring.
Originally, giraffes had short necks.
Today most giraffes have long necks. (used twice)

9. All the finches on the Galápagos Islands

   a. are unrelated but descended from a common ancestor.
   b. are descended from a common ancestor, and therefore related.
   c. rarely compete for the same food source.
   d. Both a and c are correct.
   e. Both b and c are correct.

10. Evolution is considered a

    a. hypothesis because it is supported by data from the fossil record.
    b. hypothesis because it is supported by multiple types of data.
    c. theory because it is supported by data from the fossil record.
    d. theory because it is supported by multiple types of data.

11. Catastrophists were not able to explain

    a. multiple extinctions.
    b. the replacement of one group of organisms by another.
    c. successive changes that link groups of organisms in the fossil record.
    d. More than one of these are correct.

For questions 12–15, match the description with the type of evidence for evolution it supports, as listed in the key. Answers can be used more than once.

Key:

    a. biogeographical
    b. anatomical
    c. biochemical

12. The genetic code is the same for all organisms.

13. The human kneebone and spine were derived from ancestral structures that supported four-legged animals.

14. The South American continent lacks rabbits even though the environment is quite suitable.

15. The amino acid sequence of human hemoglobin is more similar to that of rhesus monkeys than to that of mice.

16. Fossils that serve as transitional links allow scientists to
    a. determine how prehistoric animals interacted with each other.
    b. deduce the order in which various groups of animals arose.
    c. relate climate change to evolutionary trends.
    d. determine why evolutionary changes occur.

17. Among vertebrates, the flipper of a dolphin and the fin of a tuna are
    a. homologous structures.
    b. homogeneous structures.
    c. analogous structures.
    d. reciprocal structures.

18. Which of these pairs is mismatched?
    a. Charles Darwin—natural selection
    b. Cuvier—series of catastrophes explains the fossil record
    c. Lamarck—uniformitarianism
    d. All of these are correct.

19. According to the inheritance of acquired characteristics hypothesis,
    a. if a man loses his hand, then his children will also be missing a hand.
    b. changes in phenotype are passed on by way of the genotype to the next generation.
    c. organisms are able to bring about a change in their phenotype.
    d. evolution is striving toward particular traits.
    e. All of these are correct.

20. Organisms
    a. compete with other members of their species.
    b. differ in fitness.
    c. are adapted to their environment.
    d. are related by descent from common ancestors.
    e. All of these are correct.

21. DNA nucleotide differences between organisms
    a. indicate how closely related organisms are.
    b. indicate that evolution occurs.
    c. explain why there are phenotypic differences.
    d. are to be expected.
    e. All of these are correct.

22. The fossil record offers direct evidence for common descent because you can
    a. see that the types of fossils change over time.
    b. sometimes find common ancestors.
    c. trace the ancestry of a particular group.
    d. trace the biological history of living things.
    e. All of these are correct.

Go to www.mhhe.com/maderessentials for more quiz questions.

Bioethical Issue

The term "theory" in science is reserved for those ideas that scientists have found to be all-encompassing because they are based on data collected in a number of different fields. Evolution is a scientific theory. So is the cell theory, which says that all organisms are composed of cells, and so is the atomic theory, which says that all matter is composed of atoms. No one argues that schools should teach alternatives to the cell theory or the atomic theory. Yet, confusion reigns over the use of the expression "the theory of evolution."

No wonder most scientists in our country are dismayed when state legislatures or school boards rule that teachers must put forward a variety of "theories" on the origin of life, including one that runs contrary to the mass of data that supports the theory of evolution. An organization in California called the Institute for Creation Research advocates that students be taught an "intelligent-design theory," which says that DNA could never have arisen without the involvement of an "intelligent agent," and that gaps in the fossil record mean that species arose fully developed with no antecedents.

Since no purely religious ideas can be taught in the schools, the advocates for an intelligent-design theory are careful not to mention the Bible or any strictly religious ideas. Still, teachers who have a solid scientific background do not feel comfortable teaching an intelligent-design theory because it does not meet the test of a scientific theory. Science is based on hypotheses that have been tested by observation and/or experimentation. A scientific theory has stood the test of time—that is, no hypotheses have been supported by observation and/or experimentation that run contrary to the theory. On the contrary, the Theory of Evolution is supported by data collected in such wide-ranging fields as development, anatomy, geology, and biochemistry.

The polls consistently show that nearly half of all Americans prefer to believe the Old Testament account of creation. That, of course, is their right. But should schools be required to teach an intelligent-design theory that traces its roots back to the Old Testament and is not supported by observation and experimentation?

Understanding the Terms

adaptation 221
analogous structure 226
artificial selection 222
biogeography 219
evolution 215
fitness 222
fossil record 224
fossils 217
homologous structure 226
natural selection 221
speciation 220
strata 217
uniformitarianism 219
vestigial structure 226

Match the terms to these definitions:

a. _____ Structure that is similar in different types of organisms explainable by a common ancestry.

b. _____ Lyell's theory that geological changes occur slowly and at a uniform rate.

c. _____ Darwin's proposed mechanism for evolutionary change.

d. _____ In an organism, remains of a nonfunctional structure that was once functional in an ancestor of the organism.

e. _____ Trait that helps an organism be better adapted to its environment.

Evolution on a Small Scale

CHAPTER 15

Male competition for mates can influence evolution.

Female choice of showy mates can influence evolution.

You might think that fitness means keeping in shape, but to an evolutionist it means having more fertile offspring than other individuals. Think about it—only if an animal reproduces can that animal's genes be passed on and become prevalent in the next generation. Adaptation to the environment increases the chance of reproducing, but so does *sexual selection,* which occurs because of an advantage that helps an animal get a mate.

Do male competition and female choice also influence human evolution?

▶ ▶ ▶

Such advantages as increased size, resplendent feathers, evolution of horns, or enlarged canines help males fight for and attract females. Females, in turn, must choose carefully. Perhaps to a female, the showier male is healthier or more appealing. If so, the same characteristics will be advantageous for her sons! Sexual selection increases the chances of reproducing, but can shorten the life span. A large, showy male that fights a lot most likely doesn't live as long as a small, inconspicuous male.

Do male competition and female choice occur among humans? Some think so. They point out that human males tend to be larger and more aggressive than females, and that statistically males have a shorter life span. Also, wealthy and successful males are more apt to be attractive to women, and sometimes older men marry younger women who are still fertile, thereby increasing their own fitness.

Sexual selection is one of the factors that influences evolution on a small scale (microevolution), which is the subject of this chapter.

15.1 Microevolution

Thus far, it may seem to you as though individuals evolve, but this is not the case. As evolution occurs, genetic and therefore phenotypic changes occur within a population. A **population** is all the members of a single species occupying a particular area and reproducing with one another. **Microevolution** pertains to evolutionary changes within a population.

Darwin stressed that the members of a population vary as in Figure 15.1, but he did not know how variations come about and how they are transmitted. Today, DNA sequencing in a number of plants and animals shows that each gene in sexually reproducing organisms has many alleles. The reshuffling of alleles during sexual reproduction can result in a range of phenotypes among the members of a population (see Chapter 10). Even so, as we shall see, sexual reproduction in and of itself cannot bring about evolution. Evolution is influenced by several other circumstantial factors.

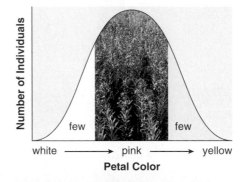

Figure 15.1 Population of lupine flowers.

Among the members of a flowering plant population, petal color can be a continuous variation, so that there is a range of phenotypes. Few individuals have the extreme phenotypes, white petals and yellow petals, and most individuals have the intermediate phenotype, pink petals. A graph of the phenotypes results in the bell-shaped curve shown above a photo of the population.

Evolution in a Genetic Context

It was not until the 1930s that population geneticists were able to apply the principles of genetics to populations and thereafter to develop a way to recognize when evolution has occurred. In population genetics, the various alleles at all the gene loci in all individuals make up the **gene pool** of the population.

To simplify matters, we will describe the gene pool of a population in terms of gene frequencies, assuming just two alleles per gene locus. Suppose that in a *Drosophila* population, 36% of the flies are homozygous dominant for long wings, 48% are heterozygous, and 16% are homozygous recessive for short wings. Therefore, in a population of 100 individuals, we have

36 *LL*, 48 *Ll*, and 16 *ll*

What is the number of the *L* allele and the *l* allele in the population?

Number of *L* alleles:			**Number of *l* alleles:**		
LL	(2 *L* × 36)	= 72	*LL*	(0 *l*)	= 0
Ll	(1 *L* × 48)	= 48	*Ll*	(1 *l* × 48)	= 48
ll	(0 *L*)	= 0	*ll*	(2 *l* × 32)	= 32
		120 *L*			80 *l*

To determine the frequency of each allele, calculate its percentage from the total number of alleles in the population. In each case, for the dominant allele *L*, 120/200 = 0.6; for the recessive allele *l*, 80/200 = 0.4. The sperm and eggs produced by this population will also contain these alleles in these frequencies. Assuming random mating (all possible gametes have an equal chance to combine with any other), we can calculate the ratio of genotypes in the next generation by using a Punnett square, as you first learned in Chapter 10.

There is an important difference between a Punnett square that represents a cross between individuals and the following one. Below, the sperm and eggs are those produced by the members of a population— not those produced by a single male and female. As you can see, the frequency of the allele in the next generation is the product of the frequencies of the parental generation. The results of the Punnett square indicate that the frequency for each allele in the next generation is still 0.6 for *L* and 0.4 for *l*:

		sperm	
		0.6 *L*	0.4 *l*
eggs	0.6 *L*	0.36 *LL*	0.24 *Ll*
	0.4 *l*	0.24 *Ll*	0.16 *ll*

Genotype frequencies:
0.36 *LL* + 0.48 *Ll* + 0.16 *ll*

Figure 15.2   Freckles.

A dominant allele causes freckles—so why doesn't everyone have freckles? The Hardy-Weinberg principle, which states that sexual reproduction in and of itself doesn't change allele frequencies, explains why dominant alleles don't become more prevalent with each generation.

Therefore, sexual reproduction alone cannot bring about a change in allele frequencies. Also, the dominant allele need not increase from one generation to the next. Dominance does not cause an allele to become common (Fig. 15.2). Microevolution requires that allele frequencies within a population change, and influences aside from sexual reproduction are required to bring this about.

The potential constancy, or equilibrium state, of gene pool frequencies was independently recognized in 1908 by

Check Your Progress

How does a population geneticist know that evolution has occurred?

Answer: Evolution has occurred if allele frequencies within a population have changed from one generation to the next.

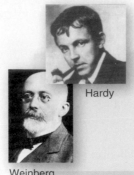

Hardy

Weinberg

$$p^2 + 2pq + q^2$$

p^2 = frequency of homozygous dominant individuals

p = frequency of dominant allele

q^2 = frequency of homozygous recessive individuals

q = frequency of recessive allele

$2pq$ = frequency of heterozygous individuals

Realize that $p + q = 1$ (There are only 2 alleles.)

$p^2 + 2pq + q^2 = 1$ (These are the only genotypes.)

Figure 15.3   **Calculating gene pool frequencies using the Hardy-Weinberg equation.**

Example

An investigator has determined by inspection that 16% of a human population has a recessive trait. Using this information, we can complete all the genotype and allele frequencies for the population, provided the conditions for Hardy-Weinberg equilibrium are met.

Given: q^2 = 16% = 0.16 are homozygous recessive individuals

Therefore, $q = \sqrt{0.16} = 0.4$ = frequency of recessive allele
$p = 1.0 - 0.4 = 0.6$ = frequency of dominant allele
$p^2 = (0.6)(0.6) = 0.36 = 36\%$ are homozygous dominant individuals
$2pq = 2(0.6)(0.4) = 0.48 = 48\%$ are heterozygous individuals

} 84% have the dominant phenotype

or
$= 1.00 - 0.52 = 0.48$

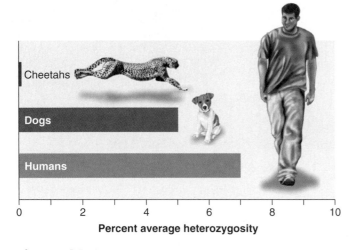

Cheetahs

Dogs

Humans

| 0 | 2 | 4 | 6 | 8 | 10 |

Percent average heterozygosity

Figure 15.4   **Random mating.**

These data suggest that heterozygosity is so infrequent in cheetahs, dogs, and humans that the chance of random mating resulting in evolution is slim. Cheetahs went through a bottleneck (see page 237), and this reduced their heterozygosity to almost zero. Dogs have been so inbred by humans that each breed is practically a separate species. Humans also have limited heterozygosity; on the average only 65 pairs of alleles out of 1,000 are heterozygous, and all the rest are homozygous.

G. H. Hardy, an English mathematician, and W. Weinberg, a German physician. They used the binomial equation ($p^2 + 2pq + q^2$) to calculate the genotypic and allele frequencies of a population. Figure 15.3 shows how this is done. The Hardy-Weinberg principle states that an equilibrium of allele frequencies in a gene pool, calculated by using the expression $p^2 + 2pq + q^2$, will remain in effect in each succeeding generation of a sexually reproducing population as long as five conditions are met:

1. No mutations: Allelic changes do not occur, or changes in one direction are balanced by changes in the opposite direction.
2. No gene flow: Migration of alleles into or out of the population does not occur.
3. Random mating: Individuals pair by chance and not according to their genotypes or phenotypes (Fig. 15.4).
4. No genetic drift: The population is very large, and changes in allele frequencies due to chance alone are insignificant.
5. No selection: No selective agent favors one genotype over another.

Often these conditions are rarely, if ever, met and allele frequencies in the gene pool of a population do change from one generation to the next. Therefore, microevolution does occur. The significance of the Hardy-Weinberg principle is that it tells us what factors cause evolution—those that violate the conditions listed. Evolution can be detected by noting any deviation from a Hardy-Weinberg equilibrium of allele frequencies in the gene pool of a population.

For a change in allele frequencies to be subject to natural selection, it must result in a change of phenotype frequencies. **Industrial melanism** provides us with an example (Fig. 15.5). Before soot was introduced into the air due to industry, the original peppered moth population in Great Britain included only

10% dark-colored moths. When dark-colored moths rest on light trunks, they are seen and eaten by predatory birds. With the advent of industry, the trunks of trees darkened, and the light-colored moths became visible and were eaten. The birds acted as a selective agent, and microevolution occurred; the last observed generation of peppered moths had 80% dark-colored moths.

Causes of Microevolution

Any conditions that deviate from the list of conditions for allelic equilibrium cause evolutionary change. Thus, these five conditions can cause a divergence from the Hardy-Weinberg equilibrium: mutation, gene flow, nonrandom mating, genetic drift, and natural selection.

Genetic Mutations

Mutations are the raw material for evolutionary change. Without mutations, there could be no new variations among members of a population. Prokaryotes reproduce asexually and therefore are dependent on mutations alone to introduce variations. All mutations that occur and result in phenotypic differences can be tested by the environment. However, in sexually reproducing organisms, mutations, if recessive, do not immediately affect the phenotype.

In a changing environment, even a seemingly harmful mutation that results in a phenotypic difference can be the source of an adaptive variation. For example, the water flea *Daphnia* ordinarily thrives at temperatures around 20°C, but a mutation exists that requires *Daphnia* to live at temperatures between 25°C and 30°C. The adaptive value of this mutation is entirely dependent on environmental conditions.

If a trait is polygenic, with many alleles for each gene locus, certain combinations of these alleles might be more adaptive than others in a particular environment. The most favorable genotype may not occur until just the right combination of alleles are grouped in a single individual. For example, if alleles K and T are present in one parent, and alleles Q and R are present in the other parent, an offspring could inherit all of these alleles, and the result might be a protein with a different structure that enables the offspring to tolerate a warmer environment than its parents did.

a.

b.

Figure 15.5 Industrial melanism and microevolution.

Microevolution has occurred when there is a change in gene pool frequencies—in this case, due to natural selection. **a.** When birds cannot see light-colored moths on light tree trunks, the light-colored phenotype is more frequent in the population. **b.** When birds cannot see dark-colored moths on dark tree trunks, the dark-colored phenotype is more frequent in the population. Therefore, when trees become sooty due to pollution, the percentage of the dark-colored phenotype increases.

Gene Flow

Gene flow, also called *gene migration,* is the movement of alleles between populations by migration of breeding individuals. Gene flow can increase the variation within a population by introducing novel alleles that were produced by mutation in another population. Constant gene flow can occur between adjacent animal populations due to the migration of organisms. Continued gene flow makes gene pools similar and reduces the possibility of allele frequency differences between populations now and in the future. Indeed, gene flow among populations can prevent speciation from occurring. Due to gene flow, the snake populations featured in Figure 15.6 are subspecies instead of separate species.

Nonrandom Mating

Random mating occurs when individuals pair by chance and not according to their genotypes or phenotypes. Inbreeding, or mating between relatives to a greater extent than by chance, is an example of **nonrandom mating.** Inbreeding does not change allele frequencies, but it does gradually increase the proportion of homozygotes, because the homozygotes that result must necessarily produce only homozygotes.

    Assortative mating occurs when individuals tend to mate with those that have the same phenotype with respect to a certain characteristic. For example, in humans, tall people tend to mate with each other. Assortative mating causes the population to subdivide into two phenotypic classes, between which gene exchange is reduced. Homozygotes for the gene loci that control the trait in question increase in frequency, and heterozygotes for these loci decrease in frequency.

    Sexual selection favors characteristics that increase the likelihood of obtaining mates, and in this way it promotes nonrandom mating. In most species, males that compete best for access to females and/or have a phenotype that pleases females are more apt to mate and have increased fitness (see the introduction to this chapter.)

Figure 15.6 **Gene flow.**

Each rat snake represents a separate population of snakes. Because the populations are adjacent to one another, interbreeding occurs, and so does gene flow between the populations. This keeps their gene pools somewhat similar, and each of these populations is considered a subspecies of the species *Elaphe obsoleta.* Therefore, each has a three-part name.

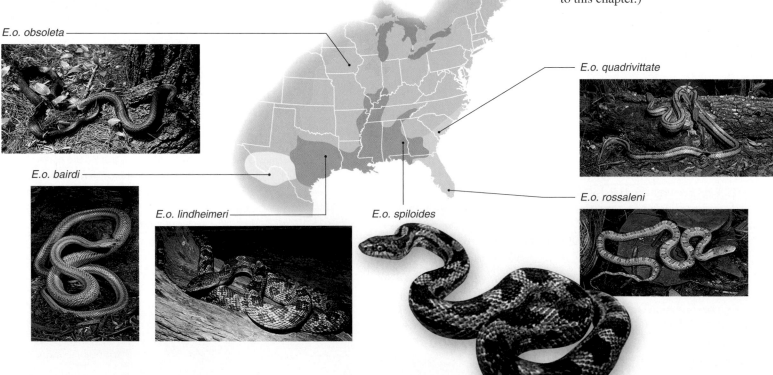

E.o. obsoleta

E.o. quadrivittate

E.o. bairdi

E.o. rossaleni

E.o. lindheimeri

E.o. spiloides

Genetic Drift

Genetic drift refers to changes in the allele frequencies of a gene pool due to chance. Although genetic drift occurs in both large and small populations, a larger population is expected to suffer less of a sampling error than a smaller population. Suppose you had a large bag containing 1,000 green balls and 1,000 blue balls, and you randomly drew 10%, or 200, of the balls. Because there is a large number of balls of each color in the bag, you can reasonably expect to draw 100 green balls and 100 blue balls, or at least a ratio close to this. It is extremely unlikely that you would draw 200 green or 200 blue balls. But suppose you had a bag containing only 10 green balls and 10 blue balls, and you drew 10%, or only 2 balls. You could easily draw two green balls or two blue balls, or one of each color.

When a population is small, there is a greater chance that some rare genotype might not participate at all in the production of the next generation. Suppose in a small population of frogs, certain frogs by chance do not pass on their traits. Certainly, the next generation will have a change in allele frequencies (Fig. 15.7). When genetic drift leads to a loss of one or more alleles, other alleles over time become *fixed* in the population.

In an experiment involving brown eye color, each of 107 *Drosophila* populations was kept in its own culture bottle. Every bottle contained eight heterozygous flies of each sex. There were no homozygous recessive or homozygous dominant flies. From the many offspring, the experimenter chose at random eight males and eight females. This action, which represented genetic drift, continued for 19 generations. By the nineteenth generation, 25% of the populations contained only homozygous recessive flies, and 25% contained only homozygous dominant flies with the allele for brown eye color.

Genetic drift is a random process, and therefore it is not likely to produce the same results in several populations. In California, there are a number of cypress groves, each a separate population. The phenotypes within each grove are more similar to one another than they are to the phenotypes in the other groves. Some groves have longitudinally shaped trees, and others have pyramidally shaped trees. The bark is rough in some colonies and smooth in others. The leaves are gray to bright green or bluish, and the cones are small or large. Because the environmental conditions are similar for all the groves and no correlation has been found between phenotype and environment across groves, scientists hypothesize that these variations among populations are due to genetic drift.

Bottleneck Effect Sometimes a species is subjected to near extinction because of a natural disaster (e.g., earthquake or fire) or because of overharvesting and habitat loss. It is as though most of the population has stayed behind and only a few survivors have passed through the neck of a bottle (Fig. 15.8). This so-called **bottleneck effect** prevents the majority of genotypes from participating in the production of the next generation.

The extreme genetic similarity found in cheetahs is believed to be due to a bottleneck. In a study of 47 different enzymes, each of which can occur in several different forms in other types of cats, all the cheetahs studied had exactly the same form. This demonstrates that genetic drift can cause certain alleles to be lost from a population. Exactly what caused the cheetah bottleneck is not known. It is speculated that perhaps cheetahs

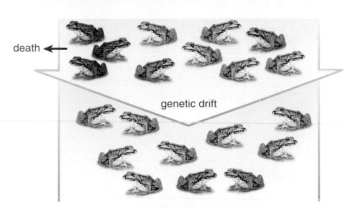

Figure 15.7   Genetic drift.

Genetic drift occurs when by chance only certain members of a population (in this case, green frogs) reproduce and pass on their genes to the next generation. The allele frequencies of the next generation's gene pool may be markedly different from those of the previous generation.

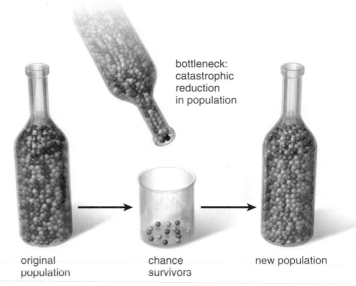

Figure 15.8   Bottleneck effect.

A bottleneck effect may occur when a catastrophic reduction in a population occurs, such as after a major epidemic or a hurricane. For example, a parent population contains roughly equal numbers of genotypes represented by blue, yellow, purple, green, and red marbles. The chance survivors of the catastrophe have genotypes represented by blue and yellow, resulting in a new population with altered gene pool frequencies.

Check Your Progress

1. Explain how gene flow alters allele frequencies.
2. Explain why assortative mating is a type of nonrandom mating.
3. Explain why genetic drift is more likely to happen in a small population.

Answers: 1. Individuals add and remove alleles when they move into and out of populations. 2. When individuals choose a phenotype like their own, mating is not random. 3. The chance that a rare genotype might not participate in passing alleles to the next generation is greater.

Figure 15.9   Founder effect.

A member of the founding population of Amish in Pennsylvania had a recessive allele for a rare kind of dwarfism linked with polydactylism. The percentage of the Amish population now carrying this allele is much higher compared to that of the general population.

were slaughtered by nineteenth-century cattle farmers protecting their herds, or were captured by Egyptians as pets 4,000 years ago, or were decimated by a mass extinction tens of thousands of years ago. Today, cheetahs suffer from relative infertility because of the intense inbreeding that occurred after the bottleneck.

Founder Effect   The **founder effect** is an example of genetic drift in which rare alleles, or combinations of alleles, occur at a higher frequency in a population isolated from the general population. After all, founding individuals contain only a fraction of the total genetic diversity of the original gene pool. Which particular alleles are carried by the founders is dictated by chance alone. The Amish of Lancaster County, Pennsylvania, are an isolated group that was founded by German settlers. Today, as many as one in 14 individuals carries a recessive allele that causes an unusual form of dwarfism (affecting only the lower arms and legs) and polydactylism (extra fingers) (Fig. 15.9). In most populations, only one in 1,000 individuals has this allele.

15.2   Natural Selection

Natural selection is the process that results in adaptation of a population to the biotic (living) and abiotic (nonliving) environments. In the biotic environment, organisms acquire resources through competition, predation, and parasitism. The abiotic environment includes weather conditions, dependent chiefly upon temperatures and precipitation. Charles Darwin, the father of modern evolutionary theory, became convinced that species evolve with time and suggested natural selection as the mechanism for adaptation to the environment (see Chapter 14). In Table 15.1, Darwin's hypothesis of natural selection is stated in a way that is consistent with modern genetics.

As a result of natural selection, the most *fit* individuals become more prevalent in a population, and this way a population changes over time. The most fit individuals are those that reproduce more than others because they are adapted to the environment.

Types of Selection

Most of the traits on which natural selection acts are polygenic and controlled by more than one pair of alleles located at different gene loci. Such traits have a range of phenotypes, the frequency distribution of which usually resembles a bell-shaped curve, as shown in Figures 15.10–15.12.

Check Your Progress

Describe the evolutionary consequences of natural selection.

Answer: Natural selection results in the adaptation of a population to biotic and abiotic environments.

Table 15.1	Natural Selection

Evolution by natural selection requires:

1. Variation. The members of a population differ from one another.

2. Inheritance. Many of these differences are heritable genetic differences.

3. Degrees of adaptiveness. Some of these genetic differences affect how well an organism is adapted to its environment.

4. Degrees of successful reproduction. Individuals that are better adapted to their environment are more likely to reproduce, and their fertile offspring will make up a greater proportion of the next generation.

Three types of natural selection are possible for any particular trait: directional selection, stabilizing selection, and disruptive selection.

Directional Selection

Directional selection occurs when an extreme phenotype is favored and the distribution curve shifts in that direction. Such a shift can occur when a population is adapting to a changing environment.

Industrial melanism, discussed earlier and depicted in Figure 15.5, is an example of directional selection. Drug resistance is another. As you may know, widespread use of antibiotics and pesticides results in a wide distribution of bacteria and insects that are resistant to these chemicals. When an antibiotic is administered, some bacteria may survive because they are genetically resistant to the antibiotic. These are the bacteria that are likely to pass on their genes to the next generation. As a result, the number of resistant bacteria keeps increasing. Drug-resistant strains of bacteria that cause tuberculosis have now become a serious threat to the health of people worldwide.

Another example of directional selection is the human struggle against malaria, a disease caused by an infection of the liver and the red blood cells. The *Anopheles* mosquito transmits the disease-causing protozoan *Plasmodium vivax* from person to person. In the early 1960s, international health authorities thought malaria would soon be eradicated. A new drug, chloroquine, seemed effective against *Plasmodium*, and spraying of DDT (an insecticide) had reduced the mosquito population. But by the mid-1960s, *Plasmodium* was showing signs of chloroquine resistance, and worse yet, mosquitoes were becoming resistant to DDT. A few drug-resistant parasites and a few DDT-resistant mosquitoes had survived and multiplied, shifting the distribution curve toward the infective success of the parasite.

The gradual increase in the size of the modern horse, *Equus*, is an example of directional selection that can be correlated with a change in the environment—in this case, from forest conditions to grassland conditions (Fig. 15.10). Even so, as discussed previously, the evolution of the horse should not be viewed as a straight line of descent because we know of many side branches that became extinct.

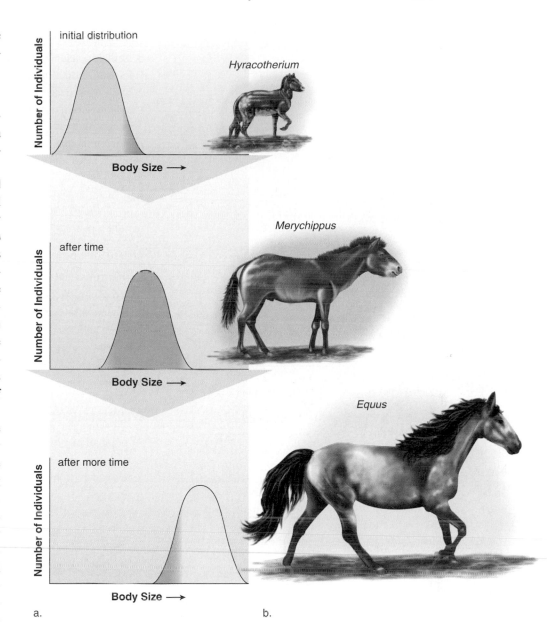

Figure 15.10   Directional selection.

a. Natural selection favors one extreme phenotype, and the distribution curve shifts. **b.** *Equus,* the modern-day horse, evolved from *Hyracotherium,* which was about the size of a dog. This small animal could have hidden among trees, and it had low-crowned teeth for browsing. When grasslands began to replace forests, the ancestors of *Equus* may have been subject to selection pressure for the development of strength, intelligence, speed, and durable grinding teeth. Larger animals were stronger and more successful in combat, those with larger skulls and brains had better sensory processing, those with longer legs and better-developed hooves could escape enemies, and animals with durable teeth could feed more efficiently on grass. Animals with these characteristics tended to produce more offspring.

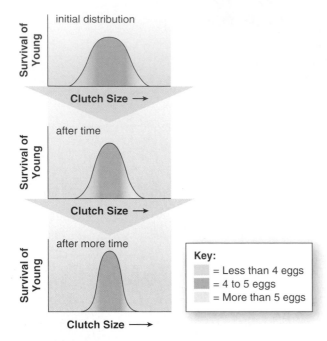

Figure 15.11 **Stabilizing selection.**

Stabilizing selection occurs when natural selection favors the intermediate phenotype over the extremes. For example, Swiss starling birds that lay four to five eggs (usual clutch size) have more young survive than those that lay less than four eggs or more than five eggs.

Stabilizing Selection

Stabilizing selection occurs when an intermediate phenotype is favored. With stabilizing selection, extreme phenotypes are selected against, and individuals near the average are favored. This is the most common form of selection because the average individual is well-adapted to its environment.

As an example, consider that when Swiss starlings lay four to five eggs, more young survive than when the female lays more or less than this number (Fig. 15.11). Genes determining physiological characteristics, such as the production of yolk, and behavioral characteristics, such as how long the female will mate, are involved in determining clutch size. In humans, birth weight ranges from 0.89 to 4.9 kilograms. But most babies have an intermediate birth weight—the weight that favors survival.

Disruptive Selection

In **disruptive selection,** two or more extreme phenotypes are favored over any intermediate phenotype. For example, British land snails (*Cepaea nemoralis*) are found in low-vegetation areas (grass fields and hedgerows) and in forests. In low-vegetation areas, thrushes feed mainly on snails with dark shells that lack light bands, and in forest areas, they feed mainly on snails with light-banded shells. Therefore, these two distinctly different phenotypes are found in the population (Fig. 15.12).

Maintenance of Variations

A population always shows some genotypic variation. The maintenance of variation is beneficial because populations with limited variation may not be able to adapt to new conditions and may become extinct. How can variation be maintained in spite of selection constantly working to reduce it?

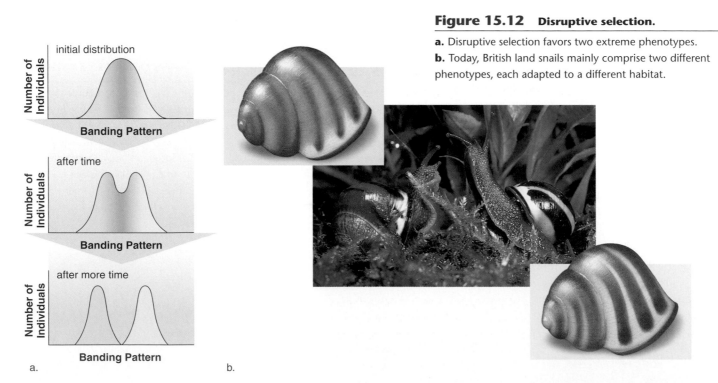

Figure 15.12 **Disruptive selection.**

a. Disruptive selection favors two extreme phenotypes.
b. Today, British land snails mainly comprise two different phenotypes, each adapted to a different habitat.

Check Your Progress

1. Contrast directional selection with stabilizing selection.
2. Describe the effect of disruptive selection.

Answers: 1. Directional selection occurs when one extreme phenotype is favored; stabilizing selection occurs when the intermediate phenotype is favored. *2.* Since the intermediate phenotype is selected against, two distinctly different populations eventually develop at the phenotypic extremes.

First, we must remember that the forces that promote variation are still at work: Mutation still creates new alleles, and recombination still recombines these alleles during gametogenesis and fertilization. Second, gene flow might still occur. If the receiving population is small and is mostly homozygous, gene flow can be a significant source of new alleles. Finally, natural selection favors certain phenotypes, but the other types may still remain in reduced frequency. And disruptive selection even promotes differences in form, called **polymorphism,** in a population. Diploidy and the heterozygote also help because they maintain recessive alleles in the gene pool.

Diploidy and the Heterozygote

Only alleles that are exposed (cause a phenotypic difference) are subject to natural selection. In diploid organisms, this fact makes the heterozygote a potential protector of recessive alleles that might otherwise be weeded out of the gene pool. Because the heterozygote remains in a population, so does the possibility of the recessive phenotype, which might have greater fitness in a changing environment. When natural selection favors the ratio of two or more phenotypes in generation after generation, it is called *balanced polymorphism.* Sickle cell disease offers an example of balanced polymorphism.

Sickle Cell Disease Individuals with sickle cell disease have the genotype Hb^SHb^S and tend to die at an early age due to hemorrhaging and organ destruction. Those who are heterozygous and have sickle cell trait (Hb^AHb^S) are better off because their red blood cells usually become sickle-shaped only when the oxygen content of the environment is low. Ordinarily, those with a normal genotype (Hb^AHb^A) are the most fit.

Geneticists studying the distribution of sickle cell disease in Africa have found that the recessive allele (Hb^S) has a higher frequency (0.2 to as high as 0.4 in a few areas) in regions where the disease malaria is also prevalent (Fig. 15.13). What is the connection between higher frequency of the recessive allele and malaria? Malaria is caused by a parasite that lives in and destroys the red blood cells of the normal homozygote (Hb^AHb^A). However, the parasite is unable to live in the red blood cells of the heterozygote (Hb^AHb^S) because the infection causes the red blood cells to become sickle-shaped. Sickle-shaped red blood cells lose potassium, and this causes the parasite to die. In an environment where malaria is prevalent, the heterozygote is favored. Each of the homozygotes is selected against but is maintained because the heterozygote is favored

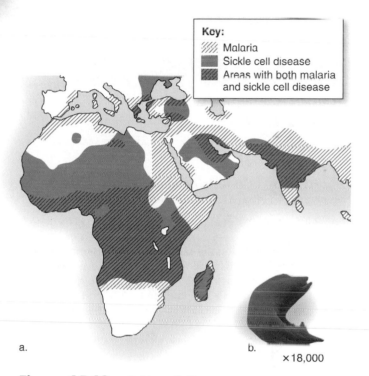

a.

b.
×18,000

Key:
- Malaria
- Sickle cell disease
- Areas with both malaria and sickle cell disease

Figure 15.13 **Sickle cell disease.**

a. Hash marks show the areas where malaria was prevalent in Africa, the Middle East, southern Europe, and southern Asia in 1920, before eradication programs began; shown in orange are the areas where sickle cell disease most often occurred. The overlap of these two distributions suggested a causal connection. **b.** Micrograph of a sickled red blood cell.

in parts of Africa subject to malaria. Table 15.2 summarizes the effects of the three possible genotypes.

Table 15.2	Sickle Cell Disease	
Genotype	**Phenotype**	**Result**
$Hb^A Hb^A$	Normal	Dies due to malarial infection
$Hb^A Hb^S$	Sickle cell trait	Lives due to protection from both
$Hb^S Hb^S$	Sickle cell disease	Dies due to sickle cell disease

---

THE CHAPTER IN REVIEW

Summary

15.1   Microevolution

Evolution in a Genetic Context
Microevolution involves several elements:

- All the various genes of a population make up its gene pool.
- Hardy-Weinberg equilibrium is present when gene pool allele frequencies remain the same from generation to generation. Certain conditions have to be met to achieve an equilibrium.
- The conditions are (1) no mutations, (2) no gene flow, (3) random mating, (4) no genetic drift, and (5) no selection. Since these conditions are rarely met, a change in gene pool frequencies is likely.
- When gene pool frequencies change, microevolution has occurred. Deviations from a Hardy-Weinberg equilibrium allow us to determine when evolution has taken place.

Causes of Microevolution
Microevolution is caused by five conditions:

- Mutations are the raw material for evolutionary change. Certain genotypic variations may be of evolutionary significance, only if the environment changes. Recombinations help bring about adaptive genotypes.
- Gene flow occurs when a breeding individual (in animals) migrates to another population or when gametes and seeds (in plants) are carried into another population. Constant gene flow between two populations causes their gene pools to become similar.
- Nonrandom mating occurs when relatives mate (inbreeding) or when assortative mating takes place. Both of these cause an increase in homozygotes. Sexual selection occurs when a characteristic that increases the chances of mating is favored.
- Genetic drift occurs when allele frequencies are altered by chance—that is, by sampling error. Genetic drift is particularly evident after a bottleneck, when severe inbreeding occurs, or when founders start a new population.
- Natural selection (see Section 15.2).

15.2   Natural Selection

Today we believe that adaptation comes about because the more fit individuals, who reproduce more than others, are adapted to the environment.

Types of Selection
Most of the traits of evolutionary significance are polygenic, and a range of phenotypes in a population result in a bell-shaped curve. Three types of selection occur:

Directional Selection   The curve shifts in one direction, as when dark-colored peppered moths become prevalent in polluted areas.

Stabilizing Selection   The peak of the curve increases, as when most human babies have the intermediate birth weight. Babies that are very small or very large are less fit than those of intermediate weight.

Disruptive Selection   The curve has two peaks, as when British land snails vary because a wide geographic range causes selection to vary.

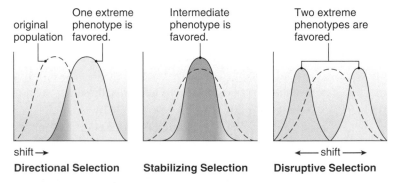

Directional Selection   Stabilizing Selection   Disruptive Selection

Maintenance of Variations
Despite constant natural selection, variation is maintained because:

- Mutations and recombination still occur; gene flow among small populations can introduce new alleles; and natural selection still occurs.
- In sexually reproducing diploid organisms, the heterozygote acts as a repository for recessive alleles whose frequency is low. In sickle cell disease, the heterozygote is more fit in areas where malaria occurs, and therefore both homozygotes are maintained in the population.

Thinking Scientifically

Cystic fibrosis (CF) is the most common serious genetic disorder in Caucasians. People with CF have an average life span of 25 years. The disease is expressed in individuals who are homozygous recessive for the cystic fibrosis transmembrane regulator (*CFTR*) gene. They are not able to make a functional CFTR protein, which causes their cells to accumulate chloride ions. This results in the formation of a thick mucus in the lungs, leading to frequent lung infections. In addition, secretory ducts are blocked, causing nutritional problems. Since this is a serious and typically fatal disorder, why do you suppose natural selection has not eliminated (or at least dramatically reduced) the defective form of the gene from the human population?

Testing Yourself

Choose the best answer for each question.

1. A population consists of 48 *AA*, 54 *Aa*, and 22 *aa* individuals. What is the frequency of the *A* allele?

   a. 0.60
   b. 0.40
   c. 0.62
   d. 0.42
   e. 0.58

2. Which of the following is the binomial equation?

   a. $2p^2 + 2pq + 2q^2$
   b. $p^2 + pq + q^2$
   c. $2p^2 + pq + 2q^2$
   d. $p^2 + 2 pq + q^2$

For questions 3 and 4, consider that about 70% of white North Americans can taste the chemical phenylthiocarbamide. The ability to taste is due to the dominant allele *T*. Nontasters are *tt*. Assume this population is in Hardy-Weinberg equilibrium.

3. What is the frequency of *t*?

   a. 0.30
   b. 0.70
   c. 0.55
   d. 0.09
   e. 0.60

4. What is the frequency of heterozygous tasters?

   a. 0.495
   b. 0.21
   c. 0.42
   d. 0.2475
   e. 0.45

5. Typically, mutations are immediately expressed and tested by the environment in

   a. prokaryotes.
   b. eukaryotes.
   c. prokaryotes and eukaryotes.
   d. neither prokaryotes nor eukaryotes.

6. The offspring of better-adapted individuals are expected to make up a larger proportion of the next generation. The most likely explanation is

   a. mutations and nonrandom mating.
   b. gene flow and genetic drift.
   c. mutations and natural selection.
   d. mutations and genetic drift.

7. The Northern elephant seal went through a severe population decline as a result of hunting in the late 1800s. The population has rebounded but is now homozygous for nearly every gene studied. This is an example of

   a. negative assortative mating.
   b. migration.
   c. mutation.
   d. a bottleneck.
   e. disruptive selection.

8. Which of the following generally results in a net gain in genetic variability?

   a. genetic drift
   b. mutation
   c. natural selection
   d. bottleneck
   e. More than one of these are correct.

For questions 9–15, indicate the effect of each of the conditions of the Hardy-Weinberg principle on genotype and allele frequencies. Each answer may be used more than once.

Key:

   a. alters genotype and allele frequencies
   b. alters genotype frequency only
   c. alters allele frequency only
   d. does not alter genotype or allele frequency

9. Mutation

10. Gene flow

11. Inbreeding

12. Assortative mating

13. Genetic drift

14. Bottleneck

15. Natural selection

16. A small, reproductively isolated religious sect called the Dunkers was established by 27 families that came to the United States from Germany 200 years ago. The frequencies for blood group alleles in this population differ significantly from those in the general U.S. population. This is an example of

   a. negative assortative mating.
   b. natural selection.
   c. founder effect.
   d. bottleneck effect.
   e. gene flow.

17. Assuming a Hardy-Weinberg equilibrium, 21% of a population is homozygous dominant, 50% is heterozygous, and 29% is homozygous recessive. What percentage of the next generation is predicted to be homozygous recessive?

   a. 21%
   b. 50%
   c. 29%
   d. 42%
   e. 58%

18. When a population is small, there is a greater chance of

   a. gene flow.
   b. genetic drift.
   c. natural selection.
   d. mutations occurring.
   e. sexual selection.

19. Which of the following is not expected to help maintain genetic variability?

   a. gene flow
   b. mutation
   c. recombination
   d. disruptive selection
   e. genetic drift

20. The sickle cell allele is maintained in regions where malaria is prevalent because

   a. the allele confers resistance to the parasite.
   b. gene flow is high in those regions.
   c. disruptive selection is occurring.
   d. genetic drift randomly selects for the allele.

21. Complete the following by drawing a curve in the other two graphs to show the effect of directional selection.

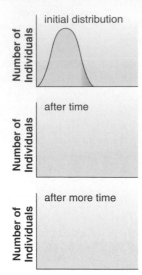

22. Complete the following by drawing a curve in the other two graphs to show the effect of stabilizing selection.

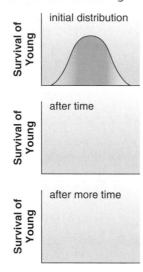

23. Complete the following by drawing curves in the other two graphs to show the effect of disruptive selection.

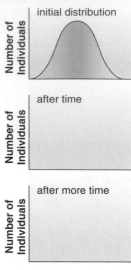

Go to www.mhhe.com/maderessentials for more quiz questions.

Bioethical Issue

The highly regarded population geneticist Sir Ronald Aylmer Fisher published a book in 1930 entitled *The Genetical Theory of Natural Selection.* In the book, he claims that civilizations fail because the people with the highest level of fitness (those at the top of the societal ladder) do not reproduce as often as those in the less affluent classes. He suggests that high-income families be paid to have children in order to improve the population. In other words, Fischer is suggesting that we carry out selection within the human population.

Can you envision any scenario in which our society should encourage reproduction by the most-fit individuals and discourage it by the least-fit? If so, how should it be done? What characteristics would be appropriate to select for or against?

Understanding the Terms

assortative mating   236
bottleneck effect   237
directional selection   239
disruptive selection   240
founder effect   238
gene flow   236
gene pool   233
genetic drift   237
industrial melanism   234

microevolution   232
mutation   235
natural selection   238
nonrandom mating   236
polymorphism   241
population   232
sexual selection   236
stabilizing selection   240

Match the terms to these definitions:

a. _____ Frequencies of all the alleles of all the genes in all individuals in a population.

b. _____ Mating between phenotypically similar males and females.

c. _____ Causes nonrandom mating by favoring characteristics that increase the likelihood of obtaining a mate.

d. _____ Sharp decline in the number of individuals in a population due to severe selection pressure or a natural calamity.

Evolution on a Large Scale

The common green iguana, a strong swimmer native to South America, could be ancestral to other types of iguanas.

The rhinoceros iguana, with bumps on its snout, lives to the northeast of South America, only on Hispaniola.

The unique marine iguana feeds at sea and lives to the west of South America, only in the Galápagos Islands.

Iguanas are a type of lizard known for a very long tail, which can be up to three times their body length. The green iguana, the second largest lizard in the world, is common throughout the South American continent. Its claws enable it to climb trees, where it feeds on succulent leaves and soft fruits along riverbanks. It is an excellent swimmer and uses rivers to travel to new feeding areas.

▶ ▶ ▶

Islands to the west and northeast of South America don't have green iguanas. Instead, the Galápagos Islands to the west have one type of land iguana and one type of marine iguana. Marine iguanas, black in color, are unique to the Galápagos Islands—they occur no place else on planet Earth. These air-breathing reptiles can dive to a depth exceeding 10 meters (33 feet) and remain submerged for more than 30 minutes. They use their claws to cling to the bottom rocks while feeding on algae.

Hispaniola,[1] an island to the northeast of South America, has a type of iguana called the rhinoceros iguana, which has three horny bumps on its snout and is dark brown to black in color. This iguana lives mainly in dry forests with rocky, limestone habitats and feeds on a wide variety of plants.

The Theory of Evolution can be used to explain the occurrence of the unique iguanas on the Galápagos Islands and Hispaniola. It can be hypothesized that ancestral green iguanas swam or hitchhiked on floating driftwood from South America to the Galápagos to the west and to Hispaniola to the northeast. After arrival, a green iguana population, cut off from the others of its species, evolved into these new and different species. Speciation is the topic of this chapter.

16.1 Macroevolution

Chapter 15 considered evolution on a small scale—that is, microevolution. In this chapter, we turn our attention to evolution on a large scale—that is, **macroevolution.** The history of life on Earth is a part of macroevolution. Macroevolution requires **speciation,** the splitting of one species into two or more new species. Speciation involves the gene pool changes that we studied in Chapter 15.

Species originate, adapt to their environments, and then may become extinct (Fig. 16.1). Without the origin and extinction of species, life on Earth would not have an ever-changing history.

Defining Species

Before we consider the origin of species, we first need to define a species. Appearance is not always a good criterion. The members of different species can look quite similar, while the members of a single species can be diverse in appearance. The *biological species concept* offers a way to define a species that does not depend on appearance: The members of a species interbreed and have a shared gene pool, and each species is reproductively isolated from every other species. For example, the flycatchers in Figure 16.2 are members of separate species because they do not interbreed in nature.

According to the biological species concept, gene flow occurs between the populations of a species but not between populations of different species. The red maple and the sugar maple are found over a wide geographical range in

---

[1] Hispaniola is divided into Haiti and the Dominican Republic.

Figure 16.1   Dinosaurs.

Triceratops (*left*) and *Tyrannosaurus rex* (*right*) were dinosaurs that lived about 65 million years ago when flowering plants were already flourishing. The prevailing forms of life on Earth change as new species come into existence, flourish, and then decline or become extinct.

the eastern half of the United States, and each species is made up of many populations. However, the members of each species' population rarely hybridize in nature. Therefore, these two types of plants are separate species. In contrast, the human species has many populations that certainly differ in physical appearance (Fig. 16.3). We know, however, that all humans belong to one species because the members of these populations can produce fertile offspring.

The biological species concept is useful, as we shall see, but even so, it has its limitations. For example, it applies only to sexually reproducing organisms and cannot apply to asexually reproducing organisms. Then, too, sexually reproducing organisms are not always as reproductively isolated as we would expect. Some North American orioles live in the western half of the continent and some in the eastern half of the continent. Yet even the two most genetically distant oriole species, as recognized by analyzing their mitochondrial DNA, will hybridize where they meet in the middle of the continent.

There are other definitions of species aside from the biological definition. Later in this chapter, we will define species as a category of classification below the rank of genus. Species in the same genus share a recent common ancestor. A **common ancestor** is an ancestor for two or more different groups. Your grandmother is the common ancestor for you and your cousins. Similarly, there is a common ancestor for all species of roses.

Acadian flycatcher,
Empidonax virescens

Traill's flycatcher,
Empidonax trailli

least flycatcher,
Empidonax minimus

Figure 16.2   Three species of flycatchers.

Although these flycatcher species are nearly identical in appearance, we know they are separate species because they are reproductively isolated—the members of each species reproduce only with one another. Each species has a characteristic song and its own particular habitat during the mating season as well.

a.   b.

Figure 16.3   Human populations.

The Massai of East Africa (**a**) and the Eskimos that live near the Arctic Circle (**b**) are both members of the species *Homo sapiens* because Massai can reproduce with Eskimos and produce fertile offspring.

Reproductive Barriers

As mentioned, for two species to be separate, they must be reproductively isolated—that is, gene flow must not occur between them. Reproductive barriers are isolating mechanisms that prevent successful reproduction from occurring (Fig. 16.4). In evolution, reproduction is successful only when it produces fertile offspring.

Prezygotic (before the formation of a zygote) **isolating mechanisms** are those that prevent reproduction attempts and make it unlikely that fertilization will be successful if mating is attempted. Habitat isolation, temporal isolation, behavioral isolation, mechanical isolation, and gamete isolation make it highly unlikely that particular genotypes will contribute to the gene pool of a population.

Habitat isolation When two species occupy different habitats, even within the same geographic range, they are less likely to meet and attempt to reproduce. This is one of the reasons that the flycatchers in Figure 16.2 do not mate, and that the red maple and sugar maple do not exchange pollen. In tropical rain forests, many animal species are restricted to a particular level of the forest canopy, and in this way they are isolated from similar species.

Temporal isolation Two species can live in the same locale, but if each reproduces at a different time of year, they do not attempt to mate. For example, *Reticulitermes hageni* and *R. virginicus* are two species of termites. The former has mating flights in March through May, whereas the latter mates in the fall and winter months.

Behavioral isolation Many animal species have courtship patterns that allow males and females to recognize one another (Fig. 16.5). Male fireflies are recognized by females of their species by the pattern of their flashings; similarly, male crickets are recognized by females of their species by their chirping. Many males recognize females of their species by sensing chemical signals called pheromones. For example, female gypsy moths

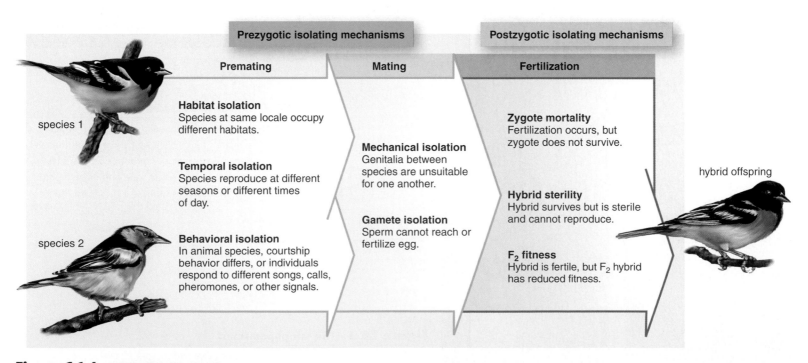

Figure 16.4   Reproductive barriers.

Prezygotic isolating mechanisms prevent mating attempts or a successful outcome should mating take place. No zygote is ever formed. Postzygotic isolating mechanisms prevent offspring from reproducing—that is, the hybrids are unable to breed successfully.

secrete chemicals from special abdominal glands. These chemicals are detected downwind by receptors on the antennae of males.

Mechanical isolation When animal genitalia or plant floral structures are incompatible, reproduction cannot occur. Inaccessibility of pollen to certain pollinators can prevent cross-fertilization in plants, and the sexes of many insect species have genitalia that do not match, or other characteristics that make mating impossible. For example, male dragonflies have claspers that are suitable for holding only the females of their own species.

Gamete isolation Even if the gametes of two different species meet, they may not fuse to become a zygote. In animals, the sperm of one species may not be able to survive in the reproductive tract of another species, or the egg may have receptors only for sperm of its species. In plants, only certain types of pollen grains can germinate so that sperm successfully reach the egg.

Postzygotic (after the formation of a zygote) **isolating mechanisms** prevent hybrid offspring (reproductive product of two different species) from developing or breeding, even if reproduction attempts have been successful.

Zygote mortality A hybrid zygote may not be viable, and so it dies. A zygote with two different chromosome sets may fail to go through mitosis properly, or the developing embryo may receive incompatible instructions from the maternal and paternal genes so that it cannot continue to exist.

Hybrid sterility The hybrid zygote may develop into a sterile adult. As is well known, a cross between a horse and a donkey produces a mule, which is usually sterile—it cannot reproduce (Fig. 16.6). Sterility of hybrids generally results from complications in meiosis that lead to an inability to produce viable gametes. A cross between a cabbage and a radish produces offspring that cannot form gametes, even though the diploid number is 18, an even number, most likely because the cabbage chromosomes and the radish chromosomes could not align during meiosis.

F_2 *fitness* If hybrids can reproduce, their offspring are unable to reproduce. In some cases, mules are fertile, but their offspring (the F_2 generation) are not fertile.

Figure 16.5   Prezygotic isolating mechanism.

An elaborate courtship display allows the blue-footed boobies of the Galápagos Islands to select a mate. The male lifts his feet in a ritualized manner that shows off their bright blue color.

horse   donkey

mating

fertilization

mule (hybrid)

Usually mules cannot reproduce. If an offspring does result, it cannot reproduce.

Figure 16.6   Postzygotic isolating mechanism.

Mules are horse-donkey hybrids. Mules are infertile due to a difference in the chromosomes inherited from their parents.

Check Your Progress

1. Define a species according to the biological species concept.
2. Describe the limitations of the biological species concept.
3. Contrast prezygotic isolating mechanisms with postzygotic isolating mechanisms.

Answers: 1. A biological species is a group of organisms that can interbreed. The members of this species are reproductively isolated from members of other species. 2. The biological species concept cannot be applied to asexually reproducing organisms. In addition, reproductive isolation may not be complete, so some breeding between species may occur. 3. Prezygotic isolating mechanisms do not allow zygotes to be formed. Postzygotic mechanisms allow zygotes to be formed, but the zygotes either die or become offspring that are infertile.

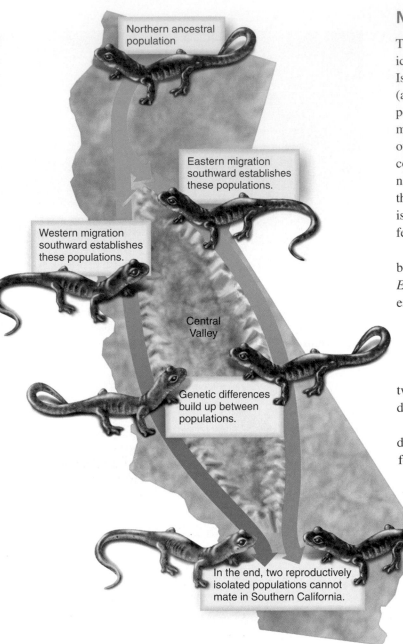

Figure 16.7   Allopatric speciation.

In this example of allopatric speciation, the Central Valley of California is separating a range of populations descended from the same northern ancestral species. The limited contact between the populations on the west and those on the east allow genetic changes to build up so that two populations, both living in the south, do not reproduce with one another and therefore can be designated as separate species.

Models of Speciation

The introduction to this chapter suggests that the green iguana of South America may be the common ancestor for both the marine iguana on the Galápagos Islands (to the west of South America) and the rhinoceros iguana on Hispaniola (an island to the northeast of South America). If so, how could it have happened? Green iguanas are strong swimmers, so by chance a few could have migrated to these islands, where they formed populations separate from each other and from the parent population back in South America. Each population continued on its own evolutionary path as new mutations, genetic drift, and natural selection occurred. Eventually, reproductive isolation developed and there were three species of iguanas. A speciation model based on geographic isolation of populations is called **allopatric speciation** because *allo* means different and *patria* loosely means homeland.

Figure 16.7 features an example of allopatric speciation, which has been extensively studied in California. Members of an ancestral population of *Ensatina* salamanders existing in the Pacific Northwest migrated southward, establishing a range of populations. Each population was exposed to its own selective pressures along the coastal mountains and along the Sierra Nevada mountains. Due to the barrier created by the Central Valley of California, no gene flow occurred between the eastern populations and the western populations. Genetic differences increased from north to south, resulting in two distinct forms of *Ensatina* salamanders in southern California that differ dramatically in color and interbreed only rarely.

With **sympatric speciation,** a population develops into two or more reproductively isolated groups without prior geographic isolation. The best evidence for this type of speciation is found among plants, where it can occur by means of *polyploidy,* an increase in the number of sets of chromosomes to 3*n* or above. The presence of sex chromosomes makes it difficult for polyploidy speciation to occur in animals. In plants, hybridization between two species can be followed by a doubling of the chromosome number. Such polyploid plants are reproductively isolated by a postzygotic mechanism; they can reproduce successfully only with other similar polyploids, and backcrosses with their parents are sterile. Therefore, three species instead of two species result. Figure 16.8 shows that the parents of our present-day bread wheat had 28 and 14

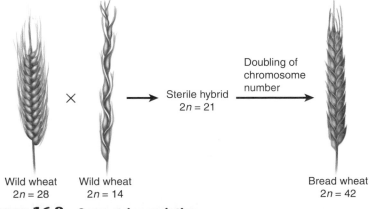

Figure 16.8   Sympatric speciation.

In this example of sympatric speciation, two populations of wild wheat hybridized many years ago. The hybrid is sterile, but chromosome doubling allowed some plants to reproduce. These plants became today's bread wheat.

chromosomes, respectively. The hybrid with 21 chromosomes is sterile, but polyploid bread wheat with 42 chromosomes is fertile because the chromosomes can align during meiosis.

Adaptive Radiation

One of the best examples of speciation through adaptive radiation is provided by the finches on the Galápagos Islands, which are often called Darwin's finches because Darwin first realized their significance as an example of how evolution works. During **adaptive radiation,** many new species evolve from a single ancestral species. The many species of finches that live on the Galápagos Islands are believed to be descended from a single type of ancestral finch from the mainland (Fig. 16.9). The populations on the various islands were subjected to the founder effect involving genetic drift, genetic mutations, and the process of natural selection. Because of natural selection, each population became adapted to a particular habitat on its island. In time, the various populations became so genotypically different that now, when by chance they reside on the same island, they do not interbreed and are therefore separate species. There is evidence that the finches use beak shape to recognize members of the same species during courtship. Rejection of suitors with the wrong type of beak is a behavioral prezygotic isolating mechanism.

Similarly, inhabiting the Hawaiian Islands is a wide variety of honeycreepers, all descended from a common goldfinchlike ancestor that arrived from Asia or North America about 5 million years ago. Today, honeycreepers have a range of beak sizes and shapes for feeding on various food sources, including seeds, fruits, flowers, and insects.

Figure 16.9   Darwin's finches.

Each of Darwin's finches is adapted to gathering and eating a different type of food. Tree finches have beaks largely adapted to eating insects and, at times, plants. Ground finches have beaks adapted to eating the flesh of the prickly-pear cactus or different-sized seeds.

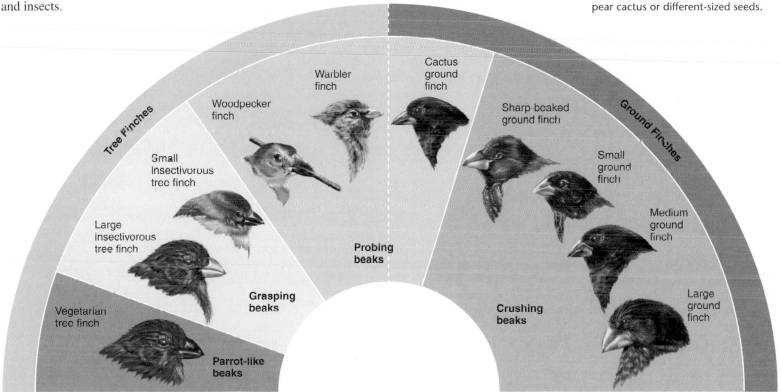

Table 16.1	The Geological Timescale: Major Divisions of Geological Time and Some of the Major Evolutionary Events That Occurred

Era	Period	Epoch	MYA	Plant and Animal Life
Cenozoic*		Holocene	0–0.01	AGE OF HUMAN CIVILIZATION; Destruction of tropical rain forests accelerates extinctions.
		SIGNIFICANT MAMMALIAN EXTINCTION		
	Neogene	Pleistocene	0.01–2	Modern humans appear; modern plants spread and diversify.
		Pliocene	2–6	First hominids appear; modern angiosperms flourish.
		Miocene	6–24	Apelike mammals, grazing mammals, and insects flourish; grasslands spread; and forests contract.
		Oligocene	24–37	Monkeylike primates appear; modern angiosperms appear.
	Paleogene	Eocene	37–58	All modern orders of mammals are present; subtropical forests flourish.
		Paleocene	58–65	Primates, herbivores, carnivores, insectivores are present; angiosperms diversify.
	MASS EXTINCTION: DINOSAURS AND MOST REPTILES			
Mesozoic	Cretaceous		65–144	Placental mammals and modern insects appear; angiosperms spread and conifers persist.
	Jurassic		144–208	Dinosaurs flourish; birds and angiosperms appear.
	MASS EXTINCTION			
	Triassic		208–250	First mammals and dinosaurs appear; forests of conifers and cycads dominate land; corals and molluscs dominate seas.
	MASS EXTINCTION			
Paleozoic	Permian		250–286	Reptiles diversify; amphibians decline; and gymnosperms diversify.
	Carboniferous		286–360	Amphibians diversify; reptiles appear; and insects diversify. Age of great coal-forming forests.
	MASS EXTINCTION			
	Devonian		360–408	Jawed fishes diversify; insects and amphibians appear; seedless vascular plants diversify and seed plants appear.
	Silurian		408–438	First jawed fishes and seedless vascular plants appear.
	MASS EXTINCTION			
	Ordovician		438–510	Invertebrates spread and diversify; jawless fishes appear; nonvascular plants appear on land.
	Cambrian		510–543	Marine invertebrates with skeletons are dominant and invade land, and marine algae flourish.
Precambrian time			600	Oldest soft-bodied invertebrate fossils
			1,400–700	Protists evolve and diversify.
			2,000	Oldest eukaryotic fossils
			2,500	O_2 accumulates in atmosphere
			3,500	Oldest known fossils (prokaryotes)
			4,500	Earth forms.

* Many authorities divide the Cenozoic era into the Tertiary period (contains Paleocene, Eocene, Oligocene, Miocene, and Pliocene) and the Quaternary period (contains Pleistocene and Holocene).

16.2 The History of Species

The history of the origin and extinction of species on Earth is best discovered by studying the fossil record. **Fossils** are the traces and remains of past life or any other direct evidence of past life. Traces include trails, footprints, burrows, worm casts, or even preserved droppings. Most fossils consist of the remains of hard parts, such as shells, bones, or teeth, because these are usually not consumed or destroyed (Fig. 16.10). **Paleontology** is the science of discovering and studying the fossil record and, from it, making decisions about the history of species.

The Geological Timescale

Because life-forms have evolved over time, the strata (layers of sedimentary rock, see Fig. 14.2a) of the Earth's crust contain different fossils. By studying the strata and the fossils they contain, geologists have been able to construct a geological timescale (Table 16.1). It divides the history of life on Earth into eras, then periods, and then epochs, and describes the types of fossils common to each of these divisions of time. Notice that in the geological timescale, only the periods of the Cenozoic era are divided into epochs, meaning that more attention is given to the evolution of primates and flowering plants than to the earlier evolving organisms. Despite an epoch assigned to modern civilization, humans have only been around about .04% of the history of life.

In contrast, prokaryotes existed alone for some 2 billion years before the eukaryotic cell and multicellularity arose during Precambrian time. Some prokaryotes became the first photosynthesizers to add oxygen to the atmosphere. The presence of oxygen may have spurred the evolution of the eukaryotic cell and multicellularity during the Precambrian. All major groups of animals evolved during what is sometimes called the Cambrian explosion. The fossil record for Precambrian time is meager, but the fossil record for the Cambrian period is rich. The evolution of the invertebrate external skeleton accounts for this increase in the number of fossils. Perhaps this skeleton which impedes the uptake of oxygen couldn't evolve until oxygen was plentiful. Or perhaps the external skeleton was merely a defense against predation.

The origin of life on land is another interesting topic. During the Paleozoic era, plants are present on land before animals. Nonvascular plants preceded vascular plants, and among these, cone-bearing plants (gymnosperms) preceded flowering plants (angiosperms). Among vertebrates, the fishes are aquatic, and the amphibians invaded land. The reptiles (e.g., dinosaurs) are ancestral to both birds and mammals. The number of species in the world has continued to increase until the present time, despite the occurrence of five mass extinctions and one significant mammalian extinction during the history of life on Earth.

The Pace of Speciation

Many evolutionists believe as Darwin did that evolutionary changes occur gradually. In other words, they support a *gradualistic model* to explain the pace of evolution. Speciation probably occurs after populations become isolated, with

Figure 16.10 Fossils.

a. A fern leaf from 245 million years ago (MYA) remains because it was buried in sediment that hardened to rock. **b.** This midge (40 MYA) became embedded in amber (hardened resin from a tree).
c. The shells of ammonites (135 MYA) left a mold filled in by minerals.
d. Most fossils, such as this early insectivore mammal (47 MYA) are remains of hard parts because they do not decay as the soft parts do. **e.** These tree trunks (190 MYA) are petrified because minerals have replaced their soft parts. **f.** Signs of ancient life, such as these dinosaur footprints (135 MYA) are traces of past life.

Time

Change

a. Gradualistic model

Time

Change

b. Punctuated equilibrium model

Figure 16.11   Pace of evolution.

a. According to the gradualistic model, new species evolve slowly from an ancestral species. **b.** According to the punctuated equilibrium model, new species appear suddenly and then remain largely unchanged until they become extinct.

each group continuing slowly on its own evolutionary pathway. These evolutionists often show the evolutionary history of groups of organisms by drawing the type of **evolutionary tree** shown in Figure 16.11*a*. Note that in this diagram, an ancestral species has given rise to two separate species, represented by a slow change in plumage color. The gradualistic model suggests that it is difficult to indicate when speciation has occurred because there would be so many transitional links. In some cases, it has been possible to trace the evolution of a group of organisms by finding transitional links.

More often, however, species appear quite suddenly in the fossil record, and then they remain essentially unchanged phenotypically until they undergo extinction. Some paleontologists have therefore developed a *punctuated equilibrium model* to explain the pace of evolution. The model says that a period of equilibrium (no change) is punctuated (interrupted) by speciation. Figure 16.11*b* shows the type of diagram paleontologists prefer to use when representing the history of evolution over time. This model suggests that transitional links are less likely to become fossils and less likely to be found. Speciation most likely involves only an isolated subpopulation at one locale. Only when this new subpopulation expands and replaces existing species is it apt to show up in the fossil record.

The differences between these two models are subtle, especially when we consider that the "sudden" appearance of a new species in the fossil record could represent many thousands of years because geological time is measured in millions of years.

Mass Extinctions of Species

As we have noted, most species exist for only a limited amount of time (measured in millions of years), and then they die out (become extinct). **Mass extinctions** are the disappearance of a large number of species or a higher taxonomic group within a relatively short period of time. The geological timescale in Table 16.1 shows the occurrence of five mass extinctions: at the ends of the Ordovician, Devonian, Permian, Triassic, and Cretaceous periods. Also, there was a significant mammalian extinction at the end of the Pleistocene epoch. While many factors contribute to mass extinctions, two possible types of events (continental drift and meteorite impacts) are believed to be particularly significant.

Continental drift—the movement of continents—is believed to have contributed to several extinctions. You may have noticed that the coastlines of several continents are mirror images of each other. For example, the outline of the west coast of Africa matches that of the east coast of South America. Also, the same geological structures are found in many of the areas where the continents at some time in the past touched. A single mountain range runs through South America, Antarctica, and Australia, for example. But the mountain range is no longer continuous because the continents have drifted apart. The reason the continents drift is explained by a branch of geology known as *plate tectonics,* which is based on the fact that the Earth's crust is fragmented into slablike plates that float on a lower, hot mantle layer (Fig. 16.12). The continents and the ocean basins are a part of these rigid plates, which move like conveyor belts.

The loss of habitats is a significant cause of extinctions, and continental drift can lead to massive habitat changes. We know that 250 million years ago (MYA), at the time of the Permian mass extinction, all the Earth's continents came together to form one large continent called Pangaea (Fig. 16.13*a*). The result was dramatic environmental change that profoundly affected life. Marine organisms suffered as the oceans were joined, and the amount of shoreline, where many marine organisms live, was reduced. They didn't recover until some continents

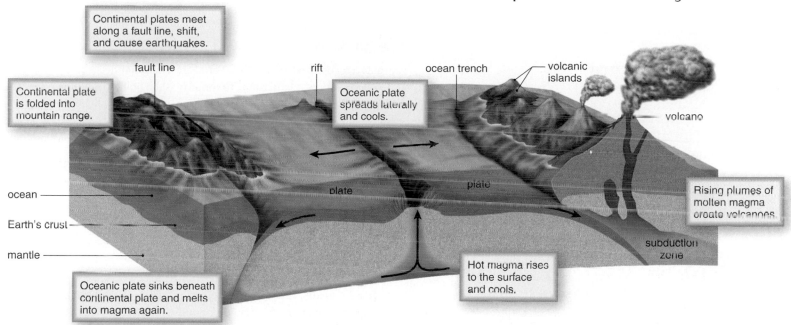

Continental plates meet along a fault line, shift, and cause earthquakes.

fault line

rift

ocean trench

volcanic islands

Continental plate is folded into mountain range.

Oceanic plate spreads laterally and cools.

volcano

ocean

plate

plate

Rising plumes of molten magma create volcanoes.

Earth's crust

subduction zone

mantle

Oceanic plate sinks beneath continental plate and melts into magma again.

Hot magma rises to the surface and cools.

Figure 16.12   Plate tectonics.

The Earth's surface is divided into several solid tectonic plates floating on the fluid magma beneath them. At rifts in the ocean floor, two plates gradually separate as fresh magma wells up and cools, enlarging the plates. Mountains, including volcanoes, are raised where one plate pushes beneath another at subduction zones. Where two plates slowly grind past each other at a fault line, tension builds up, which is released occasionally in the form of earthquakes.

drifted away from the poles and warmth returned (Fig. 16.13b). Terrestrial organisms were affected as well because the amount of interior land, which tends to have a drier and more erratic climate, increased. Immense glaciers developing at the poles withdrew water from the oceans and chilled even once-tropical land.

The other event that is believed to have contributed to mass extinctions is the impact of a meteorite as it crashed into the Earth. The result of a large meteorite striking Earth could have been similar to that of a worldwide atomic bomb explosion: A cloud of dust would have mushroomed into the atmosphere, blocking out the sun and causing plants to freeze and die. This type of event has been proposed as a primary cause of the Cretaceous extinction that saw the demise of the dinosaurs. Cretaceous clay contains an abnormally high level of iridium, an element that is rare in the Earth's crust but more common in meteorites. A layer of soot has been identified in the strata alongside the iridium, and a huge crater that could have been caused by a meteorite was found in the Caribbean–Gulf of Mexico region on the Yucatán peninsula.

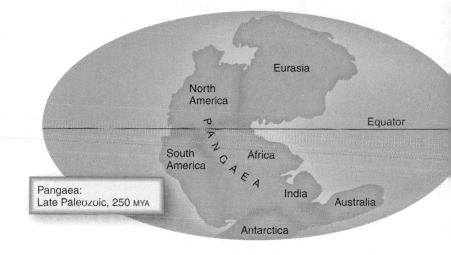

Eurasia

North America

Equator

South America

Africa

India

Australia

Pangaea: Late Paleozoic, 250 MYA

Antarctica

a.

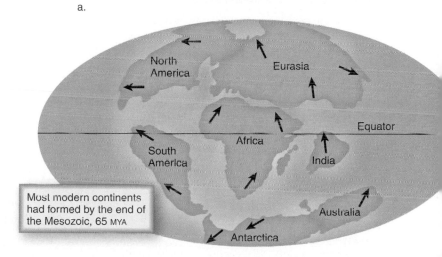

North America

Eurasia

Equator

Africa

South America

India

Most modern continents had formed by the end of the Mesozoic, 65 MYA

Australia

Antarctica

b.

Figure 16.13   Continental drift.

a. About 250 MYA, all the continents were joined into a supercontinent called Pangaea. **b.** By 65 MYA, all the continents had begun to separate. This process is continuing today. North America and Europe are separating at a rate of about 2 centimeters per year.

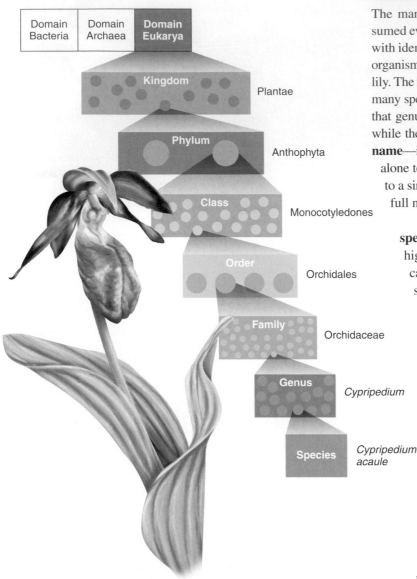

Figure 16.14 Taxonomy hierarchy.

A domain is the most inclusive of the classification categories. The plant kingdom is in the domain Eukarya. In the plant kingdom are several phyla, each represented here by lavender circles. The phylum Anthophyta has only two classes (the monocots and eudicots). The class Monocotyledones encompasses many orders. In the order Orchidales are many families; in the family Orchidaceae are many genera; and in the genus *Cypripedium* are many species—for example, *Cypripedium acaule*. (This illustration is diagrammatic and doesn't necessarily show the correct number of subcategories.)

16.3 Classification of Species

The many recognized species on Earth are classified according to their presumed evolutionary relationship. **Taxonomy** is the branch of biology concerned with identifying, naming, and classifying organisms. Classification begins when organisms are given a scientific name. For example, *Lilium buibiferum* is a type of lily. The first word, *Lilium,* is the genus in a classification category that can contain many species. The second word, the **specific epithet,** refers to one species within that genus. Notice that the scientific name is in italics; the genus is capitalized, while the specific epithet is not. The species is designated by the full **binomial name**—in this case, *Lilium buibiferum.* However, the genus name can be used alone to refer to a group of related species. Also, the genus can be abbreviated to a single letter if used with the specific epithet (e.g., *L. buibiferum*) and if the full name has been given previously.

Today, taxonomists use the following categories of classification: **species, genus, family, order, class, phylum,** and **kingdom.** Recently, a higher taxonomic category, the **domain,** has been added to this list. There can be several species within a genus, several genera within a family, and so forth. In this hierarchy, the higher the category, the more inclusive it is (Fig. 16.14). Also, the organisms that fill a particular classification category are distinguishable from other organisms by sharing a structural, chromosomal, or molecular feature that distinguishes one group from another. Organisms in the same domain have general characteristics in common; those in the same species have quite specific characteristics in common.

In most cases, each category of classification can be subdivided into three additional categories, as in superorder, order, suborder, and infraorder. Considering these, there are more than 30 categories of classification.

Classification and Phylogeny

Taxonomy and classification are a part of the broader field of **systematics,** which is the study of the diversity of organisms at all levels of organization. One goal of systematics is to determine **phylogeny,** the evolutionary history of a group of organisms. When we say that two species (or genera, families, etc.) are closely related, we mean they share a recent **common ancestor.**

Using **traditional systematics,** Figure 16.15 shows how the classification of groups of organisms allows us to construct a **phylogenetic tree,** a diagram that indicates common ancestors and lines of descent (lineages). In order to classify organisms and to construct a phylogenetic tree, it is necessary to determine the characteristics, or simply, characters, of the various groups, called **taxa** (sing., taxon). A *primitive character* is present in the common ancestor and in all taxa belonging to that group. A *derived character* is found only in a particular line of descent. Whether a character is primitive or derived depends on the group being considered. For example, a primitive character for primates may be a derived character for mammals. Different lineages diverging from a common ancestor are expected to have different derived characters. Thus, all the animals in the family Cervidae have antlers, but they are highly branched in red deer and palmate (having the shape of a hand) in reindeer.

Tracing Phylogeny

Systematists gather all sorts of data in order to discover the evolutionary relationships between species. They rely heavily on a combination of data from the fossil record, homology, and molecular data to determine the correct sequence of common ancestors in any particular group of organisms. If you can determine the common ancestors in the history of life, you know how evolution occurred and can classify organisms accordingly.

Homology *Homology* is character similarity that stems from having a common ancestor. Comparative anatomy, including embryological evidence, provides information regarding homology. **Homologous structures** are related to each other through common descent. The forelimbs of vertebrates are homologous because they contain the same bones organized in the same general way as in a common ancestor (see Fig. 14.14). However, a horse has but a single digit and toe (the hoof), while a bat has four lengthened digits that support its membranous wings.

Deciphering homology is sometimes difficult because of convergent evolution and parallel evolution. **Convergent evolution** is the acquisition of the same or similar characters in distantly related lines of descent. Similarity due to convergence is termed **analogy.** The wings of an insect and the wings of a bat are analogous. You may recall from Chapter 14 that **analogous structures** have the same function in different groups, but organisms with these structures do not have a common ancestor. Both cactuses and spurges are adapted similarly to a hot, dry environment, and both are succulent, spiny, flowering plants. However, the details of their flower structure indicate that these plants are not closely related. **Parallel evolution** is the acquisition of the same or a similar character in two or more related lineages without it being present in a common ancestor. A similar banding pattern is found in several species of moths, for example. It is sometimes difficult to tell if features are primitive, derived, convergent, or parallel.

Molecular Data Speciation occurs when mutations bring about changes in the base-pair sequences of DNA. Systematists, therefore, assume that the more

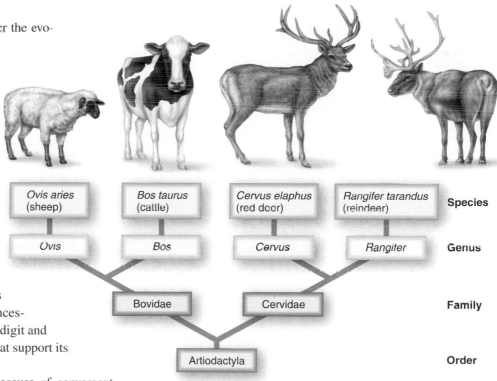

| *Ovis aries* (sheep) | *Bos taurus* (cattle) | *Cervus elaphus* (red deer) | *Rangifer tarandus* (reindeer) | **Species** |

| *Ovis* | *Bos* | *Cervus* | *Rangifer* | **Genus** |

| Bovidae | Cervidae | **Family** |

| Artiodactyla | **Order** |

a.

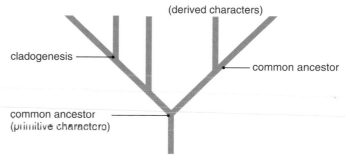

(derived characters)

cladogenesis

common ancestor

common ancestor
(primitive characters)

b.

Figure 16.15 Classification and phylogeny.

a. The classification and phylogenetic tree for a group of organisms is ideally constructed to reflect their phylogenetic history. A species is most closely related to other species in the same genus, more distantly related to species in other genera of the same family, and so forth, on through order, class, phylum, and kingdom.
b. A diagrammatic tree showing the location of common ancestors, which are said to have primitive (ancestral) characters. Following cladogenesis (a divergence), each line of descent has its own derived (evolved) characters.

Check Your Progress

1. List the categories of classification, in order, from the largest category to the smallest.
2. Contrast homologous structures with analogous structures.
3. What is a phylogenetic tree?

Answers: **1.** Domain, kingdom, phylum, class, order, family, genus, species. **2.** Homologous structures are similar in anatomy because they are derived from a common ancestor. Analogous structures have the same function in different groups and do not share a common ancestry. **3.** A phylogenetic tree is a diagram that indicates common ancestors and lines of descent.

closely species are related, the fewer changes will be found in DNA base-pair sequences. Because molecular data are straightforward and numerical, they can sometimes sort out relationships obscured by inconsequential anatomical variations or convergence. Software breakthroughs have made it possible to analyze nucleotide sequences quickly and accurately using a computer. Also, these analyses are available to anyone doing comparative studies through the Internet, so each investigator doesn't have to start from scratch. The combination of accuracy and availability of past data has made molecular systematics a standard way to study the relatedness of groups of organisms today.

All cells have ribosomes, which are essential for protein synthesis. Further, the genes that code for ribosomal RNA (rRNA) have changed very slowly during evolution because any drastic changes lead to malfunctioning cells. Therefore, it is believed that comparative rRNA sequencing provides a reliable indicator of the similarity between organisms. Ribosomal RNA sequencing helped investigators conclude that all living things can be divided into the three domains.

One study involving DNA differences produced the data shown in Figure 16.16. Although the data suggest that chimpanzees are more closely related to humans than to other apes, in most classifications, humans and chimpanzees are placed in different families; humans are in the family Hominidae, and chimpan-

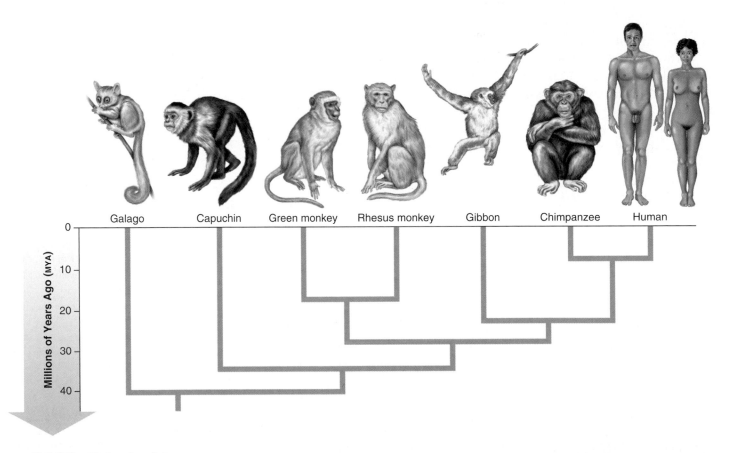

Figure 16.16   Molecular data.

The relationships of certain primate species are based on a study of their genomes. The length of the branches indicates the relative number of DNA base-pair differences between the groups. These data, along with knowledge of the fossil record for one divergence, make it possible to suggest a date for the other divergences in the tree.

zees are in the family Pongidae. In contrast, the rhesus monkey and the green monkey, which have more numerous DNA differences, are placed in the same family (Cercopithecidae). To be consistent with the data, shouldn't humans and chimpanzees also be in the same family? *Traditional systematists,* in particular, believe that since humans are markedly different from chimpanzees because of adaptation to a different environment, it is justifiable to place humans in a separate family.

Cladistic Systematics

Cladistics is a school of systematics that was introduced in 1950 and has become increasingly influential. The goal is to produce *testable* hypotheses about the evolutionary relationships among a series of organisms. Cladistics depends on distinguishing primitive from shared derived characters to classify organisms and arrange taxa in a type of diagram called a *cladogram.* A cladogram traces the evolutionary history of the group being studied. Let's see how it works.

The first step when constructing a cladogram is to draw up a table that summarizes the characters of the taxa being compared (Fig. 16.17a). Although this table concerns anatomical characters only, cladists use all types of data, including molecular data, to arrive at evolutionary relationships. At least one, but preferably several, species are studied as an *outgroup,* a group that is not part of the study group. In this example, lancelets are in the outgroup, and selected vertebrates (eels, newts, snakes, and lizards) are in the study group. Any character found in both the outgroup and the study group is a shared primitive character presumed to have been present in an ancestor common to both the outgroup and the study group. Lancelets (in the outgroup) have a supporting rod called a notochord as an embryo, and so do all vertebrates (in the study group). Therefore, a notochord is a primitive character shared by all taxa in Figure 16.17a.

Cladists are always guided by the principle of *parsimony,* which states that the least number of assumptions is the most probable. Thus, they construct a cladogram that minimizes the number of assumed evolutionary changes, or leaves the fewest number of derived characters unexplained. Therefore, any character in the table found in scattered taxa (in this case, four bony limbs and a long cylindrical body) is not used to construct the cladogram because we would have to assume that these characters evolved more than once among the taxa of the study group. The other characters are designated as shared derived characters—that is, they are homologies shared by only certain taxa of the study group.

The Cladogram

In a cladogram, a *clade* is an evolutionary branch that includes a common ancestor together with all its descendant species. A clade includes the taxa that have shared derived characters. The cladogram in Figure 16.17b has three clades in the study group that differ in size because the first includes the other two, and so forth. Based on Figure 16.17a, the common ancestor at the root of the tree has one primitive character—a notochord when an embryo—and all taxa in the cladogram share this character. Figure 16.17a also tells us that all the taxa in the study group belong to a clade that has vertebrae; only newts, snakes, and lizards are in a clade that has lungs and a three-chambered heart; and only snakes and lizards are in a clade that has an amniotic egg and internal fertilization. Therefore, this is the sequence in which these characters must have evolved during the evolutionary history of vertebrates.

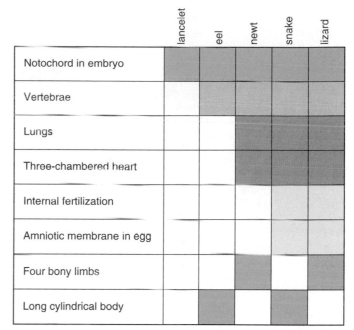

a.

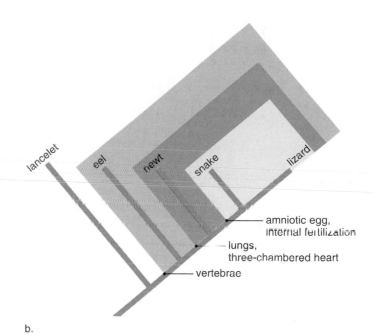

b.

Figure 16.17 Constructing a cladogram.

a. First, a table is drawn up that lists characters for all the taxa. An examination of the table shows which characters are primitive (notochord aqua) and which are derived (lavender, orange, and yellow). The shared derived characters distinguish the taxa. **b.** In a cladogram, the shared derived characters are sequenced in the order they evolved and are used to define clades. A clade contains a common ancestor and all the species that share the same derived characters (homologies). Four bony limbs and a long cylindrical body were not used in constructing the cladogram because they are in scattered taxa.

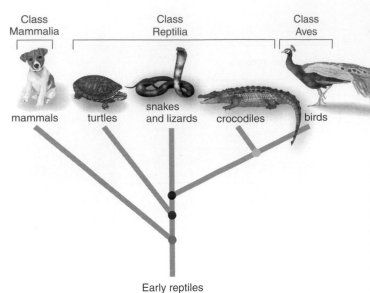

Early reptiles

a. Traditional systematics: phylogenetic tree

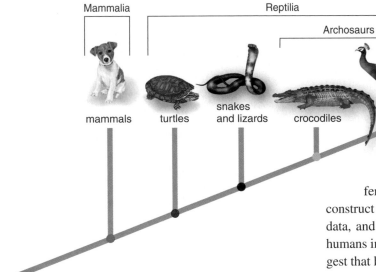

b. Cladistic systematics: cladogram

Figure 16.18 **Traditional versus cladistic view of reptilian phylogeny.**

Traditionalists are more likely to use value judgments than are cladists. **a.** Although crocodiles and birds share many characteristics, traditionalists place birds in a different class because the evolution of feathers is such a significant event in the history of animals. **b.** Cladists make no value judgments and place crocodiles and birds in the same class on the basis of shared derived characters. The colors show that the placement of common ancestors is the same in both diagrams.

A cladogram is objective—it lists the characters used to construct the cladogram. Cladists regard a cladogram as a hypothesis that can be tested and either corroborated or refuted on the basis of additional data. Therefore, they consider a cladogram to be more scientific than a traditional phylogenetic tree.

Traditionalists Versus Cladists

Traditional systematists use a wider range of information than do cladists, and they make value judgments about which characters are more indicative of phylogeny. Traditionalists believe that organisms adapted to a new environment need not be classified with the common ancestor from which they evolved. In the traditional phylogenetic tree shown in Figure 16.18*a*, birds and mammals are placed in different classes because it is quite obvious to the most casual observer that mammals (having hair and mammary glands) and birds (having feathers) are quite different in appearance from reptiles (having scaly skin). The traditionalist tree shows that mammals and birds have a different reptilian ancestor.

Cladists prefer the cladogram in Figure 16.18*b*. All the animals shown are in one clade because they all evolved from a common ancestor that laid eggs. Mammals can be placed in their own class because they alone have hair and mammary glands and three middle ear bones. Not so for reptiles: The only thing that dinosaurs, crocodiles, snakes, lizards, and turtles have in common is that they are not birds or mammals. On the other hand, crocodiles and birds share common derived anatomical and behavioral characters not present in snakes and lizards. On this basis, cladists place crocodiles and birds in the same taxon. They agree that birds and crocodiles are adapted to different ways of life, but this observation is not the type of data used to construct a cladogram. Most traditionalists today regard adaptations as valuable data, and they place birds in a separate taxon from reptiles. They also place humans in a different family from all the apes, even though molecular data suggest that humans shared a common ancestor with chimpanzees.

Check Your Progress

1. Describe how cladists classify organisms.
2. Explain how cladists employ the principle of parsimony.
3. Contrast traditionalists with cladists.

Answers: 1. Cladistics uses only shared derived characters to determine that several taxa share a common ancestor and belong to a clade. 2. Cladists use the minimum number of assumptions to construct a cladogram. 3. Traditionalists use a wider range of information than do cladists and make value judgments about the importance of certain characters.

Classification Systems

Classification systems change over time. Until recently, most biologists utilized the **five-kingdom system** of classification, which contains kingdoms for the plants, animals, fungi, protists, and monerans. Organisms were placed into these kingdoms based on type of cell (prokaryotic or eukaryotic), level of organization (unicellular or multicellular), and type of nutrition. In the five-kingdom system, the monerans were distinguished by their structure—they were prokaryotic (lack a membrane-bounded nucleus)—whereas the organisms in the other kingdoms were eukaryotic (have a membrane-bounded nucleus). The kingdom Monera contained all prokaryotes, which according to the fossil record evolved first.

Within the past ten years, new information has called into question the five-kingdom system of classification. Molecular and cellular data suggest that there are two groups of prokaryotes—the bacteria and the archaea—and that these groups are fundamentally different from each other. In fact, they are so different that they should be assigned to separate domains, a category of classification that is higher than the kingdom category. The bacteria arose first, and they were followed by the archaea and then the eukarya (Fig. 16.19). The archaea and eukarya are more closely related to each other than either is to the bacteria. Systematists, using the **three-domain system** of classification, are in the process of sorting out what kingdoms belong within **domain Bacteria** and **domain Archaea. Domain Eukarya** contains kingdoms for protists, animals, fungi, and plants. Later in this text, we will study the individual kingdoms that occur within the domain Eukarya. The protists do not share one common ancestor, and some suggest that the kingdom should be divided into many different kingdoms. How many kingdoms is still a matter of debate among systematists, illustrating that taxonomy can be subjective until data are available and an objective conclusion is acceptable to most systematists.

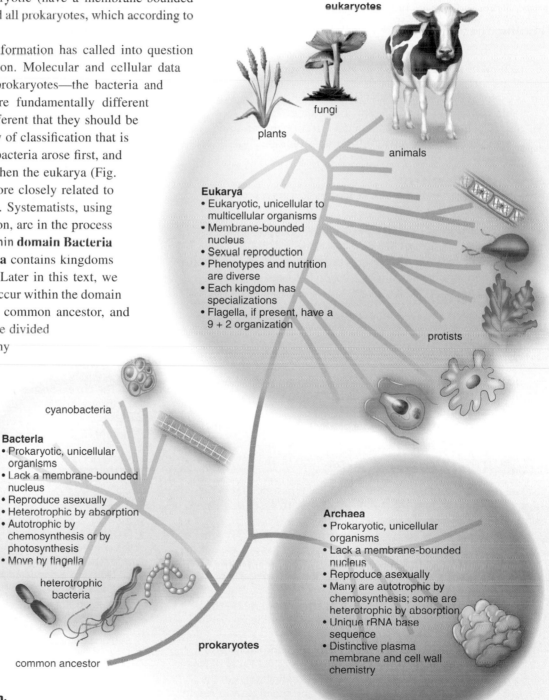

Eukarya
- Eukaryotic, unicellular to multicellular organisms
- Membrane-bounded nucleus
- Sexual reproduction
- Phenotypes and nutrition are diverse
- Each kingdom has specializations
- Flagella, if present, have a 9 + 2 organization

eukaryotes

plants

fungi

animals

protists

Bacteria
- Prokaryotic, unicellular organisms
- Lack a membrane-bounded nucleus
- Reproduce asexually
- Heterotrophic by absorption
- Autotrophic by chemosynthesis or by photosynthesis
- Move by flagella

cyanobacteria

heterotrophic bacteria

Archaea
- Prokaryotic, unicellular organisms
- Lack a membrane-bounded nucleus
- Reproduce asexually
- Many are autotrophic by chemosynthesis; some are heterotrophic by absorption
- Unique rRNA base sequence
- Distinctive plasma membrane and cell wall chemistry

prokaryotes

common ancestor

Figure 16.19 Three-domain system.

In this system, the prokaryotes are in the domains Bacteria and Archaea. The eukaryotes are in the domain Eukarya, which contains four kingdoms for the protists, animals, fungi, and plants.

Summary

16.1 Macroevolution

Macroevolution is evolution on a large scale because it considers the history of life on Earth. Macroevolution begins with speciation, the origin of new species. Without speciation and the extinction of species, life on Earth would not have a history.

Defining Species
The biological concept of species

- recognizes a species by its inability to reproduce with members of another group.
- is useful because species can look similar, and members of the same species can have different appearances.
- has its limitations. Hybridization does occur between some species, and it is only relevant to sexually reproducing organisms.

Reproductive Barriers
Prezygotic and postzygotic barriers keep species from reproducing with one another.

- Prezygotic isolating mechanisms involve habitat isolation, temporal isolation, and behavioral isolation.
- Postzygotic isolating mechanisms prevent hybrid offspring from developing or breeding if reproduction has been successful.

Models of Speciation
Allopatric speciation and sympatric speciation are two models of speciation.

- The allopatric speciation model proposes that a geographic barrier keeps groups of populations apart. Meanwhile, prezygotic and postzygotic isolating mechanisms develop, and these prevent successful reproduction if these two groups come into future contact.
- The sympatric speciation model proposes that a geographic barrier is not required.

16.2 The History of Species

The fossil record, as outlined by the geological timescale, traces the history of life in broad terms. It has been possible to absolutely date fossils by using radioactive dating techniques.

The fossil record

- can be used to support a gradualistic model: Two groups of organisms arise from an ancestral species and gradually become two different species.
- can be used to support a punctuated equilibrium model: A period of equilibrium (no change) is interrupted by speciation within a relatively short period of time.
- shows that at least five mass extinctions and one significant mammalian extinction have occurred during the history of life on Earth. Two major contributors to mass extinctions are the loss of habitats due to continental drift and the disastrous results from a meteorite impacting Earth.

16.3 Classification of Species

Systematics

- assigns every species a scientific name, which immediately indicates its genus. From that beginning, species are also assigned to a family, order, class, phylum, kingdom, and domain according to their molecular and structural characters.
- attempts to determine phylogeny, the evolutionary history of a group of organisms. Systematics relies on the fossil record, homology, and molecular data to determine relationships among organisms.
- has different schools. **Cladists** use shared primitive and shared derived characters to construct cladograms. In a cladogram, a clade consisting of a common ancestor and all the species derived from a common ancestor, the species have shared derived characteristics. **Traditionalists** are not opposed to making value judgments (particularly about adaptations to new environments) in constructing phylogenetic trees, but cladists use only objective data when constructing cladograms.

Classification Systems

- The **five-kingdom system** recognized the kingdoms for the monerans, protists, animals, fungi, and plants.

Five-Kingdom System

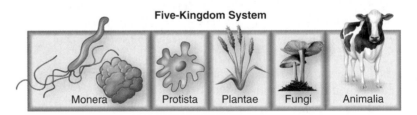

Monera   Protista   Plantae   Fungi   Animalia

- The **three-domain system** uses molecular data to designate three evolutionary domains: Bacteria, Archaea, and Eukarya:

Three-Domain System

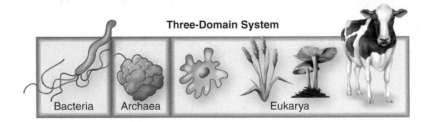

Bacteria   Archaea   Eukarya

- Domains Bacteria and Archaea contain prokaryotes.
- Domain Eukarya contains kingdoms for the protists, animals, fungi, and plants.

Thinking Scientifically

1. Use the data from the following table to fill in the phylogenetic tree for vascular plants. Plants with vascular tissue have transport tissue.

	Ferns	Conifers	Ginkgos	Monocots	Eudicots
vascular tissue	X	X	X	X	X
produce seeds		X	X	X	X
naked seeds		X	X		
needle-like leaves		X			
fan-shaped leaves			X		
enclosed seeds				X	X
one embryonic leaf				X	
two embryonic leaves					X

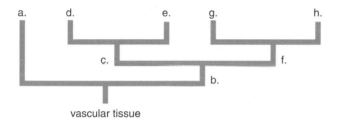

2. The Hawaiian Islands are located thousands of kilometers from any mainland. Each island arose from the sea bottom and was colonized by plants and animals that drifted in on ocean currents or winds. Each island has a unique environment in which its inhabitants have evolved. Consequently, most of the plant and animal species on the islands do not exist anywhere else in the world.

   In contrast, on the islands of the Florida Keys, there are no unique or indigenous species. All of the species on those islands also exist on the mainland. Suggest an explanation for these two different patterns of speciation.

Testing Yourself

Choose the best answer for each question.

1. A biological species
   a. always looks different from other species.
   b. always has a different chromosome number from that of other species.
   c. is reproductively isolated from other species.
   d. never occupies the same niche as other species.

For questions 2–9, indicate the type of isolating mechanism described in each scenario.

Key:
   a. habitat isolation
   b. temporal isolation
   c. behavioral isolation
   d. mechanical isolation
   e. gamete isolation
   f. zygote mortality
   g. hybrid sterility
   h. low F$_2$ fitness

2. Males of one species do not recognize the courtship behaviors of females of another species.
3. One species reproduces at a different time than another species.
4. A cross between two species produces a zygote that always dies.
5. Two species do not interbreed because they occupy different areas.
6. A hybrid between two species produces gametes that are not viable.
7. Two species of plants do not mate with each other because they are visited by different pollinators.
8. The sperm of one species cannot survive in the reproductive tract of another species.
9. The offspring of two hybrid individuals exhibit poor vigor.
10. Complete the following diagram illustrating allopatric speciation by using these phrases: genetic changes (used twice), geographic barrier, species 1, species 2, species 3.

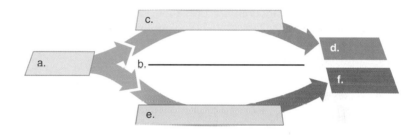

11. Transitional links are least likely to be found if evolution proceeds according to the
    a. gradualistic model.
    b. punctuated equilibrium model.

12. Which is the scientific name of an organism?
    a. *Rosa rugosa*
    b. *Rosa*
    c. *rugosa*
    d. *rugosa rugosa*
    e. Both a and d are correct.

13. Which of these statements best pertains to taxonomy? Species
    a. always have three-part names, such as *Homo sapiens sapiens*.
    b. are always reproductively isolated from other species.
    c. always share the most recent common ancestor.
    d. always look exactly alike.
    e. Both c and d are correct.

14. Which of the following groups are domains? Choose more than one answer if correct.
    a. bacteria
    b. archaea
    c. eukarya
    d. animals
    e. plants

15. Which pair is mismatched?
    a. homology—character similarity due to a common ancestor
    b. molecular data—DNA strands match
    c. fossil record—bones and teeth
    d. homology—functions always differ
    e. molecular data—DNA and RNA data

16. One benefit of the fossil record is
    a. that hard parts are more likely to fossilize.
    b. that fossils can be dated.
    c. its completeness.
    d. that fossils congregate in one place.
    e. All of these are correct.

17. The discovery of common ancestors in the fossil record, the presence of homologies, and molecular data similarities help scientists decide
    a. how to classify organisms.
    b. the proper cladogram.
    c. how to construct phylogenetic trees.
    d. how evolution occurred.
    e. All of these are correct.

18. In cladistics,
    a. a clade must contain the common ancestor plus all its descendants.
    b. derived characters help construct cladograms.
    c. data for the cladogram are presented.
    d. the species in a clade share homologous structures.
    e. All of these are correct.

19. Answer these questions about the following cladogram.
    a. This cladogram contains how many clades? How are they designated in the diagram?
    b. What character is shared by all animals in the study group? What characters are shared by only snakes and lizards?
    c. Which animals share a recent common ancestor? How do you know?

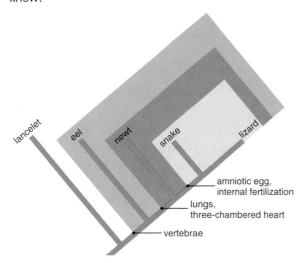

20. Allopatric, but not sympatric, speciation requires
    a. reproductive isolation.
    b. geographic isolation.
    c. spontaneous differences in males and females.
    d. prior hybridization.
    e. rapid rate of mutation.

21. Which kingdom is mismatched?
    a. Protista—domain Bacteria
    b. Protista—multicellular algae
    c. Plantae—flowers and mosses
    d. Animalia—arthropods and humans
    e. Fungi—molds and mushrooms

Go to www.mhhe.com/maderessentials for more quiz questions.

Bioethical Issue

Paleoanthropologists often use cladistics to create phylogenetic trees based on bone and tooth measurements. According to developmental biologists, though, bone shape and arrangement are influenced by both the environment and complex genetic interactions. Therefore, some evolutionary biologists are now suggesting that skeletal anatomy cannot be used to produce cladograms that make biological sense. Cladograms based on skull and tooth measurements do not coincide with those based on DNA base-pair sequence similarities. In order to determine the most valuable traits for cladistic analyses, researchers need to determine the molecular pathways involved in the development of the human body. This is an area that would require millions of dollars of research funding.

   Would you support the use of government funding for research into the genetics and biology of body development, for the purpose of better understanding our phylogenetic tree? If not, would you support the idea if the research had the potential to advance our knowledge of related medical problems, such as osteoporosis?

Understanding the Terms

adaptive radiation   251
allopatric speciation   250
analogous structure   257
analogy   257
binomial name   256
cladistics   259
class   256
common ancestor   247, 256
convergent evolution   257
domain   256
domain Archaea   261
domain Bacteria   261
domain Eukarya   261
evolutionary tree   254
family   256
five-kingdom system   261
fossil   253
genus   256
homologous structure   257
kingdom   256
macroevolution   246

mass extinction   254
order   256
paleontology   253
parallel evolution   257
phylogenetic tree   256
phylogeny   256
phylum   256
postzygotic isolating
   mechanism   249
prezygotic isolating
   mechanism   248
speciation   246
species   256
specific epithet   256
sympatric speciation   250
systematics   256
taxon (pl., taxa)   256
taxonomy   256
three-domain system   261
traditional systematics   256

Match the terms to these definitions:

a. _____ Evolution of several new species from a single ancestral species.

b. _____ Science of discovering and studying the fossil record.

c. _____ Disappearance of a large number of species in a short period of time.

d. _____ Scientific name that includes genus and species.

e. _____ Diagram that tells the evolutionary history of a group of organisms.

f. _____ Most inclusive of the classification categories in the three-domain system.

g. _____ Anatomical structure indicative of relatedness through common descent.

The First Forms of Life

CHAPTER

17

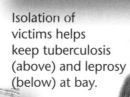

Isolation of victims helps keep tuberculosis (above) and leprosy (below) at bay.

Fear of polio (above) and AIDS (right) can cause us to modify our behavior.

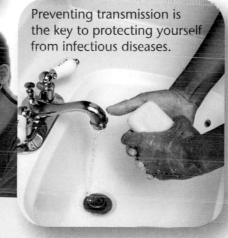

Preventing transmission is the key to protecting yourself from infectious diseases.

AIDS, polio, tuberculosis, gonorrhea, SARS, leprosy, malaria. *Microbes,* named for their microscopic size, cause these diseases. Disease-causing microbes, discussed in this chapter, include tiny viruses, some bacteria, and certain protists. Although microscopic size unifies the microbes, their structures differ. Viruses are noncellular, bacteria are prokaryotes, and protists are eukaryotes. Viruses always cause disease, but only some bacteria and a few protists do. Among ▶ ▶ ▶

the protists, algae rarely cause disease because they usually make their own food. The parasitic way of life allows the disease-causing bacteria and protozoans (protists) to get their food and reproduce. Viruses don't need food—after entering a cell, they cause the cell to make hundreds of viruses at one time.

We didn't always know that microbes cause disease. The germ theory of disease didn't take hold until the twentieth century. After that, rapid progress was made in identifying which microbes cause which diseases and how they are transmitted. For example, we now know that AIDS and gonorrhea are sexually transmitted, tuberculosis and SARS are contracted by way of the respiratory tract, and the *Anopheles* mosquito spreads malaria. Each microbe lives in a different part of the body: HIV replicates in just certain types of white blood cells, tuberculosis usually centers in the lungs, and the protozoan that causes malaria reproduces in red blood cells.

This chapter provides basic information about microbes and the various ways they reproduce. Such knowledge enables us not only to fight the diseases microbes sometimes cause, but also to take advantage of the many benefits they provide.

17.1 The Viruses

The study of **viruses** has contributed much to our understanding of disease, genetics, and even the characteristics of living things. Their contribution is surprising because viruses are not included in the classification of organisms. They are noncellular, while organisms are always cellular. Also, they are amazingly small; at 0.2 microns, a virus is only about one-fifth the size of a bacterium.

Each type of virus always has at least two parts: an outer **capsid,** composed of protein subunits, and an inner core of nucleic acid—either DNA or RNA. The adenovirus shown in Figure 17.1 also has spikes (formed from a glycoprotein) that are involved in attaching the virus to the host cell. In some animal viruses, the capsid is surrounded by an outer membranous envelope with glycoprotein spikes. The envelope is actually a piece of the host's plasma membrane that also contains proteins produced by the virus. The interior of a virus can contain various enzymes such as the polymerases, which are needed to produce viral DNA and/or RNA. The viral genome has at most several hundred genes; by contrast, a human cell contains thousands of genes.

Should viruses be considered living? Both scientific and philosophical debates have raged with regard to this question. Viruses are *obligate intracellular parasites* because they can only reproduce inside a living cell. Outside a living cell, viruses can be stored as chemicals or even synthesized in the laboratory from chemicals. Still, viruses do have a genome that mutates and functions to direct their reproduction when inside a cell.

Viral Reproduction

Viruses are specific to a particular host cell because a portion of the capsid (often a spike) adheres in a lock-and-key manner to a specific molecule (called a receptor) on the host cell's outer surface. Once inside, the viral genome *takes over the metabolic machinery of the host cell*. In large measure, the virus uses

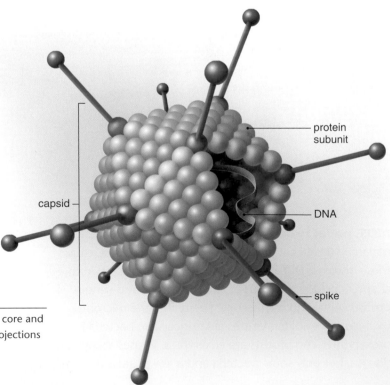

Figure 17.1   Adenovirus anatomy.

Typical of viruses, adenoviruses have a nucleic acid core and a coat of protein, the so-called capsid. Note the projections called spikes—they help a virus enter a cell.

the host's enzymes, ribosomes, transfer RNA (tRNA), and ATP for its reproduction.

Reproduction of Bacteriophages

A **bacteriophage,** or simply *phage,* is a virus that reproduces in a bacterium. Bacteriophages are named in different ways; the one shown in Figure 17.2 is called lambda (λ). When phage λ reproduces, it can undergo the lytic cycle or the lysogenic cycle (Fig. 17.3). The **lytic cycle** may be divided into five stages: attachment, penetration, biosynthesis, maturation, and release. During *attachment,* the capsid combines with a receptor in the bacterial cell wall. During *penetration,* a viral enzyme digests away part of the cell wall, and viral DNA is injected into the bacterial cell. *Biosynthesis* of viral components begins after the virus inactivates host genes not necessary to viral replication. The machinery of the host cell then carries out (1) viral DNA replication and (2) production of multiple copies of the capsid protein subunits. During *maturation,* viral DNA and capsids assemble to produce several hundred viral particles. Lysozyme, an enzyme coded for by a viral gene, disrupts the cell wall, and the *release* of phage particles occurs. The bacterial cell dies as a result.

In the **lysogenic cycle,** the infected bacterium does not immediately produce phage, but may do so sometime in the future. In the meantime, the phage is *latent*—not actively reproducing. Following attachment and penetration, *integration* occurs: Viral DNA becomes incorporated into bacterial DNA with no destruction of host DNA. While latent, the viral DNA is called a *prophage.* The prophage is replicated along with the host DNA, and all subsequent cells, called lysogenic cells, carry a copy of the prophage. Certain environmental factors, such as ultraviolet radiation, can induce the prophage to enter the lytic stage of biosynthesis, followed by maturation and release.

The herpesviruses, which cause cold sores, genital herpes, and chickenpox in humans, are examples of infections that remain latent much of the time. Herpesviruses linger in spinal ganglia until stress, excessive sunlight, or some other stimulus causes them to undergo the lytic cycle. HIV (human immunodeficiency virus), the cause of AIDS, remains relatively latent in lymphocytes, slowly releasing new viruses.

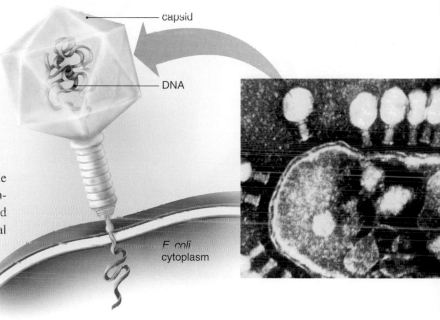

Figure 17.2   Bacteriophage lambda (λ).

The micrograph shows many viral particles attached to a bacteriophage and the blow-up shows how DNA from a virus enters a bacterium.

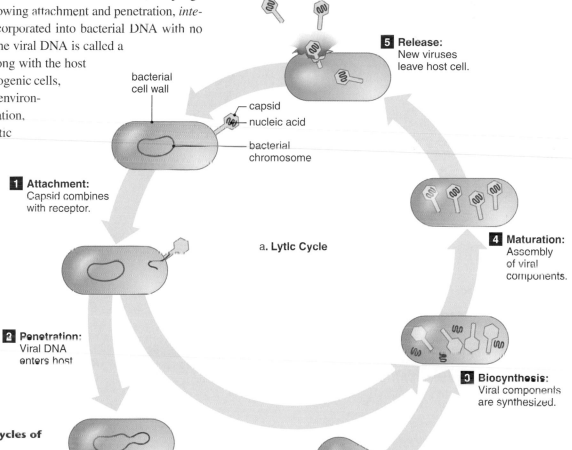

1 **Attachment:** Capsid combines with receptor.

2 **Penetration:** Viral DNA enters host

a. **Lytic Cycle**

b. **Lysogenic Cycle**

Integration

Cloning of viral DNA

5 **Release:** New viruses leave host cell.

capsid
nucleic acid
bacterial chromosome

bacterial cell wall

4 **Maturation:** Assembly of viral components.

3 **Biosynthesis:** Viral components are synthesized.

prophage

Figure 17.3   Lytic and lysogenic cycles of a virus called lambda.

a. In the lytic cycle, viral particles escape when the cell is lysed (broken open). **b.** In the lysogenic cycle, viral DNA is integrated into host DNA. At some time in the future, the lysogenic cycle can be followed by the last three steps of the lytic cycle.

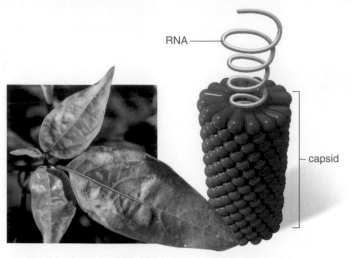

Figure 17.4 **Infected tobacco plant.**

Mottling and a distorted leaf shape are typical of a viral infection in plants.

Plant Viruses

Crops and garden plants are also subject to viral infections. Plant viruses tend to enter through damaged tissues and then move about the plant through *plasmodesmata,* cytoplasmic strands that extend between plant cell walls. The best-studied plant virus is tobacco mosaic virus, a long, rod-shaped virus with only one type of protein subunit in its capsid (Fig. 17.4). Not all viruses are deadly, but over time, they often debilitate a plant.

Viruses are passed from one plant to another by insects, and by gardening tools, which move sap from one plant to another. Viral particles are also transmitted by way of seeds and pollen. Unfortunately, no chemical can control viral diseases. Until recently, the only way to control viral diseases was to destroy symptomatic plants and to control the insect vector, if there is one. Now that bioengineering is routine, it is possible to transfer genes conferring disease resistance between plants. One of the most successful examples to date is the creation of papaya plants resistant to papaya ring spot virus (PRSV) in Hawaii. One transgenic line is now completely resistant to PRSV.

Animal Viruses

Viruses that cause diseases in animals, including humans, reproduce in a manner similar to that of bacteriophages. However, there are modifications. In particular, some, but not all, animal viruses have an outer membranous envelope beyond their capsid. After attachment to a receptor in the plasma membrane, viruses with an envelope either fuse with the plasma membrane or enter by endocytosis. After an enveloped virus enters, uncoating follows—that is, the capsid is removed. Once the viral genome, either DNA or RNA, is free of its covering, biosynthesis plus the other steps then proceed, as described in Figure 17.5. While a naked animal virus exits the host cell in the same manner as does a bacteriophage, those with an envelope bud from the cell. During budding, the virus picks up its envelope, consisting mainly of lipids and proteins, from the host plasma membrane. Spikes, those portions of the envelope that allow the virus to enter a host cell, are coded for by viral genes.

Retroviruses

Retroviruses are RNA animal viruses that have a DNA stage. Figure 17.5 illustrates the reproduction of HIV, the retrovirus that causes AIDS. A retrovirus contains a special enzyme called reverse transcriptase, which carries out transcription of RNA to cDNA. The DNA is called cDNA because it is a DNA copy of the viral genome. Following replication of the single strand, the resulting double-stranded DNA is integrated into the host genome. The viral DNA remains in the host genome and is replicated when

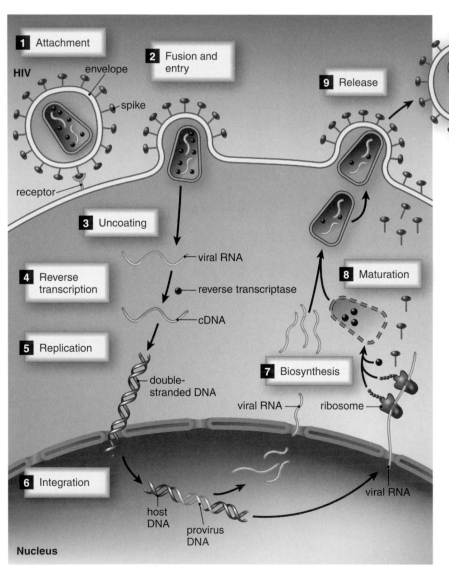

Figure 17.5 **Reproduction of HIV.**

HIV, the virus that causes AIDS, goes through the steps noted in the boxes. Because HIV is a retrovirus with an RNA genome, the enzyme reverse transcriptase is utilized to produce a DNA copy of the genome, called cDNA. After cDNA undergoes replication, DNA integrates its genome into host DNA.

host DNA is replicated. When and if this DNA is transcribed, new viruses are produced by the steps we have already cited: biosynthesis, maturation, and release. Being an animal virus with an envelope, HIV buds from the cell.

Emerging Viruses

HIV is an **emerging virus,** the causative agent of a disease that only recently has arisen and infected people. Other examples of emerging viruses are West Nile virus, SARS virus, hantavirus, Ebola virus, and avian influenza (bird flu) virus (Fig. 17.6).

Infectious diseases emerge in several different ways. In some cases, the virus is simply transported from one location to another. The West Nile virus is making headlines because it changed its range: It was transported into the United States and is taking hold in bird and mosquito populations. Severe acute respiratory syndrome (SARS) was clearly transported from Asia to Toronto, Canada. A world in which you can begin your day in Bangkok and end it in Los Angeles is a world in which disease can spread at an unprecedented rate.

Other factors can also cause infectious viruses to emerge. Viruses are well known for their high mutation rates. Some of these mutations affect the structure of the spikes, so a virus that previously could only infect a particular animal species can now also infect the human species. For example, the diseases AIDS and Ebola fever are caused by viruses that at one time infected only monkeys and apes. SARS is another example of a mutant virus that most likely jumped species. A related virus was isolated from the palm civet, a catlike carnivore sold for food in China. Wild ducks are resistant to avian influenza viruses that can spread from them to chickens, which increases the likelihood the disease, often called bird flu, will spread to humans. Another possible way a virus could emerge is by a change in the mode of transmission. What would happen if a little-known virus were suddenly able to be transmitted like the common cold?

Drug Control of Human Viral Diseases

Because viruses reproduce using the metabolic machinery of the cell, it has been difficult to develop antiviral drugs. If a patient has not been vaccinated against the agent, he or she is usually told to simply let the virus run its course. However, some antiviral drugs are available. Most antiviral compounds, such as ribavirin and acyclovir, are structurally similar to nucleotides, and therefore they interfere with viral genome synthesis. Compounds related to acyclovir are commonly used to suppress herpes outbreaks. HIV is treated with antiviral compounds specific to a retrovirus. The well-publicized drug AZT and others block reverse transcriptase. And HIV protease inhibitors block enzymes required for the maturation of viral proteins.

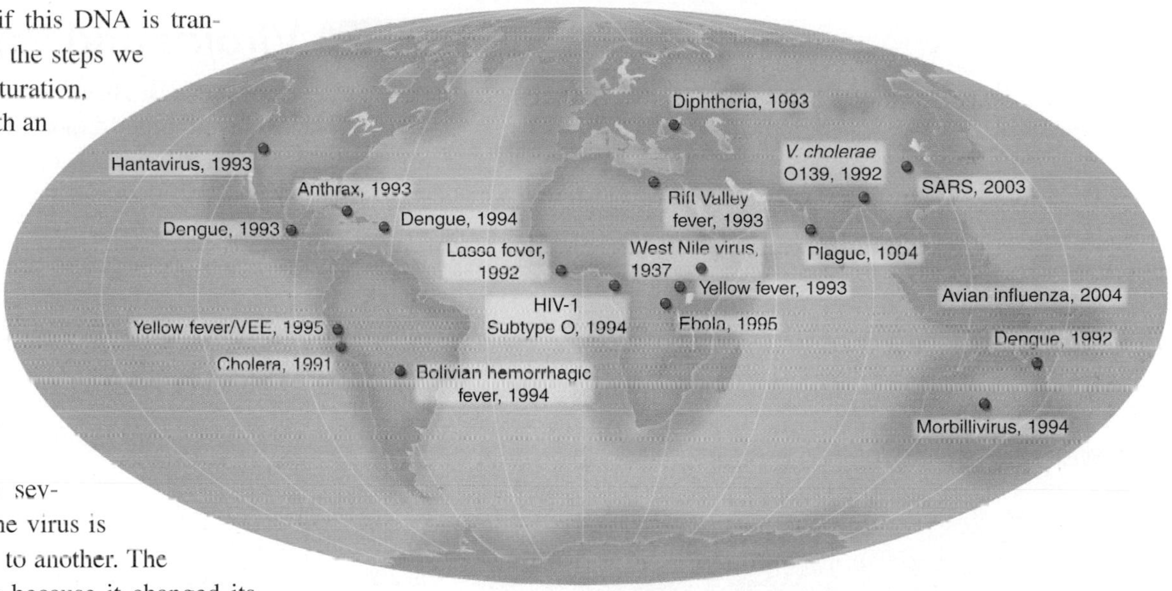

Figure 17.6 **Emerging diseases.**

Emerging diseases, such as those noted here according to their country of origin, are new or demonstrate increased prevalence. These disease-causing agents may have acquired new virulence factors, or environmental factors may have encouraged their spread to an increased number of hosts. Avian influenza (bird flu) is a new emerging disease.

Check Your Progress

1. All viruses contain what two components?
2. Contrast a lytic with a lysogenic viral life cycle.
3. List ways in which viruses can move among plants.
4. What happens when animal viruses bud from the host cell?
5. Describe the significance of reverse transcriptase for retroviruses.

Answers: 1. All viruses are composed of an outer capsid made of protein and an inner core of DNA or RNA. 2. In the lytic cycle, the bacterium is used to immediately produce more virus particles. In the lysogenic cycle, the phage is latent, but it may at some point be stimulated to enter the lytic cycle. 3. Insect vectors, gardening tools, seeds, and pollen. 4. They acquire an outer envelope, which in part consists of the host plasma membrane. 5. Retroviruses have an RNA genome. They use the enzyme reverse transcriptase to make a DNA copy of their genome. When this replicates, the double helix can integrate into the host cell genome and become latent.

Figure 17.7   Kuru.

Kuru is a fatal prion disease that infected the Fore (pronounced for-ay), a tribe of the remote highlands of Papua New Guinea, prior to the 1950s. Following their death, family members were ritualistically cooked and eaten, with the closest female relatives and children usually consuming the brain, the organ most likely to pass on kuru. A misshapen protein is the causative agent of all prion diseases, whether they occur in animals (e.g., mad cow disease) or in humans (e.g., kuru, CJD). The inset shows a diseased monkey brain.

17.2   Viroids and Prions

About a dozen crop diseases have been attributed not to viruses but to **viroids,** which are naked strands of RNA (not covered by a capsid). Like viruses, though, viroids direct the cell to produce more viroids.

Some diseases in humans have been attributed to **prions,** a term coined for proteinaceous infectious particles. The discovery of prions began when it was observed that members of a primitive tribe in the highlands of Papua New Guinea died from a disease called kuru (meaning trembling with fear) after participating in the cannibalistic practice of eating a deceased person's brain (Fig. 17.7). The causative agent was smaller than a virus—it was a rogue protein. It appears that a normal protein changes shape so that its polypeptide chain is in a different configuration. The result is a fatal prion and a neurodegenerative disorder.

It is believed that a prion can interact with normal proteins to turn them over to the "dark side," but the mechanism is unclear. The process has been best studied in a disease called scrapie, which attacks sheep. Other prion diseases include the popularized mad cow disease, human maladies such as Creutzfeldt-Jakob syndrome (CJD), and a variety of chronic wasting syndromes in animals.

Check Your Progress

1. **Contrast a viroid with a prion.**
2. **Prions cause disease. Why then can proteins that become prions be found in healthy brains?**

Answers: 1. A viroid is composed of naked RNA, while a prion is composed of protein. 2. Prions are derived from normal proteins that have become prions due to a change in shape.

17.3   The Prokaryotes

What was the first cellular life on Earth like? What features in modern organisms are the most primitive? The answers to these questions will probably come from a study of prokaryotes. These cells harbor no nucleus to contain their genome. They have no sealed compartments and no membrane-bounded organelles to perform specific functions. Yet these structurally simple, unicellular organisms are endowed with far greater metabolic capabilities than more structurally complex organisms; some of them need neither air nor organic matter to survive. The two types of **prokaryotes** are the bacteria and the archaea.

Bacteria

Bacteria are the most diverse and prevalent organisms on Earth. Billions of bacteria exist in nearly every square meter of soil, water, and air. They also make a home on your skin and in your intestines. Although tens of thousands of different bacteria have been identified, this is likely only a very small fraction of living bacteria. Less than 1% of bacteria in the soil can be grown on agar plates in the laboratory. Molecular genetic techniques are now being used to discover the extent of bacterial diversity.

General Biology of Bacteria

Bacterial Structure Bacteria have a number of different shapes; however, most bacteria are spheres, called cocci, or rods, called bacilli, or spirals, called spirilla (Fig. 17.8). Intermediate forms between a sphere and rod are coccobacilli. A spirillum can be slightly curved (called a vibrio) or highly coiled (called a spirochete). Many bacteria grow as single cells, but some form doublets (the diplococci and diplobacilli). Others form filaments (chains) as do the streptococci, the cause of strep throat. A third growth habit resembles a bunch of grapes, as in the staphylococci, a cause of food poisoning.

Figure 4.4 on page 51 reviews the structure of a bacterium. A bacterium, being a prokaryote, has no nucleus. A single, closed circle of double-stranded DNA constitutes the chromosome, which occurs in a limited area of the cell called the **nucleoid.** In some cases, extrachromosomal DNA molecules called **plasmids** are also found in bacteria.

Bacteria have ribosomes but not membrane-bounded organelles such as mitochondria and chloroplasts. Those that are photosynthetic have thylakoid membranes, but these are not enclosed by another membrane. Motile bacteria generally use **flagella** for locomotion, but never cilia. The bacterial flagellum is not structured like a eukaryotic flagellum (Fig. 17.9). It has a filament composed of three strands of the protein flagellin wound in a helix. The filament is inserted into a hook that is anchored by a basal body. The 360° rotation causes the bacterium to spin as it moves forward and backward.

Bacteria have an outer cell wall strengthened not by cellulose, but by **peptidoglycan,** a complex molecule containing a unique amino disaccharide and peptide fragments. The cell wall prevents bacteria from bursting or collapsing due to osmotic changes. Parasitic bacteria are further protected from host defenses by a polysaccharide capsule that surrounds the cell wall.

Bacterial Reproduction Bacteria (and Archaea) reproduce asexually by means of **binary fission.** The single, circular chromosome replicates, and then the two copies separate as the cell enlarges. The newly formed plasma membrane and cell wall partition the two new cells, with a chromosome in

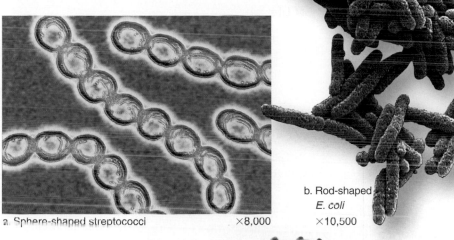

b. Rod-shaped
E. coli

a. Sphere-shaped streptococci ×8,000 ×10,500

c. Spirochete, *T. pallidum*

Figure 17.8 Shapes of bacteria.

a. Streptococci, which exist as chains of cocci, cause a number of illnesses, including strep throat. **b.** *Escherichia coli*, which lives in your intestine, is a rod-shaped bacillus. **c.** *Treponema pallidum*, the cause of syphilis, is a spirochete.

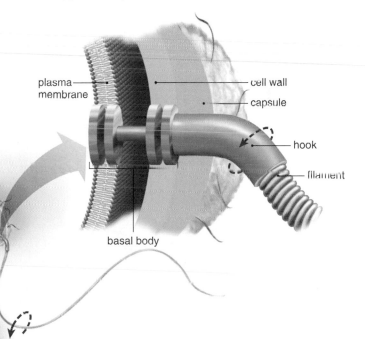

plasma membrane — cell wall

capsule

hook

filament

basal body

Figure 17.9 Flagella.

Each flagellum of a bacterium contains a basal body, a hook, and a filament. Arrows here indicate that the hook (and filament) rotates 360°. Flagella rotating counterclockwise result in forward motion of the bacterium; a clockwise rotation causes the bacterium to move backward.

Check Your Progress

1. List the two types of prokaryotes.
2. List the three shapes of bacteria.
3. Describe the chromosome of a bacterium.

Answers: 1. Bacteria and archaea. 2. Spherical (cocci), rod (bacilli), and spiral (spirillum). 3. A single, closed circle of double-stranded DNA.

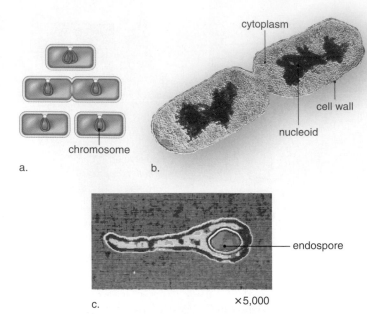

a.

b.

cytoplasm

cell wall

nucleoid

chromosome

c.
×5,000

endospore

Figure 17.10 **Bacterial reproduction and survival.**

a, b. When conditions are favorable to growth, prokaryotes divide to multiply. **c.** Formation of endospores allows bacteria to survive unfavorable environmental conditions.

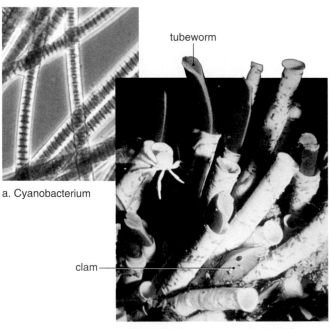

a. Cyanobacterium

tubeworm

clam

b. Hosts for chemoautotrophic bacteria

Figure 17.11 **Bacterial autotrophs.**

a. Cyanobacteria are photoautotrophs that photosynthesize in the same manner as plants—they split water and release oxygen.
b. Certain chemoautotrophic bacteria live inside tubeworms, where they produce organic compounds without the need of sunlight. In this way they help support ecosystems at hydrothermal vents deep in the ocean.

each one (Fig. 17.10*a,b*). Mitosis, which requires the formation of a spindle apparatus, does not occur in prokaryotes. Binary fission turns one cell into two cells, two cells into four cells, four cells into eight cells, and continuing on until billions of cells have been produced.

In eukaryotes, genetic recombination occurs as a result of sexual reproduction. Sexual reproduction does not occur among prokaryotes, but three means of genetic recombination have been observed in bacteria. **Conjugation** occurs between bacteria when a donor cell passes DNA to a recipient cell by way of tubes, called *sex pili,* which temporarily join the two cells. Conjugation takes place only between bacteria in the same or closely related species. **Transformation** occurs when a bacterium picks up (from the surroundings) free pieces of DNA secreted by live prokaryotes or released by dead prokaryotes. During **transduction,** bacteriophages carry portions of bacterial DNA from one cell to another. Plasmids, which sometimes carry genes for resistance to antibiotics, can be transferred between infectious bacteria by any of these means.

When faced with unfavorable environmental conditions, some bacteria form **endospores** (Fig. 17.10*c*). A portion of the cytoplasm and a copy of the chromosome dehydrate and are then encased by three heavy, protective spore coats. The rest of the bacterial cell deteriorates, and the endospore is released. Spores survive in the harshest of environments—desert heat and desiccation, boiling temperatures, polar ice, and extreme ultraviolet radiation. They also survive for very long periods. When anthrax spores 1,300 years old germinate, they can still cause a severe infection (usually seen in cattle and sheep). Humans also fear a deadly but uncommon type of food poisoning called botulism, which is caused by the germination of endospores inside cans of food. Spore formation is not a means of reproduction, but it does allow survival and dispersal of bacteria to new places.

Bacterial Nutrition Eukaryotes are limited in their metabolic capabilities. For example, plants are **photoautotrophs** (many times called autotrophs or photosynthesizers) that can only perform oxygenic photosynthesis: They depend on solar energy to split water and energize electrons for the reduction of carbon dioxide. Among bacteria, the **cyanobacteria** are also photoautotrophs and do the same. The cyanobacteria may well represent the oldest lineage of oxygenic organisms (Fig. 17.11*a*). Some fossil cyanobacteria have been dated at 3.7 billion years old. Many cyanobacteria are capable of fixing atmospheric nitrogen and reducing it to an organic form. Therefore, they need only minerals, air, sunlight, and water for growth.

Other bacterial photosynthesizers don't release oxygen because they take electrons from a source other than water; some split hydrogen sulfide (H_2S) and release sulfur (S) in marshes where they live anaerobically.

The **chemoautotrophs** (many times called chemosynthesizers) don't use solar energy at all. They reduce carbon dioxide using energetic electrons derived from inorganic molecules such as ammonia or hydrogen gas. Electrons can also be extracted from certain minerals, such as iron. Some chemoautotrophs oxidize sulfur compounds spewing from deep-sea vents 2.5 kilometers below sea level. The organic compounds they produce support the growth of communities of organisms found at vents, where only darkness prevails (Fig. 17.11*b*).

Most bacteria are **chemoheterotrophs** (often referred to as simply heterotrophs) and, like animals, take in organic nutrients that they use as a source of energy and building blocks to synthesize macromolecules. Unlike animals, they are **saprotrophs** that send enzymes into the environment and decompose almost any large organic molecule to smaller ones that are absorbable. There is probably no natural organic molecule that cannot be digested by at least one bacterial species. Bacteria play a critical role in recycling matter and making inorganic molecules available to photosynthesizers.

Heterotrophic bacteria may be either free-living or **symbiotic,** meaning that they form (1) mutualistic (both partners benefit), (2) commensalistic (one partner benefits, the other is not harmed), or (3) parasitic (one partner benefits, the other is harmed) relationships. Mutualistic bacteria that live in human intestines release vitamins K and B_{12}, which we can use to help produce blood components. In the stomachs of cows and goats, special mutualistic prokaryotes digest cellulose, enabling these animals to feed on grass.

Commensalism often occurs when one population modifies the environment in such a way that a second population benefits. Obligate anaerobes can live in our intestines only because the bacterium *Escherichia coli* uses up the available oxygen. The parasitic bacteria cause diseases, as discussed on the next page.

Environmental and Medical Importance of Bacteria

Bacteria in the Environment   For an ecosystem to sustain its populations, the chemical elements available to living things must eventually be recycled. A fixed and limited amount of elements are available to living things—the rest are either buried too deep in the Earth's crust or present in forms that are not usable. All living organisms, including producers, consumers, and decomposers, are involved in the important process of cycling elements to sustain life (see Fig. 1.4). *Bacteria are decomposers* that digest dead organic remains and return inorganic nutrients to producers. Without the work of decomposers, life would soon come to a halt.

While decomposing, bacteria perform reactions needed for biogeochemical cycling, such as for the carbon and nitrogen cycles. Let's examine how bacteria participate in the nitrogen cycle. Plants are unable to fix atmospheric nitrogen (N_2), but they need a source of ammonia or nitrate in order to produce proteins. Bacteria in the soil can fix atmospheric nitrogen and/or change nitrogen compounds into forms that plants can use. In addition, mutualistic bacteria live in the root nodules of soybean, clover, and alfalfa plants where they reduce atmospheric nitrogen to ammonia, which is used by plants (Fig. 17.12). Without the work of bacteria, nitrogen would not be available for plants to produce proteins, nor available to animals that feed on plants or other animals.

Bioremediation is the biological cleanup of an environment that contains harmful chemicals called *pollutants*. To help correct the situation, the vast ability of bacteria to break down almost any substance, including sewage, is being exploited (Fig. 17.13a). People have added thousands of tons of slowly degradable pesticides and herbicides, nonbiodegradable detergents, and plastics to the environment. Strains of bacteria are being developed specifically for cleaning up these particular types of pollutants. Some strains have been used to remove Agent Orange, a potent herbicide, from soil samples, and dual cultures of two types of bacteria have been shown to degrade PCBs, chemicals formerly used as coolants and lubricants. The occasional oil spill spoils beaches and kills wildlife. The ability of bacteria to degrade petroleum has been improved by biotechnology and the addition of a growth-promoting fertilizer (Fig. 17.13b). Without the fertilizer, the lack of nitrogen and phosphate in seawater limits their growth.

Figure 17.12   **Nodules of a legume.**

Although some free-living bacteria carry on nitrogen fixation, those of the genus *Rhizobium* invade the roots of legumes, with the resultant formation of nodules. Here the bacteria convert atmospheric nitrogen to an organic nitrogen that the plant can use. These are nodules on the roots of a soybean plant.

a.

b.

after   before

Figure 17.13   **Bioremediation.**

a. Bacteria have been used for many years in sewage treatment plants to break down human wastes. **b.** Increasingly, the ability of bacteria to break down pollutants is being researched and enhanced. Here technicians are relying on bacteria to clean up an oil spill.

Figure 17.14 **Bacteria in food processing.**

Bacteria and fungi (see Chapter 18) are used to help produce food products. Fungal fermentation results in carbon dioxide and alcohol, which help produce bread and wine, respectfully. Bacterial fermentation results in acids that give some types of cheeses their characteristic taste.

Bacteria in Food Science and Biotechnology   A wide variety of food products are created through the action of bacteria (Fig. 17.14). Under anaerobic conditions, bacteria carry out fermentation, which results in a variety of alcohols and acids. One of these acids is lactate, a product that pickles cucumbers, curdles milk into cheese, and gives these foods their characteristic tangy flavor. Other bacterial fermentations can produce flavor compounds, such as the propionic acid in Swiss cheese. Bacterial fermentation is also useful in the manufacture of such products as vitamins and antibiotics—in fact, most antibiotics known today were discovered in soil bacteria.

As you know, biotechnology can be used to alter the genome and the products generated by bacterial cultures. Bacteria can be genetically engineered to produce medically important products such as insulin, human growth hormone, and vaccines against a number of human diseases. The natural ability of bacteria to perform all manner of reactions is also enhanced through biotechnology.

Bacterial Diseases in Humans   Microbes that can cause disease are called **pathogens.** Pathogens are able to (1) produce a toxin, and/or (2) adhere to surfaces, and sometimes (3) invade organs or cells.

Toxins are small organic molecules, or small pieces of protein or parts of the bacterial cell wall that are released when bacteria die. Toxins are poisonous, and bacteria that produce a toxin usually cause serious diseases. In almost all cases, the growth of the microbes themselves does not cause disease; the toxins they release cause disease. When someone steps on a rusty nail, bacteria may be introduced deep into damaged tissue. The damaged area does not have good blood flow and can become anaerobic. *Clostridium tetani*, the cause of tetanus, proliferates under these conditions. The bacteria never leave the site of the wound, but the tetanus toxin they produce does move throughout the body. This toxin prevents the relaxation of muscles. In time, the body contorts because all the muscles have contracted. Eventually, the person suffocates.

Adhesion factors allow a pathogen to bind to certain cells, and this determines which organs or cells of the body will be its host. *Shigella dysenteriae* produces a toxin, but it is also able to stick to the intestinal wall, which makes it a more life-threatening cause of dysentery. Also, invasive mechanisms that give a pathogen the ability to move through tissues and into the bloodstream result in a more medically significant disease than if it were localized. Usually a person can recover from food poisoning caused by *Salmonella*. But some strains of *Salmonella* have virulence factors that allow the bacteria to penetrate the lining of the colon and move beyond this organ. Typhoid fever, a life-threatening disease, can then result.

Because bacteria are cells in their own right, a number of antibiotic compounds are active against bacteria and are widely prescribed. Most antibacterial compounds fall within two classes: those that inhibit protein biosynthesis and those that inhibit cell wall biosynthesis. Erythromycin and tetracyclines can inhibit bacterial protein biosynthesis because bacterial ribosomes function somewhat differently than eukaryotic ribosomes. Cell wall biosynthesis inhibitors generally block the formation of peptidoglycan, the substance found in the bacterial cell wall. Penicillin, ampicillin, and fluoroquinolone (such as Cipro) inhibit bacterial cell wall biosynthesis without harming animal cells.

Check Your Progress

1. Explain how bacteria reproduce asexually.
2. List the three ways in which genetic recombination can occur in bacteria.
3. Describe the function of endospores in bacteria.
4. Contrast photoautotrophs with chemoautotrophs.
5. Describe a significant role of bacteria in the nitrogen cycle that is beneficial to plants and animals.
6. Describe the modes of action of the two major classes of antibiotics.

Answers: 1. The single chromosome copies itself, and the two copies are pulled apart as the cell enlarges. The cell is then partitioned into two cells, with a chromosome in each one. 2. Conjugation, transformation, and transduction. 3. Endospores allow bacteria to survive harsh conditions and disperse to new places. 4. Photoautotrophs use energy from the sun to produce sugars, while chemoautotrophs use high-energy electrons from chemicals. 5. Bacteria in nodules and the soil change nitrogen to a form plants can use. Animals acquire organic nitrogen from plants or from animals that have fed on plants. 6. One class inhibits cell wall biosynthesis, while the other inhibits protein biosynthesis.

One problem with antibiotic therapy has been increasing bacterial resistance to antibiotics. When penicillin was first introduced, less than 3% of *Staphylococcus aureus* strains were resistant to it. Now, due to natural selection, 90% or more are resistant.

Archaea

As previously discussed in Chapter 16, scientists currently propose that the tree of life contains three domains: Archaea, Bacteria, and Eukarya. Because archaea and some bacteria are found in extreme environments (hot springs, thermal vents, salt basins), they may have diverged from a common ancestor relatively soon after life began. Later, the eukarya are believed to have split off from the archaeal line of descent. Archaea and eukarya share some of the same ribosomal proteins (not found in bacteria), initiate transcription in the same manner, and have similar types of tRNA.

Structure and Function

The plasma membranes of archaea contain unusual lipids that allow them to function at high temperatures. The archaea have also evolved diverse cell wall types, which facilitate their survival under extreme conditions. The cell walls of archaea do not contain peptidoglycan as do the cell walls of bacteria. In some archaea, the cell wall is largely composed of polysaccharides, and in others, the wall is pure protein. A few have no cell wall.

Archaea have retained primitive and unique forms of metabolism. The ability to form methane, is one type of metabolism that is performed only by some archaea, called methanogens. Most archaea are chemoautotrophs (see page 272), and few are photosynthetic. This suggests that chemoautotrophy predated photoautotrophy during the evolution of prokaryotes. Archaea are sometimes mutualistic or even commensalistic, but none are parasitic—that is, archaea are not known to cause infectious diseases.

Types of Archaea

Archaea are often discussed in terms of their unique habitats. The **methanogens** (methane makers) are found in anaerobic environments, such as in swamps, marshes, and the intestinal tracts of animals. They couple the production of methane (CH_4) from hydrogen gas (H_2) and carbon dioxide to the formation of ATP (Fig. 17.15). This methane, which is also called *biogas,* is released into the atmosphere where it contributes to the greenhouse effect and global warming. About 65% of the methane in our atmosphere is produced by methanogenic archaea.

The **halophiles** require high salt concentrations for growth (usually 12–15%; by contrast, the ocean is about 3.5% salt). Halophiles have been isolated from highly saline environments such as the Great Salt Lake in Utah, the Dead Sea, solar salt ponds, and hypersaline soils (Fig. 17.16). These archaea have evolved a number of mechanisms to survive in high-salt environments. They depend on a pigment related to the rhodopsin in our eyes to absorb light energy for the purpose of pumping chloride and another similar type pigment for the purpose of synthesizing ATP.

A third major type of archaea are the **thermoacidophiles** (Fig. 17.17). These archaea are isolated from extremely hot, acidic environments such as hot springs, geysers, submarine thermal vents, and around volcanoes. They reduce sulfur to sulfides and survive best at temperatures above 80°C; some can even grow at 105°C (remember that water boils at 100°C)! Metabolism of sulfides results in acidic sulfates and these bacteria grow best at pH 1 to 2.

Figure 17.15   Methanogen habitat and structure.

a. A swamp where methanogens live. **b.** Micrograph of *Methanosarcina mazei,* a methanogen.

Figure 17.16   Halophile habitat and structure.

a. Great Salt Lake, Utah, where halophiles live. **b.** Micrograph of *Halobacterium salinarium,* a halophile.

Figure 17.17   Thermoacidophile habitat and structure.

a. Boiling springs and geysers in Yellowstone National Park where thermoacidophiles live. **b.** Micrograph of *Sulfolobus acidocaldarius,* a thermoacidophile. (The dark central region has yet to be identified and could be an artifact of staining.)

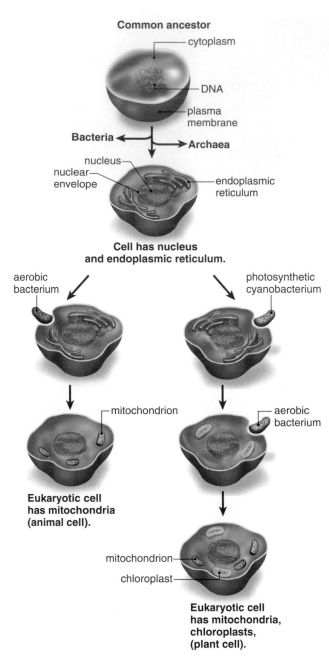

Figure 17.18 Evolution of the eukaryotic cell.

Invagination of the plasma membrane could account for the formation of the nucleus and certain other organelles. The endosymbiotic hypothesis suggests that mitochondria and chloroplasts are derived from prokaryotes that were taken up by a much larger eukaryotic cell.

17.4 The Protists

Like ancient creatures from another planet, the **protists** inhabit the oceans and other watery environments of the world. Their morphological diversity is their most outstanding feature—unicellular diatoms are encrusted in silica "hatboxes"; dinoflagellates have plates of armor; and ciliates shaped like slippers have complex structures.

Many protists are unicellular, but all are eukaryotes with a nucleus and a wide range of organelles. It is widely held that the organelles of eukaryotic cells arose from close symbiotic associations between bacteria and primitive eukaryotes (Fig. 17.18). This so-called *endosymbiotic hypothesis* is supported by the presence of double membranes around mitochondria and chloroplasts. Also, these organelles have their own genomes, although incomplete, and their ribosomal genes point to bacterial origins. The mitochondria appear closely related to certain bacteria, and the chloroplasts are most closely related to cyanobacteria.

To explain the diversity of protists, we can well imagine that once the eukaryotic cell arose, it provided the opportunity for many different lineages to begin. Some unicellular protists have organelles not seen in other eukaryotes. For example, food is digested in food vacuoles, and excess water is expelled when contractile vacuoles discharge their contents.

Protists also possibly give us insight into the evolution of a multicellular organism with differentiated tissues. Some protists are a colony of single cells, with certain cells specialized to produce eggs and sperm, and others are multicellular, with tissues specialized for various purposes. Perhaps the first type of organization preceded the second in a progression toward multicellular organisms.

General Biology of Protists

The complexity and diversity of protists make it difficult to classify them. The variety of protists is so great that it's been suggested they could be split into more than a dozen kingdoms. Due to limited space, this text groups the phyla according to modes of nutrition.

Traditionally, the term **algae** means aquatic photosynthesizer. At one time, botanists classified algae as plants because they contain chlorophyll *a* and carry on photosynthesis. In aquatic environments, algae are a part of the phytoplankton, photosynthesizers that lie suspended in the water. They are producers, which serve as a source of food for other organisms and pour oxygen into the environment. In terrestrial systems, algae are found in soils, on rocks, and in trees. One type of alga is a symbiote of animals called corals, which depend on them for food as they build the coral reefs of the world. Others partner with fungi in lichens capable of living in harsh terrestrial environments.

The definition of a **protozoan** as a unicellular chemoheterotroph explains why protozoans were originally classified with the animals. Often a protozoan has some form of locomotion, either by flagella, pseudopods, or cilia. In aquatic environments, protozoans are a part of the zooplankton, suspended microscopic heterotrophs that serve as a food source for animals. While most are free-living, some protozoans are human pathogens, often causing diseases of the blood. In many cases, their complex life cycles inhibit the development of suitable treatments.

There is still one other group of protists: the slime molds and water molds. These creatures are chemoheterotrophs, but the slime molds ingest their food in the same manner as the protozoan called an amoeba, while the water molds are saprotrophic, like fungi.

Check Your Progress

List the features of mitochondria and chloroplasts that support the endosymbiotic hypothesis.

Answer: Double membrane, the presence of DNA, and ribosomal genes similar to those of bacteria.

Algae

The green alga *Chlamydomonas* serves as our model for algal structure (Fig. 17.19). The most conspicuous organelle in the algal cell is the chloroplast. Algal chloroplasts share many features with those of plants, and the two groups likely share a common origin; for example, the photosynthetic pigments are housed in thylakoid membranes. Not surprisingly, then, algae perform photosynthesis in the same manner as plants. Pyrenoids are organelles found in algae that are active in starch storage and metabolism. Vacuoles are seen in algae, along with mitochondria. Algae generally have a cell wall, and many produce a slime layer that can be harvested and used for food processing. Some algae are nonmotile, while others possess flagella.

Algae can reproduce asexually or sexually in most cases. Asexual reproduction can occur by binary fission, as in bacteria. Some proliferate by forming flagellated spores called zoospores, while others simply fragment, with each fragment becoming a progeny alga. Sexual reproduction generally requires the formation of gametes that combine to form a zygote.

One traditional way to classify algae is based on the color of the pigments in their chloroplasts: **green algae, red algae, golden-brown algae,** and **brown algae** (Fig. 17.20). The green algae are most closely related to plants, and they are commonly represented by three species: *Chlamydomonas;* colonial *Volvox,* a large, hollow sphere with dozens to hundreds of cells; and *Spirogyra,* a filamentous alga in which the chloroplasts form a green spiral ribbon. The coralline red algae deposit calcium carbonate in their cell walls and contribute to the formation of coral reefs. The golden-brown algae are represented by the diatoms, which have a hatbox structure with each half an elaborate shell composed of silica. The green algae, red algae, and brown algae include the multicellular seaweeds.

The Protozoans

Protozoans are unicellular, but their cells are very complex. Many of the functions we normally associate with the organs of multicellular organisms are performed by organelles in

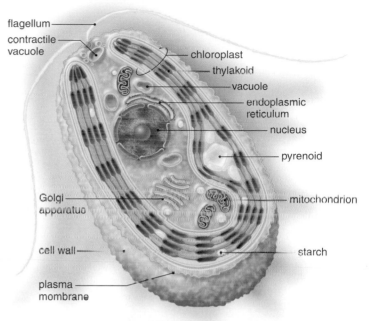

Figure 17.19   *Chlamydomonas.*

Chlamydomonas, a green alga, has the organelles and other possible structures typical of a motile algal cell.

Figure 17.20   **Algal diversity.**

a. *Acetabularia,* a unicellular green alga; **b.** *Bossiella,* a coralline red alga; **c.** *Fucus,* a brown alga; **d.** *Licmorpha,* a stalked diatom; **e.** *Chlamydomonas,* a unicellular green alga; **f.** *Volvox,* a colonial green alga with daughter colonies inside; **g.** *Spirogyra,* a filamentous green alga undergoing conjugation to form zygotes; **h.** *Ceratium,* an armored dinoflagellate; **i.** *Macrocystis,* a brown alga; **j.** *Sargassum,* a brown alga.

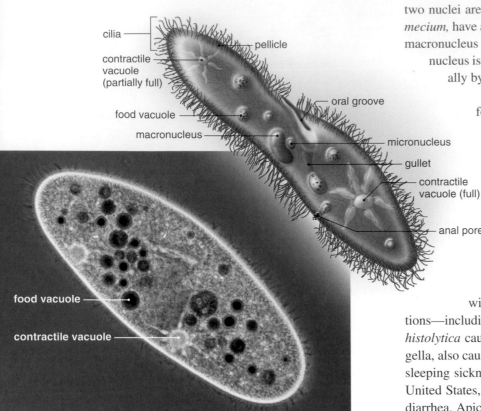

protozoans. Some protozoans have more than one nucleus. In some cases, the two nuclei are identical in size and function. Other protozoans, such as *Paramecium,* have a large macronucleus and a small micronucleus (Fig. 17.21). The macronucleus produces mRNA and directs metabolic functions. The micronucleus is important for reproduction. Protozoans usually reproduce asexually by binary fission.

Protozoans are heterotrophic, and many feed by engulfing food particles. Phagocytic vacuoles act as their "stomachs" into which digestive enzymes and acid are added. Secretory vacuoles release enzymes that may enhance any pathogenicity. Contractile vacuoles permit osmoregulation, particularly in fresh water.

The various protozoans shown in Figure 17.22 illustrate that they are usually motile. The **ciliates,** so named because they move by cilia, are represented by *Paramecium,* and the **amoeboids** by amoebas, which move by cytoplasmic extensions called **pseudopods.** The **radiolarians** and **foraminiferans** are two types of marine amoeboids with calcium carbonate skeletons important in limestone formations—including the White Cliffs of Dover in Britain. The amoeba *Entamoeba histolytica* causes amoebic dysentery. The **zooflagellates,** which move by flagella, also cause diseases. In this group, a **trypanosome** is the cause of African sleeping sickness, a blood disease that cuts off circulation to the brain. In the United States, *Giardia lamblia* contaminates water supplies and causes severe diarrhea. Apicomplexans, commonly called **sporozoans** because they produce spores, are unlike other protozoan groups because they are not motile. One genus, *Plasmodium,* causes malaria, the most widespread and dangerous protozoan disease. Malaria is transmitted by a mosquito. Toxoplasmosis, another protozoan disease, is commonly transmitted by cats.

Figure 17.21 **A paramecium.**

A paramecium is a ciliate, a type of complex protozoan that moves by cilia.

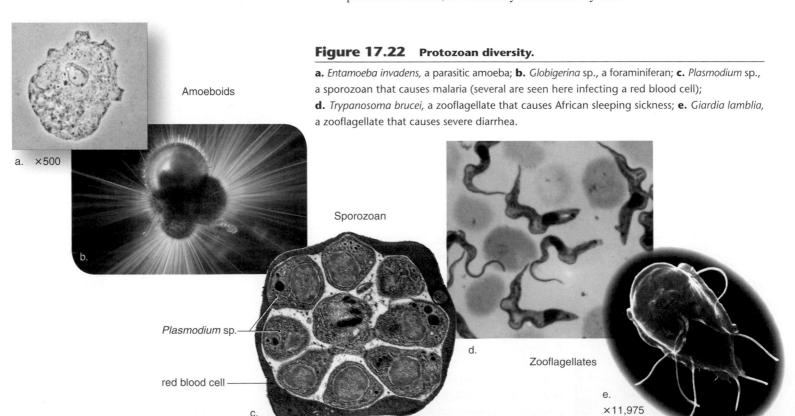

Figure 17.22 **Protozoan diversity.**

a. *Entamoeba invadens,* a parasitic amoeba; **b.** *Globigerina* sp., a foraminiferan; **c.** *Plasmodium* sp., a sporozoan that causes malaria (several are seen here infecting a red blood cell); **d.** *Trypanosoma brucei,* a zooflagellate that causes African sleeping sickness; **e.** *Giardia lamblia,* a zooflagellate that causes severe diarrhea.

Slime Molds and Water Molds

In forests and woodlands, **slime molds** feed on, and therefore help dispose of, dead plant material. They also feed on bacteria, keeping their population sizes under control. **Water molds** decompose remains but are also significant parasites of plants and animals in ecosystems. Slime molds and water molds were once classified as fungi, but unlike fungi, all have flagellated cells at some time during their life cycles. Only the water molds have a cell wall, but it contains cellulose, not the chitin of fungal cell walls. Both slime molds and water molds form spores, each a small, single-celled reproductive body capable of becoming a new organism. The spores of water molds are flagellated, but those of slime molds are windblown.

Although grouped together here, slime molds and water molds may be more closely related to the amoeboids than to each other. Indeed, the vegetative state of the slime molds is mobile and amoeboid, and like the amoeboids, they ingest their food by phagocytosis.

Plasmodial Slime Molds   Usually, plasmodial slime molds exist as a plasmodium—a diploid, multinucleated, cytoplasmic mass enveloped by a slime sheath that creeps along, phagocytizing decaying plant material in a forest or agricultural field. At times unfavorable to growth, such as during a drought, the plasmodium develops many sporangia. A sporangium is a reproductive structure that produces spores resistant to dry conditions. When favorable moist conditions return, the spores germinate, releasing a flagellated cell or an amoeboid cell. Eventually, two of them fuse to form a zygote that feeds and grows, producing a multinucleated plasmodium once again. Figure 17.23 shows the life cycle of a plasmodial slime mold.

Check Your Progress

1. Compare and contrast algae with protozoans.
2. List the major features of protozoans.
3. Describe the significance of slime molds in terrestrial ecosystems.

Answers: 1. Both algae and protozoans are protists, but algae are autotrophs, while protozoans are heterotrophs. 2. Protozoans are unicellular, may have more than one nucleus, reproduce asexually via binary fission, and are phagocytes. 3. Slime molds feed on bacteria and decaying plant material.

Plasmodium phagocytizes food

Sporangia produce spores

young plasmodium

mature plasmodium

sporangia formation begins

zygote

FERTILIZATION

diploid (2n)

haploid (n)

young sporangium

MEIOSIS

fusion

amoeboid cells

flagellated cells

germinating spore

spores

mature sporangium

Figure 17.23   Plasmodial slime molds.

As long as conditions are favorable, plasmodial slime molds exist as multinucleated diploid plasmodium that creeps along the forest floor phagocytizing organic remains. When conditions become unfavorable, sporangia form where meiosis produces haploid spores, structures that can survive unfavorable times. Spores germinate to release independent haploid cells. Fusion of these cells produces a zygote that becomes a mature plasmodium once again.

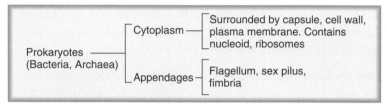

THE CHAPTER IN REVIEW

Summary

Viruses are noncellular particles.

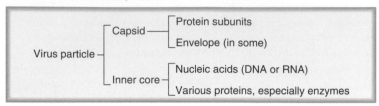

Structure

- Viruses have at least two parts: an outer capsid composed of protein subunits and an inner core of nucleic acid, either DNA or RNA.
- Animal viruses are either naked (no outer envelope) or they have an outer membranous envelope.

Reproduction

Viruses are obligate intracellular parasites that can reproduce only inside living cells. Bacteriophages can have a lytic or lysogenic life cycle.

The **lytic cycle** consists of these steps:

- Attachment
- Penetration
- Biosynthesis
- Maturation
- Release

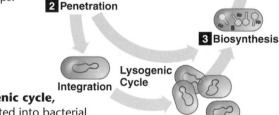

5 Release

1 Attachment

Lytic Cycle   **4 Maturation**

2 Penetration

3 Biosynthesis

Integration   **Lysogenic Cycle**

In the **lysogenic cycle,** viral DNA is integrated into bacterial DNA for an indefinite period of time, but it can undergo the last three steps of the lytic cycle at any time.

Plant Viruses

Crops and garden plants are subject to viral infections. Not all viruses are deadly, but over time they debilitate a plant.

Animal Viruses

The reproductive cycle in animal viruses has the same steps as in a bacteriophage, with modifications if the virus has an envelope, in which case:

- Fusion or endocytosis brings virus into cell.
- Uncoating is needed to free the genome from the capsid.
- Budding releases the viral particles from the cell.

HIV, the AIDS virus, is an RNA retrovirus. These viruses have an enzyme, reverse transcriptase, which carries out reverse transcription. This produces cDNA, which replicates, forming a double helix that becomes integrated into host DNA.

Drug Control of Human Viral Diseases

Antiviral drugs are structurally similar to a nucleotide and interfere with viral genome synthesis. HIV protease inhibitors block enzymes required for maturation of viral proteins.

Viroids are naked (not covered by a capsid) strands of RNA that can cause disease.

Prions are protein molecules that have a misshapen tertiary structure. Prions cause such diseases as CJD in humans and mad cow disease in cattle when they cause other proteins of their own kind to also become misshapen.

The bacteria and archaea are prokaryotes. Prokaryotes lack a nucleus and most of the other cytoplasmic organelles found in eukaryotic cells.

Prokaryotes (Bacteria, Archaea)

- Cytoplasm — Surrounded by capsule, cell wall, plasma membrane. Contains nucleoid, ribosomes
- Appendages — Flagellum, sex pilus, fimbria

General Biology of Bacteria

Bacterial Structure

- Structure can be rods (bacilli), spheres (cocci), and curved or spiral (vibrio, spirillum, or spirochete).
- Single, closed circle of double-stranded DNA (chromosome) is in a nucleoid.
- Flagellum is unique and rotates, causing the organism to spin.
- Cell wall contains peptidoglycan.

Bacterial Reproduction and Survival

- Reproduction is asexual by binary fission.
- Genetic recombination occurs by means of conjugation, transformation, and transduction.
- Endospore formation allows bacteria to survive an unfavorable environment.
- Endospores are extremely resistant to destruction; the genetic material can thereby survive unfavorable conditions.

Bacterial Nutrition

- Bacteria can be autotrophic. Cyanobacteria are photoautotrophs—they photosynthesize, as do plants. Chemoautotrophs oxidize inorganic compounds, such as

COMPARISON OF VIRUSES AND PROKARYOTES		
Characteristic of Life	**Viruses**	**Prokaryotes**
Consist of cell	No	Yes
Metabolize	No	Yes
Respond to stimuli	No	Yes
Multiply	Yes (always inside living cell)	Yes (usually independently)
Evolve	Yes	Yes

hydrogen gas, hydrogen sulfide, and ammonia, to acquire energy to make their own food. Chemoautotrophs are chemosynthesizers that support communities at deep-sea vents.
- Like animals, most bacteria are chemoheterotrophs (heterotrophs), but they are saprotrophic decomposers. Many heterotrophic prokaryotes are symbiotic. The mutualistic nitrogen-fixing bacteria live in nodules on the roots of legumes.

Environmental and Medical Importance of Bacteria

Bacteria in the Environment
- As decomposers, bacteria keep inorganic nutrients cycling in ecosystems.
- The reactions they perform keep the nitrogen cycle going.
- Bacteria play an important role in bioremediation.

Bacteria in Food Science and Biotechnology
- Bacterial fermentations are important in the production of foods.
- Genetic engineering allows bacteria to produce medically important products.

Bacteria Diseases in Humans
- Bacterial pathogens that can cause diseases are able to: (1) produce a toxin, (2) adhere to surfaces, and (3) sometimes invade organs or cells.
- Indiscriminate antibiotic therapy has led to bacterial resistance to some antibiotics.

Archaea

The archaea (domain Archaea) are a second type of prokaryote. Some characteristics of archaea are:

- They appear to be more closely related to the eukarya than to the bacteria.
- They do not have peptidoglycan in their cell walls, as do the bacteria, and they share more biochemical characteristics with the eukarya than do bacteria.
- They are well known for living under harsh conditions, such as anaerobic marshes (methanogens), salty lakes (halophiles), and sulfur springs (thermoacidophiles).

17.4   The Protists

General Biology of Protists

- Protists are eukaryotes. Endosymbiotic events may account for the presence of mitochondria and chloroplasts in eukaryotic cells. Some are multicellular with differentiated tissues.
- Protists have great ecological importance because in largely aquatic environments they are the producers (algae) or sources (algae and protozoans) of food for other organisms.

Algae

- Algae possess chlorophylls. They store reserve food as starch and have cell walls, as do plants.
- Algae are divided according to their pigments, being either green, brown, golden-brown, or red. Many are unicellular, but the green, red, and brown algae include multicellular seaweeds.

Protozons

- Protozoans are heterotrophic and usually motile by means of cilia (paramecia), pseudopods (amoebas), or flagella (zooflagellates)

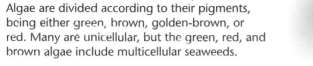

- Sporozoans cause diseases, including malaria, the most serious protozoan disease.
- A zooflagellates (trypanosome) causes African sleeping sickness; *Giardia* causes severe diarrhea.

Slime Molds and Water Molds
- Slime molds are decomposers in forests and woodlands.
- Water molds have cell walls of cellulose, which distinguish them from fungi.

Thinking Scientifically

1. The bacterium *Agrobacterium tumefasciens* is a plant pathogen. When it infects a plant, it transforms the plant's genome with a set of bacterial genes. These genes code for a bacterial food supply and for proteins that encourage plant cell proliferation, providing a home for the bacteria. Plant genetic engineers have taken advantage of this property to introduce foreign genes into plants. Suggest how they might do this without creating diseased plants.

2. If a plant is genetically engineered to produce the capsid protein of a virus, it becomes resistant to that virus. However, these transgenic plants are not resistant if you infect them with the RNA of the virus. How, and when in the infection process, do you suppose this resistance mechanism is acting?

Testing Yourself

Choose the best answer for each question.

1. A virus contains
   a. a cell wall.
   b. a plasma membrane.
   c. nucleic acid.
   d. cytoplasm.
   e. More than one of these are correct.

2. Label the parts of a virus in the following illustration.

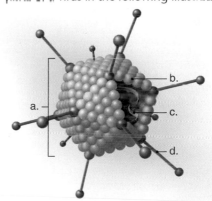

3. The five stages of the lytic cycle occur in this order:
   a. penetration, attachment, release, maturation, biosynthesis
   b. attachment, penetration, release, biosynthesis, maturation
   c. biosynthesis, attachment, penetration, maturation, release
   d. attachment, penetration, biosynthesis, maturation, release
   e. penetration, biosynthesis, attachment, maturation, release

4. Animal viruses
   a. contain both DNA and RNA.
   b. sometimes have an envelope.
   c. sometimes infect bacteria.
   d. do not reproduce inside cells.

5. The enzyme that is unique to retroviruses is

   a. reverse transcriptase.   c. DNA gyrase.
   b. DNA polymerase.   d. RNA polymerase.

6. Which of these is mismatched?

   a. amoeboids—pseudopods   c. algae—variously colored
   b. sporozoans—disease agents   d. slime molds—trypanosomes

7. Which of the following statements about viroids is false?

   a. They are composed of naked RNA.
   b. They cause plant diseases.
   c. They have a broader range of hosts than most viruses.
   d. They die once they reproduce.

8. Prion proteins cause disease when they

   a. enlarge in size.   c. change shape.
   b. break into small pieces.   d. interact with DNA.

9. Bacterial cells contain

   a. ribosomes.   d. vacuoles.
   b. nuclei.   e. More than one of these are
   c. mitochondria.   correct.

10. The primary producers at deep-sea vents are

    a. heterotrophs.   c. chemoautotrophs.
    b. symbionts.   d. photoautotrophs.

For questions 11–15, determine which type of organism is being described. Each answer in the key may be used more than once.

Key:

   a. bacteria
   b. archaea
   c. both bacteria and archaea
   d. neither bacteria nor archaea

11. Peptidoglycan in cell wall.

12. Methanogens.

13. Sometimes parasitic.

14. Contain a nucleus.

15. Plasma membrane contains lipids.

16. Unlike plants, algae contain

    a. chloroplasts.   d. pyrenoids.
    b. vacuoles.   e. cell walls.
    c. mitochondria.

17. Label the parts of a paramecium in the following illustration.

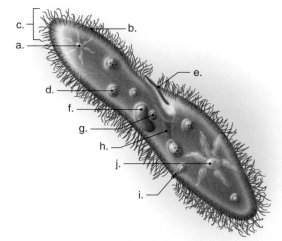

Go to www.mhhe.com/maderessentials for more quiz questions.

Bioethical Issue

The Food and Drug Administration (FDA) estimates that physicians annually write 50 million unnecessary prescriptions for antibiotics to treat viral infections. Of course, these antibiotics are ineffective against viruses. Furthermore, the frequent exposure of bacteria to these drugs has resulted in the development of numerous strains with antibiotic resistance. Doctors may prescribe antibiotics because patients demand them. In addition, if the cause of the illness is not known, it is safer to prescribe an antibiotic that turns out to be ineffective than to withhold the antibiotic when it would have helped. The FDA has, therefore, initiated a new policy requiring manufacturers to label antibiotics with precautions about their misuse.

Do you think this type of labeling information will educate consumers enough to significantly reduce the inappropriate use of antibiotics? Physicians already know about the dangers of antibiotic resistance, but may prescribe them anyway. This is probably due, in large part, to the fact that they are liable for erring on the side of withholding an antibiotic but not for inappropriately prescribing one. Should physicians be held more accountable for prescribing antibiotics? If so, how?

Understanding the Terms

Match the terms to these definitions:

a. _____ Phage life cycle in which the infected bacterium does not immediately produce phage.

b. _____ RNA animal virus with a DNA stage.

c. _____ Proteinaceous infectious particle.

d. _____ Type of molecule found in a bacterial cell wall.

e. _____ Method of asexual reproduction in bacteria.

Land Environment: Plants and Fungi

CHAPTER 18

Without plants, there would be no animals.

Twelve plants, wheat among them, stand between humans and starvation.

Cotton is a textile fiber that accounts for about half of all the fabric sold in the world.

Have you thanked a green plant today? Without plants, we and our civilized life could not exist. Plants provide us with our food, either directly or indirectly, and with the oxygen we breathe. Oxygen also rises into the stratosphere and becomes the ozone shield that absorbs ultraviolet rays and makes life on land possible.

▶ ▶ ▶

The bulk of the human diet comes from just 12 plants. Wheat is associated with Europe, corn with the Americas, and rice with the Far East. Other significant food crops include white and sweet potatoes, cassava, soybeans, common beans, sugarcane, sugar beets, coconuts, and bananas.

Eons ago, the bodies of plants became the coal we now burn to produce much of our electricity. The wood of trees is also commonly used as fuel. Then, too, fermentation of plant materials produces alcohol, which can be used directly to fuel automobiles or as a gasoline additive.

Much of our clothing comes from plants. People still like the feel of cotton next to their skin, while linen from flax makes a stronger cloth. Cotton is used in all sorts of other cloth products, such as towels, sheets, and upholstered furniture. Even the manufacture of rayon depends on cellulose from plant cell walls.

Thousands of other household products come from plants. Wood is used to make furniture as well as the houses that contain the furniture. Jute is used for rope, carpet, insulation, and burlap bags. Coconut oil is found in soaps, shampoos, and suntan lotions. Even toothpastes contain such plant flavorings as peppermint and spearmint. Suffice it to say, it would be impossible to mention all the products from plants that we depend on for our daily needs.

18.1 Onto Land

Plants are a diverse group that range in size and structure from the smallest aquatic duckweed to the largest giant sequoia. Plants, unlike algae, are adapted for living on land. Still, several types of evidence support the conclusion that plants are related, through evolution, to green algae. Both green algae and plants (1) contain chlorophylls *a* and *b* and various accessory pigments, (2) store excess carbohydrates as starch, and (3) have cellulose in their cell walls. Additionally, in recent years biochemists have compared the sequences of DNA bases coding for ribosomal RNA between organisms. The results suggest that plants are most closely related to a group of green algae encrusted with calcium carbonate and, therefore, known as stoneworts, perhaps those in the genus *Chara*. The split between green algae and plants may have occurred about 500 million years ago (MYA).

Chara

Today, plants are classified into four groups (Fig. 18.1). It is possible to associate each group with one of four evolutionary events, all of them adaptations to a land existence. First, all the groups of plants protect and nourish a multicellular embryo within the body of the plant (Fig. 18.1*a*). Since green algae do not protect the embryo as all plants do, this may be the first evolved feature that separates plants from green algae.

a. In mosses, the embryo is protected by a special structure.

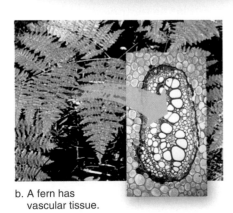

b. A fern has vascular tissue.

c. In a pine tree, seeds disperse offspring.

d. In flowering plants, the seeds are enclosed in fruits.

Figure 18.1   **Representatives of the four major groups of plants.**

The next event was most likely the evolution of **vascular tissue.** Vascular tissue is essential for the transport of water and solutes from the roots to the leaves of terrestrial plants (Fig. 18.1*b*). The fossil record indicates that vascular plants evolved about 420 MYA.

The third event was the evolution of seeds. Two groups of plants produce seeds (Fig. 18.1*c,d*). A **seed** is composed of an embryo and stored organic nutrients within a protective coat (see Fig. 18.10). When you plant a seed, a plant of the next generation emerges; in other words, seeds disperse offspring. Seeds are highly resistant structures well suited to protect a plant embryo from drought, and to some extent from predators, until conditions are favorable for germination. The first seed plants appear in the fossil record about 360 MYA.

The fourth event was the evolution of the **flower,** a reproductive structure (Fig. 18.1*d*). Flowers attract pollinators such as insects, and they also give rise to fruits that cover seeds. Plants with flowers may have evolved about 160 MYA.

Figure 18.2 traces the evolutionary history of plants and will serve as a backdrop as we discuss the four major groups of plants in the pages that follow.

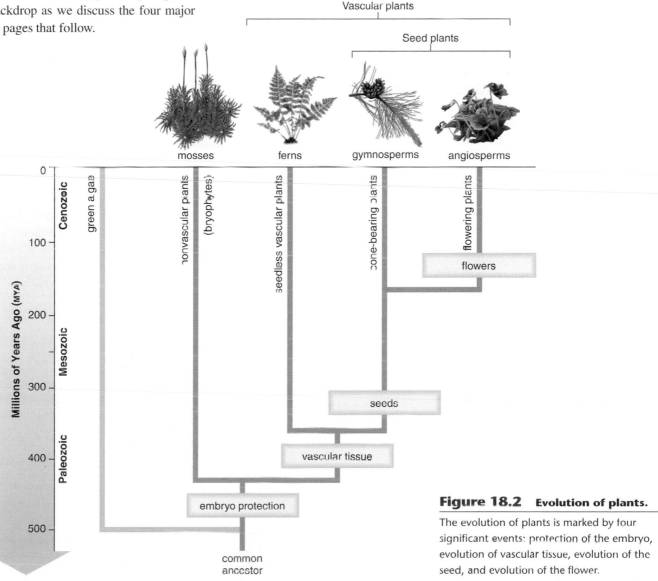

Figure 18.2 Evolution of plants.

The evolution of plants is marked by four significant events: protection of the embryo, evolution of vascular tissue, evolution of the seed, and evolution of the flower.

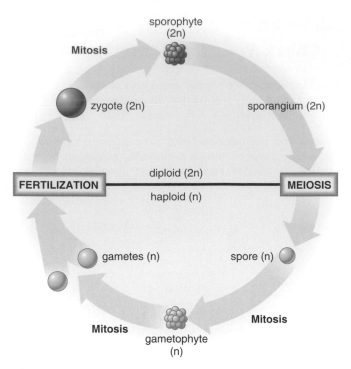

Figure 18.3   **Alternation of generations.**

Alternation of Generations

The life cycle of plants is quite different from that of animals. All plants undergo an **alternation of generations,** which means that each type of plant exists in two forms (Fig. 18.3). One form is the diploid sporophyte, and the other is the haploid gametophyte. The **sporophyte** (2n) is named for its production of haploid spores by meiosis. A **spore** is a reproductive cell that develops into a new organism without the need to fuse with another reproductive cell. In the plant life cycle, a spore undergoes mitosis and becomes a gametophyte. The **gametophyte** (n) is named for its production of gametes. A sperm and egg fuse, forming a zygote that undergoes mitosis and becomes the sporophyte.

You'll want to make two observations in Figure 18.3. First note that, in plants, meiosis produces haploid spores. This is consistent with the realization that the sporophyte is the diploid generation and spores are haploid. Second, note that mitosis occurs as a spore becomes a gametophyte, and it also occurs as a zygote becomes a sporophyte. Mitosis must happen in both places in order to have two generations.

The Dominant Generation

When you bring a plant to mind, you're probably thinking of the dominant generation—the generation that is larger, lasts longer, and is the generation we recognize as the plant. In nonvascular plants, the gametophyte is the dominant generation, and in the other groups of plants, the sporophyte is the dominant generation. In the history of plants, *only the sporophyte evolves vascular tissue;* therefore, the trend toward sporophyte dominance is an adaptation to life on land. As the sporophyte gains in dominance, the gametophyte becomes microscopic. It also becomes dependent upon the sporophyte, the generation best adapted to a dry land environment.

Note the appearance of the generations in Figure 18.4. In mosses (**bryophytes**), the gametophyte is much larger than the sporophyte. In ferns, the gametophyte is a small, independent, heart-shaped structure. In contrast, the female gametophyte of cone-bearing plants (**gymnosperm**) and flowering plants (**angiosperm**) is microscopic—it is retained within the body of the sporophyte plant. This protects the female gametophyte

---

Check Your Progress

1. Explain what is meant by alternation of generations.
2. What generation is dominant among mosses? In other groups of plants? Give an explanation for this trend of generation dominance among plants.

Answers: 1. Every plant exists in two forms. The diploid sporophyte produces haploid spores by meiosis. The haploid gametophyte produces gametes. 2. The gametophyte is dominant among mosses, and the sporophyte is dominant among the other groups of plants. The sporophyte evolved vascular tissue, and vascular tissue is a distinct advantage in a land environment.

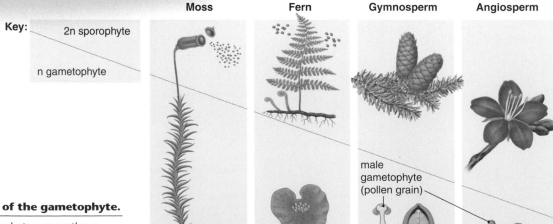

Key:
2n sporophyte
n gametophyte

Moss   Fern   Gymnosperm   Angiosperm

male gametophyte (pollen grain)

female gametophyte

Figure 18.4   **Reduction in the size of the gametophyte.**
Notice the reduction in the size of the gametophyte among these representatives of today's plants. This trend occurred as plants became adapted for life on land.

from drying out. Also, the male gametophyte of seed plants lies within a pollen grain. Pollen grains have strong protective walls and are transported by wind, insects, or birds to reach the egg. In the life cycle of seed plants, the spores, the gametes, and the zygote are protected from drying out in the land environment.

18.2 Diversity of Plants

In the evolution of plants, the nonvascular plants evolved before the vascular plants.

Nonvascular Plants

The **nonvascular plants,** often called the *bryophytes,* do not have true roots, stems, and leaves—which, by definition, must contain vascular tissue. In bryophytes, the gametophyte is the dominant generation.

The most familiar bryophytes are the liverworts and mosses, which are low-lying plants. In bryophytes, the gametophyte consists of leafy shoots, which produce the gametes (Fig. 18.5). Flagellated sperm swim in a film of water to reach an egg. After a sperm fertilizes an egg, the resulting zygote becomes an embryo that develops into a sporophyte. The sporophyte is attached to, and derives its nourishment from, the photosynthetic gametophyte. The sporophyte produces windblown spores, an adaptation to life on land.

The common name of several plants implies that they are mosses when they are not. Irish moss is an edible red alga that grows in leathery tufts along northern seacoasts. Reindeer moss, a lichen, is the dietary mainstay of reindeer and caribou in northern lands. Club mosses, discussed later in this chapter, are vascular plants, and Spanish moss, which hangs in grayish clusters from trees in the southeastern United States, is a flowering plant of the pineapple family.

Adaptations and Uses of Bryophytes

The lack of vascular tissue and the presence of swimming sperm largely account for the short height of bryophytes such as mosses. Still, mosses can be found from the Antarctic through the tropics to parts of the Arctic. Although most mosses prefer damp, shaded locations in the temperate zone, some survive in deserts, and others inhabit bogs and streams. In forests, they frequently form a mat that covers the ground and rotting logs. In dry environments, they may become shriveled, turn brown, and look completely dead. As soon as it rains, the plant becomes green and resumes metabolic activity. Mosses are even better than flowering plants at living on stone walls, on fences, and in shady cracks of hot, exposed rocks. When bryophytes colonize bare rock, the rock degrades to soil that can be used for their own growth and the growth of other organisms.

In areas such as bogs, where the ground is wet and acidic, dead mosses, especially *Sphagnum,* do not decay. The accumulated moss, called peat or bog moss, can be used as fuel. Peat moss is also commercially important in another way. Because it has special nonliving cells that can absorb moisture, peat moss is often used in gardens to improve the water-holding capacity of the soil. The so-called copper mosses live only in the vicinity of copper, and can serve as an indicator of copper deposits.

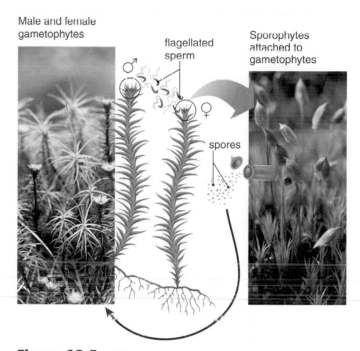

Figure 18.5   Mosses.

In bryophytes, exemplified by mosses, the gametophyte is the dominant generation. Sperm swim from male shoots to female shoots, and the zygote develops into an attached sporophyte. The sporophyte produces haploid spores that develop into gametophyte shoots.

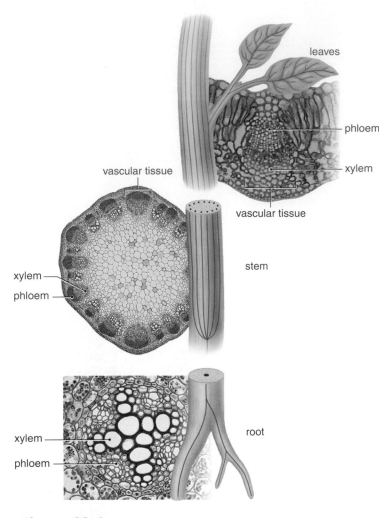

Figure 18.6 **Vascular tissue.**

The roots, stems, and leaves of vascular plants, such as this flowering plant, have vascular tissue (xylem and phloem). Xylem transports water and minerals; phloem transports organic compounds.

Vascular Plants

All the other plants we will study are **vascular plants.** The vascular plants usually have *true* roots, stems, and leaves. The roots absorb water from the soil, and the stem conducts water to the leaves. The leaves are fully covered by a waxy cuticle except where it is interrupted by stomata, little pores for gas exchange, the opening and closing of which can be regulated to control water loss.

Vascular tissue consists of **xylem,** which conducts water and minerals up from the roots, and **phloem,** which conducts organic nutrients from one part of a plant to another (Fig. 18.6). The walls of conducting cells in xylem are strengthened by **lignin,** an organic compound that makes them stronger, more waterproof, and resistant to attack by parasites and predators. Only because of strong cell walls can plants reach great heights.

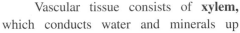

Seedless Vascular Plants

The seedless vascular plants include club mosses, horsetails, and ferns. Today club mosses are represented in temperate forests by ground pines, so called because of their limited height and superficial resemblance to a pine tree. Horsetails are taller and have a stem that is interrupted by whorls of slender, green branches. Horsetails live in moist habitats about the globe. This category of plants has vascular tissue but does not produce seeds. Instead, windblown spores disperse the plant to new locations.

Before the seed plants evolved, the seedless vascular plants were as tall as today's trees, and they dominated the swamp forests of the Carboniferous period (Fig. 18.7). A large number of these plants died but did not decompose completely. Instead, they were compressed to form the coal that we still mine and burn today. (Oil has a similar origin, but it most likely formed in marine sedimentary rocks and included animal remains.)

Figure 18.7 **The Carboniferous period.**

Growing in the swamp forests of the Carboniferous period were treelike club mosses (*left*), treelike horsetails (*right*), and lower, fernlike foliage (*left*). When the trees fell, they were covered by water and did not decompose completely. Sediment built up and turned to rock, the pressure of which caused the organic material to become coal, a fossil fuel that helps run our industrialized society today.

Ferns

Ferns are a widespread group of plants that are well known for their attractiveness. The large and conspicuous leaves of ferns, called **fronds,** are commonly divided into leaflets. The royal fern has fronds that stand about 1.8 meters tall; those of the hart's tongue fern are straplike and leathery; and those of the maidenhair fern are broad, with subdivided leaflets (Fig. 18.8).

Adaptations and Uses of Ferns Ferns are most often found in a moist environment because the water-dependent gametophyte, which lacks vascular tissue, is separate from the sporophyte. Also, flagellated sperm require an outside source of moisture in which to swim to the eggs (Fig. 18.9). Once established, some ferns, such as the bracken fern, can spread into drier areas because their rhizomes, which grow horizontally in the soil, produce new plants.

At first it may seem that ferns do not have much economic value, but they are frequently used by florists in decorative bouquets and as ornamental plants in the home and garden. Wood from tropical tree ferns is often used as a building material because it resists decay, particularly by termites. Ferns, especially the ostrich fern, are used as food, and in the northeastern United States, many restaurants feature fiddleheads (that season's first growth) as a special treat. Ferns also have medicinal value; many Native Americans use ferns as an astringent during childbirth to stop bleeding, and the maidenhair fern is the source of a cold medicine.

maidenhair fern

royal fern

hart's tongue fern

Figure 18.8 **Diversity of ferns.**

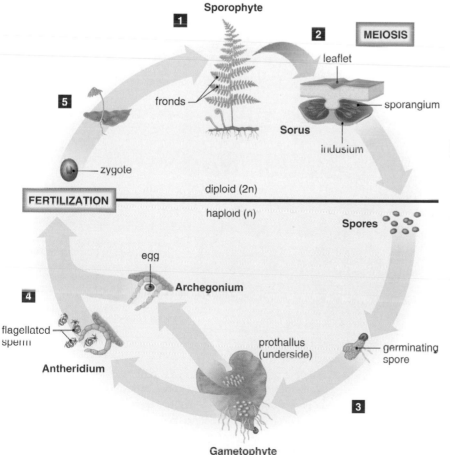

Figure 18.9 **Fern life cycle.**

1 The sporophyte is dominant in ferns. **2** In the fern shown here, sori are on the underside of a fern. Each sorus, protected by an indusium, contains sporangia, in which meiosis occurs and spores are produced and then released. **3** A spore germinates into a prothallus (the gametophyte), which has sperm-bearing and egg-bearing structures on its underside. **4** Fertilization takes place when moisture is present, because the flagellated sperm must swim in a film of water to the egg. **5** The resulting zygote begins its development inside an archegonium, and eventually a young sporophyte becomes visible. The young sporophyte develops, and fronds appear.

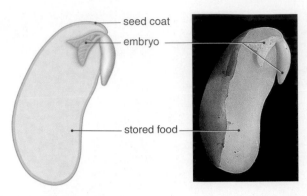

Figure 18.10   Seed anatomy.

A split bean seed showing the seed coat, embryo, and stored food (cotyledon).

Check Your Progress

1. What structure in seed plants becomes a seed?
2. What are the components of a seed?
3. What is the difference between pollination and fertilization?

Answers: 1. Ovule. 2. A seed contains an embryo, stored food, and a seed coat. 3. Pollination is the delivery of the male gametophyte to the region of the female gametophyte. Fertilization occurs when the sperm from the male gametophyte unites with the egg in the female gametophyte.

Seed Plants

Seed plants are the most plentiful plants in the biosphere today. Most trees are seed plants, and so are almost all your garden plants. The major parts of a seed are shown in Figure 18.10. The seed coat and stored food protect the embryo and allow it to survive harsh conditions during a period of dormancy (arrested state), until environmental conditions become favorable for growth. Seeds can even remain dormant for hundreds of years. When a seed germinates, the stored food is a source of nutrients for the growing seedling. The survival value of seeds largely accounts for the dominance of seed plants today.

Seed plants have two types of spores and produce two kinds of gametophytes—male and female. The gametophytes are microscopic and consist of just a few cells. **Pollen grains** are drought resistant male gametophytes. **Pollination** occurs when a pollen grain is brought to the vicinity of the female gametophyte by wind or a pollinator. Later, sperm move toward the female gametophyte through a growing pollen tube, and fertilization occurs. *Note that no external water is needed to accomplish fertilization.* The whole male gametophyte, rather than just the sperm as in seedless plants, moves to the female gametophyte.

A female gametophyte develops within an **ovule,** which eventually becomes a seed (Fig. 18.11). The embryo in a seed will be the sporophyte plant. In gymnosperms, the ovules are not completely enclosed by sporophyte tissue at the time of pollination. In angiosperms, the ovules are completely enclosed within diploid sporophyte tissue (an ovary), which becomes a fruit. We shall see that the flower has many advantages, one of which is the production of fruit.

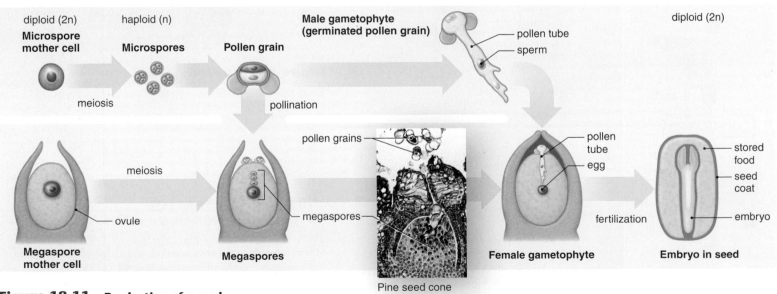

Figure 18.11   Production of a seed.

Development of the male gametophyte (*above*) begins when a microspore mother cell undergoes meiosis to produce microspores, each of which becomes a pollen grain. Development of the female gametophyte (*below*) begins in an ovule, where a megaspore mother cell undergoes meiosis to produce megaspores, only one of which will undergo mitosis to become the female gametophyte. During pollination, the pollen grain is carried to the vicinity of the ovule. The pollen grain germinates, and a nonflagellated sperm travels in a pollen tube to the egg produced by the female gametophyte. Following fertilization, the zygote becomes the embryo; tissue within the ovule becomes the stored food, and the ovule wall becomes the seed coat.

Gymnosperms

In **gymnosperms,** ovules and seeds are exposed on the surface of a cone scale (modified leaf). The term gymnosperm means "naked seed." Ancient gymnosperms, including cycads, were present in the swamp forests of the Carboniferous period (Fig. 18.12). Of these, the conifers have become a dominant plant group today.

Conifers

Pines, spruces, firs, cedars, hemlocks, redwoods, cypresses, and others are all **conifers** (Fig. 18.13). The name *conifer* signifies plants that bear **cones,** but other types of gymnosperms are also cone-bearing. The coastal redwood, a conifer native to northwestern California and southwestern Oregon, is the tallest living vascular plant; it may attain nearly 100 meters in height. Another conifer, the bristlecone pine of the White Mountains of California, is the oldest living tree—one specimen is 4,900 years of age.

Adaptations and Uses of Pine Trees   Vast areas of northern temperate regions are covered in evergreen coniferous forests. The tough, needlelike leaves of pines conserve water because they have a thick cuticle and recessed stomata. This type of leaf helps them live in areas where frozen topsoil makes it difficult for the roots to obtain plentiful water.

   A substance called resin protects leaves and other parts of the tree from insect and fungal attacks. The resin of certain pines is harvested; the liquid portion, called turpentine, is a paint thinner, while the solid portion is used on stringed instruments. The wood of pines is used extensively in construction, and vast forests of pines are planted for this purpose. The wood consists primarily of xylem tissue that lacks some of the more rigid cell types found in flowering trees. Therefore, it is considered a "soft" rather than a "hard" wood.

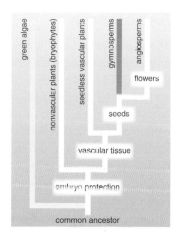

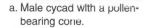

a. Male cycad with a pollen-bearing cone.

b. Female cycad with a seed-bearing cone.

Figure 18.12   Cycads.

Cycads are an ancient group of gymnosperms that are threatened today because they are slow growing. Herbivorous dinosaurs of the Mesozoic era most likely fed on cycad seeds.

fleshy seed cones (juniper berries)

pollen cones

seed cones

b.

c.

a.

Figure 18.13   Conifers.

a. Pine trees are the most common of the conifers. The pollen cones (male) are smaller than the seed cones (female) and produce plentiful pollen. A cluster of pollen cones may produce more than one million pollen grains. Other conifers include (**b**) the spruces, which make beautiful Christmas trees, and (**c**) junipers, which have fleshy seed cones.

Figure 18.14 *Amborella trichopoda.*

Molecular data suggest this plant is most closely related to the first flowering plants.

Angiosperms

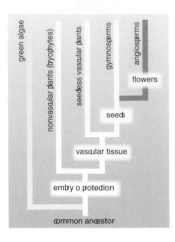

The **angiosperms** (meaning covered seeds) are an exceptionally large and successful group of plants, with 240,000 known species—six times the number of species of all other plant groups combined. Angiosperms, also called the flowering plants, live in all sorts of habitats, from fresh water to desert, and from the frigid north to the torrid tropics. They range in size from the tiny, almost microscopic duckweed to *Eucalyptus* trees over 100 meters tall. Most plants in your garden produce flowers, and therefore are angiosperms. In northern climates, the trees that lose their leaves are flowering plants. In subtropical and tropical climates, flowering trees as well as gymnosperms tend to keep their leaves all year.

Although the first fossils of angiosperms are no older than about 135 million years, the angiosperms probably arose much earlier. Indirect evidence suggests the possible ancestors of angiosperms may have originated as long ago as 160 MYA. To help solve the mystery of their origin, botanists have turned to DNA comparisons to find a living plant that is most closely related to the first angiosperms. Their data point to *Amborella trichopoda* as having the oldest lineage among today's angiosperms (Fig. 18.14). This small shrub, which has small, cream-colored flowers, lives only on the island of New Caledonia in the South Pacific.

The Flower

Most flowers have certain parts in common despite their dissimilar appearances. The flower parts, called **sepals, petals, stamens,** and **carpels,** occur in whorls (circles) (Fig. 18.15). The sepals, collectively called the **calyx,** protect the flower bud before it opens. The sepals may drop off or may be colored like the petals. Usually, however, sepals are green and remain in place. The petals, collectively called the **corolla,** are quite diverse in size, shape, and color. The petals often attract a particular pollinator. Each stamen consists of a stalk called a **filament** and an **anther,** where pollen is produced in pollen sacs. In most flowers, the anther is positioned where the pollen can be carried away by wind or a pollinator. One or more carpels is at the center of a flower. A carpel has three major regions: ovary, style, and stigma. The swollen base is the **ovary,** which contains from one to hundreds of ovules. The **style** elevates the **stigma,** which is sticky or otherwise adapted for the reception of pollen grains. Glands located in the region of the ovary produce nectar, a nutrient that is gathered by pollinators as they go from flower to flower.

Check Your Progress

1. List the components of the stamen. Where is pollen formed?
2. List the components of the carpel. Which part becomes a seed? The fruit?

Answers: 1. The stamen contains the anther and the filament. Pollen forms in the pollen sacs of the anther. 2. The carpel contains the stigma, style, and ovary. An ovule in the ovary becomes a seed, and the ovary becomes the fruit.

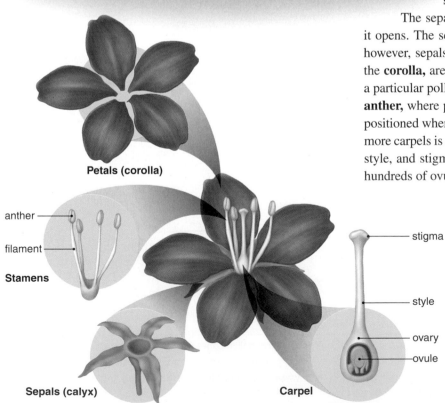

Petals (corolla)

anther

filament

Stamens

stigma

style

ovary

ovule

Sepals (calyx)

Carpel

Figure 18.15 **Generalized flower.**

A flower has four main parts: sepals, petals, stamens, and carpels. A stamen has a filament and an anther. A carpel has an ovary, a style, and a stigma. The ovary contains ovules.

Flowering Plant Life Cycle

In angiosperms, the flower produces seeds enclosed by fruit. The ovary of a carpel contains several ovules, and each of these eventually holds an egg-bearing female gametophyte called the embryo sac. During pollination, a pollen grain is transported by various means from the anther of a stamen to the stigma of a carpel, where it germinates. The **pollen tube** carries the two sperm into a small opening of an ovule. During **double fertilization,** one sperm unites with an egg nucleus forming a diploid zygote, and the other sperm unites with two other nuclei forming a triploid (3n) **endosperm** (Fig. 18.16). In angiosperms, the endosperm is the stored food.

Ultimately, the ovule becomes a seed that contains a sporophyte embryo. In some seeds, the endosperm is absorbed by the seed leaves, called **cotyledons,** whereas in other seeds, endosperm is digested as the seed germinates. When you open a peanut, the two halves are the cotyledons. If you look closely, you will see the embryo between the cotyledons. A **fruit** is derived from an ovary and possibly accessory parts of the flower. Some fruits (e.g., apple) provide a fleshy covering for their seeds, and other fruits provide a dry covering (e.g., pea pod, shell of peanut).

Figure 18.16 Life cycle of flowering eudicot.

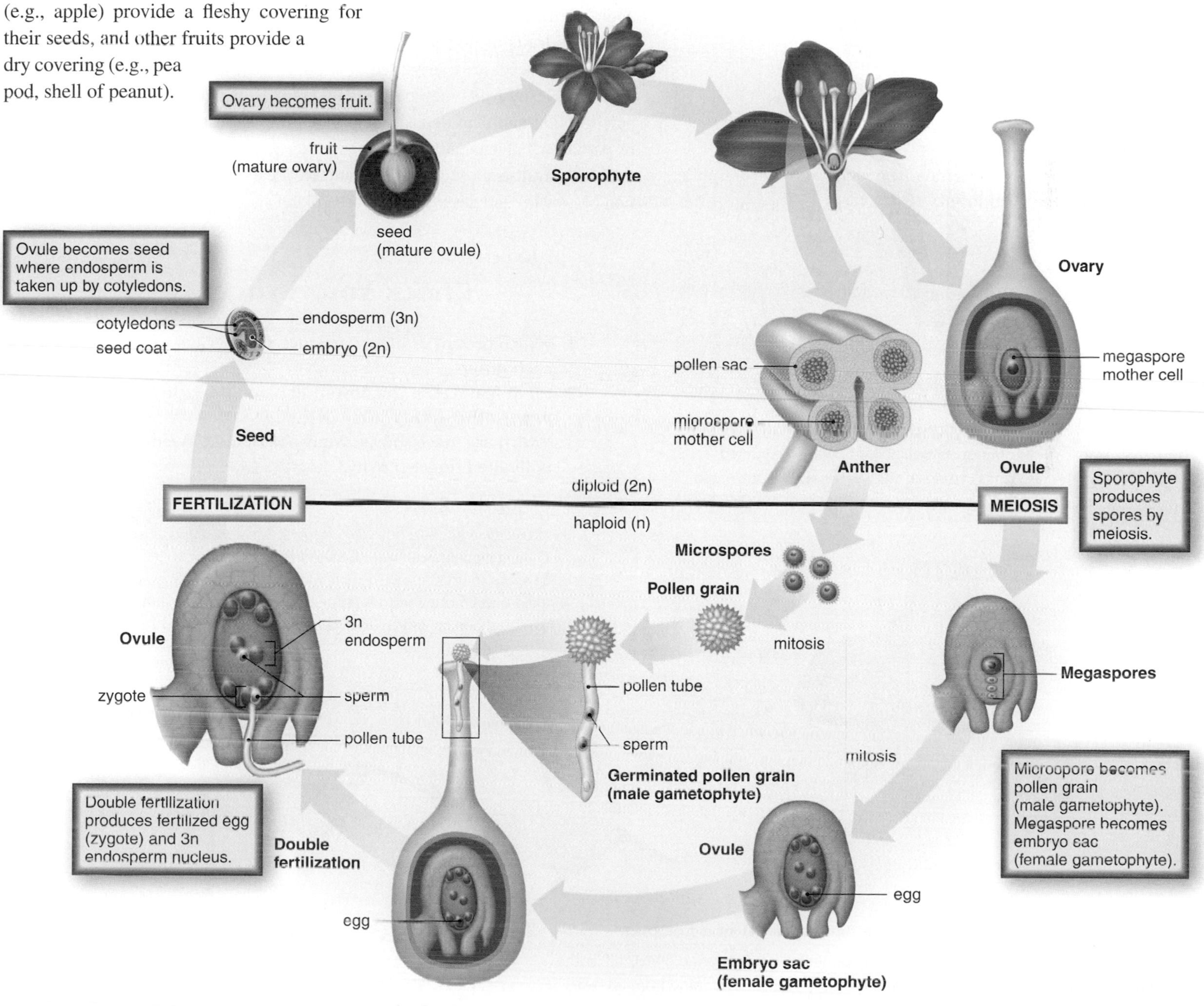

Adaptations and Uses of Angiosperms

Successful completion of sexual reproduction in angiosperms requires the effective dispersal of pollen and then seeds. Adaptations have resulted in various means of dispersal of pollen and seeds. Wind-pollinated flowers are usually not showy, whereas insect-pollinated flowers and bird-pollinated flowers are often colorful (Fig. 18.17*a–c*). Night-blooming flowers attract nocturnal mammals or insects; these flowers are usually aromatic and white or cream-colored (Fig. 18.17*d*). Although some flowers disperse their pollen by wind, many are adapted to attract specific pollinators, such as bees, wasps, flies, butterflies, moths, and even bats, which carry only particular pollen from flower to flower. For example, bee-pollinated flowers are usually blue or yellow and have ultraviolet shadings that lead the pollinator to the location of nectar at the base of the flower. In turn, the mouthparts of bees are fused into a long tube that is able to obtain nectar from this location. Today, there are some 240,000 species of flowering plants and over 900,000 species of insects. This diversity suggests that the success of angiosperms has contributed to the success of insects and vice versa.

The fruits of flowers protect and aid in the dispersal of seeds. Dispersal occurs when seeds are transported by wind, gravity, water, and animals to another location. Fleshy fruits may be eaten by animals, which transport the seeds to a new location and then deposit them when they defecate. Because animals live in particular habitats and/or have particular migration patterns, they are apt to deliver the fruit-enclosed seeds to a suitable location for seed germination (initiation of growth) and development of the plant.

Figure 18.17 Pollinators.

a. A bee-pollinated flower is a color other than red (bees can't see red). **b.** Butterfly-pollinated flowers are wide, allowing the butterfly to land. **c.** Hummingbird-pollinated flowers are curved back, allowing the bird's beak to reach the nectar. **d.** Bat-pollinated flowers are large and sturdy and able to withstand rough treatment.

Check Your Progress

1. Which groups of plants produce seeds? Give examples of each group.
2. What features of the flowering plant life cycle are not found in any other group?
3. Which are more showy, wind-pollinated flowers or animal-pollinated flowers? Why?

Answers: 1. Gymnosperms (cone-bearing, such as cycads and pine trees) and angiosperms (flowering plants, such as fruit trees and garden plants) produce seeds. 2. Presence of an ovary leads to production of seeds enclosed by a fruit. Animals are often used as pollinators. 3. Animal-pollinated flowers are showy, and in different ways, such as color and fragrance, attract their particular pollinators.

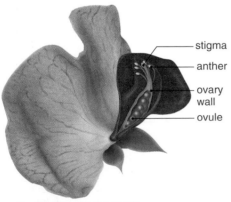

stigma
anther
ovary wall
ovule

Ovules become the seeds.

a.

b.   c.

Ovary becomes the fruit.

d.

Figure 18.18 Pea flower and the development of a pea pod.

a. Pea flower. **b.** Pea flower still has its petals soon after fertilization. **c.** Petals fall and the pea pod is quite noticeable. **d.** Pea pod is clearly visible developing from an ovary.

Economic Benefits of Plants

One of the primary economic benefits of plants is the use of their fruits as food. Botanists use the term "fruit" in a much broader way than do laypeople. Among the foods mentioned in the introduction, you would have no trouble recognizing a banana as a fruit. A coconut is also a fruit, as are the grains (corn, wheat, rice) and the pods that contain beans or peas (Fig. 18.18). Cotton is derived from the cotton boll, a fruit containing seeds with seed hairs that become textile fibers used to make cloth.

Other economic benefits of plants include foods and commercial products made from roots, stems, and leaves. Cassava and sweet potatoes are edible roots, while white potatoes are the tubers of underground stems. Most furniture and paper are made from the wood of a tree trunk (Fig. 18.19). Also, the many chemicals produced by plants make up 50% of all pharmaceuticals and various other types of products we can use. The cancer drug taxol originally came from the bark of the Pacific yew tree. Today, plants are even bioengineered to produce certain substances of interest.

Indirectly, the economic benefits of plants are often dependent on pollinators. Only if pollination occurs can these plants produce a fruit and propagate themselves. In recent years, the populations of bees and other pollinators have been declining worldwide, principally due to a parasitic mite but also due to the widespread use of pesticides. Consequently, some plants are endangered because they have lost their normal pollinators. Because of our dependence on flowering plants, we should protect pollinators!

Figure 18.19 Wooden furniture.

Maple, walnut, and mahogany furniture is made from the wood of these trees.

Ecological Benefits of Plants

The ecological benefits of flowering plants are so important that we could not exist without them. Plants produce food for themselves and directly or indirectly for all other organisms in the biosphere. The oxygen they produce is also used by themselves and all organisms that carry out cellular respiration.

Forests are an important part of the water cycle and the carbon cycle. The roots of trees in particular hold soil in place and absorb water that returns to the atmosphere. Without these functions of trees and other plants, rainwater runs off and contributes to flooding. The absorption of carbon dioxide by plants lessens the amount in the atmosphere. CO_2 in the atmosphere contributes to global warming because it and other gases trap heat near the Earth. The burning of tropical rain forests is double trouble for global warming because it adds CO_2 to the atmosphere and removes trees that otherwise would absorb CO_2. Some plants can also be used to clean up toxic messes. For example, poplar, mustard, and mulberry species take up lead, uranium, and other pollutants from the soil.

In addition to all the other uses of plants, we should not forget their aesthetic value. Almost everyone prefers to vacation in a natural setting and enjoys the sight of trees and flowers in the yard and in the house.

Check Your Progress

1. What are some economic benefits of flowering plants?
2. What are some ecological benefits of flowering plants?

Answers: 1. Use of fruits as food, such as beans and rice; use of fruits as textile fibers, such as cotton; use of wood as a building material for construction and furniture; and use of chemicals flowering plants produce as pharmaceuticals, such as taxol. 2. Flowering plants are producers in ecosystems; forests help keep ecological cycles functioning and keep the environment healthy.

18.3 The Fungi

Asked whether **fungi** are animal or vegetable, most would choose vegetable. But this would be wrong because fungi do not have chloroplasts, and they can't photosynthesize. Then, too, fungi are not animals, even though they are chemoheterotrophs like animals. Animals ingest their food, but fungi release digestive enzymes into their immediate environment and then absorb the products of digestion. Also, animals are motile, but fungi are nonmotile and do not have flagella at any stage in their life cycle. The fungal life cycle differs from that of both animals and plants because fungi produce windblown spores during both an asexual and a sexual life cycle.

Table 18.1 contrasts fungi with plants and animals. The many unique features of fungi indicate that although fungi are multicellular eukaryotes (except for the unicellular yeasts), they are not closely related to any other group of organisms. DNA sequence data suggest fungi are distantly related to animals rather than plants.

General Biology of a Fungus

Representative fungi are shown in Figure 18.20. All parts of a fungus are composed of *hyphae* (sing., hypha), which are thin filaments of cells. The hyphae are packed closely together to form a complex structure, such as a mushroom. However, the main body of a fungus is not the mushroom or the puffball or the morel; these are just temporary reproductive structures. Mold shows best that the main body of a fungus is a mass of hyphae called a **mycelium** (Fig. 18.21). The mycelium penetrates the soil, the wood, or the bread in which the fungus is

Figure 18.20   Diversity of fungi.

a. Scarlet hood, an inedible mushroom. **b.** Spores exploding from a puffball. **c.** Common bread mold. **d.** A morel, an edible fungus.

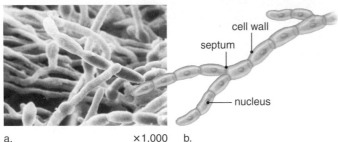

cell wall
septum
nucleus

a.   ×1,000   b.

Figure 18.21   Body of a fungus.

a. The body of a fungus is called a mycelium. **b.** A mycelium contains many individual chains of cells, and each chain is called a hypha.

Check Your Progress

1. Compare and contrast nutrition in fungi with that in animals.
2. Describe the relationship between a mycelium and hyphae.

Answers: 1. Both fungi and animals are chemoheterotrophs, but animals ingest their food, while fungi absorb organic nutrients from the environment. 2. A mycelium is a mass of hyphae (filaments of cells).

Table 18.1	How Fungi Differ from Plants and Animals		
Feature	**Fungi**	**Plants**	**Animals**
Nutrition	Chemoheterotrophic by absorption	Photosynthetic	Chemoheterotrophic by ingestion
Movement	Nonmotile	Nonmotile	Motile
Body	Mycelium of hyphae	Specialized tissues/organs	Specialized tissues/organs
Adult chromosome number	Haploid	Haploid/diploid	Diploid
Cell wall	Composed of chitin	Composed of cellulose	No cell wall
Reproduction	Spores	Spores/gametes	Gametes

growing. Mycelia in the soil can become large enough to cover acres, making them the largest creatures on Earth and earning them the label "the humongous fungus among us."

Fungal cells typically have thick cell walls, but unlike plants, the fungal cell walls do not contain cellulose. They are made of another polysaccharide in which the glucose monomers contain amino groups (amino sugars) and form a polymer called chitin. This polymer is also the major structural component of the exoskeleton of insects and arthropods, such as lobsters and crabs. The walls dividing the cells of a hypha have pores that allow the cytoplasm to pass from one cell to the other along the length of the hypha. The hyphae give the mycelium quite a large surface area, which facilitates the ability of the mycelium to absorb nutrients. Hyphae move toward a food source by growing at their tips, and the hyphae of a mycelium absorb and then pass nutrients on to the growing tips.

Black Bread Mold

Multicellular organisms are characterized by specialized cells. Black bread mold demonstrates that the hyphae of a fungus may be specialized for various purposes (Fig. 18.22). In this fungus, horizontal hyphae exist on the surface of the bread; other hyphae grow into the bread, anchoring the mycelium and carrying out digestion; and some form stalks that bear sporangia. A *sporangium* is a structure that produces spores.

The mycelia of two different mating types are featured in the center and bottom of Figure 18.22. During asexual reproduction, each mycelium produces sporangia, where spore formation occurs. Spores are resistant to environmental damage, and they are often made in large numbers. Fungal spores are windblown, a distinct advantage for a nonmotile organism living on land. The spores germinate into new mycelia without going through any developmental stages, another feature that distinguishes fungi from animals.

Sexual reproduction in fungi involves the conjugation of hyphae from different mating types (usually designated + and −). In black bread mold, the tips of + and − hyphae join, the nuclei fuse, and a thick-walled zygospore results. The zygospore undergoes a period of dormancy before it germinates, producing sporangia. Meiosis occurs within the sporangia, producing spores of both mating types. The spores, which are dispersed by air currents, give rise to new mycelia. In fungi, only the zygote is diploid, and all other stages of the life cycle are haploid, a distinct difference from animals.

Figure 18.22   Life cycle of black bread mold.

During asexual reproduction, sporangia produce asexual spores (purple and blue arrows). During sexual reproduction, two hyphae tips fuse, and then two nuclei fuse, forming a zygote that develops a thick, resistant wall (zygospore). When conditions are favorable, the zygospore germinates, and meiosis within a sporangium produces windblown spores (yellow arrows).

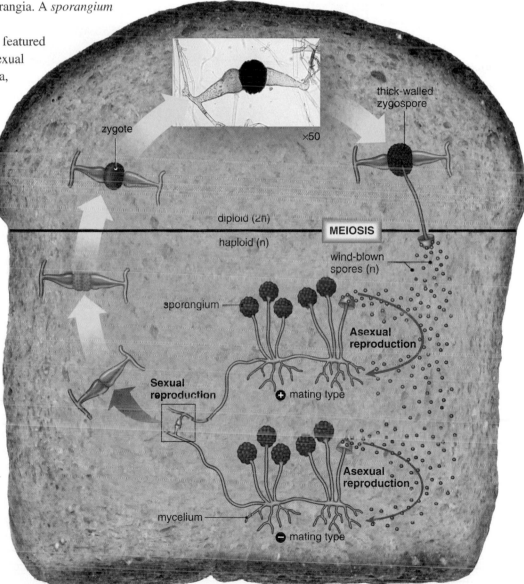

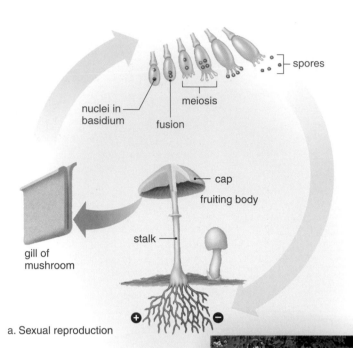

a. Sexual reproduction

Figure 18.23 **Sexual reproduction produces mushrooms.**

a. Fusion of + and − hyphae tips results in hyphae that form the mushroom (a fruiting body). The nuclei fuse in clublike structures attached to the gills of a mushroom, and meiosis produces spores. **b.** A circle of mushrooms is called a fairy ring.

Figure 18.24 **Carnivorous fungus.**

Fungi like this oyster fungus, a type of bracket fungus, grow on trees because they can digest cellulose and lignin. If this fungus meets a roundworm in the process, it immobilizes the worm and digests it also. The worm is a source of nitrogen for the fungus.

Mushroom

When you eat a mushroom, you are eating a *fruiting body,* whose function is to produce spores. In mushrooms, when the tips of + and − hyphae fuse, the haploid nuclei do not fuse immediately. Instead, so-called dikaryotic (two nuclei) hyphae form a mushroom consisting of a stalk and a cap. Club-shaped structures called basidia (sing., basidium) project from the gills located on the underside of the cap (Fig. 18.23*a*). Fusion of nuclei inside these structures is followed by meiosis and production of windblown spores. Each cap of a mushroom produces tens of thousands of spores.

In the soil, a mycelium is absorbing nutrients and may form a ring. Therefore, when the weather turns warm and rain is plentiful, the mycelium may put forth a circle of mushrooms on your lawn. At one time, people believed this ring indicated where fairies had joined hands and danced the night before (Fig. 18.23*b*).

Ecological Benefits of Fungi

Most fungi are **saprotrophs** that decompose the remains of plants, animals, and microbes in the soil. Fungal enzymes can degrade cellulose and even lignin in the woody parts of plants. That is why fungi so often grow on dead trees. This also means that fungi can be used to remove excess lignin from paper pulp. Ordinarily, lignin is difficult to extract from pulp and ends up being a pollutant once it is removed.

Along with the bacteria, which are also decomposers, fungi play an indispensable role in the environment by returning inorganic nutrients to photosynthesizers. Many people take advantage of the activities of bacteria and fungi by composting their food scraps or yard waste. When a gardener makes a compost pile and provides good conditions for decomposition to occur, the result is a dark, crumbly material that makes excellent fertilizer. And while the material may smell bad as decomposition is occurring, the finished compost looks and smells like rich, moist earth.

Some fungi eat animals that they encounter as they feed on their usual meals of dead organic remains. For example, the oyster fungus excretes a substance that anesthetizes any nematodes (roundworms) feeding on it (Fig. 18.24). After the worms become inactive, the fungal hyphae penetrate and digest their bodies, absorbing the nutrients. Other fungi snare, trap, or fire projectiles into nematodes and other small animals before digesting them. The animals serve as a source of nitrogen for the fungus.

Check Your Progress

1. Contrast sexual reproduction in black bread mold with sexual reproduction in a mushroom.
2. How do fungi contribute to ecological cycling?

Answers: 1. In black bread mold, meiosis produces spores in sporangia. In a mushroom, meiosis produces spores in clublike structures attached to the gills. 2. When fungi secrete enzymes that digest the remains of plants and animals in the environment, inorganic nutrients are returned to plants.

Mutualistic Relationships

In a mutualistic relationship, two different species live together and help each other out. **Lichens** are a mutualistic association between a particular fungus and cyanobacteria or green algae (Fig. 18.25). The fungal partner is efficient at acquiring nutrients and moisture, and therefore lichens can survive in poor soils, as well as on rocks with no soil. The organic acids given off by fungi release from rocks the minerals that can be used by the photosynthetic partner. Lichens are ecologically important because they produce organic matter and create new soil, allowing plants to invade the area.

Lichens occur in three varieties: compact crustose lichens, often seen on bare rocks or tree bark; shrublike fruticose lichens; and leaflike foliose lichens. Regardless, the body of a lichen has three layers. The fungal hyphae form a thin, tough upper layer and a loosely packed lower layer. These layers shield the photosynthetic cells in the middle layer. Specialized fungal hyphae that penetrate or envelop the photosynthetic cells transfer organic nutrients to the rest of the mycelium. The fungus not only provides minerals and water to the photosynthesizer, but also offers protection from predation and desiccation. Lichens can reproduce asexually by releasing fragments that contain hyphae and an algal cell. At first, the relationship between the fungi and algae was likely a parasite-and-host interaction. Over evolutionary time, the relationship apparently became more mutually beneficial, although how to test this hypothesis is a matter of debate at the present time.

Mycorrhizal fungi form mutualistic relationships with the roots of most plants, helping them grow more successfully in dry or poor soils, particularly those deficient in inorganic nutrients (Fig. 18.26). The fungal hyphae greatly increase the surface area from which the plant can absorb water and nutrients. It has been found beneficial to encourage the growth of mycorrhizal fungi when restoring lands damaged by strip mining or chemical pollution.

Mycorrhizal fungi may live on the outside of roots, enter between root cells, or penetrate root cells. The fungus and plant cells exchange nutrients, with the fungus bringing water and minerals to the plant and the plant providing organic carbon to the fungus. Early plant fossils indicate that the relationship between fungi and plant roots is an ancient one, and therefore it may have helped plants adapt to life on dry land. The general public is not familiar with mycorrhizal fungi, but a few people relish truffles, a mycorrhizal fungus that grows in oak and beech forests. Truffles are considered a gourmet delicacy.

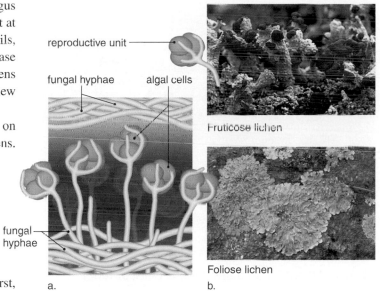

reproductive unit

fungal hyphae — algal cells

fungal hyphae

Fruticose lichen

Foliose lichen

a.   b.

Figure 18.25 Lichens.

a. Morphology of a lichen. **b.** Examples of lichens.

Figure 18.26 Plant growth experiment.

A soybean plant without mycorrhizal fungi grows poorly compared to two others infected with different strains of mycorrhizal fungi.

Check Your Progress

1. What organisms make up a lichen? Why is the relationship mutualistic?
2. What is the ecological importance of mycorrhizal fungi?

Answers: 1. A lichen is composed of a fungus and a cyanobacterium or a green alga. Each partner gives to the other. The fungus provides protection, water, and minerals, and the photosynthesizer provides organic nutrients. 2. Mycorrhizal fungi allow plants to colonize degraded land, and this helps restore the area.

Figure 18.27 **Commercial importance of fungi.**

Delicacies such as various types of mushrooms are fungi, and some of our other favorite foods require the participation of fungi to produce them. Significant medicines have also come from fungi.

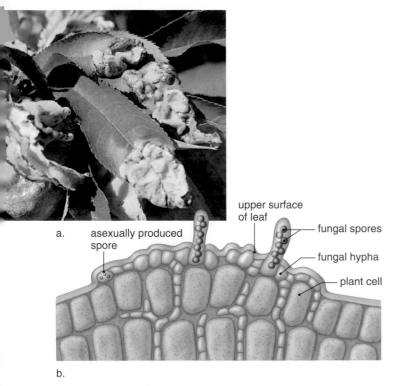

a. asexually produced spore

upper surface of leaf

fungal spores

fungal hypha

plant cell

b.

Figure 18.28 **Plant fungal disease.**

Peach leaf curl causes lesions on the leaf by invading the leaf, as shown in (**a**) the photo and (**b**) the diagram.

Economic Benefits of Fungi

Fungi help us produce medicines and many types of foods (Fig. 18.27). The mold *Penicillium* was the original source of penicillin, a breakthrough antibiotic that lead to an important class of cillin antibiotics. Cillin antibiotics have saved millions of lives.

Yeast fermentation is utilized to make bread, beer, wine, and distilled spirits. During fermentation, yeast produce the carbon dioxide that causes dough to rise when gas pockets are preserved as the bread bakes. Ethanol, also a product of yeast fermentation, is desired for the making of alcoholic beverages. Other types of fungal fermentations contribute to the manufacture of various cheeses and soy sauce from soybeans. Another commercial application of interest is the use of fungi to soften the centers of certain candies.

Fungi as Food

In the United States, the consumption of mushrooms has been steadily increasing. In 2001, total consumption of all mushrooms totalled 1.13 billion pounds—21% greater than in 1991. In addition to adding taste and texture to soups, salads, and omelets, and being used in stir-fry and on salads, mushrooms are an excellent low-calorie meat substitute with great nutritional value and lots of vitamins. Although there are thousands of mushroom varieties in the world, the white button mushroom, *Agaricus bisporus,* dominates the U.S. market. However, in recent years, sales of brown-colored variants have surged in popularity and have been one of the fastest-growing segments of the mushroom industry. Portabella is a marketing name used by the mushroom industry for the more flavorful brown strains of *A. bisporus*. The mushroom is brown because it is allowed to open, exposing the mature gills with their brown spores; crimini is the same brown strain, but it is not allowed to open before it is harvested. Non-*Agaricus* varieties, especially shiitake and oyster, have slowly gained in popularity over the past decade. Shiitake is touted for lowering cholesterol levels and having antitumor and antiviral properties.

Fungi as Disease-Causing Organisms

Fungi cause diseases in both plants and humans.

Fungi and Plant Diseases

Fungal pathogens, which usually gain access to plants by way of the stomata or a wound, are a major concern for farmers. Serious crop losses occur each year due to fungal disease. As much as a third of the world's rice crop is destroyed each year by rice blast disease. Corn smut is a major problem in the midwestern United States. Various rusts attack grains, and leaf curl is a disease of fruit trees (Fig. 18.28).

The life cycle of rusts may be particularly complex, since it requires two different host species to complete the cycle. Black stem rust of wheat uses barberry bushes as an alternate host. Eradication of barberry bushes in areas where wheat is grown helps control this rust. Fungicides are regularly applied to crops to limit the negative effects of fungal pathogens. Wheat rust can also be controlled by producing new and resistant strains of wheat.

Fungi and Human Diseases

As is well known, certain mushrooms are poisonous. The ergot fungus that grows on grain can result in ergotism when a person eats contaminated bread. Ergotism is characterized by hysteria, convulsions, and sometimes death.

Mycoses are diseases caused by fungi. Mycoses have three possible levels of invasion: Cutaneous mycoses only affect the epidermis; subcutaneous mycoses affect deeper skin layers; and systemic mycoses spread their effects throughout the body by traveling in the bloodstream. Fungal diseases that can be contracted from the environment include ringworm from soil, rose gardener's disease from thorns, Chicago disease from old buildings, and basketweaver's disease from grass cuttings. Opportunistic fungal infections now seen in AIDS patients stem from fungi that are always present in the body but take the opportunity to cause disease when the immune system becomes weakened.

Candida albicans causes the widest variety of fungal infections. Disease occurs when antibacterial treatments kill off the microflora community, allowing *Candida* to proliferate. Vaginal *Candida* infections are commonly called "yeast infections" in women. Oral thrush is a *Candida* infection of the mouth common in newborns and AIDS patients (Fig. 18.29*a*). In individuals with inadequate immune systems, *Candida* can move throughout the body, causing a systemic infection that can damage the heart, brain, and other organs.

Ringworm is a group of related diseases caused, for the most part, by fungi in the genus *Tinea*. Ringworm is a cutaneous infection that does not penetrate the skin. The fungal colony grows outward, forming a ring of inflammation. The center of the lesion begins to heal, giving the lesion its characteristic appearance, a red ring surrounding an area of healed skin (Fig. 18.29*b*). Athlete's foot is a form of tinea that affects the foot, mainly causing itching and peeling of the skin between the toes (Fig. 18.29*c*).

The vast majority of people living in the midwestern United States have been infected with *Histoplasma capsulatum*. This common soil fungus, often associated with bird droppings, leads in most cases to a mild "fungal flu." However, about 3,000 of these cases develop into a severe disease. About 50 persons die each year from histoplasmosis when the fungus grows within cells of the immune system. Lesions are formed in the lungs, leaving calcifications that are visible in X-ray images and resemble those of tuberculosis.

Because fungi are more closely related to animals than to bacteria, it is hard to design an antibiotic against fungi that does not also harm humans. Thus, researchers exploit any biochemical differences they can discover between humans and fungi.

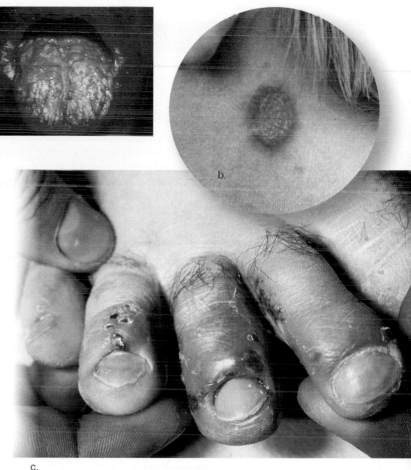

c.

Figure 18.29 Human fungal diseases.

a. Thrush, or oral candidiasis, is characterized by the formation of white patches on the tongue. **b.** Ringworm and (**c**) athlete's foot are caused by *Tinea* spp.

Check Your Progress

1. List the ways in which humans use fungi.
2. Why is a white button mushroom white and a portabella mushroom brown?
3. What are some common fungal diseases of humans? When is a fungal disease opportunistic?

Answers: 1. Fungi are used to process food and to make various types of medicines. Mushrooms are an important source of food. 2. A white button mushroom has not been allowed to mature. A portabella mushroom is mature, and spores are present. 3. Common fungal diseases include *Candida* (yeast) infections, ringworm, athlete's foot, and histoplasmosis. An opportunistic fungal disease occurs in a person with a weakened immune system.

THE CHAPTER IN REVIEW

Summary

18.1 Onto Land

Plants probably evolved from a multicellular, freshwater green alga about 500 MYA. Whereas algae are adapted to life in the water, plants are adapted to living on land. During the evolution of plants, four significant events are associated with adaptation to a land existence:

| 1. Protection of embryo | 2. Evolution of vascular tissue |
| 3. Evolution of seeds | 4. Evolution of flower |

Alternation of Generations

All plants have an alternation of generations life cycle, in which each type of plant exists in two forms, the sporophyte (2n) and the gametophyte (n):

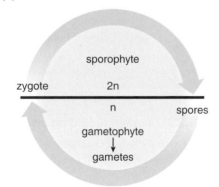

sporophyte
2n
zygote
n   spores
gametophyte
↓
gametes

18.2 Diversity of Plants

Nonvascular Plants

Bryophytes, represented by the mosses, are plants with the following characteristics:

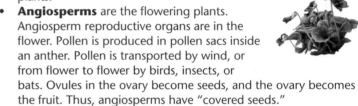

sporophyte

gametophyte

- There is no well-developed vascular tissue.
- The gametophyte is dominant, and flagellated sperm swim in external moisture to the egg.
- The sporophyte is dependent on the gametophyte.
- Windblown spores disperse the gametophyte.

Vascular Plants

In vascular plants, the dominant sporophyte has two kinds of well-defined conducting tissues. Xylem is specialized to conduct water and dissolved minerals, and phloem is specialized to conduct organic nutrients and hormones.

Seedless Vascular Plants The seedless vascular plants grew to enormous sizes during the Carboniferous period. Today, they include club mosses (e.g., ground pines), horsetails, and ferns.

Ferns best represent seedless vascular plants. They have the following characteristics:

- The sporophyte generation is dominant and produces windblown spores.
- The gametophyte is separate and independent. Flagellated sperm swim in external moisture to the egg.

sporophyte

Seed Plants Seed plants have male and female gametophytes. Gametophytes are reduced in size. The female gametophyte is retained within an ovule, and the male gametophyte is the mature pollen grain. The ovule becomes the seed, which contains a sporophyte embryo, food, and a seed coat.

- **Gymnosperms** are cone-bearing plants, represented by the pine tree. They have "naked seeds" because the seeds are not enclosed by fruit, as are those of flowering plants.

- **Angiosperms** are the flowering plants. Angiosperm reproductive organs are in the flower. Pollen is produced in pollen sacs inside an anther. Pollen is transported by wind, or from flower to flower by birds, insects, or bats. Ovules in the ovary become seeds, and the ovary becomes the fruit. Thus, angiosperms have "covered seeds."

18.3 The Fungi

Fungi include unicellular yeasts and multicellular mushrooms and molds.

General Biology of a Fungus

Fungi are neither animals nor plants.

- The body of a fungus is composed of thin filaments of cells, called hyphae, that form a mass called a mycelium.
- The cell wall contains chitin.
- Fungi produce windblown spores during both asexual and sexual reproduction.

Ecological Benefits of Fungi

- Fungi are saprotrophs that carry on external digestion. As decomposers, fungi keep ecological cycles going in the biosphere.
- Lichens (fungi plus cyanobacteria or green algae) are primary colonizers in poor soils or on rocks.
- Mycorrhizal fungi grow on or in plant roots and help the plant absorb minerals and water.

Economic Benefits of Fungi

- Fungi help produce foods and medicines, as well as serving as a source of food themselves.

Fungi as Disease-Causing Organisms

- Fungal pathogens of plants include blasts, smuts, and rusts that attack crops of great economic importance, such as rice and wheat.
- Human diseases caused by fungi include thrush, ringworm, and histoplasmosis.

Thinking Scientifically

1. Bare-root pine tree seedlings transplanted into open fields often grow very slowly. However, pine seedlings grow much more vigorously if they are dug from their native environment and then transplanted into a field, as long as some of the original soil is retained on the seedlings. Why is it so important to retain some native soil on the seedlings?

2. Some orchids produce flowers that have modified petals resembling female wasps. Male wasps emerge from their pupal stage a week earlier than females. The male wasps are attracted to the flowers even though they are not rewarded with nectar. Why do you suppose the male wasps visit the flowers? Why is this an especially effective pollen dispersal strategy?

Testing Yourself

Choose the best answer for each question.

1. Which of the following is not a plant adaptation to land?
   a. recirculation of water
   b. protection of embryo in maternal tissue
   c. development of flowers
   d. creation of vascular tissue
   e. seed production

2. Plant spores are
   a. haploid and genetically different from each other.
   b. haploid and genetically identical to each other.
   c. diploid and genetically different from each other.
   d. diploid and genetically identical to each other.

3. Label the parts of the generalized plant life cycle in the following illustration.

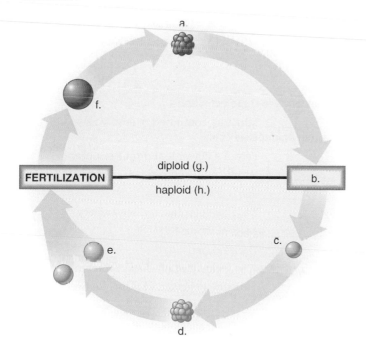

For questions 4–10, identify the group(s) to which each feature belongs. Each answer in the key may be used more than once. Each question may have more than one answer.

Key:
   a. nonvascular plants
   b. ferns
   c. gymnosperms
   d. angiosperms

4. Produce lignin in cell walls.
5. Exhibit alternation of generations.
6. Produce seeds.
7. Lack true roots, stems, and leaves.
8. Produce ovules that are not completely surrounded by sporophyte tissue.
9. Produce swimming sperm.
10. Produce flowers.
11. Label the parts of the flower in the following illustration.

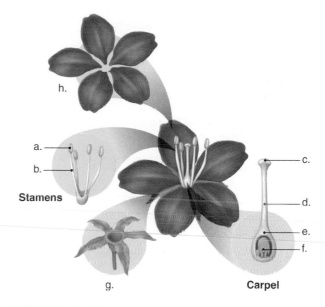

12. A fruit is derived from
    a. the corolla.
    b. an ovary.
    c. the stamen.
    d. an ovule.
    e. the calyx.

13. Which of the following statements about fungi is false?
    a. Most fungi are multicellular.
    b. Fungal cell walls are composed of cellulose.
    c. Fungi are nonmotile.
    d. Fungi digest their food before ingesting it.
    e. Fungi have haploid and diploid stages in their sexual life cycle.

14. The spores produced by a fruiting body are
    a. haploid and genetically different from each other.
    b. haploid and genetically identical to each other.
    c. diploid and genetically different from each other.
    d. diploid and genetically identical to each other.

15. Which of the following is a true statement?
    a. People don't eat mushrooms because they might be poisonous.
    b. Penicillin is derived from a fungus.
    c. The alcohol from yeast fermentation makes bread rise.
    d. Fungi are prokaryotes like bacteria.
    e. Fungi ingest their food in the same way animals do.

16. A fungal spore
    a. contains an embryonic organism.
    b. germinates directly into an organism.
    c. is always windblown.
    d. is most often diploid.
    e. Both b and c are correct.

17. Lichens
    a. can live on bare rocks.
    b. need a nitrogen source in order to live.
    c. are parasitic on trees.
    d. can reproduce asexually.
    e. Both a and d are correct.

18. Mycorrhizal fungi
    a. are a type of lichen.
    b. help plants gather solar energy.
    c. help plants gather inorganic nutrients.
    d. All of these are correct.

19. *Tinea* infections are generally found
    a. in the eye.
    b. in cartilage.
    c. on the skin.
    d. in the vagina.

20. Which of the following diseases is (are) caused by *Candida*?
    a. oral thrush
    b. athlete's foot
    c. vaginal yeast infection
    d. ringworm
    e. Both a and c are correct.

21. The gametophyte is the dominant generation in
    a. ferns.
    b. mosses.
    c. gymnosperms.
    d. angiosperms.
    e. More than one of these are correct.

22. A seed is a mature
    a. embryo.
    b. ovule.
    c. ovary.
    d. pollen grain.

23. A mushroom is like a plant because it
    a. is a multicellular eukaryote.
    b. produces spores.
    c. is adapted to a land environment.
    d. is photosynthetic.
    e. All but d are correct.

Go to www.mhhe.com/maderessentials for more quiz questions.

Bioethical Issue

Bristlecone pines are among the oldest trees on Earth. In 1964, a graduate student working in the southwestern United States took core samples from several trees to determine their age. One tree was found to be over 4,000 years old. When the student's coring tool broke, the U.S. Forest Service gave him permission to cut down the tree in order to accurately determine its age. The tree was found to be 4,862 years old—the oldest known living creature on Earth.

Did the student do anything wrong, scientifically or morally, considering that he was given permission to cut down the tree? Who should be responsible for protecting unique trees like the bristlecone pine?

Understanding the Terms

alternation of generations 286	mycelium 296
angiosperm 286, 292	mycorrhizal fungi 299
anther 292	nonvascular plant 287
bryophyte 286	ovary 292
calyx 292	ovule 290
carpel 292	petal 292
cone 291	phloem 288
conifer 291	pollen grain 290
corolla 292	pollen tube 293
cotyledon 293	pollination 290
double fertilization 293	saprotroph 298
endosperm 293	seed 285
fern 289	sepal 292
filament 292	spore 286
flower 285	sporophyte 286
frond 289	stamen 292
fruit 293	stigma 292
fungus (pl., fungi) 296	style 292
gametophyte 286	vascular plant 288
gymnosperm 286, 291	vascular tissue 285
lichen 299	xylem 288
lignin 288	

Match the terms to these definitions:

a. _____ Structure composed of embryo and stored food within a protective coat.

b. _____ Part of a flower that produces pollen.

c. _____ Water-conducting vascular tissue.

d. _____ Fern leaves.

e. _____ Male gametophytes in seed plants.

f. _____ Mass of fungal filaments.

g. _____ Decomposer.

h. _____ Generation that produces spores in a plant.

i. _____ Structure in flowering plants derived from the ovary.

Both Water and Land: Animals

Both a cat and a duckbill platypus, which has a bill like a duck, are mammals.

Both a gull, which flies in the air, and a penguin, which doesn't, are birds.

Classification isn't all that easy. A duckbill platypus has webbed feet and a paddlelike tail for swimming. It uses its bill to find and collect food. Millions of tiny, jelly-filled pits lining the toothless bill are electroreceptors that detect tiny electrical charges given off by the platypus's prey. It eats worms, crustaceans, fish, tadpoles, and adult frogs, which it grinds with two hornlike plates at the back of its bill. To reproduce, the duckbill platypus lays hard-shelled eggs.

▶ ▶ ▶

Somewhat surprisingly, the duckbill platypus is classified as a mammal, the same as a cat. The two characteristics of a mammal are hair and mammary glands, and the duckbill platypus has both. Thick hair covers its body, except for its bill and feet. Female platypuses produce milk from mammary glands, but because they have no nipples, the young lick the milk from their fur.

It's clear to us that gulls are birds. After all, they fly in the air. But penguins don't fly. When on land, they walk around on webbed hind feet with a flipper hanging at each side. Penguins are excellent divers and spend most of their time at sea fishing for fish, squid, and small crustaceans called krill. The Emperor penguin has been known to dive to a depth of more than 1,500 feet and stay down for over 18 minutes. Penguins' solid bones—not hollow like those of other birds—help them stay underwater.

So what makes a penguin a bird? Like the gull, the attractive coat of a penguin is made of feathers, and its flipper, built like a bird's wing, allows it to "fly" through the water. Penguins lay hard-shelled eggs and raise their chicks on land. A penguin might look like a mammal from a distance, but it is clearly a bird!

The content of this chapter will help you learn to classify some of the other major groups of animals.

19.1 Evolution of Animals

Animals can be contrasted with plants and fungi. Like both of these, animals are multicellular eukaryotes, but unlike plants, which make their food through photosynthesis, they are chemoheterotrophs and must acquire nutrients from an external source. So must fungi, but fungi digest their food externally and absorb the breakdown products. Animals ingest (eat) their food and digest it internally.

Animals usually carry on sexual reproduction and begin life only as a fertilized diploid egg. From this starting point, they undergo a series of developmental stages to produce an organism that has specialized tissues, usually within organs that carry on specific functions. Two types of tissues in particular—muscles and nerves—characterize animals. The presence of these tissues allows an animal to exhibit motility and a variety of flexible movements. The evolution of these tissues enables many types of animals to search actively for their food and to prey on other organisms. Coordinated movements also allow animals to seek mates, shelter, and a suitable climate—behaviors that have resulted in the vast diversity of animals. The more than 30 animal phyla we recognize today are believed to have evolved from a single ancestor.

Figure 19.1 illustrates the characteristics of a complex animal using the frog as an example. A frog goes through a number of embryonic stages

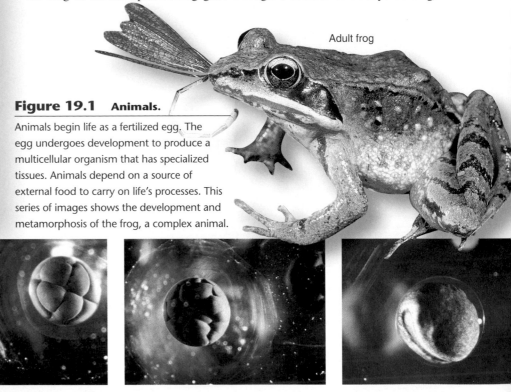

Adult frog

Figure 19.1   Animals.

Animals begin life as a fertilized egg. The egg undergoes development to produce a multicellular organism that has specialized tissues. Animals depend on a source of external food to carry on life's processes. This series of images shows the development and metamorphosis of the frog, a complex animal.

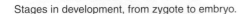

Stages in development, from zygote to embryo.

Stages in metamorphosis, from hatching to tadpole.

to become a larval form (the tadpole) with specialized organs, including muscular and nervous systems that enable it to swim. A larva is an immature stage that typically lives in a different habitat and feeds on different foods than the adult. By means of a change in body form called *metamorphosis*, the larva, which only swims, turns into a sexually mature adult frog that swims and hops. The aquatic tadpole lives on plankton, and the terrestrial adult typically feeds on insects and worms. A large African bullfrog will try to eat just about anything, including other frogs, as well as small fish, reptiles, and mammals.

The Evolutionary History of Animals

Animals are believed to have evolved from a protistan ancestor over 600 MYA, but the exact time of their appearance is not known. Figure 19.2 shows that the seas teemed with strange-looking invertebrates (animals without an **endoskeleton** [internal skeleton] of bone or cartilage) during the Cambrian period. Animal life was sparse before this period, and then suddenly representatives of all of today's animal phyla appear in the fossil record within some 10 million years. The appearance of this great variety of animals is so dramatic that the event is called the **Cambrian explosion.**

Paleontologists ask, "Why are animal fossils suddenly prevalent during this period?" It could be that only during the Cambrian period did animals evolve a protective **exoskeleton** (external skeleton), hard portions able to survive the forces that are apt to destroy fossils. Perhaps oxygen gas, pumped into the atmosphere by cyanobacteria and algae, was in sufficient quantity to permit aquatic animals to acquire oxygen, even though they were covered by hard outer skeletons. Then, too, predation may have played a role: Skeletons may have evolved during the Cambrian period because skeletons help protect animals from other animals.

Check Your Progress

1. Compare and contrast nutrition in animals with nutrition in fungi.
2. Compare and contrast nutrition in animals with nutrition in plants.

Answers: 1. Both animals and fungi are chemoheterotrophs. Fungi digest their food externally, while animals digest their food internally. 2. Animals are chemoheterotrophs, while plants are photosynthesizers.

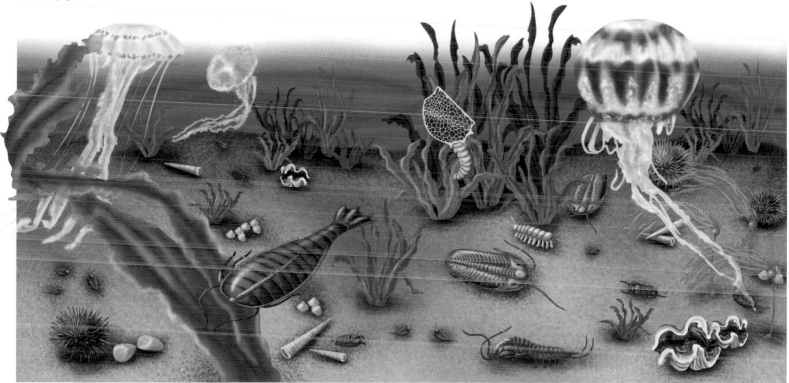

Figure 19.2 **Sea life in the Cambrian period.**

The animals depicted here are found as fossils in the Burgess Shale, a formation in the Rocky Mountains of British Columbia. Some lineages represented by these animals are still evolving today; others have become extinct.

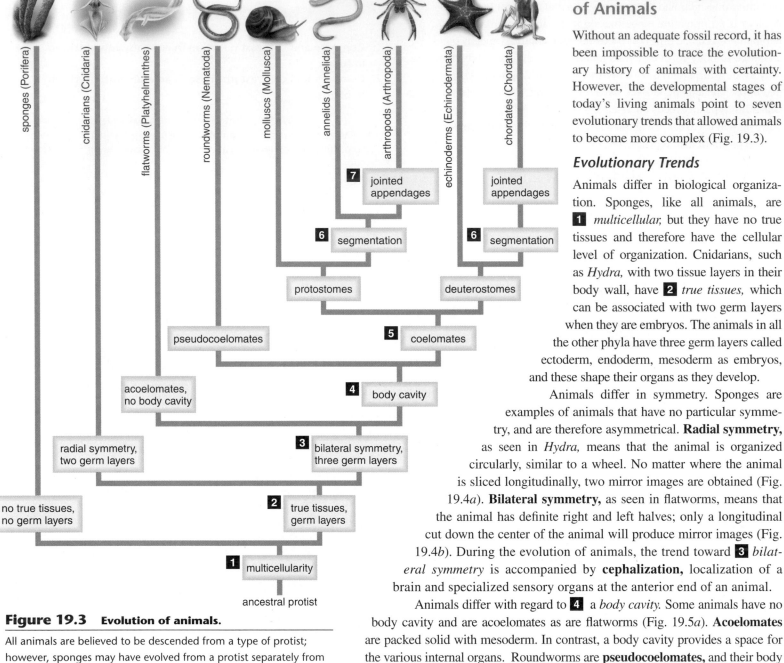

Figure 19.3   Evolution of animals.

All animals are believed to be descended from a type of protist; however, sponges may have evolved from a protist separately from the rest of the animals.

The Evolutionary Tree of Animals

Without an adequate fossil record, it has been impossible to trace the evolutionary history of animals with certainty. However, the developmental stages of today's living animals point to seven evolutionary trends that allowed animals to become more complex (Fig. 19.3).

Evolutionary Trends

Animals differ in biological organization. Sponges, like all animals, are **1** *multicellular,* but they have no true tissues and therefore have the cellular level of organization. Cnidarians, such as *Hydra,* with two tissue layers in their body wall, have **2** *true tissues,* which can be associated with two germ layers when they are embryos. The animals in all the other phyla have three germ layers called ectoderm, endoderm, mesoderm as embryos, and these shape their organs as they develop.

Animals differ in symmetry. Sponges are examples of animals that have no particular symmetry, and are therefore asymmetrical. **Radial symmetry,** as seen in *Hydra,* means that the animal is organized circularly, similar to a wheel. No matter where the animal is sliced longitudinally, two mirror images are obtained (Fig. 19.4*a*). **Bilateral symmetry,** as seen in flatworms, means that the animal has definite right and left halves; only a longitudinal cut down the center of the animal will produce mirror images (Fig. 19.4*b*). During the evolution of animals, the trend toward **3** *bilateral symmetry* is accompanied by **cephalization,** localization of a brain and specialized sensory organs at the anterior end of an animal.

Animals differ with regard to **4** a *body cavity.* Some animals have no body cavity and are acoelomates as are flatworms (Fig. 19.5*a*). **Acoelomates** are packed solid with mesoderm. In contrast, a body cavity provides a space for the various internal organs. Roundworms are **pseudocoelomates,** and their body

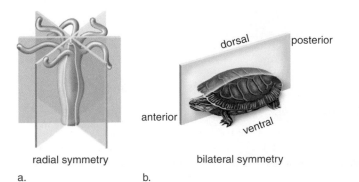

radial symmetry

a.

bilateral symmetry

b.

Figure 19.4   Radial versus bilateral symmetry.

a. With radial symmetry, two mirror images are obtained no matter how the animal is sliced longitudinally. Radially symmetrical animals tend to stay in one place and reach out in all directions to get their food. **b.** With bilateral symmetry, mirror images are obtained only if the animal is sliced down the middle. Bilaterally symmetrical animals tend to actively go after their food.

cavity is incompletely lined by mesoderm—that is, a layer of mesoderm exists beneath the body wall but not around the gut (Fig. 19.5b). Most animals are **5** coelomates in which the body cavity is completely lined with mesoderm (Fig. 19.5c). In animals with a coelom, mesentery, which is composed of strings of mesoderm, supports the internal organs. In coelomate animals, such as earthworms, lobsters, and humans, the mesoderm can interact not only with the ectoderm but also with the endoderm. Therefore, body movements are freer because the outer wall can move independently of the organs, and the organs have the space to become more complex. In animals without a skeleton, a coelom even acts as a so-called hydrostatic skeleton.

Animals can be nonsegmented or **6** *segmented*. **Segmentation** is the repetition of body parts along the length of the body. Annelids, such as an earthworm; arthropods, such as a lobster; and chordates, such as ourselves, are segmented. To illustrate your segmentation, run your hand along your backbone, which is composed of a series of vertebrae. Segmentation leads to specialization of parts because the various segments can become differentiated for specific purposes.

Two groups of animals, the arthropods and chordates, have **7** *jointed appendages,* which are particularly useful for movement on land. A lobster has a jointed exoskeleton, while a human has a jointed endoskeleton.

Molecular Data

Modern phylogenetic investigations take into account molecular data, primarily nucleotide sequences, when classifying animals. It is assumed that the more closely related two organisms are, the more nucleotide sequences they will have in common. An evolutionary tree based on molecular data is quite different from the one based only on anatomical characteristics that we have been discussing.

In the traditional tree, the protostomes are restricted to three phyla which have a true coelom (see Fig. 19.3). So, for example, flatworms, which are acoelomate, are not protostomes, whereas segmented worms are protostomes because they have a coelom, and their development follows a particular pattern.

Figure 19.6 shows an evolutionary tree based on molecular data. These data suggest that many more animal phyla should be designated protostomes because their rRNA sequences are so similar. However, the protostomes are divided into two groups. One group has a particular type of immature stage, the trochophore larva, as reflected in their name, lophotrochozoa. (The *lopho* stands for a feeding apparatus called a lophophore, which is seen in some animal phyla we are not studying.) The other group contains the roundworms and the arthropods. Both of these types of animals shed their outer covering as they grow; therefore, they are called ecdysozoa, which means molting animals.

Notice, too, that segmentation doesn't play a defining role in the phylogenetic tree based on molecular data. In the traditional tree, the segmented worms (e.g., earthworm) are placed close to the arthropods, which are also segmented. In the new tree, the segmented worms are lophotrochozoa, and the arthropods are ecdysozoa.

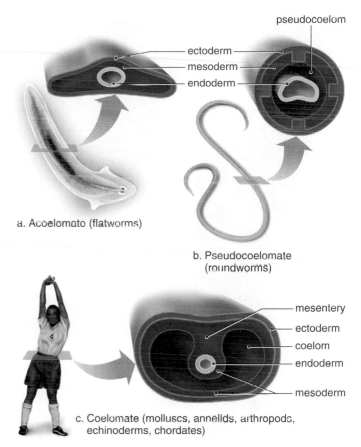

Figure 19.5 Type of body cavity.

a. Flatworms don't have a body cavity; they are acoelomates, and mesoderm fills the space between ectoderm and endoderm.
b. Roundworms are pseudocoelomates; they have a body cavity, and mesoderm lies next to the ectoderm but not the endoderm. **c.** Other animals are coelomates, and mesoderm lines the entire body cavity.

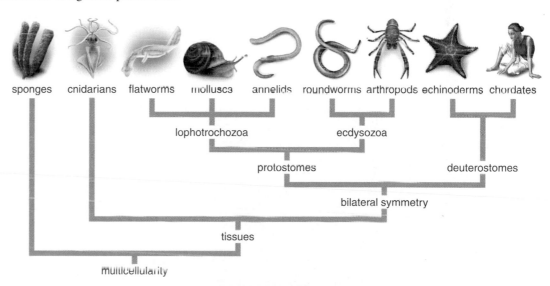

Figure 19.6 Proposed evolutionary tree of animals.

This proposed evolutionary tree is based on RNA sequence data. Compared to the traditional tree, many more phyla are considered protostomes and they are divided into two unique groups. The annelids (segmented worms, such as the earthworm) are placed in one group, and the arthropods are placed in another, even though both are segmented animals.

Body wall

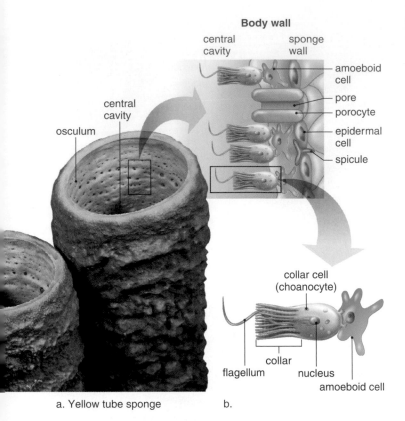

a. Yellow tube sponge   b.

Figure 19.7   Sponge anatomy.

Water enters a sponge through pores and circulates past collar cells before exiting at the mouth. Collar cells digest small particles that become trapped by their appendages, flagella in the microvilli of their collar. Amoeboid cells transport nutrients from cell to cell.

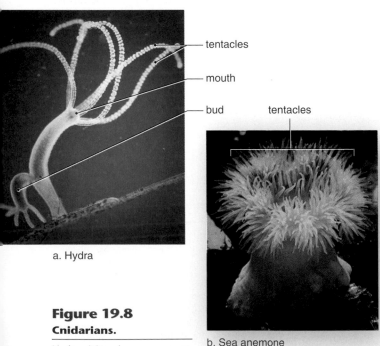

a. Hydra

Figure 19.8
Cnidarians.

Hydras (**a**) and sea anemones (**b**) are solitary polyps that use tentacles laden with stinging cells to capture their food.

b. Sea anemone

19.2   Introducing the Invertebrates

This section describes the invertebrate animals—the sponges, cnidarians, flatworms, and roundworms. Several key evolutionary trends are seen among these phyla (see Fig. 19.3).

Sponges: Multicellularity

Sponges are placed in phylum Porifera because their saclike bodies are perforated by many pores (Fig. 19.7a). Sponges are aquatic, largely marine animals that vary greatly in size, shape, and color. Sponges are multicellular but lack organized tissues. Therefore, *sponges have the cellular level of organization.* Evolutionists believe that sponges are outside the mainstream of the rest of animal evolution.

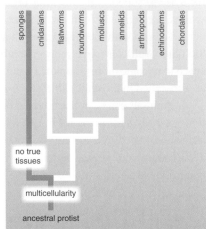

The interior of a sponge is lined with flagellated cells called *collar cells,* or choanocytes. The beating of the flagella produces water currents that flow through the pores into the central cavity and out through the osculum, the upper opening of the body. Even a simple sponge only 10 centimeters tall is estimated to filter as much as 100 liters of water each day. It takes this much water to supply the needs of the sponge. A sponge is a sedentary **filter feeder,** an organism that filters its food from the water by means of a straining device—in this case, the pores of the walls and the microvilli making up the collar of collar cells (Fig. 19.7b). Microscopic food particles that pass between the microvilli are engulfed by the collar cells and digested by them in food vacuoles.

Sponges can reproduce both asexually and sexually. They reproduce asexually by fragmentation or by *budding.* During budding, a small protuberance appears and gradually increases in size until a complete organism forms. Budding produces colonies of sponges that can become quite large. During sexual reproduction, eggs and sperm are released into the central cavity, and the zygote develops into a flagellated larva that may swim to a new location. If the cells of a sponge are mechanically separated, they will reassemble into a complete and functioning organism! Like many less specialized organisms, sponges are also capable of regeneration, or growth of a whole from a small part.

Some sponges have an endoskeleton composed of *spicules,* small needle-shaped structures with one to six rays. Most sponges have fibers of spongin, a modified form of collagen; a bath sponge is the dried spongin skeleton from which all living tissue has been removed. Today, however, commercial "sponges" are usually synthetic.

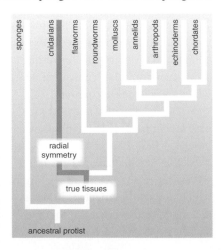

Cnidarians: True Tissues

Cnidarians (phylum Cnidaria) are an ancient group of invertebrates with a rich fossil record. Cnidarians are radially symmetrical and capture their prey with a ring of tentacles that bear specialized stinging cells, called cnidocytes (Fig. 19.8). Each cnidocyte has a capsule called a **nematocyst,** containing a long, spirally coiled, hollow thread. When the trigger of the cnidocyte is touched, the nematocyst is discharged. Some nematocysts merely trap a prey or predator; others have spines that penetrate and inject paralyzing toxins before the prey is captured and drawn into the gastrovascular cavity. Most cnidarians live in the sea, though there are a few freshwater species.

During development, cnidarians have two germ layers (ectoderm and endoderm), and as adults *cnidarians have the tissue level of organization.* Two basic body forms are seen among cnidarians—the polyp and the medusa. The mouth of a polyp is directed upward from the substrate, while the mouth of the medusa is directed downward. A medusa has much jellylike packing material and is commonly called a "jellyfish." Polyps are tubular and generally attached to a rock (Fig. 19.9).

Cnidarians, as well as other marine animals, have been the source of medicines, particularly drugs that counter inflammation.

Flatworms: Bilateral Symmetry

Flatworms (phylum Platyhelminthes) have *bilateral symmetry.* Like all the other animal phyla we will study, they also have three germ layers. However, the flatworms have no coelom—they are acoelomates. The presence of mesoderm in addition to ectoderm and endoderm gives bulk to the animal and leads to greater complexity.

Free-living flatworms, called **planarians,** have several body systems (Fig. 19.10), including a ladderlike nervous system. A small anterior brain and two lateral nerve cords are joined by cross-branches called transverse nerves. Planarians exhibit cephalization; aside from a brain, the "head" end has light-sensitive organs (the eyespots) and chemosensitive organs located on the auricles. Their three muscle layers—an outer circular layer, an inner longitudinal layer, and a diagonal layer—allow for varied movement. Their ciliated epidermis allows planarians to glide along a film of mucus.

Figure 19.9   More cnidarians.

a. The Portuguese man-of-war is a colony of polyp and medusa types of individuals. One polyp becomes a gas-filled float, and the other polyps are specialized for feeding. **b.** The calcium carbonate skeletons of corals form the coral reefs, areas of biological diversity. **c.** In jellyfishes, the medusa is the primary stage of the life cycle, and a polyp form remains quite small and inconspicuous.

a. Portuguese man-of-war

b. Cup coral   c. Jellyfish

Figure 19.10   Anatomy of a planarian.

a. This drawing shows that flatworms are bilaterally symmetrical and have a head region with eyespots. **b.** The excretory system with flame cells is shown in detail. **c.** The nervous system has a ladderlike appearance. **d.** The reproductive system (shown in brown) has both male and female organs, and the digestive system (shown in pink) has a single opening. When the pharynx is extended as shown in (**a**), a planarian sucks food up into a gastrovascular cavity that branches throughout its body.

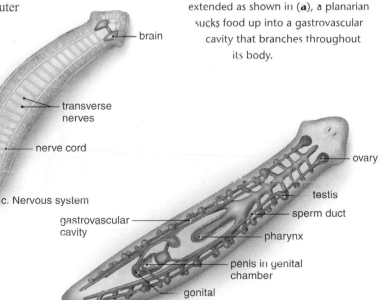

brain

transverse nerves

nerve cord

c. Nervous system

eyespots

auricle

pharynx extended through mouth

a. External appearance

excretory canal

flame cells

b. Excretory system

gastrovascular cavity

ovary

testis

sperm duct

pharynx

penis in genital chamber

gonital pore

d. Reproductive and digestive systems

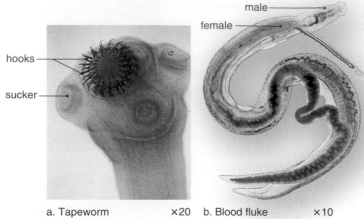

a. Tapeworm   ×20   b. Blood fluke   ×10

Figure 19.11   Parasitic flatworms.

a. The anterior end of a tapeworm has hooks and suckers for attachment to the intestinal wall. **b.** Sexes are separate in blood flukes, which cause schistosomiasis.

The animal captures food by wrapping itself around the prey, entangling it in slime, and pinning it down. Then the planarian extends a muscular pharynx, and by a sucking motion tears up and swallows its food. The pharynx leads into a three-branched gastrovascular cavity where digestion occurs. The digestive tract is incomplete because it has only one opening.

Planarians are **hermaphrodites,** meaning that they possess both male and female sex organs. The worms practice cross-fertilization: The penis of one is inserted into the genital pore of the other, and a reciprocal transfer of sperm takes place. The fertilized eggs hatch in 2–3 weeks as tiny worms.

The parasitic flatworms belong to two classes: the tapeworms and the flukes. As adults, tapeworms are endoparasites (internal parasites) of various vertebrates, including humans (Fig. 19.11a). They vary in size from a few millimeters to nearly 20 meters. Tapeworms have a well-developed anterior region called the scolex, which bears hooks and suckers for attachment to the intestinal wall of the host. Behind the scolex, a series of reproductive units called proglottids contain a full set of female and male sex organs. After fertilization, the organs within a proglottid disintegrate, and it becomes filled with mature eggs. The eggs, or else the mature proglottids, are eliminated in the feces of the host.

Flukes are all endoparasites of various vertebrates. The anterior end of these animals has an oral sucker and at least one other sucker used for attachment to the host. Flukes are usually named for the organ they inhabit; for example, there are blood, liver, and lung flukes. The blood fluke (*Schistosoma* spp.) occurs predominantly in the Middle East, Asia, and Africa. Nearly 800,000 persons die each year from infection by this fluke, called schistosomiasis. Adults are small (approximately 2.5 centimeters long) and may live for years in their human hosts (Fig. 19.11b).

Roundworms: Pseudocoelomates

Roundworms (phylum Nematoda) possess two anatomical features not seen in animals discussed previously: a *body cavity* and a *complete digestive tract* (Fig. 19.12). The body cavity is a pseudocoelom and is incompletely lined with mesoderm (see Fig. 19.5). The digestive tract is complete because it has both a mouth and an anus. Worms

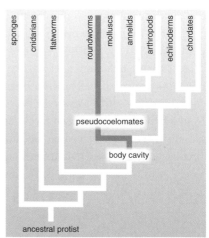

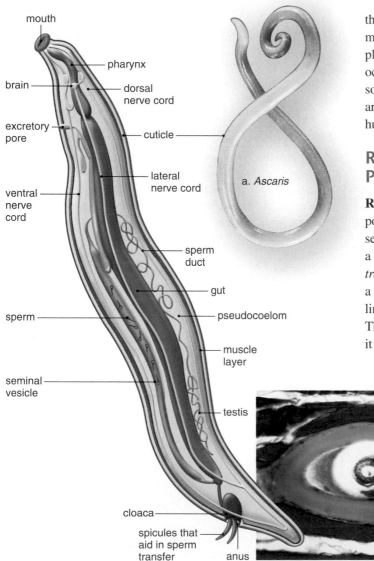

a. *Ascaris*

b. *Ascaris* anatomy   c. *Trichinella*   ×124

Figure 19.12   Roundworm anatomy.

a. The roundworm *Ascaris*. **b.** Roundworms such as *Ascaris* have a pseudocoelom and a complete digestive tract with a mouth and an anus. The sexes are separate; this is a male roundworm. **c.** The larvae of the roundworm *Trichinella* penetrate striated muscle fibers, where they coil in a sheath formed from the muscle fiber.

in general do not have a skeleton, but the fluid-filled pseudocoelom supports muscle contraction and enhances flexibility.

The roundworms are nonsegmented, meaning that they have a smooth outside body wall. Roundworms are generally colorless and less than 5 centimeters in length, and they occur almost everywhere—in the sea, in fresh water, and in the soil—in such numbers that thousands of them can be found in a small area. Many are free-living and feed on algae, fungi, microscopic animals, dead organisms, and plant juices, causing great agricultural damage. Parasitic roundworms live anaerobically in every type of animal and many plants. Several parasitic roundworms infect humans.

A female *Ascaris,* a human parasite, is very prolific, producing over 200,000 eggs daily. The eggs are eliminated with host feces, and under the right conditions they can develop into a worm within two weeks. The eggs enter the body via uncooked vegetables, soiled fingers, or ingested fecal material, and hatch in the intestines. The juveniles make their way into the cardiovascular system and are carried to the heart and lungs. From the lungs, the larvae travel up the trachea, where they are swallowed and eventually reach the intestines. There the larvae mature and begin feeding on intestinal contents.

Trichinosis is a fairly serious human infection rarely seen in the United States. The female trichina worm burrows into the wall of the host's small intestine; here she deposits live larvae that are carried by the bloodstream to the skeletal muscles, where they encyst. Once the adults are in the small intestine, digestive disorders, fatigue, and fever occur. After the larvae encyst, the symptoms include aching joints, muscle pain, and itchy skin. Humans catch the disease when they eat infected meat.

Elephantiasis is caused by a roundworm called a filarial worm, which utilizes mosquitoes as a secondary host. Because the adult worms reside in lymphatic vessels, fluid return is impeded, and the limbs of an infected human can swell to an enormous size, even resembling those of an elephant. When a mosquito bites an infected person, it can transport larvae to a new host.

Other roundworm infections are more common in the United States. Children frequently acquire pinworm infection, and hookworm is seen in the southern states, as well as worldwide. A hookworm infection can be very debilitating because the worms attach to the intestinal wall and feed on blood. Good hygiene, proper disposal of sewage, thorough cooking of meat, and regular deworming of pets usually protect people from parasitic roundworms.

19.3   Protostomes and Deuterostomes Compared

Protostomes and deuterostomes can be distinguished on the basis of embryological development. In this text, we will discuss molluscs, annelids, and arthropods as representative protostomes, and echinoderms and chordates as deuterostomes. The first embryonic opening in both protostomes and deuterostomes is called the blastopore. In **protostomes** (*proto,* before), the blastopore becomes the mouth, and in **deuterostomes,** it becomes the anus. The coelom develops differently in protostomes and deuterostomes (Fig. 19.13).

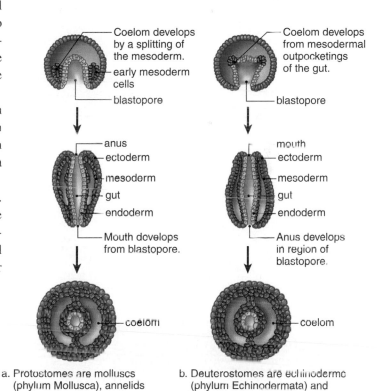

a. Protostomes are molluscs (phylum Mollusca), annelids (phylum Annelida), and arthropods (phylum Arthropoda).

b. Deuterostomes are echinoderms (phylum Echinodermata) and chordates (phylum Chordata).

Figure 19.13   Development in protostomes and deuterostomes.

a. In protostomes, the mesoderm splits, creating a coelom; the blastopore becomes the mouth. **b.** In deuterostomes, the primitive gut outpockets to form the coelom; the blastopore becomes the anus.

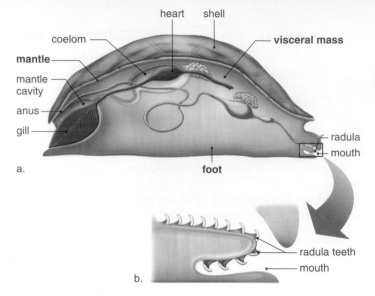

a.

b.

Figure 19.14   Body plan of molluscs.

Molluscs have a three-part body: **a.** a muscular foot, a visceral mass, and a mantle. **b.** In the mouth, the radula is a tonguelike organ that bears rows of tiny, backward-pointing teeth.

Land snail

Nudibranch

Octopus

Chambered nautilus

Scallop

Mussels

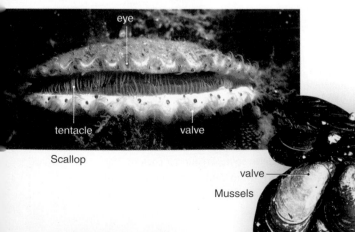

Molluscs, Annelids, and Arthropods

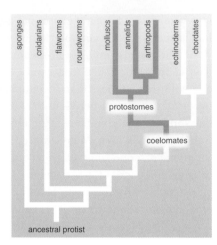

Traditionally, molluscs, annelids, and arthropods have been grouped together as protostomes.

Molluscs

Despite being a very large and diversified group, all molluscs (phylum Mollusca) have a body composed of at least three distinct parts: (1) The *visceral mass* is the soft-bodied portion that contains internal organs; (2) the *foot* is the strong, muscular portion used for locomotion; (3) the *mantle* is a membranous or sometimes muscular covering that envelops, but does not completely enclose, the visceral mass. In addition, the *mantle cavity* is the space between the two folds of the mantle. The mantle may secrete an exoskeleton called a *shell.* Another feature often present is a rasping, tonguelike *radula,* an organ that bears many rows of teeth and is used to obtain food (Fig. 19.14).

In **gastropods** (meaning stomach-footed), including nudibranchs, conchs, and snails, the foot is ventrally flattened, and the animal moves by muscle contractions that pass along the foot (Fig. 19.15, *top*). Many gastropods are herbivores that use their radulas to scrape food from surfaces. Others are carnivores, using their radulas to bore through surfaces such as bivalve shells to obtain food. In snails that are terrestrial, the mantle is richly supplied with blood vessels and functions as a lung.

In **cephalopods** (meaning head-footed), including octopuses, squids, and nautiluses, the foot has evolved into tentacles about the head (Fig. 19.15, *middle*). The tentacles seize prey, and then a powerful beak and a radula tear it apart. Cephalopods possess well-developed nervous systems and complex sensory organs. The large brain is formed from a fusion of ganglia, and nerves leaving the brain supply various parts of the body. An especially large pair of nerves controls the rapid contraction of the mantle, allowing these animals to move quickly by jet propulsion of water. Rapid movement and the secretion of a brown or black pigment from an ink gland help cephalopods escape their enemies. In the squid and octopus, a well-developed eye resembles that of vertebrates, having a cornea, lens, retina, and iris. Octopuses have no shell, and squids have only a remnant of one concealed beneath the skin. Scientific experiments reveal that octopuses in particular are highly intelligent.

Clams, oysters, scallops, and mussels are called **bivalves** because of the two parts to their shells (Fig. 19.15, *bottom*). They have a muscular foot that projects ventrally from the shell. In a clam, such as the freshwater clam, the calcium carbonate shell has an inner layer of mother-of-pearl. The clam is a filter feeder. Food particles and water enter the mantle cavity by way of the incurrent siphon, a

Figure 19.15   Three classes of molluscs.

Top row: Snails (*left*) and nudibranchs (*right*) are gastropods. Middle row: Octopuses (*left*) and nautiluses (*right*) are cephalopods. Bottom row: Scallops (*left*) and mussels (*right*) are bivalves.

posterior opening between the two valves. Mucous secretions cause smaller particles to adhere to the gills, and ciliary action sweeps them toward the mouth.

Molluscs have some economic importance as a source of food and pearls. If a foreign body is placed between the mantle and the shell of a clam, concentric layers of shell are deposited about the particle to form a pearl.

Annelids: Segmented Worms

Annelids (phylum Annelida) are *segmented,* as can be seen externally by the rings that encircle the body of an earthworm. Partitions called septa divide the well-developed, fluid-filled coelom, which is used as a hydrostatic skeleton to facilitate movement. In annelids, the complete digestive tract body plan has led to specialization of parts (Fig. 19.16). For example, the digestive system may include a pharynx, esophagus, crop, gizzard, intestine, and accessory glands. Annelids have an extensive closed circulatory system with blood vessels that run the length of the body and branch to every segment. The nervous system consists of a brain connected to a ventral nerve cord, with ganglia in each segment. The excretory system consists of nephridia in most segments. A **nephridium** is a tubule that collects waste material and excretes it through an opening in the body wall.

Most annelids are polychaetes (having many setae per segment) that live in marine environments. Setae are bristles that anchor the worm or help it move. The clam worm *Nereis* is a predator. It preys on crustaceans and other small animals, capturing them with a pair of strong, chitinous jaws that extend with a part of the pharynx (Fig. 19.17a). In support of its predatory way of life, *Nereis* has a well-defined head region, with eyes and other sense organs. Other polychaetes are sedentary (sessile) tube worms, with tentacles that form a funnel-shaped fan. Water currents created by the action of cilia trap food particles that are directed toward the mouth (Fig. 19.17b).

The oligochaetes (few setae per segment) include the earthworms. Earthworms do not have a well-developed head, and they reside in soil where there is adequate moisture to keep the body wall moist for gas exchange. They are scavengers that feed on leaves or any other organic matter, living or dead, that can conveniently be taken into the mouth along with dirt.

Leeches have no setae but have the same body plan as other annelids. Most are found in fresh water, but some are marine or even terrestrial. The medicinal leech can be as long as 20 centimeters, but most leeches are much shorter. A leech has two *suckers,* a small one around the mouth and a large posterior one. While some leeches are free-living, most are fluid feeders that penetrate the surface of an animal by using a proboscis or their jaws, and then suck in fluids with their powerful pharynx. Leeches are able to keep blood flowing and prevent clotting by means of a substance in their saliva known as *hirudin,* a powerful anticoagulant. This secretion has added to their potential usefulness in the field of medicine today (Fig. 19.17c).

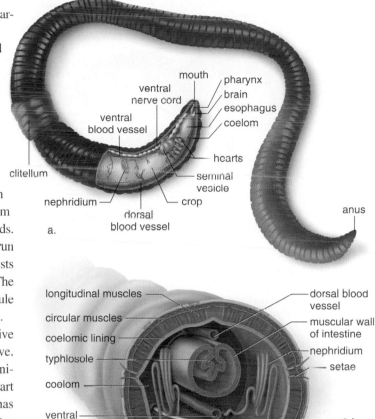

a.

Figure 19.16 Earthworm anatomy.

a. Internal anatomy of the anterior part of an earthworm. **b.** Cross section of an earthworm.

Figure 19.17 Other annelids.

A polychaete can be predacious like this marine clam worm (**a**) or live in a tube like this fan worm called the Christmas tree worm (**b**). **c.** The medicinal leech, also an annelid, is sometimes used to remove blood from tissues after surgery.

a. Clam worm

b. Christmas tree worm

c. Medicinal leech

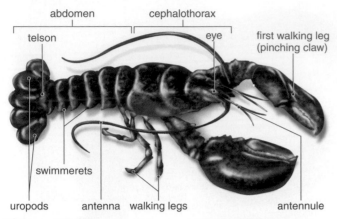

Figure 19.18 **Exoskeleton and jointed appendages.**

Arthropods, such as a lobster, have various appendages attached to the head region of a cephalothorax and five pairs of walking legs attached to the thorax region. Appendages called swimmerets, used in reproduction and swimming, are attached to the abdomen. The uropods and telson make up a fan-shaped tail.

Figure 19.19 **Monarch butterfly metamorphosis.**

a. A caterpillar (larva) eats and grows. **b.** After the larva goes through several molts, it builds a cocoon around itself and becomes a pupa. **c.** Inside the pupa, the larva undergoes changes in organ structure to become an adult. **d, e.** The adult butterfly emerges from the cocoon and reproduces, and the cycle begins again.

Arthropods: Jointed Appendages

Arthropods (phylum Arthropoda) are extremely diverse. Over one million species have been discovered and described, but some experts suggest that as many as 30 million arthropods may exist—most of them insects. The success of arthropods can be attributed to the following six characteristics:

1. *Jointed appendages.* Basically hollow tubes moved by muscles, jointed appendages have become adapted to different means of locomotion, food gathering, and reproduction (Fig. 19.18). These modifications account for much of the diversity of arthropods.
2. *Exoskeleton.* A rigid but jointed exoskeleton is composed primarily of **chitin,** a strong, flexible, nitrogenous polysaccharide. The exoskeleton serves many functions, including protection, prevention of desiccation, attachment for muscles, and locomotion.

   Because an exoskeleton is hard and nonexpandable, arthropods must undergo **molting,** or shedding of the exoskeleton, as they grow larger. During molting, arthropods are vulnerable and are attacked by many predators.
3. *Segmentation.* In many species, the repeating units of the body are called segments. Each has a pair of jointed appendages. In others, the segments are fused into a head, thorax, and abdomen.
4. *Well-developed nervous system.* Arthropods have a brain and a ventral nerve cord. The head bears various types of sense organs, including compound and simple eyes. Many arthropods also have well-developed touch, smell, taste, balance, and hearing capabilities. Arthropods display many complex behaviors and communication skills.
5. *Variety of respiratory organs.* Marine forms utilize gills; terrestrial forms have book lungs (e.g., spiders) or air tubes called tracheae. Tracheae serve as a rapid way to transport oxygen directly to the cells. The circulatory system is open, with the dorsal heart pumping blood into various sinuses throughout the body.
6. *Reduced competition through metamorphosis.* Many arthropods undergo a change in form and physiology as a larva becomes an adult. Metamorphosis allows the larva to have a different lifestyle than the adult (Fig. 19.19). For example, larval crabs live among and feed on plankton, while adult crabs are bottom dwellers that catch live prey or scavenge dead organic matter. Among insects such as butterflies, the caterpillar feeds on leafy vegetation, while the adult feeds on nectar.

Crustaceans, a name derived from their hard, crusty exoskeleton, are a group of largely marine arthropods that include barnacles, shrimps, lobsters, and crabs (Fig. 19.20). There are also some freshwater crustaceans, including the crayfish, and some terrestrial ones, including the sowbug, or pillbug. Although crustacean anatomy is extremely diverse, the head usually bears a pair of compound eyes and five pairs of appendages. The first two pairs of appendages, called antennae and antennules, respectively, lie in front of the mouth and have sensory functions. The other three pairs are mouthparts used in feeding. In a lobster, the thorax bears five pairs of walking legs. The first walking leg is a pinching claw. The *gills* are situated above the walking legs. The head and thorax are fused into a cephalothorax, which is covered on the top and sides by a nonsegmented carapace. The abdominal segments, which are largely muscular, are equipped with swimmerets, small paddlelike structures. The last two segments bear the uropods and the telson, which make up a fan-shaped tail to propel the lobster backward (see Fig. 19.18).

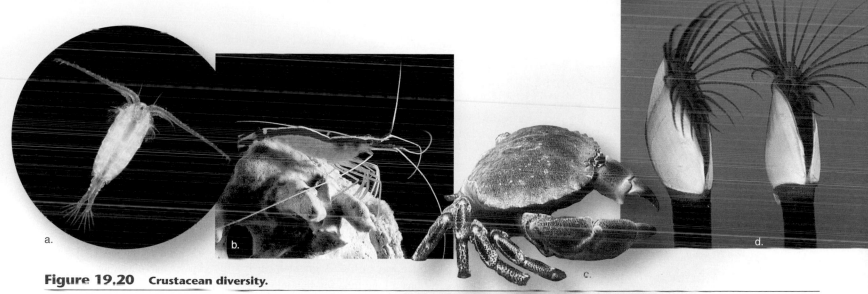

a.

b.

c.

d.

Figure 19.20 Crustacean diversity.

a. A copepod uses its long antennae for floating and its feathery maxillae for filter feeding. Shrimp (**b**) and crabs (**c**) are decapods—they have five pairs of walking legs. Shrimp resemble crayfish more closely than they do crabs, which have a reduced abdomen. Marine shrimp feed on copepods. **d.** The gooseneck barnacle is attached to an object by a long stalk. Barnacles have no abdomen and a reduced head; the thoracic legs project through a shell to filter feed. Barnacles often live on man-made objects such as ships, buoys, and cables.

Crustaceans play a vital role in the food chain. Tiny crustaceans known as krill are a major source of food for baleen whales, sea birds, and seals. Countries such as Japan are harvesting krill for human use. Copepods and other small crustaceans are primary consumers in marine and aquatic ecosystems. Many species of lobsters, crabs, and shrimp are important in the seafood industry. Some barnacles are destructive to wharfs, piers, and boats.

Among arthropods, the **arachnids** include spiders, scorpions, ticks, mites, and harvestmen ("daddy longlegs") (Fig. 19.21*a*). Spiders have a narrow waist that separates the cephalothorax from the abdomen. Some spiders inject poisons into their prey and digest their food externally before sucking it into the stomach. Spiders use silk threads for all sorts of purposes, from lining their nests to catching prey. The internal organs of spiders also show how they are adapted to a terrestrial way of life. Malpighian tubules work in conjunction with rectal glands to reabsorb ions and water before a relatively dry nitrogenous waste (uric acid) is excreted. Invaginations of the inner body wall form lamellae ("pages") of spiders' so-called book lungs.

Scorpions are the oldest terrestrial arthropods (Fig. 19.21*b*). Today, they occur in the tropics, subtropics, and temperate regions of North America. They are nocturnal and spend most of the day hidden under a log or a rock. Ticks and mites are parasites. Ticks suck the blood of vertebrates and sometimes transmit diseases, such as Rocky Mountain spotted fever or Lyme disease. Chiggers, the larvae of certain mites, feed on the skin of vertebrates.

The horseshoe crab is grouped with the arachnids because the first pair of appendages are pincerlike structures used for feeding and defense. Horseshoe crabs have pedipalps, which they use as feeding and sensory structures, and four pairs of walking legs (Fig. 19.21*c*).

Centipedes (Fig. 19.21*d*), with a pair of appendages on every segment, are carnivorous, while millipedes (Fig. 19.21*e*), with two pairs of legs on most segments, are herbivorous. These animals are grouped with the insects because they have similar head appendages.

a.

b.

d.

e.

c.

Figure 19.21 More arthropods.

a. The black widow spider is a poisonous spider that spins a web. **b.** Scorpions have large pincers in front and along abdomen, which ends with a stinger containing venom. **c.** Horseshoe crabs are common along the North American east coast. **d.** A centipede has a pair of appendages on almost every segment. **e.** A millipede has two pairs of legs on most segments.

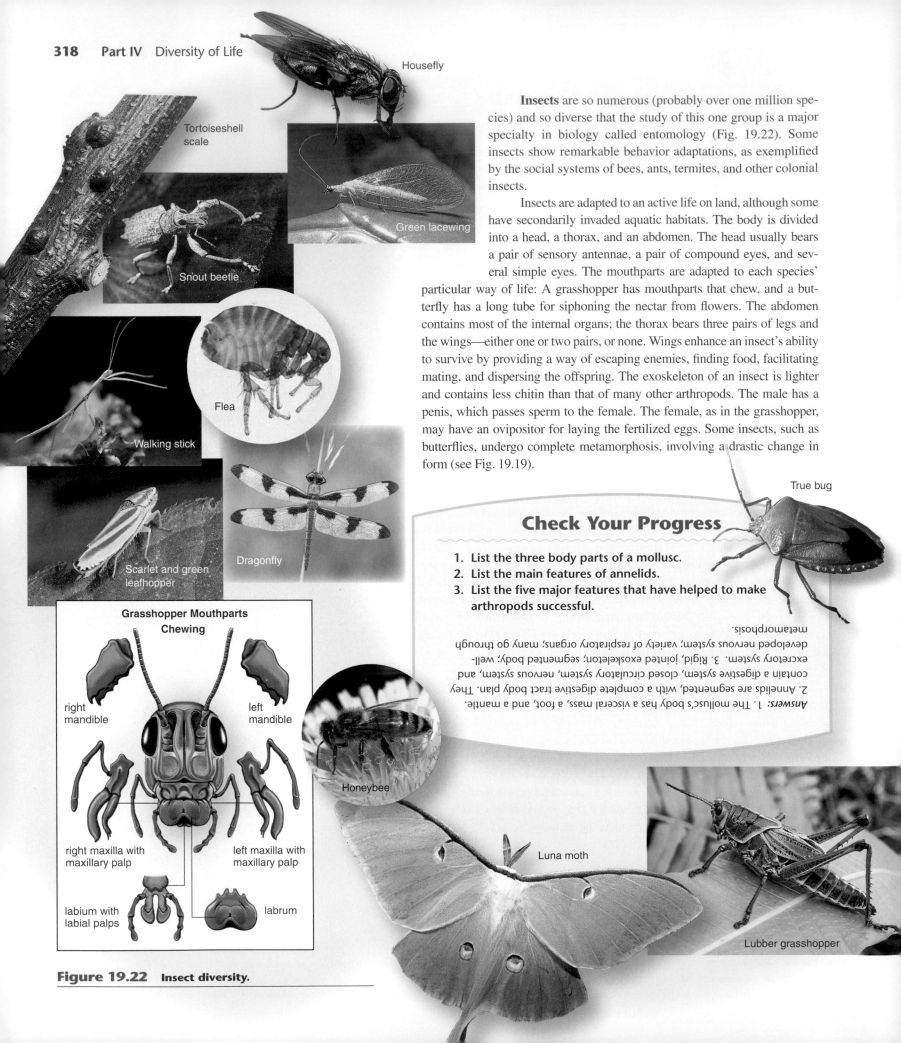

Housefly

Tortoiseshell scale

Green lacewing

Snout beetle

Flea

Walking stick

Scarlet and green leafhopper

Dragonfly

Insects are so numerous (probably over one million species) and so diverse that the study of this one group is a major specialty in biology called entomology (Fig. 19.22). Some insects show remarkable behavior adaptations, as exemplified by the social systems of bees, ants, termites, and other colonial insects.

Insects are adapted to an active life on land, although some have secondarily invaded aquatic habitats. The body is divided into a head, a thorax, and an abdomen. The head usually bears a pair of sensory antennae, a pair of compound eyes, and several simple eyes. The mouthparts are adapted to each species' particular way of life: A grasshopper has mouthparts that chew, and a butterfly has a long tube for siphoning the nectar from flowers. The abdomen contains most of the internal organs; the thorax bears three pairs of legs and the wings—either one or two pairs, or none. Wings enhance an insect's ability to survive by providing a way of escaping enemies, finding food, facilitating mating, and dispersing the offspring. The exoskeleton of an insect is lighter and contains less chitin than that of many other arthropods. The male has a penis, which passes sperm to the female. The female, as in the grasshopper, may have an ovipositor for laying the fertilized eggs. Some insects, such as butterflies, undergo complete metamorphosis, involving a drastic change in form (see Fig. 19.19).

True bug

Check Your Progress

1. List the three body parts of a mollusc.
2. List the main features of annelids.
3. List the five major features that have helped to make arthropods successful.

Answers: 1. The mollusc's body has a visceral mass, a foot, and a mantle. 2. Annelids are segmented, with a complete digestive tract body plan. They contain a digestive system, closed circulatory system, nervous system, and excretory system. 3. Rigid, jointed exoskeleton; segmented body; well-developed nervous system; variety of respiratory organs; many go through metamorphosis.

Grasshopper Mouthparts
Chewing

right mandible

left mandible

right maxilla with maxillary palp

left maxilla with maxillary palp

labium with labial palps

labrum

Honeybee

Luna moth

Lubber grasshopper

Figure 19.22 **Insect diversity.**

19.5   Echinoderms and Chordates

The deuterostomes (see page 313) include the echinoderms and the chordates.

Echinoderms

It may seem surprising that **echino-derms** (phylum Echinodermata) lack features associated with vertebrates such as humans, and yet are related to chordates (Fig. 19.23). For example, the echinoderms are often radially, not bilaterally, symmetrical. Their larva is a free-swimming filter feeder with bilateral symmetry, but it metamorphoses into a radially symmetrical adult. Also, adult echinoderms do not have a head, brain, or segmentation. The nervous system consists of nerves in a ring around the mouth extending outward radially. Nevertheless, they are classified as deuterostomes as are the chordates.

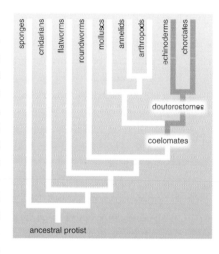

Echinoderm locomotion depends on a water vascular system. In the sea star, water enters this system through a sieve plate. Eventually it is pumped into many tube feet, expanding them. When the foot touches a surface, the center withdraws, producing suction that causes the foot to adhere to the surface. By alternating the expansion and contraction of its many tube feet, a sea star moves slowly along.

Echinoderms don't have a complex respiratory, excretory, or circulatory system. Fluids within the coelomic cavity and a water vascular system carry out many of these functions. For example, gas exchange occurs across the skin gills and the tube feet. Nitrogenous wastes diffuse through the coelomic fluid and the body wall.

In ecosystems, most echinoderms feed variously on organic matter in the sea or substratum, but sea stars prey upon crustaceans, molluscs, and other invertebrates. From the human perspective, sea stars cause extensive economic loss because they consume oysters and clams before they can be harvested. Fishes and sea otters eat echinoderms, and scientists favor echinoderms for embryological research.

Figure 19.23   Echinoderm diversity.

a. Anatomy of a sea star. **b.** Sea lilies are immobile, but feather stars can move about. They usually cling to coral or sponges where they feed on plankton. **c.** Sea cucumbers have a long, leathery body that resembles a cucumber, except for the feeding tentacles about the mouth. **d.** Brittle stars have a central disk from which long, flexible arms radiate. **e.** Sea urchins and sand dollars have spines for locomotion, defense, and burrowing.

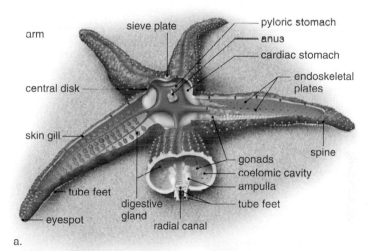

a.

b. Sea lily (above), feather star (right)

c. Sea cucumber

d. Brittle star

e. Sea urchin (left), sand dollar (right)

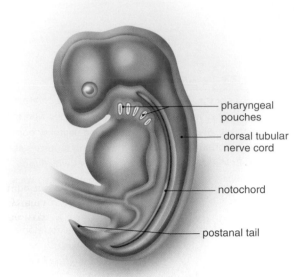

- pharyngeal pouches
- dorsal tubular nerve cord
- notochord
- postanal tail

Figure 19.24   The four chordate characteristics.

a.

b.

Figure 19.25   Invertebrate chordates.

a. Tunicates (sea squirts) have numerous gill slits, the only chordate characteristic that remains in the adult. **b.** Lancelets have all four chordate characteristics as adults.

Chordates

To be classified as a **chordate** (phylum Chordata), an animal must at some time during its life history have the four characteristics depicted in Figure 19.24 and listed here:

1. A *dorsal supporting rod,* called a **notochord.** The notochord is located just below the nerve cord toward the back (i.e., dorsal). **Vertebrates** have an endoskeleton of cartilage or bone, including a vertebral column that has replaced the notochord during development.

2. A *dorsal tubular nerve cord.* Tubular means that the cord contains a canal filled with fluid. In vertebrates, the nerve cord is protected by the vertebrae. Therefore, it is called the spinal cord because the vertebrae form the spine.

3. *Pharyngeal pouches.* These structures are seen only during embryonic development in most vertebrates. In the invertebrate chordates, the fishes, and some amphibian larvae, the pharyngeal pouches become functioning gills. Water passing into the mouth and the pharynx goes through the gill slits, which are supported by gill arches. In terrestrial vertebrates that breathe with lungs, the pouches are modified for various purposes. In humans, the first pair of pouches become the auditory tubes. The second pair become the tonsils, while the third and fourth pairs become the thymus gland and the parathyroids.

4. A *tail.* Because the tail extends beyond the anus, it is called a postanal tail.

The Invertebrate Chordates

There are a few invertebrate chordates in which the notochord is never replaced by the vertebral column. **Tunicates** (subphylum Urochordata) live on the ocean floor and take their name from a tunic that makes the adults look like thick-walled, squat sacs. They are also called sea squirts because they squirt water from one of their siphons when disturbed (Fig. 19.25*a*). The tunicate larva is bilaterally symmetrical and has the four chordate characteristics. Metamorphosis produces the sessile adult in which numerous cilia move water into the pharynx and out numerous gill slits, the only chordate characteristic that remains in the adult.

Lancelets (subphylum Cephalochordata) are marine chordates only a few centimeters long. They have the appearance of a lancet—a small, two-edged surgical knife (Fig. 19.25*b*). Lancelets are found in the shallow water along most coasts, where they usually lie partly buried in sandy or muddy substrates with only their anterior mouth and gill apparatus exposed. They feed on microscopic particles filtered out of the constant stream of water that enters the mouth and exits through the gill slits. Lancelets retain the four chordate characteristics as adults. In addition, segmentation is present, as witnessed by the fact that the muscles are segmentally arranged and the dorsal tubular nerve cord has periodic branches.

Evolutionary Trends Among the Chordates

Figure 19.26 depicts the evolutionary tree of the chordates and previews the animal groups we will discuss in the remainder of this chapter. The figure also lists at least one main evolutionary trend that distinguishes each group of animals from the preceding ones. The tunicates and lancelets are invertebrate chordates; they don't have vertebrae. The vertebrates are the fishes, amphibians, reptiles, birds, and mammals. Fishes with an endoskeleton of cartilage and some bone in their scales were the first to have jaws. Early bony fishes had lungs. Amphibians were the first group to clearly have jointed appendages and to invade land. However, the fleshy appendages of lobe-finned fishes from the Devonian era contained bones homologous to those of terrestrial vertebrates. These fishes are believed to be ancestral to the amphibians. Reptiles, birds, and mammals have a means of reproduction suitable to land. During development, an amnion and other extraembryonic membranes are present. These membranes carry out all the functions needed to support the embryo as it develops into a young offspring, capable of feeding on its own

Check Your Progress

1. What are unique features of echinoderms?
2. List the four distinguishing features of chordates.

Answers: 1. Echinoderms have radial symmetry as adults and a water vascular system with tube feet for locomotion. 2. Chordates have a notochord, dorsal tubular nerve cord, pharyngeal pouches, and postanal tail.

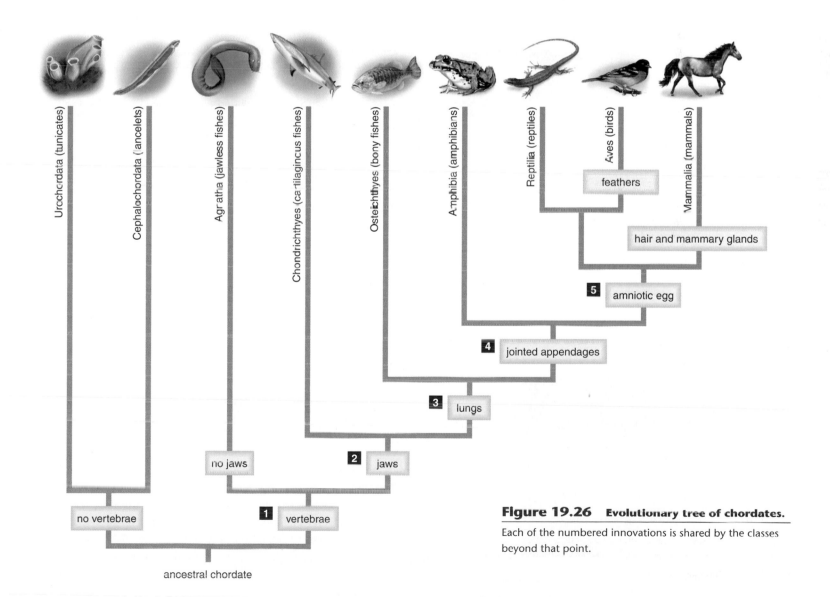

Figure 19.26   Evolutionary tree of chordates.

Each of the numbered innovations is shared by the classes beyond that point.

ancestral chordate

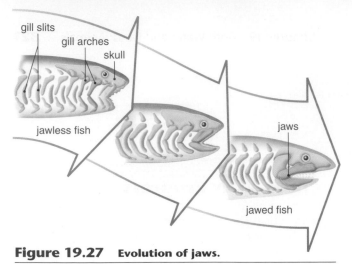

Figure 19.27   Evolution of jaws.

Jaws evolved from the anterior gill arches of ancient jawless fishes.

**Figure 19.28
Diversity of fishes.**

a. The lamprey is a jawless fish. Note the toothed oral disk. **b.** The shark is a cartilaginous fish. **c.** The soldierfish, a bony fish, has the typical appearance of a ray-finned fish.

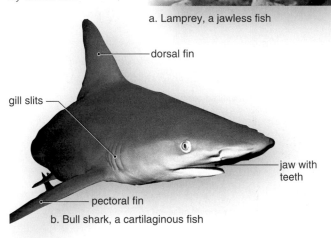

a. Lamprey, a jawless fish

b. Bull shark, a cartilaginous fish

c. Soldierfish, a bony fish

Fishes: First Jaws and Lungs

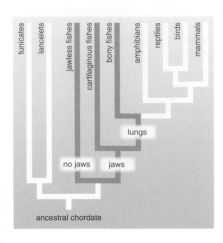

The first vertebrates were jawless fishes, which wiggled through the water and sucked up food from the ocean floor. Today, there are three living classes of fishes: jawless fishes, cartilaginous fishes, and bony fishes. The two latter groups have *jaws,* tooth-bearing bones of the head. Jaws are believed to have evolved from the first pair of gill arches, structures that ordinarily support gills (Fig. 19.27). The presence of jaws permits a predatory way of life.

Living representatives of the **jawless fishes** (class Agnatha) are cylindrical and up to a meter long. They have smooth, scaleless skin and no jaws or paired fins. The two groups of living jawless fishes are *hagfishes* and *lampreys* (Fig. 19.28*a*). The hagfishes are scavengers, feeding mainly on dead fishes, while some lampreys are parasitic. When parasitic, the round mouth of the lamprey serves as a sucker. The lamprey attaches itself to another fish and taps into its circulatory system. Water cannot move in through the lamprey's mouth and out over the gills, as is common in all other fishes. Instead, water moves in and out through the gill openings.

Cartilaginous fishes (class Chondrichthyes) are the sharks (Fig. 19.28*b*), the rays, and the skates, which have skeletons of cartilage instead of bone. The small dogfish shark is often dissected in biology laboratories. One of the most dangerous sharks inhabiting both tropical and temperate waters is the hammerhead shark. The largest sharks, the whale sharks, feed on small fishes and marine invertebrates and do not attack humans. Skates and rays are rather flat fishes that live partly buried in the sand and feed on mussels and clams.

Three well-developed senses enable sharks and rays to detect their prey: (1) They have the ability to sense electric currents in water—even those generated by the muscle movements of animals; (2) they, and all other types of fishes, have a lateral line system, a series of cells that lie within canals along both sides of the body and can sense pressure caused by a fish or another animal swimming nearby; (3) they have a keen sense of smell; the part of the brain associated with this sense is twice as large as the other parts. Sharks can detect about one drop of blood in 115 liters (25 gallons) of water.

Bony fishes (class Osteichthyes) are by far the most numerous and diverse of all the vertebrates. Most of the bony fishes we eat, such as perch, trout, salmon, and haddock, are **ray-finned fishes** (Fig. 19.28*c*). Their fins, which are used in balancing and propelling the body, are thin and supported by bony spikes. Ray-finned fishes have various ways of life. Some, such as herring, are filter feeders; others, such as trout, are opportunists; and still others, such as piranhas and barracudas, are predaceous carnivores.

Ray-finned fishes have a swim bladder, which usually serves as a buoyancy organ. By secreting gases into the bladder or by absorbing gases from it, these fishes can change their density, and thus go up or down in the water. The streamlined shape, fins, and muscle action of ray-finned fishes are all suited to locomotion in the water. Their skin is covered by bony scales that protect the body but do not prevent water loss. When ray-finned fishes respire, the gills are kept continuously moist by the passage of water through the mouth and out the gill slits. As the water passes over the gills, oxygen is absorbed by the blood, and carbon dioxide is given off. Ray-finned fishes have a single-circuit circulatory system. The heart is a simple pump, and the blood flows through the chambers, including a nondivided atrium and ventricle, to the gills. Oxygenated blood leaves the gills and goes to the body proper, eventually returning to the heart for recirculation.

Another type of bony fish, called the **lobe-finned fishes,** evolved into the amphibians (Fig. 19.29). These fishes not only had fleshy appendages that could be adapted to land locomotion, but most also had a **lung,** which was used for respiration.

Amphibians: Jointed Vertebrate Limbs

Amphibians (class Amphibia), whose class name means living on both land and in the water, are represented today by frogs, toads, newts, and salamanders. Aside from *jointed limbs* (Fig. 19.29*b*), amphibians have other features not seen in bony fishes: eyelids for keeping their eyes moist, ears adapted to picking up sound waves, and a voice-producing larynx. The brain is larger than that of a fish. Adult

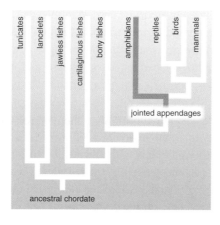

amphibians usually have small lungs. Air enters the mouth by way of nostrils, and when the floor of the mouth is raised, air is forced into the relatively small lungs. Respiration is supplemented by gas exchange through the smooth, moist, and glandular skin. The amphibian heart has only three chambers, compared to the four of mammals. Mixed blood is sent to all parts of the body; some is sent to the skin, where it is further oxygenated.

Most members of this group lead an amphibious life—that is, the larval stage lives in the water, and the adult stage lives on the land (see Fig. 19.1). However, the adult usually returns to the water to reproduce.

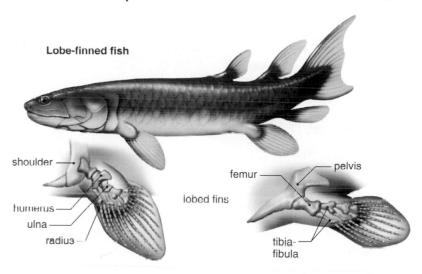

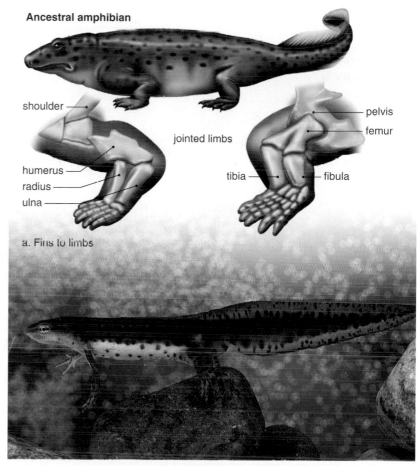

a. Fins to limbs

b

Figure 19.29 **Evolution of amphibians.**
a. A lobe-finned fish compared with an amphibian. A shift in the position of the bones in the forelimbs and hind limbs lifts and supports the body. **b.** Newt (shown here) and frogs (see Fig. 19.1) are types of living amphibians.

Check Your Progress

1. **Explain the evolutionary significance of the development of a jaw.**
2. **Among amphibians, what features in particular are adaptations to life on land?**

Answers: 1. The jaw allowed vertebrates to become predators. 2. Amphibians have jointed limbs, eyelids for keeping their eyes moist, ears for picking up sound waves, a voice-producing larynx, and usually lungs.

Figure 19.30 **Reptiles.**

Snakes wave a forked tongue in the air to collect chemical molecules, which are then brought back into the mouth and delivered to an organ in the floor of the mouth. Analyzed chemicals help the snake trail its prey, recognize a predator, or find a mate.

Reptiles: Amniotic Egg

Reptiles (class Reptilia) diversified and were most abundant between 245 and 66 MYA. These animals included the mammal-like reptiles, the ancestors of today's living mammals, and the dinosaurs, which became extinct. Some dinosaurs are remembered for their great size. *Brachiosaurus,* a herbivore, was about 23 meters (75 feet) long and about 17 meters (56 feet) tall. *Tyrannosaurus rex,* a carnivore, was 5 meters (16 feet) tall when standing on its hind legs. The bipedal stance of some reptiles was preadaptive for the evolution of wings in birds.

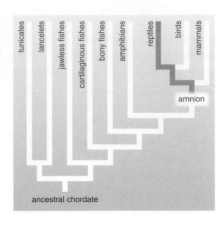

The reptiles living today are mainly turtles, crocodiles, snakes, and lizards. The body is covered with hard, keratinized scales, which protect the animal from desiccation and from predators. Another adaptation for a land existence is the manner in which snakes use their tongue as a sense organ (Fig. 19.30). Reptiles have well-developed lungs enclosed by a protective rib cage. The heart has four chambers, but the septum that divides the two halves is incomplete in certain species; therefore, there is some exchange of O_2-rich and O_2-poor blood.

Perhaps the most outstanding adaptation of the reptiles is that they have a means of reproduction suitable to a land existence. The penis of the male passes sperm directly to the female. Fertilization is internal, and the female lays leathery, flexible, shelled eggs. The *amniote egg* made development on land possible and eliminated the need for a swimming larval stage during development. The amniote egg provides the developing embryo with atmospheric oxygen, food, and water; removes nitrogenous wastes; and protects the embryo from drying out and from mechanical injury. This is accomplished by the presence of extraembryonic membranes (Fig. 19.31).

Fishes, amphibians, and reptiles are **ectothermic,** meaning that their body temperature matches the temperature of the external environment. If it is cold externally, they are cold internally; if it is hot externally, they are hot internally. Reptiles regulate their body temperatures by exposing themselves to the sun if they need warmth or by hiding in the shadows if they need cooling off.

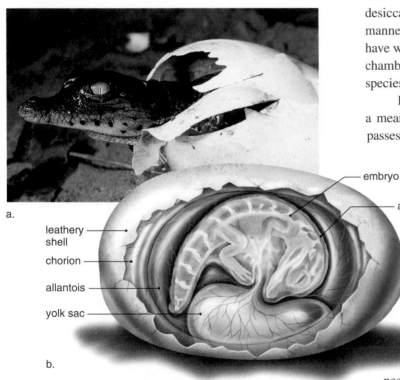

a.

b.

Figure 19.31 **Reproduction on land.**

a. A baby American crocodile hatching out of its shell. Note that the shell is leathery and flexible, not brittle like a bird's egg. **b.** Inside the egg, the embryo is surrounded by extraembryonic membranes. The chorion aids gas exchange, the yolk sac provides nutrients, the allantois stores waste, and the amnion encloses a fluid that prevents drying out and provides protection.

Birds: Feathers

Birds (class Aves) are characterized by the presence of *feathers,* which are modified reptilian scales. (Perhaps you have noticed the scales on the legs of a chicken.) However, birds lay a hard-shelled amniote egg rather than the leathery egg of reptiles. The exact history of birds is still in dispute, but gathering evidence indicates that birds are closely related to bipedal dinosaurs and that they should be classified as such.

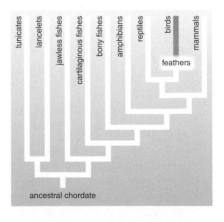

Nearly every anatomical feature of a bird can be related to its ability to fly (Fig. 19.32). The forelimbs are modified as wings. The hollow, very light bones are laced with air cavities. A horny beak has replaced jaws equipped with teeth, and a slender neck connects the head to a rounded, compact torso. Respiration is efficient since the lobular lungs form anterior and posterior air sacs. The presence of these sacs means that the air moves one way through the lungs, and gases are continuously exchanged across respiratory tissues. Another benefit of air sacs is that they lighten the body and aid flying.

Birds have a four-chambered heart that completely separates O_2-rich blood from O_2-poor blood. Birds are **endotherms** and generate internal heat. Many endotherms can use metabolic heat to maintain a constant internal temperature. This may be associated with their efficient nervous, respiratory, and circulatory systems. Also, their feathers provide insulation. Birds have no bladder and excrete uric acid in a semidry state.

Birds have particularly acute vision and well-developed brains. Their muscle reflexes are excellent. These adaptations are suited to flight. An enlarged portion of the brain seems to be the area responsible for instinctive behavior. A ritualized courtship often precedes mating. Many newly hatched birds require parental care before they are able to fly away and seek food for themselves. A remarkable aspect of bird behavior is the seasonal migration of many species over very long distances. Birds navigate by day and night, whether it's sunny or cloudy, by using the sun and stars and even the Earth's magnetic field to guide them.

Traditionally, the classification of birds was particularly based on beak and foot types (Fig. 19.33), and to some extent on habitat and behavior. The various orders include birds of prey with notched beaks and sharp talons; shorebirds with long, slender, probing bills and long, stiltlike legs; woodpeckers with sharp, chisel-like bills and grasping feet; waterfowl with webbed toes and broad bills; penguins with wings modified as flippers; and songbirds with perching feet. Now genetics is used to determine relationships among birds.

upstroke

downstroke

Figure 19.32   Bird flight.

Birds fly by flapping their wings. Bird flight requires an airstream and a powerful wing downstroke for lift, a force at right angles to the airstream.

a.

Check Your Progress

1. **Describe how reptiles are adapted to reproduce on land.**
2. **List the anatomical features of birds that help them fly.**

Answers: **1.** Fertilization is internal in reptiles and results in the production of a leathery amniote egg. **2.** Birds' forelimbs are modified as wings; their bones are usually hollow with air cavities; air sacs involved in respiration also lighten the body; and they have a well-developed nervous system and muscles.

Figure 19.33   Bird beaks.

a. A cardinal's beak allows it to crack tough seeds. **b.** A bald eagle's beak allows it to tear prey apart. **c.** A flamingo's beak strains food from the water with bristles that fringe the mandibles.

b.

c.

Mammals: Hair and Mammary Glands

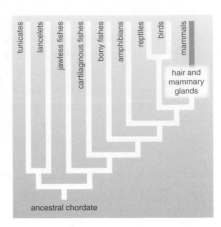

The first **mammals** were small, about the size of mice. During all the time the dinosaurs flourished (165 MYA), mammals were a minor group that changed little. Some of the earliest mammalian groups are still represented today by the monotremes and marsupials, but they are not abundant. The placental mammals that evolved later went on to live in many habitats.

The two chief characteristics of mammals (class Mammalia) are hair and milk-producing mammary glands. Almost all mammals are endotherms and generate internal heat. Many of the adaptations of mammals are related to temperature control. Hair, for example, provides insulation against heat loss and allows mammals to be active, even in cold weather.

Mammary glands enable females to feed (nurse) their young without leaving them to find food. Nursing also creates a bond between mother and offspring that helps ensure parental care while the young are helpless. In most mammals, the young are born alive after a period of development in the uterus, a part of the female reproductive system. Internal development shelters the young and allows the female to move actively about while the young are maturing.

Monotremes are mammals that, like birds, have a *cloaca,* a terminal region of the digestive tract serving as a common chamber for feces, excretory wastes, and sex cells. They also lay hard-shelled amniote eggs. They are represented by the spiny anteater and the duckbill platypus, both of which live in Australia. The female duckbill platypus lays her eggs in a burrow in the ground. She incubates the eggs, and after hatching, the young lick up milk that seeps from mammary glands on her abdomen. The spiny anteater has a pouch on the belly side formed by swollen mammary glands and longitudinal muscle (Fig. 19.34*a*). Hatching takes place in this pouch, and the young remain there for about 53 days. Then they stay in a burrow, where the mother periodically visits and nurses them.

The young of **marsupials** begin their development inside the female's body, but they are born in a very immature condition. Newborns crawl up into a pouch on their mother's abdomen. Inside the pouch, they attach to the nipples

a.

Figure 19.34 **Monotremes and marsupials.**

a. The spiny anteater is a monotreme that lives in Australia. **b.** The opossum is the only marsupial in the United States. The Virginia opossum is found in a variety of habitats. **c.** The koala is an Australian marsupial that lives in trees.

b.

c.

of mammary glands and continue to develop. Frequently, more are born than can be accommodated by the number of nipples, and it's "first come, first served."

The Virginia opossum is the only marsupial that occurs north of Mexico (Fig. 19.34*b*). In Australia, however, marsupials underwent adaptive radiation for several million years without competition. Thus, marsupial mammals are now found mainly in Australia, with some in Central and South America as well. Among the herbivorous marsupials, koalas are tree-climbing browsers (Fig. 19.34*c*), and kangaroos are grazers. The Tasmanian wolf or tiger, thought to be extinct, was a carnivorous marsupial about the size of a collie dog.

The vast majority of living mammals are **placental mammals** (Fig. 19.35). In these mammals, the extraembryonic membranes of the reptilian egg (see Fig. 19.31) have been modified for internal development within the uterus of the female. The chorion contributes to the fetal portion of the placenta, while a part of the uterine wall contributes to the maternal portion. Here, nutrients, oxygen, and waste are exchanged between fetal and maternal blood.

Mammals are adapted to life on land and have limbs that allow them to move rapidly. In fact, an evaluation of mammalian features leads us to the obvious conclusion that they lead active lives. The brain is well developed; the lungs are expanded not only by the action of the rib cage but also by the contraction of the diaphragm, a horizontal muscle that divides the thoracic cavity from the abdominal cavity; and the heart has four chambers. The internal temperature is constant, and hair, when abundant, helps insulate the body.

The mammalian brain is enlarged due to the expansion of the cerebral hemispheres that control the rest of the brain. The brain is not fully developed until after birth, and young learn to take care of themselves during a period of dependency on their parents.

Mammals can be distinguished by their methods of obtaining food and their mode of locomotion. For example, bats have membranous wings supported by digits; horses have long, hoofed legs; and whales have paddlelike forelimbs. The specific shape and size of the teeth may be associated with whether the mammal is a herbivore (eats vegetation), a carnivore (eats meat), or an omnivore (eats both meat and vegetation). For example, mice have continuously growing incisors; horses have large, grinding molars; and dogs have long canine teeth.

Figure 19.35   Placental mammals.

Placental mammals have adapted to various ways of life. **a.** Deer are herbivores that live in forests. **b.** Lions are carnivores on the African plain. **c.** Monkeys inhabit tropical forests. **d.** Whales are sea-dwelling placental mammals.

d.

a.

b.

c.

a.

b.

Figure 19.38   *Australopithecus afarensis.*

a. A reconstruction of Lucy on display at the St. Louis Zoo.
b. These fossilized footprints occur in ash from a volcanic eruption some 3.7 MYA. The larger footprints are double, and a third, smaller individual was walking to the side. (A female holding the hand of a youngster may have been walking in the footprints of a male.) The footprints suggest that *A. afarensis* walked bipedally.

Check Your Progress

Contrast the australopithecines with *Homo habilis.*

Answer: Australopithecines were apelike above the waist and humanlike below the waist. *Homo habilis* had a larger brain and smaller cheek teeth than the australopithecines, and was able to use stone tools and possibly speech.

Australopithecines

It's possible that one of the **australopithecines,** a group of hominids that evolved and diversified in Africa about 4 MYA, is a direct ancestor of humans. More than 20 years ago, a team led by Donald Johanson unearthed nearly 250 fossils of a hominid called *Australopithecus afarensis.* A now-famous female skeleton dated at 3.18 MYA is known worldwide by its field name, Lucy. (The name derives from the Beatles song "Lucy in the Sky with Diamonds.") Although her brain was quite small (400 cc), the shapes and relative proportions of Lucy's limbs indicate that she stood upright and walked bipedally (Fig. 19.38*a*). Even better evidence of bipedal locomotion comes from a trail of footprints dated about 3.7 MYA. The larger prints are double, as though a smaller-sized individual was stepping in the footfalls of another—and there are additional small prints off to the side, within hand-holding distance (Fig. 19.38*b*).

Since the australopithecines were apelike above the waist (small brain) and humanlike below the waist (walked erect), it seems that human characteristics did not evolve all at one time. The term **mosaic evolution** is applied when different body parts change at different rates and therefore at different times.

Homo habilis

Homo habilis, dated between 2.0 and 1.9 MYA, may be ancestral to modern humans. Some of these fossils have a brain size as large as 775 cc, which is about 45% larger than the brain of *A. afarensis.* The cheek teeth are smaller than those of any of the australopithecines. Therefore, it is likely that this early *Homo* was omnivorous and ate meat in addition to plant material. Bones at the campsites of *H. habilis* bear cut marks, indicating that they used tools to strip the meat from bones.

The stone tools made by *H. habilis,* whose name means "handyman," are rather crude (see Fig. 19.37). It's possible that these are the cores from which these hominids took flakes sharp enough to scrape away hide, cut tendons, and easily remove meat from bones.

The skulls of *H. habilis* suggest that the portions of the brain associated with speech areas were enlarged. We can speculate that the ability to speak may have led to hunting cooperatively. Some members of the group may have remained plant gatherers, and if so, both hunters and gatherers most likely ate together and shared their food. In this way, society and culture could have begun. **Culture,** which encompasses human behavior and products (such as technology and the arts), is dependent upon the capacity to speak and transmit knowledge. We can further speculate that the advantages of a culture to *H. habilis* may have hastened the extinction of the australopithecines.

Homo erectus

Homo erectus and like fossils are found in Africa, Asia, and Europe and dated between 1.9 and 0.3 MYA. Although all fossils assigned the name *H. erectus* are similar in appearance, there is enough discrepancy to suggest that several different species have been included in this group. Compared to *H. habilis,* *H. erectus* had a larger brain (about 1,000 cc) and a flatter face. The recovery of an almost complete skeleton of a 10-year-old boy indicates that *H. erectus* was much taller than the hominids discussed thus far. Males were 1.8 meters tall (about 6 feet), and females were 1.55 meters (approaching 5 feet). Indeed, these hominids were erect and most likely had a striding gait like that of modern humans. The robust and most likely heavily muscled skeleton still retained some australopith-ecine features. Even so, the size of the birth canal indicates that infants were born in an immature state that required an extended period of care.

It is believed that *H. erectus* first appeared in Africa and then migrated into Asia and Europe. At one time, the migration was thought to have occurred about 1 MYA, but recently *H. erectus* fossil remains in Java and the Republic of Georgia have been dated at 1.9 and 1.6 MYA, respectively. These remains push the evolution of *H. erectus* in Africa to an earlier date than has yet been determined. In any case, such an extensive population movement is a first in the history of humankind and a tribute to the intellectual and physical skills of the species.

H. erectus was the first hominid to use fire, and it also fashioned more advanced tools than earlier *Homo* species. These hominids used heavy, teardrop-shaped axes and cleavers, as well as flakes that were probably used for cut-ting and scraping. Some investigators believe *H. erectus* was a systematic hunter that brought kills to the same site over and over again. In one location, researchers have found over 40,000 bones and 2,647 stones. These sites could have been "home bases" where social interaction occurred and a prolonged childhood allowed time for learning. Perhaps a language evolved and a culture more like our own developed.

Evolution of Modern Humans

Most researchers believe that ***Homo sapiens*** (modern humans) evolved from *H. erectus,* but they differ as to the details. The hypothesis that *H. sapiens* evolved separately from *H. erectus* in Asia, Africa, and Europe is called the **multiregional continuity hypothesis** (Fig. 19.39*a*). This hypothesis proposes that evolution to modern humans was essentially similar in several different places. If so, each region should show a continuity of its own anatomical char-acteristics from the time when *H. erectus* first arrived.

Opponents argue that it seems highly unlikely that evolution would have produced essentially the same result in these different places. They suggest, instead, the **out-of-Africa hypothesis,** which proposes that *H. sapiens* evolved from *H. erectus* only in Africa, and thereafter *H. sapiens* migrated to Europe and Asia about 100,000 years BP (before present) (Fig. 19.39*b*). If so, fossils dated 200,000 BP and 100,000 BP are expected to be markedly different from each other.

According to which hypothesis would modern humans be most genetically alike? The multiregional continuity hypothesis states that human populations have been evolving separately for a long time; therefore, genetic differences would be expected between groups. According to the out-of-Africa hypothesis, we are all descended from a few individuals from about 100,000 years BP. Therefore, the out-of-Africa hypothesis suggests that we are more genetically similar.

A few years ago, a study attempted to show that all the people of Europe (and the world, for that matter) have essentially the same mitochondrial DNA. Called the "mitochondrial Eve" hypothesis by the press (note that this is a mis-nomer because no single ancestor is proposed), the statistics that calculated the date of the African migration were found to be flawed. Still, the raw data—which indicate a close genetic relationship among all Europeans—support the out-of-Africa hypothesis.

These opposing hypotheses have sparked many other innovative studies. The final conclusions are still being determined.

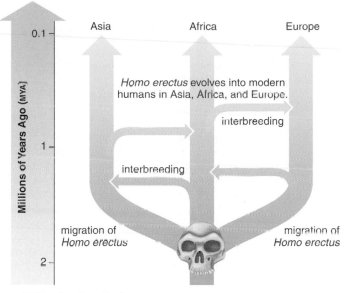

a. Multiregional continuity

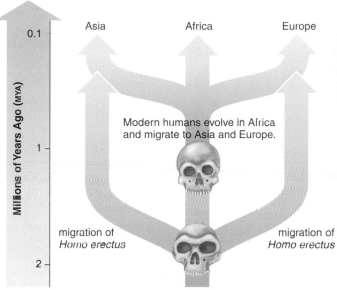

b. Out of Africa

Figure 19.39   Evolution of modern humans.

a. The multiregional continuity hypothesis proposes that *Homo sapiens* evolved separately in at least three different places: Asia, Africa, and Europe. Therefore, continuity of genotypes and phenotypes is expected in each region, but not between regions. **b.** The out-of-Africa hypothesis proposes that *Homo sapiens* evolved only in Africa; then this species migrated and supplanted populations of *Homo* in Asia and Europe about 100,000 years ago.

Neandertals

Neandertals (*H. neandertalensis*) take their name from Germany's Neander Valley, where one of the first Neandertal skeletons, dated some 200,000 years ago, was discovered. It's possible that the Neandertals were already present in Eurasia when modern humans represented by Cro-Magnons (discussed next) arrived on the scene. In that case, competition with the Cro-Magnons may have made the Neandertals become extinct. The Neandertals had massive brow ridges, and their nose, jaws, and teeth protruded far forward. The forehead was low and sloping, and the lower jaw lacked a chin. The Neandertals were heavily muscled, especially in the shoulders and neck. The bones of their limbs were shorter and thicker than those of modern humans. The Neandertals lived in Europe and Asia during the last Ice Age, and their sturdy build could have helped conserve heat.

The Neandertals give evidence of being culturally advanced. Most lived in caves, but those living in the open may have built houses. They manufactured a variety of stone tools, including spear points, which could have been used for hunting, and scrapers and knives, which would have helped with food preparation. They most likely successfully hunted bears, woolly mammoths, rhinoceroses, reindeer, and other contemporary animals. They buried their dead with flowers and tools and may have had a religion.

Cro-Magnons

Cro-Magnons are the oldest fossils to be designated *H. sapiens*. Cro-Magnons, who are named after a fossil location in France, had a thoroughly modern appearance. They made advanced tools, including compound tools, as when stone flakes were fitted to a wooden handle. They may have been the first to make knife-like blades and to throw spears, enabling them to kill animals from a distance. They were such accomplished hunters that some researchers believe they were responsible for the extinction of many larger mammals, such as the giant sloth, the mammoth, the saber-toothed tiger, and the giant ox, during the late Pleistocene epoch.

Cro-Magnons hunted cooperatively, and were perhaps the first to have a language. They are believed to have lived in small groups, with the men hunting by day while the women remained at home with the children, gathering and processing food items. Probably, the women also were engaged in maintenance tasks. The Cro-Magnon culture included art. They sculpted small figurines out of reindeer bones and antlers. They also painted beautiful drawings of animals, some of which have survived on cave walls in Spain and France (Fig. 19.40).

Figure 19.40 Cro-Magnons.

Cro-Magnon people are the first to be designated *Homo sapiens*. Their tool-making ability and other cultural attributes, including their artistic talents, are legendary.

THE CHAPTER IN REVIEW

Summary

19.1   Evolution of Animals

Animals are motile, multicellular heterotrophs that ingest their food.

Evolutionary History of Animals

Animals representative of today's phyla appear in the fossil record during the Cambrian period. This burst of diversification is called the Cambrian explosion.

Evolutionary Tree of Animals

An evolutionary tree, based on seven innovations, depicts the seven evolutionary trends that occurred as animals evolved:

- Multicellularity
- True tissues
- Bilateral symmetry
- Body cavity
- Coelomates
- Segmentation
- Jointed appendages

19.2   Introducing the Invertebrates

Sponges are multicellular (lack tissues) and asymmetrical.

Cnidarians have two tissue layers, are radially symmetrical, have a saclike digestive cavity, and possess stinging cells and nematocysts.

Flatworms have ectoderm, endoderm, and mesoderm but no coelom. They are bilaterally symmetrical and have a saclike digestive cavity.

Roundworms have a pseudocoelom and a complete digestive tract.

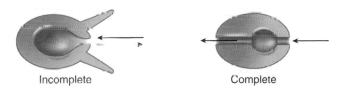

Incomplete   Complete

19.3   Protostomes and Deuterostomes Compared

On the basis of embryological data:

- Molluscs, annelids, and arthropods are protostomes; the first embryonic opening becomes the mouth, and the coelom develops by a splitting of mesoderm.

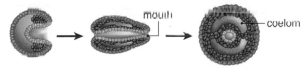

Development in protostomes

- Echinoderms and chordates are deuterostomes; the second embryonic opening becomes the mouth, and the coelom forms by outpocketing of the primitive gut.

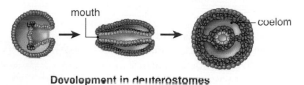

Development in deuterostomes

19.4   Molluscs, Annelids, and Arthropods

Molluscs

The body of a mollusc typically contains a visceral mass, a mantle, and a foot.

Gastropods   Snails, representatives of this group, have a flat foot, a one-part shell, and a mantle cavity that carries on gas exchange.

Cephalopods   Octopuses and squids display marked cephalization, move rapidly by jet propulsion, and have a closed circulatory system.

Bivalves   Bivalves such as clams have a hatchet foot and a two-part shell, and are filter feeders.

Annelids: Segmented Worms

Annelids are segmented worms; segmentation is seen both externally and internally.

Marine worms   Polychaetes are worms that have many setae. A clam worm is a predaceous marine worm with a defined head region.

Earthworms   Earthworms are oligochaetes that scavenge for food in the soil and do not have a well-defined head region.

Leeches   Leeches are annelids that feed by sucking blood.

Arthropods: Jointed Appendages

Arthropods are the most varied and numerous of animals. Their success is largely attributable to a flexible exoskeleton and specialized body regions.

Crustaceans have a head that bears compound eyes, antennae, and mouthparts. Five pairs of walking legs are present.

Arachnids have four pairs of walking legs attached to a cephalothorax. Horseshoe crabs are marine arachnids. Spiders live on land and spin silk, which they use to capture prey as well as for other purposes.

Insects have three pairs of legs attached to the thorax. Insects have adaptations to a terrestrial life, such as wings for flying.

19.5   Echinoderms and Chordates

Echinoderms

Echinoderms have radial symmetry as adults (not as larvae) and endoskeletal spines. Typical echinoderms have tiny skin gills, a central nerve ring with branches, and a water vascular system for locomotion, as exemplified by the sea star.

Chordates

Chordates (tunicates, lancelets, and vertebrates) have a notochord, a dorsal tubular nerve cord, pharyngeal pouches, and a postanal tail at some time in their life history.

Invertebrate Chordates   Adult tunicates lack chordate characteristics except gill slits, but adult lancelets have the four chordate characteristics and show obvious segmentation.

Vertebrate Chordates   Vertebrate chordates include the fishes, amphibians, reptiles, birds, and mammals.

Fishes
- The first vertebrates, represented by hagfishes and lampreys, lacked jaws and fins.
- Cartilaginous fishes, represented by sharks and rays, have jaws and a skeleton made of cartilage.
- Bony fishes have jaws and fins supported by bony spikes; the bony fishes include those that are ray-finned and a few that are lobe-finned. Some of the lobe-finned fishes have lungs.

Amphibians (frogs, toads, newts, and salamanders) evolved from the lobe-finned fishes and have two pairs of jointed vertebrate limbs. Frogs usually return to the water to reproduce; frog tadpoles metamorphose into terrestrial adults.

Reptiles (today's snakes, lizards, turtles, and crocodiles) lay a shelled egg, which contains extraembryonic membranes, including an amnion that allows them to reproduce on land.

Birds are feathered, which helps them maintain a constant body temperature. They are adapted for flight; their bones are hollow with air cavities; lungs form air sacs that allow one-way ventilation; and they have well-developed sense organs.

Mammals have hair and mammary glands. The former helps them maintain a constant body temperature, and the latter allows them to nurse their young.
- Monotremes lay eggs.
- Marsupials have a pouch in which the newborn matures.
- Placental mammals, which are far more varied and numerous, retain offspring inside the uterus until birth.

19.6   Human Evolution

Arboreal

Primates are mammals adapted to living in trees. During the evolution of primates, various groups diverged in a particular sequence from the line of descent leading to today's primates. Prosimians (tarsiers and lemurs) diverged first, followed by the monkeys, then the apes, and then humans. Molecular biologists tell us we are most closely related to the African apes, with which we share a common ancestor about 7 MYA.

Evolution of Hominids

Human evolution occurred in Africa.

Bipedal

- The most famous australopithecine is Lucy (3.18 MYA) whose brain was small but who walked bipedally.

Tool Use

- *Homo habilis*, present about 2 MYA, is certain to have made tools.

Brain Size Increased
- *Homo erectus*, with a brain capacity of 1,000 cc and a striding gait, was the first to migrate out of Africa.

Evolution of Modern Humans

Two contradicting hypotheses have been suggested about the origin of modern humans:

Multiregional Continuity Hypothesis   Modern humans originated separately in Asia, in Europe, and in Africa.

Out-of-Africa Hypothesis   Modern humans originated in Africa and, after migrating into Europe and Asia, replaced the other *Homo* species (including the Neandertals) found there. Cro-Magnon was the first hominid to be considered *Homo sapiens*.

Thinking Scientifically

1. Recently, three fossil skulls of *Homo sapiens,* dating to about 160,000 years BP, were discovered in eastern Africa. These skulls fill a gap between the 100,000-year-old *H. sapiens* skulls found in Africa and Israel and 500,000-year-old skulls of archaic *H. sapiens* found in Ethiopia. Supporters of the out-of-Africa hypothesis argue that this discovery strengthens their position by documenting the succession of human ancestors from 6 MYA through this most recent group. Does this finding prove that the out-of-Africa hypothesis is correct? If not, what fossil evidence might yet be found to support the multiregional continuity hypothesis?

2. Think of the animals in this chapter that are radially symmetrical (cnidarians, many adult echinoderms). How is their lifestyle different from that of bilaterally symmetrical animals? How does their body plan complement their lifestyle?

Testing Yourself

Choose the best answer for each question.

1. Which of the following is the least likely explanation for the Cambrian explosion?
   a. The development of skeletons may have provided protection against predators.
   b. Mutation rates were higher than normal.
   c. Oxygen levels were on the rise.
   d. Skeletons survived as fossils.

2. In the following diagram, label the major regions of the bodies of acoelomates, pseudocoelomates, and coelomates.

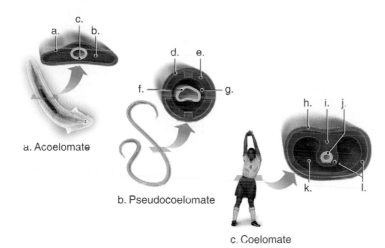

a. Acoelomate

b. Pseudocoelomate

c. Coelomate

3. Sponges are ancestors of
   a. cnidarians.
   b. protostomes.
   c. deuterostomes.
   d. None of these are correct.

For questions 4–10, identify the group(s) to which each feature belongs. Each answer may be used more than once. Each question may have more than one answer.

Key:
   a. flatworms
   b. roundworms
   c. molluscs
   d. annelids
   e. arthropods
   f. echinoderms
   g. chordates

4. Most have an endoskeleton.
5. Typically bilaterally symmetrical.
6. Nonsegmented.
7. Lack any kind of coelom.
8. Chitin exoskeleton.
9. Contain a mantle.
10. Move by pumping water.
11. Insects have wings and three pairs of legs attached to the
    a. abdomen.
    b. thorax.
    c. head.
    d. midsection.
12. Cartilaginous fishes detect their prey by sensing
    a. electrical currents.
    b. odors.
    c. pressure changes.
    d. More than one of these are correct.
    e. All of these are correct.

13. Which of the following is not a feature of mammals?
    a. hair
    b. milk-producing glands
    c. ectothermic
    d. four-chambered heart
    e. diaphragm to help expand lungs
14. This species was probably the first to have the use of fire.
    a. *Homo habilis*
    b. *Homo erectus*
    c. *Homo sapiens*
    d. *Australopithecus robustus*
    e. *Australopithecus afarensis*
15. Which of these is matched correctly?
    a. *Australopithecus afarensis*—bipedal but small brain
    b. *Homo habilis*—made tools and migrated often
    c. *Homo erectus*—had fire and migrated out of Europe to Africa
    d. *Homo sapiens*—projecting face and had culture
    e. Both a and c are correct.
16. Cnidarians are considered to be organized at the tissue level because they contain
    a. ectoderm and endoderm.
    b. ectoderm.
    c. ectoderm and mesoderm.
    d. endoderm and mesoderm.
    e. mesoderm.
17. A cnidocyte is
    a. a digestive cell.
    b. a reproductive cell.
    c. a stinging cell.
    d. an excretory cell.
18. Which of the following is not a feature of a coelomate?
    a. radial symmetry
    b. three germ layers
    c. complete digestive tract body plan
    d. organ level of organization
19. A mollusc's shell is secreted by the
    a. foot.
    b. head.
    c. visceral mass.
    d. mantle.
20. The type of mollusc that produces tentacles is a
    a. gastropod.
    b. bivalve.
    c. univalve.
    d. cephalopod.
21. A feature of annelids is
    a. a segmented body.
    b. acoelomate.
    c. a sac body plan.
    d. radial symmetry.
22. Which of the following is not a feature of an insect?
    a. compound eyes
    b. eight legs
    c. antennae
    d. an exoskeleton
    e. jointed legs

23. Which of the following undergoes metamorphosis?
    a. grasshopper
    b. crayfish
    c. earthworm
    d. More than one of these are correct.

24. Which of the following is not an arachnid?
    a. spider
    b. tick
    c. mite
    d. scorpion
    e. beetle

25. Unlike bony fishes, amphibians have
    a. ears.
    b. jaws.
    c. a circulatory system.
    d. a heart.

26. Which of the following is not an adaptation for flight in birds?
    a. air sacs
    b. modified forelimbs
    c. bones with air cavities
    d. acute vision
    e. well-developed bladder

27. Examples of monotremes include the
    a. spiny anteater and duckbill platypus.
    b. opossum and koala.
    c. badger and skunk.
    d. porcupine and armadillo.

28. Mammals are distinguished based on
    a. size and hair type.
    b. mode of reproduction.
    c. number of limbs and method of caring for young.
    d. number of mammary glands and number of offspring.

29. The first humanlike feature to evolve in the hominids was
    a. a large brain.
    b. massive jaws.
    c. a slender body.
    d. bipedal locomotion.

30. In *H. habilis*, enlargement of the portions of the brain associated with speech probably led to
    a. cooperative hunting.
    b. the sharing of food.
    c. the development of culture.
    d. All of these are correct.

31. Mitochondrial DNA data support which hypothesis for the evolution of humans?
    a. multiregional continuity
    b. out-of-Africa

32. The first *Homo* species to use art appears to be
    a. *H. neandertalensis*.
    b. *H. erectus*.
    c. *H. habilis*.
    d. *H. sapiens*.

Go to www.mhhe.com/maderessentials for more quiz questions.

Bioethical Issue

People who approve of laboratory research involving animals point out that even today it would be difficult to develop new vaccines and medicines against infectious diseases, new surgical techniques for saving human lives, or new treatments for spinal cord injuries without the use of animals. Even so, most scientists favor what are now called the "three Rs": (1) replacement of animals by in vitro, or test-tube, methods whenever possible; (2) reduction of the number of animals used in experiments; and (3) refinement of experiments to cause less suffering to animals.

F. Barbara Orlans of the Kennedy Institute of Ethics at Georgetown University says, "It is possible to be both pro research and pro reform." She feels that animal activists need to accept that sometimes animal research is beneficial to humans, and all scientists need to consider the ethical dilemmas that arise when animals are used for laboratory research. Do you approve of this compromise?

Understanding the Terms

Match the terms to these definitions:

a. _____ Localization of the brain and sensory organs at the anterior end of an animal.

b. _____ Body cavity enclosed by mesoderm.

c. _____ Possesses both male and female sex organs.

d. _____ Longitudinal cut through the body at any point gives two equal halves.

e. _____ Generate internal heat.

Plant Anatomy and Growth

The world's tallest living tree, a coastal redwood, is taller than the Statue of Liberty.

Fertilizer contains the minerals plants need to grow.

Carnivorous plants have evolved an unusual way to get the nutrients they need.

OUTLINE

If you want to see an interesting plant called the pitcher plant, take a walk in an acidic bog. Pitcher plants like to grow in areas where nitrogen and phosphorus, which are required nutrients for plants, are lacking. What's their secret? Pitcher plants feed on animals, particularly insects, to get the nutrients they need. That's right—like a few other types of plants, pitcher plants are carnivorous.

▶ ▶ ▶

A pitcher plant is named for its leaves, which are shaped like a container we call a pitcher. The pitcher attracts flying insects because it has a scent, is brightly colored, and provides nectar that insects can eat. When an insect lands on the lip of the pitcher, a slippery substance and downward-pointing hairs encourage it to slide into a pit. There, juices secreted by the leaf begin digesting the insect. The plant then absorbs these nutrients into its tissues. Although the pitcher is designed to attract insects, animals as large as rats have been found in pitcher plants! The unique adaptations of pitcher plants allow them to thrive in environments where other plants don't have a chance of surviving.

Some animals are able to turn the tables on the pitcher plant by taking advantage of the liquid they provide. Larvae of mosquitoes and flies have been known to develop safely inside a pitcher plant.

Carnivorous plants such as the pitcher plant are fairly unusual in the plant world, but so are many others, such as the coastal redwoods—the tallest trees on planet Earth. In this chapter, you will learn about the organs and structures of flowering plants, the nutrition they need, and how they transport nutrients within plant tissues.

20.1 Plant Organs

A flowering plant, whether a cactus, a daisy, or an apple tree, has a root system and a shoot system (Fig. 20.1). The **root system** simply consists of the roots, while the **shoot system** consists of the stem, the leaves, and the organs of sexual reproduction. The three vegetative organs—the leaf, the stem, and the root—perform functions that allow a plant to live and grow. The flower, which functions during reproduction, is discussed in Chapter 21.

Activity at the **terminal bud,** the bud at the tip of a main shoot or a branch shoot, causes the shoot to increase in length, while activity at the tip of the root causes the root system to increase in length. This is called *primary growth*, and it would be equivalent to you increasing in height by growing from your head and your feet!

Leaves

Leaves are usually the chief organs of photosynthesis, and as such they require a supply of solar energy, carbon dioxide, and water. Broad and thin foliage leaves have a maximum surface area for the collection of solar energy and the absorption of carbon dioxide. Leaves receive water from the root system by way of vascular tissue that terminates in the leaves. Without carrying on photosynthesis, a plant would not be able to grow.

The wide portion of a foliage leaf is called the **blade.** The **petiole** is a stalk that attaches the blade to the stem. Not all leaves are foliage leaves. Some are specialized to protect the plant, attach to objects (tendrils), or

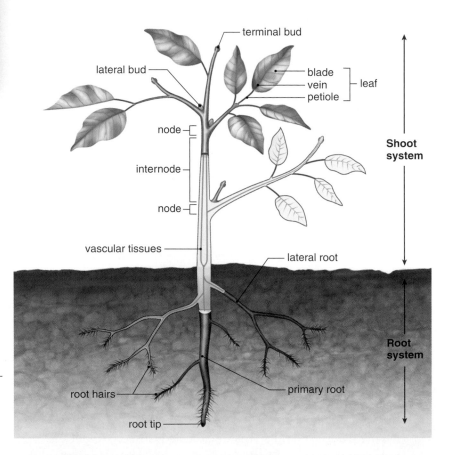

Figure 20.1   The body of a plant.

A plant has a root system, which extends below ground, and a shoot system, which is composed of the stem and leaves. Activity at the terminal bud and the root tip allows a plant to increase in length, a process called primary growth.

even capture insects (Fig. 20.2). Some trees, called *evergreens,* retain their leaves, and others, called **deciduous,** lose their leaves during a particular season of the year.

Stems

A **stem** is the main axis of a plant. Terminal bud, also called an apical bud, activity produces new leaves and tissues during primary growth. Lateral (side) branches grow from a **lateral bud,** also called an axillary bud, located at the angle where a leaf joins the stem. A **node** occurs where a leaf or leaves are attached to the stem, and an **internode** is the region between the nodes (see Fig. 20.1). Nodes and internodes are used to identify the stem, even if it happens to be an underground stem. In some plants, the nodes of horizontal stems asexually produce new plants.

In addition to supporting the leaves, a stem houses vascular tissue that transports water and minerals from the roots through the stem to the leaves and transports the products of photosynthesis, usually in the opposite direction.

Stems may have functions other than transport. In some plants (e.g., cactus), the stem is the primary photosynthetic organ and serves as a water reservoir (Fig 20.2*a*).

Roots

Just as plants are diverse, so are roots (Fig. 20.3). All root systems support the plant by anchoring it in the soil, as well as absorbing water and minerals from the soil for the entire plant. As a rule of thumb, the root system is at least equivalent in size and extent to its shoot system. Therefore, an apple tree has a much larger root system than a corn plant. Also, the extent of a root system depends on the environment. A single corn plant may have roots as deep as 2.5 meters (m), while a mesquite tree that lives in the desert may have roots that penetrate to a depth of 20 m.

The cylindrical shape of roots and their slimy surface allow them to penetrate the soil as they grow and permit water to be absorbed from all sides. The absorptive capacity of a root is also increased by its many **root hairs,** located in a special zone near the root tip. Root hairs are so numerous that they increase the absorptive surface of a root tremendously. It has been estimated that a single rye plant has about 14 billion root-hair cells, and if placed end to end, the root hairs would stretch 10,626 kilometers. Root-hair cells are constantly being replaced, so this same rye plant forms about 100 million new root-hair cells every day. You probably know that a plant yanked out of the soil will not fare well when transplanted; this is because small lateral roots and root hairs are torn off. Transplantation is more apt to be successful if you take a part of the surrounding soil along with the plant, leaving as many of the lateral roots and the root hairs intact as possible.

Figure 20.3*a* shows the diversity of roots. A carrot plant has one main taproot, which stores the products of photosynthesis. Grass has fibrous roots that cling to the soil (Fig. 20.3*b*), and corn plants have prop roots that grow from the stem for better support (Fig. 20.3*c*). **Perennial plants** are ones that are able to regrow next season because their roots survive, even though the shoot system dies back.

a. Cactus

b. Cucumber

c. Venus's flytrap

Figure 20.2
Leaf diversity.

a. The spines of a cactus protect the fleshy stem from herbivores. **b.** The tendrils of a cucumber are leaves modified to attach the plant to a support. **c.** Venus's flytrap leaves serve as the trap for unsuspecting insects. After digesting this fly, the plant will absorb the nutrients released.

Figure 20.3 Root diversity.

a. A taproot has one main root. **b.** A fibrous root has many slender roots with no main root. **c.** Prop roots are organized for support.

a. Taproot

b. Fibrous root system

c. Prop roots

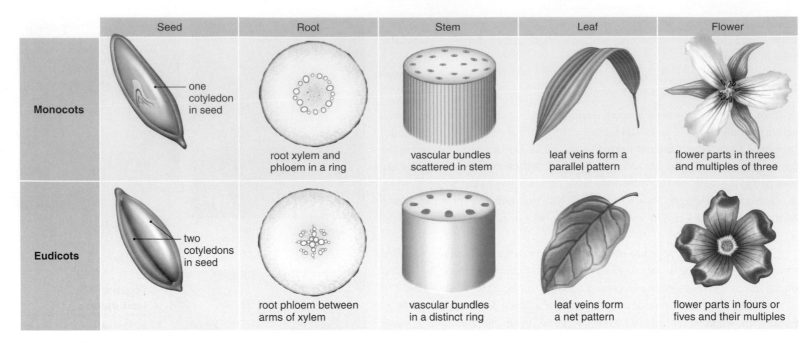

	Seed	Root	Stem	Leaf	Flower
Monocots	one cotyledon in seed	root xylem and phloem in a ring	vascular bundles scattered in stem	leaf veins form a parallel pattern	flower parts in threes and multiples of three
Eudicots	two cotyledons in seed	root phloem between arms of xylem	vascular bundles in a distinct ring	leaf veins form a net pattern	flower parts in fours or fives and their multiples

Figure 20.4   Flowering plants are either monocots or eudicots.

Five features are typically used to distinguish monocots from eudicots: number of cotyledons; arrangement of vascular tissue in roots, stems, and leaves; and number of flower parts.

Check Your Progress

1. List the three vegetative organs in a plant.
2. What is the function of leaves, and what are some possible leaf specializations?
3. List three functions of roots.
4. List the major differences between monocots and eudicots.

Answers: 1. Vegetative organs are the leaves, stem, and root. 2. Sites of photosynthesis; some leaves also store water, attach to objects, and capture insects. 3. Anchorage, absorb water and minerals, and store products of photosynthesis. 4. Monocots: embryo with single cotyledon; xylem and phloem in a ring in the root; scattered vascular bundles in the stem; parallel leaf veins; flower parts in multiples of three. Eudicots: embryo with two cotyledons; phloem located between arms of xylem in the root; vascular bundles in a ring in the stem; netted leaf veins; flower parts in multiples of fours or fives.

Monocot Versus Eudicot Plants

Figure 20.4 shows how flowering plants can be divided into two major groups. One of these differences concerns the **cotyledons,** which are embryonic leaves present in seeds. The cotyledons wither after the first true leaves appear. Some embryos have one cotyledon, and these plants are known as monocotyledons, or **monocots.** Other embryos have two cotyledons, and these plants are known as eudicotyledons, or **eudicots** (true dicots). The cotyledons of eudicots supply nutrients for seedlings, but the cotyledons of monocots store some nutrients and act as a transfer tissue for nutrients stored elsewhere.

Although the distinction between monocots and eudicots may seem of limited importance, it can be associated with many differences in their structure. Vascular plants contain two main types of transport tissue, the **xylem** for water and minerals and the **phloem** for organic nutrients. In a sense, xylem and phloem are to plants what veins and arteries are to animals. In the monocot root, transport tissue, also called vascular tissue, occurs in a ring, but in the eudicot, phloem is located between the arms of the xylem. In a stem, transport tissue occurs in the **vascular bundles.** The vascular bundles are scattered in the monocot stem, and they occur in a ring in the eudicot stems.

In a leaf, transport tissue forms **leaf veins.** In monocots, the veins are parallel, while in eudicots, the leaf veins form a net pattern. Monocots and eudicots also have different numbers of flower parts, and this difference will be discussed further in Chapter 21.

The eudicots are the larger group and include some of our most familiar flowering plants—from dandelions to oak trees. The monocots include grasses, lilies, orchids, and palm trees, among others. Some of our most significant food sources are monocots, including rice, wheat, and corn.

20.2 Plant Tissues and Cells

Unlike humans, flowering plants grow their entire life because they have **meristem** (embryonic) tissue, composed of cells that divide. **Apical meristem** is located in the terminal bud of the shoot system and in the root tip. When apical meristem cells divide, one of the daughter cells remains a meristem cell, and the other differentiates into one of the three types of specialized tissues of a plant:

1. **Epidermal tissue** forms the outer protective covering of a plant.
2. **Ground tissue** fills the interior of a plant and helps carry out the functions of a particular organ.
3. **Vascular tissue** transports water and nutrients in a plant and provides support.

Another type of meristem tissue is associated with vascular tissue and is called **vascular cambium.** When the meristem cells of vascular cambium divide, they give rise to new vascular tissue, which causes a plant to increase in girth (called *secondary growth*).

Epidermis and Ground Tissue

The entire body of a plant is covered by a layer of **epidermis** composed of closely packed epidermal cells. The walls of epidermal cells that are exposed to air are covered with a waxy **cuticle** to minimize water loss. The cuticle also protects against bacteria and other organisms that might cause disease. As mentioned in Section 20.1, roots have root hairs, long slender projections of epidermal cells (Fig. 20.5a). These hairs increase the surface area of the root for absorption of water and minerals; they also help anchor the plant firmly in place. In leaves, the epidermis often contains **stomata** (sing., stoma). A stoma is a small opening surrounded by two guard cells (Fig. 20.5b). When the stomata are open, gas exchange and water loss occur.

In the trunk of a tree, the epidermis is replaced by cork, which is a part of bark (Fig. 20.5c). New cork cells are made by a meristem called cork cambium. As the new cork cells mature, they increase slightly in volume, and their walls become encrusted with *suberin,* a lipid material, so that they are waterproof and chemically inert. These nonliving cells protect the plant and make it resistant to attack by fungi, bacteria, and animals.

Ground tissue forms the bulk of leaves, stems, and roots. Ground tissue contains three types of cells (Fig. 20.6). **Parenchyma** cells are the least specialized of the cell types and are found in all the organs of a plant. They may contain chloroplasts and carry on photosynthesis, or they may contain colorless plastids that store the products of photosynthesis. **Collenchyma** cells are like parenchyma cells except they have thicker primary walls. The thickness is uneven and usually involves the corners of the cell. Collenchyma cells often form bundles just beneath the epidermis and give flexible support to immature regions of a plant body. The familiar strands in celery stalks are composed mostly of collenchyma cells. **Sclerenchyma** cells have thick secondary cell walls impregnated with *lignin,* which makes plant cell walls tough and hard. If we compare a cell wall to reinforced concrete, cellulose fibrils would play the role of steel rods, and lignin would be analogous to the cement. Most sclerenchyma cells are nonliving; their primary function is to support the mature regions of a plant. The long fibers in plants, composed of strings of sclerenchyma, make them useful for a number of purposes. For example, cotton and flax fibers can be woven into cloth, and hemp fibers can make strong rope.

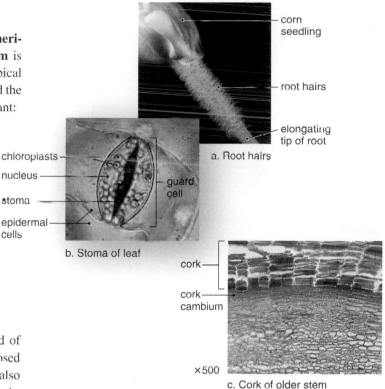

corn seedling

root hairs

elongating tip of root

a. Root hairs

chloroplasts

nucleus

stoma

guard cell

epidermal cells

b. Stoma of leaf

cork

cork cambium

×500

c. Cork of older stem

Figure 20.5 **Modifications of epidermal tissue.**

a. Root epidermis has root hairs, which help to absorb water. **b.** Leaf epidermis contains stomata (sing., stoma) for gas exchange. **c.** Cork, which is a part of bark, replaces epidermis in older, woody stems.

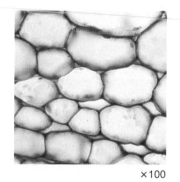

×100

a. Parenchyma cells

×340

b. Collenchyma cells

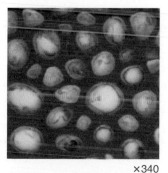

×340

c. Sclerenchyma cells

Figure 20.6 **Ground tissue cells.**

a. Parenchyma cells are the least specialized of the plant cells.
b. Collenchyma cells are much thicker and more irregular than parenchyma cells.
c. Sclerenchyma cells have very thick walls and are nonliving— their only function is to give strong support.

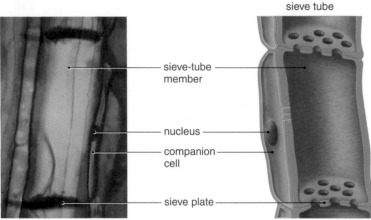

a.

b.

Figure 20.7   Vascular tissue.

a. Photomicrograph of xylem vascular tissue, with (*left*) drawing of a vessel (composed of vessel elements) and (*right*) drawing of tracheids. **b.** Photomicrograph of phloem vascular tissue, with drawing of a sieve tube. Each sieve-tube member has a companion cell.

Check Your Progress

1. List the three types of tissue in a plant and the cells that make up these tissues.
2. Compare and contrast xylem with phloem.
3. What is the importance of the leaf tissue called mesophyll to a plant?

Answers: 1. Epidermal tissue: epidermal cells; ground tissue: parenchyma, collenchyma, and sclerenchyma cells; vascular tissue: xylem (vessel elements and tracheids) and phloem (sieve-tube members). 2. Xylem transports water and minerals from roots to leaves. Phloem transports organic compounds throughout the plant. 3. Photosynthesis, which produces organic food for a plant, occurs in the mesophyll.

Vascular Tissue

The two types of transport tissue, more properly called vascular tissue, have different functions. Xylem transports water and minerals from the roots to the leaves; phloem transports sugar, in the form of sucrose, and other organic compounds, such as hormones, often from the leaves to the roots. Xylem contains two types of conducting cells: vessel elements and tracheids (Fig. 20.7*a*). Both types of conducting cells are hollow and nonliving, but the **vessel elements** are larger, have perforated end walls, and are arranged to form a continuous pipeline for water and mineral transport. The end walls and side walls of **tracheids** have pits, depressions that allow water to move from one tracheid to another.

The conducting cells of phloem are **sieve-tube members,** arranged to form a continuous sieve tube (Fig. 20.7*b*). Sieve-tube members contain cytoplasm but no nuclei. The term *sieve* refers to a cluster of pores in the end walls, collectively known as a sieve plate. Each sieve-tube member has a companion cell, which does have a nucleus. The two are connected by numerous plasmodesmata, and the nucleus of the companion cell may control and maintain the life of both cells. The companion cells are also believed to be involved in the transport function of phloem.

Vascular tissue, consisting of both xylem and phloem, extends from the root through the stem to the leaves and vice versa (see Fig. 20.1). In the root, the vascular tissue is located in a central cylinder; in the stem, it is in vascular bundles; and in the leaves, it is found in leaf veins.

20.3   Organization of Leaves

Figure 20.8 shows that leaves can be organized variously. In simple leaves, the blade is not divided as it is in compound leaves. Notice that it is possible to tell from the placement of the lateral bud whether you are looking at several individual leaves or one compound leaf.

Figure 20.9 shows a typical eudicot leaf of a temperate-zone plant in cross section. At the top and bottom are layers of epidermal tissue that often bear protective hairs and/or glands that secrete irritating substances. These appendages discourage insects from eating leaves. The epidermis characteristically has an outer, waxy cuticle that helps keep the leaf from drying out. The cuticle prevents loss of water and also prevents gas exchange because it is not gas permeable.

The interior of a leaf is composed of **mesophyll,** the tissue that carries on photosynthesis. Notice how leaf veins terminate at the mesophyll. This termination is important because the leaf veins transport water and minerals to a leaf and transport the product of photosynthesis, a sugar, away from the leaf. Mesophyll has two distinct regions: **palisade mesophyll,** containing elongated cells, and **spongy mesophyll,** containing irregular cells bounded by air spaces. The loosely packed arrangement of the cells in the spongy layer increases the amount of surface area for gas exchange and water loss.

The lower epidermis of eudicot leaves and both surfaces of monocot leaves contain stomata. Carbon dioxide for photosynthesis enters a leaf at the stomata, and the by-product of photosynthesis, oxygen, exits a leaf at the stomata. Also, water brought into a leaf by leaf veins evaporates from spongy mesophyll and exits at the stomata.

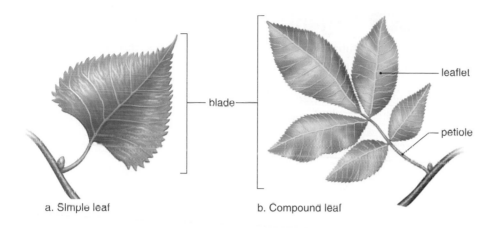

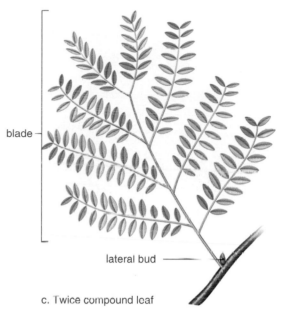

Figure 20.8 **Simple versus compound leaves.**

a. The cottonwood tree has a simple leaf. **b.** The shagbark hickory has a compound leaf. **c.** The honey locust has a twice compound leaf. A lateral bud appears where a leaf attaches to the stem.

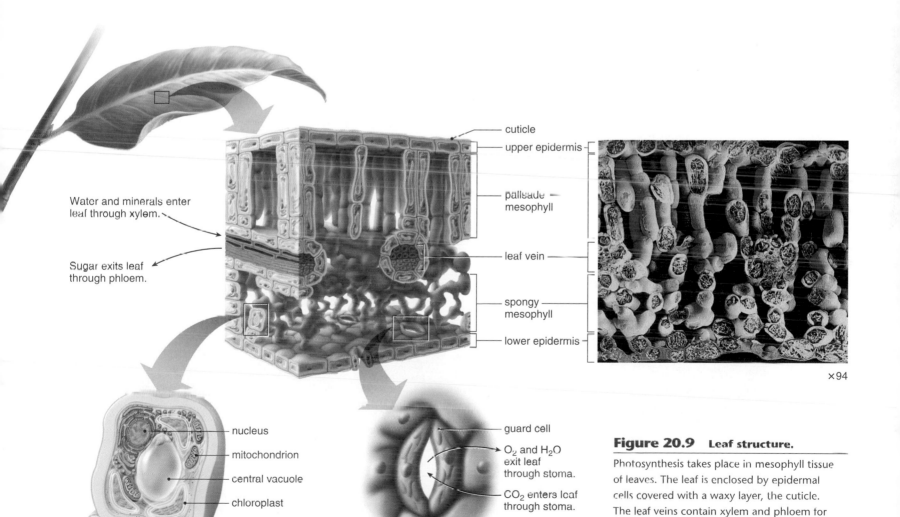

Water and minerals enter leaf through xylem.

Sugar exits leaf through phloem.

cuticle
upper epidermis
palisade mesophyll
leaf vein
spongy mesophyll
lower epidermis

×94

nucleus
mitochondrion
central vacuole
chloroplast

Leaf cell

guard cell
O_2 and H_2O exit leaf through stoma.
CO_2 enters leaf through stoma.
epidermal cell

Stoma and guard cells

Figure 20.9 **Leaf structure.**

Photosynthesis takes place in mesophyll tissue of leaves. The leaf is enclosed by epidermal cells covered with a waxy layer, the cuticle. The leaf veins contain xylem and phloem for the transport of water and solutes. A stoma is an opening in the epidermis that permits the exchange of gases.

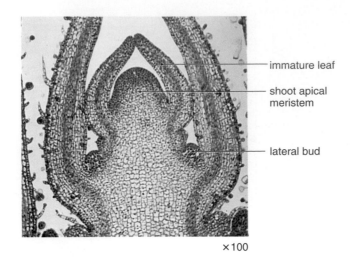

×100

Figure 20.10 Shoot tip.

The terminal bud contains the apical meristem that produces leaves and also cells that become the tissues of stems.

Figure 20.11 Nonwoody (herbaceous) stems.

a. In eudicot stems, the vascular bundles are arranged in a ring around well-defined ground tissue called pith. **b.** In monocot stems, the vascular bundles are scattered within the ground tissue.

20.4 Organization of Stems

During the primary growth of stems, the shoot apical meristem in the terminal bud produces new cells that elongate and then differentiate to become the tissues and organs of the shoot system, namely the leaves and the stem. The shoot apical meristem is protected by the immature leaves that envelop it. Also, note the lateral buds from which lateral (side) branches develop (Fig. 20.10).

Nonwoody Stems

A stem that experiences only primary growth is nonwoody. Plants that have nonwoody stems, such as zinnias and daisies, are termed **herbaceous** plants. As with leaves, the outermost tissue of a herbaceous stem is the epidermis, which is covered by a waxy cuticle to prevent water loss. Herbaceous stems have distinctive vascular bundles, where xylem and phloem are found. In each bundle, xylem is typically found toward the inside of the stem, and phloem is found toward the outside.

In the herbaceous eudicot stem, the vascular bundles are arranged in a distinct ring that separates the cortex from the central pith, which stores water and the products of photosynthesis. The cortex is sometimes green and carries on photosynthesis. In the herbaceous monocot stem, the vascular bundles are scattered throughout the stem. The cortex is a narrow band of cells beneath the epidermis, and ground tissue occupies the rest of the stem. Figure 20.11 contrasts herbaceous eudicot and monocot stems.

The vascular tissue of a stem supports the growth of the shoot system. The sclerenchyma cells of vascular tissue and the strong walls of vessel members and tracheids help support the shoot system as it increases in length. Further, the vascular bundles bring water and minerals to the leaf veins, which distribute them to the mesophyll cells of leaves, the primary organs of photosynthesis in most plants. The vascular bundles also distribute the products of photosynthesis to the root system and to any immature leaves that are not yet photosynthesizing.

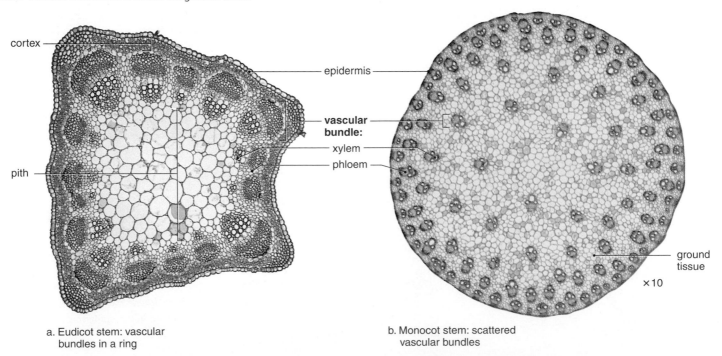

a. Eudicot stem: vascular bundles in a ring

b. Monocot stem: scattered vascular bundles

Woody Stems

Plants with woody stems experience both primary and secondary growth. Secondary growth, which occurs only in trees and shrubs, increases the girth of stems, branches, and roots. Secondary growth occurs because of a change in the location and activity of vascular cambium, which, as you may recall from Section 20.2, is a type of meristem tissue. In herbaceous plants, vascular cambium is present between the xylem and phloem of each vascular bundle. In woody plants, the vascular cambium forms a ring of meristem that divides parallel to the surface of the plant and produces new xylem and phloem each year. Eventually, a woody eudicot stem has three distinct areas: the bark, the wood, and the pith. Vascular cambium occurs between the bark and the wood (Fig. 20.12).

Bark

The **bark** of a tree contains cork, cork cambium, cortex, and phloem. It is very harmful to remove the bark of a tree, because without phloem, organic nutrients cannot be transported. Although secondary phloem is produced each year by vascular cambium, it does not build up in the same manner as xylem.

Cork cambium is located beneath the epidermis. When cork cambium first begins to divide, it produces tissue that disrupts the epidermis and replaces it with cork cells. **Cork cells** are impregnated with suberin, a waxy layer that makes them waterproof but also causes them to die. This is a protective strategy, because now the stem is less edible. In a woody stem, gas exchange is impeded, except at *lenticels,* which are pockets of loosely arranged cork cells not impregnated with suberin.

Wood

Wood is secondary xylem that builds up year after year, thereby increasing the girth of trees. In trees that have a growing season, vascular cambium is dormant during the winter. In the spring, when moisture is plentiful and leaves require much water for growth, the secondary xylem contains wide vessels with thin walls. In this so-called *spring wood,* wide vessels transport sufficient water to the growing leaves. Later in the season, moisture is scarce, and the wood at this time, called *summer wood,* has a lower proportion of vessels. Strength is required because the tree is growing larger, and summer wood contains numerous thick-walled tracheids. At the end of the growing season, just before the cambium becomes dormant again, only heavy fibers with especially thick secondary walls may develop. When the trunk of a tree has spring wood followed by summer wood, the two together make up one year's growth, or an **annual ring** (Fig. 20.12). You can tell the age of a tree by counting the annual rings.

Wood is one of the most useful and versatile materials known. It is used for building structures and for making furniture, and in addition to its functionality, its beauty gives us pleasure. Much wood is also used for heating and cooking, as well as for producing paper, chemicals, and pharmaceuticals. The resins derived from wood also have commercial importance.

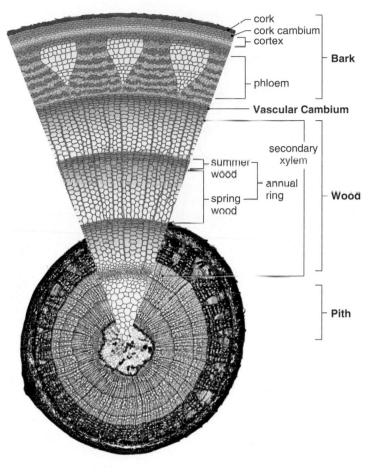

Figure 20.12 **Organization of a woody stem.**

In a woody stem, vascular cambium produces new secondary phloem and secondary xylem each year. Secondary xylem builds up to form wood consisting of annual rings. Counting the annual rings tells you that this woody stem is three years old.

Check Your Progress

1. What transport tissues are in a vascular bundle?
2. How are vascular bundles arranged in monocot stems? In eudicot stems?
3. Contrast primary growth with secondary growth.
4. List the components of bark.
5. Relate spring wood and summer wood to annual rings.

Answers: 1. A vascular bundle contains xylem and phloem. 2. Vascular bundles are scattered in monocot stems and form a ring in eudicot stems. 3. Primary growth is growth in length and is nonwoody; secondary growth is growth in girth and is woody. 4. Bark is composed of cork, cork cambium, cortex, and phloem. 5. An annual ring is composed of one year's growth of wood—one layer of spring wood followed by one layer of summer wood.

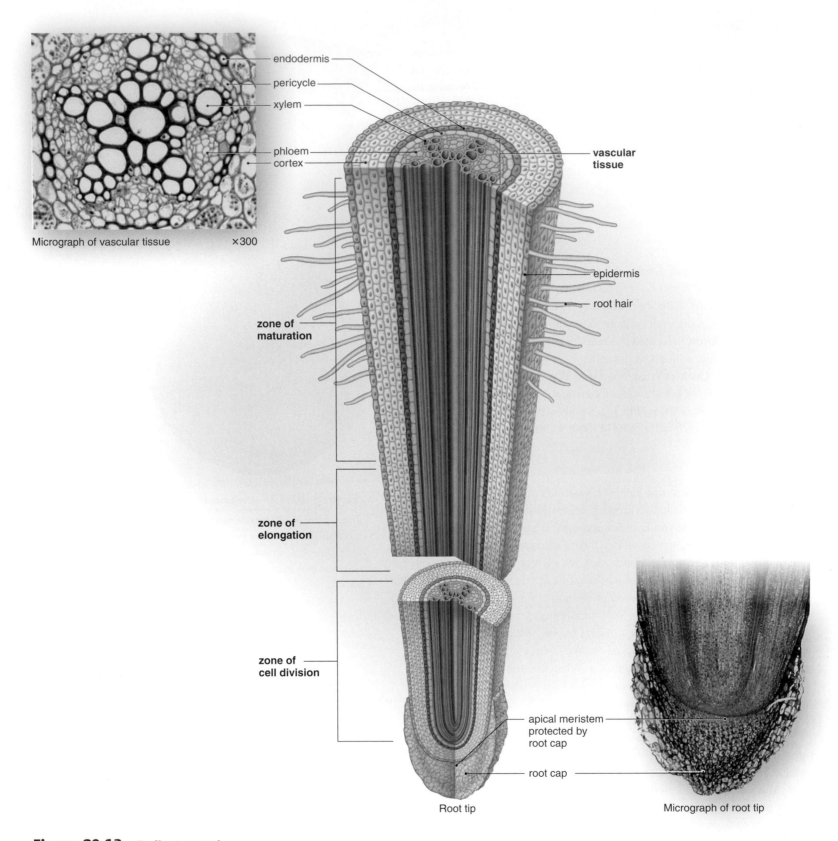

Micrograph of vascular tissue ×300

endodermis

pericycle

xylem

phloem

cortex

vascular tissue

epidermis

root hair

zone of maturation

zone of elongation

zone of cell division

apical meristem protected by root cap

root cap

Root tip

Micrograph of root tip

Figure 20.13 Eudicot root tip.

The root tip is divided into three zones, best seen in a longitudinal section such as this. In the eudicot, xylem is typically star-shaped, and phloem lies between the points of the star. Water and minerals must eventually pass through the cytoplasm of endodermal cells in order to enter xylem; endodermal cells regulate the passage of minerals into xylem.

20.5 Organization of Roots

Figure 20.13, a longitudinal section of a eudicot root, reveals zones where cells are in various stages of differentiation as primary growth occurs. The apical meristem of a root is in the zone of a cell division, protected by a **root cap.** When this apical meristem divides, it provides cells to the zone of elongation above. In the zone of elongation, the cells lengthen as they become specialized. The zone of maturation, which contains fully differentiated cells, is recognizable because many of its epidermal cells bear root hairs.

Tissues of a Eudicot Root

Figure 20.13 also shows a cross section of a root at the region of maturation. The following specialized tissues are identifiable:

Epidermis The epidermis, which forms the outer layer of the root, consists of only a single layer of largely thin-walled, rectangular cells. In the zone of maturation, many epidermal cells have root hairs.

Cortex Large, thin-walled parenchyma cells make up the **cortex** of the root. The cells contain starch granules, and the cortex may function in food storage.

Endodermis The **endodermis** is a single layer of rectangular cells that fit snugly together. Further, a layer of impermeable material on all but two sides forces water and minerals to pass through endodermal cells. In this way, the endodermis regulates the entrance of minerals into the vascular tissue of the root.

Pericycle The **pericycle** is the first layer of cells inside the endodermis. It has retained its capacity to divide and can start the development of branch, or lateral, roots (Fig. 20.14).

Vascular tissue Xylem appears star-shaped in eudicot roots because several arms of tissue radiate from a common center (see Fig. 20.13). The phloem is found in separate regions between the arms of the xylem.

Organization of Monocot Roots

Monocot roots have the same growth zones as eudicot roots, but they do not undergo secondary growth as many eudicot roots do. Also, the organization of their tissues is slightly different. The ground tissue of a monocot, called the **pith,** is centrally located and surrounded by a vascular ring composed of alternating xylem and phloem bundles. Monocot roots also have pericycle, endodermis, cortex, and epidermis.

Comparison with Stems

Both roots and stems have apical meristem regions that account for their primary growth. The branching of a stem occurs at a lateral bud, whereas the branching of a root occurs some distance from apical meristem and is initiated by the pericycle. Eudicot roots, like woody stems, experience secondary growth. Vascular cambium located between the arms of the xylem and the phloem produces secondary xylem and phloem. As secondary xylem builds up, a root grows in girth.

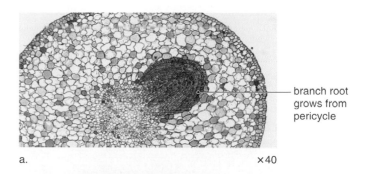

a. ×40

branch root grows from pericycle

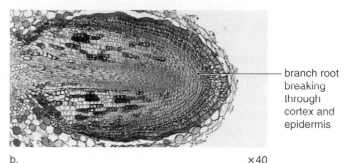

b. ×40

branch root breaking through cortex and epidermis

Figure 20.14 **Branching of eudicot root.**

These cross sections of a willow show the origination (**a**) and growth (**b**) of a branch root from the pericycle.

Check Your Progress

1. Describe the relationship between the root apical meristem and the root cap.
2. List the functions of the cortex, the endodermis, and the pericycle in a root.
3. Compare and contrast roots with stems.

Answers: 1. The root apical meristem is located at the tip of the root and is covered by the root cap. 2. Cortex: food storage; endodermis: control of mineral uptake; pericycle: formation of branch roots. 3. Both roots and stems have apical meristems and are capable of branching. Stems branch at lateral buds, while root branches originate in the pericycle.

solution lacks nitrogen   complete nutrient solution

Figure 20.15   Nitrogen deficiency.

This experiment shows that sunflower plants respond poorly if their growth medium lacks nitrogen.

— root

— root nodule

a.

b.

Figure 20.16   Adaptations of roots for mineral uptake.

a. Nitrogen-fixing bacteria live in nodules on the roots of plants, particularly legumes. **b.** Rough lemon plants with mycorrhizal fungi (*right*) grow much better than plants without mycorrhizal fungi (*left*).

20.6 Plant Nutrition

Plant nutrition is remarkable to us because plants require only inorganic nutrients, and from these they make all the organic compounds that compose their bodies. Of course, they require carbon, hydrogen, and oxygen, which they can acquire from carbon dioxide and water, but they also need elements that are absorbed as minerals. An element is termed an essential nutrient if a plant cannot live without it. The essential nutrients are divided into *macronutrients* and *micronutrients,* according to their relative concentrations in plant tissue. The following diagram and slogan show which are the macronutrients and which are the micronutrients:

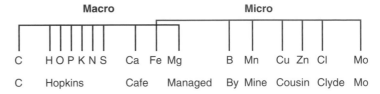

All of these elements play vital roles in plant cells. For example, nitrogen and phosphorus are required to make nucleic acids, and nitrogen and sulfur are important parts of amino acids. Figure 20.15 shows how an insufficient supply of the macronutrient nitrogen can affect a plant. The micronutrients are often cofactors for enzymes in various metabolic pathways.

It's interesting to observe that humans make use of a plant's ability to take minerals from the soil. For example, we depend on plants for supplies of calcium to build our bones and teeth, and for iron to help carry oxygen to our cells. Minerals such as copper and zinc, which we acquire by eating plants, are also cofactors for our own enzymes.

Adaptations of Roots for Mineral Uptake

Minerals enter a plant at its root system, and two mutualistic relationships assist roots in fulfilling this function. The air all about us contains about 78% nitrogen (N_2), but plants can't make use of it. Most plants depend on bacteria in the soil to *fix* nitrogen—that is, the bacteria change atmospheric nitrogen (N_2) to nitrate (NO_3^-) or ammonium (NH_4^+), both of which plants can take up and use. Some plants, such as legumes, have roots colonized by bacteria that are able to take up atmospheric nitrogen and reduce it to a form suitable for incorporation into organic compounds (Fig. 20.16*a*). The bacteria live in **root nodules,** and the plant supplies the bacteria with carbohydrates; the bacteria in turn furnish the plant with nitrogen compounds.

Another mutualistic relationship involves fungi and almost all plant roots (Fig. 20.16*b*). This association is called a **mycorrhizal association,** literally fungal roots. The hyphae of the fungus increase the surface area available for water uptake and break down organic matter, releasing inorganic nutrients that the plant can use. In return, the root furnishes the fungus with sugars and amino acids. Plants are extremely dependent on mycorrhizal fungi. For example, orchid seeds, which are quite small and contain limited nutrients, do not germinate until a mycorrhizal fungus has invaded their cells. The visual appearance of plants that do not have adequate nitrogen convinces us how important these associations are (see Fig. 20.15).

Check Your Progress

1. Explain the significance of nitrogen-fixing bacteria in the soil to plants.
2. Explain how both partners benefit from a mycorrhizal association.

Answers: 1. The bacteria convert atmospheric nitrogen to nitrate or ammonium, which can be taken up by plant roots. 2. The fungus obtains sugars and amino acids from the plant. The plant obtains inorganic nutrients and water from the fungus.

20.7 Transport of Nutrients

Water and Mineral Transport in Xylem

Water and minerals are taken up by root hairs at the same time, and they are both transported within xylem. The vessel elements, assisted by the tracheids, constitute a continuous system for water and mineral transport in a plant (see Fig. 20.7). In other words, a pipeline for water transport exists from the roots through the stem to the leaves. People have long wondered how plants, especially very tall trees, lift water to the leaves against gravity. Water enters root cells by osmosis, and sure enough, this does create a positive pressure called root pressure. But this pressure is not nearly enough to account for the movement of water and minerals all the way to the leaves.

The mechanism by which water and minerals travel up the xylem is called the **cohesion-tension model.** To understand how it works, we have to turn our attention to the anatomy of leaves and the properties of water. A leaf, as you know, has small openings called stomata, and the stomata open to air spaces in the leaf. Dry air passing across the leaves causes water to evaporate from the surface of spongy mesophyll cells. Water vapor in the air spaces of a leaf exits a leaf at the stomata, a phenomenon called **transpiration.** The evaporation of water creates a *tension* that is sufficient to pull a water column up from the roots to the leaves (Fig. 20.17). Why? Because of the properties of water. Water molecules have hydrogen atoms bonded to one another, and they tend to stay together—if one moves up, they all move up. This tendency of water molecules to cling together is called the *cohesion property* of water. Also, water molecules tend to adhere to the sides of the xylem vessels. This *adhesion property* of water gives the water column extra strength and prevents it from slipping back.

The total amount of water a plant loses through transpiration over a long period of time is surprisingly large. At least 90% of the water taken up by roots is eventually lost at the leaves. A single corn plant loses somewhere between 135 and 200 liters of water through transpiration during a growing season. As transpiration occurs, the water column is pulled upward, first within the leaf, then from the stem, and finally from the roots.

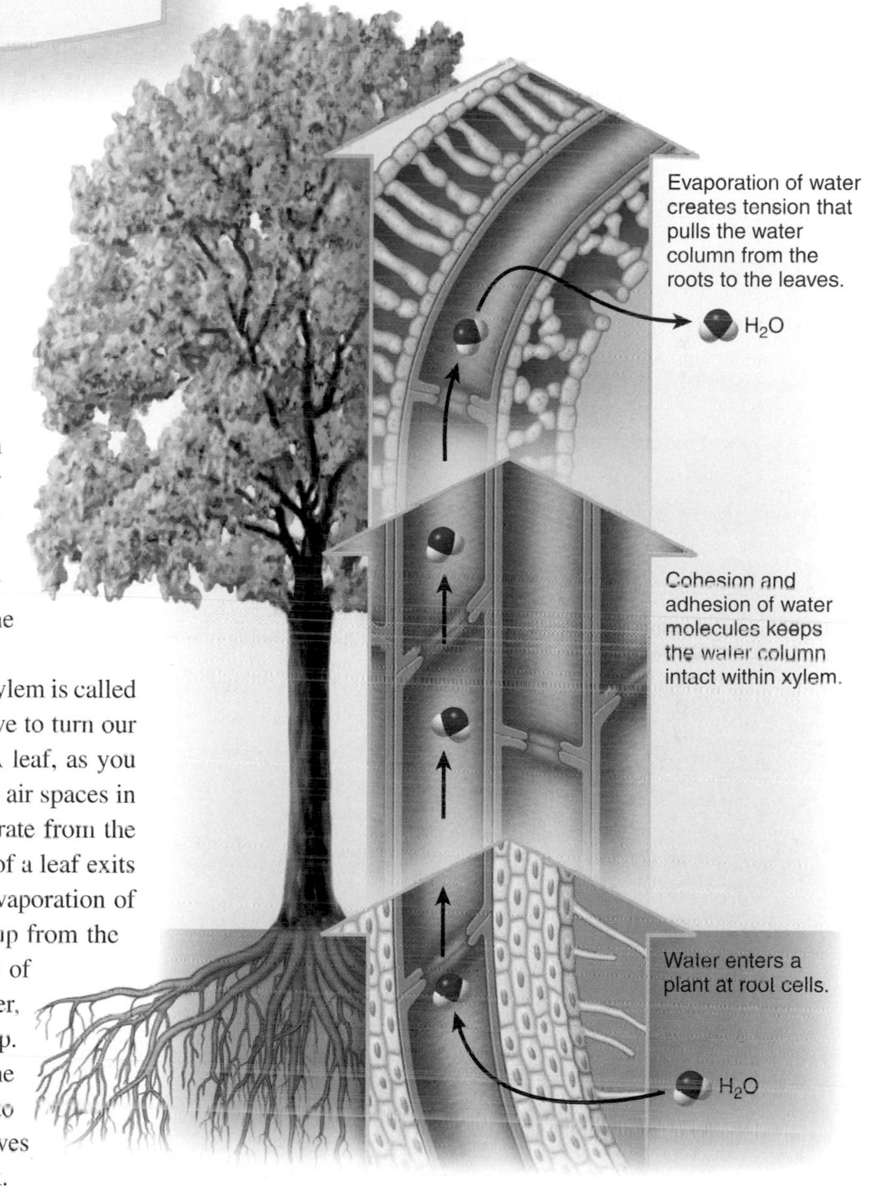

Evaporation of water creates tension that pulls the water column from the roots to the leaves.

H_2O

Cohesion and adhesion of water molecules keeps the water column intact within xylem.

Water enters a plant at root cells.

H_2O

Figure 20.17   Cohesion-tension model of xylem transport.

How does water rise to the top of tall trees? Xylem vessels are water-filled pipelines from the roots to the leaves. When water evaporates from spongy mesophyll into the air spaces of leaves, this water column is pulled upward due to the cohesion of water molecules with one another and the adhesion of water molecules to the sides of the vessels.

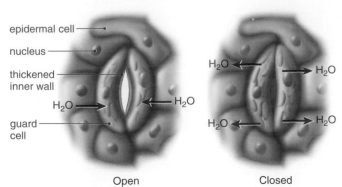

epidermal cell

nucleus

thickened
inner wall

H_2O

guard
cell

H_2O H_2O

H_2O H_2O

Open   Closed

Figure 20.18   Opening and closing of stomata.

Stomata open when water enters guard cells and turgor pressure
increases. Stomata close when water exits guard cells and turgor
pressure is lost.

Opening and Closing of Stomata

Stomata are able to open and close because of changes in turgor pressure within
their guard cells (Fig. 20.18). When water enters the guard cells, turgor pressure
is created; when water leaves the guard cells, turgor pressure is lost. For transpiration to occur, the stomata must stay open. Still, when a plant is under stress
and about to wilt from lack of water, the stomata close. Now the plant is unable
to take up carbon dioxide from the air, and photosynthesis ceases.

In plants kept in the dark, stomata open and close on a 24-hour basis, just
as if they were responding to the presence of sunlight in the daytime and the
absence of sunlight at night. Such circadian rhythms (behaviors that occur every
24 hours) are areas of intense investigation at this time.

Organic Nutrient Transport in Phloem

Phloem transports organic nutrients from stored supplies in roots, or from
mature leaves that are photosynthesizing, to all other parts of a plant, including
young leaves that have not yet reached their full photosynthetic potential and
flowers in the process of making seeds and fruits. Nutrients are transported to
any part of a plant, including roots, when they lack a ready supply.

Just as xylem is continuous
throughout a plant, so is phloem. In
phloem, the sieve-tube members align
end to end, and plasmodesmata extend
through sieve plates from one sieve-tube
member to the other, allowing direct
cytoplasmic exchange. Sieve tubes,
therefore, form a continuous pathway
for transport of organic nutrients, primarily the sugar sucrose.

The mechanism by which nutrients
are transported through phloem is called
the **pressure-flow model** (Fig. 20.19). The
process begins during the growing season
when photosynthesizing leaves are making sugar, and are therefore a *source* of
sugar. This sugar is actively transported
into sieve tubes, and water follows passively by osmosis. The buildup of water
within sieve tubes creates a positive pressure that starts a flow of phloem contents.
The other parts of a plant, such as roots
that are not photosynthesizing, are a *sink*
for sugar, because they require a supply
of sugar. Here, sugar is actively transported out of phloem, and water follows
passively by osmosis. In this way, phloem
contents continue to flow from source
(e.g., leaves) to a sink (e.g., root).

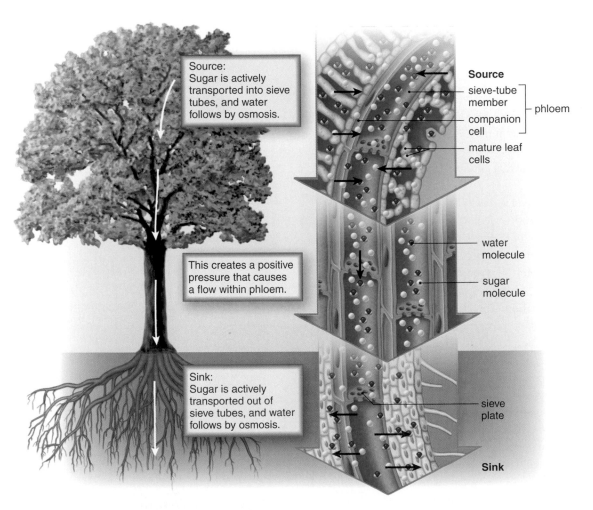

Source:
Sugar is actively
transported into sieve
tubes, and water
follows by osmosis.

This creates a positive
pressure that causes
a flow within phloem.

Sink:
Sugar is actively
transported out of
sieve tubes, and water
follows by osmosis.

Source

sieve-tube
member

companion
cell

phloem

mature leaf
cells

water
molecule

sugar
molecule

sieve
plate

Sink

Figure 20.19   Pressure-flow model of phloem transport.

Sugar and water enter sieve-tube members at a source. This creates a positive pressure, which causes phloem
contents to flow. Sieve-tube members form a continuous pipeline from a source to a sink, where sugar and water
exit sieve-tube members.

The experiment described in Figure 20.20 supports the pressure-flow model. Two bulbs within a beaker of distilled water are connected by a glass tube. The glass tube is analogous to phloem. The first bulb contains solute at a higher concentration than the second bulb. Each bulb is bounded by a differentially permeable membrane. Distilled water flows into the first bulb because it has the higher solute concentration. The entrance of water creates *pressure,* and causes water to flow toward the second bulb. This pressure not only drives water toward the second bulb, but it also provides enough force for water to move out through the membrane of the second bulb—even though the second bulb contains a higher concentration of solute than the distilled water.

The beauty of the pressure-flow model is that it can account for the flow of sugar in either direction, depending on which is the source and which is the sink. For example, in the spring, recently formed leaves can be a sink, and roots that have held stored sugar over the winter can be a source. Thus, organic nutrients will flow from the roots to the leaves, instead of the other way around. Maple sugar is made from the sap of maple trees flowing upward in the spring.

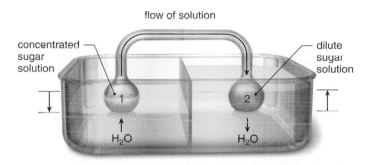

Figure 20.20   Experiment supporting the pressure-flow model.

When a concentrated sugar solution is placed in the first bulb, water enters and causes a flow of both water and sugar to the second bulb. The pressure is sufficient to cause water to exit the second bulb, even though the surrounding water contains no sugar. As water enters the first bulb, water level decreases *(on left);* as water exits the second bulb, water level increases *(on right).*

Check Your Progress

1. Explain why water is under tension in stems.
2. Explain the significance of the cohesion and adhesion properties of water to water transport.
3. Explain how sugars move from source to sink in a plant.

Answers: 1. Water is pulled upward in stems via evaporation from leaf surfaces, so it is under tension. 2. When water molecules are pulled upward during transpiration, their cohesiveness allows them to cling together, creating a continuous column. Adhesion allows them to cling to the sides of xylem vessels, so the column of water does not slip down. 3. Sugars build up in sieve tubes at sources, creating pressure as water flows in as well. The pressure is relieved at the other end when sugars and water are removed at the sink.

THE CHAPTER IN REVIEW

Summary

20.1 Plant Organs

A flowering plant has a root system and a shoot system. Both grow at their tips.

Leaves carry on photosynthesis.

Stems support leaves, conduct materials to and from roots and leaves, and help store plant products.

Roots anchor a plant, absorb water and minerals, and store the products of photosynthesis.

Monocot Versus Eudicot Plants

Flowering plants are divided into eudicots and monocots according to:

- Number of cotyledons in the seed.
- Arrangement of vascular tissue in leaves, stems, and roots.
- Number of flower parts:

monocot flower   eudicot flower

20.2 Plant Tissues and Cells

Plants have three types of specialized tissue:

Epidermal tissue is composed of only epidermal cells.

Ground tissue contains:

- Parenchyma cells, which are thin-walled and capable of photosynthesis when they contain chloroplasts.
- Collenchyma cells, which have thicker walls for flexible support.
- Sclerenchyma cells, which are hollow, nonliving support cells with secondary walls.

Vascular tissue consists of:

- Xylem, which contains vessels composed of vessel elements and tracheids. Xylem transports water and minerals.
- Phloem, which contains sieve tubes composed of sieve-tube members, each of which has a companion cell. Phloem transports organic nutrients.

xylem

phloem

20.3 Organization of Leaves

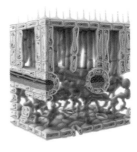

- A leaf has an upper epidermis, with stomata mostly in the lower epidermis.
- Stomata allow water vapor and oxygen to escape and carbon dioxide to enter the leaf.
- Mesophyll (palisade and spongy) forms the body of a leaf and carries on photosynthesis.

20.4 Organization of Stems

- Primary growth of a stem is due to the activity of the shoot apical meristem, which is protected within a terminal bud.
- Nonwoody eudicots have epidermis, cortex tissue, vascular bundles in a ring, and an inner pith.
- Monocot stems have scattered vascular bundles.

eudicot stem   monocot stem

Woody Stems

- Secondary growth of a woody stem is due to vascular cambium, which produces new xylem and phloem every year.
- Cork replaces epidermis in woody plants.
- A woody stem has bark (contains cork, cork cambium, cortex, and phloem). Wood contains annual rings of xylem.

20.5 Organization of Roots

A eudicot root tip has three zones:

- Zone of cell division (contains root apical meristem) protected by the root cap.
- Zone of elongation.
- Zone of maturation (has root hairs).

In a eudicot root:

- Xylem appears star-shaped.
- Phloem is in separate regions between the arms of the xylem.

20.6 Plant Nutrition

Plants need only inorganic nutrients to make all the organic compounds that make up their bodies. Some nutrients are essential, being either macronutrients or micronutrients.

Adaptations of Roots for Mineral Uptake

Some roots have nodules where bacteria fix nitrogen and produce forms that plants can use. Most roots have mycorrhizal fungi.

20.7 Transport of Nutrients

Water and Mineral Transport in Xylem

The cohesion-tension model of xylem transport states that transpiration (evaporation of water from spongy mesophyll at stomata) creates tension, which pulls water upward in xylem. This method works only because water molecules are cohesive and adhesive.

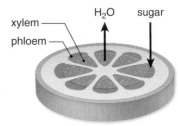

xylem
phloem

H$_2$O   sugar

Organic Nutrient Transport in Phloem

The pressure-flow model of phloem transport states that sugar is actively transported into phloem at a source, and water follows by osmosis. The resulting increase in pressure creates a flow, which moves water and sugar to a sink.

Thinking Scientifically

1. Legumes must be able to distinguish between nitrogen-fixing bacteria and all other bacteria in the soil in order to develop nodules. How do you suppose they can "perceive" differences among the diverse species of bacteria in the soil and engulf only those that will be beneficial?

2. *Welwitschia* is a genus of plant that lives in the Namib and Mossamedes deserts in Africa. Annual rainfall averages only 2.5 centimeters (1 inch) per year. Contrary to what you might expect, the leaves of *Welwitschia* plants contain a large number of stomata (22,000 per square centimeter). Can you suggest an adaptive explanation for the abundance of stomata on these desert plants?

Testing Yourself

Choose the best answer for each question.

1. It is possible to distinguish between stems and roots by looking for

   a. petioles on stems.
   b. petioles on roots.
   c. nodes on stems.
   d. nodes on roots.

2. In contrast to monocots, eudicots

   a. have parallel veins.
   b. produce root xylem and phloem in a ring.
   c. use endosperm as the major source of nutrients for seedlings.
   d. have flower parts in multiples of threes.
   e. produce a ring of vascular bundles in the stem.

3. Leaves require stomata because

   a. epidermal cells are sealed by a cuticle.
   b. vascular tissue is not near the epidermis.
   c. water cannot move across plant cell walls.
   d. sclerenchyma cells contain lignin.

4. Flexible support in immature regions of a plant is provided by

   a. vascular tissue.
   b. collenchyma cells.
   c. sclerenchyma cells.
   d. parenchyma cells.
   e. epidermal tissue.

5. Label the parts of a plant in the following diagram.

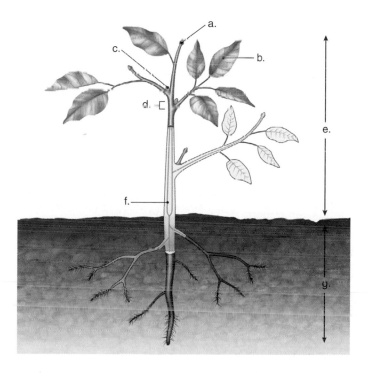

For questions 6–10, identify the plant organ that contains each feature. Each answer may be used more than once. Each question may have more than one answer.

Key:

   a. nonwoody stem
   b. woody stem
   c. root (nonwoody)
   d. leaf

6. Mesophyll

7. Vascular bundles

8. Ground tissue

9. Vascular cambium

10. Cuticle

11. During secondary growth, a tree adds more xylem and phloem through the activity of the

    a. apical meristems.
    b. vascular cambium.
    c. intercalary meristems.
    d. cork cambium.

12. Root nodules are important because they

    a. encourage the growth of mycorrhizal fungi.
    b. represent areas of extensive branch root growth.
    c. contain nitrogen-fixing bacteria.
    d. provide extra oxygen to the plant root system.

13. In contrast to transpiration, the transport of organic nutrients in the phloem

    a. requires energy input from the plant.
    b. always flows in one direction.
    c. results from tension.
    d. does not require living cells.

14. Label the parts of a leaf in the following illustration.

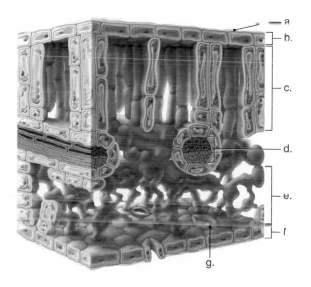

15. Label the parts of the eudicot stem in the illustration following.

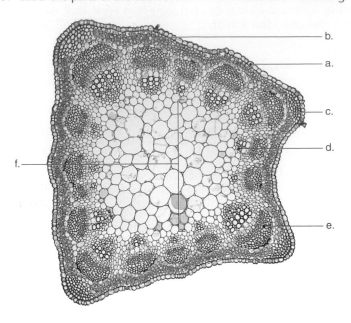

16. Label the parts of the root in the following illustration.

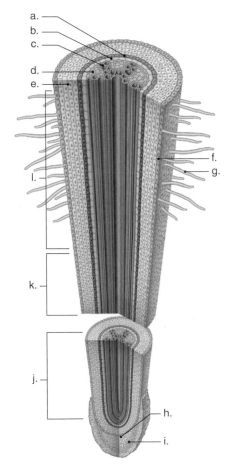

Go to www.mhhe.com/maderessentials for more quiz questions.

Bioethical Issue

Wood is a renewable resource used for everything from constructing homes to making the paper on which this page is printed. However, some sources of wood are, for all practical purposes, nonrenewable. The old-growth forests of the Pacific Northwest are threatened by logging and will probably never be restored. Recently, the Bush administration eased restrictions on the logging of old-growth forests by allowing forest managers to begin logging without completing an inventory of rare plants and animals in the area. Opponents of this new policy say that it will double the amount of federal land on which logging occurs and will result in the extinction of hundreds of species. Representatives of the timber industry, on the other hand, applaud the change because previously they were required to spend years surveying the land before they could log it. Do you agree with the new policy?

Understanding the Terms

annual ring 345	palisade mesophyll 342
apical meristem 341	parenchyma 341
bark 345	perennial plant 339
blade 338	pericycle 347
cohesion-tension model 349	petiole 338
collenchyma 341	phloem 340
cork cambium 345	pith 347
cork cell 345	pressure-flow model 350
cortex 347	root cap 347
cotyledon 340	root hair 339
cuticle 341	root nodule 348
deciduous 339	root system 338
endodermis 347	sclerenchyma 341
epidermal tissue 341	shoot system 338
epidermis 341	sieve-tube member 342
eudicot 340	spongy mesophyll 342
ground tissue 341	stem 339
herbaceous 344	stomata (sing., stoma) 341
internode 339	terminal bud 338
lateral bud 339	tracheid 342
leaf 338	transpiration 349
leaf vein 340	vascular bundle 340
meristem 341	vascular cambium 341
mesophyll 342	vascular tissue 341
monocot 340	vessel element 342
mycorrhizal association 348	wood 345
node 339	xylem 340

Match the terms to these definitions:

a. _____ A plant that dies back and then regrows during the next season.

b. _____ Vascular tissue that conducts organic nutrients and contains sieve-tube members.

c. _____ Structures in a leaf that allow water vapor and oxygen to escape and carbon dioxide to enter.

d. _____ Secondary xylem that builds up year after year and becomes annual rings.

e. _____ Undifferentiated embryonic tissue in the active growth regions of plants.

Plant Responses and Reproduction

CHAPTER 21

Length of continuous darkness, not light, controls flowering.

A hormone causes plants to bend toward the light.

Some plants have their own patents.

Issuing patents for plants is not a new practice. The first plant to be patented was a peach tree in 1932, and since then numerous plants have been patented. In fact, the United States Patent and Trademark Office has a special patent class just for asexually reproducing plants.

Why might you patent a plant? Plant breeders develop plants that are more beautiful, more useful, or more resistant ▶ ▶ ▶

to disease. A patent gives the plant breeder exclusive rights to sell a particular plant for 20 years. One of the more controversial plant patents is the one for Roundup Ready® plants. These plants have been genetically modified to be resistant to Roundup®, a herbicide widely used to kill weeds that compete with plants for nutrients and water. The benefit of Roundup Ready wheat, cotton, or soybeans is that after fields are sprayed with Roundup, the crop survives but the weeds are destroyed. This method of fighting weeds saves time and labor compared to traditional means of controlling weeds. Although Roundup Ready crops have these benefits, what about possible risks? The biggest drawback to Roundup Ready crops is that their pollen could possibly transmit resistant genes to other plants, weeds in particular, and then these weeds would not be controlled by Roundup. A catastrophe of great magnitude could be in the making.

In this chapter, you will learn about other ways plants of commercial importance have been genetically engineered. But first, the chapter introduces the plant hormones involved in plant responses and flowering. It also reviews sexual and asexual reproduction in flowering plants in more detail than those topics were covered in Chapter 18.

21.1 Responses in Flowering Plants

Plants usually respond to environmental stimuli such as light, gravity, and seasonal changes by altering their pattern of growth in some way. Hormones help control these responses.

Plant Hormones

Plant hormones are small organic molecules produced by the plant that serve as chemical signals between cells and tissues. Currently, the five commonly recognized groups of plant hormones are auxins, gibberellins, cytokinins, abscisic acid, and ethylene. Other chemicals produced commercially, some of which differ only slightly from the natural hormones, also affect the growth of plants. These and the naturally occurring hormones are sometimes grouped together and called plant growth regulators.

Plant hormones bring about a physiological response in target cells after binding to a specific receptor protein in the plasma membrane.

Auxins

The most common naturally occurring **auxin** is indoleacetic acid (IAA). It is produced in the shoot apical meristem and is found in young leaves and in flowers and fruits. Therefore, you would expect auxin to affect many aspects of plant growth and development, and it does. Figure 21.1 shows how the hormone auxin brings about elongation of a cell, a necessary step toward differentiation and maturation of a plant cell.

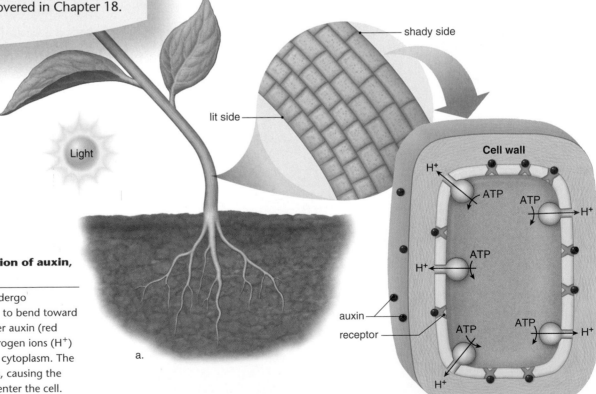

Figure 21.1 Mode of action of auxin, a plant hormone.

a. Plant cells on the shady side undergo elongation, and this causes a stem to bend toward the light. **b.** Elongation occurs after auxin (red balls) binds to a receptor and hydrogen ions (H⁺) are actively transported out of the cytoplasm. The resulting acidity activates enzymes, causing the cell wall to weaken, and water to enter the cell. The cell then elongates.

Effects of Auxin   Auxin brings about **apical dominance,** the inhibition of lateral bud growth by the presence of the shoot tip (Fig. 21.2). Release from apical dominance occurs when pruning removes the shoot tip. Then, the lateral buds grow and the plant takes on a bushier appearance. Apical dominance is believed to be the result of downward transport of auxin produced in the apical meristem. When IAA is applied to the stump, apical dominance is restored.

The application of a weak solution of auxin to a cutting from a woody stem causes roots to develop more quickly than they would otherwise. Auxin production by seeds also promotes the growth of fruit. As long as auxin is concentrated in leaves or fruits rather than in the stem, leaves and fruits do not fall off. Therefore, trees can be sprayed with auxin to keep mature fruit from falling to the ground. Auxin is also involved in phototropism, in which stems bend toward a light source, as well as gravitropism, in which roots curve downward and stems curve upward in response to gravity.

How Auxins Work   When a plant is exposed to unidirectional light, auxin moves to the shady side, where it binds to receptors and activates an ATP-driven pump that transports hydrogen ions (H$^+$) out of the cell (see Fig. 21.1). The acidic environment weakens cellulose fibrils, and activated enzymes further degrade the cell wall. Water now enters the cell, and the resulting increase in turgor pressure causes the cells on the shady side to elongate and the stem to bend toward the light.

Gibberellins

Gibberellins were discovered in 1926 when a Japanese scientist was investigating a fungal disease of rice plants called "foolish seedling disease." The plants elongated too quickly, causing the stem to weaken and the plant to collapse. The fungus infecting the plants produced an excess of a chemical called gibberellin, named after the fungus *Gibberella fujikuroi.* It wasn't until 1956 that a form of gibberellin now known as gibberellic acid was isolated from a flowering plant rather than from a fungus. Sources of gibberellin in flowering plant parts are young leaves, roots, embryos, seeds, and fruits. Gibberellins are growth-promoting hormones that bring about elongation of the cells. We know of about 70 gibberellins, and they differ chemically only slightly. The most common of these is gibberellic acid, GA$_3$ (the subscript designation distinguishes it from other gibberellins).

Effects of Gibberellins   When gibberellins are applied externally to plants, the most obvious effect is stem elongation between the nodes (Fig. 21.3). Gibberellins can cause dwarf plants to grow, cabbage plants to become as much as 2 meters tall, and bush beans to become pole beans.

Dormancy is a period during which a plant or a seed does not grow, even though conditions may be favorable for growth. The dormancy of seeds and buds can be broken by applying gibberellins. Research with barley seeds has shown how GA$_3$ influences the germination of seeds. Endosperm is the tissue that serves as food for the embryo and seedling as they undergo development. Barley seeds have a large, starchy endosperm that must be broken down into sugars to provide energy for the embryo to grow. It is hypothesized that after GA$_3$ attaches to a receptor in the plasma membrane, calcium ions (Ca^{2+}) combine with a protein. This complex is believed to activate the gene that codes for amylase. Amylase then acts on starch to release sugars as a source of energy for seed germination.

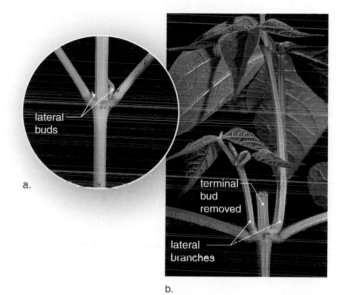

Figure 21.2   Apical dominance.

a. Lateral bud growth is inhibited when a plant retains its terminal bud. **b.** When the terminal bud is removed, lateral branches develop and the plant is bushier.

Figure 21.3   Effect of gibberellins.

The plant on the right was treated with gibberellins; the plant on the left was not treated. Gibberellins are often used to promote stem elongation in economically important plants, but the exact mode of action remains unclear.

Cytokinins

Cytokinins were discovered as a result of attempts to grow plant tissue and organs in culture vessels in the 1940s. It was found that cell division occurs when coconut milk (a liquid endosperm) and yeast extract are added to the culture medium. Although the effective agent or agents could not be isolated, they were collectively called cytokinins because, as you may recall, cytokinesis means division of the cytoplasm. A naturally occurring cytokinin was not isolated until 1967. Because it came from the kernels of maize (*Zea*), it was called zeatin.

Effects of Cytokinins The cytokinins promote cell division. They are derivatives of adenine, one of the purine bases in DNA and RNA. Cytokinins have been isolated from actively dividing tissues of roots and also in seeds and fruits. A synthetic cytokinin, called kinetin, also promotes cell division.

It has been found that **senescence** (aging) of leaves can be prevented by the application of cytokinins. When a plant organ, such as a leaf, loses its natural color, it is most likely undergoing senescence. During senescence, large molecules within the leaf are broken down and transported to other parts of the plant. Senescence does not always affect the entire plant at once; for example, as some plants grow taller, they naturally lose their lower leaves. Not only can cytokinins prevent the death of leaves, but they can also initiate leaf growth. Lateral buds begin to grow despite apical dominance when cytokinin is applied to them.

Researchers are well aware that the *ratio of auxin to cytokinin* and the acidity of the culture medium determine whether a plant tissue forms an undifferentiated mass, called a callus, or differentiates to form roots, vegetative shoots, leaves, or floral shoots. Some reports suggest that chemicals called oligosaccharins (chemical fragments released from the cell wall) are effective in directing differentiation. Perhaps the reception of auxin and cytokinins, which leads to the activation of enzymes, releases these fragments from the cell wall.

Abscisic Acid

Abscisic acid (ABA) is produced by any "green tissue" with chloroplasts, monocot endosperm, and roots. It was once believed that abscisic acid functioned in **abscission,** the dropping of leaves, fruits, and flowers from a plant. But although the external application of abscisic acid promotes abscission, this hormone is no longer believed to function naturally in this process. Instead, the hormone ethylene, discussed next, is thought to bring about abscission.

Effects of Abscisic Acid Abscisic acid is sometimes called the stress hormone because it initiates and maintains seed and bud dormancy and brings about the closure of stomata. Dormancy has begun when a plant stops growing and prepares for adverse conditions (even though conditions at the time are favorable for growth). For example, it is believed that abscisic acid moves from leaves to vegetative buds in the fall, and thereafter these buds are converted to winter buds. A winter bud is covered by thick, hardened scales (Fig. 21.4*a*). A reduction in the level of abscisic acid and an increase in the level of gibberellins are believed to break seed and bud dormancy. Then seeds germinate, and buds send forth leaves.

Abscisic acid brings about the closing of stomata when a plant is under water stress (Fig. 21.4*b*). In some unknown way, abscisic acid causes potassium ions (K^+) to leave guard cells. Thereafter, the guard cells lose water, and a stoma closes.

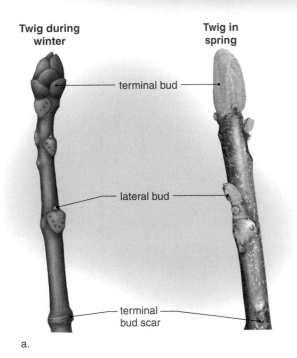

Twig during winter — terminal bud, lateral bud, terminal bud scar

Twig in spring — terminal bud

a.

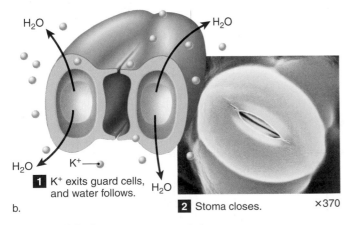

H_2O H_2O

H_2O K^+ **1** K^+ exits guard cells, and water follows. H_2O

2 Stoma closes. ×370

b.

Figure 21.4 Effects of abscisic acid.

a. Abscisic acid encourages the formation of winter buds (left), and a reduction in the amount of abscisic acid breaks bud dormancy (right). **b.** Abscisic acid also brings about the closing of a stoma by influencing the movement of potassium ions (K^+) out of guard cells.

Ethylene

Ethylene is a gas that can move freely in the air. Like the other hormones studied, ethylene works with other hormones to bring about certain effects.

Effects of Ethylene Ethylene is involved in abscission. Low levels of auxin and perhaps gibberellin, as compared to the levels in the stem, probably initiate abscission. But once the process of abscission has begun, ethylene stimulates certain enzymes, such as cellulase, which cause leaf, fruit, or flower drop (Fig. 21.5a,b). Cellulase hydrolyzes cellulose in plant cell walls.

In the early 1900s, it was common practice to prepare citrus fruits for market by placing them in a room with a kerosene stove. Only later did researchers realize that ethylene, an incomplete combustion product of kerosene, was ripening the fruit (Fig. 21.5c). Because it is a gas, ethylene can act from a distance. A barrel of ripening apples can induce ripening in a bunch of bananas, even if they are in different containers. If a plant is wounded due to physical damage or infection, ethylene is released at the wound site. This is why one rotten apple spoils the whole barrel.

Table 21.1 summarizes the effects of the five groups of plant hormones.

a. No abscission　　　　b. Abscission

c.

Figure 21.5 Functions of ethylene.

a. Normally, there is no abscission when a holly twig is placed under a glass jar for a week. **b.** When an ethylene-producing ripe apple is also under the jar, abscission of the holly leaves occurs. **c.** Ethylene causes fruits to ripen, making them luscious to eat.

Check Your Progress

1. Describe the effects of cytokinins on plant growth.
2. Describe the effects of abscisic acid on plant growth.
3. Describe the effects of ethylene on plant growth.

Answers: 1. Cytokinins promote cell division in actively growing parts of a plant. They also prevent senescence and initiate leaf growth. 2. Abscisic acid initiates and maintains seed and bud dormancy. It also causes stomata to close when a plant is water-stressed. 3. Ethylene stimulates abscission of leaves, fruits, and flowers. It also enhances fruit ripening.

Table 21.1	Functions of the Major Plant Hormones	
Hormone	**Functions**	**Where Produced or Found in Plant**
Auxins	Maintain apical dominance; involved in phototropism and gravitropism; promote growth of roots in tissue culture; prevent leaf and fruit drop.	Apical meristems, immature leaves
Gibberellins	Promote stem elongation between nodes; break seed and bud dormancy and influence germination of seeds.	Apical meristem, immature leaves, seeds
Cytokinins	Promote cell division; prevent senescence; along with auxin, promote differentiation leading to roots, shoots, leaves, or floral shoots.	Root apical meristem
Abscisic acid	Initiates and maintains seed and bud dormancy; promotes formation of winter buds; promotes closure of stomata.	Endosperm, roots, tissue with chloroplasts
Ethylene	Promotes abscission (leaf, fruit, or flower drop); promotes ripening of fruit.	Aging parts of plant and ripening fruit; apical meristem and nodes of stem

Figure 21.6   Positive phototropism.

The stem of a plant curves toward the light. This response is due to the accumulation of auxin on the shady side of the stem.

Figure 21.7   Negative gravitropism.

The stem of a plant curves away from the direction of gravity 24 hours after the plant was placed on its side. This response is due to the accumulation of auxin on the lower side of the stem.

Environmental Stimuli and Plant Responses

Plant responses are strongly influenced by such environmental stimuli as light, day length, gravity, and touch. The ability of a plant to respond to environmental signals fosters the survival of the plant and the species in a particular environment.

Plant responses to environmental signals can be rapid, as when stomata open in the presence of light, or they can take some time, as when a plant flowers in season. Despite their variety, most plant responses to environmental signals are due to growth and sometimes differentiation, brought about at least in part by particular hormones.

Plant Tropisms

Plant growth toward or away from a directional stimulus is called a **tropism.** Tropisms are due to differential growth—one side of an organ elongates faster than the other, and the result is a curving toward or away from the stimulus. The following two well-known tropisms were each named for the stimulus that causes the response:

> Phototropism: a movement in response to a light stimulus
> Gravitropism: a movement in response to gravity

Growth toward a stimulus is called a positive tropism, and growth away from a stimulus is called a negative tropism. Figure 21.6 illustrates positive **phototropism**—stems curve toward the light. Figure 21.7 illustrates negative **gravitropism**—stems curve away from the direction of gravity. Roots, of course, exhibit positive gravitropism.

The role of auxin in the positive phototropism of stems has been studied for quite some time. Because blue light in particular causes phototropism to occur, it is believed that a yellow pigment related to the vitamin riboflavin acts as a photoreceptor for light. Following reception, auxin migrates from the bright side to the shady side of a stem. The cells on that side elongate faster than those on the bright side, causing the stem to curve toward the light (see Fig. 21.1). Negative gravitropism of stems occurs because auxin moves to the lower part of a stem when a plant is placed on its side.

Photoperiodism

Flowering is a striking response in angiosperms to environmental seasonal changes. In some plants, flowering occurs according to the **photoperiod,** which is the ratio of the length of day to the length of night over a 24-hour period. Plants can be divided into three groups:

1. **Short-day plants/long-night plants** flower when the day length is shorter and the night is longer than a definite length of time called the *critical length*. (Examples are cocklebur, poinsettia, and chrysanthemum.)
2. **Long-day plants/short-night plants** flower when the day length is longer and the night is shorter than a critical length. (Examples are wheat, barley, clover, and spinach.)
3. **Day-neutral plants** do not depend on day/night length for flowering. (Examples are tomato and cucumber.)

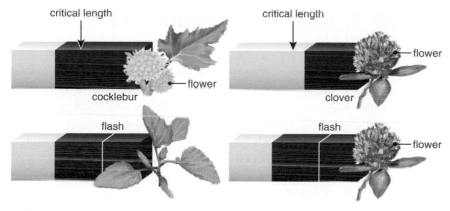

critical length

critical length

—flower

—flower

cocklebur

clover

flash

flash

—flower

a. Short-day (long-night) plants

b. Long-day (short-night) plant

Figure 21.8 Length of darkness controls flowering.
a. The cocklebur flowers when days are short and nights are long (*top*). If a long night is interrupted by a flash of light, the cocklebur will not flower (*bottom*). **b.** Clover flowers when days are long and nights are short (*top*). If a long night is interrupted by a flash of light, clover will still flower (*bottom*). Therefore, we can conclude that the length of continuous darkness controls flowering.

Further, we should note that both long-day plants and short-day plants can have the same critical length. Figure 21.8 illustrates that the cocklebur and the clover have the same critical length. The cocklebur flowers when the day is shorter (night is longer) than 8.5 hours, and clover flowers when the day is longer (night is shorter) than 8.5 hours.

Experiments have shown that the *length of continuous darkness, not light, controls flowering.* For example, the cocklebur will not flower if a suitable length of darkness is interrupted by a flash of light. On the other hand, clover will flower when an unsuitable length of darkness is interrupted by a flash of light. (Interrupting the light period with darkness has no effect on flowering.) Nurseries use these kinds of data to make all kinds of flowers available throughout the year (Fig. 21.9).

Phytochrome and Plant Flowering

If flowering is dependent on night length, plants must have some way to detect these periods. This appears to be the role of **phytochrome,** a blue-green leaf pigment. The proportion of red light to far-red light determines the particular form of phytochrome. P_{fr} is apt to be present in plant leaves during the day, but P_{fr} is converted to P_r as night approaches. There is also a slow metabolic replacement of P_{fr} by P_r during the night. Is this the timing device that tells the plant the length of darkness? Perhaps phytochrome, rather than a hormone, is involved in a signaling pathway that results in flowering. Although researchers have been looking for a flowering hormone for many years, as yet one has not been discovered.

Other Functions of Phytochrome

Apparently, the presence of P_{fr} indicates to some seeds that sunlight is present and conditions are favorable for germination. Such seeds must be only partly covered with soil when planted. Phytochrome may also affect leaf expansion and stem branching. In the absence of P_{fr}, stems elongate as a way to reach sunlight. Seedlings that are grown in the dark *etiolate* —that is, the stem increases in length, and the leaves remain small. Once the seedling is exposed to sunlight and P_{fr} is present, the seedling begins to grow normally—the leaves expand, and the stem branches. It appears that P_{fr} binds to regulatory proteins in the cytoplasm, and the complex migrates to the nucleus, where it binds to particular genes.

Figure 21.9 Flowering.
Nurseries know how to regulate the photoperiod so that many types of flowers are available year-round for special occasions.

Check Your Progress

1. Compare and contrast phototropism with gravitropism.
2. Contrast short-day, long-day, and day-neutral plants.
3. List the functions of phytochrome in plants.

Answers: 1. Both are differential growth responses. Phototropism is a response to light, while gravitropism is a response to gravity. 2. Short-day plants flower when the night is longer than a critical length; long-day plants flower when the night is shorter than a critical length; and day-neutral plants do not depend on day/night length to stimulate flowering. 3. Phytochrome functions in detection of photoperiod, stimulation of seed germination, and development of normal stems and leaves.

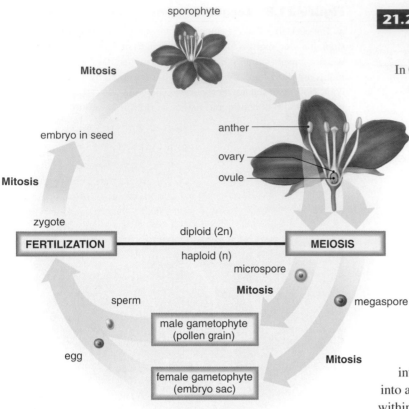

Figure 21.10 Alternation of generations in flowering plants.

In flowering plants, there are two types of spores and two gametophytes, male and female. Flowering plants are adapted to a land existence: The spores, the gametophytes, and the zygote are protected from drying out, in large part by the sporophyte.

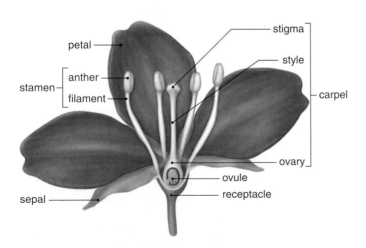

Figure 21.11 Anatomy of a flower.

A complete flower has all flower parts: sepals, petals, stamens, and at least one carpel.

21.2 Sexual Reproduction in Flowering Plants

In Chapter 18, we noted that plants have two multicellular stages in their life cycle, and therefore their life cycle is called an **alternation of generations.** In this life cycle, depicted in Figure 18.3, a diploid sporophyte alternates with a haploid gametophyte:

The **sporophyte** (2n) produces haploid spores by meiosis. The spores develop into gametophytes.

The **gametophytes** (n) produce gametes. Upon fertilization, the cycle returns to the 2n sporophyte.

Overview of the Plant Life Cycle

Flowering plants have an alternation of generations life cycle, but with the modifications shown in Figure 21.10. First, we will give an overview of the flowering plant life cycle, and then we will discuss the life cycle in more depth. In flowering plants, the sporophyte is dominant, and it is the generation that bears flowers. The flower is the reproductive structure of angiosperms. The flower produces two types of spores: microspores and megaspores. A **microspore** develops into a male gametophyte, which is a pollen grain. A **megaspore** develops into a female gametophyte, the embryo sac, which is microscopic and retained within the flower.

A pollen grain is either windblown or carried by an animal to the vicinity of the embryo sac. At maturity, a pollen grain contains two nonflagellated sperm. The embryo sac contains an egg.

A pollen grain develops a pollen tube, and the sperm move down the pollen tube to the embryo sac. After a sperm fertilizes an egg, the zygote becomes an embryo, still within the flower. The structure that houses the embryo develops into a seed. The seed also contains stored food and is surrounded by a seed coat. The seeds are enclosed by a fruit, which aids in dispersing the seeds. When a seed germinates, a new sporophyte emerges and develops into the dominant sporophyte.

As you learned in Chapter 18, the life cycle of flowering plants is well adapted to a land existence. No external water is needed to transport the pollen grain to the embryo sac, or to enable the sperm to reach the egg. All stages of the life cycle are protected from drying out.

Flowers

The **flower** is a unique reproductive structure found only in angiosperms (Fig. 21.11). Flowers produce the spores and protect the gametophytes. They often attract pollinators that aid in transporting pollen from plant to plant. Flowers also produce the fruits that enclose the seeds. The success of angiosperms, with over 240,000 species, is largely attributable to the evolution of the flower.

In monocots, flower parts occur in threes and multiples of three; in eudicots, flower parts are in fours or fives and multiples of four or five (Fig. 21.12).

A typical flower has four whorls of modified leaves attached to a **receptacle** at the end of a flower stalk.

sepal

carpel

stamens

petal

a. Daylily, a monocot

carpel

stamens

petal

b. Wild geranium, a eudicot

Figure 21.12   Monocot versus eudicot flowers.

a. Monocots, such as daylilies, have flower parts in threes. In particular, note the three petals. **b.** Geraniums are eudicots. They have flower parts in fours or fives; note the five petals of this flower.

1. The **sepals,** which are the most leaflike of all the flower parts, are usually green, and they protect the bud as the flower develops.
2. An open flower also has a whorl of **petals,** whose color accounts for the attractiveness of many flowers. The size, the shape, and the color of petals are attractive to a specific pollinator. Wind-pollinated flowers may have no petals at all.
3. **Stamens** are the "male" portion of the flower. Each stamen has two parts: the **anther,** a saclike container, and the **filament,** a slender stalk. Pollen grains develop from the microspores produced in the anther.
4. At the very center of a flower is the **carpel,** a vaselike structure that represents the "female" portion of the flower. A carpel usually has three parts: the **stigma,** an enlarged sticky knob; the **style,** a slender stalk; and the **ovary,** an enlarged base that encloses one or more **ovules.** The ovule becomes the seed, and the ovary becomes the fruit.

A flower can have a single carpel or multiple carpels. Sometimes several carpels are fused into a single structure, in which case the ovary has several chambers, each of which contains ovules. A carpel usually contains a number of ovules, which play a significant role in the production of megaspores, female gametophytes, and finally seeds.

Not all flowers have sepals, petals, stamens, and a carpel. Those that do are said to be *complete,* and those that do not are said to be *incomplete.* Flowers that have both stamens and carpels are called bisexual flowers; those with only stamens or only carpels are unisexual flowers. If staminate flowers and carpellate flowers are on one plant, the plant is called *monoecious* (Fig. 21.13). If staminate and carpellate flowers occur on separate plants, the plant is called *dioecious.* Holly trees are dioecious, and if red berries are a priority, it is necessary to acquire a plant with staminate flowers and another plant with carpellate flowers.

a. Staminate flowers

b. Carpellate flowers

Figure 21.13   Corn plants are monoecious.

A single corn plant has clusters of staminate flowers (**a**) and carpellate flowers (**b**). Staminate flowers produce the pollen that is carried by wind to the carpellate flowers, where an ear of corn develops.

Check Your Progress

1. Briefly describe the two generations in the life cycle of plants.
2. How are the microspore and the megaspore alike, and what does each become in the life cycle of a flowering plant?

Answers: 1. A plant's life cycle alternates between a multicellular diploid sporophyte and a multicellular haploid gametophyte. 2. Both are haploid products of meiosis. The microspore develops into the male gametophyte, while the megaspore develops into the female gametophyte.

From Spores to Fertilization

Now that we have some acquaintance with the flowering plant life cycle, we will examine it in more detail. As you know, seed plants produce two types of spores: **Microspores** become male gametophytes, the mature pollen grain, and the **megaspore** becomes the embryo sac, the female gametophyte. Just exactly where does this happen, and how do these events contribute to the life cycle of flowering plants?

Microspores develop into pollen grains in the anthers of stamens (Fig. 21.14). A **pollen grain** is at first an immature **male gametophyte** that consists of two cells. The larger cell will eventually produce a pollen tube. The smaller cell divides, either now or later, to become two sperm. This is why the stamen is called the "male" portion of the flower.

Figure 21.14   Life cycle of a flowering plant.

Follow the development of a microspore into a germinated pollen grain that contains sperm. Also follow the development of a megaspore into the female gametophyte that contains an egg. Note the occurrence of double fertilization, in which one sperm from the pollen tube fertilizes the egg, producing a zygote, and the other joins with two other female gametophyte cells to become endosperm. Now the ovule becomes a seed.

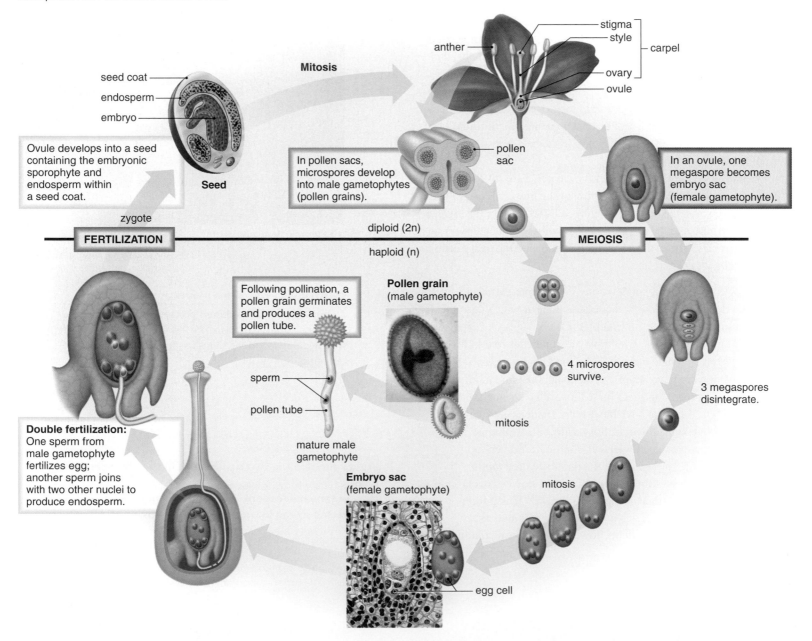

Pollen grains are distinctive to the particular plant (Fig. 21.15), and **pollination** is simply the transfer of pollen from the anther to the stigma of a carpel. Plants often have adaptations that foster *cross-pollination,* which occurs when the pollen landing on the stigma is from a different plant of the same species. For example, the carpels may mature only after the anthers have released their pollen. Cross-pollination may also be brought about with the assistance of an animal pollinator. If a pollinator such as a bee goes from flower to flower of only one type of plant, cross-pollination is more likely to occur in an efficient manner. The secretion of nectar is one way that insects are attracted to plants, and over time, certain pollinators have become adapted to reach the nectar of only one type of flower. In the process, pollen is inadvertently picked up and taken to another plant of the same type. Plants attract particular pollinators in still other ways. For example, through the evolutionary process, orchids of the genus *Ophrys* have flowers that look like female wasps. Males of that species pick up pollen when they attempt to copulate with these flowers!

Figure 21.14 also allows you to follow the development of the megaspore. In an ovule, within the ovary of a carpel, one megaspore develops into a seven-celled embryo sac. The **embryo sac** is the **female gametophyte,** and one of these cells is an egg cell. This is why the carpel is the "female" portion of the flower. Fertilization of the egg occurs after a pollen grain lands on the stigma of a carpel and develops a pollen tube. A pollen grain that has germinated and produced a pollen tube is the mature male gametophyte (Fig. 21.14, *middle*). A pollen tube contains two sperm. Once it reaches the ovule, **double fertilization** occurs. One sperm unites with the egg, forming a 2n zygote. The other sperm unites with two nuclei centrally placed in the embryo sac, forming a 3n endosperm cell.

Check Your Progress

1. What specific part of a flower produces male gametophytes? Female gametophyte?
2. Contrast the development and structure of the male gametophyte and the female gametophyte.
3. What are the products of double fertilization in angiosperms?

Answers: 1. Male gametophytes are produced in the anther of the stamen. The female gametophyte is produced in an ovule within the ovary of the carpel. 2. Each microspore produces a two-celled pollen grain. Later, one cell produces two sperm. Following pollination, the pollen grain germinates and produces a pollen tube. One of the four megaspores produces a seven-celled female gametophyte, called the embryo sac, within the ovule. 3. When one sperm fertilizes the egg, a zygote results. When the second sperm joins with two other nuclei of the embryo sac, the endosperm results.

b. ×85

c. ×1230

Figure 21.15   Pollen.

a. Pollen grains are so distinctive that a paleontologist can use fossilized pollen to date the appearance of a plant in a particular area. Pollen grains can become fossils because their strong walls are resistant to chemical and mechanical damage. **b.** Pollen grains of Canadian goldenrod. **c.** Pollen grains of a pussy willow.

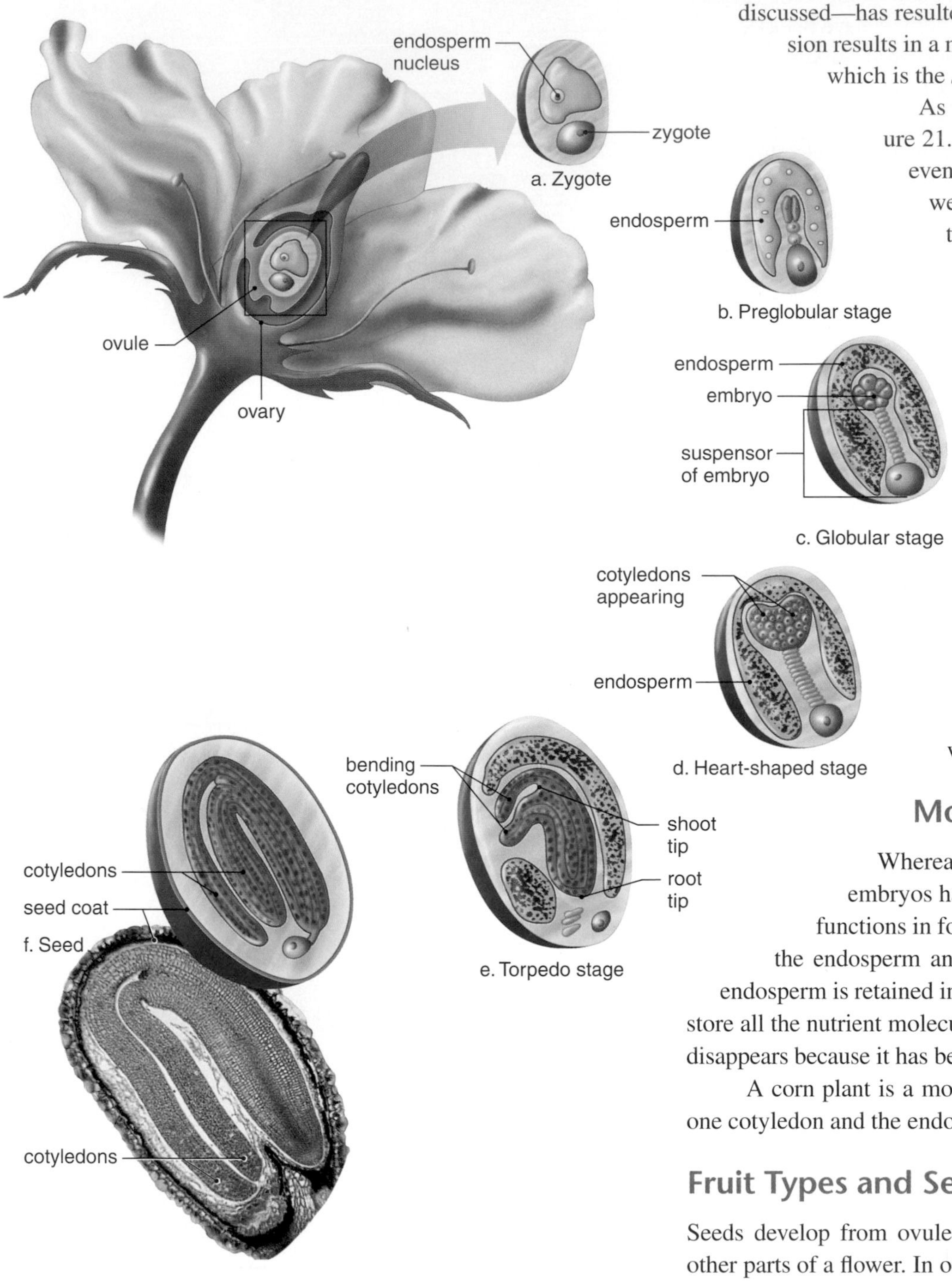

Development of the Seed in a Eudicot

It is now possible to account for the three parts of a seed. The ovule wall will become a protective covering called the *seed coat.* Double fertilization—just discussed—has resulted in an endosperm nucleus and a zygote. Cell division results in a multicellular *embryo* and a multicellular endosperm, which is the *stored food* of a seed.

As the embryo passes through the stages noted in Figure 21.16, tissue specialization occurs as evidenced by the eventual appearance of the plant axis (Fig. 21.16e), as well as the shoot tip and root tip, each of which contains apical meristem. When the seed germinates, the shoot tip produces a shoot system, and the root tip produces a root system.

Notice also that as development proceeds, the endosperm reduces in size, and the cotyledons, which are a part of the embryo, increase in size. **Cotyledons** are embryonic leaves, present in seeds. Cotyledons wither when the first *true leaves* grow and become functional. In many eudicots, the cotyledons absorb the developing endosperm and become large and fleshy. The food stored by the cotyledons will nourish the embryo when it resumes growth. The common garden bean is a good example of a eudicot seed with large cotyledons and no endosperm (see Fig. 21.19). A **seed** contains the embryo and stored food within a seed coat.

Monocots Versus Eudicots

Whereas eudicot embryos have two cotyledons, monocot embryos have only one cotyledon. In monocots, the cotyledon functions in food storage, and it also absorbs food molecules from the endosperm and passes them to the embryo. In other words, the endosperm is retained in monocot seeds. In eudicots, the cotyledons usually store all the nutrient molecules that the embryo uses. Therefore, the endosperm disappears because it has been taken up by the two cotyledons.

A corn plant is a monocot; consequently, in a corn kernel, there is only one cotyledon and the endosperm is present (see Fig. 21.20).

Fruit Types and Seed Dispersal

Seeds develop from ovules, and fruits develop from ovaries and sometimes other parts of a flower. In other words, flowering plants have seeds enclosed by a **fruit** (Fig. 21.17).

Fruits can be dry or fleshy. *Dry fruits* are generally a dull color with a thin and dry ovary wall so that the food is largely confined to the seeds. In grains such as wheat, corn, and rice, the fruit looks like a seed. Nuts (e.g., walnuts,

Figure 21.16 **Development of the seed in a eudicot.** Development begins with (**a**) the zygote and ends with (**f**) the seed. As the embryo develops, it progresses through the stages noted from (**b**) to (**e**).

Figure 21.17   Examples of fruits.

Some fruits are dry, as in walnuts and peas. Some fruits are fleshy, as in peaches and apples. To a botanist, any plant product derived from an ovary plus perhaps other flower parts is a fruit.

pecans) have a hard outer shell covering a single seed. A legume, such as a pea, has a several-seeded fruit that splits open to release the seeds. A legume illustrates that what we call a vegetable may actually be a fruit.

In contrast to dry fruits, *fleshy fruits* are usually juicy and brightly colored. A drupe (e.g., peach, cherry, olive) is a "stone fruit"—the outer part of the ovary wall is fleshy, but there is an inner stony layer. Inside the stony layer is the seed. A berry, such as a tomato, contains many seeds. An apple is a pome, in which a dry ovary covers the seeds, and the fleshy part is derived from the receptacle of the flower. A strawberry is an interesting fruit because the flesh is derived from the receptacle, and what appear to be the seeds are actually dry fruits!

Dispersal of Seeds

For plants to be widely distributed, their seeds have to be dispersed—that is, distributed preferably long distances from the parent plant.

Plants have various means to ensure that dispersal takes place. The hooks and spines of clover, burr, and cocklebur fruits attach to the fur of animals and the clothing of humans. Birds and mammals sometimes eat fruits, including the seeds, which then pass out of the digestive tract with the feces some distance from the parent plant. Squirrels and other animals gather seeds and fruits, which they bury some distance away. Humans greatly assist in the dispersal of grains, which are normally windblown (Fig. 21.18).

The fruit of the coconut palm, which can be dispersed by ocean currents, may land many hundreds of kilometers away from the parent plant. Some plants have fruits with trapped air or seeds with inflated sacs that help them float in water. Many seeds are dispersed by wind. Woolly hairs, plumes, and wings are all adaptations for this type of dispersal. The seeds of an orchid are so small and light that they need no special adaptation to carry them far away. The somewhat heavier dandelion fruit uses a tiny "parachute" for dispersal. The winged fruit of a maple tree, which contains two seeds, has been known to travel up to 10 kilometers from its parent. A touch-me-not plant has seed pods that swell as they mature. When the pods finally burst, the ripe seeds are hurled out.

Figure 21.18   Humans help disperse seeds and fruits.

When farmers plant their crops, they are dispersing seeds sometimes enclosed by fruits.

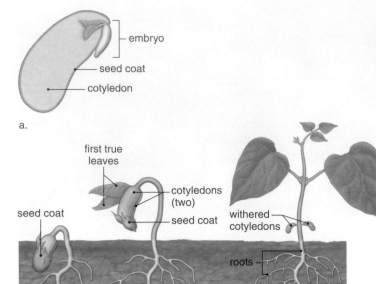

a.

b.

Figure 21.19 Common garden bean, a eudicot.

a. Seed structure. **b.** Germination and development of the seedling. Notice that there are two cotyledons and that the leaves are net-veined.

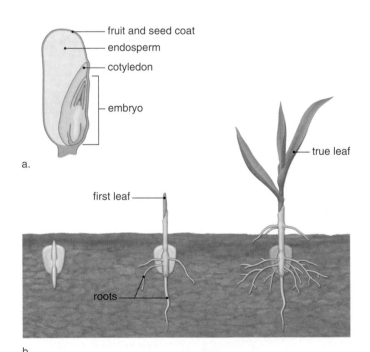

a.

b.

Figure 21.20 Corn, a monocot.

a. Corn kernel structure. **b.** Germination and development of the seedling. Notice that there is one cotyledon and that the leaves are parallel-veined.

Germination of Seeds

Following dispersal, seeds may **germinate.** If so, they begin to grow so that a seedling appears. Germination doesn't usually take place until there is sufficient water, warmth, and oxygen to sustain growth. In deserts, germination does not occur until there is adequate moisture. These requirements help ensure that seeds do not germinate until the most favorable growing season has arrived.

Some seeds do not germinate until they have been dormant for a period of time. For seeds, dormancy is the time during which no growth occurs, even though conditions may be favorable for growth. In the temperate zone, seeds often have to be exposed to a period of cold weather before dormancy is broken. Fleshy fruits (e.g., apples, pears, oranges, and tomatoes) contain inhibitors so that germination does not occur until the seeds are removed and washed. Aside from water, bacterial action and even fire can act on the seed coat, allowing it to become permeable to water. The uptake of water causes the seed coat to burst.

Eudicot Versus Monocot Germination

If the two cotyledons of a bean seed are parted, you can see the cotyledons and a rudimentary plant with immature leaves. As the eudicot seedling emerges from the soil, the shoot is hook-shaped to protect the immature leaves as they start to grow. The cotyledons shrivel up as the true leaves of the plant begin photosynthesizing (Fig. 21.19). A corn kernel is actually a fruit, and therefore the outer covering is the fruit and seed coat combined (Fig. 21.20). Inside is the single cotyledon. Also, both the immature leaves and the root are covered by sheaths. The sheaths are discarded when the seedling begins growing, and the immature leaves become the first true leaves of the corn plant.

Check Your Progress

1. Identify the origin of each of the three parts of a seed.
2. What are cotyledons, and what role do they play?
3. How does a dry fruit differ from a fleshy fruit? Which would you expect to be windblown—the fruit or the seeds? Which would you expect to be eaten by animals—the fruit or the seeds? Why?
4. In general, what conditions do seeds require to germinate? Contrast how a eudicot seedling and a monocot seedling protect the first true leaves.

Answers: 1. The embryo is derived from the zygote; the stored food is derived from the endosperm; and the seed coat is derived from the ovule wall. *2.* Cotyledons are embryonic leaves that are present in seeds. They wither when the first true leaves grow and become functional. Cotyledons store nutrients derived from endosperm (in eudicots), or both store and take nutrients from the endosperm (in monocots) *3.* Dry fruits, with a dull, thin, and dry covering derived from the ovary, are more apt to be windblown. Fleshy fruits, with a juicy covering derived from the ovary and possibly other parts of the flower, are more apt to be eaten by animals. *4.* Seeds require water, warmth, and oxygen in order to germinate. Eudicot seedlings have a hook shape, and monocot seedlings have a sheath to protect the first true leaves.

21.3 Asexual Reproduction in Flowering Plants

Because plants contain nondifferentiated meristem tissue, they routinely reproduce asexually by vegetative propagation. In asexual reproduction, there is only one parent, instead of two as in sexual reproduction. For example, complete strawberry plants can grow from the nodes of stolons, and iris plants can grow from the nodes of rhizomes (Fig. 21.21). White, or Irish, potatoes are actually portions of underground stems, and each eye is a bud that will produce a new potato plant if it is planted with a portion of the swollen tuber. Sweet potatoes are modified roots; they can be propagated by planting sections of the root. You may have noticed that the roots of some fruit trees, such as cherry and apple trees, produce "suckers," small plants that can be used to grow new trees.

There are several ways that seedless fruits can arise, but usually some sort of stimulation causes the plant to produce fruit, even though fertilization never occurred. In the seedless watermelon, which is triploid and cannot go through meiosis to form sperm and eggs, pollination is used as a stimulus to cause the plants to form fruit. Often, hormones, such as auxins, gibberellins, or cytokinins, are used to stimulate seedless fruit formation. Seedless grapes actually contain seeds, but they lack an embryo—the embryo is routinely aborted by this type of grape—and the grapes are sprayed with gibberellin to increase their size.

Propagation of Plants in Tissue Culture

Tissue culture is the growth of a tissue in an artificial liquid or solid culture medium. Plant cells are **totipotent,** which means that each plant cell can become an entire plant. This has led to the commercial production of somatic embryos in tissue culture (Fig. 21.22). Adult cells are digested to release *protoplasts,* which

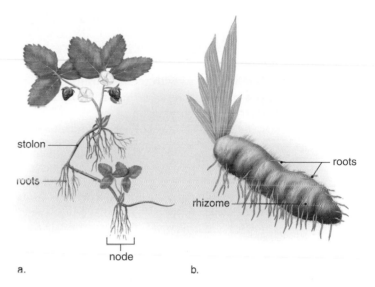

stolon

roots

roots

rhizome

node

a.

b.

Figure 21.21 Vegetative propagation.

a. A strawberry plant has an aboveground horizontal stem called a stolon. Every other node produces a new plant. **b.** An iris has an underground horizontal stem called a rhizome, whose nodes produce new plants.

a. Protoplasts (naked cells)

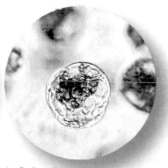

b. Cell wall regeneration

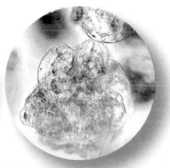

c. Aggregates of cells

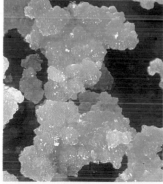

d. Callus (undifferentiated mass)

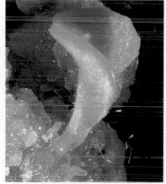

e. Somatic embryo

f. Plantlet

Figure 21.22 Tissue culture.

a. When plant cell walls are removed by digestive enzyme action, cells without cell walls, called protoplasts, result. **b.** Cell walls regenerate, and cell division begins. **c.** Cell division produces aggregates of cells. **d.** An undifferentiated mass, called a callus, develops. **e.** From the callus, somatic embryos such as this one appear. **f.** The embryos develop into plantlets that can be transferred to soil for growth into adult plants.

are plant cells without cell walls. The cell then goes through the stages shown in Figure 21.22. Thousands and even millions of somatic embryos can be produced in tanks called bioreactors. The embryos can be encapsulated in protective hydrated gel (called artificial seeds) and shipped anywhere. This method of production is utilized for certain vegetables, such as tomato, celery, and asparagus, and for ornamental plants, such as lilies, begonias, and African violets. Plants generated from somatic embryos vary somewhat because of mutations that arise during the production process. These so-called *somaclonal variations* are another way to produce new plants with desirable traits.

Instead of using mature plant tissues as the starting material, you can use meristem tissue as a source of plant cells. In this case, the end result is the production of *clonal plants* that do have the same traits (Fig. 21.23). If the correct proportions of hormones are added to the liquid medium, many new shoots will develop from a single shoot tip. When these are removed, more shoots form. Another advantage to producing identical plants from meristem tissue is that the plants will be virus-free. (The presence of plant viruses weakens plants and makes them less productive.)

Anther culture is a technique in which mature anthers are cultured in a medium containing vitamins and growth regulators. The haploid cells within the pollen grains divide, producing proembryos consisting of as many as 20 to 40 cells. Finally, the pollen grains rupture, releasing haploid embryos. The experimenter can then generate a haploid plant, or chemical agents can be added that encourage chromosomal doubling. After chromosomal doubling, the resulting plants are diploid, and the homologous chromosomes carry the same genes. Anther culture is a direct way to produce plants that are certain to have the same characteristics.

The culturing of plant tissues has also led to a technique called *cell suspension culture.* Rapidly growing calluses are cut into small pieces and shaken in a liquid nutrient medium so that single cells or small clumps of cells break off and form a suspension. These cells produce the same chemicals as the entire plant. For example, cell suspension cultures of *Cinchona ledgeriana* produce quinine, and those of *Digitalis lanata* produce digitoxin, both of which are useful drugs for humans.

Genetic Engineering of Plants

Traditionally, **hybridization,** the crossing of different varieties of plants or even species, was used to produce plants with desirable traits. Hybridization, followed by vegetative propagation of the mature plants, generated a large number of identical plants with these traits. Today, it is possible to directly alter the genes of organisms, and in that way produce new varieties with desirable traits.

Tissue Culture and Genetic Engineering

Genetic engineering can be done utilizing protoplasts (plant cells lacking cell walls) in tissue culture (see Fig. 21.22). A foreign piece of DNA isolated from any type of organism—plant, animal, or bacteria—is placed in the tissue culture medium. High-voltage electrical pulses can then be used to create pores in the plasma membrane so that the foreign gene enters the cells. In one such procedure, a gene for the production of the firefly enzyme luciferase was inserted into tobacco protoplasts, and the adult plants glowed when sprayed with the substrate luciferin.

Figure 21.23 **Whole plants from meristem tissue.**

Each of these flasks contains meristem tissue and a growth medium. In the presence of light and with slow rotation of the flasks, a new plant will appear in a few weeks.

Unfortunately, the regeneration of cereal grains from protoplasts has been difficult. As a result, other methods are used to introduce DNA into plant cells with intact cell walls. In one technique, foreign DNA is inserted into the plasmid of the bacterium *Agrobacterium,* which normally infects plant cells. When a bacterium with the recombinant plasmid infects a plant, the plasmid enters the cells of the plant.

Today, it is possible to use a gene gun to bombard a callus with microscopic DNA-coated metal particles. Then, genetically altered somatic embryos develop into genetically altered adult plants. Many plants, including corn and wheat varieties, have been genetically engineered by this method. Such plants are called **genetically modified plants (GMPs),** or **transgenic plants,** because they carry a foreign gene and have new and different traits. Figure 21.24 shows two types of transgenic plants.

Agricultural Plants with Improved Traits

Corn and cotton plants, in addition to soybean and potato plants, have been engineered utilizing single gene transfers to be resistant to either herbicides or insect pests. Some corn and cotton plants have been developed that are resistant to both insects and herbicides. In 2001, transgenic crops were planted on more than 72 million acres worldwide, and the acreage is expected to triple in about five years. If crops are resistant to a broad-spectrum herbicide and weeds are not, then the herbicide can be used to kill the weeds. When herbicide-resistant plants were planted, weeds were easily controlled, less tillage was needed, and soil erosion was minimized. However, wildlife and native plants may decline when increased amounts of herbicides are used.

Some citizens are concerned about GMPs because of possible effects on their health. This concern has been prompted by limited laboratory data suggesting that some people may be allergic to GMPs (Fig. 21.25). Some ecologists are concerned about the effect GMPs may have on the environment. It's possible that beneficial insects feeding on the GMPs might be harmed by them, or that GMPs might pass their resistant genes to certain weeds, which would then be difficult to control. Some feel that drastic consequences might occur in the future, even though no life-threatening effects of GMPs have so far been recorded.

In the meantime, researchers are trying to produce more types of GMPs (Fig. 21.26). A salt-tolerant tomato will soon be field-tested. First, scientists identified a gene coding for a channel protein that transports Na^+ in a vacuole, preventing it from interfering with plant metabolism. Then the scientists used the gene to engineer tomato plants that maximally produce the channel protein. The GMPs thrived when watered with a salty solution. Irrigation, even with fresh water, inevitably leads to salinization of the soil, which reduces crop yields. Salt-tolerant crops would increase yield on such land. Salt-tolerant and also drought- and cold-tolerant cereals, rice, and sugarcane might help provide enough food for the growing world population.

a. Herbicide-resistant soybean plant

b. Nonresistant potato plant

c. Pest-resistant potato plant

Figure 21.24   Transgenic plants.

a. These soybean plants have been genetically modified to be resistant to herbicides. Nonresistant potato plants (**b**) can be genetically modified to be resistant to pests (**c**).

Figure 21.25   Regulation of genetically modified plants.

Activists protest the present regulation of genetically modified plants, which they believe to be too permissive.

Transgenic Crops of the Future

Improved Agricultural Traits

Herbicide resistant	Wheat, rice, sugar beets, canola
Salt tolerant	Cereals, rice, sugarcane, canola
Drought tolerant	Cereals, rice, sugarcane
Cold tolerant	Cereals, rice, sugarcane
Improved yield	Cereals, rice, corn, cotton
Modified wood pulp	Trees
Disease protected	Wheat, corn, potatoes

Improved Food-Quality Traits

Fatty acid/oil content	Corn, soybeans
Protein/starch content	Cereals, potatoes, soybeans, rice, corn
Amino acid content	Corn, soybeans

a. Desirable traits

b. Salt intolerant   Salt tolerant

Figure 21.26 Genetically modified crops of the future.

a. Genetically modified crops of the future include those with improved agricultural traits or food quality. **b.** A salt-tolerant tomato plant has been engineered. The plant to the left does poorly when watered with a salty solution, but the engineered plant to the right is tolerant of the solution. The development of salt-tolerant crops would increase food production in the future.

Potato blight is the most serious potato disease in the world. About 150 years ago, it was responsible for the Irish potato famine, which caused the death of millions of people. By placing a gene from a naturally blight-resistant wild potato into a farmed variety, researchers have now created potato plants that are no longer vulnerable to a range of blight strains.

Some progress has also been made in increasing the food quality of crops. Soybeans have been developed that mainly produce monounsaturated fatty acids, a change that may improve human health. These altered plants also produce acids that can be used as hardeners in paints and plastics. The necessary genes were taken from other plants and transferred into the soybean DNA.

Other types of genetically engineered plants are expected to increase productivity. Stomata might be altered to take in more carbon dioxide or to lose less water. A team of Japanese scientists is working on introducing the C_4 photosynthetic capability into rice. Unlike C_3 plants, C_4 plants do well in hot, dry weather (see Chapter 6). These modifications would require a more complete reengineering of plant cells than the single-gene transfers that have been done so far.

Commercial Products

Single-gene transfers also have allowed plants to produce various products, including human hormones, clotting factors, and antibodies. One type of antibody made by corn can deliver radioisotopes to tumor cells, and another made by soybeans may be developed to treat genital herpes. The tobacco mosaic virus has been used as a vector to introduce a human gene into adult tobacco plants in the field. (Note that this technology bypasses the need for tissue culture completely.) Tens of grams of α-galactosidase, an enzyme needed for the treatment of a human lysosome storage disease, were harvested per acre of tobacco plants. And it only took 30 days to get tobacco plants to produce antibodies to treat non-Hodgkin's lymphoma after being sprayed with a genetically engineered virus.

Check Your Progress

1. List the techniques used to promote uptake of DNA by plant cells in tissue culture.
2. List the potential problems with genetically modified plants (GMPs).
3. List the potential benefits of GMPs.

Answers: 1. Techniques to promote DNA uptake include electrical pulses, *Agrobacterium* infection, and the gene gun. 2. People might be allergic to GMPs, and GMPs could harm the environment. 3. GMPs may be able to tolerate high salt, drought, and cold; able to resist diseases and pests; have improved nutritional quality; and be modified to make products of medical importance to humans.

Summary

21.1 Responses in Flowering Plants

Plants respond to environmental stimuli such as light, gravity, and seasonal changes.

Plant Hormones
Plant hormones lead to physiological changes within the cell. The five commonly recognized groups of plant hormones are:

- Auxins: apical dominance, two types of tropisms—phototropism and gravitropism, growth of roots
- Gibberellins: promote stem elongation, break seed dormancy
- Cytokinins: promote cell division, prevent senescence of leaves, influence differentiation of plant tissues
- Abscisic acid: initiates and maintains seed and bud dormancy, closing of stomata
- Ethylene: causes abscission of leaves, fruits, and flowers; ripens fruits

Plant Responses to Environmental Stimuli
Environmental signals play a significant role in plant growth and development.

Plant Tropisms   Tropisms are growth responses toward or away from unidirectional stimuli.

Phototropism: Response to a light
Gravitropism: Response to gravity

- Auxin is responsible for the negative gravitropism exhibited by stems. Stems grow upward opposite the direction of gravity.

Photoperiodism   Flowering is a response to seasonal changes, namely length of the night.

- Short-day plants flower when nights are longer than a critical length.
- Long-day plants flower when nights are shorter than a critical length.
- Some plants are day/night-neutral.

Phytochrome and Plant Flowering   Phytochrome is a plant pigment that responds to daylight. Functions of phytochrome in plant cells include:

- Brings about flowering
- Encourages germination
- Influences leaf expansion
- Affects stem branching

21.2 Sexual Reproduction in Flowering Plants

The life cycle of flowering plants is adapted to a land existence.

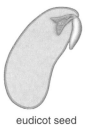

- Flowering plants have an alternation of generations life cycle with separate male and female gametophytes.
- Pollen grain is the male gametophyte
- Female gametophyte, located in the ovule of a flower, produces an egg
- Double fertilization; zygote and endosperm result

Development of Eudicot Seed
The zygote undergoes a series of stages to become an embryo. In eudicots, the embryo has two cotyledons, which absorb the endosperm. In monocots, the embryo has a single cotyledon and also endosperm. Aside from the embryo and stored food, a seed has a seed coat.

eudicot seed   monocot seed

Fruit Types and Seed Dispersal
A fruit is a mature, ripened ovary and may also be composed of other flower parts. Some fruits are dry (e.g., nuts, legumes), and some are fleshy (e.g., apples, peaches). In general, fruits aid the dispersal of seeds. Following dispersal, a seed germinates.

21.3 Asexual Reproduction in Flowering Plants

Many flowering plants reproduce asexually.

- Nodes located on stems (either aboveground or underground) give rise to entire plants.
- Roots produce new shoots.

Propagation of Plants in Tissue Culture
The production of clonal plants utilizing tissue culture is now a commercial venture. Plant cells in tissue culture produce chemicals of medical importance.

Genetic Engineering of Plants
The practice of plant tissue culture facilitates genetic engineering to produce plants that have improved agricultural or food-quality traits. Plants can also be engineered to produce chemicals of use to humans.

Thinking Scientifically

1. In late November every year, florists ship truckloads of poinsettia plants to stores around the country. Typically, the plants are individually wrapped in plastic sleeves. If the plants remain in the sleeves for too long during shipping and storage, their leaves begin to curl under and eventually fall off. What plant hormone do you suppose causes this response? How do you suppose the plastic sleeves affect the response?

2. Snow buttercups (*Ranunculus adoneus*) live in alpine regions and produce sun-tracking flowers. The flowers face east in the morning to absorb the sun's warmth in order to attract pollinators and speed the growth of fertilized ovules. The flowers track the sun all day, continually bending to face the sun. How might you determine whether the flowers or stems are responsible for sun tracking? Assuming you have determined that the stem is responsible, how would you determine which region of the stem follows the sun? How might you determine whether auxin causes this growth response?

Testing Yourself

Choose the best answer for each question. For questions 1–8, identify the plant hormone in the key that is associated with each phenomenon. Each answer may be used more than once.

Key:

    a. auxin
    b. gibberellin
    c. cytokinin
    d. abscisic acid
    e. ethylene

1. Initiates and maintains seed and bud dormancy.
2. Stimulates root development on cuttings.
3. Capable of moving from plant to plant in the air.
4. Promotes stem elongation.
5. Responsible for apical dominance.
6. Stimulates leaf, fruit, and flower drop.
7. Overcomes seed and bud dormancy.
8. Delays senescence.
9. Stigma is to carpel as anther is to
    a. sepal.
    b. stamen.
    c. ovary.
    d. style.
10. _____ always promotes cell division.
    a. Auxin
    b. Phytochrome
    c. Cytokinin
    d. None of these are correct.

11. The embryo of a flowering plant can be found in the
    a. pollen.
    b. anther.
    c. microspore.
    d. seed.

12. Which is not a plant hormone?
    a. auxin
    b. cytokinin
    c. gibberellin
    d. All of these are plant hormones.

13. Label the parts of the flowering plant life cycle in the following illustration.

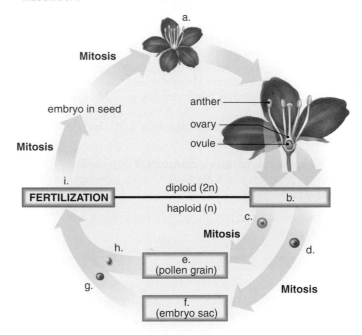

14. The term totipotent means
    a. that each plant cell can become an entire plant.
    b. hormones control all plant growth.
    c. all cells develop from the same tissue.
    d. None of these are correct.

15. Nondifferentiated meristem allows for _____ in _____.
    a. elongation, meiosis
    b. sexual reproduction, vegetative propagation
    c. asexual reproduction, tissue culture
    d. asexual reproduction, meiosis

16. Double fertilization refers to the formation of a _____ and a _____.
    a. zygote, zygote
    b. zygote, pollen grain
    c. zygote, megaspore
    d. zygote, endosperm

17. Which is the correct order of the following events:
    (1) megaspore becomes embryo sac, (2) embryo formed,
    (3) double fertilization, (4) meiosis?

    a. 1, 2, 3, 4
    b. 4, 1, 3, 2
    c. 4, 3, 2, 1
    d. 2, 3, 4, 1

18. In the absence of abscisic acid, plants may have difficulty

    a. forming winter buds.
    b. closing the stomata.
    c. Both a and b are correct.
    d. Neither a nor b is correct.

19. Phytochrome plays a role in

    a. flowering.
    b. stem growth.
    c. leaf growth.
    d. All of these are correct.

20. A plant requiring a dark period of at least 14 hours will

    a. flower if a 14-hour night is interrupted by a flash of light.
    b. not flower if a 14-hour night is interrupted by a flash of light.
    c. not flower if the days are 14 hours long.
    d. Both b and c are correct.

21. Short-day plants

    a. are the same as long-day plants.
    b. are apt to flower in the fall.
    c. do not have a critical photoperiod.
    d. will not flower if a short day is interrupted by bright light.
    e. All of these are correct.

22. Which of these is a correct statement?

    a. Both stems and roots show positive gravitropism.
    b. Both stems and roots show negative gravitropism.
    c. Only stems show positive gravitropism.
    d. Only roots show positive gravitropism.

23. Which of the following plant hormones causes plants to grow in an upright position?

    a. auxin
    b. gibberellins
    c. cytokinins
    d. abscisic acid
    e. ethylene

24. The function of the flower is to _____, and the function of fruit is to _____.

    a. produce fruit; provide food for humans
    b. aid in seed dispersal; attract pollinators
    c. attract pollinators; assist in seed dispersal
    d. produce the ovule; produce the ovary

25. The megaspore is similar to the microspore in that both

    a. have the diploid number of chromosomes.
    b. become an embryo sac.
    c. become a gametophyte that produces a gamete.
    d. are necessary to seed production.
    e. Both c and d are correct.

26. Unlike taxis in animals, tropism in plants

    a. is a response to the environment.
    b. may be stimulated by light.
    c. is a differential growth response.
    d. requires the perception of an environmental cue.

27. Lettuce seeds require light for germination. Assuming that all the seeds are viable, what percent germination would you expect from seeds exposed to red light, then to far-red light, and then to red light again?

    a. 0%
    b. 33%
    c. 50%
    d. 67%
    e. 100%

28. Which of the following is not a component of the carpel?

    a. stigma
    b. filament
    c. ovary
    d. ovule
    e. style

29. Label the parts of the flower in the following diagram.

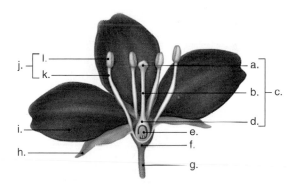

30. The two sperm cells in a pollen grain are

    a. haploid and genetically different from each other.
    b. haploid and genetically identical to each other.
    c. diploid and genetically different from each other.
    d. diploid and genetically identical to each other.

31. In contrast to eudicots, monocot

    a. embryos have two cotyledons.
    b. seeds contain endosperm.
    c. cotyledons store food.
    d. embryos undergo a heart stage of development.

32. Genetically modified plants can be created using genes from
    a. plants.
    b. animals.
    c. bacteria.
    d. fungi.
    e. All of these are correct.

33. In an accessory fruit, such as an apple, the bulk of the fruit is from the
    a. ovary.
    b. style.
    c. pollen.
    d. receptacle.

34. Label the parts of the eudicot seed and seedling in the following diagram.

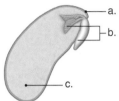

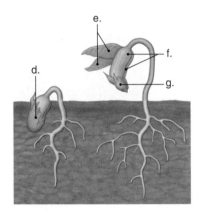

35. Plant biotechnology can lead to
    a. increased crop production.
    b. disease-resistant plants.
    c. treatment of human disease.
    d. All of these are correct.

36. The calyx is composed of
    a. sepals.
    b. petals.
    c. anthers.
    d. ovaries.
    e. stigmas.

37. A seed is a mature
    a. embryo.
    b. ovule.
    c. ovary.
    d. pollen grain.

Go to www.mhhe.com/maderessentials for more quiz questions.

Bioethical Issue

Witchweed is a serious parasitic weed that destroys 40% of Africa's cereal crop annually. Because it is intimately associated with its host (the cereal crop), witchweed is difficult to selectively destroy with herbicides. One control strategy is to create genetically modified cereals with herbicide resistance. Then, herbicides will kill the weed without harming the crop. Herbicide-resistant sorghum was created to help solve the witchweed problem. However, scientists discovered that the herbicide resistance gene could be carried via pollen into Johnson grass, a relative of sorghum and a serious weed in the United States. Farmers would have a new challenge if Johnson grass could no longer be effectively controlled with herbicides. Consequently, efforts to create herbicide-resistant sorghum were put on hold.

Do you think farmers in Africa should be able to grow herbicide-resistant sorghum, allowing them to control witchweed? Or, should the production of herbicide-resistant sorghum be banned worldwide in order to avoid the risk of introducing the herbicide resistance gene into Johnson grass?

Understanding the Terms

abscisic acid (ABA)   358
abscission   358
alternation of generations   362
anther   363
apical dominance   357
auxin   356
carpel   363
cotyledon   366
cytokinin   358
day-neutral plant   360
dormancy   357
double fertilization   365
embryo sac   365
ethylene   359
female gametophyte   365
filament   363
flower   362
fruit   366
gametophyte   362
genetically modified plant
   (GMP)   371
germinate   368
gibberellin   357
gravitropism   360
hybridization   370

long-day plant   360
male gametophyte   364
megaspore   362, 364
microspore   362, 364
ovary   363
ovule   363
petal   363
photoperiod   360
phototropism   360
phytochrome   361
pollen grain   364
pollination   365
receptacle   362
seed   366
senescence   358
sepal   363
short-day plant   360
sporophyte   362
stamen   363
stigma   363
style   363
tissue culture   369
totipotent   369
transgenic plant   371
tropism   360

Match the terms to these definitions:

a. _____ Plant hormone involved in gravitropism and phototropism.

b. _____ Stress hormone in plants.

c. _____ Ratio of the day length to the night length.

d. _____ Pigment responsible for detection of the length of night.

e. _____ Product of meiosis that develops into a male gametophyte.

Being Organized and Steady

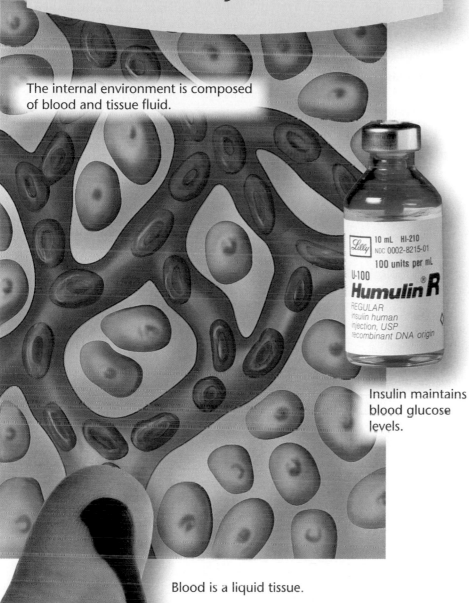

The internal environment is composed of blood and tissue fluid.

Insulin maintains blood glucose levels.

Blood is a liquid tissue.

To keep you healthy, your body is always maintaining homeostasis, the relative constancy of the internal environment. Your cells are bathed by tissue fluid, which continually exchanges substances with the blood. Therefore, the internal environment consists of blood and tissue fluid.

How might homeostasis be maintained? Let's take an example. Despite extreme fluctuations in our diet, the blood ▶ ▶ ▶

glucose (sugar) concentration usually stays within a normal range, day in and day out. This is primarily due to the work of two hormones, called insulin and glucagon. Just after you have eaten, the blood glucose level usually rises. In response, your pancreas releases insulin, which causes the body's cells, including those of the liver, to take up glucose from the blood. Now the blood glucose level lowers. The liver ordinarily stores glucose, so if your blood runs out of glucose before you eat again, the other hormone, glucagon, causes the liver to release sugar into the blood.

If by chance the liver runs out of stored glucose, it will convert other molecules such as amino acids from protein to glucose in order to maintain the blood glucose level constant. This might happen if you have diabetes mellitus, and your pancreas doesn't produce enough insulin or if you are on a low-carbohydrate diet. The inability of the pancreas to produce insulin or the inability of the liver to maintain the glucose constant are serious conditions that require good medical care. If homeostasis cannot be maintained, death can occur due to either an extremely high or extremely low blood glucose level. Of all the organs in the body, the brain requires a constant supply of glucose.

Understanding the structure and function of the various types of tissues in the body is of critical importance to understanding the functions of various organ systems. In this chapter, you will learn about basic tissue types, and how they come together to form organs and organ systems. In addition, you will learn how the body maintains homeostasis.

22.1 The Body's Organization

In Part One of this text, you studied the general structure and function of a plant cell and an animal cell. Now it is time to consider that an animal's body is composed of specialized cells that perform particular functions. Your body contains trillions of cells, of which there are a hundred different types. Cells of the same type occur within a tissue. A **tissue** is a group of similar cells performing a similar function. An organ contains different types of tissues arranged in a certain fashion. In other words, the structure and function of an organ are dependent on the tissues it contains. That's why it is sometimes said that tissues, not organs, are the structural and functional units of the body. Several organs are found within an organ system, and the organs of a system work together to perform necessary functions for the organism.

Let's take an example (Fig. 22.1). In humans, the urinary system contains these organs: kidneys, ureters, bladder, and urethra. The kidneys remove waste molecules from the blood and produce urine. The tubular shape of the ureters is suitable for passing urine to the bladder, which can store urine because it is expandable. Urine passes out of the body by way of the tubular urethra. The functions of the urinary system—to produce, store, and rid the body of metabolic wastes—are dependent on its organs, which in turn are dependent on the tissues making up these organs.

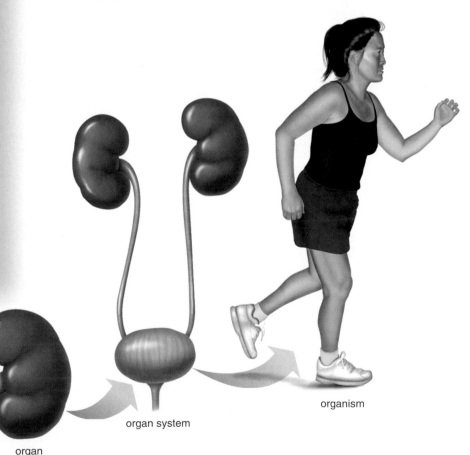

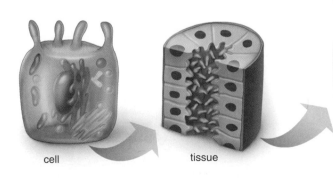

organism

organ system

organ

cell

tissue

Figure 22.1 **Levels of biological organization.**

A tissue is composed of specialized cells all having the same structure and performing the same functions. An organ is composed of those types of tissues that help it perform particular functions. An organ system contains several organs and has functions necessary to the continued existence of an organism.

Other examples also show that the work of an organ is dependent on its tissues. In the digestive system, the intestine absorbs nutrients. The cells of the tissue lining the lumen (cavity) of the small intestine have microvilli that increase the available surface area for absorption. Muscle cells, within muscle tissue, shorten when they contract because they have intracellular components that move past one another. Nerve cells, within nervous tissue, have long, slender projections that carry impulses to distant body parts. The biological axiom that "structure suits function" and vice versa begins with the specialized cells within a tissue. And thereafter, this truism applies also to organs and organ systems.

From the many different types of animal cells, biologists have been able to categorize tissues into just four major types:

Epithelial tissue (epithelium) covers body surfaces and lines body cavities.
Connective tissue binds and supports body parts.
Muscular tissue moves the body and its parts.
Nervous tissue receives stimuli and conducts nerve impulses.

Except for nervous tissue, each type of tissue is subdivided into even more types (Fig. 22.2). We look at each of these in the sections that follow to see how their particular structure allows them to perform their functions.

Check Your Progress

1. List the levels of biological organization, and explain how they are related to each other.
2. List the four major types of tissues in animals.

Answers: 1. An organism contains organ systems, which are composed of groups of organs that work together. An organ is composed of tissues, which determine its structure and function. Each tissue is composed of cells. 2. The four tissue types are epithelial, connective, muscular, and nervous.

Figure 22.2 Types of vertebrate tissues.

The four classes of vertebrate tissues are epithelial (pink), connective (blue), muscular (tan), and nervous (yellow).

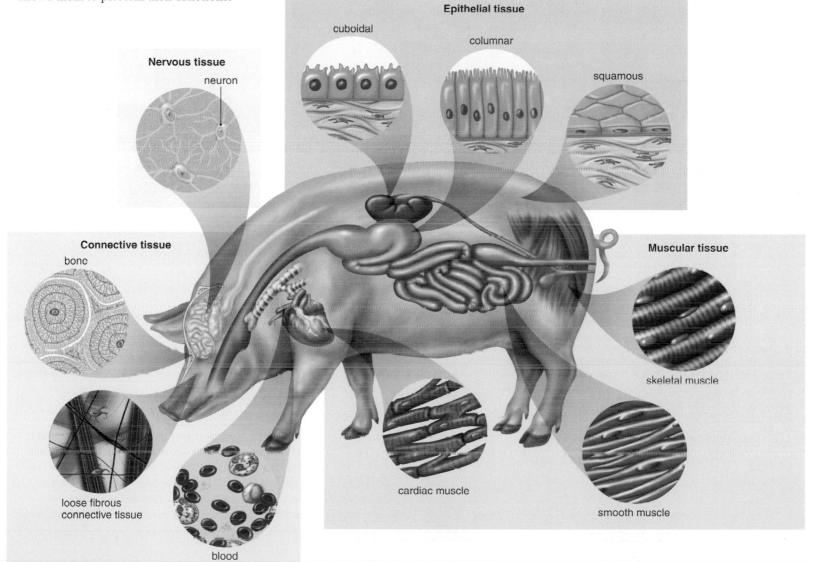

Nervous tissue
neuron

Epithelial tissue
cuboidal
columnar
squamous

Connective tissue
bone
loose fibrous connective tissue
blood

Muscular tissue
skeletal muscle
cardiac muscle
smooth muscle

Epithelial Tissue Protects

Epithelial tissue, also called *epithelium,* forms the external and internal linings of many organs and covers the surface of the body. Therefore, for a substance to enter or exit the body at the digestive tract, the lungs, or the genital tract, it must cross an epithelial tissue. Epithelial cells adhere to one another, but are generally only one cell thick. This characteristic enables them to fulfill a protective function and yet allows substances to pass through to a tissue beneath them.

The cells of epithelial tissue differ in shape (Fig. 22.3). Squamous epithelium, such as that lining the air spaces of the lungs and blood vessels, is composed of flattened cells. Cuboidal epithelium, which lines the **lumen** (cavity) of the kidney tubules, contains cube-shaped cells. Columnar epithelium has cells resembling rectangular pillars or columns, with nuclei usually located near the bottom of each cell. Columnar epithelium is found lining the lumen of the digestive tract.

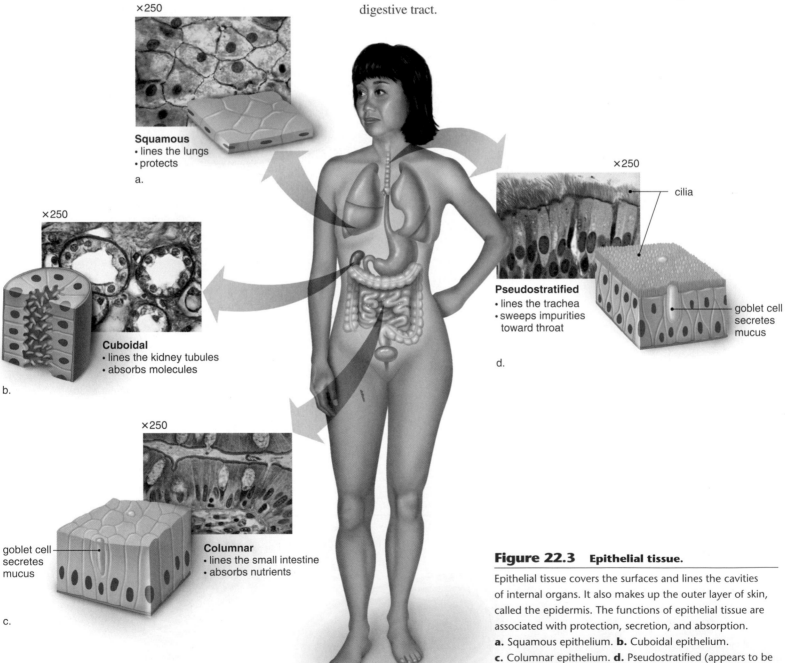

×250

Squamous
• lines the lungs
• protects

a.

×250

Cuboidal
• lines the kidney tubules
• absorbs molecules

b.

×250

goblet cell secretes mucus

Columnar
• lines the small intestine
• absorbs nutrients

c.

×250

cilia

Pseudostratified
• lines the trachea
• sweeps impurities toward throat

goblet cell secretes mucus

d.

Figure 22.3 **Epithelial tissue.**

Epithelial tissue covers the surfaces and lines the cavities of internal organs. It also makes up the outer layer of skin, called the epidermis. The functions of epithelial tissue are associated with protection, secretion, and absorption. **a.** Squamous epithelium. **b.** Cuboidal epithelium. **c.** Columnar epithelium. **d.** Pseudostratified (appears to be layered but is not) epithelium.

×250

Stratified
• epidermis of skin
• protects

Figure 22.4   **Skin.**

The outer portion of skin, called the epidermis, is a stratified epithelium. The many layers of tightly packed cells reinforced by the protein keratin make the skin protective against water loss and pathogen invasion. New epidermal cells arise in the most inner layer and are shed at the outer layer. In reptiles, such as this gila monster, the epidermis forms scales, which are simply projections hardened with keratin. When the epidermis is shed, the scales have to be replaced. Pigmented epidermal cells account for coloration of the skin.

One or more epithelial cells are the primary components of glands, which produce products and secrete them by exocytosis. During secretion, the contents of a vesicle are released when the vesicle fuses with the plasma membrane. Some glands have a duct, and others are ductless. Each mucus-secreting goblet cell within the lining of the digestive tract is a single-celled gland that produces a protective mucus (see Fig. 22.3). Pseudostratified epithelium lines the trachea (windpipe) (Fig. 22.3*d*). Along the trachea, mucus traps foreign particles, and the upward motion of cilia carries the mucus to the back of the throat, where it may be either swallowed or expelled. Smoking can cause a change in mucus secretion and inhibit ciliary action, and the result is an inflammatory condition called chronic bronchitis.

The outer region of the skin, called the epidermis, is a stratified (layered) epithelium in which the cells have been reinforced by keratin, a protein that provides strength (Fig. 22.4). A stratified epithelium allows skin to protect the body from injury, drying out, and possible pathogen (virus and bacterium) invasion.

The lining of the urinary bladder is a transitional epithelium, whose structure suits its function. When the walls of the bladder are relaxed, the transitional epithelium consists of several layers of cuboidal cells. When the bladder is distended with urine, the epithelium stretches, and the outer cells take on a squamous appearance. The cells are able to slide in relation to each other, while at the same time forming a barrier that prevents any part of urine from diffusing into other regions of the body.

Epithelial tissue is well known for constantly replacing its cells. This feature is particularly useful along the digestive tract, where rough food particles and enzymes can damage the lining. The liver, which is composed of cells of epithelial origin, can regenerate whole portions removed during surgery. But there is a price to pay for the ability of epithelial tissue to constantly divide—it is more likely to become cancerous than are other tissues. Cancers of epithelial tissue within the digestive tract, lungs, and breast are called carcinomas.

Check Your Progress

1. Describe the composition and activity of a gland.
2. Explain why cancers are more likely to develop in epithelial tissue than in other types of tissue.

Answers: 1. A gland is usually composed primarily of epithelium. The epithelial cells produce and secrete a product, such as mucus. 2. Epithelial tissue is constantly dividing, so errors that might occur during cell division and result in cancer are more likely to occur.

Connective Tissue Connects and Supports

The many different types of **connective tissue** are all involved in binding organs together and providing support and protection. As a rule, connective tissue cells are widely separated by a **matrix,** a noncellular material that varies from solid to semifluid to fluid. The matrix usually has fibers, notably collagen fibers. Collagen is the most common protein in the human body, which gives you some idea of how prevalent connective tissue is.

Loose Fibrous and Related Connective Tissues

Let's consider **loose fibrous connective tissue** first and then compare the other types to it (see Fig. 22.2). This tissue occurs beneath an epithelium and connects it to other tissues within an organ. It also forms a protective covering for many internal organs, such as muscles, blood vessels, and nerves. Its cells are called **fibroblasts** because they produce a matrix that contains fibers, including collagen fibers and elastic fibers. The presence of loose fibrous connective tissue in the walls of lungs and arteries allows these organs to expand.

Adipose tissue is a type of loose connective tissue in which the fibroblasts enlarge and store fat, and there is limited matrix (Fig. 22.5). Adipose tissue is located beneath the skin and around organs such as the heart and kidneys.

Compared to loose fibrous connective tissue, **dense fibrous connective tissue** contains more collagen fibers, and they are packed closely together (Fig. 22.5). This type of tissue has more specific functions than does loose fibrous connective tissue. For example, dense fibrous connective tissue is found in **tendons,** which connect muscles to bones, and in **ligaments,** which connect bones to other bones at joints.

Figure 22.5   Types of connective tissue.

The human knee provides examples of most types of connective tissue.

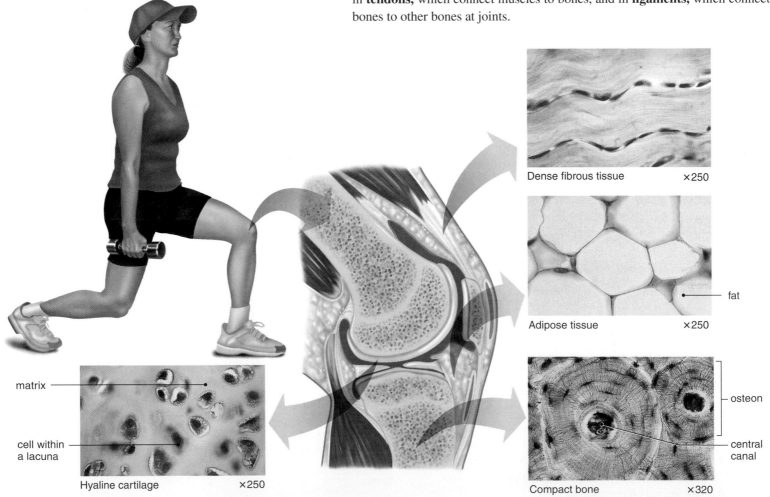

matrix

cell within a lacuna

Hyaline cartilage   ×250

Dense fibrous tissue   ×250

fat

Adipose tissue   ×250

osteon

central canal

Compact bone   ×320

In **cartilage,** the cells lie in small chambers called **lacunae,** separated by a matrix that is solid yet flexible. **Hyaline cartilage,** the most common type of cartilage, contains only very fine collagen fibers (Fig. 22.5). The matrix has a white, translucent appearance. Hyaline cartilage is found in the nose and at the ends of the long bones and the ribs, and it forms rings in the walls of respiratory passages. The human fetal skeleton is also made of this type of cartilage, which is later replaced by bone. Cartilaginous fishes, such as sharks, have a cartilaginous skeleton throughout their lives.

Bone is the most rigid connective tissue (Fig. 22.5). It consists of an extremely hard matrix of inorganic salts, notably calcium salts, deposited around collagen fibers. The inorganic salts give bone rigidity, and the protein fibers provide elasticity and strength, much as steel rods do in reinforced concrete. **Compact bone,** the most common type, consists of cylindrical structural units called osteons. The central canal of each osteon is surrounded by rings of hard matrix. Bone cells are located in spaces called lacunae between the rings of matrix. Blood vessels in the central canal carry nutrients that allow bone to renew itself.

Blood

Blood is composed of several types of cells suspended in a liquid matrix called plasma. Blood is unlike other types of connective tissue in that the matrix (i.e., plasma) is not made by the cells (Fig. 22.6). Some people do not classify blood as connective tissue; instead, they suggest a separate tissue category called vascular tissue.

Blood serves the body well. It transports nutrients and oxygen to cells and removes their wastes. It helps distribute heat and also plays a role in fluid, ion, and pH balance. Also, various components of blood help protect us from disease, and blood's ability to clot prevents fluid loss.

Blood contains two types of cells. **Red blood cells** are small, biconcave, disk-shaped cells without nuclei. The presence of the red pigment hemoglobin makes the cells red, and in turn, makes the blood red. Hemoglobin combines with oxygen, and in this way red blood cells transport oxygen.

White blood cells may be distinguished from red blood cells by the fact that they are usually larger, have a nucleus, and without staining would appear translucent. White blood cells fight infection in two primary ways. Some white blood cells are phagocytic and engulf infectious pathogens, while other white blood cells produce antibodies, molecules that combine with foreign substances to inactivate them.

Platelets are another component of blood, but they are not complete cells; rather, they are fragments of giant cells present only in bone marrow. When a blood vessel is damaged, platelets form a plug that seals the vessel, and injured tissues release molecules that help the clotting process.

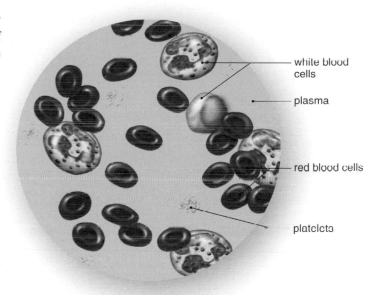

white blood cells

plasma

red blood cells

platelets

Figure 22.6 Blood, a liquid tissue.

Blood is classified as connective tissue because the cells are separated by a matrix—plasma. Plasma, the liquid portion of blood, usually contains several types of cells. The components of blood are red blood cells, white blood cells, and platelets (which are actually fragments of larger cells).

Check Your Progress

1. Compare and contrast loose fibrous connective tissue with dense fibrous connective tissue.
2. Contrast the function of tendons with that of ligaments.
3. Contrast the structure and function of red blood cells with those of white blood cells.

Answers: 1. Both loose and dense fibrous connective tissue are involved in binding organs together and providing support and protection. Loose fibrous connective tissue contains fewer collagen fibers, and these are less closely packed than those in dense fibrous connective tissue. 2. Tendons connect muscles to bones, while ligaments connect bones to other bones. 3. Red blood cells are smaller and lack a nucleus. The presence of hemoglobin for carrying oxygen makes them red. White blood cells contain a nucleus and without staining would appear translucent. They play a role in fighting infection.

Muscular Tissue Moves the Body

Muscular tissue and nervous tissue account for the ability of animals and their parts to move. **Muscle tissue** is also sometimes called *contractile tissue* because it contains contractile protein filaments, called actin and myosin filaments, that interact to produce movement. The three types of vertebrate muscles are skeletal, cardiac, and smooth.

Skeletal muscle, also called voluntary muscle, is attached by tendons to the bones of the skeleton, and when it contracts, bones move. Contraction of skeletal muscle is under voluntary control and occurs faster than in the other muscle types. The cells of skeletal muscle, called fibers, are cylindrical and quite long—sometimes they run the length of the muscle (Fig. 22.7a). They arise during development when several cells fuse, resulting in one fiber with multiple nuclei. The nuclei are located at the periphery of the cell, just inside the plasma membrane. The fibers have alternating light and dark bands that give them a **striated** appearance. These bands are due to the placement of actin filaments and myosin filaments in the cell.

Cardiac muscle is found only in the walls of the heart, and its contraction pumps blood and accounts for the heartbeat. Like skeletal muscle, cardiac muscle has striations, but the contraction of the heart is autorhythmic and involuntary (Fig. 22.7b). Cardiac muscle cells also differ from skeletal muscle cells in that they have a single, centrally placed nucleus. The cells are branched and seemingly fused one with the other, and the heart appears to be composed of one large interconnecting mass of muscle cells. Actually, cardiac muscle cells are separate and individual, but they are bound end to end at **intercalated disks,** areas where folded plasma membranes allow the contraction impulse to spread from one cell to the other.

Smooth muscle is so named because the cells lack striations. The spindle-shaped cells form layers in which the thick middle portion of one cell is opposite the thin ends of adjacent cells. Consequently, the nuclei form an irregular pattern in the tissue (Fig. 22.7c). Smooth muscle is not under voluntary control, and therefore is said to be involuntary. Smooth muscle is also sometimes called *visceral* muscle because it is found in the walls of the viscera (intestine, stomach, and other internal organs) and blood vessels. Smooth muscle contracts more slowly than skeletal muscle but can remain contracted for a longer time. When the smooth muscle of the intestine contracts, food moves along its lumen (central cavity). When the smooth muscle of the blood vessels contracts, blood vessels constrict, helping to raise blood pressure.

Figure 22.7 **Types of muscular tissue.**

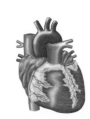
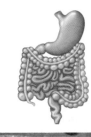

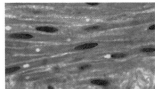
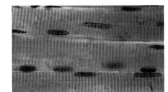

×250 ×100 ×100

a. **Skeletal muscle**
  • has striated, tubular, multinucleated fibers.
  • is usually attached to skeleton.
  • is voluntary.

b. **Cardiac muscle**
  • has striated, branched, uninucleated fibers.
  • occurs in walls of heart.
  • is involuntary.

c. **Smooth muscle**
  • has spindle-shaped, nonstriated, uninucleated fibers.
  • occurs in walls of internal organs.
  • is involuntary.

Nervous Tissue Communicates

Nervous tissue coordinates body parts and allows an animal to respond to the environment. The nervous system depends on (1) sensory input, (2) integration of data, and (3) motor output to carry out its functions. Nerves conduct impulses from sensory receptors to the spinal cord and the brain, where integration occurs. The phenomenon called sensation occurs only in the brain, however. Nerves then conduct nerve impulses away from the spinal cord and brain to the muscles and glands, causing them to contract and secrete, respectively. In this way, a coordinated response to both internal and external stimuli is achieved.

A nerve cell is called a **neuron.** Every neuron has three parts: dendrites, a cell body, and an axon (Fig. 22.8). A *dendrite* is an extension that conducts signals toward the cell body. The *cell body* contains the major concentration of the cytoplasm and the nucleus of the neuron. An *axon* is an extension that conducts nerve impulses. The brain and spinal cord contain many neurons, whereas **nerves** contain only axons of neurons. The dendrites and cell bodies of these neurons are located in the spinal cord or brain, depending on whether it is a spinal nerve or a cranial nerve. A nerve is like a land-based telephone trunk cable, because just like telephone wires, each axon is a communication channel independent of the others.

In addition to neurons, nervous tissue contains **neuroglia,** cells that support and nourish neurons. They outnumber neurons nine to one and take up more than half the volume of the brain. Although their primary function is support, research is currently being conducted to determine how much neuroglia directly contribute to brain function. Schwann cells are the type of neuroglia that encircle long nerve fibers within nerves, forming a protective myelin sheath.

dendrite
cell body
nucleus
nucleus of Schwann cell
axon
impulse
myelin sheath

×60

Nervous tissue
• Brain; spinal cord; nerves
• Conduction of nerve impulses

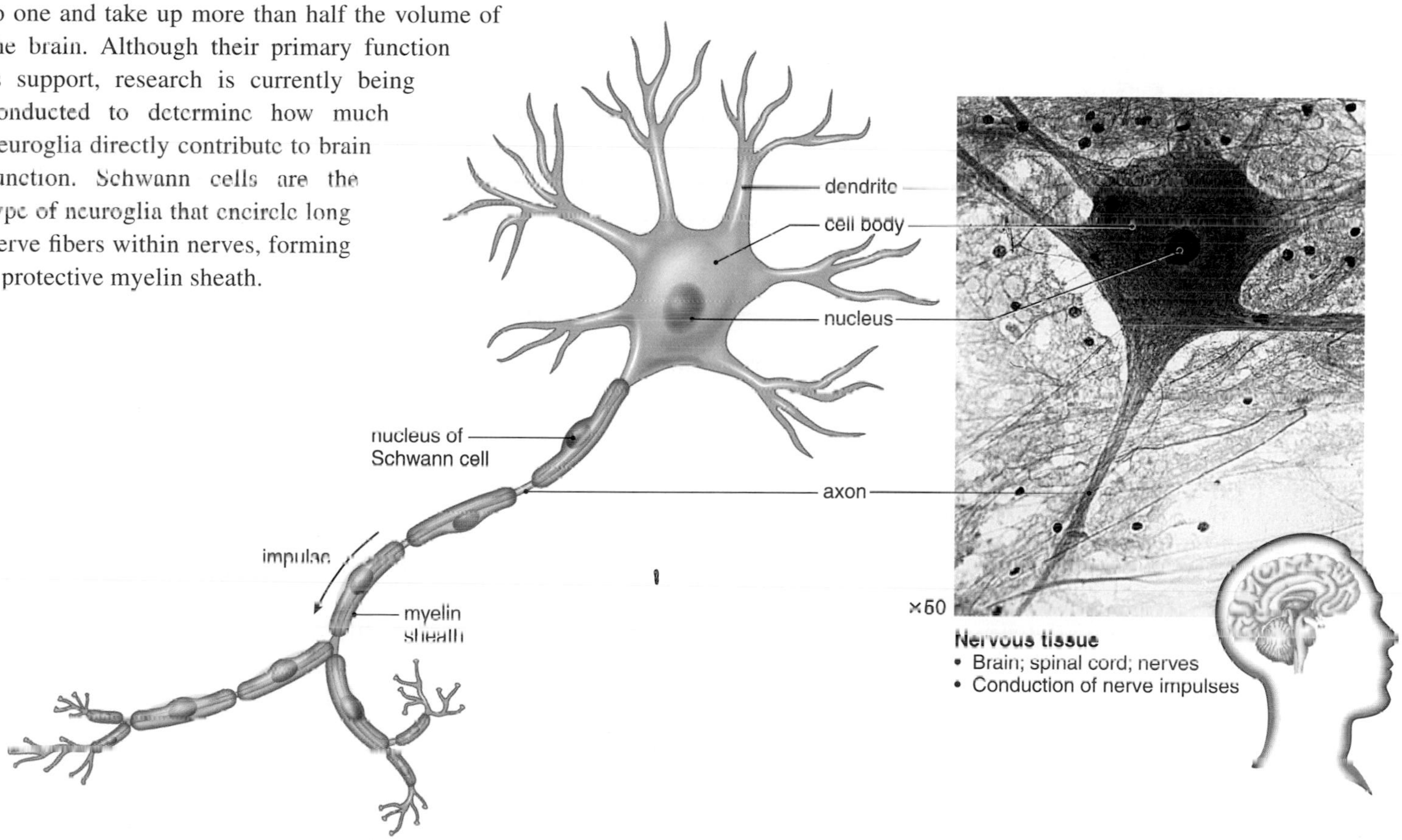

Figure 22.8 Nervous tissue.

Neurons are surrounded by neuroglia, such as Schwann cells, which envelop axons and form the myelin sheath. Only axons conduct nerve impulses.

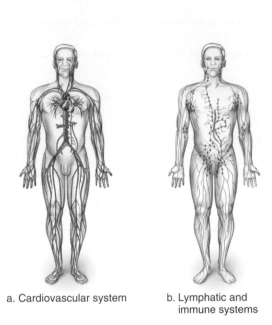

a. Cardiovascular system   b. Lymphatic and
immune systems

Figure 22.9 **The body's transport systems.**

Aside from transport, the cardiovascular and lymphatic systems are involved in defense against disease.

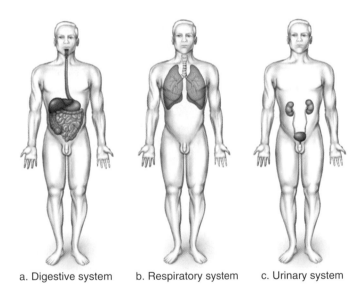

a. Digestive system   b. Respiratory system   c. Urinary system

Figure 22.10 **The body's maintenance systems.**

The digestive, respiratory, and urinary systems keep the internal environment constant.

22.2 Organs and Organ Systems

Organs are composed of a number of tissues, and the structure and function of an organ are dependent on the tissues that make it up. Since an organ has many types of tissues, it can perform a function that none of the tissues can do alone. To take an example, the function of the bladder is to store urine and to expel it when convenient. But this function is dependent on the individual tissues making up the bladder. The epithelium lining the bladder helps the organ store urine by stretching while still preventing urine from leaking into the body proper. The muscles of the bladder propel the urine forward so that it is voided from the body. Similarly, the functions of an organ system are dependent on its organs. The function of the urinary system is to produce urine, store it, and then transport it out of the body. The kidneys produce urine, the bladder stores it, and various tubes transport it from the kidneys to the bladder and also out of the body.

This text divides the systems of the body into those involved in (1) transport of fluids about the body, (2) maintenance of the body, (3) control of the body's systems, (4) sensory input and motor output, and (5) reproduction. All these systems have functions that contribute to homeostasis, the relative constancy of the internal environment (see Section 22.3).

Transport

The **cardiovascular system** (Fig. 22.9a) consists of the heart and the blood vessels that carry blood throughout the body. The body's cells are surrounded by a liquid called **tissue fluid;** blood transports nutrients and oxygen to tissue fluid for the cells, and removes waste molecules excreted by cells from the tissue fluid. Recall that the internal environment of the body consists of the blood within the blood vessels and the tissue fluid that surrounds the cells.

The **lymphatic system** (Fig. 22.9b) consists of lymphatic vessels, lymph, lymph nodes, and other lymphatic organs. Lymphatic vessels absorb fat from the digestive system and collect excess tissue fluid, which is returned to the blood in the cardiovascular system.

These two systems are also involved in defense against disease. Certain blood cells in the lymph and blood are part of an **immune system,** which specifically protects the body from disease.

Maintenance of the Body

Three systems (digestive, respiratory, and urinary) add substances to and/or remove substances from the blood. If the composition of the blood remains constant, so does that of the tissue fluid.

The **digestive system** (Fig. 22.10a) consists of the various organs along the digestive tract together with the associated organs, such as teeth, salivary glands, the liver, and the pancreas. This system receives food and digests it into nutrient molecules that enter the blood.

The **respiratory system** (Fig. 22.10b) consists of the lungs and the tubes that take air to and from the lungs. The respiratory system brings oxygen into the body and takes carbon dioxide out of the body through the lungs. It also exchanges gases with the blood.

The **urinary system** (Fig. 22.10c) contains the kidneys and the urinary bladder along with tubes that transport urine. This system rids blood of wastes and also helps regulate the fluid level and chemical content of the blood.

Control

The **nervous system** (Fig. 22.11*a*) consists of the brain, spinal cord, and associated nerves. The nerves conduct nerve impulses from receptors to the brain and spinal cord. They also conduct nerve impulses from the brain and spinal cord to the muscles and glands, allowing us to respond to both external and internal stimuli. (Sensory receptors and sense organs are sometimes considered a part of the nervous system.)

The **endocrine system** (Fig. 22.11*b*) consists of the hormonal glands, which secrete chemicals that serve as messengers between body parts. Both the nervous and endocrine systems coordinate and regulate the functions of the body's other systems. The endocrine system also helps maintain the proper functioning of the male and female reproductive organs.

Sensory Input and Motor Output

The **integumentary system** (Fig. 22.12*a*) consists of the skin and its accessory structures. The sensory receptors in the skin, and in organs such as the eyes and ears, are sensitive to certain external stimuli and communicate with the brain and spinal cord by way of nerve fibers. These messages may cause the brain to bring about a response to a stimulus.

The **skeletal system** (Fig. 22.12*b*) and the **muscular system** (Fig. 22.12*c*) enable the body and its parts to move as a result of motor output. The skeleton, as a whole, serves as a place of attachment for the skeletal muscles. Contraction of muscles in the muscular system actually accounts for movement of body parts.

These systems, along with the integumentary system, also protect and support the body. The skeletal system, consisting of the bones of the skeleton, protects body parts. For example, the skull forms a protective encasement for the brain, as does the rib cage for the heart and lungs. The skin and muscles assist in this endeavor because they are exterior to the bones.

Reproduction

The **reproductive system** (Fig. 22.13) involves different organs in the male and female. The male reproductive system consists of the testes, other glands, and various ducts that conduct semen to and through the penis. The testes produce sex cells called sperm. The female reproductive system consists of the ovaries, oviducts, uterus, vagina, and external genitals. The ovaries produce sex cells called eggs. When a sperm fertilizes an egg, an offspring begins development.

Check Your Progress

1. Compare and contrast the function of the cardiovascular system with that of the lymphatic system.
2. Describe the common function of the nervous and endocrine systems.

Answers: 1. Both are transport systems. The cardiovascular system carries blood throughout the body, while the lymphatic system absorbs fat from the digestive system and collects excess tissue fluid and returns it to the blood. 2. Both the nervous and endocrine systems coordinate and regulate the functions of other systems.

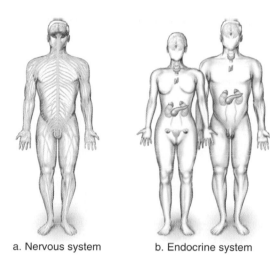

a. Nervous system   b. Endocrine system

Figure 22.11 **The control systems.**

The nervous and endocrine systems coordinate the other systems of the body.

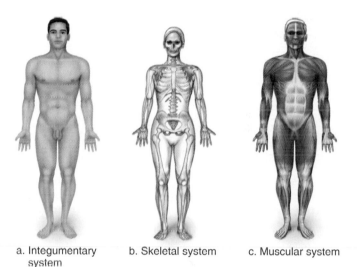

a. Integumentary system   b. Skeletal system   c. Muscular system

Figure 22.12 **Sensory input and motor output.**

Sensory receptors in the skin (integumentary system) and the sense organs send input to the skeletal and muscular systems, the control systems that cause a response to stimuli.

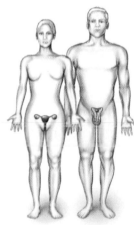

Reproductive system

Figure 22.13 **The reproductive system.**

The reproductive organs in the male and female differ. The reproductive system functions to ensure survival of the species.

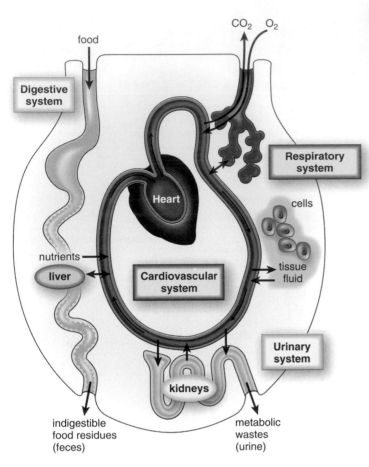

Figure 22.14   Constancy of the internal environment (blood and tissue fluid).

The respiratory system exchanges gases with the external environment and with the blood. The digestive system takes in food and adds nutrients to the blood, and the urinary system removes metabolic wastes from the blood and excretes them. When blood exchanges nutrients and oxygen for carbon dioxide and other wastes with tissue fluid, the composition of the tissue fluid stays within normal limits.

Check Your Progress

1. Why are blood and tissue fluid considered the internal environment?
2. How do the digestive, urinary, and respiratory systems contribute to homeostasis?

Answers: 1. Cells are bathed in a liquid environment called tissue fluid, and tissue fluid exchanges substances with the blood. Therefore, both are considered the internal environment of the body. 2. The digestive system adds nutrients to the blood, the urinary system removes metabolic wastes from the blood, and the respiratory system takes away carbon dioxide and adds oxygen from the blood.

22.3   Homeostasis

It is likely you are most familiar with the use of the word *environment* to mean the external environment. There is much concern these days about our external environment because of the need to keep pollution to a minimum in order to maintain the health of ecosystems and the organisms, including ourselves, that live in them. Your body, however, is composed of many cells, and therefore its internal environment is the environment of the cells. Cells live in a liquid environment called tissue fluid. Tissue fluid is constantly renewed by exchanges with the blood (Fig. 22.14). Therefore, blood and tissue fluid are the internal environment of the body. Tissue fluid remains relatively constant only as long as blood composition remains near normal levels. Relatively constant means that the composition of both tissue fluid and blood usually falls within a certain range of normality.

All the systems of the body contribute to maintaining **homeostasis,** the relative constancy of the internal environment. The cardiovascular system conducts blood to and away from capillaries, where exchange occurs. The heart pumps the blood, and thereby keeps it moving toward the capillaries. Red blood cells transport oxygen and participate in the transport of carbon dioxide. White blood cells fight infection, and platelets participate in the clotting process. The lymphatic system is accessory to the cardiovascular system. Lymphatic capillaries collect excess tissue fluid and return it via lymphatic vessels to the cardiovascular system. Lymph nodes help purify lymph and keep it free of pathogens.

The digestive system takes in and digests food, providing nutrient molecules that enter the blood to replace those that are constantly being used by the body cells. The respiratory system removes carbon dioxide from and adds oxygen to the blood. The chief regulators of blood composition are the kidneys and the liver. Urine formation by the kidneys is extremely critical to the body, not only because it rids the body of metabolic wastes, but also because the kidneys carefully regulate blood volume, salt balance, and pH. The liver, among other functions, regulates the glucose concentration of the blood. Immediately after glucose enters the blood, the liver removes the excess for storage as glycogen. Later, the glycogen is broken down to replace the glucose that was used by body cells. In this way, the glucose composition of the blood remains constant. The liver also removes toxic chemicals, such as ingested alcohol and other drugs. The liver makes urea, a nitrogenous end product of protein metabolism.

The nervous system and the endocrine system regulate the other systems of the body. They work together to control body systems so that homeostasis is maintained. In negative feedback mechanisms involving the nervous system, sensory receptors send nerve impulses to control centers in the brain, which then direct effectors to become active. Effectors can be muscles or glands. Muscles bring about an immediate change. Endocrine glands secrete hormones that bring about a slower, more lasting change that keeps the internal environment relatively stable.

Because of homeostasis, even though external conditions may change dramatically, internal conditions stay within a narrow range. For example, the temperature of the body is maintained near 37°C (97° to 99°F), even if the surrounding temperature is lower or higher. If you eat acidic foods, the pH of your blood still stays about 7.4, and even if you eat a candy bar, the amount of sugar in your blood remains at just about 0.1%.

Negative Feedback

Negative feedback is the primary homeostatic mechanism that allows the body to keep the internal environment constant. A negative feedback mechanism has at least two components: a sensor and a control center (Fig. 22.15). The sensor detects a change in the internal environment (a stimulus); the control center initiates an effect that brings conditions back to normal again. Now, the sensor is no longer activated. In other words, a negative feedback mechanism is present when the output of the system dampens the original stimulus.

Let's take a simple example. When the pancreas detects that the blood glucose level is too high, it secretes insulin, a hormone that causes cells to take up glucose. Now the blood sugar level returns to normal, and the pancreas is no longer stimulated to secrete insulin.

When conditions exceed their limits and feedback mechanisms cannot compensate, illness results. For example, if the pancreas is unable to produce insulin, the blood sugar level becomes dangerously high, and the individual can become seriously ill. The study of homeostatic mechanisms is therefore medically important.

Mechanical Example

A home heating system is often used to illustrate how a more complicated negative feedback mechanism works (Fig. 22.16). You set the thermostat at, say, 20°C (68°F). This is the *set point*. The thermostat contains a thermometer, a sensor that detects when the room temperature is above or below the set point. The thermostat also contains a control center; it turns the furnace off when the room is warm and turns it on when the room is cool. When the furnace is off, the room cools a bit, and when the furnace is on, the room warms a bit. In other words, a negative feedback system results in fluctuation above and below the set point.

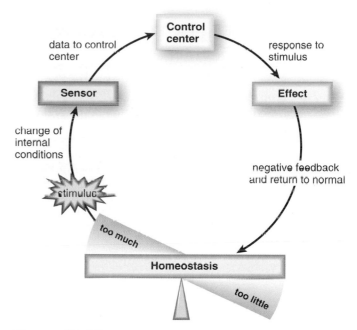

Figure 22.15 Negative feedback mechanism.

This diagram shows how the basic elements of a negative feedback mechanism work.

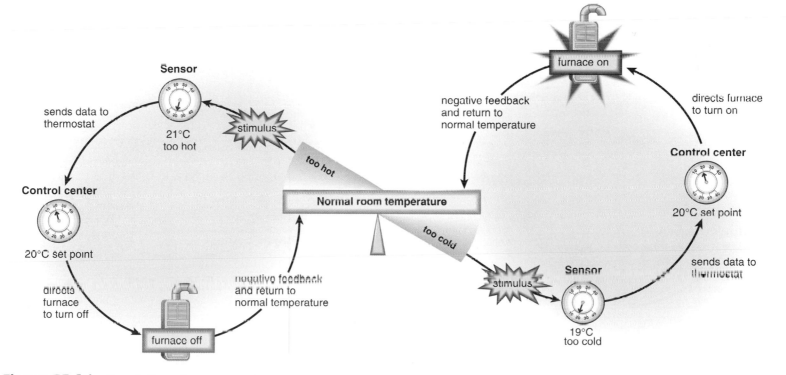

Figure 22.16 Regulation of room temperature.

This diagram shows how room temperature is returned to normal when the room becomes too hot. A contrary cycle in which the furnace turns on and gives off heat returns the room temperature to normal when the room becomes too cold.

Check Your Progress

1. Describe the components of a negative feedback mechanism.
2. Describe how the control center in the brain responds to a drop in internal temperature.
3. Describe how the control center in the brain responds to a rise in internal temperature.

Answers: 1. A sensor detects a change in the internal environment; the control center initiates an effect that brings conditions back to normal and the sensor is no longer activated. 2. First, the control center brings about constriction of the blood vessels of the skin. If the temperature continues to drop, shivering begins in order to generate heat. 3. Blood vessels of the skin dilate, and sweat glands are activated.

Human Example: Regulation of Body Temperature

The thermostat for body temperature is located in a part of the brain called the hypothalamus. When the core body temperature falls below normal, the control center directs (via nerve impulses) the blood vessels of the skin to constrict (Fig. 22.17). This action conserves heat. If the core body temperature falls even lower, the control center sends nerve impulses to the skeletal muscles, and shivering occurs. Shivering generates heat, and gradually body temperature rises to 37°C (98.6°F). When the temperature rises to normal, the control center is inactivated.

When the body temperature is higher than normal, the control center directs the blood vessels of the skin to dilate. More blood is then able to flow near the surface of the body, where heat can be lost to the environment. In addition, the nervous system activates the sweat glands, and the evaporation of sweat helps lower body temperature. Gradually, body temperature decreases to 37°C (98.6°F).

Notice that a negative feedback mechanism prevents change in the same direction; body temperature does not get warmer and warmer, because warmth brings changes that decrease body temperature. Also, body temperature does not get colder and colder, because a body temperature below normal causes changes that bring body temperature up.

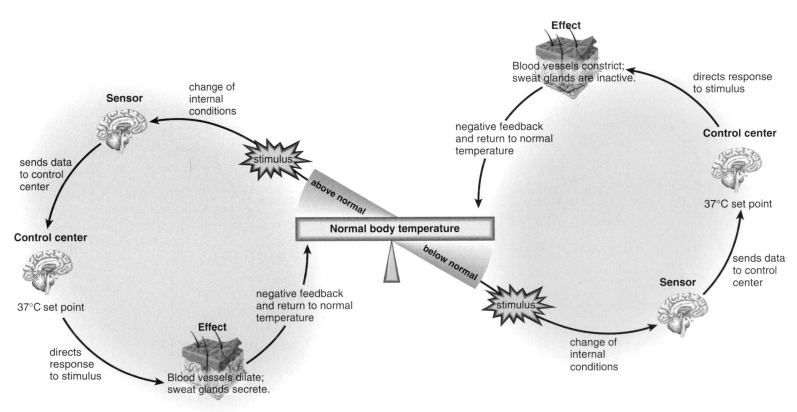

Figure 22.17   Regulation of body temperature.

This diagram shows how normal body temperature is maintained by a negative feedback system.

Summary

22.1   The Body's Organization

Levels of animal organization:

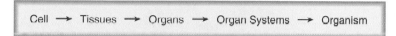

Cell → Tissues → Organs → Organ Systems → Organism

Four groups of vertebrate tissues:

| Epithelial | Connective | Muscular | Nervous |

Epithelial Tissue Protects
Epithelial tissue covers the body and lines its cavities.

- Types of epithelial tissue are: squamous, cuboidal, and columnar.
- Epithelial cells sometimes form glands that secrete either into ducts or into the blood.

Connective Tissue Connects and Supports
In connective tissue, cells are separated by a matrix that contains fibers (e.g., collagen fibers). The four types are:

- Loose fibrous connective tissue, including adipose tissue
- Dense fibrous connective tissue (tendons and ligaments)
- Cartilage and bone (matrix for cartilage is more flexible than that for bone)
- Blood (matrix is a liquid called plasma)

Muscular Tissue Moves the Body
Muscular tissue is of three types: skeletal, cardiac, smooth

- Both skeletal muscle (attached to bones) and cardiac muscle (forms the heart wall) are striated.
- Both cardiac muscle (wall of heart) and smooth muscle (wall of internal organs) are involuntary.

Nervous Tissue Communicates

- Nervous tissue is composed of neurons and several types of neuroglia.
- Each neuron has dendrites, a cell body, and an axon. Axons conduct nerve impulses.

22.2   Organs and Organ Systems

Organs make up organ systems, which are summarized in the table below.

22.3   Homeostasis

Homeostasis is the relative constancy of the internal environment. The internal environment consists of blood and tissue fluid. Due to exchange of nutrients and wastes with the blood, tissue fluid stays constant:

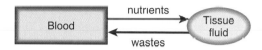

Blood →(nutrients)→ Tissue fluid
Blood ←(wastes)← Tissue fluid

All organ systems contribute to the relative constancy of tissue fluid and blood.

- The cardiovascular system transports nutrients to cells and wastes from cells.
- The lymphatic system absorbs excess tissue fluid and functions in immunity.
- The digestive system takes in food and adds nutrients to the blood.
- The respiratory system carries out gas exchange with the external environment and the blood.
- The urinary system (i.e., the kidneys) removes metabolic wastes and regulates the pH and salt content of the blood.
- The nervous system and endocrine system regulate the other systems.

Negative Feedback
Negative feedback mechanisms keep the internal environment relatively stable. When a sensor detects a change above or below a set point, a control center brings about an effect that reverses the change and brings conditions back to normal again. Examples include:

- Regulation of blood glucose level by insulin.
- Regulation of room temperature by a thermostat and furnace.
- Regulation of body temperature by the brain and sweat glands.

ORGAN SYSTEMS	
Transport Cardiovascular (heart and blood vessels) Lymphatic (lymphatic vessels)	*Sensory* Integumentary (skin)
Maintenance Digestive (e.g., stomach, intestines) Respiratory (tubes and lungs) Urinary (tubes and kidneys)	*Motor* Skeletal (bones and cartilage) Muscular (muscles)
Control Nervous (brain, spinal cord, and nerves) Endocrine (glands)	*Reproduction* Reproductive (tubes and testes in males; tubes and ovaries in females)

Thinking Scientifically

1. Patients with muscular dystrophy often develop heart disease. However, heart disease in these patients may not be detected until it has progressed past the point at which treatment options are effective. New research indicates that the link between muscular dystrophy and heart disease is so strong that muscular dystrophy patients should be screened for heart disease early, so that treatment may be begun. What do you suppose is the link between muscular dystrophy and heart disease?

2. According to a recent study, African Americans are 34% more likely to die from a liver transplant than are white Americans. What might be potential causes for this differential survival rate? How would you test your hypotheses?

Testing Yourself

Choose the best answer for each question.

1. Label the levels of biological organization in the following illustration.

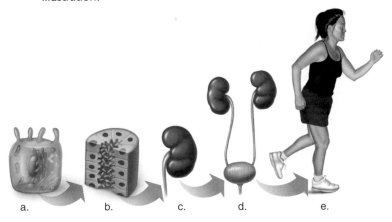

a.   b.   c.   d.   e.

2. Which of the following is not a feature of epidermis?

   a. contains keratin for strength
   b. stratified
   c. contains a matrix
   d. composed of epithelial cells

For questions 3–9, identify the tissue in the key that matches the description. Each answer may be used more than once. Each question may have more than one answer.

Key:
   a. epithelial tissue
   b. connective tissue
   c. muscular tissue
   d. nervous tissue

3. Contains cells that are separated by a matrix.

4. Has a protective role.

5. Covers the body surface and forms linings of organs.

6. Composed of a group of cells with a similar function.

7. Capable of replacing its cells.

8. Allows animals to respond to their environment.

9. Contains actin and myosin filaments.

10. Label the blood smear in the following illustration.

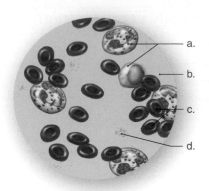

a.

b.

c.

d.

11. White blood cells fight infection by

    a. producing antibodies.
    b. producing toxins.
    c. engulfing pathogens.
    d. More than one of these are correct.
    e. All of these are correct.

12. Which of the following systems does not add or remove substances from the blood?

    a. digestive
    b. cardiovascular
    c. urinary
    d. respiratory

13. Identify the type of muscle associated with each of the following illustrations.

   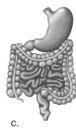

a.   b.   c.

14. The major function of tissue fluid is to

    a. provide nutrients and oxygen to cells and remove wastes.
    b. keep cells hydrated.
    c. prevent cells from touching each other.
    d. provide flexibility by allowing cells to slide over each other.

15. Label the parts of a neuron in the following illustration.

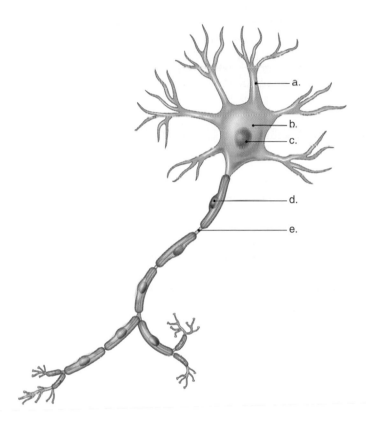

16. Which of the following is a function of skin?

    a. temperature regulation
    b. protection against water loss
    c. collection of sensory input
    d. protection from invading pathogens
    e. All of these are correct.

17. Endocrine glands secrete

    a. tissue fluid.
    b. wastes.
    c. toxins.
    d. hormones.

18. When a human being is cold, the blood vessels

    a. dilate, and the sweat glands are inactive.
    b. dilate, and the sweat glands are active.
    c. constrict, and the sweat glands are inactive.
    d. constrict, and the sweat glands are active.
    e. contract so that shivering occurs.

19. The loss of _____ will prevent proper blood clotting.

    a. red blood cells
    b. white blood cells
    c. platelets
    d. oxygen

20. Blood is a(n) _____ tissue because it has a _____

    a. connective, gap junction
    b. muscle, matrix
    c. epithelial, gap junction
    d. connective, matrix

21. Skeletal muscle is

    a. striated.
    b. under voluntary control.
    c. multinucleated.
    d. All of these are correct.

22. A reduction in red blood cells would cause problems with

    a. fighting infection.
    b. carrying oxygen.
    c. blood clotting.
    d. None of these are correct.

23. Which choice is true of both cardiac and skeletal muscle?

    a. striated
    b. single nucleus per cell
    c. multinucleated cells
    d. involuntary control

24. The skeletal system functions in

    a. blood cell production.
    b. mineral storage.
    c. movement.
    d. All of these are correct.

25. The correct order for a negative feedback mechanism is:

    a. sensory detection, control center, effect brings about change in environment.
    b. control center, sensory detection, effect brings about change in environment.
    c. sensory detection, control center, effect causes no change in environment.
    d. None of these are correct.

26. Which of the following is an example of negative feedback?

    a. Air conditioning goes off when room temperature lowers.
    b. Insulin decreases blood sugar levels after eating a meal.
    c. Heart rate increases when blood pressure drops.
    d. All of these are examples of negative feedback.

27. Which of these is not a type of epithelial tissue?

    a. cuboidal and columnar
    b. bone and cartilage
    c. squamous
    d. pseudostratified
    e. All of these are epithelial tissue.

28. Chemical messengers in the body are secreted by the

    a. integumentary system.
    b. nervous system.
    c. lymphatic system.
    d. digestive system.
    e. endocrine system.

29. Urine formation is critical for the body because it
    a. removes metabolic wastes.
    b. helps regulate blood volume.
    c. helps regulate salts.
    d. helps regulate pH.
    e. All of these are correct.

30. Which of the following act as slow effectors in the negative feedback system?
    a. muscles
    b. epidermal cells
    c. endocrine glands
    d. red blood cells
    e. senses

31. Which tissue is more apt to line a lumen?
    a. epithelial tissue
    b. connective tissue
    c. nervous tissue
    d. muscular tissue
    e. only smooth muscle

32. Tendons and ligaments
    a. are connective tissue.
    b. are associated with the bones.
    c. are found in vertebrates.
    d. contain collagen.
    e. All of these are correct.

33. Which tissue has cells in lacunae?
    a. epithelial tissue
    b. cartilage
    c. bone
    d. smooth muscle
    e. Both b and c are correct.

34. Cardiac muscle is
    a. striated.
    b. involuntary.
    c. smooth.
    d. composed of many fibers fused together.
    e. Both a and b are correct.

35. Which of these components of blood fight infection?
    a. red blood cells
    b. white blood cells
    c. platelets
    d. hydrogen ions
    e. All of these are correct.

36. Which of these body systems contribute to homeostasis?
    a. digestive and urinary systems
    b. respiratory and nervous systems
    c. nervous and endocrine systems
    d. immune and cardiovascular systems
    e. All of these are correct.

Go to www.mhhe.com/maderessentials for more quiz questions.

Bioethical Issue

Despite widespread efforts to convince people of the need for organ donors, supply continues to lag far behind demand. One proposed strategy to bring supply and demand into better balance is to develop an "insurance" program for organs. In this program, participants would pay the "premium" by promising to donate their organs at death. They, in turn, would receive priority for transplants as their "benefit." To avoid the problem of too many high-risk people applying for this type of insurance, a medical exam would be required so that only people with a normal risk of requiring a transplant would be accepted. Do you suppose this system would be an improvement over the current system in which organ donation is voluntary, with no tangible benefit? Can you think of any other strategy that would be more effective for increasing organ donations?

Understanding the Terms

blood 383	lymphatic system 386
bone 383	matrix 382
cardiac muscle 384	muscular system 387
cardiovascular system 386	muscle tissue 384
cartilage 383	negative feedback 389
compact bone 383	nerve 385
connective tissue 382	nervous system 387
dense fibrous connective tissue 382	nervous tissue 385
	neuroglia 385
digestive system 386	neuron 385
endocrine system 387	platelet 383
epithelial tissue 380	red blood cell 383
fibroblast 382	reproductive system 387
homeostasis 388	respiratory system 386
hyaline cartilage 383	skeletal muscle 384
immune system 386	skeletal system 387
integumentary system 387	smooth (visceral) muscle 384
intercalated disks 384	striated 384
lacuna 383	tendon 382
ligament 382	tissue 378
loose fibrous connective tissue 382	tissue fluid 386
	urinary system 386
lumen 380	white blood cell 383

Match the terms to these definitions:

a. _____ Connect bones to bones at joints.

b. _____ The most rigid connective tissue.

c. _____ Fragments of cells that seal damaged blood vessels.

d. _____ Called voluntary muscle; attached to bones by tendons.

e. _____ Characterized by alternating light and dark bands.

f. _____ Nervous tissue cells that support and nourish neurons.

g. _____ The organ system that consists of skin and its accessory structures.

h. _____ The system that consists of hormonal glands.

i. _____ The relative constancy of the body's internal environment.

j. _____ The homeostatic mechanism that is composed of a sensor, a control center, and an effect.

The Transport Systems

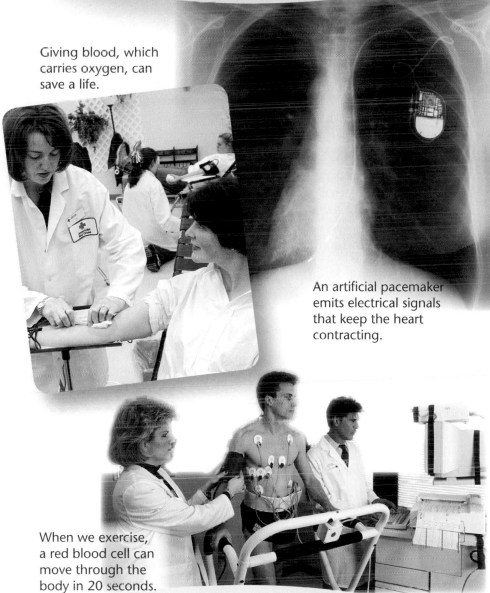

Giving blood, which carries oxygen, can save a life.

An artificial pacemaker emits electrical signals that keep the heart contracting.

When we exercise, a red blood cell can move through the body in 20 seconds.

The beat of the heart is under involuntary control, so we rarely consider the complex series of events that occurs when the heart moves blood to the lungs and the body tissues. Ordinarily, contraction of the heart is controlled by its natural pacemaker (referred to as the SA node), which automatically triggers the heart's rhythmic beating. At rest, the heart contracts about 70 times per minute, and each contraction moves the entire blood volume of approximately 4–6 liters

▶ ▶ ▶

through the system once every minute. The nervous system gets involved when the body requires an increased amount of oxygen, as when you exercise vigorously. A particular nerve releases an adrenaline-like chemical in order to increase the activity of the SA node and therefore the heart rate. This action of the nervous system is mimicked whenever adrenaline is injected directly into the heart muscle, as is sometimes shown in the movies.

When damage occurs to the SA node, physicians often implant an artificial pacemaker. The heart responds to electrical signals generated by the SA node, and the artificial pacemaker emits electrical signals at the proper rate in order to duplicate the activity of a normal SA node. It is amazing to think that a simple electrical device can keep the entire cardiovascular system running.

The cardiovascular system is one of the body's transport systems. It must move blood to the tissues at a rate appropriate to meet the oxygen and nutrient demands of the cells. In this chapter, you will learn how the cardiovascular system works and how its functioning can be adjusted to meet the body's needs. Some of the consequences of a malfunctioning system are also discussed.

23.1 Open and Closed Circulatory Systems

In some animals, the body plan makes a circulatory system unnecessary (Fig. 23.1). In a hydra, cells are either part of an external layer, or they line the gastrovascular cavity. In either case, each cell is exposed to water and can independently exchange gases and rid itself of wastes. The cells that line the gastrovascular cavity are specialized to carry out digestion. They pass nutrient molecules to other cells by diffusion. In a planarian, a trilobed gastrovascular cavity branches throughout the small, flattened body. No cell is very far from one of the digestive branches, so nutrient molecules can diffuse from cell to cell. Similarly, diffusion meets the respiratory and excretory needs of the cells.

Some other animals, such as nematodes and echinoderms, rely on movement of fluid within a coelom to circulate substances.

Open Circulatory Systems

More complex animals have a circulatory system in which a pumping **heart** moves a fluid into blood vessels. The grasshopper is an example of an invertebrate animal that has an **open circulatory system** (Fig. 23.2a). A tubular heart pumps a fluid called hemolymph through a network of channels and cavities in the body. Eventually, hemolymph (a combination of blood and tissue fluid) drains back to the heart. When the heart contracts, openings called ostia (sing., ostium) are closed;

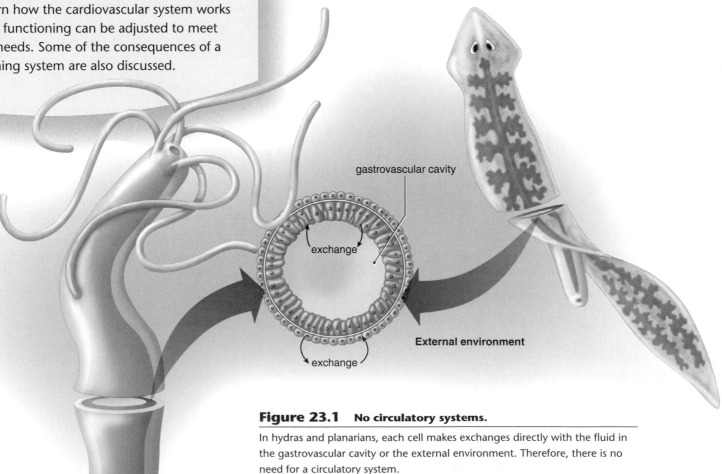

gastrovascular cavity

exchange

External environment

exchange

Figure 23.1   No circulatory systems.

In hydras and planarians, each cell makes exchanges directly with the fluid in the gastrovascular cavity or the external environment. Therefore, there is no need for a circulatory system.

when the heart relaxes, the hemolymph is sucked back into the heart by way of the ostia. The hemolymph of a grasshopper is colorless because it does not contain hemoglobin or any other respiratory pigment that combines with and carries oxygen. Oxygen is taken to cells, and carbon dioxide is removed from them by way of air tubes, called tracheae, which are found throughout the body. The tracheae provide efficient transport and delivery of respiratory gases, while at the same time restricting water loss.

Closed Circulatory Systems

All vertebrates and some invertebrates have a **closed circulatory system** (Fig. 23.2b), which is more properly called a **cardiovascular system** because it consists of a strong muscular heart and blood vessels. In humans, the heart has two receiving chambers, called atria (sing., atrium), and two pumping chambers, called ventricles. There are three kinds of vessels: arteries, which carry blood away from the heart; capillaries, which exchange materials with tissue fluid; and veins, which return blood to the heart. Blood is always contained within these vessels and never runs freely into the body unless an injury occurs.

As blood passes through capillaries, the pressure of blood forces some water out of the blood and into the tissue fluid. Some of this fluid returns directly to a capillary, and some is picked up by lymphatic capillaries in the vicinity. The fluid, now called lymph, is returned to the cardiovascular system by lymphatic vessels. The function of the lymphatic system is discussed in Section 23.2.

Figure 23.2   Open versus closed circulatory systems.

a. The grasshopper has an open circulatory system. Hemolymph freely bathes the internal organs. The heart, a pump, keeps the hemolymph moving, but an open system probably could not rapidly supply oxygen to wing muscles. These muscles receive oxygen directly from tracheae (air tubes). **b.** Vertebrates and some invertebrates have a closed circulatory system. The heart pumps blood into the arteries, which take blood away from the heart to the capillaries where exchange takes place. Veins then return blood to the heart.

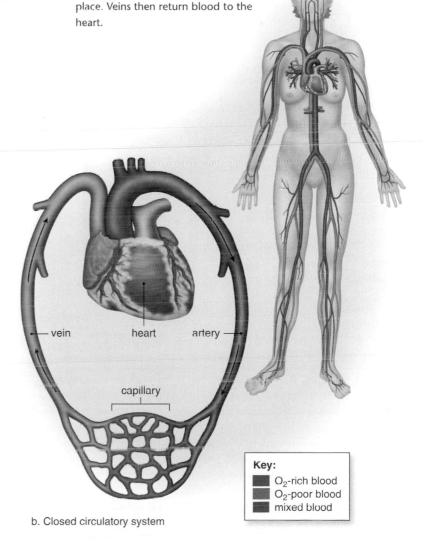

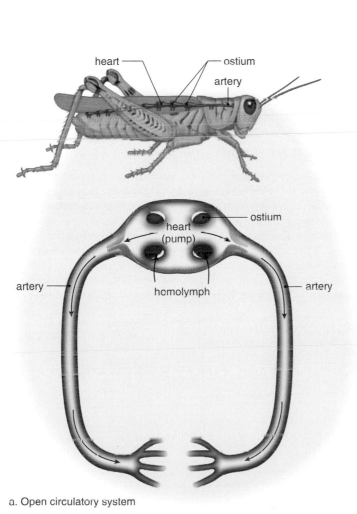

a. Open circulatory system

b. Closed circulatory system

Key:
- O$_2$-rich blood
- O$_2$-poor blood
- mixed blood

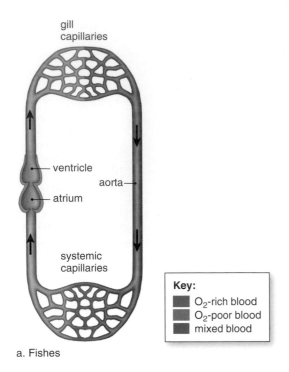

a. Fishes

Key:
- ■ O$_2$-rich blood
- ■ O$_2$-poor blood
- ■ mixed blood

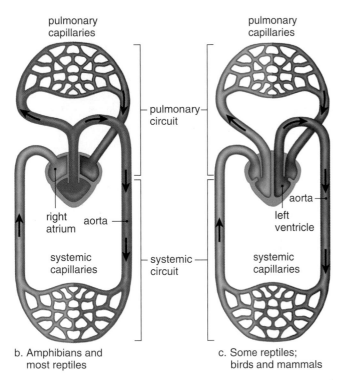

b. Amphibians and most reptiles

c. Some reptiles; birds and mammals

Figure 23.3 **Comparison of circulatory circuits in vertebrates.**

a. In a fish, the blood moves in a single circuit. The heart has a single atrium and ventricle, and it pumps the blood into the gill region, where gas exchange takes place. Blood pressure created by the pumping of the heart is dissipated after the blood passes through the gill capillaries. **b.** Amphibians and most reptiles have a two-circuit system in which the heart pumps blood to both the lungs and the body itself. There is a single ventricle, and some mixing of O$_2$-rich and O$_2$-poor blood takes place. **c.** The pulmonary and systemic circuits are completely separate in crocodiles (a reptile) and in birds and mammals. The right side pumps blood to the lungs, and the left side pumps blood to the body proper.

Comparison of Circulatory Pathways

Two different types of circulatory pathways are seen among vertebrate animals. In fishes, blood follows a one-circuit (single-loop) pathway through the body. The heart has a single atrium and a single ventricle (Fig. 23.3*a*). The pumping action of the ventricle sends blood under pressure to the gills, where gas exchange occurs. After passing through the gills, blood is under reduced pressure and flow. However, this single circulatory loop has advantages in that the gill capillaries receive O$_2$-poor blood and the systemic capillaries receive O$_2$-rich blood.

As a result of evolutionary changes, the other vertebrates have a two-circuit (double-loop) circulatory pathway. The heart pumps blood to the tissues, called a **systemic circuit,** and also pumps blood to the lungs, called a **pulmonary circuit.** This double pumping action is seen in terrestrial animals that utilize lungs to breathe air.

In amphibians, the heart has two atria, but only a single ventricle (Fig. 23.3*b*), and some mixing of O$_2$-rich and O$_2$-poor blood does occur. The same holds true for most reptiles, except that the ventricle has a partial septum. The hearts of crocodiles, which are reptiles, and those of all birds and mammals are divided into right and left halves (Fig. 23.3*c*). The right ventricle pumps blood to the lungs, and the left ventricle, which is larger than the right ventricle, pumps blood to the rest of the body. This arrangement provides adequate blood pressure for both the pulmonary and systemic circuits.

Check Your Progress

1. Compare and contrast an open circulatory system with a closed circulatory system.
2. List and describe the functions of the three types of vessels in a cardiovascular system.
3. Contrast a one-circuit circulatory pathway with a two-circuit pathway.

Answers: 1. Both use a heart to pump fluid. An open system pumps hemolymph through channels and cavities. The hemolymph eventually drains back to the heart. A closed system pumps blood through vessels that carry blood both away from and back to the heart. 2. Arteries carry blood away from the heart, capillaries exchange their contents with tissue fluid, and veins return blood back to the heart. 3. The one-circuit pathway utilizes a heart with one atrium and one ventricle to send blood to the gill capillaries and then to the systemic capillaries in a single loop. The two-circuit pathway pumps blood to both the pulmonary and systemic capillaries simultaneously.

23.2 Transport in Humans

In the human cardiovascular system, like that of other vertebrates, the heart pumps blood into blood vessels that take it to capillaries where exchanges take place. In the lungs, carbon dioxide is exchanged for oxygen, and in the tissues, nutrients and oxygen are exchanged for carbon dioxide and other wastes. These exchanges in the lungs and tissues are so important that if the heart stops pumping, death results.

The Human Heart

In humans, the heart is a double pump: The right side of the heart pumps O_2-poor blood to the lungs, and the left side of the heart pumps O_2-rich blood to the tissues (Fig. 23.4). (The "right side of the heart" refers to how the heart is positioned in your body and not to the right side of the diagram.) The heart can be a double pump because a **septum** separates the right from the left side. Further, the septum is complete and prevents O_2-poor blood from mixing with O_2-rich blood.

Each side of the heart has two chambers. The upper, thin-walled chambers are called atria (sing., **atrium**), and they receive blood. The lower chambers are the thick-walled **ventricles,** which pump the blood away from the heart.

Valves occur between the atria and the ventricles, and between the ventricles and attached vessels. Because these valves close after the blood moves through, they keep the blood moving in the correct direction. The valves between the atria and ventricles are called the **atrioventricular valves,** and the valves between the ventricles and their attached vessels are called **semilunar valves** because their cusps look like half moons.

The right atrium receives blood from attached veins (the **venae cavae**) that are returning O_2-poor blood to the heart from the tissues. After the blood passes through an atrioventricular valve (also called the *tricuspid valve*), the right ventricle pumps it through the *pulmonary semilunar valve* into the **pulmonary trunk and arteries** that take it to the lungs. The **pulmonary veins** bring O_2-rich blood to the left atrium. After this blood passes through an atrioventricular valve (also called the *bicuspid valve*), the left ventricle pumps it through the *aortic semilunar valve* into the **aorta,** which takes it to the tissues.

Like mechanical valves, the heart valves are sometimes leaky; they may not close properly, and there is a backflow of blood. A **heart murmur** is often due to leaky atrioventricular valves, which allow blood to pass back into the atria after they have closed. Rheumatic fever is a bacterial infection that begins in the throat and spreads throughout the body. The bacteria attack various organs, including the heart valves. When damage is severe, the valve can be replaced with a synthetic valve or one taken from a pig's heart.

Another observation is in order. Some people associate O_2-poor blood with all veins and O_2-rich blood with all arteries, but this idea is incorrect: Pulmonary arteries and pulmonary veins are just the reverse. That's why the pulmonary arteries are colored blue and the pulmonary veins are colored red in Figure 23.4. The correct definitions are that an **artery** is a vessel that takes blood away from the heart, and a **vein** is a vessel that takes blood to the heart.

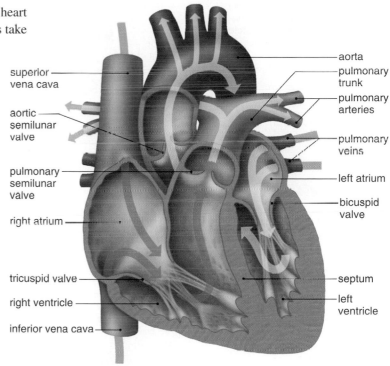

Figure 23.4   Heart anatomy and path of blood through the heart.

The right side of the heart receives and pumps O_2-poor blood to the lungs, and therefore is colored blue. The left side of the heart receives and pumps O_2-rich blood to tissues.

Check Your Progress

1. Contrast atria with ventricles.
2. Trace the path of blood through the heart from the venae cavae to the aorta.
3. What is the purpose of the heart valves?

Answers: 1. Atria are located in the upper part of the heart, are thin-walled, and receive blood. Ventricles, located in the lower part of the heart, are thick-walled, and pump blood. 2. From the body: venae cavae, right atrium, tricuspid valve, right ventricle, pulmonary semilunar valve, pulmonary trunk and arteries. From the lungs: pulmonary veins, left atrium, bicuspid valve, left ventricle, aortic semilunar valve, aorta. 3. Heart valves prevent backward flow, and therefore keep the blood moving in the right direction.

Time	Atria	Ventricles
0.15 sec	Systole	Diastole
0.30 sec	Diastole	Systole
0.40 sec	Diastole	Diastole

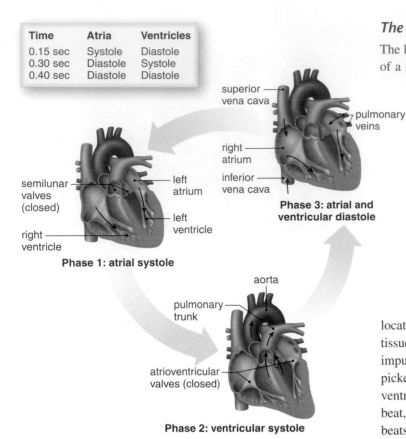

Phase 1: atrial systole

Phase 2: ventricular systole

Phase 3: atrial and ventricular diastole

Figure 23.5   The heartbeat.

A heartbeat is a cycle of events. Phase 1: The atria contract and pass blood to the ventricles. Phase 2: The ventricles contract and blood moves into the attached arteries. Phase 3: Both the atria and ventricles relax while the heart fills with blood.

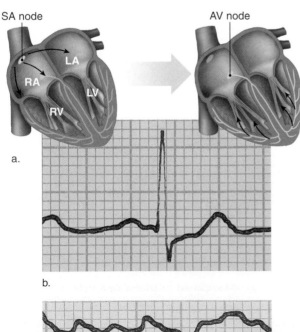

a.

b.

c.

The Heartbeat

The heart's pumping action, known as the heartbeat or **cardiac cycle,** consists of a series of events: First the atria contract, then the ventricles contract, and then they both rest. Figure 23.5 lists and depicts the events of a heartbeat, using the term **systole** to mean contraction and **diastole** to mean relaxation. When the heart beats, the familiar lub-dub sound is caused by the closing of the heart valves. The longer and lower-pitched *lub* occurs when the atrioventricular valves close, and the shorter and sharper *dub* is heard when the semilunar valves close. The **pulse** is a wave effect that passes down the walls of the arterial blood vessels when the aorta expands and then recoils following the ventricular systole. Because there is one pulse per ventricular systole, the pulse rate can be used to determine the heart rate. The heart beats about 70 times per minute, although a normal adult heart rate can vary from 60 to 100 beats per minute.

The beat of the heart is regular because it has an intrinsic pacemaker, called the **SA (sinoatrial) node.** The nodal tissue of the heart, located in two regions of the atrial wall, is a unique type of cardiac muscle tissue. Every 0.85 seconds, the SA node automatically sends out an excitation impulse that causes the atria to contract (Fig. 23.6*a*). When this impulse is picked up by the **AV (atrioventricular) node,** it passes to fibers that cause the ventricles to contract. If the SA node fails to work properly, the ventricles still beat, due to impulses generated by the AV node, but the beat is slower (40–60 beats per minute). To correct this condition, it is possible to implant an artificial pacemaker, which automatically gives an electrical stimulus to the heart every 0.85 seconds.

Although the beat of the heart is intrinsic, it is regulated by the nervous system and various hormones. Activities such as yoga and meditation lead to activation of the vagus nerve, which slows the heart rate. Exercise or anxiety leads to release of the hormones norepinephrine and epinephrine, which cause it to speed up.

An **electrocardiogram (ECG)** is a recording of the electrical changes that occur in the wall of the heart during a cardiac cycle. Body fluids contain ions that conduct electrical currents, and therefore the electrical changes in heart muscle can be detected on the skin's surface. When an electrocardiogram is being taken, electrodes placed on the skin are connected by wires to an instrument that detects these electrical changes (Fig. 23.6*b*). Various types of abnormalities can be detected by an electrocardiogram. One of these, called ventricular fibrillation, is due to uncoordinated contraction of the ventricles (Fig. 23.6*c*). Ventricular fibrillation is of special interest because it can be caused by an injury or a drug overdose. It is the most common cause of sudden cardiac death in a seemingly healthy person. Once the ventricles are fibrillating, they must be defibrillated by applying a strong electrical current for a short period of time. Then the SA node may be able to reestablish a coordinated beat.

Figure 23.6   Control of the heartbeat.

a. The beat of the heart occurs regularly because the SA node (called the pacemaker) automatically sends out an impulse that causes the atria (RA, LA) to contract and is picked up by the AV node. Thereafter, the ventricles (RV, LV) contract. **b.** An electrocardiogram records the electrical changes that occur as the heart beats. The large spike is associated with ventricular activation. **c.** Ventricular fibrillation produces an irregular ECG.

Figure 23.7   Blood vessels.

a. Arteries have well-developed walls with a thick middle layer of elastic fibers and smooth muscle.
b. Capillary walls are composed of an epithelium only one cell thick. **c.** Veins have flabby walls, particularly because the middle layer is not as thick as in arteries. Veins have valves, which point toward the heart.

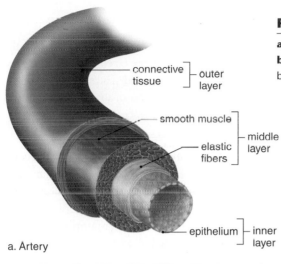

connective tissue — outer layer

smooth muscle
elastic fibers — middle layer

epithelium — inner layer

a. Artery

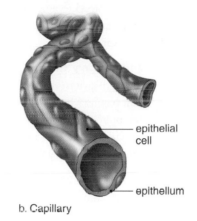

epithelial cell

epithellum

b. Capillary

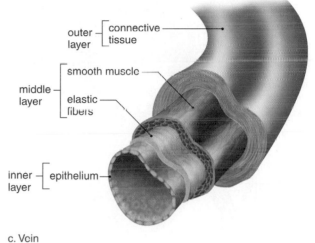

outer layer — connective tissue

smooth muscle
middle layer — elastic fibers

inner layer — epithelium

c. Vein

Blood Vessels

Arteries transport blood away from the heart. When the heart contracts, blood is sent under pressure into the arteries; thus, **blood pressure** accounts for the flow of blood in the arteries. Arteries have a much thicker wall than veins because of a well-developed middle layer composed of smooth muscle and elastic fibers. The elastic fibers allow arteries to expand and accommodate the sudden increase in blood volume that results after each heartbeat. The smooth muscle strengthens the wall and prevents overexpansion (Fig. 23.7a).

Arteries branch into arterioles. **Arterioles** are small arteries just visible to the naked eye, and their diameter can be regulated by the nervous system, depending on the needs of the body. When arterioles are dilated, more blood flows through them, and when they are constricted, less blood flows. The constriction of arterioles can also raise blood pressure.

Arterioles branch into **capillaries,** which are extremely narrow, microscopic tubes with a wall composed of only epithelium, often called endothelium because it's inside the other layers (Fig. 23.7b). Capillaries, which are usually so narrow that red blood cells pass through in single file, allow exchange of nutrient and waste molecules across their thin walls. Capillary beds (many capillaries interconnected) are so prevalent that in humans, all cells are within 60–80 micrometers (μm) of a capillary. The capillaries are so extensive that blood pressure drops and blood flows only slowly along. The entrance to a capillary bed is controlled by bands of muscle called precapillary sphincters. During muscular exercise, the sphincters relax, and the capillary beds of the muscles are open. Also, after an animal has eaten, the capillary beds in the digestive tract are open.

Venules and **veins** collect blood from the capillary beds and return it to the heart. First the venules drain the blood from the capillaries, and then they join to form a vein. Blood pressure is much reduced by the time blood reaches the veins. The walls of veins are much thinner and their diameter wider than those of arteries (Fig. 23.7c). This allows skeletal muscle contraction to push on the veins, forcing the blood past a valve (Fig. 23.8). Valves within the veins point, or open, toward the heart, preventing a backflow of blood when they close. When inhalation occurs, the thoracic pressure falls and abdominal pressure rises as the chest expands. This action also aids the flow of venous blood back to the heart because blood flows in the direction of reduced pressure.

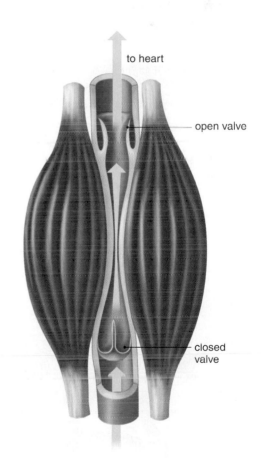

to heart

open valve

closed valve

Figure 23.8   Movement of blood in a vein.

Pressure on the walls of a vein, exerted by skeletal muscles, increases blood pressure within the vein and forces a valve open. Closure of the valves prevents the blood from flowing in the opposite direction.

The Pulmonary and Systemic Circuits

The human cardiovascular system includes two major circular pathways, the pulmonary circuit and the systemic circuit. The **pulmonary circuit** moves blood to and from the lungs, where O_2-poor becomes O_2-rich blood. The **systemic circuit** moves blood to and from the body proper. The function of the systemic circuit is to serve the needs of the body's cells. Figure 23.9 traces the path of blood in both circuits.

Pulmonary Circuit O_2-poor blood from all regions of the body collects in the right atrium and then passes into the right ventricle (RV). The pulmonary circuit begins when the RV pumps blood to the lungs via the pulmonary trunk and the pulmonary arteries. As blood passes through pulmonary capillaries, carbon dioxide is given off and oxygen is picked up. O_2-rich blood returns to the heart via the pulmonary veins. Pulmonary veins enter the left atrium.

Systemic Circuit O_2-rich blood enters the left atrium from the lungs and passes into the left ventricle (LV). The systemic circuit begins when the LV pumps the blood into the **aorta.** Arteries branching from the aorta carry blood to all areas and organs of the body, where it passes through capillaries and collects in veins. Veins converge on the **venae cavae,** which return the O_2-poor blood to the right atrium. In the systemic circuit, arteries contain O_2-rich blood

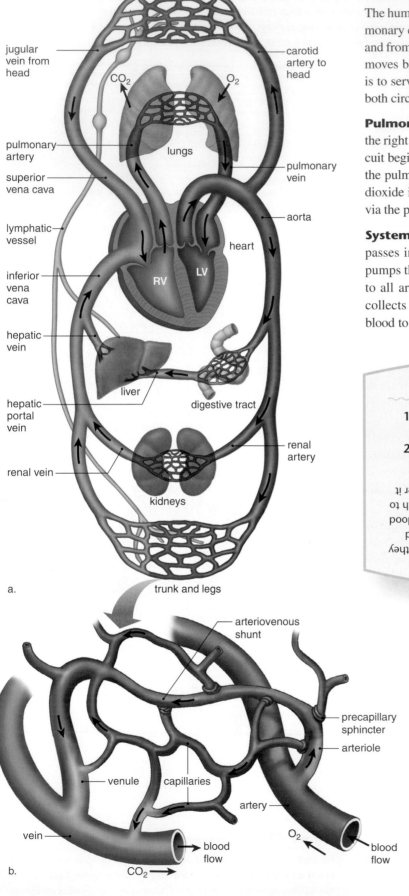

a.

b.

Check Your Progress

1. **Explain what happens during a heartbeat and what makes the familiar *lub-dub* sounds.**
2. **Outline the flow of blood through blood vessels.**

Answers: 1. First the atria contract, then the ventricles contract, and then they both rest. The *lub* sound occurs when the atrioventricular valves close, and the *dub* sound occurs when the semilunar valves close. 2. Arteries carry blood away from the heart to small arteries called arterioles. The arterioles branch to form small capillaries. Venules collect blood from the capillaries and deliver it to veins, which transport it back to the heart.

Figure 23.9   Path of blood.

a. Overview of cardiovascular system. When tracing blood from the right to the left side of the heart in the pulmonary circuit, you must include the pulmonary vessels. When tracing blood from the digestive tract to the right atrium in the systemic circuit, you must include the hepatic portal vein, the hepatic vein, and the inferior vena cava. **b.** To move from an artery to a vein, blood must move through a capillary bed where exchanges occur between blood and tissue fluid. When precapillary sphincters shut down a capillary bed, blood moves through a thoroughfare vessel (called an arteriovenous shunt) from arteriole to venule.

and have a bright red color, and veins contain O_2-poor blood and appear dull red or, when viewed through the skin, blue.

A **portal system** begins and ends in capillaries. For example, the hepatic portal takes blood from the intestines to the liver. The liver, an organ of homeostasis, modifies substances absorbed by the intestines and monitors the normal composition of the blood. The hepatic vein leaves the liver and enters the inferior vena cava.

Lymphatic System

The **lymphatic system** consists of lymphatic vessels and various lymphatic organs. Just now we are interested in the lymphatic vessels that take up excess tissue fluid and return it to cardiovascular veins in the shoulders, namely the subclavian veins (Fig. 23.10). The lymphatic vessels also take up fat in the form of lipoproteins at the digestive tract and transport it to these veins. As you will see in Chapter 26, the lymphatic system works with the immune system to help defend the body against disease.

Lymphatic vessels are quite extensive; most regions of the body are richly supplied with lymphatic capillaries. The construction of the larger lymphatic vessels is similar to that of cardiovascular veins, including the presence of valves. Also, the movement of lymph within these vessels is dependent upon skeletal muscle contraction. When the muscles contract, the lymph is squeezed past a valve that closes, preventing the lymph from flowing backward.

The lymphatic system is a one-way system that begins at the lymphatic capillaries. These capillaries take up fluid that has diffused from and not been reabsorbed by the blood capillaries. Once tissue fluid enters the lymphatic vessels, it is called **lymph.** The lymphatic capillaries join to form larger lymphatic vessels that merge before entering one of two ducts: the thoracic duct or the right lymphatic duct. The thoracic duct is much larger than the right lymphatic duct. It serves the lower limbs, the abdomen, the left arm, and the left side of both the head and the neck. The right lymphatic duct serves the right arm, the right side of both the head and the neck, and the right thoracic area. The lymphatic ducts enter the subclavian veins.

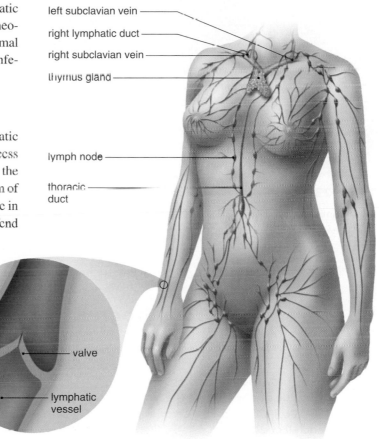

Figure 23.10   Lymphatic vessels.

Lymphatic vessels drain excess fluid from the tissues and return it to the cardiovascular system. The enlargement shows that lymphatic vessels, like cardiovascular veins, have valves to prevent backward flow. Lymph nodes filter lymph and remove impurities.

Cardiovascular Disorders

Cardiovascular disease is the leading cause of untimely death in Western countries. In the United States, it is estimated that about 20% of the population suffers from **hypertension,** which is high blood pressure. Normal blood pressure is 120/80 mm Hg. The top number is called *systolic* blood pressure because it is due to the contraction of the ventricles, and the bottom number is called *diastolic* blood pressure because it is due to the resting of the ventricles. Hypertension occurs when blood pressure readings are higher than these numbers—say, 160/100. Hypertension is sometimes called a silent killer because it may not be detected until a stroke or heart attack occurs.

Inheritance and lifestyle contribute to hypertension. For example, hypertension is often seen in individuals who have atherosclerosis (formerly called arteriosclerosis), an accumulation of soft masses of fatty materials, particularly cholesterol, beneath the inner linings of arteries

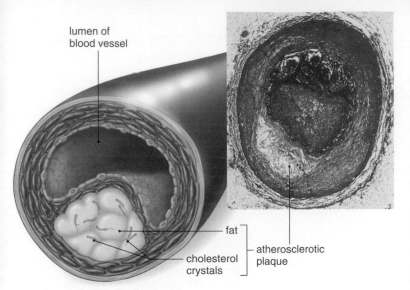

lumen of
blood vessel

fat

cholesterol
crystals

atherosclerotic
plaque

Figure 23.11   Plaque.

Plaque is an irregular accumulation of cholesterol and fat that closes off blood vessels. Of special concern is the closure of coronary arteries.

(Fig. 23.11). Such deposits, called **plaque,** tend to protrude into the lumen of the vessel, interfering with the flow of blood. Atherosclerosis begins in early adulthood and develops progressively through middle age, but symptoms may not appear until an individual is 50 or older. To prevent its onset and development, the American Heart Association and other organizations recommend a diet low in saturated fat and cholesterol and rich in fruits and vegetables. Smoking, drug or alcohol abuse, obesity, and lack of exercise contribute to the risk of atherosclerosis.

Plaque can cause a clot to form on the irregular arterial wall. As long as the clot remains stationary, it is called a *thrombus*, but when and if it dislodges and moves along with the blood, it is called an *embolus*. If *thromboembolism* is not treated, serious health problems can result. A cardiovascular accident, also called a **stroke,** often occurs when a small cranial arteriole bursts or is blocked by an embolus. Lack of oxygen causes a portion of the brain to die, and paralysis or death can result. A person is sometimes forewarned of a stroke by a feeling of numbness in the hands or the face, difficulty in speaking, or temporary blindness in one eye. If a coronary artery becomes completely blocked due to thromboembolism, a heart attack can result.

The coronary arteries bring O_2-rich blood from the aorta to capillaries in the wall of the heart, and the cardiac veins return O_2-poor blood from the capillaries to the right ventricle. If the coronary arteries narrow due to cardiovascular disease, the individual may first suffer from angina pectoris, chest pain that is often accompanied by a radiating pain in the left arm. When a coronary artery is completely blocked, a portion of the heart muscle dies due to a lack of oxygen. This is known as a **heart attack.** Two surgical procedures are frequently performed to correct a blockage or facilitate blood flow. In a coronary bypass operation, a portion of a blood vessel from another part of the body is sutured from the aorta to the coronary artery, past the point of obstruction (Fig. 23.12*a*). Now blood flows normally again from the aorta to the wall of the heart. In balloon angioplasty, a plastic tube is threaded through an artery to the blockage, and a balloon attached to the end of the tube is inflated to break through the blockage. A stent is often used to keep the vessel open (Fig. 23.12*b*).

Check Your Progress

1. The blockage of which vessels in the systemic circuit lead to a heart attack? Describe the two surgical procedures used to correct blockage of these arteries.
2. Describe the function of the lymphatic system.
3. What conditions might occur as a result of hypertension and plaque?

Answers: 1. The coronary arteries that serve the heart. In balloon angioplasty, a balloon is inserted into the artery and inflated to break the blockage. In a coronary bypass operation, a piece of a blood vessel from another part of the body is attached to the artery so that blood flows through the new section rather than through the blocked area. 2. The lymphatic system takes up excess tissue fluid and returns it to cardiovascular veins. 3. Thromboembolism, stroke, heart attack.

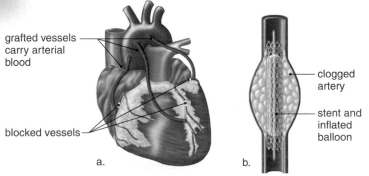

grafted vessels
carry arterial
blood

clogged
artery

stent and
inflated
balloon

blocked vessels

a.   b.

Figure 23.12   Treatment for clogged coronary arteries.

(*left*) Many heart procedures, such as coronary bypass and stent insertion, can be performed using robotic surgery techniques. **a.** During a coronary bypass operation, blood vessels (usually veins from the leg) are stitched to the heart, taking blood past the region of obstruction. **b.** During stenting, a cylinder of expandable metal mesh is positioned inside the coronary artery by using a catheter. Then, a balloon is inflated so that the stent expands and opens the artery.

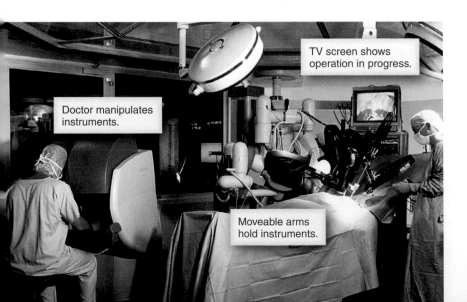

TV screen shows
operation in progress.

Doctor manipulates
instruments.

Moveable arms
hold instruments.

23.3 Blood: A Transport Medium

The **blood** of mammals helps maintain homeostasis. Blood's numerous functions include the following:

1. Transports substances to and from the capillaries, where exchanges with tissue fluid take place.
2. Helps defend the body against invasion by pathogens (e.g., disease-causing viruses and bacteria).
3. Helps regulate body temperature.
4. Forms clots, preventing a potentially life-threatening loss of blood.

In humans, blood has two main portions: the liquid portion, called plasma, and the formed elements, consisting of various cells and platelets (Fig. 23.13).

Plasma

Plasma is composed mostly of water (90–92%) and proteins (7–8%), but it also contains smaller quantities of many types of molecules, including nutrients, wastes, and salts. The salts and proteins are involved in buffering the blood, effectively keeping the pH near 7.4. They also maintain the blood's osmotic pressure so that water has an automatic tendency to enter blood capillaries. Several plasma proteins are involved in blood clotting, and others transport large organic molecules in the blood. Albumin, the most plentiful of the plasma proteins, transports bilirubin, a breakdown product of hemoglobin. Lipoproteins transport cholesterol.

Formed Elements

The **formed elements** are red blood cells, white blood cells, and platelets. Among the formed elements, **red blood cells,** also called erythrocytes, transport oxygen. Red blood cells are small, biconcave disks that at maturity lack a nucleus and contain the respiratory pigment hemoglobin (Fig. 23.14). There are 6 million red blood cells per mm^3 of whole blood, and each one of these cells contains about 250 million hemoglobin molecules. **Hemoglobin** contains iron, which combines loosely with oxygen; in this way, red blood cells transport oxygen. If the number of red blood cells is insufficient, or if the cells do not have enough hemoglobin, the individual suffers from anemia and has a tired, run-down feeling.

Red blood cells are manufactured continuously within certain bones, namely the skull, the ribs, the vertebrae, and the ends of the long bones. The hormone *erythropoietin* stimulates the production of red blood cells. The kidneys produce erythropoietin when they act on a precursor made by the liver. Now available as a drug, erythropoietin is helpful to persons with anemia and has also been abused by athletes to enhance their performance.

Before they are released from the bone marrow into the blood, red blood cells lose their nuclei and begin to synthesize hemoglobin. After living about 120 days, they are destroyed chiefly in the liver and the spleen, where they are engulfed by large phagocytic cells. When red blood cells are destroyed, hemoglobin is released. The iron is recovered and returned to the red bone marrow for reuse. Other portions of the molecules (i.e., heme) undergo chemical degradation and are excreted by the liver as bile pigments in the bile. The bile pigments are primarily responsible for the color of feces.

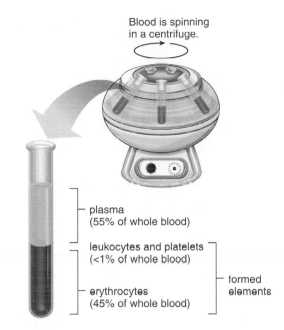

Blood is spinning in a centrifuge.

plasma
(55% of whole blood)

leukocytes and platelets
(<1% of whole blood)

erythrocytes
(45% of whole blood)

formed elements

Figure 23.13 **Components of blood.**

When blood settles in a test tube without clotting, it is apparent that it is composed of plasma and formed elements.

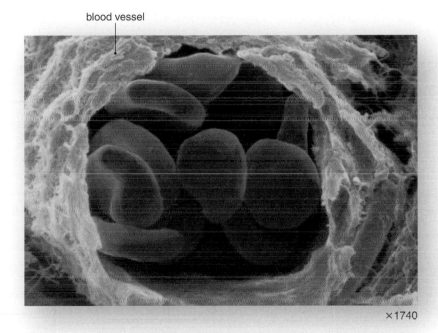

blood vessel

×1740

Figure 23.14 **Micrograph of red blood cells.**

Red blood cells contain hemoglobin, a red pigment that accounts for the red color of blood.

Formed Elements	Function
Red Blood Cells (erythrocytes) 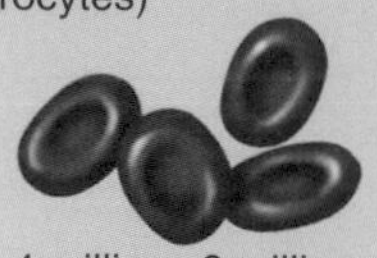 4 million–6 million per mm³ blood	Transport O_2 and help transport CO_2.
White Blood Cells (leukocytes) 5,000–11,000 per mm³ blood	Fight infection.
Granular leukocytes • Neutrophils 40–70%	Phagocytize pathogens and cellular debris.
• Eosinophils 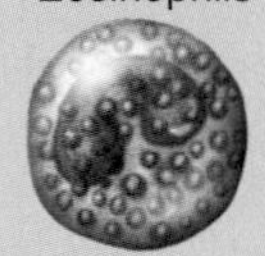 1–4%	Use granule contents to digest large pathogens, such as worms, and reduce inflammation.
• Basophils 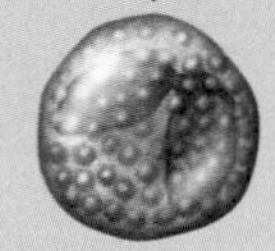 0–1%	Promote blood flow to injured tissues and the inflammatory response.
Agranular leukocytes • Lymphocytes 20–45%	Responsible for specific immunity; B cells produce antibodies; T cells destroy cancer and virus-infected cells.
• Monocytes 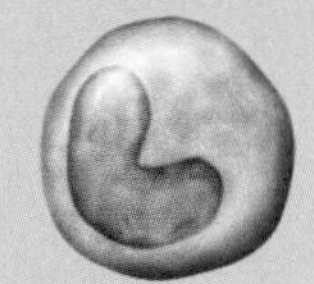 4–8%	Become macrophages that phagocytize pathogens and cellular debris.
Platelets (thrombocytes)  150,000–300,000 per mm³ blood	Aid clotting.

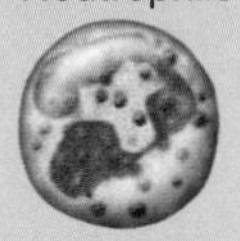

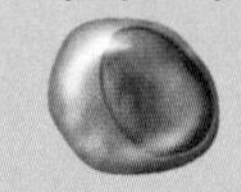

• Appearance with Wright's stain.

Figure 23.15 Formed elements.

White blood cells are quite varied and have different functions that are all associated with defense of the body against infections.

White blood cells, also called leukocytes, help fight infections. White blood cells differ from red blood cells in that they are usually larger and have a nucleus, they lack hemoglobin, and without staining, they appear translucent. Figure 23.15 shows the appearance of the various types of white blood cells. With staining, white blood cells appear light blue unless they have granules that bind with certain stains. The white blood cells that have granules also have a lobed nucleus. The agranular leukocytes have no granules and a spherical or indented nucleus. There are approximately 5,000–11,000 white blood cells per mm³ of blood. Growth factors are available to increase the production of all white blood cells, and these are helpful to people with low immunity, such as AIDS patients.

Red blood cells are confined to the blood, but white blood cells are able to squeeze between the cells of a capillary wall. Therefore, they are found in tissue fluid and lymph and in lymphatic organs. When an infection is present, white blood cells greatly increase in number. Many white blood cells live only a few days—they probably die while engaging pathogens. Others live months or even years.

When microorganisms enter the body due to an injury, the response is called an *inflammatory response* because swelling, reddening, heat, and pain occur at the injured site. Damaged tissue releases kinins, which dilate capillaries, and histamines, which increase capillary permeability. White blood cells called **neutrophils,** which are amoeboid, squeeze through the capillary wall and enter the tissue fluid, where they phagocytize foreign material. White blood cells called **monocytes** appear and are transformed into **macrophages,** large phagocytizing cells that release white blood cell growth factors. Soon the number of white blood cells increases explosively. A thick, yellowish fluid called pus contains a large proportion of dead white blood cells that have fought the infection.

Lymphocytes, another type of white blood cell, also play an important role in fighting infection. Certain lymphocytes called T cells attack infected cells that contain viruses. Other lymphocytes, called B cells, produce antibodies. Each B cell produces just one type of antibody, which is specific for one type of antigen. An **antigen,** which is most often a protein but sometimes a polysaccharide, causes the body to produce an **antibody** because the antigen doesn't belong to the body. Antigens are present in the outer covering of parasites or in their toxins. When antibodies combine with antigens, the complex is often phagocytized by a macrophage. An individual is actively immune when a large number of B cells are all producing the antibody needed for a particular infection.

Platelets and Blood Clotting

Platelets (also called thrombocytes) result from fragmentation of certain large cells, called megakaryocytes, in the red bone marrow. Platelets are produced at a rate of 200 billion a day, and the blood contains 150,000–300,000 per mm³. These formed elements are involved in blood clotting, or coagulation.

Blood contains at least 12 clotting factors that participate in the formation of a blood clot. *Hemophilia* is an inherited clotting disorder in which the

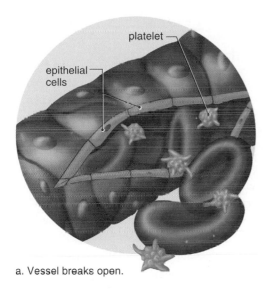

a. Vessel breaks open.

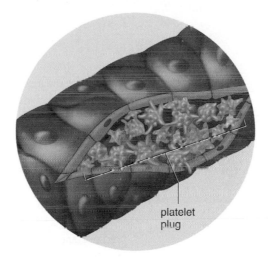

b. Platelet plug forms.

Figure 23.16   Blood clotting.

a. When a capillary is injured, blood begins leaking out. **b.** Platelets congregate to form a platelet plug, and this temporarily seals the leak. **c.** Platelets and damaged tissue cells release an activator that sets in motion a series of reactions that end with (**d**) a blood clot.

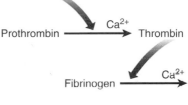

Platelets and damaged tissue cells release Prothrombin activator

Prothrombin $\xrightarrow{Ca^{2+}}$ Thrombin

Fibrinogen $\xrightarrow{Ca^{2+}}$ Fibrin threads (Red blood cells are trapped among fibrin threads.)

c. Clotting occurs.

liver is unable to produce one of the clotting factors. The slightest bump can cause the affected person to bleed into the joints, and this leads to degeneration of the joints. Bleeding into muscles can lead to nerve damage and muscular atrophy. The most frequent cause of death due to hemophilia is bleeding into the brain.

Prothrombin and fibrinogen, two proteins involved in blood clotting, are manufactured and deposited in blood by the liver. Vitamin K, found in green vegetables and also formed by intestinal bacteria, is necessary for the production of prothrombin, and if by chance this vitamin is missing from the diet, hemorrhagic disorders can develop.

A series of reactions leads to the formation of a blood clot (Fig. 23.16a–c). When a blood vessel in the body is damaged, platelets clump at the site of the puncture and form a plug that temporarily seals the leak. Platelets and the injured tissues release a clotting factor called prothrombin activator that converts prothrombin to thrombin. This reaction requires calcium ions (Ca^{2+}). **Thrombin,** in turn, acts as an enzyme that severs two short amino acid chains from each fibrinogen molecule. These activated fragments then join end to end, forming long threads of fibrin. Fibrin threads wind around the platelet plug in the damaged area of the blood vessel and provide the framework for the clot. Red blood cells also are trapped within the fibrin threads; these cells make a clot appear red (Fig. 23.16d). Clot retraction follows, during which the clot gets smaller as platelets contract. A fluid called serum is squeezed from the clot. A fibrin clot is present only temporarily. As soon as blood vessel repair is initiated, an enzyme called plasmin destroys the fibrin network and restores the fluidity of plasma.

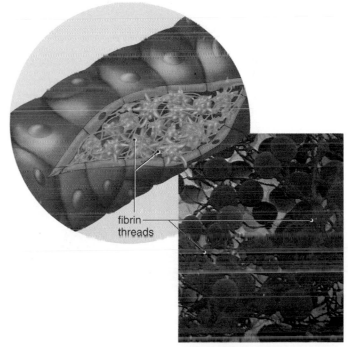

fibrin threads

d. Blood clot is present.   × 1430

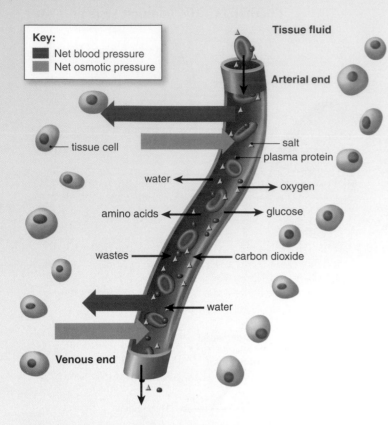

Key:
- ■ Net blood pressure
- ■ Net osmotic pressure

Tissue fluid

Arterial end

tissue cell

salt

plasma protein

water

oxygen

amino acids

glucose

wastes

carbon dioxide

water

Venous end

Figure 23.17 **Capillary exchange.**

A capillary, illustrating the exchanges that take place and the forces that aid the process. At the arterial end of a capillary (*top*), the blood pressure is higher than the osmotic pressure; therefore, water tends to leave the bloodstream. In the midsection, molecules, including oxygen and carbon dioxide, follow their concentration gradients. At the venous end of a capillary (*bottom*), the osmotic pressure is higher than the blood pressure; therefore, water tends to enter the bloodstream. Notice that the red blood cells and the plasma proteins are too large to exit a capillary.

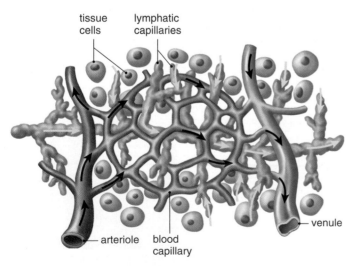

tissue cells

lymphatic capillaries

venule

arteriole

blood capillary

Figure 23.18 **Lymphatic capillary bed.**

A lymphatic capillary bed lies near a blood capillary bed. The heavy black arrows show the flow of blood. The yellow arrows show that lymph is formed when lymphatic capillaries take up excess tissue fluid.

Capillary Exchange in the Tissues

In Figure 23.1, we noted that in some animals, exchanges are carried out by each cell individually because there is no cardiovascular system. When an animal does have a cardiovascular system, tissue fluid (the fluid between the cells) makes exchanges with blood within a capillary. Notice in Figure 23.17 that amino acids, oxygen, and glucose exit a capillary and enter tissue fluid to be used by cells. On the other hand, carbon dioxide and wastes exit tissue fluid and enter a capillary to be taken away and excreted from the body. A chief purpose of the cardiovascular system is to take blood to the capillaries where exchange occurs. Without this exchange, homeostasis is not maintained, and the cells of the body perish.

Figure 23.17 illustrates certain mechanics of capillary exchange. Blood pressure and osmotic pressure are two opposing forces at work along the length of a capillary. Blood pressure is caused by the beating of the heart, while osmotic pressure is due to the salt and protein content of the blood. Blood pressure holds sway at the arterial end of a capillary and water exits. Blood pressure is reduced by the time blood reaches the venous end of a capillary, and osmotic pressure now causes water to enter. Midway between the arterial and venous ends of a capillary, blood pressure pretty much equals osmotic pressure, and passive diffusion alone causes nutrients to exit and wastes to enter. This works because tissue fluid always has the lesser amount of nutrients and the greater amount of wastes. After all, cells use nutrients and thereby create wastes.

The exchange of water at a capillary is not exact, and the result is always excess tissue fluid. Excess tissue fluid is collected by lymphatic capillaries, and in this way it becomes lymph (Fig. 23.18). Lymph contains all the components of plasma except much lesser amounts of protein. Lymph is returned to the cardiovascular system when the major lymphatic vessels enter the subclavian veins in the shoulder region.

In addition to nutrients and wastes, blood distributes heat to body parts. When you are warm, many capillaries that serve the skin are open, and your face is flushed. This helps rid the body of excess heat. When you are cold, skin capillaries close, conserving heat.

Check Your Progress

1. List the functions of blood.
2. Contrast red blood cells with white blood cells.
3. Describe how a blood clot forms.

Answers: 1. Blood transports substances to and from the capillaries, defends against pathogen invasion, helps regulate body temperature, and forms clots to prevent excessive blood loss. 2. Red blood cells are smaller, lack a nucleus, contain hemoglobin, and are red in color. White blood cells are larger, have a nucleus, do not contain hemoglobin, and are translucent in appearance. 3. Platelets accumulate at the site of injury and release a clotting factor that results in the synthesis of thrombin. Thrombin synthesizes fibrin threads that provide a framework for the clot.

THE CHAPTER IN REVIEW

Summary

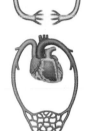

23.1 Open and Closed Circulatory Systems

- Some invertebrates do not have a transport system because their body plan allows each cell to exchange molecules with the external environment.
- Other invertebrates do have a transport system. Some have an open circulatory system, and some have a closed circulatory system.
- All vertebrates have a closed circulatory system in which arteries carry blood away from the heart to capillaries, where exchange takes place, and veins carry blood to the heart.

Comparison of Vertebrate Circulatory Pathways

- Fishes have a one-circuit pathway because the heart, with a single atrium and ventricle, pumps blood only to the gills.
- Other vertebrates have both pulmonary and systemic circuits. Amphibians have two atria but a single ventricle. Birds and mammals, including humans, have a heart with two atria and two ventricles, in which O_2-rich blood is always separate from O_2-poor blood.

23.2 Transport in Humans

The cardiovascular system consists of the heart and the blood vessels.

The Human Heart

The heart has a right and a left side. Each side has an atrium and a ventricle. Valves keep the blood moving in the correct direction.

Right Side of the Heart The atrium receives O_2-poor blood from the tissues, and the ventricle pumps it to the lungs.

Left Side of the Heart The atrium receives O_2-rich blood from the lungs, and the ventricle pumps it to the tissues.

Heartbeat During a heartbeat, first the atria contract and then the ventricles contract. The heart sounds, lub-dub, are caused by the closing of valves.

SA Node The SA node (pacemaker) causes the two atria to contract. The SA node also stimulates the AV node.

AV Node The AV node causes the two ventricles to contract.

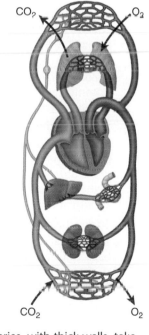

Transport Through Blood Vessels Arteries, with thick walls, take blood away from the heart to arterioles, which take blood to capillaries that have walls composed only of epithelial cells. Venules take blood from capillaries and merge to form veins, which have thinner walls than arteries have.

- Blood pressure, created by the beat of the heart, accounts for the flow of blood in the arteries.
- Skeletal muscle contraction is largely responsible for the flow of blood in the veins, which have valves preventing backward flow.

Blood Has Two Circuits

- In the pulmonary circuit, blood travels to and from the lungs.
- In the systemic circuit, the aorta divides into blood vessels that serve the body's cells. Venae cavae return O_2-poor blood to the heart.

Cardiovascular Disorders
Hypertension and atherosclerosis are two conditions that lead to heart attack and stroke. Following a heart-healthy diet, getting regular exercise, maintaining a proper weight, and not smoking cigarettes are protective against these conditions.

The Lymphatic System
The lymphatic system is a one-way system that consists of lymphatic vessels and lymphatic organs. The lymphatic vessels receive fat at the digestive tract and excess tissue fluid at blood capillaries, and carry these to the subclavian veins (cardiovascular veins in the shoulders).

23.3 Blood: A Transport Medium

Blood has two main parts: plasma and formed elements.

Plasma
Plasma contains mostly water (90–92%) and proteins (7–8%), but also nutrients, wastes, and salts.

- The proteins and salts buffer the blood and maintain its osmotic pressure.
- The proteins also have specific functions, such as participating in blood clotting and transporting molecules.

Formed Elements
The red blood cells contain hemoglobin and function in oxygen transport. Defense against disease depends on the various types of white blood cells:

Red blood cells

- Neutrophils and monocytes are phagocytic and are especially responsible for the inflammatory response.
- Lymphocytes are involved in the development of specific immunity to disease.

White blood cells

The platelets and two plasma proteins, prothrombin and fibrinogen, function in blood clotting, an enzymatic process that results in fibrin threads.

Platelets

Capillary Exchange in the Tissues
Capillary exchange in the tissues helps keep the internal environment constant. When blood reaches a capillary,

- Water moves out at the arterial end due to blood pressure.
- Water moves in at the venous end due to osmotic pressure.
- Nutrients diffuse out of and wastes diffuse into the capillary between the arterial end and the venous end.
- Lymphatic capillaries in the vicinity pick up excess tissue fluid and return it to cardiovascular veins.

Thinking Scientifically

1. Explain why the evolution of the four-chambered heart was critical for the development of an endothermic lifestyle—the generation of internal heat—in birds and mammals.

2. Provide a physiological explanation for the benefit gained by athletes who train at high altitudes.

Testing Yourself

Choose the best answer for each question.

1. In insects with an open circulatory system, oxygen is taken to cells by

   a. blood.
   b. hemolymph.
   c. tracheae.
   d. capillaries.

2. Label the components of the cardiovascular system in the following diagram.

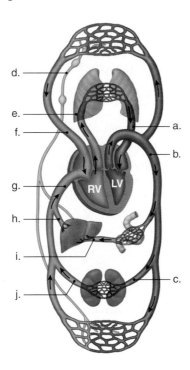

3. Which of the following statements is true?

   a. Arteries carry blood away from the heart, and veins carry blood to the heart.
   b. Arteries carry blood to the heart, and veins carry blood away from the heart.
   c. Arteries carry O_2-rich blood, and veins carry O_2-poor blood.
   d. Arteries carry O_2-poor blood, and veins carry O_2-rich blood.

4. Label the following diagram of the heart.

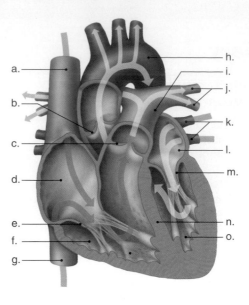

5. In humans, blood returning to the heart from the lungs returns to

   a. the right ventricle.
   b. the right atrium.
   c. the left ventricle.
   d. the left atrium
   e. both the right and left sides of the heart.

6. An artificial pacemaker replaces the effect of the

   a. vagus nerve.
   b. SA (sinoatrial) node.
   c. AV (atrioventricular) node.
   d. ventricular systole.

7. Which of the following lists the events of the cardiac cycle in the correct order?

   a. contraction of atria, rest, contraction of ventricles
   b. contraction of ventricles, rest, contraction of atria
   c. contraction of atria, contraction of ventricles, rest
   d. contraction of ventricles, contraction of atria, rest

8. The "lub," the first heart sound, is produced by the closing of

   a. the aortic semilunar valve.
   b. the pulmonary semilunar valve.
   c. the atrioventricular (tricuspid) valve.
   d. the atrioventricular (bicuspid) valve.
   e. both atrioventricular valves.

9. An electrocardiogram measures

   a. chemical signals in the brain and heart.
   b. electrical activity in the brain and heart.
   c. chemical signals in the heart.
   d. electrical changes in the wall of the heart.

10. Place the following blood vessels in order, from largest to smallest in diameter.

    a. arterioles, capillaries, arteries
    b. arteries, arterioles, capillaries
    c. capillaries, arteries, arterioles
    d. arterioles, arteries, capillaries
    e. arteries, capillaries, arterioles

11. The lymphatic system

    a. returns excess tissue fluid to cardiovascular veins.
    b. is found in limited regions of the body.
    c. relies most extensively on diffusion to move lymph
    d. is a two-way system.
    e. More than one of these are correct.

For questions 12–15, identify the cardiovascular disorder in the key that matches the description. Each answer may be used more than once. Each question may have more than one answer.

Key:
    a. hypertension
    b. atherosclerosis
    c. stroke
    d. heart attack

12. May not be detected until after a stroke or heart attack.

13. Sometimes called a silent killer.

14. Results from the accumulation of plaque.

15. May be caused by an embolism.

16. Exchange of the gases oxygen and carbon dioxide occurs across the _____ of the _____.

    a. veins, lungs
    b. capillaries, tissues
    c. arteries, tissues
    d. All of these are correct.

17. The average heart rate is about _____ beats per minute.

    a. 100
    b. 45
    c. 60
    d. 70

18. If the SA node fails and the AV node takes over, the result will be a heart rate that is

    a. slower.
    b. faster.
    c. the same.
    d. None of these are correct.

19. Which association is incorrect?

    a. white blood cells—infection fighting
    b. red blood cells—blood clotting
    c. plasma—water, nutrients, and wastes
    d. red blood cells—hemoglobin
    e. platelets—blood clotting

20. In the tissues, nutrients and _____ are exchanged for _____ and other wastes.

    a. blood, oxygen
    b. oxygen, carbon dioxide
    c. hemoglobin, tissue fluid
    d. None of these are correct.

21. Lymph is formed from

    a. urine.
    b. fats.
    c. excess tissue fluid.
    d. All of these are correct.

22. A decrease in lymphocytes would result in problems associated with

    a. clotting.
    b. immunity.
    c. oxygen transportation.
    d. All of these are correct.

23. The best explanation for the slow movement of blood in capillaries is

    a. skeletal muscles press on veins, not capillaries.
    b. capillaries have much thinner walls than arteries.
    c. there are many more capillaries than arterioles.
    d. venules are not prepared to receive so much blood from the capillaries.
    e. All of these are correct.

24. Which of the following assist in the return of venous blood to the heart?

    a. valves
    b. skeletal muscle contractions
    c. respiratory movements
    d. blood flow in the direction of reduced pressure
    e. All of these are correct.

25. Water enters the venous end of capillaries because

    a. osmotic pressure is higher than blood pressure.
    b. of an osmotic pressure gradient.
    c. of higher blood pressure on the venous side.
    d. of higher blood pressure on the arterial side.
    e. of higher red blood cell concentration on the venous side.

26. One-way conduction in lymphatic vessels is aided by

    a. valves.
    b. skeletal muscle contractions.
    c. Both a and b are correct.
    d. None of these are correct.

27. Which of the following is not a function of the lymphatic system?

    a. produces blood cells
    b. returns excess tissue fluid to the blood
    c. transports lipids absorbed from the digestive system
    d. defends the body against disease

28. Red blood cells

    a. reproduce themselves by mitosis.
    b. live for several years.
    c. continually synthesize hemoglobin.
    d. are destroyed in the liver and spleen.
    e. More than one of these are correct.

29. Which of the following is not a formed element of blood?

    a. white blood cells
    b. red blood cells
    c. fibrinogen
    d. platelets

30. The last step in blood clotting

    a. is the only step that requires calcium ions.
    b. occurs outside the bloodstream.
    c. is the same as the first step.
    d. converts prothrombin to thrombin.
    e. converts fibrinogen to fibrin.

31. Hemorrhagic disorders can result when which vitamin is missing from the diet?

    a. A   d. D
    b. C   e. thiamine
    c. K

32. Water

    a. exits the bloodstream at the arterial and venous ends of a capillary.
    b. enters the bloodstream at the arterial and venous ends of a capillary.
    c. enters the bloodstream at the arterial end and exits cells at the venous end of a capillary.
    d. exits the bloodstream at the arterial end and enters cells at the venous end of a capillary.

33. In the following diagram, label arrows a.–d. as either blood pressure or osmotic pressure.

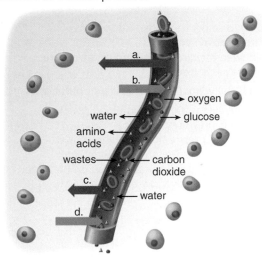

Go to www.mhhe.com/maderessentials for more quiz questions.

Bioethical Issue

In the highly competitive world of professional athletics, individuals are constantly striving to gain an edge over the competition. With six-figure salaries, international recognition, and multimillion-dollar endorsements as their incentives, some athletes resort to life-threatening measures to enhance their performance.

In recent years, various means of increasing the red blood cell count in athletes have been employed. By increasing the number of red blood cells in the body, more oxygen can be delivered to the muscles. This enhances athletic ability, especially in events requiring a high level of endurance.

In the past, blood doping was commonly used. Red blood cells (RBCs) were removed from the athlete's body and stored for several weeks. During this time, the body would release increased levels of the hormone erythropoietin to stimulate RBC production. When the RBC count returned to normal, the blood cells previously removed were injected back into the body. This was done a few days before the competition to get maximum effect. In more recent times, injections of artificially produced erythropoietin are being used to stimulate RBC production far above normal levels.

The artificial form of erythropoietin is known as epoetin alfa and is structurally identical to the naturally occurring form. Therefore, it is very difficult to determine when it is being used.

However, many athletes have paid a high price for the competitive advantage offered by epoetin alfa. The use of this hormone causes the blood to become thicker than normal. Hypertension and an increased risk of heart attack and stroke result. In recent decades, the suspected use of epoetin alfa has been linked to the deaths of numerous competitive cyclists.

Understanding the Terms

antibody   406
antigen   406
aorta   399, 402
arteriole   401
artery   399, 401
atrioventricular valve   399
atrium (pl., atria)   399
AV (atrioventricular) node   400
blood   405
blood pressure   401
capillary   401
cardiac cycle   400
cardiovascular system   397
closed circulatory system   397
diastole   400
electrocardiogram (ECG)   400
formed elements   405
heart   396
heart attack   404
heart murmur   399
hemoglobin   405
hypertension   403
lymph   403
lymphatic system   403
lymphatic vessel   403
lymphocyte   406
macrophage   406
monocyte   406
neutrophil   406
open circulatory system   396
plaque   404
plasma   405
platelet   406
portal system   403
pulmonary artery   399
pulmonary circuit   398, 402
pulmonary trunk   399
pulmonary vein   399
pulse   400
red blood cell   405
SA (sinoatrial) node   400
semilunar valve   399
septum   399
stroke   404
systemic circuit   398, 402
systole   400
thrombin   407
vein   399, 401
vena cava (pl., venae cavae)   399, 402
ventricle   399
venule   401
white blood cell   406

Match the terms to these definitions:

a. _____ Circuit that takes blood to and from the tissues.

b. _____ Upper chambers of the heart.

c. _____ Condition generally caused by leaky heart valves.

d. _____ Heart's intrinsic natural pacemaker.

e. _____ Smallest of the blood vessels.

f. _____ Major vessels in the systemic circuit.

g. _____ Tissue fluid that has entered lymphatic vessels.

h. _____ Liquid portion of blood.

i. _____ Phagocytic white blood cells derived from monocytes.

j. _____ Formed elements involved in blood clotting.

The Maintenance Systems

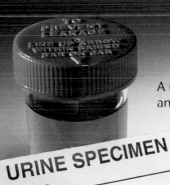

A urinalysis can involve any of 35 different tests.

URINE SPECIMEN

NAME _____

ROOM NO. _____

DATE _____ TIME _____

DOCTOR _____

INSTRUCTIONS _____

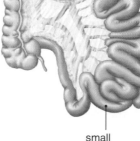

small intestine

The small intestine of humans is about 6 meters (19.7 feet) long.

The respiratory system is one of the last organ systems to develop in the human fetus.

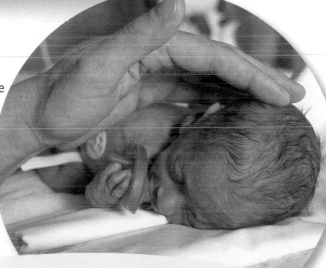

During human development, organ systems begin to function on an "as needed" basis. Within the first month of embryonic development, many organ systems (such as the cardiovascular and nervous systems) are present and functioning. However, some organ systems that appear may not be refined until much later because they are not critical to early development. When babies are born prematurely (before the 37th week of gestation),

▶ ▶ ▶

numerous health problems can occur, particularly with regard to these systems. One condition that is common in very premature babies (those born at 32 weeks or earlier) is respiratory distress syndrome (RDS). Because the lungs of a fetus do not normally function until full term (40 weeks), lung development is not complete when a baby is born prematurely. Usually, a natural surfactant that helps keep the air sacs of the lungs inflated so that gas exchange can occur is not being produced. Without treatment, RDS is often fatal because the blood is not being oxygenated. Now, an artificial surfactant is available, and premature babies are given the surfactant they lack. This treatment greatly increases the likelihood that premature babies will survive. Other medical breakthroughs are also now available to encourage the functioning of a premature baby's organ systems.

The maintenance systems have vital roles in the overall functioning of the body, and a malfunction in one of these systems can lead to problems in other systems. In this chapter, you will learn about the structure and function of the digestive, respiratory, and urinary systems, and how they support the body's other organ systems.

24.1 Digestive System

The cells of your body are bathed in tissue fluid. They acquire oxygen and nutrients and get rid of carbon dioxide and other wastes through exchanges with tissue fluid. In turn, tissue fluid makes these same exchanges with the blood. Blood is refreshed because the digestive, respiratory, and urinary systems make exchanges with the external environment. Only in this way is blood supplied with the oxygen and nutrients cells require and cleansed of waste molecules. In this first section, we consider how the digestive system contributes to homeostasis (Fig. 24.1).

Tube-Within-a-Tube Body Plan

Hydra and planarians (see Fig. 23.1) have a sac body plan—that is, the mouth, like the top of a sack, serves as both an entrance and an exit. Most other animals, such as the earthworm, have a tube-within-a-tube plan, so called because the inner tube has both an entrance and an exit—the mouth and the anus. Therefore, it is called a complete digestive tract (Fig. 24.2a). Notice that the inner tube (the digestive tract, sometimes called the alimentary canal) is separated from the outer tube (the body wall) by the **coelom.** The digestive tract of humans and other vertebrates consists of so many specialized organs that it might be hard to realize that the basic plan of vertebrates is the same as that of the earthworm (Fig. 24.2b). The digestive tract of humans exemplifies that the tube-within-a-tube plan and a complete tract result in specialization of parts.

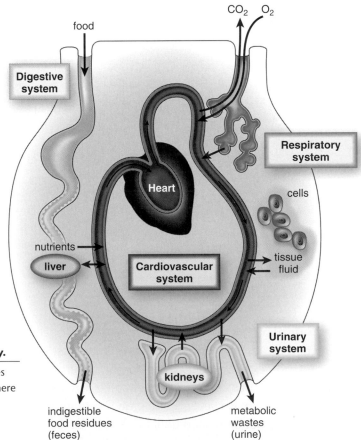

Figure 24.1 **Keeping the internal environment steady.**

The digestive system takes in food and digests it to nutrient molecules that enter the blood. The blood transports nutrients to the tissues where exchange occurs with tissue fluid. Later in this chapter we will also consider the manner in which the respiratory system and the urinary system help keep the internal environment relatively constant.

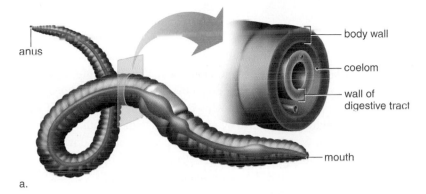

a.

Digestion of food in earthworms and humans is an extracellular process. Digestive enzymes, produced by glands in the wall of the tract or by accessory glands that lie nearby, enter the tract. Food is never found within these accessory glands, only within the tract itself.

Digestion contributes to homeostasis by providing the body's cells with the nutrients they need to continue living. A digestive tract performs the following functions:

1. Ingests food.
2. Breaks food down into small molecules that can cross plasma membranes.
3. Absorbs nutrient molecules.
4. Eliminates indigestible remains.

Mouth

In humans, the digestive system begins with the **mouth,** which chews food into pieces convenient for swallowing. Many vertebrates have teeth, an exception being birds, which lack teeth and depend on the churning of small pebbles within a gizzard to break up their food. The teeth (dentition) of mammals reflect their diet (Fig. 24.3). **Carnivores** eat meat, which is easily digestible because the cells don't have a cellulose wall. **Herbivores** eat plant material, which needs a lot of chewing and other processing to break up the cellulose walls. Humans are **omnivores;** they eat both meat and plant material. The four front teeth (top and bottom) of humans are sharp, chisel-shaped incisors used for biting. On each side of the incisors are the sharp, pointed canines used for tearing food. The premolars and molars grind and crush food. It is as though humans are carnivores in the front of their mouths and herbivores in the back.

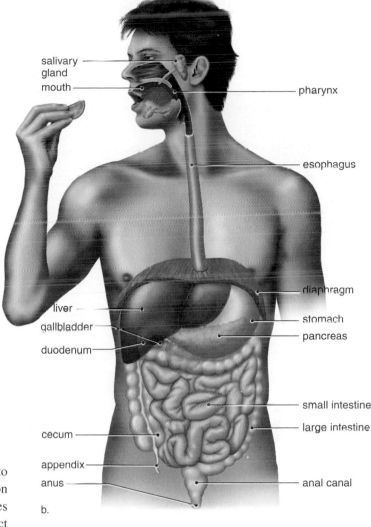

b.

Figure 24.2 Complete digestive system.

The earthworm (**a**) and humans (**b**) have complete digestive systems. A complete digestive system leads to specialization of organs along the digestive tract.

Key:
incisors	premolars
canines	molars

Figure 24.3 Dentition among mammals.

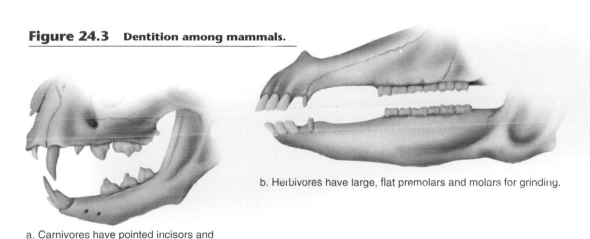

b. Herbivores have large, flat premolars and molars for grinding.

a. Carnivores have pointed incisors and enlarged canine teeth for biting and tearing.

c. The front teeth of humans are like those of carnivores, and the back teeth are like those of herbivores.

Food contains cells composed of molecules of carbohydrates, proteins, nucleic acids, and lipids. During digestion, hydrolytic enzymes break down these large molecules to smaller molecules. The process begins in the mouth. Three pairs of **salivary glands** send saliva by way of ducts to the mouth. Saliva contains the enzyme **salivary amylase,** which breaks down starch, a carbohydrate, to maltose, a disaccharide. While in the mouth, food is manipulated by a muscular tongue, which has touch and pressure receptors similar to those in the skin. Taste buds that allow us to "taste" our food are also on the tongue as well as on the surface of the mouth. The tongue mixes the chewed food with saliva and then forms this mixture into a mass called a **bolus** that is swallowed.

Swallowing

The human digestive and respiratory passages come together in the **pharynx** and then separate (Fig. 24.4a). When food is swallowed, the soft palate, the rear portion of the mouth's roof, moves back to close off the nasal cavities (Fig. 24.4b). A flap of tissue called the **epiglottis** covers the glottis, an opening into the larynx. Now the bolus must move through the pharynx into the esophagus because the air passages are blocked.

The **esophagus** is a tubular structure that takes food to the stomach, which lies below the **diaphragm,** a muscular, membranous partition that divides the **thoracic** (upper) **cavity** from the **abdominal** (lower) **cavity** of the body. When food enters the esophagus, peristalsis begins. **Peristalsis** is a rhythmic contraction that moves the contents along in tubular organs—in this case, those of the digestive tract.

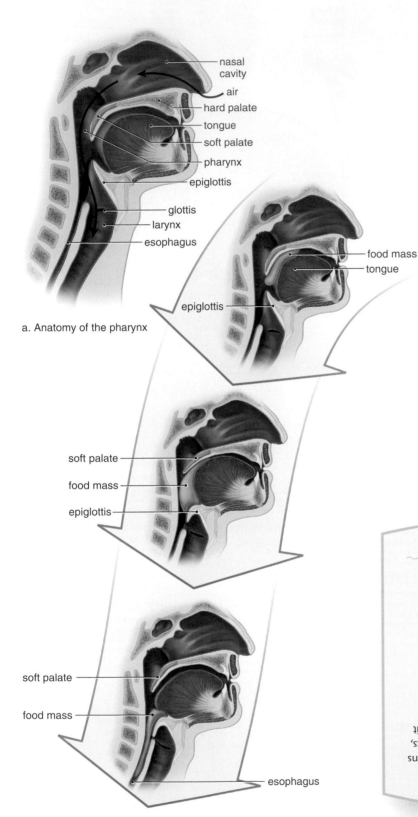

a. Anatomy of the pharynx

b. Swallowing

Figure 24.4 The human mouth, pharynx, esophagus, and larynx.

a. The palates (both hard and soft) separate the mouth from the nasal cavities. The pharynx leads to the esophagus and the larynx. **b.** When food is swallowed, the soft palate closes off the nasal cavities, and the epiglottis closes off the larynx.

Check Your Progress

1. List the three organ systems that make exchanges with the external environment and blood.
2. Compare and contrast the human digestive system with that of an earthworm.
3. List the functions of a digestive tract.
4. Explain why carnivores do not need teeth for grinding food.
5. Describe the function of peristalsis in digestion.

Answers: 1. Digestive, respiratory, and urinary systems. 2. Both are tube-within-a-tube plans, but the human system has many more specialized organs than that of an earthworm. 3. Ingest and break down food, absorb nutrients, and eliminate indigestible material. 4. Meat is easy to break down because it does not contain cellulose, so extensive chewing is not necessary. 5. Peristalsis moves materials through the digestive tract.

Stomach

The human **stomach** is a thick-walled, J-shaped organ that stretches and stores food (Fig. 24.5). It also begins the digestion of proteins, and regulates the entrance of food into the small intestine.

The wall of the stomach has deep folds, which disappear as the stomach fills to an approximate capacity of 1 liter. Therefore, humans can periodically eat relatively large meals and spend the rest of their time at other activities. But the stomach is much more than a mere storage organ, as was discovered by William Beaumont in the mid-nineteenth century. Beaumont, an American doctor, had a French Canadian patient, Alexis St. Martin, who had been shot in the stomach. When the wound healed, St. Martin was left with an opening that allowed Beaumont to look inside the stomach and to collect gastric (stomach) juices produced by gastric glands. Beaumont was able to determine that the muscular walls of the stomach contract vigorously and mix food with juices that are secreted whenever food enters the stomach. He found that gastric juice contains hydrochloric acid (HCl) and a substance, now called pepsin, that is active in digestion.

We now know that the epithelial lining of the stomach, called a mucosa, has millions of gastric glands. These gastric glands produce gastric juice containing so much hydrochloric acid that the stomach routinely has a pH of about 2. Such high acidity is usually sufficient to kill any microbes that might be in food. This low pH also promotes the activity of **pepsin**, a hydrolytic enzyme that acts on protein to produce peptides. In addition, high acidity causes *heartburn* and *gastric reflux disease* when gastric juice backs up into the esophagus.

As with the rest of the digestive tract, a thick layer of mucus protects the wall of the stomach from enzymatic action. Still, an ulcer, which is an open sore in the wall caused by the gradual destruction of tissues, may occur in some individuals. Ulcers are due to an infection by an acid-resistant bacterium, *Helicobacter pylori*, which is able to attach to the epithelial lining. Wherever the bacterium attaches, the lining stops producing mucus, and the area becomes exposed to digestive action. Then an ulcer develops.

Peristalsis pushes food along in the stomach as it does in other digestive organs (Fig. 24.5*b*). At the base of the stomach is a narrow opening controlled by a *sphincter*, a muscle that surrounds a tube and closes or opens it by contracting and relaxing. Whenever this sphincter relaxes, a small quantity of material passes through the opening into the small intestine.

Ruminants Ruminants, a type of mammal that includes cattle, sheep, goats, deer, and buffalo, are named for a part of their stomachs, the rumen (Fig. 24.6). The rumen contains symbiotic bacteria and protozoans that, unlike the mammal, can digest cellulose. After these herbivores feed on grass, it goes to the rumen, where it is broken down and then formed into small balls of cud. The cud then returns to the mouth where the animal "chews the cud." The cud may return to the rumen for a second go-round before passing through the other chambers of the stomach. The last chamber is analogous to the human stomach, being the place where protein is digested to peptides.

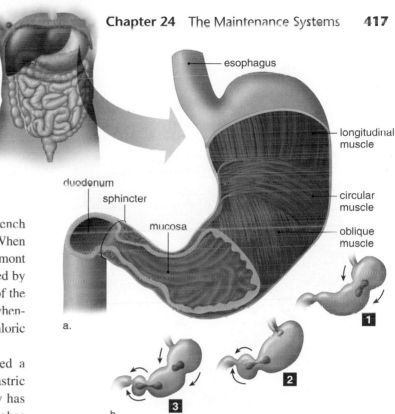

Figure 24.5   Anatomy of the human stomach.

a. The stomach has a thick wall that expands as it fills with food. The wall contains three layers of muscle, and their presence allows the stomach to churn and mix food with gastric juices. The mucosa of the stomach wall secretes mucus and contains gastric glands, which secrete gastric juices active in the digestion of protein. **b.** Peristalsis, a rhythmic contraction, occurs along the length of the digestive tract.

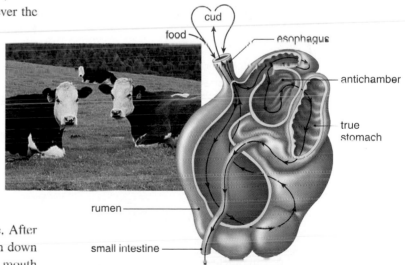

Figure 24.6   A ruminant's stomach.

Ruminants eat grass, which is made of cells with strong cellulose walls. The first chamber of a ruminant's stomach, called the rumen, contains symbiotic bacteria and protozoans that can digest cellulose. After a first pass through the rumen, the "cud" returns to the mouth where it is leisurely chewed. Then, it may return to the rumen for a second go-round of digestion before passing through to the true stomach.

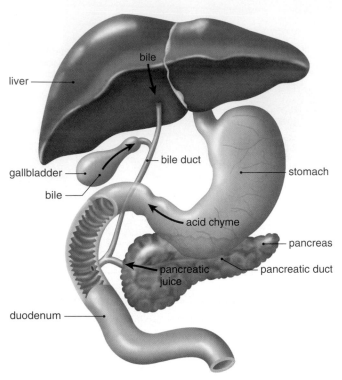

liver

bile

gallbladder

bile duct

bile

stomach

acid chyme

pancreas

pancreatic juice

pancreatic duct

duodenum

Figure 24.7 **The pancreatic and bile ducts empty into the duodenum.**

Bile, made by the liver and stored in the gallbladder, and pancreatic juice, which contains enzymes, enter the duodenum by way of ducts.

Small Intestine

Processing food in humans is more complicated than one might think. So far, the food has been chewed in the mouth and worked on by the enzyme salivary amylase, which digests starch to maltose. In addition, the digestion of protein has begun in the stomach as pepsin digests proteins to peptides. By now, the contents of the digestive tract are called **chyme.** Chyme passes from the stomach to the **small intestine,** a long, coiled tube that has two functions: (1) digestion of all the molecules in chyme, including polymers of carbohydrate, protein, nucleic acid, and fat, and (2) absorption of the nutrient molecules into the body.

The first part of the small intestine is called the duodenum. Two important accessory glands, the **liver,** the largest organ in the body, and the **pancreas,** located behind the stomach, send secretions to the duodenum by way of ducts (Fig. 24.7). The liver produces **bile,** which is stored in the **gallbladder.** Bile looks green because it contains pigments that are the products of hemoglobin breakdown. This green color is familiar to anyone who has observed how bruised tissue changes color. Hemoglobin within the bruise is breaking down into the same types of pigments found in bile. Bile also contains bile salts, which break up fat into fat droplets by a process called **emulsification.** Fat droplets mix with water and have more surface area for digestion by enzymes.

The pancreas produces pancreatic juice, which contains sodium bicarbonate ($NaHCO_3$) and digestive enzymes. $NaHCO_3$ neutralizes chyme and makes the pH of the small intestine slightly basic. The higher pH helps prevent autodigestion of the intestinal lining by pepsin and optimizes the pH for pancreatic enzymes. **Pancreatic amylase** digests starch to maltose; **trypsin** digests proteins to peptides; **lipase** digests fat droplets to glycerol and fatty acids; and **nuclease** digests nucleic acids to nucleotides.

Still more digestive enzymes are present in the small intestine. The wall of the small intestine contains fingerlike projections called **villi** (Fig. 24.8*b*). The epithelial cells of the villi produce **intestinal enzymes,** which remain attached to them. These enzymes complete the digestion of peptides and sugars. Peptides, which result from the first step in protein digestion, are digested by peptidase to amino acids. Maltose, which results from the first step in starch digestion, is digested by maltase to glucose. Other disaccharides, each of which is acted upon by a specific enzyme, are digested to simple sugars also.

Finally, these small nutrient molecules can be absorbed into the body. Our cells use these molecules as a source of energy and as building blocks to make their own macromolecules.

Absorption by Villi The wall of the small intestine is adapted to absorbing nutrient molecules because it has an extensive surface area—approximately that of a tennis court! First, the mucous membrane layer of the small intestine has circular folds that give it an almost corrugated appearance (Fig. 24.8*a*). Second, on the surface of these circular folds are the villi. Finally, the cells on the surface of the villi have minute projections called **microvilli** (Fig. 24.8*c*). If the human small intestine were simply a smooth tube, it would have to be 500 to 600 meters long to have a comparable surface area for absorption. Carnivores have a much shorter digestive tract than herbivores because meat is easier to process than plant material (Fig. 24.9).

The villi of the small intestine absorb small nutrient molecules into the body. Each villus contains an extensive network of blood capillaries and a lymphatic capillary called a **lacteal.**

Check Your Progress

1. Describe the functions of the stomach.
2. Describe the relationship between the duodenum and the liver and pancreas.

Answers: 1. The stomach stores food and continues digestion, kills microbes, and moves partially digested material into the intestines. 2. The liver and pancreas secrete bile and pancreatic juice, respectively, into the duodenum.

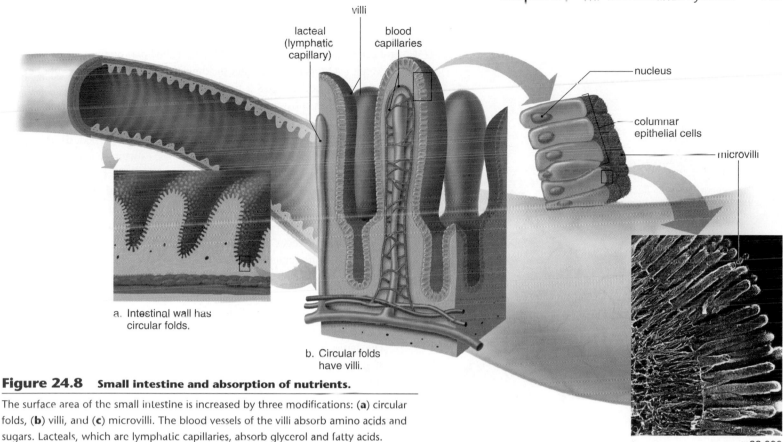

villi

lacteal
(lymphatic
capillary)

blood
capillaries

nucleus

columnar
epithelial cells

microvilli

a. Intestinal wall has
circular folds.

b. Circular folds
have villi.

×33,000
c. Cells of villi have microvilli.

Figure 24.8 **Small intestine and absorption of nutrients.**

The surface area of the small intestine is increased by three modifications: (**a**) circular folds, (**b**) villi, and (**c**) microvilli. The blood vessels of the villi absorb amino acids and sugars. Lacteals, which are lymphatic capillaries, absorb glycerol and fatty acids.

As discussed in Chapter 23, the lymphatic system is an adjunct to the cardiovascular system—its vessels carry a fluid called lymph to the cardiovascular veins. Sugars and amino acids enter the blood capillaries of a villus. In contrast, glycerol and fatty acids (digested from fats) enter the epithelial cells of the villi, and within them are joined and packaged as lipoprotein droplets, which enter a lacteal. Absorption continues until almost all nutrient molecules have been absorbed. Absorption occurs by diffusion, as well as by active transport, which requires an expenditure of cellular energy. Lymphatic vessels transport lymph to cardiovascular veins. Eventually, the bloodstream carries the nutrients absorbed by the digestive system to all the cells of the body.

Large Intestine

The word *bowel* technically means the part of the digestive tract between the stomach and the anus, but it is sometimes used to mean only the large intestine. The **large intestine** absorbs water, salts, and some vitamins. It also stores indigestible material until it is eliminated at the anus. The large intestine takes its name from its diameter rather than its length, which is shorter than that of the small intestine. The large intestine has a blind pouch below the entry of the small intestine with a small projection called the **appendix.** In humans, the appendix may play a role in fighting infections. In the condition called *appendicitis*, the appendix becomes infected and so filled with fluid that

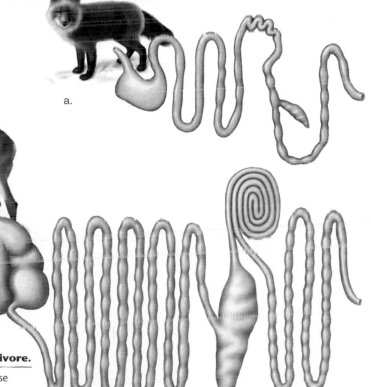

a.

b.

Figure 24.9 **Digestive tract of a carnivore compared to a ruminant herbivore.**

The digestive tract of a carnivore (**a**) is much shorter than that of a ruminant herbivore (**b**) because proteins can be more easily digested than plant matter.

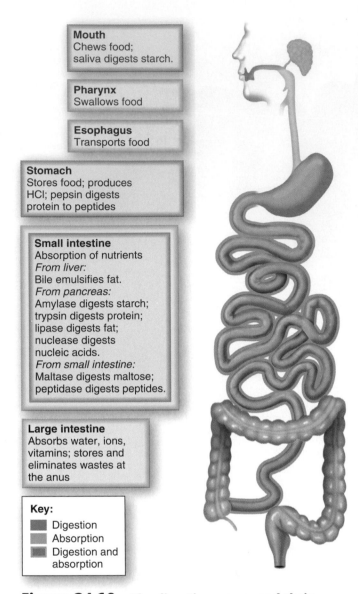

Mouth
Chews food;
saliva digests starch.

Pharynx
Swallows food

Esophagus
Transports food

Stomach
Stores food; produces
HCl; pepsin digests
protein to peptides

Small intestine
Absorption of nutrients
From liver:
Bile emulsifies fat.
From pancreas:
Amylase digests starch;
trypsin digests protein;
lipase digests fat;
nuclease digests
nucleic acids.
From small intestine:
Maltase digests maltose;
peptidase digests peptides.

Large intestine
Absorbs water, ions,
vitamins; stores and
eliminates wastes at
the anus

Key:
- ■ Digestion
- ■ Absorption
- ■ Digestion and
  absorption

Figure 24.10   The digestive organs and their functions.

A review of the processing of food to nutrient molecules and their absorption into the body.

it may burst. If an infected appendix bursts before it can be removed, a serious, generalized infection of the abdominal lining, called peritonitis, may result.

About 1.5 liters of water enter the digestive tract daily as a result of eating and drinking. An additional 8.5 liters also enter the digestive tract each day, carrying the various substances secreted by the digestive glands. About 95% of this water is absorbed by the small intestine, and much of the remaining portion is absorbed by the large intestine. If this water is not reabsorbed, **diarrhea** can occur, leading to serious dehydration and ion loss, especially in children.

The large intestine has a large population of bacteria, notably *Escherichia coli.* The bacteria break down indigestible material, and they also produce some vitamins, including vitamin K. Vitamin K is necessary for blood clotting. Digestive wastes (feces) eventually leave the body through the **anus,** the opening of the anal canal. Feces are about 75% water and 25% solid matter. Almost one-third of this solid matter is made up of intestinal bacteria. The remainder is indigestible plant material (also called fiber), fats, waste products (such as bile pigments), inorganic material, mucus, and dead cells from the intestinal lining. A diet that includes fiber adds bulk to the feces, improves regularity of elimination, and prevents *constipation.*

The large intestine (also called the colon) is subject to the development of **polyps,** which are small growths arising from the mucosa. Polyps, whether they are benign or cancerous, can be removed surgically.

Figure 24.10 reviews the digestive process.

Check Your Progress

Contrast the functions of the small intestine and the large intestine.

Answer: The small intestine digests all types of food, and it produces enzymes that complete the breakdown of food to small molecules that can be absorbed across its villi. No digestion occurs in the large intestine, but it does store digestive remains until they are eliminated. The large intestine also absorbs water, salts, and some vitamins.

Accessory Organs

The pancreas and the liver are accessory organs of digestion, along with the teeth, salivary glands, and gallbladder.

Pancreas

The pancreas (see Fig. 24.7) functions as both an endocrine gland and an exocrine gland. **Endocrine glands** are ductless and secrete their products into the blood. The pancreas is an endocrine gland when it produces and secretes the hormones **insulin** and **glucagon** into the bloodstream. **Exocrine glands** secrete into ducts. The pancreas is an exocrine gland when it produces and secretes pancreatic juice into the **duodenum** through the common bile duct.

Liver

The liver has numerous functions, including the following:

1. Detoxifies the blood by removing and metabolizing poisonous substances.

2. Produces the plasma proteins, such as albumin and fibrinogen.
3. Destroys old red blood cells and converts hemoglobin to the breakdown products in bile (bilirubin and biliverdin).
4. Produces bile, which is stored in the gallbladder before entering the small intestine, where it emulsifies fats.
5. Stores glucose as glycogen and breaks down glycogen to glucose between meals to maintain a constant glucose concentration in the blood.
6. Produces urea from amino groups and ammonia.

Blood vessels from the large and small intestines merge to form the hepatic portal vein, which leads to the liver (Fig. 24.11). The liver helps maintain the glucose concentration in blood at about 0.1% by removing excess glucose from the hepatic portal vein and storing it as glycogen. Between meals, glycogen is broken down to glucose, and glucose enters the hepatic veins. Like plant-made starch, glycogen is made up of glucose molecules, and thus it is sometimes called animal starch. If the supply of glycogen and glucose runs short, the liver converts amino acids to glucose molecules.

Amino acids contain nitrogen in the form of amino groups, whereas glucose contains only carbon, oxygen, and hydrogen. Therefore, before amino acids can be converted to glucose molecules, deamination, the removal of amino groups from amino acids, must occur. By a complex metabolic pathway, the liver converts the amino groups to urea, the most common nitrogenous (nitrogen-containing) waste product of humans. After urea is formed in the liver, it is transported by the bloodstream to the kidneys, where it is excreted.

Liver Disorders When a person is jaundiced, the skin has a yellowish tint due to an abnormally large quantity of bile pigments in the blood. In hemolytic jaundice, red blood cells are broken down in abnormally large amounts; in obstructive jaundice, the bile duct is obstructed, or the liver cells are damaged. Obstructive jaundice often occurs when crystals of cholesterol precipitate out of bile and form gallstones.

Jaundice can also result from viral infection of the liver, called *hepatitis*. Hepatitis A is most often caused by eating contaminated food. Hepatitis B and C are commonly spread by blood transfusions, kidney dialysis, and injection with unsterilized needles. These three types of hepatitis can also be spread by sexual contact.

Cirrhosis is a chronic liver disease in which the organ first becomes fatty, and later liver tissue is replaced by inactive fibrous scar tissue. Alcoholics often get cirrhosis, most likely due at least in part to the excessive amounts of alcohol the liver is forced to break down.

Regulation of Digestive Juices

Does your mouth water when you smell food cooking? Even the thought of food can sometimes cause the nervous system to order the secretion of digestive juices. The secretion of these juices is also under the influence of several peptide hormones, so called because each is a small sequence of amino acids. When you eat a meal rich in protein, the stomach wall produces a peptide hormone that enters the bloodstream and doubles back to cause the stomach to produce more gastric juices. When protein and fat are present in the small intestine, another peptide hormone made in the intestinal wall stimulates the secretion of bile and pancreatic juices. In this way, the organs of digestion regulate their own needs.

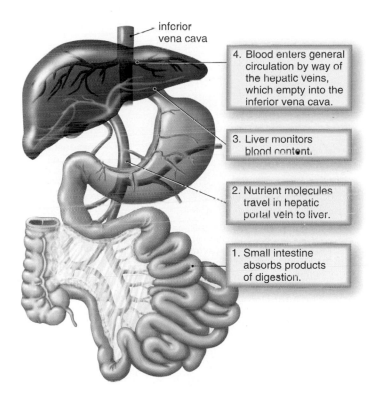

Figure 24.11 Hepatic portal system.

1. Small intestine absorbs products of digestion.
2. Nutrient molecules travel in hepatic portal vein to liver.
3. Liver monitors blood content.
4. Blood enters general circulation by way of the hepatic veins, which empty into the inferior vena cava.

inferior vena cava

The hepatic portal vein takes the products of digestion from the digestive system to the liver, where they are processed before entering hepatic veins.

Check Your Progress

List the functions of the liver.

Answer: The liver detoxifies the blood, produces plasma proteins, produces bile, stores glucose as glycogen, and produces urea.

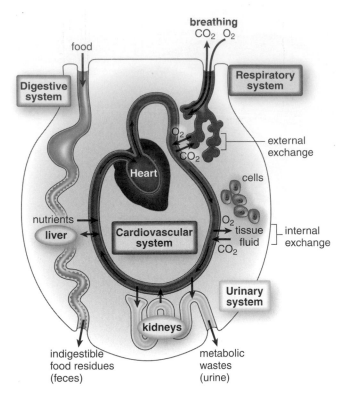

Figure 24.12   Keeping the internal environment steady.

The respiratory system functions in gas exchange. CO_2-laden blood enters the pulmonary capillaries and then CO_2 diffuses into the lungs and exits the body by way of respiratory passages. O_2-laden air enters the respiratory passages and lungs. Then O_2 diffuses into the blood at the pulmonary capillaries.

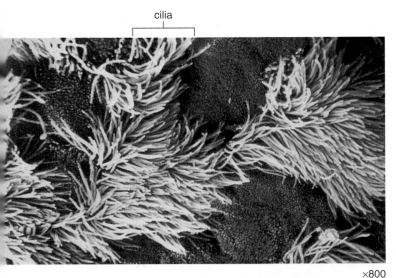

×800

Figure 24.13   Ciliated cells of the respiratory passages.

These cilia sweep impurities up away from the lungs toward the throat where they can be swallowed. Smoking first inactivates and then destroys these cilia.

24.2   Respiratory System

A respiratory system conducts oxygen-laden air to an exchange surface and conducts carbon dioxide-laden air out of the body. **Respiration** contributes to homeostasis by providing the body's cells with oxygen and removing carbon dioxide (Fig. 24.12). Respiration in complex animals requires these steps:

1. Breathing: inspiration (entrance of air into the lungs) and expiration (exit of air from the lungs).
2. External exchange of gases between the air and the blood within the lungs.
3. Internal exchange of gases between blood and tissue fluid: The body's cells exchange gases with tissue fluid.

Regardless of the particular gas-exchange surfaces of animals and the manner in which gases are delivered to the cells, in the end oxygen enters mitochondria, where cellular respiration takes place. Without the delivery of oxygen to the body's cells, ATP is not produced, and life ceases. Carbon dioxide, a waste molecule given off by cells, is a by-product of cellular respiration.

The Human Respiratory Tract

The human respiratory system includes all of the structures that conduct air in a continuous pathway to and from the **lungs** (see Fig. 24.14). As air moves through the respiratory tract, it is filtered so that it is free of debris, warmed, and humidified. By the time the air reaches the lungs, it is at body temperature and saturated with water. In the nose, hairs and cilia act as screening devices. In the respiratory passages, cilia beat upward, carrying mucus, dust, and occasional bits of food that "went down the wrong way" into the throat, where the accumulation may be swallowed or spit out (Fig. 24.13). Smoking cigarettes and cigars inactivates and eventually destroys these cilia, so that the lungs become laden with soot and debris. This is the first step toward various lung disorders.

Conversely, as air moves out of the tract, it cools and loses its moisture. As air cools, it deposits its moisture on the lining of the tract, and the nose many even drip as a result of this condensation. However, the air still retains so much moisture that on a cold day, it forms a small cloud when we breathe out.

The Upper Respiratory Tract

The upper respiratory tract consists of the nasal cavities, pharynx, and larynx (Fig. 24.14). The nose, a prominent feature of the face, is the only external portion of the respiratory system. The nose contains the **nasal cavities,** narrow canals separated from one another by a septum composed of bone and cartilage. Tears from the eyes drain into the nasal cavities by way of tear ducts. For this reason, crying produces a runny nose. The nasal cavities communicate with the **sinuses,** air-filled spaces that reduce the weight of the skull and act as resonating chambers for the voice. If the ducts leading from the sinuses become inflamed, fluid may accumulate, causing a sinus headache. The nasal cavities are separated from the mouth by a partition called the palate. The palate has two portions. Anteriorly, the hard palate is supported by bone, and posteriorly, the soft palate is not so supported.

The **pharynx** is a funnel-shaped passageway that connects the nasal cavity and mouth to the **larynx,** or voice box. The tonsils form a protective ring of lymphatic tissue at the junction of the mouth and the pharynx. *Tonsillitis* occurs when the tonsils become inflamed and enlarged. If tonsillitis occurs frequently and enlargement makes breathing difficult, the tonsils can be removed surgically, a procedure called a tonsillectomy. In the pharynx, the air passage and food passage cross because the larynx, which receives air, is anterior to the esophagus, which receives food. This arrangement may seem inefficient, since there is danger of choking if food accidentally enters the trachea, but it does have the advantage of letting you breathe through your mouth in case your nose is plugged up. In addition, it permits greater intake of air during heavy exercise, when greater gas exchange is required.

Air passes from the pharynx through the **glottis,** an opening into the larynx. The larynx is always open because it is formed by a complex of cartilages, among them the Adam's apple. At the edges of the glottis, embedded in mucous membrane, are the **vocal cords.** These flexible bands of connective tissue vibrate and produce sound when air is expelled past them through the glottis from the larynx. *Laryngitis* is an infection of the larynx with accompanying hoarseness leading to the inability to speak audibly.

Lower Respiratory Tract

The lower respiratory tract contains the respiratory tree, consisting of the trachea, bronchi, and bronchioles (Fig. 24.14). The **trachea,** commonly called the windpipe, is a tube connecting the larynx to the bronchi. The trachea is held open by a series of C-shaped, cartilaginous rings that do not completely meet in the rear. The trachea divides into two primary **bronchi,** which enter the right and left lungs. *Bronchitis* is an infection of the bronchi. As bronchitis develops, a nonproductive cough becomes a deep cough that produces mucus and perhaps pus. The deep cough of smokers indicates that they have bronchitis and that the respiratory tract is irritated. When a person stops smoking, this progression reverses, and the airways become healthy again. Chronic bronchitis is the second step toward emphysema and lung cancer caused by smoking cigarettes. Lung cancer often begins in the bronchi, and from there it spreads to the lungs.

The bronchi continue to branch until there are a great number of smaller passages called **bronchioles.** The two bronchi resemble the trachea in structure, but as the passages divide and subdivide, their walls become thinner, and rings of cartilage are no longer present. During an attack of *asthma,* the smooth muscle of the bronchioles contracts, causing constriction of the bronchioles and characteristic wheezing. Each bronchiole terminates in an elongated space enclosed by a multitude of little air pockets, or sacs, called **alveoli** (sing., alveolus), which make up the lungs (see Fig. 24.18).

Respiration in Insects While humans have one trachea, insects have many tracheae, little air tubes supported by rings of chitin that branch into every part of the body (Fig. 24.15). The tracheal system begins at spiracles, openings that perforate the insect's body wall, and ends in very fine, fluid-filled tubules, which may actually indent the plasma membranes of cells to come close to mitochondria. Ventilation is assisted by the presence of air sacs that merely assist the drawing in of air. The hemolymph in an insect does not transport oxygen, and no oxygen-carrying pigment is required.

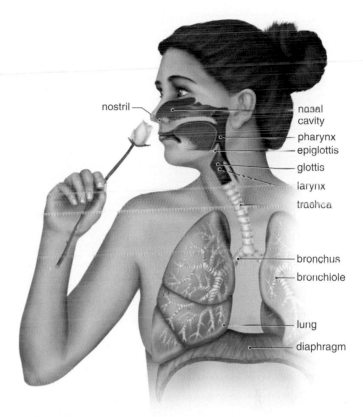

Figure 24.14 The human respiratory tract.

To reach the lungs, air moves from the nasal cavities, through the pharynx, larynx, trachea, bronchi, and finally the bronchioles, which end in the lungs.

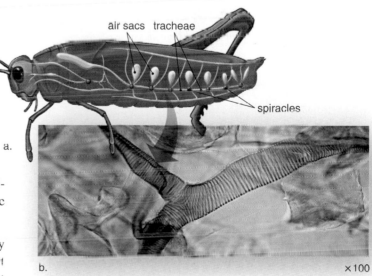

a.

b.   ×100

Figure 24.15 Tracheae of insects.

A system of air tubes extending throughout the body of an insect carries oxygen to the cells. Air enters the tracheae at openings called spiracles. From here, the air moves to smaller tubes, which take it to the cells where gas exchange takes place. **a.** Diagram of tracheae. **b.** The photomicrograph shows how the walls of the trachea are stiffened with bands of chitin.

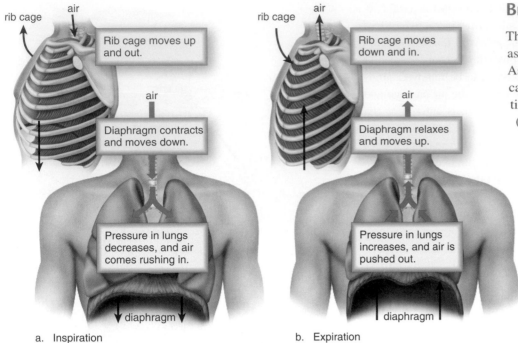

Figure 24.16 **Inspiration versus expiration.**

a. During inspiration, the thoracic cavity and lungs expand so that air is drawn in. **b.** During expiration, the thoracic cavity and lungs resume their original positions and pressures, forcing air out.

a. Inspiration

b. Expiration

Figure 24.17 **Breathing in birds.**

Two types of air sacs are attached to the lungs of birds. When a bird inhales, most of the air enters the abdominal air sacs, and when a bird exhales, air moves through the lungs to the thoracic air sacs before exiting the trachea. This one-way flow of air through the lungs allows more fresh air to be present in the lungs with each breath, and this leads to greater uptake of oxygen from one breath of air.

Breathing

The breathing mechanism of humans is the same as that of other terrestrial vertebrates except birds. As in other mammals, the volume of the thoracic cavity and lungs is increased by muscle contractions that lower the diaphragm and raise the ribs (Fig. 24.16). These movements create a negative pressure in the thoracic cavity and lungs, and air then flows into the lungs, a process called *inspiration.* It is important to realize that air comes in because the lungs have already opened up; air does not force the lungs open. When the ribs and diaphragm muscles relax, the lungs recoil, and air moves out as a result of increased pressure in the lungs, a process called *expiration.*

Because air moves in and out by the same route, some residual air is always left in the lungs of humans. In contrast, birds use a one-way ventilation mechanism (Fig. 24.17). Incoming air is carried past the lungs by a trachea that takes it to a set of abdominal air sacs. Then air passes forward through the lungs into a set of thoracic air sacs. Fresh air never mixes with used air in the lungs of birds, and thereby gas-exchange efficiency is greatly improved.

Increased concentrations of hydrogen ion (H^+) and carbon dioxide (CO_2) in the blood are the primary stimuli that increase the breathing rate in humans. The chemical content of the blood is monitored by chemoreceptors called the **aortic bodies** and **carotid bodies,** which are specialized structures in the walls of certain arteries. These receptors are very sensitive to changes in H^+ and CO_2 concentrations, but they are only minimally sensitive to a lower oxygen (O_2) concentration. Information from the chemoreceptors goes to the *breathing center* in the brain, which increases the breathing rate when concentrations of hydrogen ions and carbon dioxide rise (see Fig. 24.21). The breathing center is also directly sensitive to the chemical content of the blood, including its oxygen content.

Check Your Progress

1. List the three steps of respiration in a complex animal.
2. Contrast respiration in insects with that in humans.
3. Explain how air moves into the lungs during breathing.

Answers: 1. The three steps of respiration are breathing, external exchange, and internal exchange. 2. Insects have many trachea that branch into tubes, which deliver oxygen to the cells. Humans have a single trachea that leads to the bronchi and then to the bronchioles, which enter the lungs where gas exchange occurs. 3. Muscle contractions that lower the diaphragm and raise the ribs cause negative pressure that pulls air into the lungs.

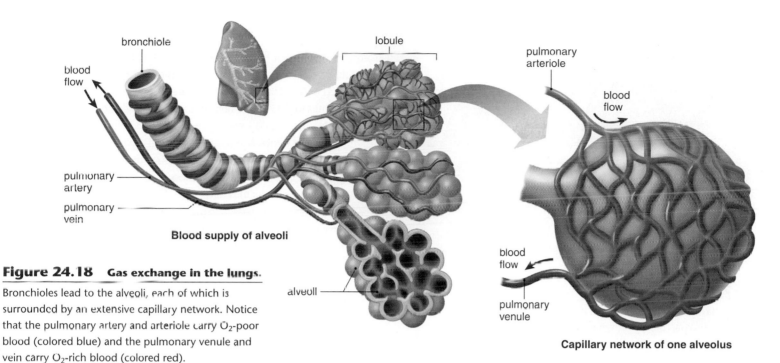

Figure 24.18   Gas exchange in the lungs.

Bronchioles lead to the alveoli, each of which is surrounded by an extensive capillary network. Notice that the pulmonary artery and arteriole carry O_2-poor blood (colored blue) and the pulmonary venule and vein carry O_2-rich blood (colored red).

Lungs and External Exchange of Gases

The lungs of humans and other mammals are more elaborately subdivided than those of amphibians and reptiles. Frogs and salamanders have a moist skin that allows them to use the surface of their body for gas exchange in addition to the lungs. It has been estimated that human lungs have a total surface area at least 50 times the skin's surface area because of the presence of alveoli (Fig. 24.18).

An alveolus, like the capillary that surrounds it, is bounded by squamous epithelium. Diffusion alone accounts for gas exchange between the alveolus and the capillary. Carbon dioxide, being more plentiful in the pulmonary venule, diffuses from a pulmonary capillary to enter an alveolus while oxygen, being more plentiful in the lungs, diffuses from an alveolus into a pulmonary capillary. The process of diffusion requires a gas-exchange region to be not only large, but also moist and thin. The alveoli are lined with surfactant, a film of lipoprotein that lowers the surface tension of water, thereby preventing the alveoli from collapsing. Some newborns, especially if premature, lack this film. Surfactant replacement therapy is now thankfully available to treat this condition.

The blood within pulmonary capillaries is indeed spread thin, and the red blood cells are pressed up against their narrow walls. The alveolar epithelium and the capillary epithelium are so close that together they are called the *respiratory membrane*. Hemoglobin in the red blood cells quickly picks up oxygen molecules as they diffuse into the blood.

Emphysema is a serious lung condition in which the walls of many alveoli have been destroyed. The lungs have less recoil and there is less surface area for gas exchange to occur.

Gills of Fish   In contrast to the lungs of terrestrial vertebrates, fish and other aquatic animals use gills as their respiratory organ (Fig. 24.19). In fish, water is drawn into the mouth and out from the pharynx across the gills. The flow of blood in gill capillaries is opposite the flow of water across the gills, and therefore the blood is always exposed to water having a higher oxygen content. In the end, about 80% to 90% of the dissolved oxygen in water is extracted.

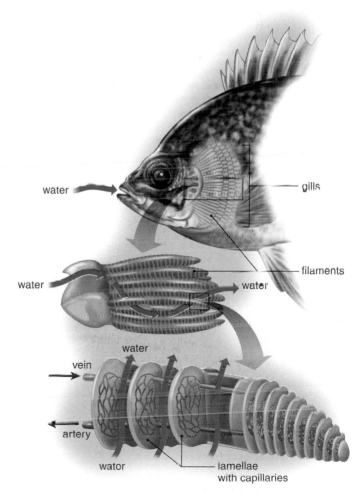

Figure 24.19   Gills of bony fishes.

Gills are finely divided into filaments, and each filament has many thin, platelike lamellae. Gases are exchanged between the capillaries inside the lamellae and the water that flows between the lamellae.

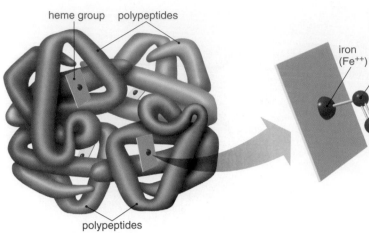

Figure 24.20   Hemoglobin.

Hemoglobin has four polypeptides, and each one is associated with a heme group. Oxygen bonds with the central iron atom of a heme group.

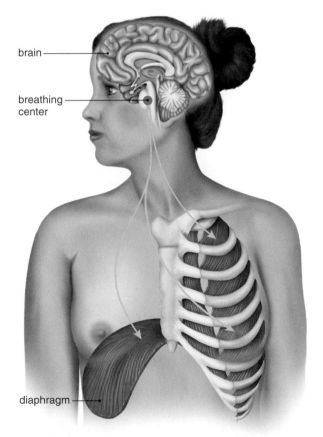

Figure 24.21   Neural control of breathing rate.

The brain regulates the breathing rate by controlling contraction of the rib cage muscles and the diaphragm. When the breathing rate increases, [H⁺] lowers as CO_2 is removed from the blood, and the pH of the blood returns to normal.

Transport and Internal Exchange of Gases

Recall from Chapter 23 that hemoglobin molecules within red blood cells carry oxygen to the body's tissues. If hemoglobin didn't transport oxygen, it's been estimated it would take three years for an oxygen molecule to move from your lungs to your toes. Each hemoglobin molecule contains four polypeptide chains, and each chain is folded around an iron-containing group called **heme.** It is actually the iron that bonds with oxygen and carries it to the tissues (Fig. 24.20). Since there are about 250 million hemoglobin molecules in each red blood cell, each cell is capable of carrying at least one billion molecules of oxygen. Hemoglobin gives up its oxygen in the tissues during internal exchange primarily because tissue fluid always has a lower oxygen concentration than blood does. This difference occurs because cells take up and utilize oxygen when they carry on cellular respiration. Another reason hemoglobin gives up oxygen is due to the warmer temperature and lower pH in the tissues, environmental conditions that are also caused by cellular respiration. When cells respire, they give off heat and carbon dioxide as by-products.

Carbon dioxide enters the blood during internal exchange because the tissue fluid always has a higher concentration of carbon dioxide. Most of the carbon dioxide is transported in the form of the **bicarbonate ion** (HCO_3^-). First, carbon dioxide combines with water, forming carbonic acid, and then this acid dissociates to a hydrogen ion (H⁺) and HCO_3^-:

$$CO_2 \;+\; H_2O \;\rightarrow\; H_2CO_3 \;\rightarrow\; H^+ \;+\; HCO_3^-$$

carbon dioxide   water   carbonic acid   hydrogen ion   bicarbonate ion

The H⁺ does cause the pH to lower, but only slightly because much of the H⁺ is absorbed by the globin portions of hemoglobin. The HCO_3^- is carried in the plasma.

What happens to the above equation in the lungs? As blood enters the pulmonary capillaries, carbon dioxide diffuses out of the blood into the alveoli. Hemoglobin gives up the H⁺ it has been carrying as this reaction occurs:

$$H^+ \;+\; HCO_3^- \;\rightarrow\; H_2CO_3 \;\rightarrow\; H_2O \;+\; CO_2$$

Now, much of the carbon dioxide diffuses out of the blood into the alveoli of the lungs. Should the blood level of H⁺ rise, the breathing center in the brain increases the breathing rate, and as more CO_2 leaves the blood, the pH of blood is corrected (Fig. 24.21).

Check Your Progress

1. Explain why oxygen diffuses into the blood from the lungs during external exchange but diffuses out of the blood into tissue fluid during internal exchange.
2. Explain how carbon dioxide is carried in the blood from the tissues to the lungs.

Answers: 1. Oxygen follows its concentration gradient; there is more oxygen in the lungs than the blood and there is more oxygen in the blood than in tissue fluid. Also, environmental conditions in the tissues causes hemoglobin to give up oxygen. 2. Carbon dioxide combines with water to form carbonic acid. Carbonic acid breaks down to the bicarbonate ion (HCO_3^-) and H⁺, which combines with hemoglobin.

24.3 Urinary System and Excretion

In complex animals, **kidneys** excrete nitrogenous wastes and are involved in regulating the water-salt balance of the body (Fig. 24.22). In addition, the mammalian kidney helps regulate the pH of blood. To summarize, these three functions can be associated with a kidney:

1. Excretion of nitrogenous wastes such as urea and uric acid.
2. Maintenance of the water-salt balance of the blood.
3. Maintenance of the acid-base balance of the blood.

In humans and other mammals, the kidneys are bean-shaped, reddish-brown organs, each about the size of a fist. They are located on either side of the vertebral column just below the diaphragm, where they are partially protected by the lower rib cage. **Urine** made by the kidneys is conducted from the body by the other organs in the urinary system. Each kidney is connected to a **ureter,** a tube that takes urine from the kidney to the **urinary bladder,** where it is stored until it is voided from the body through the single **urethra** (Fig. 24.23a). In males, the urethra passes through the penis, and in females, it opens in front of the opening of the vagina. In females, there is no connection between the genital (reproductive) and urinary systems, but there is a connection in males—that is, the urethra also carries sperm during ejaculation.

In amphibians, birds, reptiles, and some fishes, the bladder empties into the cloaca, a common chamber and outlet for the digestive, urinary, and genital tracts.

Kidneys

If a kidney is sectioned longitudinally, three major parts can be distinguished (Fig. 24.23b). The **renal cortex,** the outer region of a kidney, has a somewhat granular appearance. The **renal medulla** consists of the cone-shaped renal pyramids, which lie inside the renal cortex. The innermost part of the kidney is a hollow chamber called the **renal pelvis.** Urine collects in the renal pelvis and then is carried to the bladder by a ureter. Microscopically, each kidney is composed of about one million tiny tubules called **nephrons** (Fig. 24.23c). The nephrons of a kidney produce urine.

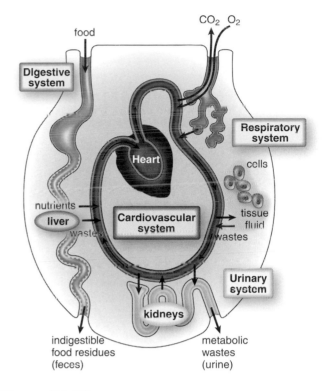

Figure 24.22   Keeping the internal environment steady.

Excretion carried out by the kidneys serves three functions: excretion of nitrogenous wastes, maintenance of water-salt balance, and maintenance of pH.

Figure 24.23   The human urinary system.

a. The human urinary system produces, stores, and expels urine from the body. **b.** The human kidney has three sections, which can be correlated with (**c**) the presence of about one million nephrons. Nephrons do the work of the kidney.

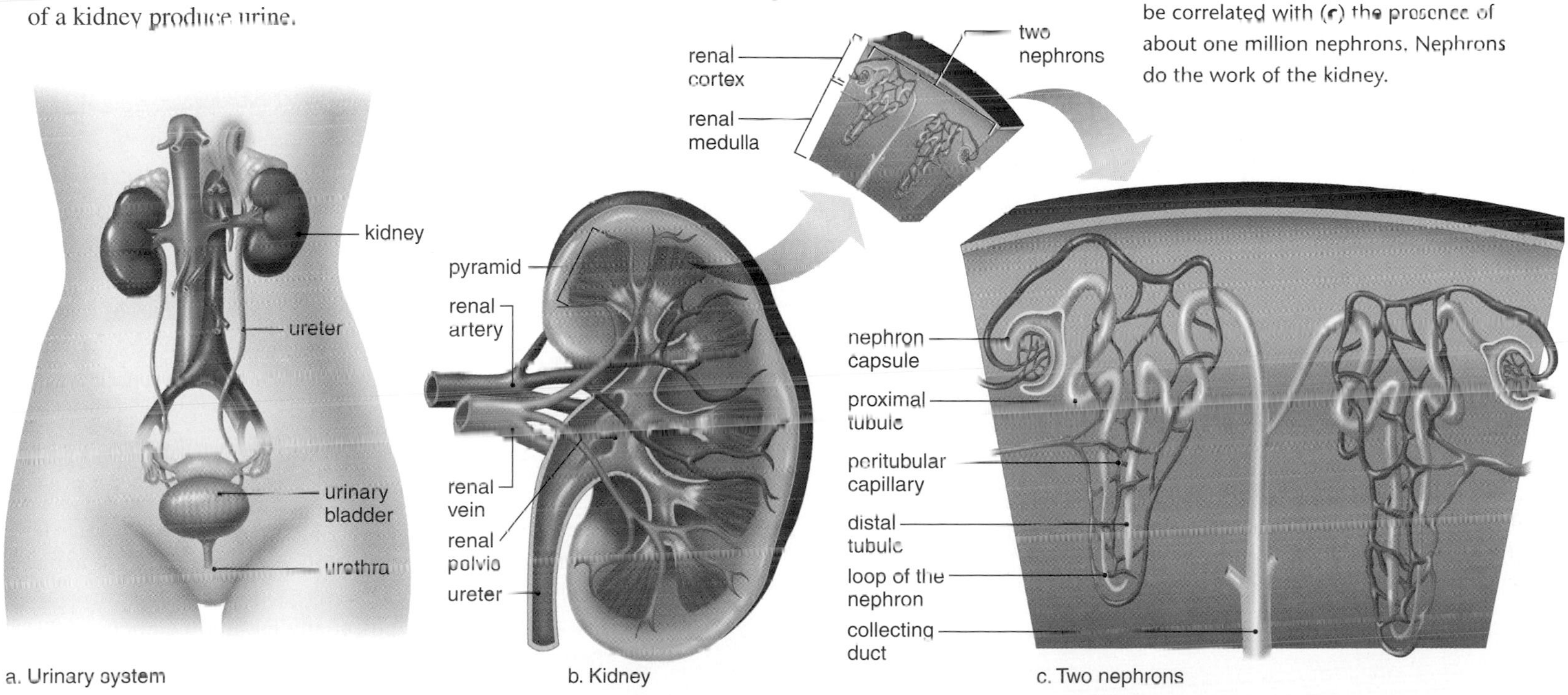

renal cortex

renal medulla

two nephrons

kidney

pyramid

renal artery

renal vein

renal pelvis

ureter

ureter

urinary bladder

urethra

nephron capsule

proximal tubule

peritubular capillary

distal tubule

loop of the nephron

collecting duct

a. Urinary system

b. Kidney

c. Two nephrons

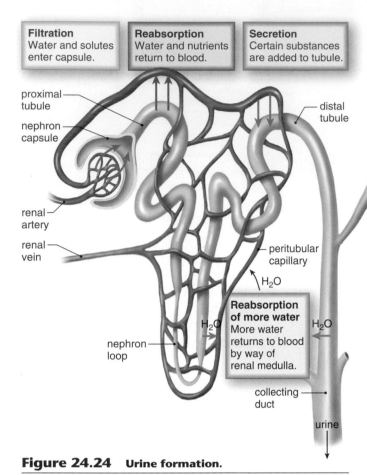

Figure 24.24 Urine formation.

Urine formation requires three steps: filtration, reabsorption, and secretion. In addition, humans are able to produce a hypertonic urine because water is reabsorbed into the blood along the length of the nephron and especially at the nephron loop.

Nephrons

Each nephron is made of several parts (Fig. 24.24). The blind end of a nephron is pushed in on itself to form the **nephron capsule.** The inner layer of the capsule is composed of specialized cells that allow easy passage of molecules. Leading from the capsule is a portion of the nephron called the **proximal tubule,** which is lined by cells with many mitochondria and tightly packed microvilli. Then comes the **nephron loop** with a descending limb and an ascending limb. This is followed by the **distal tubule.** Several distal tubules enter one **collecting duct.** A collecting duct delivers urine to the renal pelvis. The nephron loop and the collecting duct give the pyramids of the renal medulla their striped appearance (see Fig. 24.23*b*).

Each nephron has its own blood supply, and various exchanges occur between parts of the nephron and a blood capillary as urine forms.

Urine Formation

Urine formation requires three steps: filtration, reabsorption, and secretion (Fig. 24.24).

Filtration Filtration occurs whenever small substances pass through a filter and large substances are left behind. During urine formation, **filtration** is the movement of small molecules from a blood capillary to the inside of the capsule as a result of adequate blood pressure. Small molecules, such as water, nutrients, salts, and urea, move to the inside of the capsule. Plasma proteins and blood cells are too large to be part of this filtrate, so they remain in the blood.

If the composition of the filtrate were not altered in other parts of the nephron, death from loss of nutrients (starvation) and loss of water (dehydration) would quickly follow. The next step, reabsorption, helps prevent this from happening.

Reabsorption of Solutes **Reabsorption** takes place when substances from the proximal tubule move into the blood. Nutrients such as glucose and amino acids also return to the blood. This process is selective because some molecules such as glucose are both passively and actively reabsorbed. The cells of the proximal tubule have numerous microvilli, which increase the surface area, and numerous mitochondria, which supply the energy needed for active transport. However, if there is more glucose in the filtrate than there are carriers to handle it, glucose will appear in the urine. Glucose in the urine is a sign of diabetes mellitus, sometimes caused by lack of the hormone insulin.

Sodium ions (Na^+) are also actively pumped into the peritubular capillary, and then chloride ions (Cl^-) follow passively. Now water moves by osmosis from the tubule into the blood. About 60–70% of salt and water are reabsorbed at the proximal tubule.

Urea, the primary nitrogenous waste product of human metabolism, and other types of nitrogenous wastes excreted by humans are passively reabsorbed, and most remain in the filtrate.

Secretion **Secretion** refers to the transport of substances into the nephron by means other than filtration. For our purposes, secretion may be particularly associated with the distal tubule. Substances such as uric acid, hydrogen ions, ammonia, and penicillin are eliminated by secretion. The process of secretion helps rid the body of potentially harmful compounds that were not filtered into the capsule.

Regulation of Water-Salt Balance and pH   Typically, animals have some means of regulating the osmolarity of the internal environment so that the water-salt balance stays within normal limits. Insects have a unique excretory system consisting of long, thin tubules called **Malpighian tubules** attached to the gut (Fig. 24.25). Uric acid, their primary nitrogenous waste product, simply flows from the surrounding hemolymph into these tubules, and water follows a salt gradient established by active transport of potassium (K^+). Water and other useful substances are reabsorbed at the rectum, but the uric acid leaves the body at the anus. Insects that live in water or eat large quantities of moist food reabsorb little water. But insects in dry environments reabsorb most of the water and excrete a dry, semisolid mass of uric acid. Most animals can regulate the blood level of both water and salt. For example, freshwater fishes take up salt in the digestive tract and gills and produce large amounts of dilute urine. Saltwater fishes take in salt water by drinking, but then they pump ions out at the gills, and produce only small amounts of a concentrated urine.

In mammals, the long nephron loop allows secretion of a hypertonic urine (see Fig. 24.24). The ascending limb of the nephron loop pumps salt and also urea into the renal medulla, and water follows by osmosis both at the descending limb of the nephron loop and at the collecting duct. As you will see in Chapter 27, at least three hormones are involved in regulating water-salt reabsorption by the kidneys. Drinking coffee interferes with one of these hormones, and that's why coffee is a *diuretic,* a substance that causes the production of more urine.

Most mammals can also regulate the pH of the blood. The bicarbonate (HCO_3^-) buffer system of the blood and regulation of the breathing rate (see Fig. 24.21) to rid the body of CO_2 both contribute to maintaining blood pH. As helpful as these mechanisms might be, only the kidneys can excrete a wide range of acidic and basic substances. The kidneys are slower acting than the buffer/breathing mechanism, but they have a more powerful effect on pH. For the sake of simplicity, we can think of the kidneys as reabsorbing bicarbonate ions and excreting hydrogen ions as needed to maintain the normal pH of the blood (Fig. 24.26). If the blood is acidic, hydrogen ions are excreted and bicarbonate ions are reabsorbed. If the blood is basic, hydrogen ions are not excreted and bicarbonate ions are not reabsorbed. The fact that urine is most often acidic shows that usually an excess of hydrogen ions are excreted. Ammonia (NH_3) provides a means for buffering these hydrogen ions in urine: ($NH_3 + H^+ \rightarrow NH_4^+$). Ammonia (the presence of which is quite obvious in the diaper pail or kitty litter box) is produced in tubule cells by the deamination of amino acids. Phosphate provides another means of buffering hydrogen ions in urine.

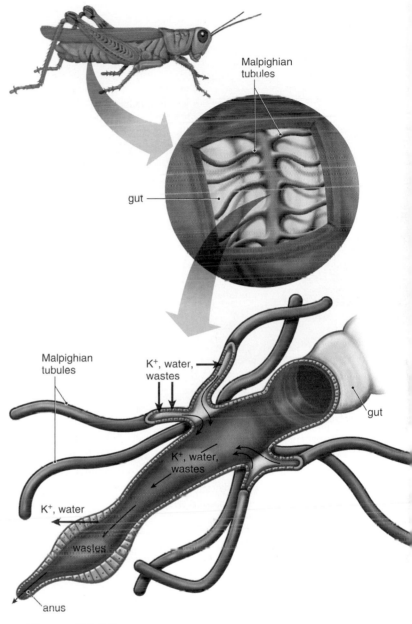

Figure 24.25   Malpighian tubules.

The Malpighian tubules of insects are attached to the gut and surrounded by the hemolymph of the open circulatory system. K^+ is secreted into these tubules, drawing in water by osmosis. Much of the water and K^+ is reabsorbed across the wall of the rectum.

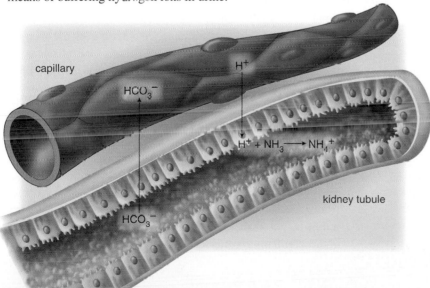

Figure 24.26   Acid-base balance.

In the kidneys, bicarbonate ions (HCO_3^-) are reabsorbed and hydrogen ions (H^+) are excreted as needed to maintain the pH of the blood. Excess hydrogen ions are buffered, for example, by ammonia (NH_3), which becomes ammonium (NH_4^+). Ammonia is produced in tubule cells by the deamination of amino acids.

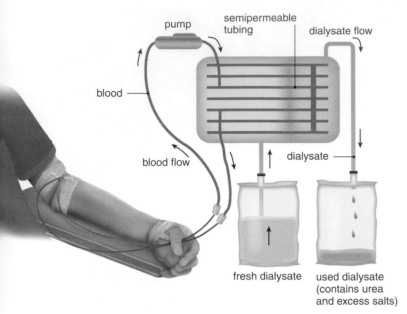

pump
semipermeable tubing
dialysate flow
blood
blood flow
dialysate
fresh dialysate
used dialysate (contains urea and excess salts)

Figure 24.27   An artificial kidney machine.

As the patient's blood is pumped through dialysis tubing, the tubing is exposed to a dialysate (dialysis solution). Wastes exit from blood into the solution because of a preestablished concentration gradient. In this way, blood is not only cleansed, but its water-salt and acid-base balances can also be adjusted.

Problems with Kidney Function

Many types of illnesses, especially diabetes, hypertension, and inherited conditions, cause progressive renal disease and renal failure. Urinary tract infections, an enlarged prostate gland, pH imbalances, or simply an intake of too much calcium can lead to kidney stones. Kidney stones form in the renal pelvis and usually pass unnoticed in the urine flow. If they grow to several centimeters and block the renal pelvis or ureter, back pressure builds up and destroys nephrons. One of the first signs of nephron damage is the presence of albumin, white blood cells, or even red blood cells in the urine, as detected by a urinalysis. If damage is so extensive that more than two-thirds of the nephrons are inoperative, urea and other waste substances accumulate in the blood. Although nitrogenous waste in the blood is a threat to homeostasis, the retention of water and salts is of even greater concern. **Edema,** fluid accumulation in the body tissues, may occur. Imbalance in the ionic composition of body fluids can lead to loss of consciousness and even to heart failure.

Hemodialysis and Kidney Replacement

Patients with renal failure can undergo **hemodialysis,** utilizing either an artificial kidney machine or **continuous ambulatory peritoneal dialysis (CAPD).** *Dialysis* is defined as the diffusion of dissolved molecules through a semipermeable membrane with pore sizes that allow only small molecules to pass through. In an artificial kidney machine, the patient's blood is passed through a membranous tube, which is in contact with a dialysis solution, or **dialysate** (Fig. 24.27). Substances more concentrated in the blood diffuse into the dialysate, and substances more concentrated in the dialysate diffuse into the blood. The dialysate is continuously replaced to maintain favorable concentration gradients. In this way, the artificial kidney can be utilized either to extract substances from the blood, including waste products or toxic chemicals and drugs, or to add substances to the blood—for example, bicarbonate ions (HCO_3^-) if the blood is acidic. In the course of a three- to six-hour hemodialysis procedure, from 50 to 250 grams of urea can be removed from a patient, which greatly exceeds the amount excreted by our kidneys within the same time frame. Therefore, a patient needs to undergo treatment only about twice a week.

CAPD is so named because the peritoneum, the epithelium that lines the abdominal cavity, is the dialysis membrane. A fresh amount of dialysate is introduced directly into the abdominal cavity from a bag that is temporarily attached to a permanently implanted plastic tube. The dialysate flows into the peritoneal cavity by gravity. Waste and salt molecules pass from the blood vessels in the abdominal wall into the dialysate before the fluid is collected four or eight hours later. The solution is drained into a bag from the abdominal cavity by gravity, and then it is discarded. One advantage of CAPD over an artificial kidney machine is that the individual can go about his or her normal activities during CAPD.

Patients with renal failure may undergo a transplant operation in which a functioning kidney from a donor is received. A person only needs one functioning kidney. As with all organ transplants, the possibility of organ rejection exists. Receiving a kidney from a close relative has the highest chance of success. The current one-year survival rate is 97% if the kidney comes from a relative, and 90% if it is from a nonrelative. In the future, it may be possible to use kidneys from pigs or kidneys created in the laboratory for transplant operations.

Check Your Progress

1. Explain how the kidneys regulate blood pH.
2. Explain how an artificial kidney machine works.

Answers: 1. The kidneys reabsorb bicarbonate ions and excrete hydrogen ions as needed to maintain blood pH. 2. The patient's blood passes through a membranous tube filled with dialysate. The dialysate is continuously replaced to maintain favorable concentration gradients so that appropriate materials diffuse into and out of the blood.

Summary

24.1   Digestive System

The digestive, respiratory, and urinary systems make exchanges with the external environment and blood, thereby supplying the blood with nutrients and oxygen and cleansing it of waste molecules. Then the blood makes exchanges with the tissue fluid, and in this way cells acquire nutrients and oxygen and rid themselves of wastes.

Tube Within a Tube Body Plan

Like the earthworm, humans have a complete digestive system that (1) ingests food; (2) breaks food down to small molecules that can cross plasma membranes; (3) absorbs these nutrient molecules; and (4) eliminates indigestible remains.

The digestive tract consists of several specialized parts:

- **Mouth:** Teeth chew the food, saliva contains salivary amylase for digesting starch, and the tongue forms a bolus for swallowing.
- **Pharynx:** The air and food passages cross in the pharynx. During swallowing, the air passage is blocked off by the soft palate and epiglottis; peristalsis begins.
- **Stomach:** The stomach expands and stores food and also churns, mixing food with the acidic gastric juices. This juice contains pepsin, an enzyme that digests protein. The stomach of ruminants has a special chamber, the rumen, where symbiotic bacteria and protozoans digest grass.
- **Small intestine:** The duodenum of the small intestine receives bile from the liver and pancreatic juice from the pancreas. Pancreatic juice contains trypsin (digests protein), lipase (digests fat), and pancreatic amylase (digests starch). The small intestine produces enzymes that finish digestion, breaking food down to small molecules that cross the villi. Amino acids and glucose enter blood capillaries. Glycerol and fatty acids are joined and packaged as lipoproteins before entering lymphatic vessels called lacteals.
- **Large intestine:** The large intestine stores the remains of digestion until they can be eliminated. It also absorbs water, salts, and some vitamins. Reduced water absorption results in diarrhea. The intake of water and fiber helps prevent constipation.

Accessory Organs

The pancreas is both an exocrine gland that produces pancreatic juice and an endocrine gland that produces the hormones insulin and glucagon. The liver produces bile, which is stored in the gallbladder. Pancreatic juice and bile enter the small intestine. The nervous system and the peptide hormones regulate the secretion of digestive juices and bile.

The hepatic portal vein carries absorbed molecules from the small intestine to the liver, an organ that performs many important functions.

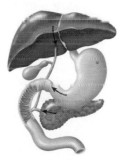

24.2   Respiratory System

A respiratory system has these functions: (1) breathing; (2) external exchange of gases; and (3) internal exchange of gases.

The respiratory tract consists of several parts:

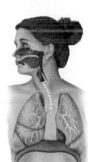

- **Nasal cavities:** In the nasal cavities of the nose, air is moistened, and hairs trap debris.
- **Pharynx:** Air crosses from front to back.
- **Larynx:** Contains the vocal cords.
- **Respiratory tree:** Consists of the trachea, bronchi, and bronchioles, which terminate in the alveoli of the lungs.

  Insects have many tracheae. A tracheal system has no need of a respiratory pigment because air is delivered directly to cells.
- **Lungs:** Contain many alveoli, air sacs surrounded by a capillary network.

Breathing

- **Inspiration:** The diaphragm lowers, and the rib cage moves up and out; the lungs expand and air rushes in.
- **Expiration:** The diaphragm relaxes and moves up; the rib cage moves down and in; pressure in the lungs increases; air is pushed out of the lungs.

Lungs and External Exchange of Gases

- Alveoli are surrounded by pulmonary capillaries.
- CO_2 diffuses from the blood into the alveoli, and O_2 diffuses into the blood from the alveoli, because of their respective concentration gradients.
- Hemoglobin takes up oxygen.

Transport and Internal Exchange of Gases

Oxygen for cellular respiration follows this path:

- Transported by the iron portion of hemoglobin.
- Exits blood at tissues by diffusion.
- Given up by hemoglobin in capillaries because cellular respiration makes tissues warmer and more acidic.

Carbon dioxide, from cellular respiration, follows this path:

- Enters blood by diffusion.
- Taken up by red blood cells and joins with water to form carbonic acid.
- Carbonic acid breaks down to H^+ and bicarbonate ion (HCO_3^-):

CO_2	$+$	H_2O	$\longrightarrow$	H_2CO_3	$\longrightarrow$	H^+	$+$	HCO_3^-
carbon dioxide		water		carbonic acid		hydrogen ion		bicarbonate ion

Bicarbonate ions are carried in plasma. H^+ combines with globin of hemoglobin.

- In the lungs, H^+ joins with bicarbonate ion to form carbonic acid, which breaks down to water and carbon dioxide:

$$H^+ \;+\; HCO_3^- \;\longrightarrow\; H_2CO_3 \;\longrightarrow\; H_2O \;+\; CO_2$$

24.3   Urinary System and Excretion

The kidneys perform the following functions:

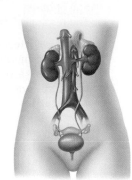

- Excrete nitrogenous wastes, such as urea and uric acid.
- Maintain the normal water-salt balance of blood.
- Maintain the acid-base balance of blood.

The urinary system consists of these parts:

- **Kidneys:** Produce urine.
- **Ureters:** Take urine to the bladder.
- **Urinary bladder:** Stores urine.
- **Urethra:** Releases urine to the outside.

Kidneys

Macroscopically, the kidneys have three parts: renal cortex, renal medulla, and renal pelvis. Microscopically, they contain the nephrons. A nephron has a nephron capsule, proximal tubule, nephron loop, and distal tubule.

Urine formation requires three steps:

- **Filtration:** Water, nutrients, and wastes move from the blood to the inside of the nephron capsule.
- **Reabsorption:** Primarily salts, water, and nutrients are reabsorbed at the proximal tubule.
- **Secretion:** Certain substances (e.g., hydrogen ions) are transported into the distal tubule from blood.

Regulation of Water-Salt Balance and pH   Animals regulate their osmolarity; examples are the Malpighian tubules of insects and the specialized organs of saltwater versus freshwater fishes. The reabsorption of water and the production of a hypertonic urine involve establishment of a solute gradient that pulls water from the descending limb of the nephron loop and from the collecting duct. The kidneys keep blood pH at about 7.4 by reabsorbing HCO_3^- and excreting H^+ as needed. Ammonia buffers H^+ in the urine.

Problems with Kidney Function   Various medical conditions, including diabetes, kidney stones, and kidney infections, can lead to renal failure. Renal failure can be treated by hemodialysis using a kidney machine or CAPD, or by a kidney transplant.

Thinking Scientifically

1. Camels can survive for days or even weeks without taking a drink. They can lose an amount of water equivalent to up to 40% of their body weight with no ill effects. This type of water loss would be lethal to any other mammal. What physiological adaptations do you suppose allow a camel to survive so long without drinking water?

2. Participants in the Atkins diet avoid carbohydrates, but eat protein-rich foods until they are satiated. Stories of success told by followers of the Atkins diet have led to the current "low-carb" craze apparent in grocery stores, restaurants, and fast-food chains. Why would a diet high in proteins lead to weight loss? Do you think the weight loss is sustainable? What might be some negative side effects of a high-protein diet?

Testing Yourself

Choose the best answer for each question.

1. Label the components of the human digestive system in the following illustration.

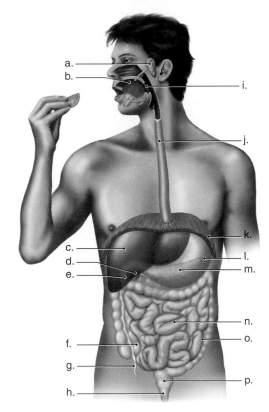

2. If plants did not have cellulose in their cell walls,
   a. an important nutrient would be missing in the diet of omnivores.
   b. the teeth of herbivores would more closely resemble those of carnivores.
   c. more birds would be carnivores.
   d. herbivores would have a longer digestive tract.

3. The stomach

   a. is lined with a thick layer of mucus.
   b. contains sphincter glands.
   c. has a pH of about 6.
   d. digests pepsin.
   e. More than one of these are correct.

4. A rumen in cows

   a. provides them with water.
   b. provides a home for symbiotic bacteria.
   c. helps them digest plant material.
   d. Both b and c are correct.

5. Which of the following is not a digestive enzyme?

   a. amylase
   b. trypsin
   c. lipase
   d. nuclease
   e. isomerase

6. Which of the following is not a function of the liver?

   a. removal of poisonous substances from the blood
   b. secretion of digestive juices
   c. production of albumin
   d. storage of glucose
   e. production of bile

7. Label the components of the human respiratory system in the following illustration.

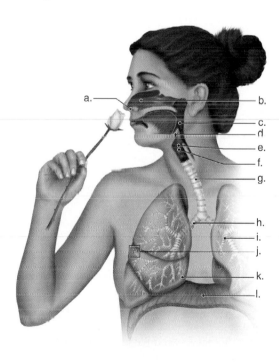

8. Infection of one of the two branches just below the trachea is called

   a. tonsillitis.
   b. meningitis.
   c. a sinus infection.
   d. bronchitis.
   e. laryngitis.

9. Birds have a higher gas-exchange capacity than other vertebrates because they

   a. have more alveoli in their lungs.
   b. have a stronger diaphragm.
   c. have a shorter trachea.
   d. never mix fresh air with used air.
   e. breathe at a more rapid rate.

10. Label the components of the human urinary system in the following illustration.

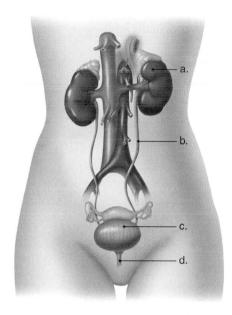

For questions 11–14, identify the kidney component in the key that matches the description.

Key:

   a. nephron capsule
   b. proximal tubule
   c. nephron loop
   d. distal tubule

11. Pumps salt into the renal medulla, and water follows by osmosis.

12. Uric acid is eliminated by secretion here.

13. Microvilli reabsorb molecules such as glucose here.

14. Filtration occurs here.

15. The presence of albumin in the urine is indicative of

   a. kidney stones.
   b. exposure to diuretics.
   c. the inability to adjust blood pH.
   d. hypertension.

16. Cleansing inhaled air involves

   a. hairs.
   b. cilia.
   c. mucus.
   d. All of these are correct.

17. Food and air both travel through the
    a. lungs.
    b. pharynx.
    c. larynx.
    d. trachea.

In questions 18–21, match the functions to the structures in the key.

Key:
    a. glottis
    b. larynx
    c. bronchi
    d. lungs

18. Passage of air into larynx.
19. Passage of air into lungs.
20. Sound production.
21. Gas exchange.
22. The use of an artificial kidney is known as
    a. filtration.
    b. excretion.
    c. hemodialysis.
    d. None of these are correct.

23. Which of the following materials would not be filtered from the blood at the nephron capsule?
    a. water
    b. urea
    c. protein
    d. nutrients
    e. salts

24. The renal medulla has a striped appearance due to the presence of which structures?
    a. nephron loop
    b. collecting duct
    c. capillaries
    d. Both a and b are correct.

Go to www.mhhe.com/maderessentials for more quiz questions.

Bioethical Issue

As a result of serious injury or illness, a person may enter a persistent vegetative state. In this condition, the person does not recognize anyone and feels no pain. Advances in medical care now allow us to use ventilators and feeding tubes to keep vegetative patients alive for years. A living will allows patients to declare that they do not wish to be maintained in this way, giving doctors the legal authority to discontinue life-sustaining treatments. But for a patient without a living will, family members are left with a difficult decision and sometimes a court battle over what they all think is best. One recent case involved a woman who was kept in a vegetative state since suffering a heart attack in 1990. If her feeding tube were removed, she would die within two weeks. Her husband asked to have the tube removed, saying that would have been his wife's wish. Her parents, on the other hand, wanted her to be kept alive. They believed she was trying to communicate with noises and facial expressions, although her doctors said those activities were simply reflexes. What would you do if you were a judge and had to determine whether the husband's or the parents' wishes should be carried out?

Understanding the Terms

Match the terms to these definitions:

a. _____ The digestive and respiratory passages come together here.

b. _____ A hydrolytic enzyme that breaks down proteins in the stomach.

c. _____ The name for the contents of the digestive tract as they move from the stomach to the intestine.

d. _____ The storage organ for bile.

e. _____ The voice box.

f. _____ The windpipe; it connects the larynx to the bronchi.

g. _____ Small air pockets in bronchioles.

h. _____ The iron-containing group in hemoglobin.

i. _____ Tiny urine-producing tubules in the kidney.

Human Nutrition

CHAPTER 25

Extremely high-protein diets may be dangerous to health.

Some vitamins taken in extreme excess can cause health problems.

Trans fats are the most harmful kind of fat in the diet.

Don't partially hydrogenate me™

Did you know that an extremely high-protein diet may eventually cause liver and kidney disease? That some vitamins can harm the body if taken in extreme excess? Or that trans fats are the worst possible kind of fat you can eat?

We hear a lot about health problems that can result from eating high-fat and high-cholesterol foods. While most individuals are aware that unsaturated fats are considered

▶ ▶ ▶

healthier than saturated fats, until recently the average person did not give much consideration to the role of trans fats in the diet. *Trans fats* are formed from liquid unsaturated fats to which hydrogen has been added to make them resemble a solid saturated fat, a process called hydrogenation. Why do manufacturers go through the trouble of creating trans fats when they could just use a natural saturated fat? The primary reason is that saturated fats must be listed on the Nutrition Facts panel of all food packages, the labeling that consumers rely on for information about a food's content. However, trans fats do not have to be listed on the Nutrition Facts panel, and thus a food product may appear to be healthier than it actually is.

Recently, the Food and Drug Administration (FDA) has decided that trans-fat content *must* be disclosed on Nutrition Facts panels beginning in 2006. Furthermore, recent media coverage has made more people aware of the dangers of diets high in trans fat, and manufacturers of products high in trans fats are considering ways to lower the trans-fat content.

In this chapter, you will learn about the components of a healthy, balanced diet as well as some of the consequences of an unhealthy one.

25.1 Nutrition

The vigilance of your immune system, the strength of your muscles and bones, the ease with which your blood circulates—all aspects of your body's functioning—depend on proper **nutrition** (Fig. 25.1). A **nutrient** is a component of food that performs a physiological function in the body. Nutrients provide us with energy, promote growth and development, and regulate cellular metabolism. They are also involved in homeostasis. For example, nutrients help maintain the fluid balance and proper pH of blood. Your body can make up for a nutrient deficiency to a degree, but eventually signs and symptoms of a **deficiency disorder** will appear. As an example, vitamin C is needed to synthesize and maintain *collagen*, the protein that holds tissues together. When the body lacks vitamin C, collagen weakens, and capillaries break easily. Gums may bleed, especially when the teeth are brushed, or tiny bruises may form under the skin when it's gently pressed. In other words, early signs of vitamin C deficiency are gums that bleed and skin that bruises easily (Fig. 25.2).

By learning about nutrition, you can improve your diet and increase the likelihood of enjoying a longer, more active and productive life. Conversely, poor diet and lack of physical activity are responsible for 400,000 deaths in the United States annually. Such lifestyle factors may soon overtake smoking as the major cause of preventable death. Everyone can benefit from learning what constitutes a poor diet versus a healthy diet so that we can choose foods that supply all the nutrients in proper balance.

Figure 25.1 **A healthy diet.**

Making the right food choices leads to better health.

Introducing the Nutrients

Several factors, including cultural and ethnic backgrounds, financial situations, environmental conditions, and psychological states, influence what we eat. A **diet** is composed of a person's typical food choices. A *balanced diet* supplies all the nutrients in the proper proportions necessary for a healthy, functioning body.

There are six classes of nutrients: carbohydrates, lipids, proteins, amino acids, minerals, vitamins, and water. An **essential nutrient** must be supplied by the diet because the body is not able to produce it or at least not in sufficient quantity to meet the body's needs. For example, certain amino acids are essential nutrients.

Carbohydrates, lipids, and proteins are called **macronutrients** because the body requires relatively large quantities of them. **Micronutrients,** such as vitamins and minerals, are needed in small quantities only. Macronutrients, not micronutrients, supply our energy needs. Although advertisements often imply that people can boost their energy levels by taking vitamin or mineral supplements, the body does not metabolize these nutrients for energy. Water does not provide energy either. Therefore, foods with high water content, such as fruits and vegetables, are usually lower in energy content than foods with less water and more macronutrient content.

Nearly every food is a mixture of nutrients. A slice of bread, for example, is about 50% carbohydrates, 35% water, 10% proteins, and 4% lipids. Vitamins and minerals make up less than 1% of the bread's nutrient content (Fig. 25.3). No single, naturally occurring food contains enough essential nutrients to meet all of our nutrient needs.

Contrary to popular belief, so-called "bad" foods or "junk foods" do have nutritional value. If a food contains water, sugar, or fat, it has nutritional value. However, such foods as sugar-sweetened soft drinks, cookies, and pastries have high amounts of fat and/or sugar in relation to their vitamin and mineral content. Therefore, these foods are more appropriately called *empty-calorie foods,* rather than junk foods. Diets that contain too many empty-calorie foods will assuredly lack enough vitamins and minerals.

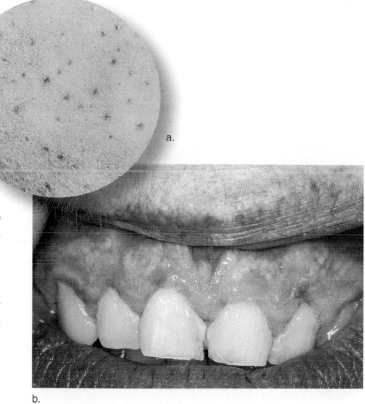

a.

b.

Figure 25.2 Vitamin C deficiency.

a. Pinpoint hemorrhages (tiny bruises that appear as red spots in the skin) are an early indication of vitamin C deficiency. **b.** Bleeding gums are another early sign of vitamin C deficiency.

Check Your Progress

1. Which classes of nutrients do not provide energy?
2. Explain why the expression "junk food" may be inappropriate.
3. Give an example of an empty-calorie food and explain why it fits this description.

Answers: 1. Vitamins, minerals, and water do not provide energy. 2. So-called "junk foods" do not contain nutrients. 3. Answers should include foods that are rich sources of simple sugars and/or fat, such as sugar-sweetened soft drinks, pastries, and cookies. These are empty-calorie foods because they provide little else besides calories.

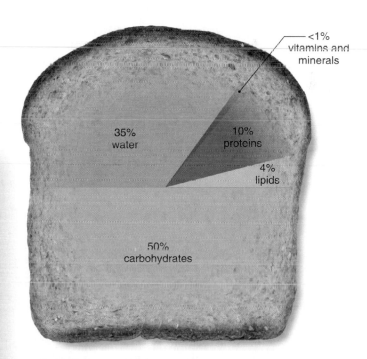

<1% vitamins and minerals

35% water

10% proteins

4% lipids

50% carbohydrates

Figure 25.3 Nutrient composition of a slice of bread.

Most foods are mixtures of nutrients.

25.2 The Classes of Nutrients

Each type of nutrient usually has more than one function in the body and can be supplied by several food sources (Fig. 25.4 and Table 25.1).

Carbohydrates

Carbohydrates are present in food as sugars, starch, and fiber. Fruits, vegetables, milk, and honey are natural sources of sugars. Glucose and fructose are monosaccharide sugars, and lactose (milk sugar) and sucrose (table sugar) are disaccharides. After absorption into the body, all sugars are converted to glucose for transport in the blood and use by cells. Glucose is the preferred direct energy source in cells.

Plants store glucose as starch, and animals store glucose as glycogen. Good sources of starch are beans, peas, cereal grains, and potatoes. Starch is digested to glucose in the digestive tract, and any excess glucose is stored as glycogen. Although other animals likewise store glucose as glycogen in liver or muscle tissue (meat), it is gone by the time an animal is eaten for food. Except for honey and milk, which contain sugars, animal foods do not contain carbohydrates.

Fiber

Fiber includes various nondigestible carbohydrates derived from plants. Food sources rich in fiber include beans, peas, nuts, fruits, and vegetables. Whole-grain products are also a good source of fiber, and are therefore more nutritious

Figure 25.4   Major food sources.

Carbohydrates, lipids, and proteins are represented.

Table 25.1	Summarizing the Classes of Nutrients		
Class of Nutrient	**Major Food Sources**	**Primary Physiological Roles**	**Macronutrient or Micronutrient**
Carbohydrates	Sugars and starches in fruits, vegetables, cereals and other grains	Glucose is metabolized for energy; fiber adds to fecal bulk, preventing constipation.	Macronutrient
Lipids	Oil, margarine, salad dressings, meat, fish, poultry, nuts, fried foods, dairy products made with whole milk	Triglycerides are metabolized for energy and stored for insulation and protection of organs. Cholesterol is used to make certain hormones, bile, and vitamin D.	Macronutrient
Proteins	Meat, fish, poultry, eggs, nuts, dried beans and peas, soybeans, cereal products, milk, cheese, yogurt	Proteins are needed for building, repairing, and maintaining tissues; synthesizing enzymes and certain hormones; and producing antibodies.	Macronutrient
Minerals	Widespread in vegetables and other foods	Regulation of energy metabolism, maintenance of fluid balance, and production of certain structures, enzymes, and hormones	Micronutrient
Vitamins	Widespread in foods; plant foods are good sources of antioxidants.	Regulation of metabolism and physiological development	Micronutrient
Water	Widespread in foods, beverages	Participant in many chemical reactions; needed for maintenance of body fluids, temperature regulation, joint lubrication, and transportation of material in cells and the body	(Not applicable)

than food products made from refined grains. During *refinement,* fiber and also vitamins and minerals are removed from grains, so that primarily starch remains. A slice of bread made from whole-wheat flour, for example, contains 3 grams of fiber; a slice of bread made from refined wheat flour contains less than a gram of fiber.

Technically, fiber is not a nutrient for humans because it cannot be digested to small molecules that enter the bloodstream. Insoluble fiber, however, adds bulk to fecal material, which stimulates movements of the large intestine, preventing constipation. Soluble fiber combines with bile acids and cholesterol in the small intestine and prevents them from being absorbed. In this way, high fiber diets may protect against heart disease. The typical American consumes only about 15 grams of fiber each day; the recommended daily intake of fiber is 25 grams for women and 38 grams for men. To increase your fiber intake, eat whole-grain foods, snack on fresh fruits and raw vegetables, and include nuts and beans in your diet (Fig. 25.5).

Can Carbohydrates Be Harmful?

If you or someone you know has lost weight by following the Atkins or South Beach diet, you may think "carbs" are unhealthy and should be avoided. According to nutritionists, however, carbohydrates should supply a large portion of your energy needs. Evidence suggests that Americans are not eating the right kind of carbohydrates. In some countries, the traditional diet is 60–70% high-fiber carbohydrates, and these people have a low incidence of the diseases that plague Americans.

Obesity is associated with diabetes type 2 and cardiovascular disease, as discussed in Section 25.3. Some nutritionists hypothesize that the high intake of foods, such as those rich in refined carbohydrates and fructose sweeteners processed from cornstarch, may be responsible for the prevalence of obesity in the United States. These foods are empty-caloric foods that provide sugars but no vitamins or minerals. Table 25.2 tells you how to reduce dietary sugars. Other nutritionists point out that consuming too much energy from any source contributes to body fat, which increases a person's risk of obesity and associated illnesses. Because many foods, such as doughnuts, cakes, pies, and cookies, are high in both refined carbohydrates and fat, it is difficult to determine which dietary component is responsible for the current epidemic of obesity among Americans.

Many people mistakenly believe children become hyperactive after eating sugar. There is no scientific basis for this belief because sucrose is broken down into glucose and fructose in the small intestine, and these sugars are absorbed into the bloodstream. The spike caused by their absorption does not last long. Excess glucose and fructose enters the liver, and fructose is converted to glucose. As you know, the liver stores glucose as glycogen, and glycogen is broken down to maintain the proper glucose level.

Lipids

Like carbohydrates, **triglycerides** (fats and oils) supply energy for cells, but **fat** is stored for the long term in the body. Fat deposits under the skin, called *subcutaneous fat,* insulate the body from cold temperatures; deeper fat deposits in the trunk protect organs against injury.

Figure 25.5 **Fiber-rich foods.**

Plants provide a good source of carbohydrates. They also provide fiber when they are not processed (refined).

Table 25.2	Reducing Dietary Sugars

TO REDUCE DIETARY SUGAR:

1. Eat fewer sweets, such as candy, soft drinks, ice cream, and pastry.

2. Eat fresh fruits or fruits canned without heavy syrup.

3. Use less sugar—white, brown, or raw—and less honey and syrups.

4. Avoid sweetened breakfast cereals.

5. Eat less jelly, jam, and preserves.

6. Drink pure fruit juices, not imitations.

7. When cooking, use spices such as cinnamon, instead of sugar, to flavor foods.

8. Do not put sugar in tea or coffee.

9. Avoid potatoes and processed foods made from refined carbohydrates, such as white bread, rice, and pasta. These foods are immediately broken down to sugar during digestion.

Table 25.3	Reducing Certain Lipids

TO REDUCE DIETARY SATURATED FAT:

1. Choose poultry, fish, or dry beans and peas as a protein source.

2. Remove skin from poultry and trim fat from red meats before cooking, and place on a rack so that fat drains off.

3. Broil, boil, or bake rather than frying.

4. Limit your intake of butter, cream, trans fats, shortening, and tropical oils (coconut and palm oils).*

5. Use herbs and spices to season vegetables instead of butter, margarine, or sauces. Use lemon juice instead of salad dressing.

6. Drink skim milk instead of whole milk, and use skim milk in cooking and baking.

TO REDUCE DIETARY CHOLESTEROL:

1. Eat white fish and poultry in preference to cheese, egg yolks, liver, and certain shellfish (shrimp and lobster).

2. Substitute egg whites for egg yolks in both cooking and eating.

3. Include soluble fiber in the diet. Oat bran, oatmeal, beans, corn, and fruits, such as apples, citrus fruits, and cranberries are high in soluble fiber.

* Although coconut and palm oils are from plant sources, they are mostly saturated fats.

Figure 25.6   Beans as food.

Beans are a good source of protein, but they don't supply all nine of the essential amino acids. To ensure a complete source of protein in the diet, they should be eaten in combination with a grain such as rice.

Nutritionists generally recommend that unsaturated rather than saturated fats be included in the diet. Two unsaturated fatty acids (alpha-linolenic and linoleic acids) are *essential dietary fatty acids*. Delayed growth and skin problems can develop when the diet lacks these essential fatty acids, which can be supplied by eating fatty fish and by including plant oils, such as canola and soybean oils, in the diet.

Animal foods such as butter, meat, whole milk, and cheeses contain saturated fats. Plant oils contain unsaturated fatty acids. Each type of oil has a particular percentage of monounsaturated and polyunsaturated fatty acids (see Fig. 3.13).

Cholesterol, a lipid, can be synthesized by the body. Cells use cholesterol to make various compounds, including bile, steroid hormones, and vitamin D. Plant foods do not contain cholesterol; only animal foods such as cheese, egg yolks, liver, and certain shellfish (shrimp and lobster) are rich in cholesterol.

Can Lipids Be Harmful?

Elevated blood cholesterol levels are associated with an increased risk of cardiovascular disease, the number one killer of Americans. A diet rich in cholesterol and saturated fats increases the risk of cardiovascular disease (see also Section 25.3). Statistical studies suggest that trans fats are even more harmful than saturated fats. Trans fats arise when unsaturated oils are hydrogenated to produce a solid fat found largely in processed foods. Trans fatty acids may reduce the function of plasma membrane receptors that clear cholesterol from the bloodstream. Trans fatty acids are found in commercially packaged foods such as cookies and crackers, in commercially fried foods such as french fries, in packaged snacks such as microwave popcorn, and in vegetable shortening and some margarines. Table 25.3 tells you how to reduce harmful lipids in the diet.

Proteins

Dietary **proteins** are digested to amino acids, which cells use to synthesize hundreds of cellular proteins. Of the 20 different amino acids, nine are *essential amino acids* that must be present in the diet. Children will not grow if their diets lack the essential amino acids. Eggs, milk products, meat, poultry, and most other foods derived from animals contain all nine essential amino acids and are "complete" or "high-quality" protein sources.

Foods derived from plants generally do not have as much protein per serving as those derived from animals, and each type of plant food generally lacks one or more of the essential amino acids. Therefore, most plant foods are "incomplete" or "low-quality" protein sources. Vegetarians, however, do not have to rely on animal sources of protein. To meet their protein needs, total vegetarians (vegans) can eat grains, beans, and nuts in various combinations (Fig. 25.6). Also, tofu, soymilk, and other foods made from processed soybeans are complete protein sources. A balanced vegetarian diet is quite possible with a little planning.

Can Proteins Be Harmful?

According to nutritionists, protein should not supply the bulk of dietary calories. The average American eats about twice as much protein as he or she needs, and some people may be on a diet that encourages the intake of proteins instead of carbohydrates as an energy source. Also, body builders, weight lifters, and other athletes may include amino acid or protein supplements in the diet because they think these supplements will increase muscle mass. However, excess amino acids are not always converted into muscle tissue. When they are used as an energy source, the liver removes the nitrogen portion (*deamination*) and uses it to form urea, which is excreted in urine. The water needed for excretion of urea can cause dehydration when a person is exercising and losing water by sweating. High-protein diets can also increase calcium loss in the urine and encourage the formation of kidney stones. Furthermore, high-protein foods often contain a high amount of fat.

Minerals

The body needs about 20 elements called **minerals** for numerous physiological roles, including regulation of biochemical reactions, maintenance of fluid balance, and incorporation into certain structures and compounds. The body contains more than 5 grams of each major mineral and less than 5 grams of each trace mineral. Table 25.4 lists the minerals and gives their functions and food sources. It also tells the health effects of too little or too much intake.

Check Your Progress

1. List three foods that are good sources of fiber.
2. What lipid sources are potentially harmful to the body?
3. Identify a food that is a complete protein source.

Answers: 1. Answers will vary, but should include plant foods only. 2. Foods rich in saturated fat, cholesterol, and trans fats are potentially harmful. 3. Answers will vary, but should be animal foods or foods made from processed soybeans.

Table 25.4	Minerals			
Mineral	**Functions**	**Food Sources**	**Conditions Caused By:**	
			Too Little	**Too Much**
Major Minerals				
Calcium (Ca^{2+})	Strong bones and teeth, nerve conduction, muscle contraction	Dairy products, leafy green vegetables	Stunted growth in children, low bone density in adults	Kidney stones; interferes with iron and zinc absorption
Phosphorus (PO_4^{3-})	Bone and soft tissue growth; part of phospholipids, ATP, and nucleic acids	Meat, dairy products, sunflower seeds, food additives	Weakness, confusion, pain in bones and joints	Low blood and bone calcium levels
Potassium (K^+)	Nerve conduction, muscle contraction	Many fruits and vegetables, bran	Paralysis, irregular heartbeat, eventual death	Vomiting, heart attack, death
Sodium (Na^+)	Nerve conduction, pH and water balance	Table salt	Lethargy, muscle cramps, loss of appetite	High blood pressure, calcium loss
Chloride (Cl^-)	Water balance	Table salt	Not likely	Vomiting, dehydration
Magnesium (Mg^{2+})	Part of various enzymes for nerve and muscle contraction, protein synthesis	Whole grains, leafy green vegetables	Muscle spasm, irregular heartbeat, convulsions, confusion, personality changes	Diarrhea
Trace Minerals				
Zinc (Zn^{2+})	Protein synthesis, wound healing, fetal development and growth, immune function	Meats, legumes, whole grains	Delayed wound healing, night blindness, diarrhea, mental lethargy	Anemia, diarrhea, vomiting, renal failure, abnormal cholesterol levels
Iron (Fe^{2+})	Hemoglobin synthesis	Whole grains, meats, prune juice	Anemia, physical and mental sluggishness	Iron toxicity disease, organ failure, eventual death
Copper (Cu^{2+})	Hemoglobin synthesis	Meat, nuts, legumes	Anemia, stunted growth in children	Damage to internal organs if not excreted
Iodine (I^-)	Thyroid hormone synthesis	Iodized table salt, seafood	Thyroid deficiency	Depressed thyroid function, anxiety
Selenium (SeO_4^{2-})	Part of antioxidant enzyme	Seafood, meats, eggs	Vascular collapse, possible cancer development	Hair and fingernail loss, discolored skin

Table 25.5 Reducing Dietary Sodium

TO REDUCE DIETARY SODIUM:

1. Use spices instead of salt to flavor foods.

2. Add little or no salt to foods at the table, and add only small amounts of salt when you cook.

3. Eat unsalted crackers, pretzels, potato chips, nuts, and popcorn.

4. Avoid hot dogs, ham, bacon, luncheon meats, smoked salmon, sardines, and anchovies.

5. Avoid processed cheese and canned or dehydrated soups.

6. Avoid brine-soaked foods, such as pickles or olives.

7. Read labels to avoid high-salt products.

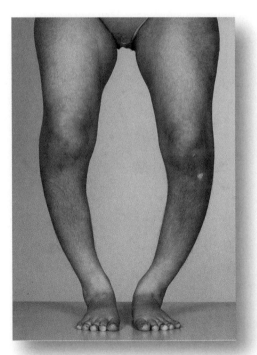

Figure 25.7   Rickets.

Bowing of bones (rickets) due to vitamin D deficiency.

Occasionally, individuals (especially women) do not receive enough iron, calcium, magnesium, or zinc in their diets. Adult females need more iron in the diet than males (18 milligrams [mg] compared to 10 mg) if they are menstruating each month. *Anemia*, characterized by a run-down feeling due to insufficient red blood cells, results when the diet lacks sufficient iron. Many people take calcium supplements as directed by a physician to counteract *osteoporosis*, a degenerative bone disease that affects an estimated one-quarter of older men and one-half of older women in the United States.

One mineral that people consume too much of is sodium. The recommended amount of sodium intake per day is 500 mg, although the average American takes in 4,000–4,700 mg each day. About one-third of the sodium we consume occurs naturally in foods; another one-third is added during commercial processing; and we add the last one-third either during home cooking or at the table in the form of table salt. Table 25.5 gives recommendations for reducing the amount of sodium in the diet.

Vitamins

Vitamins are organic compounds (other than carbohydrates, fats, and proteins including amino acids) that regulate various metabolic activities. Tables 25.6 and 25.7 list the major vitamins and some of their functions and food sources.

Although many people think vitamins can enhance health dramatically, prevent aging, and cure diseases such as arthritis and cancer, there is no scientific evidence that vitamins are "wonder drugs." However, vitamins C, E, and A are believed to defend the body against free radicals, and therefore they are termed **antioxidants.** These vitamins are especially abundant in fruits and vegetables, and so it is suggested that we eat about 4 ½ cups of fruits and vegetables per day. To achieve this goal, we should consume salad greens, raw or cooked vegetables, dried fruit, and fruit juice in addition to traditional apples and oranges and other fresh foods.

Vitamin deficiencies can lead to disorders, and even death, in humans. Although many foods in the U.S. are now enriched, or fortified, with vitamins, some individuals, especially the elderly, young children, alcoholics, and low-income people, are still at risk for vitamin deficiencies, generally as a result of poor food choices. For example, skin cells normally contain a precursor cholesterol molecule that is converted to vitamin D after UV exposure. But a vitamin D deficiency leads to a condition called rickets, in which defective mineralization of the skeleton causes bowing of the legs (Fig. 25.7). Most milk today is fortified with vitamin D, which helps prevent the occurrence of rickets. Another example is vitamin C deficiency, the effects of which are illustrated in Figure 25.2.

Water

Water comprises about 60% of an adult's body. Water participates in many chemical reactions; in addition, watery fluids lubricate joints, transport other nutrients, and help maintain body temperature. Beverages, soups, fruits, and vegetables are sources of water, and most solid foods also contain some water. The amount of *total* water (water from beverages and foods) that you need to consume depends on your physical activity level, diet, and environmental conditions. On average, men should consume about 125 ounces and women should consume about 90 ounces of total water each day. Thirst is a healthy person's best guide for meeting water needs and avoiding dehydration.

Table 25.6	Water-Soluble Vitamins			
Vitamin	**Functions**	**Food Sources**	**Conditions Caused By:**	
			Too Little	**Too Much**
Vitamin C	Antioxidant; needed for forming collagen; helps maintain capillaries, bones, and teeth	Citrus fruits, leafy green vegetables, tomatoes, potatoes, cabbage	Scurvy, delayed wound healing, infections	Gout, kidney stones, diarrhea, decreased copper
Thiamine (vitamin B_1)	Part of coenzyme needed for cellular respiration; also promotes activity of the nervous system	Whole-grain cereals, dried beans and peas, sunflower seeds, nuts	Beriberi, muscular weakness, enlarged heart	Can interfere with absorption of other vitamins
Riboflavin (vitamin B_2)	Part of coenzymes, such as FAD; aids cellular respiration, including oxidation of protein and fat	Nuts, dairy products, whole-grain cereals, poultry, leafy green vegetables	Dermatitis, blurred vision, growth failure	Unknown
Niacin (nicotinic acid)	Part of coenzymes NAD and NADP; needed for cellular respiration, including oxidation of protein and fat	Peanuts, poultry, whole-grain cereals, leafy green vegetables, beans	Pellagra, diarrhea, mental disorders	High blood sugar and uric acid, vasodilation, etc.
Folacin (folic acid)	Coenzyme needed for production of hemoglobin and formation of DNA	Dark leafy green vegetables, nuts, beans, whole-grain cereals	Megaloblastic anemia, spina bifida	May mask B_{12} deficiency
Vitamin B_6	Coenzyme needed for synthesis of hormones and hemoglobin; CNS control	Whole-grain cereals, bananas, beans, poultry, nuts, leafy green vegetables	Rarely, convulsions, vomiting, seborrhea, muscular weakness	Insomnia, neuropathy
Pantothenic acid	Part of coenzyme A needed for oxidation of carbohydrates and fats; aids in the formation of hormones and certain neurotransmitters	Nuts, beans, dark green vegetables, poultry, fruits, milk	Rarely, loss of appetite, mental depression, numbness	Unknown
Vitamin B_{12}	Complex, cobalt-containing compound; part of the coenzyme needed for synthesis of nucleic acids and myelin	Dairy products, fish, poultry, eggs, fortified cereals	Pernicious anemia	Unknown
Biotin	Coenzyme needed for metabolism of amino acids and fatty acids	Generally in foods, especially eggs	Skin rash, nausea, fatigue	Unknown

Table 25.7	Fat-Soluble Vitamins			
Vitamin	**Functions**	**Food Sources**	**Conditions Caused By:**	
			Too Little	**Too Much**
Vitamin D	A group of steroids needed for development and maintenance of bones and teeth	Milk fortified with vitamin D, fish liver oil; also made in the skin when exposed to sunlight	Rickets, bone decalcification and weakening	Calcification of soft tissues, diarrhea, possible renal damage
Vitamin E	Antioxidant that prevents oxidation of vitamin A and polyunsaturated fatty acids	Leafy green vegetables, fruits, vegetable oils, nuts, whole-grain breads and cereals	Unknown	Diarrhea, nausea, headaches, fatigue, muscle weakness
Vitamin K	Needed for synthesis of substances active in clotting of blood	Leafy green vegetables, cabbage, cauliflower	Easy bruising and bleeding	Can interfere with anticoagulant medication

Check Your Progress

1. **What two minerals are needed for strong bones and teeth?**
2. **List the vitamins that are antioxidants.**
3. **Name the nutrient in your body that is present in the largest quantity.**

Answers: 1. Calcium and phosphorus are needed for strong bones and teeth. 2. Vitamins C, E, and A (beta-carotene) are antioxidants. 3. Water is present in the largest quantity.

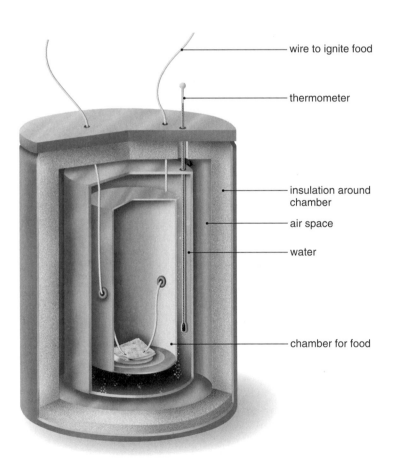

Figure 25.8   Measuring the energy content of food.

Scientists use a bomb calorimeter to determine the amount of calories supplied by a particular food.

Labels:
- wire to ignite food
- thermometer
- insulation around chamber
- air space
- water
- chamber for food

25.3 Nutrition and Health

Many serious disorders in Americans are linked to a diet that results in excess body fat. In the United States, the number of people who are overweight or obese has reached epidemic proportions. Nearly two-thirds of adult Americans have too much body fat.

Excess body fat increases the risk of diabetes type 2, cardiovascular disease, and possibly certain cancers. These conditions are among the leading causes of disability and death in the United States. Therefore, it is important for us all to stay within the recommended weight for our height.

Are You Overweight?

Medical researchers use the **body mass index (BMI)** to determine if a person is overweight or obese. On the whole, the height to which we grow is determined genetically, while our weight is influenced by other factors as well, particularly diet and lifestyle. BMI is the relationship between a person's weight and height. Here's how to calculate your BMI:

$$\frac{\text{weight (pounds)} \times 703.1}{\text{height}^2 \text{ (inches)}}$$

Example: A woman whose height is 63 inches and weighs 133 pounds has a BMI of 23.56.

$$\frac{133 \times 703.1}{63^2} \quad = \quad \frac{93512.3}{3969} \quad = \quad 23.56 \text{ (BMI)}$$

Healthy BMIs = 18.5 to 24.9
Overweight BMIs = 25.0 to 29.9
Obese BMIs = 30.0 to 39.9
Morbidly obese BMIs = 40.0 or more

Energy Intake Versus Energy Output

While genetics is a factor in being overweight, a person cannot become fat without taking in more food energy (calories) than are expended.

Energy Intake   Scientists use a bomb calorimeter to measure a food's caloric value (Fig. 25.8). A portion of food is placed inside the chamber and then ignited. As the food burns, it raises the temperature of the water that surrounds the chamber. Scientists measure the change in water temperature and calculate the number of calories in the food. The energy value of food is often reported in kilocalories (kcal). A kilocalorie is the amount of heat that raises the temperature of a liter of water by 1°C.

For practical purposes, you can estimate a food's caloric value if you know how many grams of carbohydrate, fat, protein, and alcohol it contains. Each gram of carbohydrate or protein supplies 4 kcal, and each gram of fat supplies 9 kcal. Although alcohol is not a nutrient, it is considered a food, and each gram supplies 7 kcal. Therefore, if a serving of food contains 30 grams of carbohydrate, 9 grams of fat, and 5 grams of protein, it supplies 221 kcal.

carbohydrate	30 g × 4 kcal =	120 kcal
fat	9 g × 9 kcal −	81 kcal
protein	5 g × 4 kcal =	20 kcal
Total		221 kcal

Energy Output   The body expends energy primarily for (1) metabolic functions, (2) physical activity, and (3) digesting, absorbing, and processing nutrients from food. Scientists can access a person's energy expenditure for a particular physical activity by measuring oxygen intake and carbon dioxide output during performance of the activity (Fig. 25.9).

For practical purposes, here's how to estimate your daily energy needs:

1. **Kcal needed daily for metabolic functions**   Multiply your weight in kilograms (weight in pounds divided by 2.2) times 1.0 if you are a man, and times 0.9 if you are a woman. Then multiply that number by 24.

   Example: Meghan, a woman who weighs 130 pounds (about 59 kg), would use the following steps to calculate her daily caloric use:

   0.9 kcal × 59 kg × 24 hours = approximately 1,274 kcal/day

2. **Kcal needed daily for physical activity**   Choose a multiplication factor from one of the follow categories:
   Sedentary (little or no physical activity) = .20 to .40
   Light (walk daily) = .55 to .65
   Moderate (daily vigorous exercise) = .70 to .75
   Heavy (perform physical labor/endurance training) = .80 to 1.20

   Multiply this factor times the kcal value you obtained in step 1.

   Example: Meghan performs light physical activity daily. She multiplies .55 × 1,274 kcal to determine her physical activity use, which is 701 kcal.

3. **Kcal needed for digestion, absorption, and processing of nutrients**

   Example: Meghan adds 1,274 kcal and 701 kcal to calculate her total energy needs so far. The answer is 1,975 kcal.

   Multiply this value by 0.1, and add it to the total kcal from steps 1 and 2 to get your total daily energy needs.

   Example: Meghan multiplies 1,975 kcal by 0.1 and adds that value (197.5) to 1,975 to obtain her total daily energy needs of 2,172 kcal.

   Therefore, Meghan will maintain her weight of 130 pounds if she continues to consume about 2,170 kcal a day and to perform light physical activities.

Figure 25.9   Measuring energy needed for a physical activity.

a. A sedentary job requires much less energy than (**b**) a physical job. **c.** Scientists measure oxygen intake and carbon dioxide output to determine the kcal required for a particular physical activity.

Check Your Progress

1. Ted's BMI is 32.5. According to this information, in what range is his weight?
2. Name the device used to determine the number of calories in a food.
3. Daily caloric needs are based on what three types of needs?

Answers: 1. Ted is obese. 2. A bomb calorimeter counts calories in food. 3. Daily caloric needs are based on metabolic functions, physical activity, and digestion, absorption, and processing of food.

Intake	Output	Weight Change

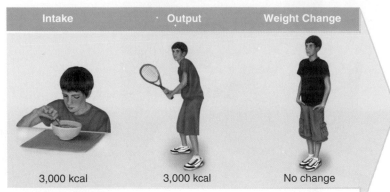

3,000 kcal 3,000 kcal No change

Energy balance (equilibrium)

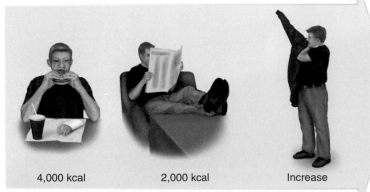

4,000 kcal 2,000 kcal Increase

Positive energy balance

2,000 kcal 3,000 kcal Decrease

Negative energy balance

Figure 25.10   How body weight changes.

These illustrations show the relationships between food intake, energy use, and weight change.

Maintaining a Healthy Weight

To maintain weight at an appropriate level, the kcal intake (eating) should not exceed the daily kcal output (metabolism + physical activity + processing food). For many Americans, this ratio is out of sync; they take in more calories than they need. The extra energy is converted to fat stored in *adipose tissue,* and they become overweight. To lose weight, an overweight person needs to lower the kcal intake and increase the kcal output in the form of physical activity. Only then does the body metabolize its stored fat for energy needs, allowing the person to lose weight. Figure 25.10 illustrates how body weight changes in relation to kcal intake and kcal output.

Dieting   Fad weight-reduction diets—high-protein, low-carb, high-fiber, and even cabbage soup diets—come and go. During the first few weeks of following a fad diet, overweight people often lose weight rapidly, because they consume fewer calories than usual, and excess body fat is metabolized for energy needs. In most cases, however, dieters become bored with eating the same foods and avoiding their favorite foods, which may be high in fat and sugars. When most dieters "go off" their diets, they regain the weight that was lost, and they often feel frustrated and angry at themselves for failing to maintain the weight loss.

There are no quick and easy solutions for losing weight. The typical fad diet is nutritionally unbalanced and difficult to follow over the long term. Weight loss and weight maintenance require making permanent lifestyle changes, such as increasing the level of physical activity and reducing portion sizes. Behavior modification allows an overweight person to lose weight safely, generally at a reasonable rate of about $1/2$ to 2 pounds per week. Once body weight is under control, it needs to be maintained by continuing to eat sensibly.

Disorders Associated with Obesity

Diabetes type 2 and cardiovascular disease are often seen in people who are obese.

Diabetes Type 2

As discussed in Chapter 27, diabetes comes in two forms, diabetes type 1 and diabetes type 2. When a person has diabetes type 1, the pancreas does not produce insulin, and the patient has to have daily insulin injections.

In contrast to diabetes type 1, children and more often adults with diabetes type 2 are usually obese and display impaired insulin production and insulin resistance. Normally, the presence of insulin causes the cells of the body to take up and metabolize glucose. In a person with insulin resistance, the body's cells fail to take up glucose even when insulin is present. Therefore, the blood glucose value exceeds the normal level, and glucose appears in the urine.

Diabetes type 2 is increasing rapidly in most industrialized countries of the world. Because diabetes type 2 is most often seen in people who are obese, dietary factors are generally believed to contribute to the development of diabetes type 2. Further, a healthy diet, increased physical activity, and weight loss have been seen to improve the ability of insulin to function properly in type 2 diabetics (Fig. 25.11). How might diet contribute to the occurrence of diabetes type 2? Simple sugars in foods, such as candy and ice cream, immediately enter the bloodstream, as do sugars from the digestion of starch within white bread and potatoes. When the blood glucose level rises rapidly, the pancreas produces an overload of insulin to bring the level under control. Chronically high insulin levels apparently lead to insulin resistance, increased fat deposition, and a high blood fatty acid level. Over the years, the body's cells become insulin resistant, and thus diabetes type 2 can occur. In addition, high fatty acid levels can lead to increased risk of cardiovascular disease.

It is well worth the effort to control diabetes type 2 because all diabetics, whether type 1 or type 2, are at risk for blindness, kidney disease, and cardiovascular disease.

Cardiovascular Disease

In the United States, cardiovascular disease, which includes hypertension, heart attack, and stroke, is among the leading causes of death. Cardiovascular disease is often due to arteries blocked by plaque, which contains saturated fats and cholesterol. Cholesterol is carried in the blood by two types of lipoproteins: low-density lipoprotein (LDL) and high-density lipoprotein (HDL). LDL is thought of as "bad" because it carries cholesterol from the liver to the cells, while HDL is thought of as "good" because it carries cholesterol from the cells to the liver, which takes it up and converts it to bile salts.

Saturated fats, including trans fats, tend to raise LDL cholesterol levels, while unsaturated fats lower LDL cholesterol levels. Beef, dairy foods, and coconut oil are rich sources of saturated fat. Foods containing partially hydrogenated oils (e.g., vegetable shortening and stick margarine) are sources of trans fats (Fig. 25.12). Unsaturated fatty acids in olive and canola oils, most nuts, and coldwater fish tend to lower LDL-cholesterol levels. Furthermore, coldwater fish (e.g., herring, sardines, tuna, and salmon) contain polyunsaturated fatty acids and especially *omega-3 unsaturated fatty acids* that can reduce the risk of cardiovascular disease. Taking fish oil supplements to obtain omega-3s is not recommended without a physician's approval, because too much of these fatty acids can interfere with normal blood clotting.

The American Heart Association recommends limiting total cholesterol intake to 300 mg per day. Eggs are very nutritious, but yolks contain about 210 mg of cholesterol. Most healthy people can eat a couple of whole eggs each week without experiencing an increase in their blood cholesterol levels. Overall, dietary saturated fats and trans fats raise LDL-cholesterol levels more than dietary cholesterol.

A physician can determine if blood lipid levels are normal. If a person's cholesterol and triglyceride levels are elevated, modifying the fat content of your diet, losing excess body fat, and exercising regularly can reduce them. If lifestyle changes do not lower blood lipid levels enough to reduce the risk of cardiovascular disease, a physician may prescribe special medications.

Figure 25.11   Exercising for good health.

Regular exercise helps prevent and control diabetes type 2.

Figure 25.12   French fries.

Most people enjoy eating french fries, but fast-food fries are apt to contain saturated fats and trans fats.

a.

b.

Figure 25.13   Eating disorders.

a. People with anorexia nervosa have a mistaken body image and think they are fat even though they are thin. **b.** Those with bulimia nervosa overeat and then purge their bodies of the food they have eaten.

Eating Disorders

People with eating disorders are dissatisfied with their body image. Social, cultural, emotional, and biological factors all contribute to the development of an eating disorder. These serious conditions can lead to malnutrition, disability, and death. Regardless of the eating disorder, early recognition and treatment are crucial. Treatment usually includes psychological counseling and antidepressant medications.

Anorexia nervosa is a severe psychological disorder characterized by an irrational fear of getting fat that results in the refusal to eat enough food to maintain a healthy body weight (Fig. 25.13*a*). A self-imposed starvation diet is often accompanied by occasional binge eating that is followed by purging and extreme physical activity to avoid weight gain. Binges usually include large amounts of high-calorie foods, and purging episodes involve self-induced vomiting and laxative abuse. About 90% of people suffering from anorexia nervosa are young women; an estimated 1 in 200 teenage girls is affected.

A person with **bulimia nervosa** binge-eats, and then purges to avoid gaining weight (Fig. 25.13*b*). The binge-purge cyclic behavior can occur several times a day. People with bulimia nervosa can be difficult to identify because their body weights are often normal and they tend to conceal their binging and purging practices. Women are more likely than men to develop bulimia; an estimated 4% of young women suffer from this condition.

Other abnormal eating practices include binge-eating disorder and muscle dysmorphia. Many obese people suffer from **binge-eating disorder,** a condition characterized by episodes of overeating that are not followed by purging. Stress, anxiety, anger, and depression can trigger food binges. A person suffering from **muscle dysmorphia** thinks his or her body is underdeveloped. Body-building activities and a preoccupation with diet and body form accompany this condition. Each day, the person may spend hours in the gym working out on muscle-strengthening equipment. Unlike anorexia nervosa and bulimia, muscle dysmorphia affects more men than women.

Check Your Progress

1. What causes a person to gain weight?
2. Why are fad diets of limited value?
3. Salmon, tuna, and other coldwater fish are good sources of certain polyunsaturated fatty acids. What are these fatty acids called?
4. Sally diets because she thinks she is fat, even though her friends tell her she is already too thin. Sally is in danger of what eating disorder?

Answers: 1. Taking in more calories than required for daily energy needs causes weight gain. 2. Typical fad diets are prone to be nutritionally unbalanced and hard to follow for the long term. 3. Omega-3 unsaturated fatty acids. 4. Sally is in danger of anorexia nervosa.

25.4 How to Plan Nutritious Meals

Planning nutritious meals and snacks involves making daily food choices based on a wide variety of information about recommended amounts of nutrients. A day's food intake should provide the proper balance of nutrients—neither too much nor too little of each nutrient. Food guides can be helpful in planning your diet. Additionally, reading the Nutrition Facts panel on packaged foods can help you choose a healthier source of nutrients.

The Food Pyramid

In 2005, the U.S. Department of Agriculture (USDA) presented an updated food guide pyramid (Fig. 25.14). The previous food guide pyramid had been criticized for not emphasizing the need for whole-grain breads and cereals, low-fat dairy products, low-sugar fruits and vegetables, and unsaturated fats such as olive and canola oil in the diet. It also did not place an emphasis on exercise as a means to control weight gain.

The new pyramid, like the old one, groups foods with similar nutritional content and ranks food groups according to dietary emphasis so the diet will be balanced. The new pyramid emphasizes foods that should be eaten often and omits foods that should not be eaten on a regular basis. Additionally, the USDA provides recommendations concerning the minimum quantity of foods in each group that should be eaten daily. The sample menu in Table 25.8 shows how the food guide pyramid can be used to plan a day's meals and snacks that supply about 2,200 kcal.

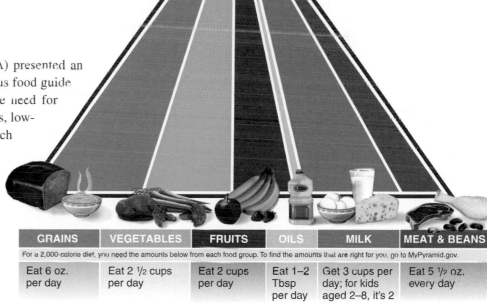

GRAINS	VEGETABLES	FRUITS	OILS	MILK	MEAT & BEANS
Eat 6 oz. per day	Eat 2 1/2 cups per day	Eat 2 cups per day	Eat 1–2 Tbsp per day	Get 3 cups per day; for kids aged 2–8, it's 2	Eat 5 1/2 oz. every day

For a 2,000-calorie diet, you need the amounts below from each food group. To find the amounts that are right for you, go to MyPyramid.gov.

Figure 25.14   The 2005 USDA food guide pyramid.

This food guide pyramid can be helpful when planning nutritious meals and snacks.

Source: Data from the U.S. Department of Agriculture.

Table 25.8		Preparing Meals and Snacks using the Food Guide Pyramid			
Breakfast	**Snack**	**Lunch**	**Snack**	**Dinner**	**Dessert**
1/2 cup oat flakes cereal	Whole-wheat muffin	Toasted American cheese sandwich (whole-grain bread)	Apple	3 oz. hamburger patty	1/2 cup frozen vanilla yogurt topped with sliced strawberries
8 oz. nonfat milk	2 Tbsp peanut butter mixed with 1 tsp sunflower seeds	1 cup vegetable salad	4 graham crackers	Whole-wheat bun	
1/2 cup blueberries	8 oz. nonfat milk	1 Tbsp olive oil and vinegar salad dressing	1/2 cup baked beans		
1 slice whole-grain toast		4 oz. tomato juice		1 cup steamed broccoli spears	
1 tsp soft margarine				1 tsp soft margarine	
4 oz. orange juice					

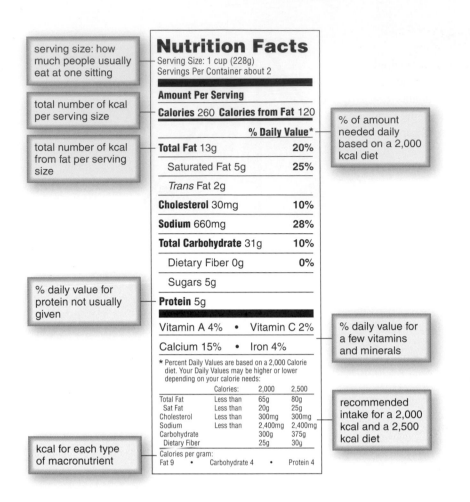

serving size: how much people usually eat at one sitting

total number of kcal per serving size

total number of kcal from fat per serving size

% daily value for protein not usually given

kcal for each type of macronutrient

% of amount needed daily based on a 2,000 kcal diet

% daily value for a few vitamins and minerals

recommended intake for a 2,000 kcal and a 2,500 kcal diet

Figure 25.15   Nutrition Facts panel.

The Nutrition Facts panel provides information about the amounts of certain nutrients and other substances in a serving of food.

Making Sense of Nutrition Labels

A Nutrition Facts panel, such as the one shown in Figure 25.15 provides specific dietary information about the product and general information about the nutrients the product contains.

Serving Size   The serving size is based on the typical serving size for the product and not necessarily what the food guide pyramid would consider a serving size. If you are comparing Calorie (kcal)[1] and other data about products of the same type, you definitely want to be sure the serving size is the same for each product.

Calories   The total number of kcal is based on the serving size. Obviously, if you eat twice the serving size, you have taken in twice the number of calories.

The bottom of the panel lists the Calories (kcal) per gram (g) of fat, carbohydrate, and protein. You can use this information to calculate the calories per serving for each type of nutrient. The total for all nutrients should agree with the figure given for Calories at the top of the panel, namely 260.

% Daily Value   The % daily value (the % of the total amount needed in a 2,000 kcal diet) is calculated by comparing the specific information about this product with the information given at the bottom of the panel. For example, the product in Figure 25.15 has a fat content of 13 g, and the total daily recommended amount is less than 65 g, so 13/65 = 20%. *These % daily values are not applicable for people who require more or less than 2,000 Calories (kcal) per day.*

A % daily value for protein is generally not given because determining % daily value would require expensive testing of the protein quality of the product by the manufacturer. Also, notice there is a % daily value for carbohydrates but not sugars because there is no daily value for sugar.

How to Use the Panel   If the serving sizes are the same, you can use Nutrition Facts panels to compare two products of the same type. For example, if you wanted to reduce your caloric intake and increase your fiber and vitamin C intakes, comparing the panels from two different food products would allow you to see which one is lowest in calories and highest in fiber and vitamin C.

Dietary Supplements

Dietary supplements are nutrients and plant products (such as herbal teas) that are used to enhance health. The U.S. government does not require dietary supplements to undergo the same safety and effectiveness testing that new prescription drugs must complete before they are approved. Therefore, many herbal products have not been tested scientifically to determine their benefits.

---

[1] Nutritionists use the term Calorie, while scientists prefer kcal.

Although people often think herbal products are safe because they are "natural," many plants, including lobelia, comfrey, and kava kava, can be poisonous.

Dietary supplements that contain nutrients can also cause harm. Most fat-soluble vitamins are stored in the body and can accumulate to toxic levels, particularly vitamins A and D. Although excesses of water-soluble vitamins can be excreted, cases involving toxic amounts of vitamins B_6, thiamin, and vitamin C have been reported. Minerals can be harmful, even deadly, when ingested in amounts that exceed the body's needs.

Healthy people can take a daily supplement that contains recommended amounts of vitamins and minerals. Some people have metabolic diseases or physical conditions that interfere with their ability to absorb or metabolize certain nutrients. These individuals may need to add certain nutrient supplements to their diet. However, people should not take high doses of dietary supplements without checking with their physicians.

The Bottom Line

While food guide pyramids differ in various ways, most nutritionists can agree that a healthy diet does the following:

- Has a moderate total fat intake and is low in saturated fat, trans fats, and cholesterol. (See Table 25.3 for help in achieving this goal.)
- Is rich in whole-grain products, vegetables, and legumes (e.g., beans and peas) as sources of complex carbohydrates and fiber.
- Is low in refined carbohydrates, such as starches and sugars. (See Table 25.2 for help in achieving this goal.)
- Is low in salt and sodium intake. (See Table 25.5 for help in achieving this goal.)
- Contains only adequate amounts of protein, largely from poultry, fish, and plant origins.
- Includes only moderate amounts of alcohol.
- Contains adequate amounts of minerals and vitamins, but avoids questionable food additives and supplements.

Check Your Progress

1. What should Jim consult to compare the caloric value and vitamin A content of canned peaches and canned pears?
2. For the product in Figure 25.15, how many calories are in a serving? How many grams of total fat are in a serving?
3. Is it always safe to take high doses of nutritional supplements?

Answers: 1. Jim should read the Nutrition Facts panels on the labels. 2. Total calories = 260; total fat = 13 grams 3. No, vitamins and minerals especially can be toxic in large doses.

THE CHAPTER IN REVIEW

Summary

25.1 | Nutrition

Nutrients perform varied physiological functions in the body. A healthy diet can lead to a longer and more active life.

Introducing the Nutrients
Nearly every food is a mixture of nutrients. Balanced diets supply nutrients in proportions necessary for health. Essential nutrients must be supplied by the diet, or else deficiency disorders result. Macronutrients (carbohydrates, lipids, and proteins) supply energy. Micronutrients (vitamins and minerals) and water do not supply energy.

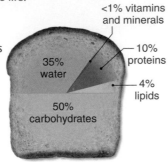

<1% vitamins and minerals
35% water
10% proteins
4% lipids
50% carbohydrates

25.2 | The Classes of Nutrients

The six classes of nutrients are carbohydrates, lipids, proteins, minerals, vitamins, and water. Their functions and food sources follow.

Carbohydrates (Macronutrient)
- Carbohydrates in the form of sugars and starch provide energy for cells. Glucose is a simple carbohydrate the body uses directly for energy.
- Sources of carbohydrates are fruits, vegetables, cereals, and other grains.
- Fiber includes nondigestible carbohydrates from plants. Fiber in the diet can prevent constipation and may protect against cardiovascular disease, diabetes, and colon cancer. Refined carbohydrates from which fiber, vitamins, and minerals have been removed are a poor dietary choice.
- Nutritionists debate whether dietary fat or refined carbohydrates are responsible for the prevalence of obesity today.

Lipids (Macronutrient)
- Triglycerides (fats and oils) supply energy and are stored for insulation and protection of organs. Alpha-linolenic and linoleic acids are essential fatty acids.
- Cholesterol is used to make bile, steroid hormones, and vitamin D. The only food sources of cholesterol are foods derived from animals. Elevated blood cholesterol levels can lead to cardiovascular disease.
- Sources of lipids include oils, fats, whole-milk dairy products, meat, fish, poultry, and nuts.
- High intake of saturated fats, trans fats, and cholesterol is harmful to health.

Proteins (Macronutrient)
- The body uses the 20 different amino acids to synthesize hundreds of proteins. Nine amino acids are essential to the diet. Most animal foods are complete sources of protein because they contain all nine essential amino acids.
- Sources of proteins include meat, fish, poultry, eggs, nuts, soybeans, and cheese.
- Healthy vegetarian diets rely on various sources of plant proteins.
- Consumption of excess proteins can be harmful because the excretion of excess urea taxes the kidneys and can lead to kidney stones.

Minerals (Micronutrient)
- Minerals regulate metabolism and are incorporated into structures and compounds.
- About 20 minerals, obtained from most foods, are needed by the body.
- Lack of the mineral calcium can lead to osteoporosis.
- Most people consume too much sodium.

Vitamins (Micronutrient)
- Vitamins, obtained from most foods, regulate metabolism and physiological development. Vitamins C, E, and A also serve as antioxidants.
- Lack of any of the vitamins can lead to certain disorders.

Water
- Water participates in chemical reactions, lubricates joints, transports other nutrients, and helps maintain body temperature.

25.3 | Nutrition and Health

Excess body fat increases the risk of diabetes type 2, cardiovascular disease, and certain cancers.

Body Mass Index
Medical researchers use the body mass index (BMI) to determine if a person has a healthy weight, is overweight, or is obese.

Energy Intake Versus Energy Output
- Energy intake: The number of daily calories (kcal) consumed is based on grams of carbohydrate, fat, and protein in the foods eaten.
- Energy output: The number of daily calories (kcal) used is based on the amounts needed for metabolic functions, physical activity, and digesting, absorbing, and processing nutrients.

Maintaining a Healthy Weight
- To maintain a healthy weight, kcal intake should not exceed kcal output. To lose weight, decrease caloric intake and increase physical activity.
- Fad diets, in general, are nutritionally unbalanced and difficult to follow over the long term.

Disorders Associated with Obesity
- Diabetes type 2: A healthy diet, increased physical activity, and weight loss improve insulin function in type 2 diabetics.
- Cardiovascular disease: Saturated fats and trans fats are associated with high blood LDL-cholesterol levels. To reduce risk of cardiovascular disease, cholesterol intake should be limited, and the diet should include sources of unsaturated, polyunsaturated, and omega-3 fatty acids to reduce cholesterol levels.

Eating Disorders
People suffering from eating disorders are dissatisfied with their body image. Anorexia nervosa and bulimia nervosa are serious psychological disturbances that can lead to malnutrition and death.

25.4 | How to Plan Nutritious Meals

Planning nutritious meals and snacks involves making healthy and informed food choices.

The Food Pyramid

Food guide pyramids group foods according to similar nutrient content. Such guides can be helpful when planning nutritious meals and snacks. The new USDA food guide pyramid also emphasizes the need for exercise to prevent weight gain.

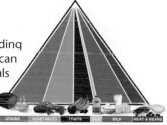

Making Sense of Nutrition Labels

The information in the Nutrition Facts panel on packaged foods can be useful when comparing foods for nutrient content, especially serving size, total calories, and % daily value.

Dietary Supplements

Dietary supplements are nutrients and plant products that are taken to enhance health. A multiple vitamin and mineral supplement that provides recommended amounts of nutrients can be taken daily, but herbal and nutritional supplements can be harmful if misused.

Nutrition Facts
Serving Size: 1 cup (228g)
Servings Per Container about 2
Amount Per Serving
Calories 260 Calories from Fat 120
% Daily Value*
Total Fat 13g — 20%
Saturated Fat 5g — 25%
Trans Fat 2g
Cholesterol 30mg — 10%
Sodium 660mg — 28%
Total Carbohydrate 31g — 10%
Dietary Fiber 0g — 0%
Sugars 5g
Protein 5g
Vitamin A 4% • Vitamin C 2%
Calcium 15% • Iron 4%

Thinking Scientifically

One reason obesity is such a pervasive problem in the United States is that serving sizes have increased dramatically in the past two decades. For example, 20 years ago, a typical soda was 6.5 ounces, while today a 20-ounce soda is considered average. A serving of french fries was 2.6 ounces in the past, but is at least 6 ounces today. Bagels have doubled in size. Complete the following table to determine the caloric values of today's servings, and then compare them.

	Soda Then 4.5 oz.	Soda Now 20 oz.	Calorie Increase
Fat (g)	0	_____	
Carbohydrate (g)	21	_____	
Protein (g)	0	_____	
Total calories	_____	_____	_____

	Fries Then 2.6 oz.	Fries Now 6 oz.	Calorie Increase
Fat (g)	11	_____	
Carbohydrate (g)	28	_____	
Protein (g)	3	_____	
Total calories	_____	_____	_____

	Bagels Then 80 oz.	Bagels Now 200 oz.	Calorie Increase
Fat (g)	0.5	_____	
Carbohydrate (g)	40	_____	
Protein (g)	6	_____	
Total calories	_____	_____	_____

Testing Yourself

Choose the best answer for each question.

1. Vitamins are considered
   a. micronutrients because they are small in size.
   b. micronutrients because they are needed in small quantities.
   c. macronutrients because they are large in size.
   d. macronutrients because they are needed in large quantities.

For questions 2–9, choose the class of nutrient from the key that matches the description. Each answer may be used more than once.

Key:
   a. carbohydrates
   b. lipids
   c. proteins
   d. minerals
   e. vitamins
   f. water

2. Preferred source of direct energy for cells.
3. Comprises the majority of body mass.
4. Not present in most animal foods.
5. Include the essential fatty acids alpha-linolenic and linoleic.
6. Include antioxidants.
7. Generally found in higher levels in animal than in plant sources.
8. An example is cholesterol.
9. Includes calcium, phosphorus, and potassium.
10. The refining of whole-grain foods removes
    a. fiber.
    b. vitamins.
    c. minerals.
    d. More than one of these are correct.
    e. All of these are correct.

11. What is the BMI (body mass index) of a person who is five feet ten inches tall and weighs 160 pounds?
    a. 31
    b. 27
    c. 23
    d. 18
    e. 12

12. A serving of graham crackers contains 24 grams of carbohydrate, 3 grams of fat, and 2 grams of protein. How many calories are in a serving?
    a. 131
    b. 236
    c. 116
    d. 261
    e. 141

13. With reference to the food pyramid, fill in each component and the recommended daily dietary amount based on a 2,000 Calorie (kcal) diet.

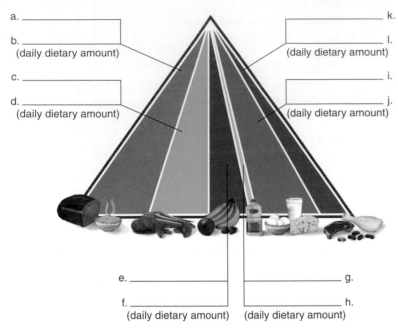

a. _____
b. _____
   (daily dietary amount)
c. _____
d. _____
   (daily dietary amount)

k. _____
l. _____
   (daily dietary amount)
i. _____
j. _____
   (daily dietary amount)

e. _____   g. _____
f. _____   h. _____
   (daily dietary amount)   (daily dietary amount)

14. You are comparing two breakfast cereals and one has fewer calories than the other per serving. What should you check before deciding that one will lead to weight loss rather than the other?

    a. the percentage of vitamins provided by each
    b. the grams of protein provided by each
    c. the grams of sugar provided by each
    d. the suggested serving size
    e. the size of the box

15. A percent daily value for sugar is not included in a nutrition label because

    a. sugar is not a nutrient.
    b. it is too difficult to determine the caloric value of sugar.
    c. there is a daily value given for carbohydrates but not for sugars.
    d. sugar quality varies from product to product.

16. The amino acids that must be consumed in the diet are called essential. Nonessential amino acids

    a. can be produced by the body.
    b. are only needed occasionally.
    c. are stored in the body until needed.
    d. can be taken in by supplements.

17. Bulimia nervosa is not characterized by

    a. a restrictive diet, but often includes excessive exercise.
    b. binge eating followed by purging.
    c. an obsession about body shape and weight.
    d. a distorted body image so that the person feels fat even when emaciated.

18. Which of these is a true statement?

    a. Technically speaking, junk foods have nutritional value.
    b. Water is not a nutrient because it contains no kcal.
    c. Even though fiber can be digested, it is not a nutrient.
    d. Cholesterol should be avoided because it plays no role in the body.
    e. A vegetarian diet is sure not to provide all the essential amino acids.

Go to www.mhhe.com/maderessentials for more quiz questions.

Bioethical Issue

Some people believe that food and drink manufacturers are at least partly to blame for the obesity epidemic in the United States. They suggest that advertising by these companies shows portion sizes that encourage excess consumption, that children are led to believe they must purchase certain products, such as soft drinks, in order to be socially accepted, and that some products that are not substitutes for meals are presented as such. Do you think the government should regulate or ban such advertising? If so, how extensive should the regulations be? If not, how might parents counter the effects of such advertising?

Understanding the Terms

Match the terms to these definitions:

a. _____ Component of food that provides a physiological function.

b. _____ Nondigestible carbohydrate derived from plants.

c. _____ Fats and oils.

d. _____ Lipid necessary for the synthesis of bile, steroid hormones, and vitamin D.

e. _____ Organic compounds, found in low levels in the body, that regulate metabolic activities.

f. _____ Compounds that defend the body against free radicals.

g. _____ Measure of height relative to weight to determine whether or not a person is overweight

h. _____ Psychological disorder in which a person refuses to eat because of an irrational fear of becoming fat.

i. _____ Eating disorder characterized by binge eating and then purging.

j. _____ Disorder in which the person performs excessive body building exercises to correct what is perceived as an underdeveloped body.

Defenses Against Disease

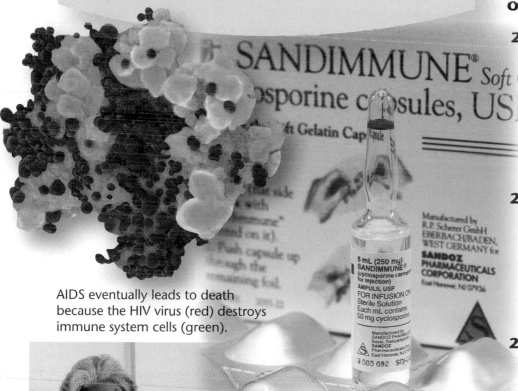

AIDS eventually leads to death because the HIV virus (red) destroys immune system cells (green).

People who have had organ transplants must take immuno-suppressive drugs for the rest of their lives.

A fever increases metabolism, enabling the immune system to fight infection more efficiently.

If you were asked the purpose of a fever, what would your answer be? Many people think the heat generated by a fever actually "burns out" the pathogens in the body, thus killing them. In reality, the heat from increased body temperature does not directly kill the pathogens, but instead increases your metabolism so that your immune system defense mechanisms can work better to fight the pathogens. In fact, an increase of 1°C in body temperature causes about a 10% ▶ ▶ ▶

OUTLINE

increase in metabolism. While a low-grade fever can be beneficial, a high-grade fever can be detrimental, as body temperature moves further away from normal. If a fever becomes too high and is not reduced, death will occur. For this reason, we often take medications to reduce fevers in case they start to elevate to potentially dangerous levels.

An interesting manipulation of body temperature is the basis for weight-loss supplements that claim to be "thermogenic." These supplements raise the body temperature slightly (as would a natural fever), increasing metabolism as a way to bring about weight loss. However, some people have suffered serious consequences, including death, after taking such supplements. For this reason, the FDA has banned one of the major thermogenic supplemental ingredients—ephedra.

The immune system has an incredible array of defenses to help keep you free of infection, as well as to protect you from cancer. Most of these defenses work remarkably well and only rarely fail. In this chapter, you will learn about how the immune system protects you on a daily basis.

26.1 Organs, Tissues, and Cells of the Immune System

The **immune system** plays an important role in keeping us healthy because it fights infections and cancer. First and foremost, the immune system contains the **lymphatic organs:** red bone marrow, the thymus gland, the lymph nodes, and the spleen (Fig. 26.1). The **tonsils,** which are located in the pharynx, and the **appendix,** which is attached to a portion of the large intestine, are well known patches of lymphatic tissue that also belong to the immune system. The lymphatic system not only contains a network of lymphatic organs and lymphatic tissues, but it also includes cells. In particular, we will be discussing the cell types mentioned in Table 26.1. You will want to refer to this table as you read the chapter.

  Immunity is the body's capability to repel foreign substances, pathogens, and cancer cells. Nonspecific immunity indiscriminately repels pathogens, while specific immunity requires that a certain antigen be present. An **antigen** is any molecule, usually a protein or carbohydrate, that stimulates the immune system.

Lymphatic Organs

Each of the lymphatic organs has a particular function in immunity, and each is rich in lymphocytes, one of the types of white blood cells.

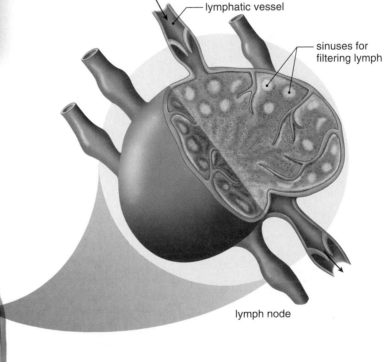

Figure 26.1 Lymphatic organs.

The red bone marrow, the thymus gland, the lymph nodes, and the spleen are lymphatic organs essential to immunity. The cells of the immune system, including lymphocytes, are found in these organs and also in the lymph of lymphatic vessels.

Red Bone Marrow

In a child, most bones have red bone marrow, but in an adult it is present only in the bones of the skull, the sternum (breastbone), the ribs, the clavicle, the pelvic bones, and the vertebral column. **Red bone marrow** produces all types of blood cells, but in this chapter we are interested in the cells listed in Table 26.1.

Lymphocytes differentiate into either **B lymphocytes (B cells)** or **T lymphocytes (T cells),** which are discussed at length later in the chapter. Bone marrow is not only the source of B lymphocytes, but also the place where B lymphocytes mature. T lymphocytes mature in the thymus gland. We shall learn that B cells produce antibodies and that T cells kill antigen-bearing cells outright.

Thymus Gland

The soft, bilobed **thymus gland** varies in size, but it is larger in children and shrinks as we get older. The thymus gland plays a role in the maturity of T lymphocytes. Immature T lymphocytes migrate from the bone marrow through the bloodstream to the thymus, and then they mature. Only about 5% of T lymphocytes ever leave the thymus. These T lymphocytes have survived a critical test: If any show the ability to react with "self" cells, they die. If they have potential to attack a foreign cell, they leave the thymus.

The thymus gland also produces thymic hormones thought to aid in the maturation of T lymphocytes.

Lymph Nodes

Lymph nodes are small, ovoid structures occurring along lymphatic vessels. Lymph nodes filter lymph and keep it free of pathogens and antigens. Lymph is filtered as it flows through a lymph node because its many sinuses (open spaces) are lined by **macrophages,** large phagocytic cells that engulf and then devour as many as a hundred pathogens and still survive (see Fig. 26.4a). Lymph nodes are also instrumental in fighting infections and cancer because they contain many lymphocytes.

Some lymph nodes are located near the surface of the body, and such nodes are named for their location. For example, inguinal nodes are in the groin, and axillary nodes are in the armpits. Physicians often feel for the presence of swollen, tender lymph nodes in the neck as evidence that the body is fighting an infection. This method is a noninvasive, preliminary way to help make a diagnosis.

Spleen

The **spleen,** which is about the size of your fist, is in the upper left abdominal cavity. The spleen's unique function is to filter the blood. This soft, spongy organ contains so-called red pulp and white pulp. Blood passing through the many sinuses in the red pulp is filtered of pathogens and debris, including worn out red blood cells, because the sinuses are lined by macrophages. The white pulp contains lymphocytes that are actively engaged in fighting infections and also cancer.

The spleen's outer capsule is relatively thin, and an infection or a severe blow can cause the spleen to burst. The spleen's functions can be fulfilled by other organs, but a person without a spleen is often slightly more susceptible to infections and may have to receive antibiotic therapy indefinitely.

Table 26.1	Immunocell Type and Function
Cell	**Function**
Macrophages	Phagocytize pathogens; inflammatory response and specific immunity
Mast cells	Release histamine, which promotes blood flow to injured tissues; inflammatory response
Neutrophils	Phagocytize pathogens; inflammatory response
Natural killer cells	Kill virus-infected and tumor cells by cell-to-cell contact
Lymphocytes	Responsible for specific immunity
B cells	Produce plasma cells and memory cells
Plasma cells	Produce specific antibodies
Memory cells	Ready to produce antibodies in the future
T cells	Regulate immune response; produce cytotoxic T cells and helper T cells
Cytotoxic T cells	Kill virus-infected and cancer cells
Helper T cells	Regulate immunity

Check Your Progress

1. List the components of the immune system.
2. List the lymphatic organs, and identify their functions.
3. What does an antibody do?

Answers: 1. The immune system is composed of lymphatic organs, tissues, and cells as well as the products of those cells, including antibodies. *2.* Red bone marrow: produces blood cells, including white blood cells; thymus: aids in maturation of T lymphocytes and testing of their ability to function; spleen: filters blood of pathogens and debris; lymph nodes: filter lymph of pathogens and debris. *3.* An antibody combines with and destroys an antigen.

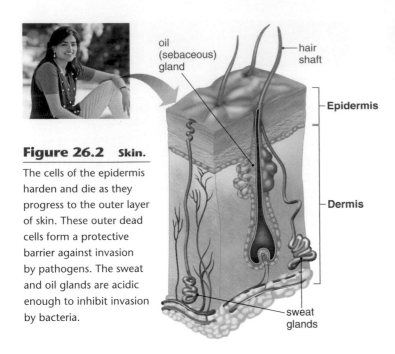

Figure 26.2 Skin.

The cells of the epidermis harden and die as they progress to the outer layer of skin. These outer dead cells form a protective barrier against invasion by pathogens. The sweat and oil glands are acidic enough to inhibit invasion by bacteria.

26.2 Nonspecific Defenses

The body has various types of nonspecific defenses that are our first line of defense against most types of infections. The nonspecific defenses are barriers to entry, the inflammatory response, the complement system, and natural killer cells.

Barriers to Entry

Skin and the mucous membranes lining the respiratory, digestive, and urinary tracts serve as mechanical barriers to entry by pathogens (Fig. 26.2). Oil gland secretions contain chemicals that weaken or kill certain bacteria on the skin. The upper respiratory tract is lined by ciliated cells that sweep mucus and trapped particles up into the throat, where they can be swallowed or expectorated (spit out). The stomach has an acidic pH, which inhibits growth or kills many types of bacteria. The various bacteria that normally reside in the intestine and other areas, such as the vagina, prevent pathogens from taking up residence.

The Inflammatory Response

Whenever tissue is damaged, a series of events occurs that is known as the **inflammatory response** (Fig. 26.3). The inflamed area has four outward signs: redness, warmth, swelling, and pain. The inflammatory response involves the first three cell types listed in Table 26.1.

When an injury occurs, damaged tissue cells and **mast cells** release chemical mediators, such as **histamine,** which cause the capillaries to dilate and become more permeable. Excess blood flow due to enlarged capillaries causes the skin to redden and become warm. Increased permeability allows proteins and fluids to escape into the tissues, resulting in swelling. The swollen area stimulates free nerve endings, causing the sensation of pain.

Neutrophils are phagocytic white blood cells that migrate to the site of injury. They are amoeboid and can change shape to squeeze through capillary walls and enter tissue fluid. When neutrophils phagocytize pathogens, they are enclosed within a vesicle. The pathogens are later destroyed by hydrolytic enzymes when this vesicle combines with a lysosome, one of the cellular organelles.

Also present are *macrophages* (Fig. 26.4*a*). As stated, lymph nodes contain macrophages, but so do some tissues, particularly connective tissues, which have resident macrophages that routinely act as scavengers, devouring old blood cells, bits of dead tissue, and other debris. During the inflammatory response, macrophages can also bring about an explosive increase in the number of leukocytes by liberating growth factors, which pass by way of blood to the red bone marrow. There, they stimulate the production and release of white blood cells, primarily neutrophils. As the infection is being overcome, some neutrophils may die. These—along with dead cells, dead bacteria, and

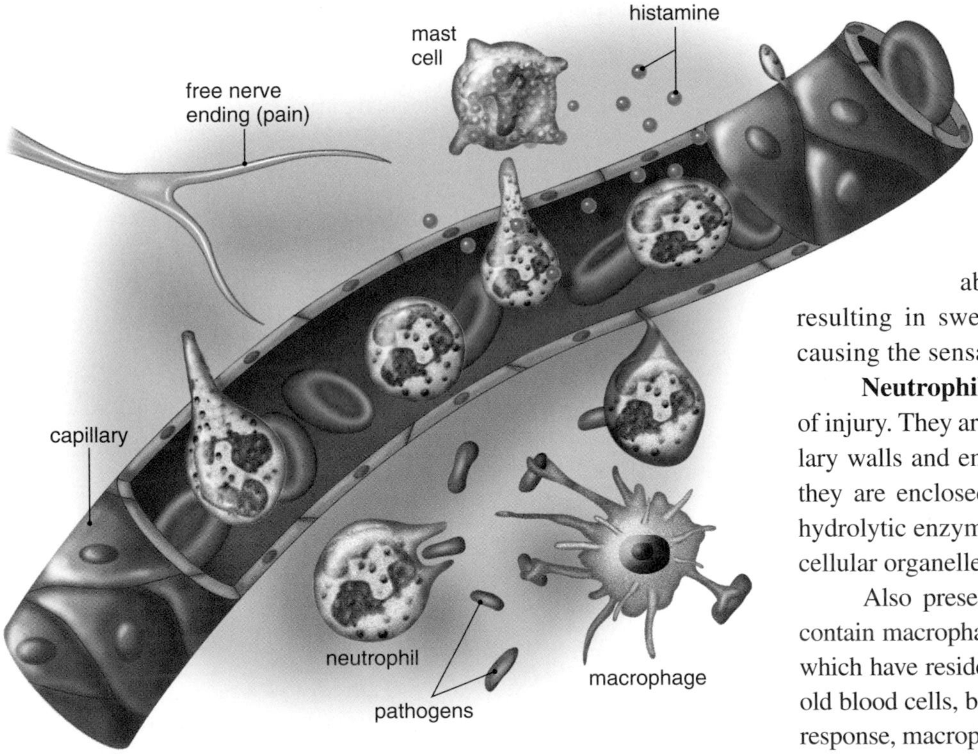

Figure 26.3 The inflammatory response.

If you cut your skin, the inflammatory response occurs immediately. As blood flow increases, the area gets red and warm. As mast cells, a type of white blood cell, release chemicals such as histamine, the capillary becomes more permeable, localized swelling occurs, and pain receptors are stimulated. Neutrophils and macrophages begin phagocytizing the bacteria.

living white blood cells—form pus, a whitish material. The presence of pus indicates that the body is trying to overcome an infection. When a blood vessel ruptures or is torn, the blood forms a clot to seal the break and prevent any further pathogen invasion.

The macrophages that have been fighting the infection move through the tissue fluid and lymph to the lymph nodes. Now lymphocytes are activated to react to the threat of an infection. Sometimes inflammation persists, and the result is chronic inflammation that is often treated by administering anti-inflammatory agents such as aspirin, ibuprofen, or cortisone. These medications act against the chemical mediators such as histamine released by the white blood cells in the area.

The Complement System

The **complement system,** often simply called complement, is composed of a number of blood plasma proteins designated by the letter C and a subscript. A limited amount of activated complement protein is needed because a domino effect occurs: As soon as a protein is activated, it in turn stimulates many more.

The complement proteins are activated when pathogens enter the body. The proteins "complement" certain immune responses, which accounts for their name. For example, they are involved in and amplify the inflammatory response because complement proteins attract phagocytes to the scene. Some complement proteins bind to the surfaces of pathogens already coated with antibodies, which ensures that the pathogens will be phagocytized by a neutrophil or macrophage.

Certain other complement proteins join to form a membrane attack complex that produces holes in the surfaces of bacteria and viruses (Fig. 26.4b). Fluids and salts then enter the pathogen to the point that it bursts.

Natural Killer Cells

Natural killer (NK) cells are large, granular lymphocytes that kill virus-infected cells and tumor (cancer) cells by cell-to-cell contact (see Table 26.1).

What makes an NK cell attack and kill a cell? The cells of your body ordinarily have "self" proteins on their surface that bind to receptors on NK cells. Sometimes virus-infected cells and cancer cells undergo alterations and lose their ability to produce self proteins. When NK cells can find no self proteins to bind to, they kill the cell by the same method as T lymphocytes do (see Fig. 26.8).

NK cells are not specific—their numbers do not increase when exposed to a particular antigen, and they have no means of "remembering" the antigen from previous contact.

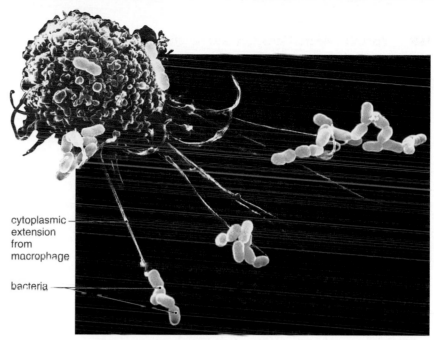

cytoplasmic extension from macrophage

bacteria

a.　　×6,000

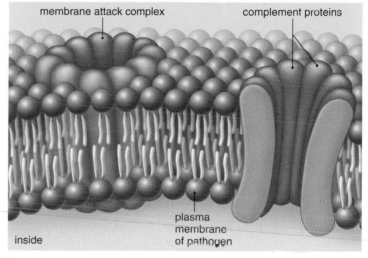

membrane attack complex　　complement proteins

plasma membrane of pathogen

inside

b.

Figure 26.4　Ways to get rid of pathogens.

a. Macrophages, the body's scavengers, engulf pathogens and chop them up inside lysosomes. **b.** Complement proteins come together and form a membrane attack complex in the surface of a pathogen. Water and salts enter, and the pathogen bursts.

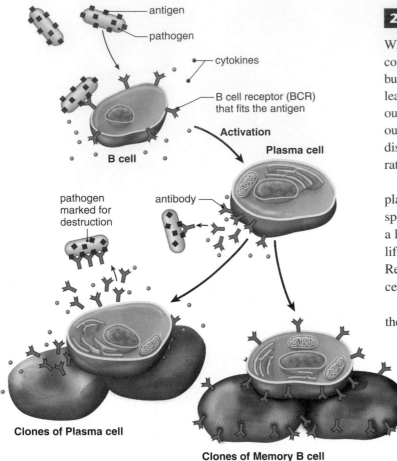

Figure 26.5   **B cell and the antibody response.**

When an antigen combines with a BCR, the B cell divides to produce plasma cells and memory B cells. Plasma cells produce antibodies but eventually disappear. Memory B cells remain in the body ready to produce the same antibody in the future.

Table 26.2	Characteristics of B Cells

- Provide an antibody response to a pathogen

- Produced and become mature in bone marrow

- Reside in lymph nodes and spleen; circulate in blood and lymph

- Directly recognize antigen and then undergo cell division

- Cell division produces antibody-secreting plasma cells as well as memory B cells

26.3 Specific Defenses

When nonspecific defenses have failed to prevent an infection, specific defenses come into play. Specific defenses respond to antigens. Pathogens have antigens, but antigens can also be part of a foreign cell or a cancer cell. When our body learns to destroy a particular antigen, we have become immune to it. Because our immune system does not ordinarily respond to the proteins on the surface of our own cells (as if they were antigens), the immune system is said to be able to distinguish "self" from "nonself." Only in this way can the immune system aid, rather than disrupt, homeostasis.

Lymphocytes are capable of recognizing an antigen because they have plasma membrane receptor proteins shaped to allow them to combine with a specific antigen. It is often said that the receptor and the antigen fit together like a lock and a key. Because we encounter a million different antigens during our lifetime, we need a great diversity of lymphocytes to protect us against them. Remarkably, diversification occurs to such an extent during the maturation process that a lymphocyte type exists for any possible antigen.

Immunity usually lasts for some time. For example, once we recover from the measles, we usually do not get the illness a second time. Immunity is primarily the result of the action of the B lymphocytes and the T lymphocytes. B lymphocytes mature in the bone marrow, and T lymphocytes mature in the thymus gland. B lymphocytes, also called B cells, give rise to plasma cells, which produce antibodies. These antibodies are secreted into the blood, lymph, and other body fluids. In contrast, T lymphocytes, also called T cells, do not produce antibodies. Some T cells regulate the immune response, and other T cells directly attack cells that bear antigens (see Table 26.1).

B Cells and the Antibody Response

Each B cell can bind only to a specific antigen—the antigen that fits the binding site of its receptor. The receptor is called a **B-cell receptor (BCR).** Some B cells never have anything to do because an antigen that fits the binding site of their type of receptor never shows up. But if an antigen does bind to a BCR, that particular B cell is activated, and it divides, producing many plasma cells and also memory B cells (Fig. 26.5). A B cell is stimulated to divide and produce plasma cells by helper T-cell secretions called cytokines, which are discussed later in this section. Plasma cells are larger than regular B cells because they have extensive rough endoplasmic reticulum for the mass production and secretion of antibodies specific to the antigen. The antibodies produced by plasma cells and secreted into the blood and lymph are identical to the BCR of the activated B cell.

Memory B cells are the means by which long-term immunity is possible. If the same antigen enters the system again, memory B cells quickly divide and give rise to more plasma cells capable of producing the correct antibodies.

Once the threat of an infection has passed, the development of new plasma cells ceases, and those present undergo apoptosis. Apoptosis is a process of programmed cell death (PCD) involving a cascade of specific cellular events leading to the death and destruction of the cell (see Fig. 8.11, p. 120).

Table 26.2 summarizes the activities of B cells.

The Function of Antibodies

Antibodies, as stated, are immunoglobulin proteins that are capable of combining with a specific antigen (Fig. 26.6). The antigen-antibody reaction can take several forms, but quite often the reaction produces complexes of antigens combined with antibodies. Such antigen-antibody complexes, sometimes called immune complexes, mark the antigens for destruction (see Fig. 26.5). An antigen-antibody complex may be engulfed by neutrophils or macrophages, or it may activate complement. Complement makes pathogens more susceptible to phagocytosis, as discussed previously.

ABO Blood Type   Most likely you know your ABO blood type—whether you have type A, B, AB, or O blood. These letters stand for antigens on your red blood cells. If you have type O blood, you don't have either antigen A or B on your red cells. Some blood types have antibodies in the plasma (Table 26.3). As an example, type O blood has both anti-A and anti-B antibodies in the plasma. You can't give a person with type O blood a transfusion from an individual with type A blood. If you do, the antibodies in the plasma will react to type A red blood cells, and agglutination will occur. **Agglutination,** the clumping of red blood cells, causes blood to stop circulating and red blood cells to burst. On the other hand, you can give type O blood to a person with any blood type because type O red blood cells bear neither A nor B antigens.

It's possible to determine who can give blood to whom based on the ABO system. However, there are other red blood cell antigens in addition to A and B that are used in typing blood. Therefore, it is best to physically put the donor's blood on a slide with the recipient's blood to observe whether the two types match (no agglutination occurs) before blood can be safely transfused from one person to another.

Table 26.3	Blood Types	
Blood Type	**Antigens on Red Blood Cells**	**Antibodies in Plasma**
A	A	Anti-B
B	B	Anti-A
AB	A,B	Neither anti-A nor anti-B
O	—	Both anti-A and anti-B

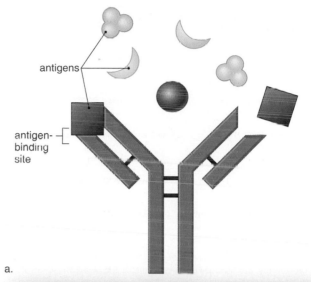

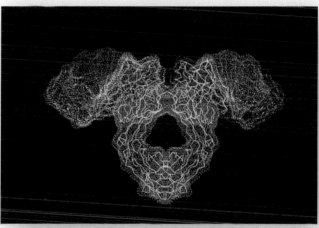

a.

b.

Figure 26.6   Why an antibody is specific.

a. During a lifetime, a person will encounter a million different antigens that differ in shape. An antibody has two variable regions that end in an antigen-binding site. The variable regions vary so much that the binding site of each antibody has a shape that will fit only one specific antigen. **b.** Computer model of an antibody molecule. The antigen combines with the two side branches.

Check Your Progress

1. Explain how antibodies are produced in response to an antigen.
2. Explain what the letters in an ABO blood type refer to.

Answers: 1. The B-cell receptor binds to the antigen and activates the B cell carrying that receptor. This causes the B cell to produce many plasma cells. The plasma cells secrete antibodies against the antigen. 2. The letters A, B, AB, and O refer to the antigens carried on red blood cells. Type A produces antigen A, type B produces antigen B, type AB produces both antigens A and B, and type O does not produce antigen A or B.

Table 26.4	Characteristics of T Cells

- Provide a cellular response to virus-infected cells and cancer cells

- Produced in bone marrow, mature in thymus gland

- Antigen must be presented in groove of an MHC[1] molecule

- Cytotoxic T cells destroy nonself protein-bearing cell

- Helper T cells secrete cytokines that control the immune response

[1] MHC = major histocompatibility complex. The MHC genes encode MHC proteins that are displayed on the cell surface and define an individual's tissue type. When tissues/organs have like MHC proteins, they are histocompatible, and transplants between individuals are possible.

T Cells and the Cellular Response

When T cells leave the thymus gland, they have unique **T-cell receptors (TCRs)** just as B cells do. Unlike B cells, however, T cells are unable to recognize an antigen without help. The antigen must be presented to them by an **antigen-presenting cell (APC),** such as a macrophage. After a macrophage phagocytizes a virus it is digested in a lysosome. An antigen from the virus is combined with an MHC "self" protein, and the complex appears on the cell surface. Then the MHC + antigen is presented to a T cell. The importance of self proteins in plasma membranes was first recognized when it was discovered that they contribute to the specificity of tissues and make it difficult to transplant tissue from one human to another.

In Figure 26.7, the antigen is represented by a red triangle, and the T cell that binds to the antigen has the specific TCR that can combine with this particular MHC + antigen. Now the T cell is activated and divides to produce more T cells. T cells that are destined to become cytotoxic T cells are activated by a macrophage that presents an antigen with an MHC I. T cells that are destined to become helper T cells are activated by a macrophage that presents an antigen with an MHC II. If the macrophage presents an antigen with an MHC II, the activated T cell will form helper T cells as in Figure 26.7.

As the illness disappears, the immune reaction wanes, and the activated T cells become susceptible to apoptosis. Apoptosis contributes to homeostasis by regulating the number of cells present in an organ, or in this case, the immune system. When apoptosis does not occur as it should, T-cell cancers (e.g., lymphomas and leukemias) can result. Also, in the thymus gland, any T cell that has the potential to destroy the body's own cells undergoes apoptosis.

Types of T Cells

The two main types of T cells are cytotoxic T cells and helper T cells. **Cytotoxic T cells** have storage vacuoles containing perforins or enzymes called granzymes. After a cytotoxic T cell binds to a virus-infected or cancer cell presenting the same antigen it has learned to recognize, it releases perforin molecules that perforate the plasma membrane, forming a pore. Cytotoxic T cells then deliver a supply of granzymes into the pore, and these cause the cell to undergo apoptosis

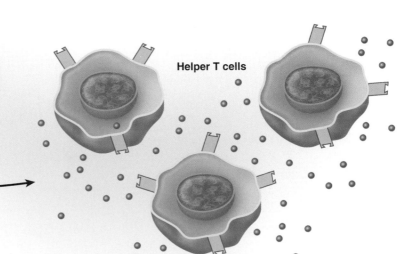

Figure 26.7 **Activation of a T cell.**

For a T cell to be activated, the antigen must be presented along with an MHC protein to the T cell by an APC, often by a macrophage. Each type of T cell bears a specific receptor, and if its TCR fits the MHC + antigen complex, it is activated to divide and produce more T cells—in this case, helper T cells, which also have this type TCR.

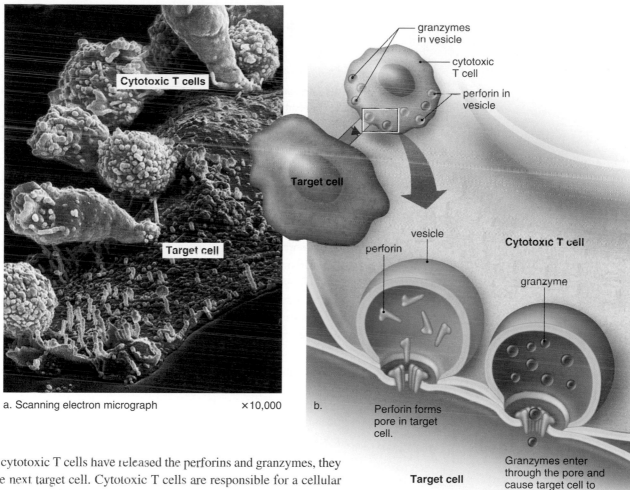

a. Scanning electron micrograph ×10,000

b.

and die. Once cytotoxic T cells have released the perforins and granzymes, they move on to the next target cell. Cytotoxic T cells are responsible for a cellular response to virus-infected and cancer cells (Fig. 26.8).

Helper T cells regulate immunity by secreting cytokines. **Cytokines** are signaling chemicals that stimulate various immune cells (e.g., macrophages, B cells, and other T cells) to perform their functions. B cells cannot be activated without T cell help. Because HIV, the virus that causes AIDS, infects helper T cells and other cells of the immune system, it inactivates the immune response and makes HIV-infected individuals susceptible to opportunistic infections. Infected macrophages serve as reservoirs for the HIV virus. AIDS is discussed in Section 26.5.

Table 26.4 summarizes the activities of T cells.

Tissue Rejection

Certain organs, such as the skin, the heart, and the kidneys, could be transplanted easily from one person to another if the body did not attempt to reject them. Rejection occurs because cytotoxic T cells and also antibodies bring about destruction of foreign tissues in the body. When rejection occurs, the immune system is correctly distinguishing between self and nonself.

Organ rejection can be controlled by carefully selecting the organ to be transplanted and administering immunosuppressive drugs. It is best if the transplanted organ has the same type of MHC proteins as those of the recipient because otherwise the transplanted organ will be antigenic to T cells. Two well-known immunosuppressive drugs, cyclosporine and tacrolimus, act by inhibiting the response of T cells to cytokines. Without cytokines, all types of immune responses are weak.

Figure 26.8 **Cytotoxic T cells and the cellular response.**

a. Scanning electron micrograph of cytotoxic T cell attacking a target cell, which is either a virus-infected or cancer cell.
b. A cytotoxic T cell attacks any cell that presents it with an MHC I + antigen it has learned to recognize. First, vesicles release perforin, which forms a pore in the target cell. Then vesicles release granzymes, which cause the cell to undergo apoptosis.

Check Your Progress

1. Compare and contrast B cells with T cells.
2. What happens to plasma cells and cytotoxic T cells when an infection has passed?

Answers: 1. Both are involved with specific defenses against disease. B cells are produced in the bone marrow and give rise to plasma cells that produce antibodies. T cells are produced in the thymus gland and when activated become either cytotoxic or helper T cells. 2. They undergo programmed cell death, called apoptosis.

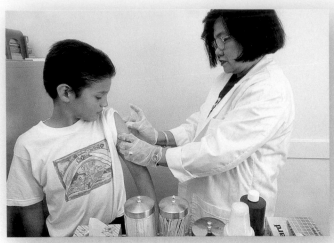

a.

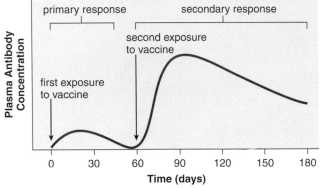

b.

Figure 26.9 **Active immunity due to immunizations.**

a. Immunization often requires more than one injection.
b. A minimal primary response occurs after the first vaccine injection, but after a second injection, the secondary response usually shows a dramatic rise in the amount of antibody present in plasma.

Check Your Progress

1. Explain why you cannot be vaccinated for most sexually transmitted diseases.
2. Describe the function of a booster shot.
3. Explain the function of gamma globulin injections to prevent illness in a patient who is unexpectedly exposed to an infectious disease.

Answers: 1. People cannot become immune to sexually transmitted diseases, so vaccines are ineffective. 2. A booster shot is an injection of the same vaccine to increase the antibody concentration to a high level. 3. Gamma globulin from someone who had the same disease will contain antibodies against the pathogen in question.

26.4 Immunizations

After you have had an infection, you are sometimes immune to getting it again. Good examples are the childhood diseases measles and mumps. Unfortunately, few sexually transmitted diseases stimulate lasting immunity; for example, a person can get gonorrhea over and over again. If lasting immunity is possible, a vaccine for the disease likely exists or can be developed. **Vaccines** are substances that usually don't cause illness, but even so, the immune system responds to them. Traditionally, vaccines are the pathogens themselves, or their products, that have been treated so that they are no longer virulent (able to cause disease). Today, it is possible to genetically engineer bacteria to mass-produce a protein from pathogens, and this protein can be used as a vaccine. This method has now produced a vaccine against hepatitis B, a virus-induced disease, and it is being used to prepare a vaccine against malaria, a protozoan-induced disease.

Immunization promotes **active immunity.** After a vaccine is given, it is possible to follow an active immune response by determining the amount of antibody present in a sample of plasma; this is called the antibody titer. After the first exposure to a vaccine, a primary response occurs. For a period of several days, no antibodies are present; then their concentration rises slowly, levels off, and gradually declines as the antibodies bind to the antigen or simply break down (Fig. 26.9). After a second exposure to the vaccine, a secondary response is expected. The concentration now rises rapidly to a level much greater than before; then it slowly declines. The second exposure is called a "booster" because it boosts the antibody concentration to a high level. The high antibody concentration now is expected to help prevent disease symptoms if the individual is exposed to the disease-causing antigen.

Active immunity is dependent upon the presence of memory B cells and possibly memory T cells that are capable of responding to lower doses of antigen. Active immunity is usually long-lasting, but a booster may be required after a certain number of years.

Although the body usually makes its own antibodies, it is sometimes possible to give an individual prepared antibodies (immunoglobulins) to combat a disease. Because these antibodies are not produced by the individual's plasma cells, this so-called **passive immunity** is temporary. For example, newborn infants are passively immune to some diseases because antibodies have crossed the placenta from the mother's blood. These antibodies soon disappear, however, so that within a few months, infants become more susceptible to infections. Breast-feeding prolongs the natural passive immunity an infant receives from the mother because antibodies are present in the mother's milk.

Even though passive immunity does not last, it is sometimes used to prevent illness in a patient who has been unexpectedly exposed to an infectious disease. Usually, the patient receives an injection of gamma globulin, a portion of blood that contains antibodies, preferably taken from an individual who has recovered from the illness. In the past, horses were immunized, and gamma globulin was taken from them to provide the needed antibodies against such diseases as diphtheria, botulism, and tetanus. Unfortunately, patients who received these antibodies became ill about 50% of the time, because the serum contained proteins that the individual's immune system recognized as foreign. This condition was called serum sickness.

26.5 Immune System Problems

The immune system usually protects us from disease because it can distinguish self from nonself. Sometimes, however, it responds in a manner that harms the body, as when individuals develop allergies or have an autoimmune disease.

Allergies

Allergies are hypersensitivities to substances in the environment, such as pollen, food, or animal hair, that ordinarily would not cause an immune reaction. The response to these antigens, called **allergens,** usually includes some unpleasant symptoms (Fig. 26.10). An allergic response is regulated by cytokines secreted by both T cells and macrophages.

Immediate allergic responses are caused by receptors attached to the plasma membrane of mast cells in the tissues. When an allergen attaches to receptors on mast cells, they release histamine and other substances that bring about the symptoms.

An immediate allergic response can occur within seconds of contact with the antigen. The symptoms can vary, but a dramatic example, anaphylactic shock, is a severe reaction characterized by a sudden and life-threatening drop in blood pressure.

Allergy shots sometimes prevent the onset of an allergic response. It has been suggested that injections of the allergen may cause the body to build up high quantities of antibodies released by plasma cells, and these combine with allergens received from the environment before they have a chance to reach the receptors located in the membranes of mast cells.

Delayed allergic responses are probably initiated by memory T cells at the site of allergen contact in the body. A classic example of a delayed allergic response is the skin test for tuberculosis (TB). When the test result is positive, the tissue where the antigen was injected becomes red and hardened. This shows that the person has been previously exposed to tubercle bacilli, the cause of TB. Contact dermatitis, which occurs when a person is allergic to poison ivy, jewelry, cosmetics, and so forth, is also an example of a delayed allergic response.

Autoimmune Diseases

When cytotoxic T cells or antibodies mistakenly attack the body's own cells as if they bear antigens, the resulting condition is known as an **autoimmune disease.** Exactly what causes autoimmune diseases is not known. However, sometimes they occur after an individual has recovered from an infection.

In the autoimmune disease *myasthenia gravis,* neuromuscular junctions do not work properly, and muscular weakness results. In *multiple sclerosis (MS),* the myelin sheath of nerve fibers breaks down, causing various neuromuscular disorders. A person with *systemic lupus erythematosus* has various symptoms prior to death due to kidney damage. In *rheumatoid arthritis,* the joints are affected (Fig. 26.11). Researchers suggest that rheumatic fever and also diabetes type 1 are autoimmune illnesses. As yet, there are no cures for autoimmune diseases, but they can be controlled with drugs.

SEM of pollen

Figure 26.10 Allergies.

When people are allergic to pollen, they develop symptoms that include watery eyes, sinus headaches, increased mucus production, labored breathing, and sneezing.

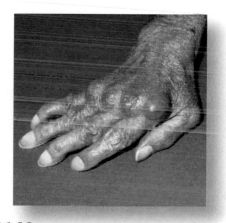

Figure 26.11 Rheumatoid arthritis.

Rheumatoid arthritis is due to recurring inflammation in skeletal joints. Complement proteins, T cells, and B cells all participate in deterioration of the joints, which eventually become immobile.

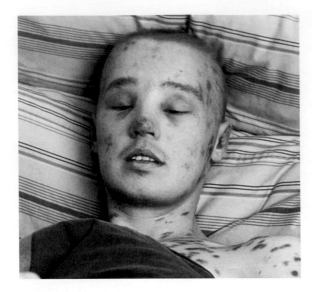

a.

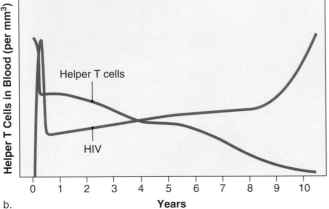

b.

Figure 26.12 **HIV infection.**

a. An HIV infection leads to AIDS, characterized by a number of associated illnesses, including a cancer that results in skin lesions.
b. At first the body produces enough helper T cells to keep the HIV infection under control, but then as the number of helper T cells declines, the HIV infection takes over.

AIDS

Understanding why AIDS patients are so sick gives us a whole new level of appreciation for the workings of a healthy immune system. An immune system ravaged by **AIDS (acquired immunodeficiency syndrome)** can no longer fight off the onslaught of viruses, fungi, and bacteria that the body encounters every day. **HIV (human immunodeficiency virus),** which causes AIDS, lives in and destroys helper T cells, which promote the activity of all the other cells in the immune system. For years, the body is able to maintain an adequate number of T cells, but finally the virus gains the upper hand. Without drug therapy, the number of T cells eventually drops from thousands to hundreds as the immune system becomes helpless (Fig. 26.12). The symptoms of AIDS begin with weight loss, chronic fever, cough, diarrhea, swollen glands, and shortness of breath, and progress to those of rare diseases. *Pneumocystic pneumonia,* a respiratory disease found in cats, and *Kaposi's sarcoma,* a very rare type of cancer, are often observed in patients with advanced AIDS. Death approaches rapidly and certainly.

HIV is transmitted by sexual contact with an infected person, including vaginal or rectal intercourse and oral/genital contact. Also, needle-sharing among intravenous drug users is high-risk behavior. Babies born to HIV-infected women may become infected before or during birth, or through breast-feeding after birth. To date, as many as 64 million people worldwide may have contracted HIV, and almost 2.2 million people have died. Male-to-male sexual contact still accounts for most new AIDS cases in the United States, but the greatest percentage of increase is now through heterosexual contact or intravenous drug use.

Advances in treatment have reduced the serious complications of an HIV infection and have prolonged life. The sooner drug therapy begins after infection, the better the chances that the immune system will not be destroyed by HIV. Also, medication must be continued indefinitely. Unfortunately, new strains of the virus have emerged that are resistant to the new drugs used for treatment. The likelihood of transmission from mother to child at birth can be lessened if the mother receives medication prior to birth and the child is delivered by cesarean section.

Many investigators are working on a vaccine for AIDS. Some are trying to develop a vaccine in the traditional way, using the entire virus. Others are working on vaccines that utilize just a single HIV protein as the vaccine. So far, no method has resulted in sufficient antibodies to keep an infection at bay. After many clinical trials, none too successful, most investigators agree that a combination of various vaccines may be the best strategy to bring about a response by both B cells and T cells.

HIV infection is preventable. Suggestions for preventing an infection are: (1) Abstain from sexual intercourse or develop a long-term monogamous (always same partner) sexual relationship with a person who is free of HIV; (2) be aware that having relations with an intravenous drug user is risky behavior; (3) avoid anal-rectal intercourse because the lining of the rectum is thin, and infected T cells easily enter the body there; (4) always use a latex condom during sexual intercourse if you do not know that your partner has been free of HIV for the past five years; (5) avoid oral sex because this can be a means of transmission; and (6) be cautious about the use of alcohol or any drug that may prevent you from being able to control your behavior.

Check Your Progress

1. Describe the relationship between allergies and allergens.
2. Contrast an immediate allergic response with a delayed allergic response.
3. Explain why AIDS patients cannot fight pathogens.

Answers: 1. Allergies are hypersensitive responses to antigens called allergens. 2. An immediate allergic response occurs within seconds of exposure to an allergen and is caused by chemicals released by mast cells. Delayed allergic responses take longer to develop and are probably initiated by memory T cells. 3. The HIV virus lives in and destroys helper T cells, which are necessary for the activity of all other immune system cells.

Summary

26.1 Organs, Tissues, and Cells of the Immune System

The immune system consists of lymphatic organs, tissues, and cells, as well as the products of these cells. The lymphatic organs are:

- Red bone marrow, where all blood cells are made and the B lymphocytes mature.
- Thymus gland, where T lymphocytes mature (see Table 26.1).
- Lymph nodes, where lymph is cleansed of pathogens and debris.
- Spleen, where blood is cleansed of pathogens and debris.
- Other organs, such as the tonsils and appendix, are patches of lymphatic tissue.

26.2 Nonspecific Defenses

Immunity involves nonspecific and specific defenses. Nonspecific defenses include:

- Barriers to entry (e.g., skin).
- Inflammatory response involves mast cells, which release histamine to increase capillary permeability resulting in redness, warmth, swelling, and pain. Neutrophils and macrophages enter tissue fluid and engulf pathogens.
- The complement system has many functions. One is to attack bacteria outright by forming a pore in the surface of a bacterium. Water and salts then enter, and the bacterium bursts.
- Natural killer cells that can tell self proteins from nonself and cause virus-infected cells to undergo apoptosis.

26.3 Specific Defenses

Specific defenses require B lymphocytes and T lymphocytes, also called B cells and T cells.

B Cells and the Antibody Response

The BCR (B-cell receptor) of each type B cell is specific to a particular antigen. When the antigen binds to a BCR, that B cell divides to produce plasma cells and memory B cells.

- Plasma cells secrete antibodies and eventually undergo apoptosis. Plasma cells are responsible for antibody response to an antigen.

Activated B cell ⟶ Plasma cells ⟶ Produce antibodies

- Memory B cells remain in the body and produce antibodies if the same antigen enters the body at a later date.

T Cells and the Cellular Response

T cells are responsible for cellular response to an infection. Each TCR (T-cell receptor) is specific to a particular antigen. For a T cell to recognize an antigen, the antigen must be presented to it, usually by a macrophage, along with an MHC protein. Thereafter, the activated T cell divides and produces either cytotoxic T cells or helper T cells.

Activated T cell ⟶ Cytotoxic T cells ⟶ Kill virus-infected cell
Activated T cell ⟶ Helper T cells ⟶ Regulate immune response

- Cytotoxic T cells kill virus-infected or cancer cells on contact because these cells bear the MHC + antigen it has learned to recognize. First, perforin is secreted, and these molecules form a pore; then granzymes are secreted and cause the cell to undergo apoptosis.
- Helper T cells produce cytokines and stimulate other immune cells.

26.4 Immunizations

Immunity occurs after an infection or a vaccination.

- A vaccine brings about immunity to a particular infection. Two injections may be required because the number of antibodies is higher after the second injection, called a booster shot. A booster shot in the future also increases the number of antibodies.

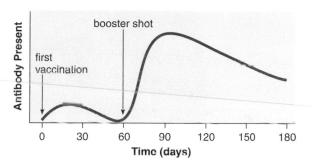

- Passive immunity (receiving preformed antibodies) is short-lived because the antibodies are administered to, not made by, the individual.

26.5 Immune System Problems

Allergies

Allergic responses to various substances can be immediate or delayed. Anaphylactic shock is a dangerous immediate allergic response.

Autoimmune Diseases

In an autoimmune disease (e.g., rheumatoid arthritis, multiple sclerosis, and perhaps diabetes type 1), the immune system mistakenly attacks the body's tissues.

AIDS

The HIV virus lives in and destroys helper T cells. Without drug treatment, the number of T cells falls off, and the individual dies from infections that are rare in healthy individuals. Sexual contact and needle-sharing transmit AIDS from person to person. Combined drug therapy has prolonged the lives of some people infected with HIV. So far, vaccine research has met with limited success. However, AIDS is a preventable disease if certain behaviors are avoided.

Thinking Scientifically

1. The human body is exposed to many antigens over a lifetime. It is estimated that the mammalian genome contains the information needed to produce a million different antibodies. Considering that there are only about 30,000 genes in the human genome, how can it make a million different antibodies?

2. The transplantation of organs from one person to another was impossible until the discovery of immunosuppressive drugs. Now, with the use of drugs such as cyclosporine, organs can be transplanted without rejection. Transplant patients must take immunosuppressive drugs for the remainder of their lives. How can a person do this and not eventually succumb to disease?

Testing Yourself

Choose the best answer for each question.

1. This lymphatic tissue is associated with the respiratory system.

   a. red bone marrow
   b. spleen
   c. tonsil
   d. appendix
   e. lymph node

2. Label the lymphatic organs in the following illustration.

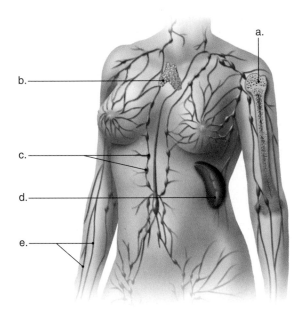

3. Which of the following is not a barrier to pathogen entry?

   a. oil gland secretions of the skin
   b. acidic pH in the stomach
   c. cilia in the upper respiratory tract
   d. nonpathogenic bacteria in the digestive tract
   e. saliva in the mouth

For questions 4–9, identify the lymphatic organ in the key that matches the description. Some answers may be used more than once.

Key:

   a. red bone marrow
   b. thymus gland
   c. lymph nodes
   d. spleen

4. Produces stem cells.

5. Located in every body cavity except the dorsal cavity.

6. Produces lymphocytes.

7. Site of maturation of T cell.

8. Contains red pulp and white pulp.

9. Shrinks with age.

10. During the inflammatory response,

    a. T cells move to the site of injury.
    b. capillaries become constricted.
    c. histamine is produced.
    d. capillaries become less permeable.
    e. More than one of these are correct.

11. Which is a lymphatic organ?

    a. spleen
    b. tonsil
    c. thymus gland
    d. All of these are correct.

12. The _____ cleanses _____, and in its absence, problems associated with _____ may occur.

    a. spleen, lymph, oxygen transport
    b. liver, lymph, infection
    c. spleen, blood, infection
    d. spleen, blood, oxygen transport

13. Which cells will phagocytize pathogens?

    a. neutrophils
    b. macrophages
    c. mast cells
    d. Both a and b are correct.

14. Complement

    a. is a general defense mechanism.
    b. is involved in the inflammatory response.
    c. is a series of proteins present in the plasma.
    d. plays a role in destroying bacteria.
    e. All of these are correct.

15. _____ release histamines.

    a. Mast cells
    b. Neutrophils
    c. Monocytes
    d. All of these are correct.

16. Which applies to T lymphocytes?

    a. mature in bone marrow
    b. mature in thymus gland
    c. Both a and b are correct.
    d. None of these are correct.

17. AIDS is caused by which of the following viruses?

    a. SIV
    b. HSV
    c. HPV
    d. AIDS
    e. HIV

18. Plasma cells are

    a. the same as memory cells.
    b. formed from blood plasma.
    c. B cells that are actively secreting antibody.
    d. inactive T cells carried in the plasma.
    e. a type of red blood cell.

19. Vaccines are associated with

    a. active immunity.
    b. long-lasting immunity.
    c. passive immunity.
    d. Both a and b are correct.

20. MHC proteins play a role in

    a. active immunity.
    b. histamine release.
    c. tissue transplantation.
    d. All of these are correct.

21. Programmed cell death is known as

    a. activation.
    b. apoptosis.
    c. clonal selection.
    d. None of these are correct.

22. Type O blood has _____ antibodies in the plasma.

    a. A
    b. B
    c. Both a and b are correct.
    d. Neither a nor b is correct.

23. Label the following diagram of skin.

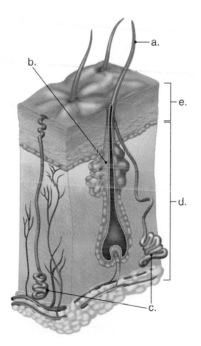

24. Which of the following is a nonspecific defense of the body?

    a. immunoglobulin
    b. B cell
    c. T cell
    d. vaccine
    e. inflammatory response

25. Antibodies combine with antigens

    a. at variable regions.
    b. at other regions
    c. only if macrophages are present.
    d. Both a and c are correct.

26. Which one of these pairs is mismatched?

    a. helper T cells —help complement react
    b. cytotoxic T cells—active in tissue rejection
    c. macrophages—activate T cells
    d. memory T cells—long-living line of T cells
    e. T cells—mature in thymus

27. An antigen-presenting cell (APC)

    a. presents antigens to T cells.
    b. secretes antibodies.
    c. marks each human cell as belonging to that particular person.
    d. secretes cytokines.

28. Which of the following would not be a participant in cell-mediated immune responses?

    a. helper T cells
    b. macrophages
    c. cytokines
    d. cytotoxic T cells
    e. plasma cells

29. Vaccines are

    a. the same as monoclonal antibodies.
    b. treated bacteria or viruses, or one of their proteins.
    c. short-lived.
    d. MHC proteins.
    e. All of these are correct.

30. Which of the following is not an example of an autoimmune disease?

    a. multiple sclerosis
    b. myasthenia gravis
    c. contact dermatitis
    d. systemic lupus erythematosus
    e. rheumatoid arthritis

31. Natural killer cells attack cells that

    a. have been determined to contain a virus.
    b. have lost their self proteins.
    c. are malformed.
    d. are dividing rapidly.

32. A person with type A blood can receive a transfusion from someone with type

    a. A blood only.
    b. A or B blood.
    c. A or AB blood.
    d. A or O blood.
    e. A, B, AB, or O blood.

33. Unlike B cells, T cells
    a. require help to recognize an antigen.
    b. are components of a specific defense response.
    c. are types of lymphocytes.
    d. contribute to homeostasis.

34. Passive immunity
    a. is permanent.
    b. may result from immunoglobulin injections.
    c. requires memory B cells.
    d. may be induced by vaccines.

35. Anaphylactic shock is an example of
    a. an immediate allergic response.
    b. an autoimmune disease.
    c. a type of serum shock.
    d. a delayed allergic response.

36. The HIV virus lives in and destroys
    a. B cells.
    b. cytotoxic T cells.
    c. helper T cells.
    d. All of these are correct.

37. AIDS patients generally die due to which of the following causes?
    a. rare diseases and infections
    b. HIV virus
    c. strokes
    d. heart attacks
    e. cancer

Go to www.mhhe.com/maderessentials for more quiz questions.

Bioethical Issue

Over 36 million people worldwide are living with AIDS. This disease is deadly without proper medical care, but can be chronic if treated. Drug companies typically charge a high price for AIDS medications because Americans and their insurance companies can afford them. However, these drugs are out of reach in many countries, such as those in Africa, where AIDS is a widespread problem. Some people argue that drug companies should use the profits from other drugs (such as those for heart disease, depression, and impotence) to make AIDS drugs affordable to those who need them. That has not happened yet. In some countries, governments have allowed companies to infringe on foreign patents held by major drug companies so that they can produce affordable AIDS drugs. Do drug companies have a moral obligation to provide low-cost AIDS drugs, even if they have to do so at a loss of revenue? Is it right for governments to ignore patent laws in order to provide their citizens with affordable drugs?

Understanding the Terms

active immunity 464
agglutination 461
AIDS (acquired immunodeficiency syndrome) 466
allergen 465
allergy 465
antibody 456
antigen 456
antigen-presenting cell (APC) 462
appendix 456
autoimmune disease 465
B-cell receptor (BCR) 460
B lymphocyte (B cell) 457
complement system 459
cytokine 463
cytotoxic T cell 462
delayed allergic response 465
helper T cell 463
histamine 458

HIV (human immunodeficiency virus) 466
immediate allergic response 465
immune system 456
immunity 456
inflammatory response 458
lymph node 457
lymphatic organs 456
macrophage 457
mast cell 458
natural killer (NK) cell 459
neutrophil 458
passive immunity 464
red bone marrow 457
spleen 457
T-cell receptor (TCR) 462
T lymphocyte (T cell) 457
thymus gland 457
tonsils 456
vaccine 464

Match the terms to these definitions:

a. _____ Region of lymphatic tissue attached to the large intestine.

b. _____ Site of blood cell production.

c. _____ Large white blood cell that phagocytizes pathogens and presents antigens to T cells.

d. _____ Chemical that causes capillaries to dilate and become more permeable.

e. _____ White blood cell produced in the bone marrow that gives rise to plasma cells.

f. _____ Clumping of red blood cells.

g. _____ Secretes granzymes into infected cells to cause apoptosis.

h. _____ Substance that doesn't cause illness, but results in immunity.

i. _____ Antigens that cause allergies.

j. _____ Condition that results when cytotoxic T cells attack the body's own cells.

The Control Systems

The nervous and endocrine systems control our behavior.

Brain tissue will die from lack of sleep faster than from starvation.

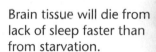

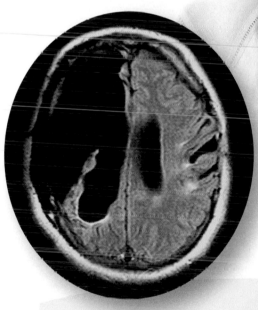

If brain tissue is removed, it is possible to live with half a brain.

Obviously, the brain's role in the human body is critical to survival. The brain controls nearly all the activities in the body, including those of the other organ systems. When particular areas of the brain malfunction, a multitude of problems, or even death, can occur. Certain conditions, such as epilepsy, inherited disease, or infection, can cause unusual electrical activity in areas of the brain, leading to severe and debilitating seizures. When the condition is serious enough ▶ ▶ ▶

OUTLINE

and cannot be controlled by medication, physicians may consider performing a surgery termed a hemispherectomy. In this procedure, a portion of the brain (anywhere from a lobe up to a complete half) is completely removed. Unbelievably, many of the functions linked to the side of the brain that was removed can be "re-mapped" to the remaining side. In fact, most children who have undergone this procedure have experienced minimal lasting side effects, other than some minor paralysis on one side of the body. Learning ability does not seem to be significantly damaged by the procedure. Such drastic surgeries have taught us that the brain is a highly adaptable organ that can recover from significant trauma.

Understanding the structure and function of the nervous system is of critical importance to understanding the functioning of other organs and organ systems in the body. In this chapter, you will learn how the central and peripheral nervous systems operate, as well as how the central nervous system communicates with the endocrine system, which in turn regulates the body's internal environment.

27.1 Nervous System

The nervous system and the endocrine system work together to regulate the activities of the other systems. Both control systems use chemical signals when they respond to changes that might threaten homeostasis, but they have different means of delivering these signals (Fig. 27.1). The nervous system quickly sends a message along a nerve fiber directly to a target organ, such as skeletal or smooth muscle. Once a chemical signal is released, the muscle brings about an appropriate response.

The endocrine system uses the blood vessels of the cardiovascular system to send a hormone as a messenger to a target organ, such as the liver. The endocrine system is slower-acting because it takes time for the hormone to move through a vessel to a target organ. Also, hormones change the metabolism of cells, and this takes time; however, the response is longer-lasting. Cellular metabolism tends to remain the same for at least a limited period of time.

Let's first turn our attention to the human nervous system. As we examine it, we will also compare it to the nervous systems of other animals.

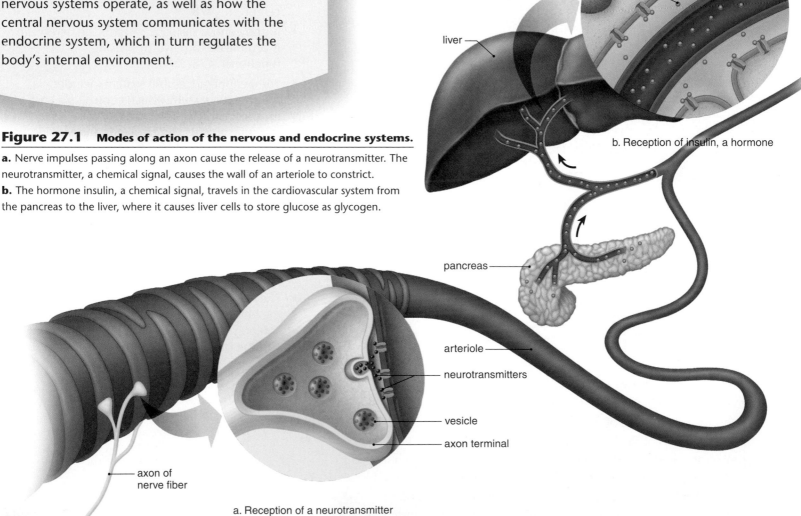

Figure 27.1 Modes of action of the nervous and endocrine systems.
a. Nerve impulses passing along an axon cause the release of a neurotransmitter. The neurotransmitter, a chemical signal, causes the wall of an arteriole to constrict.
b. The hormone insulin, a chemical signal, travels in the cardiovascular system from the pancreas to the liver, where it causes liver cells to store glucose as glycogen.

liver cell
insulin
liver
b. Reception of insulin, a hormone

pancreas

arteriole
neurotransmitters

vesicle
axon terminal

axon of nerve fiber

a. Reception of a neurotransmitter

The Human Nervous System

The nervous system is always involved in an animal's ability to move around. In fact, about the only function of the nervous system in some animals, such as planarians, is movement, particularly to feed. Planarians are predators, and the worm wraps itself around its prey, entangling it in slime and pinning it down. Then a muscular pharynx extends out the mouth, and by a sucking motion, the prey is torn up and swallowed. The bilateral symmetry of planarians is reflected in the organization of their nervous system. They have two lateral nerve cords (bundles of nerves) joined together by transverse nerves. The arrangement is called a *ladderlike* nervous system. A "brain" receives sensory information from the eyespots and sensory cells in the auricles. The two lateral nerve cords allow a rapid transfer of information from the cerebral ganglia to the posterior end, and the transverse nerves between the nerve cords keep the movement of the two sides coordinated. The nervous organization in planarians is a foreshadowing of the central and peripheral nervous systems that are found in more complex invertebrates, such as an earthworm, and in vertebrates, including humans (Fig. 27.2).

In humans, the nervous system controls the muscular system and all the other systems. The **central nervous system (CNS)** includes the brain and spinal cord, which have a central location—they lie in the midline of the body. The **peripheral nervous system (PNS)** consists of nerves that lie outside the central nervous system. The brain gives off paired cranial nerves (one on each side of the body), and the spinal cord gives off paired spinal nerves. The division between the central nervous system and the peripheral nervous system is arbitrary; the two systems work together and are connected to one another.

What makes the human nervous system more complex than the planarian system? Five trends during the evolution of the vertebrate nervous system can be identified:

- A CNS developed that is able to summarize incoming messages before ordering outgoing messages.
- Nerve cells (neurons) became specialized to send messages to the central nervous system (CNS), between neurons in the CNS, or away from the CNS.
- A brain evolved that has special centers for receiving input from various regions of the body and for directing their activity.
- The CNS became connected to all parts of the body by peripheral nerves. Therefore, the central nervous system can respond to both external and internal stimuli.
- Complex sense organs, such as the human eye and ear, arose that can detect changes in the external environment.

This chapter discusses the first four aspects of the human nervous system. Sensory input and muscle response are the topics of Chapter 28.

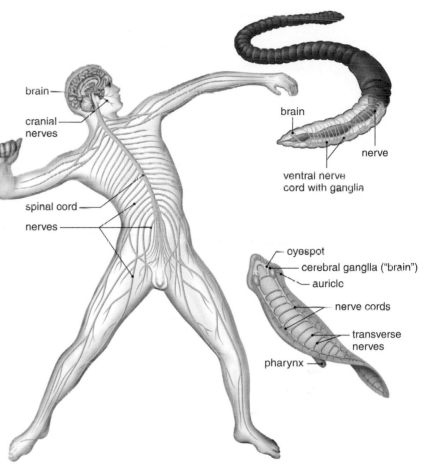

Figure 27.2 **Comparison of nervous systems.**

Invertebrates, such as a planarian and an earthworm, as well as vertebrates, such as humans, have a central nervous system (e.g., brain) and a peripheral nervous system (nerves).

brain
cranial nerves
spinal cord
nerves
brain
nerve
ventral nerve cord with ganglia
eyespot
cerebral ganglia ("brain")
auricle
nerve cords
transverse nerves
pharynx

Check Your Progress

1. **Describe the function of the nervous system in planarians, and explain how it is anatomically different from that of humans.**
2. **List the five trends toward complexity of the nervous system in animals.**

Answers: 1. The major function of the planarian nervous system is to bring about movement. The planarian nervous system is composed of two lateral nerve cords and transverse nerves that join them to form a ladderlike system. Humans have a well-developed central nervous system, which works in conjunction with the peripheral nervous system. 2. The central nervous system processes incoming messages before sending out messages; neurons are specialized for carrying messages; the brain has specialized regions for receiving information from different parts of the body; the CNS can respond to both internal and external stimuli; and sense organs detect changes in the external environment.

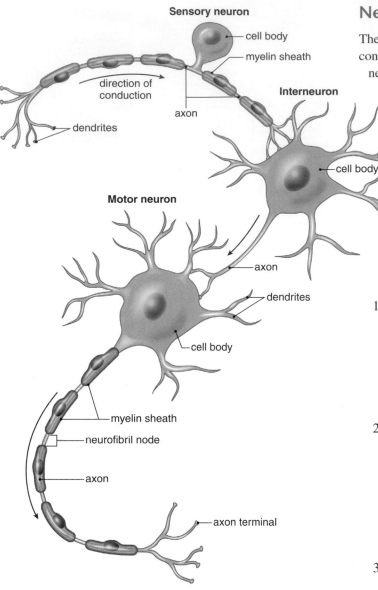

Figure 27.3   Types of neurons.

A sensory neuron, an interneuron, and a motor neuron are drawn here to show their arrangement in the body. Only axons conduct nerve impulses. In a sensory neuron, a process that extends from the cell body divides into an axon that takes nerve impulses all the way from the dendrites to the CNS. In a motor neuron and interneuron the axon extends directly from the cell body. The axon of sensory and motor neurons is covered by a myelin sheath. All long axons have a myelin sheath.

Neurons

The shape of a nerve cell, or **neuron,** is suitable to its function. The **cell body** contains the nucleus and other organelles that allow a cell to function. The neuron's many short **dendrite** nerve fibers fan out to receive signals from sensory receptors or other neurons. These signals can result in nerve impulses carried by an axon. The **axon,** a nerve fiber that is typically longer than a dendrite, is the portion of a neuron that conducts nerve impulses. An axon can reach all the way from the end of your spinal cord to the tip of your big toe or from your big toe to the spinal cord, depending on which way messages are being conducted. Long axons are covered by a white **myelin sheath** formed from the membranes of tightly spiraled cells that leave gaps called neurofibril nodes (nodes of Ranvier). Myelin sheaths account for our impression that nerves are white and glistening.

The nervous system has three types of neurons specific to its three functions (Fig. 27.3):

1. The nervous system receives sensory input. **Sensory neurons** perform this function. They take nerve impulses from sensory receptors to the CNS. The sensory receptor, which is the distal end of the axon of a sensory neuron, may be as simple as a naked nerve ending (a pain receptor), or may be built into a highly complex organ, such as the eye or ear. In any case, the axon of a sensory neuron can be quite long if the sensory receptor is far from the CNS.

2. The nervous system performs integration—in other words, the CNS sums up the input it receives from all over the body. **Interneurons** occur entirely within the CNS and take nerve impulses between various parts of the CNS. Some interneurons lie between sensory neurons and motor neurons, and some take messages from one side of the spinal cord to the other or from the brain to the spinal cord, and vice versa. They also form complex pathways in the brain where processes that account for thinking, memory, and language occur.

3. The nervous system generates motor output. **Motor neurons** take nerve impulses from the CNS to muscles or glands. Motor neurons cause muscle fibers to contract or glands to secrete, and therefore they are said to *innervate* these structures.

Check Your Progress

List the three major types of neurons and their functions.

Answer: Sensory neurons take nerve impulses from sensory receptors to the central nervous system. Interneurons carry nerve impulses between parts of the central nervous system. Motor neurons carry nerve impulses from the central nervous system to muscles or glands.

The Nerve Impulse

Like some other cellular processes, the nerve impulse is also dependent on concentration gradients. In neurons, these concentration gradients are maintained by the sodium-potassium pump. This pump actively transports sodium ions (Na^+) to the outside of the axon and actively transports potassium ions (K^+) inside. Aside from ion concentration differences across the axon's membrane, a charge difference also exists. *The inside of an axon is negative* compared to the outside. This charge difference is due in part to an unequal distribution of ions across the membrane, but is also due to the presence of large, negatively charged proteins in the axon cytoplasm.

The charge difference across the axon's membrane offers a potential for change, or an **action potential,** as the nerve impulse is also called. The nerve impulse is a rapid, short-lived, self-propagating reversal in the charge difference across the axon's membrane. Figure 27.4 shows how it works. A nerve impulse involves two types of gated channel proteins in the axon's membrane: One allows sodium (Na^+) to pass through the membrane, and the other allows potassium (K^+) to pass through the membrane. As an axon is conducting a nerve impulse, the Na^+ gates open at a particular location, and the inside of the axon becomes positive as Na^+ moves from outside the axon to the inside. The Na^+ gates close, and then the K^+ gates open. Now K^+ moves from inside the axon to outside the axon, and the charge reverses back again.

In Figure 27.4, the axon is unmyelinated, and the action potential at one locale stimulates an adjacent part of the axon's membrane to produce an action potential. In myelinated axons, an action potential at one neurofibril node causes an action potential at the next node (Fig. 27.5). This type of conduction, called **saltatory conduction,** is much faster than otherwise. In thin, unmyelinated axons, the nerve impulse travels about 1.0 m/second, and in thick, myelinated axons, the rate is more than 100 m/second. In any case, action potentials are self-propagating; each action potential generates another along the length of an axon.

The conduction of a nerve impulse (action potential) is an all-or-none event—that is, either an axon conducts a nerve impulse or it does not. The intensity of a message is determined by how many nerve impulses are generated within a given time span. An axon can conduct a volley of nerve impulses because only a small number of ions are exchanged with each impulse. As soon as an impulse has passed by each successive portion of an axon, it undergoes a short refractory period during which it is unable to conduct an impulse. During a refractory period, the sodium gates cannot yet open. This period ensures that nerve impulses travel in only one direction and do not reverse.

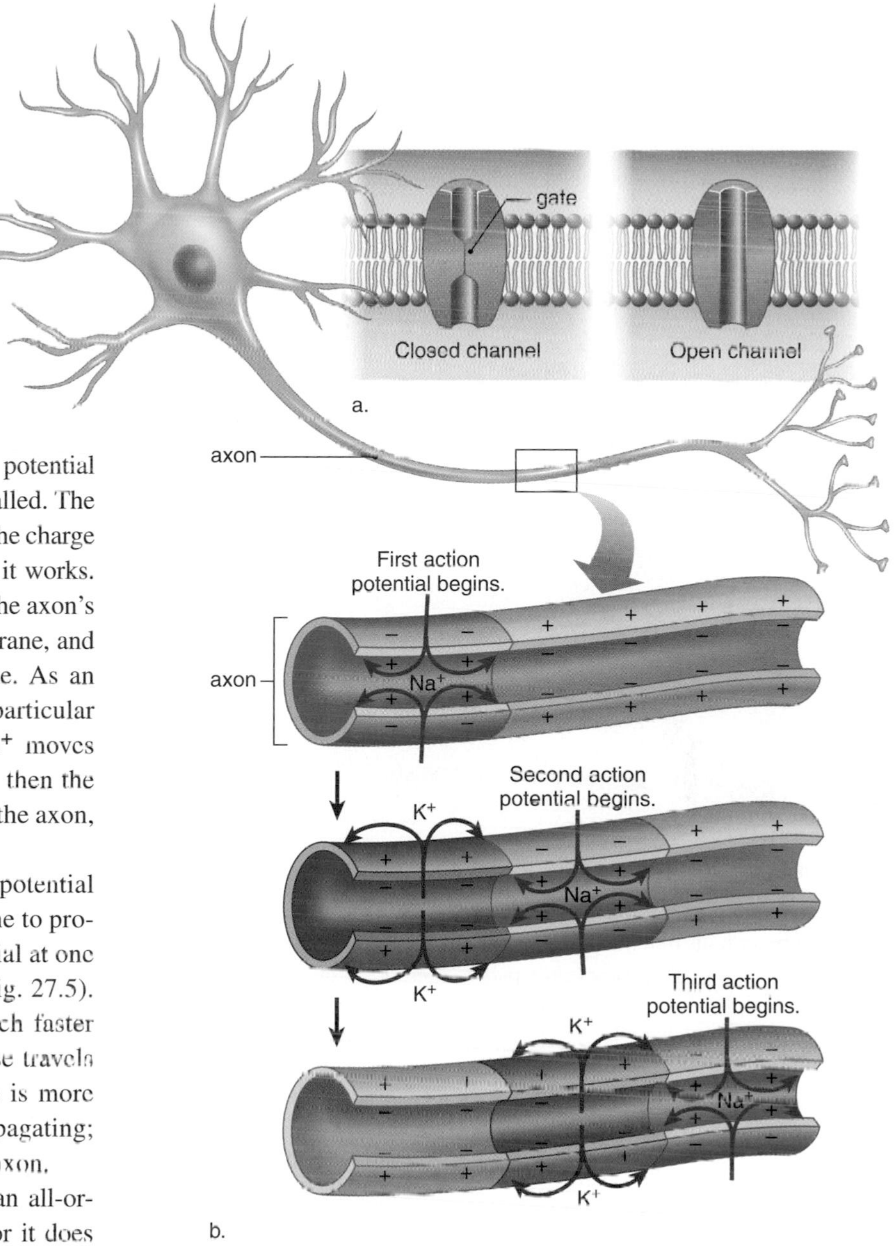

Figure 27.4   Conduction of action potentials in an unmyelinated axon.

a. Na^+ and K^+ each have their own gated channel protein by which they cross the axon's membrane. **b.** During an action potential, Na^+ enters the axon, and the charge difference between inside and outside reverses (blue); then K^+ exits, and the charge difference is restored (red). The action potential moves from section to section in an unmyelinated axon. Only a few ions are exchanged at a time.

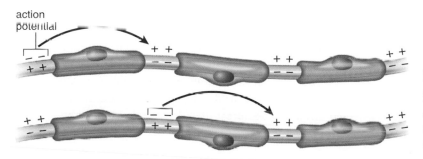

Figure 27.5   Conduction of a nerve impulse in a myelinated axon.

Action potentials can occur only at gaps in the myelin sheath called neurofibril nodes. This makes the speed of conduction much faster than in unmyelinated axons. In humans, all long axons are myelinated.

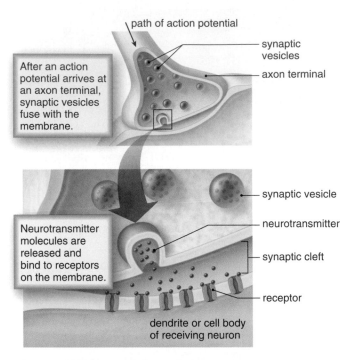

Figure 27.6 **Synapse structure and function.**

Transmission across a synapse from one neuron to another occurs when a neurotransmitter is released, diffuses across a synaptic cleft, and binds to a receptor in the plasma membrane of the next neuron. Each axon releases only one type of neurotransmitter, symbolized here by a red ball.

Labels on figure:

path of action potential

After an action potential arrives at an axon terminal, synaptic vesicles fuse with the membrane.

synaptic vesicles

axon terminal

synaptic vesicle

Neurotransmitter molecules are released and bind to receptors on the membrane.

neurotransmitter

synaptic cleft

receptor

dendrite or cell body of receiving neuron

The Synapse

Each axon has many axon terminals (Fig. 27.6). In the CNS, a terminal lies very close to the dendrite (or the cell body) of another neuron. This region of close proximity is called a **synapse.** In the PNS, when a terminal is close to a muscle cell, the region is called a neuromuscular junction. A small gap exists at a synapse, and this gap is called the **synaptic cleft.** How is it possible to excite the next neuron or a muscle cell across this gap? Transmission across a synaptic cleft is carried out by chemical signals called **neurotransmitters,** which are stored in synaptic vesicles. When nerve impulses traveling along an axon reach an axon terminal, synaptic vesicles release a neurotransmitter into the synaptic cleft. Neurotransmitter molecules diffuse across the cleft and bind to a specific receptor protein on the receiving neuron or other target cell.

Depending on the type of neurotransmitter and/or the type of receptor, the response of the receiving neuron or muscle cell can be toward excitation or toward inhibition. At least 25 different neurotransmitters have been identified, but two very well-known neurotransmitters are **acetylcholine (ACh)** and **norepinephrine (NE).**

Once a neurotransmitter has been released into a synaptic cleft and has initiated a response, the neurotransmitter is removed from the cleft. In some synapses, the receiving neuron contains enzymes that rapidly inactivate the neurotransmitter. For example, the enzyme **acetylcholinesterase (AChE)** breaks down acetylcholine. In other synapses, the original neuron rapidly reabsorbs the neurotransmitter. For example, norephinephrine is reabsorbed by the axon terminal. The short existence of neurotransmitters at a synapse prevents continuous stimulation (or inhibition) of the receiving neuron.

A single neuron has many dendrites plus the cell body, and both can have synapses with many other neurons. One thousand to ten thousand synapses per single neuron are not uncommon. Therefore, a neuron is on the receiving end of many signals. An excitatory neurotransmitter produces a potential change that drives the neuron closer to an action potential, and an inhibitory neurotransmitter produces a potential change that drives the neuron further from an action potential. Neurons integrate these incoming signals. **Integration** is the summing up of excitatory and inhibitory signals. If a neuron receives many excitatory signals (either at different synapses or at a rapid rate from one synapse), chances are the axon will transmit a nerve impulse. On the other hand, if a neuron receives both inhibitory and excitatory signals, the summing up of these signals may prohibit the axon from firing.

Drug Abuse

Many drugs that affect the nervous system act by interfering with or promoting the action of neurotransmitters. A drug can either enhance or block the release of a neurotransmitter, mimic the action of a neurotransmitter or block the receptor for it, or interfere with the removal of a neurotransmitter from a synaptic cleft. *Stimulants* are drugs that increase the likelihood of neuron excitation, and *depressants* decrease the likelihood of excitation. Increasingly, researchers believe that dopamine, a neurotransmitter in the brain, is responsible for mood. Many of the new medications developed to counter drug dependence and mental illness affect the release, reception, or breakdown of dopamine.

Drug abuse is apparent when a person takes a drug at a dose level and under circumstances that increase the potential for a harmful effect. A drug abuser often takes more of the drug than was intended. Drug abusers are apt to display a psychological and/or physical dependence on the drug. With physical dependence, formerly called "addiction," more of the drug is needed to get the same effect, and withdrawal symptoms occur when the user stops taking the drug.

Cocaine

Cocaine is an alkaloid derived from the shrub *Erythroxylon coca*. It is sold in powder form and as crack, a more potent extract. Because cocaine prevents the synaptic uptake of dopamine, the user experiences a "rush" sensation. The epinephrine-like effects of dopamine account for the state of arousal that lasts for several minutes after the rush experience.

A cocaine binge can go on for days, after which the individual suffers a crash. During the binge period, the user is hyperactive and has little desire for food or sleep but has an increased sex drive. During the crash period, the user is fatigued, depressed, and irritable, has memory and concentration problems, and displays no interest in sex.

Cocaine causes extreme physical dependence. With continued cocaine use, the body begins to make less dopamine to compensate for a seemingly excess supply. The user, therefore, experiences physical dependence withdrawal symptoms, and an intense craving for cocaine. These are indications that the person is highly dependent upon the drug.

Overdosing on cocaine can cause seizures and cardiac and respiratory arrest. It is possible that long-term cocaine abuse causes brain damage (Fig. 27.7). Babies born to addicts suffer withdrawal symptoms and may have neurological and developmental problems.

Heroin

Heroin is derived from morphine, an alkaloid of opium. Once heroin is injected into a vein (Fig. 27.8), a feeling of euphoria, along with relief of any pain, occurs within 3 to 6 minutes. Side effects can include nausea, vomiting, restlessness, anxiety, mood disorders, and respiratory and circulatory depression. Heroin binds to receptors meant for the endorphins, special neurotransmitters that kill pain and produce a feeling of tranquility. With time, the body's production of endorphins decreases. Physical dependence develops, and the euphoria originally experienced upon injection is no longer felt. Heroin withdrawal symptoms are quite severe, and infants born to women who are physically dependent also experience these withdrawal symptoms.

Marijuana

The dried flowering tops, leaves, and stems of the Indian hemp plant *Cannabis sativa* contain and are covered by a resin that is rich in THC (tetrahydrocannabinol). The names *cannabis* and *marijuana* apply to either the plant or THC. Usually, marijuana is smoked in a cigarette form called a "joint."

Recently, researchers have found that marijuana binds to a receptor for anandamide, a neurotransmitter that seems to create a feeling of peaceful contentment. The occasional marijuana user experiences a mild euphoria along with alterations in vision and judgment, which result in distortions of space and time. Motor incoordination, including the inability to speak coherently, takes place. Heavy use can result in hallucinations, anxiety, depression, rapid flow of ideas, body image distortions, paranoid reactions, and similar psychotic symptoms. Craving and difficulty in stopping usage can occur as a result of regular use.

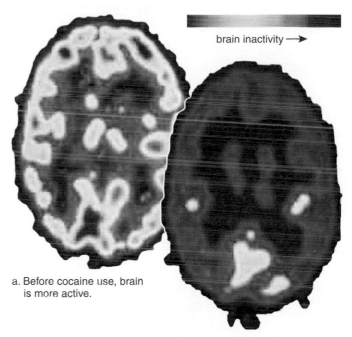

brain inactivity ⟶

a. Before cocaine use, brain is more active.

b. After cocaine use, brain is less active.

Figure 27.7   Effect of cocaine on brain.

In a cocaine user, PET scans show that (**a**) the usual activity of the brain is (**b**) reduced. The color red indicates brain tissue is active.

Figure 27.8   Drug abuse.

Blood-borne diseases such as AIDS and hepatitis B pass from one drug abuser to another when they share needles.

cerebellum

reptile (alligator)

cerebrum

optic lobe

bird (goose)

olfactory lobe

medulla

mammal (horse)

a.

Cerebrum

skull

corpus callosum

thalamus

Diencephalon —

hypothalamus

pineal gland

pituitary gland

Cerebellum

midbrain

pons

medulla oblongata

Brain stem

b.

spinal cord

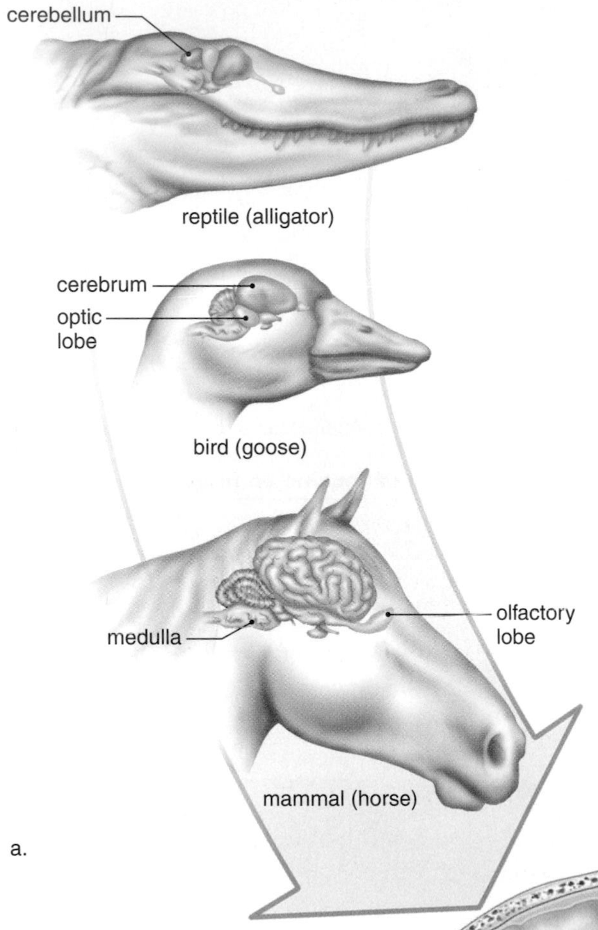

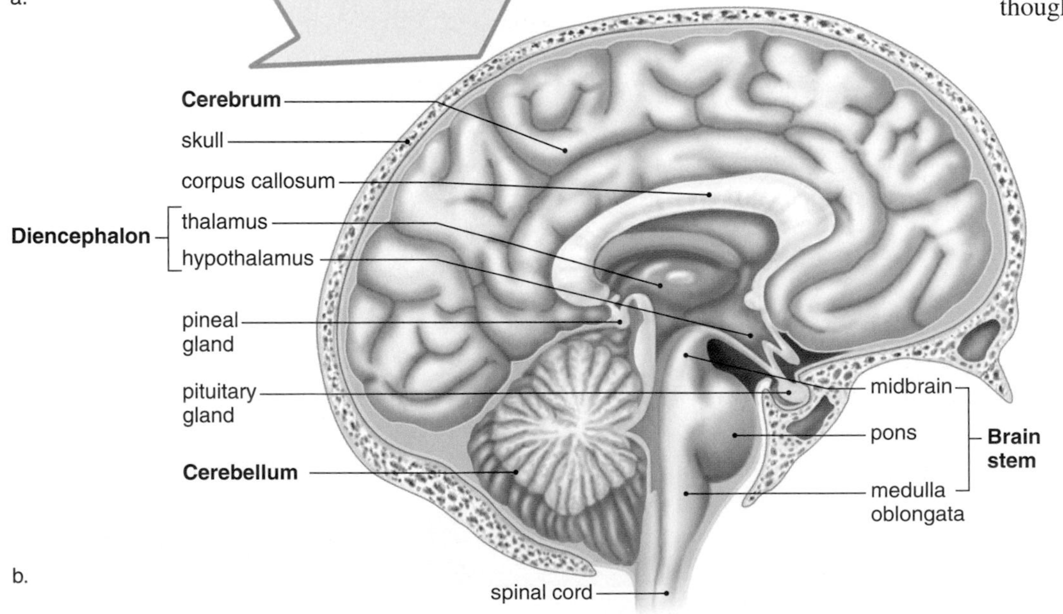

Figure 27.9 **Vertebrate brains.**

a. A comparison of reptile, bird, and mammalian brains shows that the forebrain increased in size and complexity among these animals. **b.** The human brain.

The Central Nervous System

The organization of the brains in certain vertebrates—namely, reptiles, birds, and mammals (horse and human)—is compared in Figure 27.9a. If we divide the brain into a hindbrain, midbrain, and forebrain, we can see that the forebrain is most prominent in humans. The forebrain of mammals also has an altered function in that it becomes the last depository for sensory information. This change accounts for why the forebrain carries on much of the integration for the entire nervous system before it sends out motor instructions to glands and muscles. In humans, the **spinal cord** provides a means of communication between the brain and the spinal nerves, which are a part of the PNS. (Spinal nerves leave the spinal cord and take messages to and from the skin, glands, and muscles in all areas of the body, except for the head and face.) Myelinated long fibers of interneurons in the spinal cord run together in bundles called **tracts.** These tracts connect the spinal cord to the brain. Because the tracts cross over at one point, the left side of the brain controls the right side of the body, and vice versa. Also, as discussed on page 481, the spinal cord is involved in reflex actions, which are programmed, built-in circuits that allow for protection and survival. They are present at birth and require no conscious thought to take place.

The Brain

Our discussion will center on these parts of the brain: the cerebrum, the diencephalon, the cerebellum, and the brain stem (Fig. 27.9b).

Cerebrum The cerebrum communicates with and coordinates the activities of the other parts of the brain. The cerebrum has two halves, and each half has a number of lobes, which are color coded in Figure 27.10. Most of the cerebrum is white matter where the long axons of interneurons are taking nerve impulses to and from the cerebrum. The highly convoluted outer layer of gray matter that covers the cerebrum is called the **cerebral cortex.** The cerebral cortex contains over one billion cell bodies, and it is the region of the cerebrum that accounts for sensation, voluntary movement, and higher thought processes.

Investigators have found that each part of the cerebrum has specific functions. To take an example, the **primary sensory area** located in the parietal lobe receives information from the skin, skeletal muscles, and joints. Each part of the body has its own particular receiving area. The **primary motor area,** on the other hand, is in the frontal lobe just before the small cleft that divides the frontal lobe from the parietal lobe. Voluntary commands to skeletal muscles involve the primary motor area, and the muscles in each part of the body are controlled by a certain section of the primary motor area.

The lobes of the cerebral cortex have a number of specialized centers to receive information from the sensory receptors for sight, hearing, and smell. The lobes also have *association areas* where integration occurs. The **prefrontal area,** an association area in the frontal lobe, receives information from the other association areas and uses this information to reason and plan our actions. Integration in this area accounts for our most cherished human abilities to think critically and to formulate appropriate behaviors.

Diencephalon   Beneath the cerebrum is the diencephalon, which contains the hypothalamus and the thalamus (see Fig. 27.9*b*). The **hypothalamus** is an integrating center that helps maintain homeostasis by regulating hunger, sleep, thirst, body temperature, and water balance. The hypothalamus controls the pituitary gland, and thereby serves as a link between the nervous and endocrine systems. The **thalamus** is on the receiving end for all sensory input except smell. Information from the eyes, ears, and skin arrives at the thalamus via the cranial nerves and tracts from the spinal cord. The thalamus integrates this information and sends it on to the appropriate portions of the cerebrum. The thalamus is involved in arousal of the cerebrum, and it also participates in higher mental functions such as memory and emotions.

The pineal gland, which secretes the hormone melatonin, is located in the diencephalon. Presently, there is much popular interest in melatonin because it is released at night when we are sleeping. Some researchers believe it can be used to help prevent jet lag or insomnia.

Cerebellum   The **cerebellum** has two portions that are joined by a narrow median strap. Each portion is primarily composed of white matter, which in longitudinal section has a treelike pattern (see Fig. 27.9*b*). Overlying the white matter is a thin layer of gray matter that forms a series of complex folds.

The cerebellum receives sensory input from the eyes, ears, joints, and skeletal muscles about the present position of body parts, and it also receives motor output from the cerebral cortex about where these parts should be located. After integrating this information, the cerebellum sends motor impulses by way of the brain stem to the skeletal muscles. In this way, the cerebellum maintains posture and balance. It also ensures that all of the muscles work together to produce smooth, coordinated voluntary movements. The cerebellum assists the learning of new motor skills such as playing the piano or hitting a baseball.

Brain Stem   The **brain stem,** which contains the midbrain, the pons, and the medulla oblongata, connects the rest of the brain to the spinal cord (see Fig. 27.9*b*). It contains tracts that ascend or descend between the spinal cord and higher brain centers. The **medulla oblongata** contains a number of reflex centers for regulating heartbeat, breathing, and vasoconstriction (blood pressure). It also contains the reflex centers for vomiting, coughing, sneezing, hiccuping, and swallowing. In addition, the medulla oblongata helps control various internal organs.

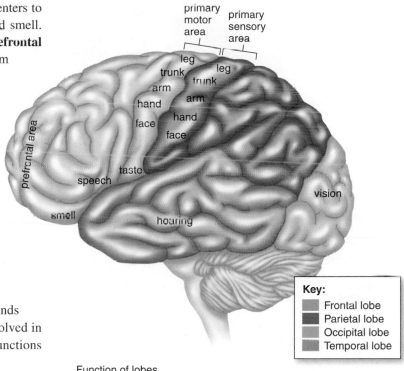

Key:
- Frontal lobe
- Parietal lobe
- Occipital lobe
- Temporal lobe

Function of lobes
- Frontal lobe—reasoning, planning, speech, movement, emotions, and problem solving
- Parietal lobe—integration of sensory input from skin and skeletal muscles, understanding speech
- Occipital lobe—seeing, perception of visual stimuli
- Temporal lobe—hearing, perception of auditory stimuli

Figure 27.10   Functional regions of the cerebral cortex.

Specific areas of the cerebral cortex receive sensory input from particular sensory receptors, integrate various types of information, or send out motor commands to particular areas of the body.

Check Your Progress

Compare and contrast the primary sensory area with the primary motor area of the cerebrum.

Answer: The primary sensory area receives information from the skin, skeletal muscles and joints. The primary motor area sends voluntary commands to skeletal muscles.

thalamus
hippocampus
hypothalamus
amygdala

Figure 27.11 **The limbic system.**

The limbic system includes the diencephalon and parts of the cerebrum. It joins higher mental functions, such as reasoning, with more primitive feelings, such as fear and pleasure. Eating is a pleasurable activity for most people.

Figure 27.12 **A nerve.**

A nerve contains the axons of many neurons.

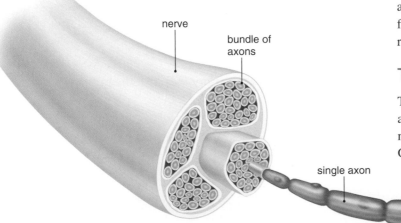

nerve
bundle of axons
single axon

The Limbic System

The **limbic system** is a complex network that includes the diencephalon and areas of the cerebrum (Fig. 27.11). The limbic system blends higher mental functions and primitive emotions into a united whole. It accounts for why activities such as sexual behavior and eating seem pleasurable and also why, say, mental stress can cause high blood pressure.

Two significant structures within the limbic system are the hippocampus and the amygdala, which are essential for learning and memory. The hippocampus, a seahorse-shaped structure that lies deep in the temporal lobe, is well situated in the brain to make the prefrontal area aware of past experiences stored in sensory association areas. The amygdala, in particular, can cause these experiences to have emotional overtones. A connection between the frontal lobe and the limbic system means that reason can keep us from acting out strong feelings.

Learning and Memory Memory is the ability to hold a thought in mind or recall events from the past, ranging from a word we learned only yesterday to an early emotional experience that has shaped our lives. Learning takes place when we retain and utilize past memories.

The prefrontal area in the frontal lobe is active during short-term memory, as when we temporarily recall a telephone number. Some telephone numbers go into long-term memory. Think of a telephone number you know by heart, and see if you can bring it to mind without also thinking about the place or person associated with that number. Most likely you cannot, because typically long-term memory is a mixture of what is called *semantic memory* (numbers, words, and so on) and *episodic memory* (persons, events, and other associations). *Skill memory* is a type of memory that can exist independent of episodic memory. Skill memory is what allows us to perform motor activities like riding a bike or playing ice hockey. A person who has **Alzheimer disease (AD)** experiences a progressive loss of memory, particularly for recent events. Gradually the person loses the ability to perform any type of daily activity and becomes bedridden. In AD patients, abnormal neurons occur, especially in the hippocampus and amygdala. Major research efforts are devoted to seeking a cure for AD.

What parts of the brain are functioning when we remember something from long ago? Our long-term memories are stored in bits and pieces throughout the sensory association areas of the cerebral cortex. The hippocampus gathers this information together for use by the prefrontal area of the frontal lobe when we remember Uncle George or our summer holiday. Why are some memories so emotionally charged? The amygdala is responsible for fear conditioning and for associating danger with sensory information received from the thalamus and the cortical sensory areas.

The Peripheral Nervous System

The peripheral nervous system (PNS) lies outside the central nervous system and contains **nerves,** which are bundles of axons (Fig. 27.12). The cell bodies of neurons are found in the CNS—that is, the brain and spinal cord—or in ganglia. Ganglia (sing., **ganglion**) are collections of cell bodies within the PNS.

Humans have 12 pairs of **cranial nerves** attached to the brain. Cranial nerves are largely concerned with the head, neck, and facial regions of the body. However, the vagus nerve is a cranial nerve that has branches not only to the pharynx and larynx, but also to most of the internal organs.

Humans have 31 pairs of **spinal nerves,** and each contains many sensory and motor axons. The dorsal root of a spinal nerve contains the axons of sensory neurons, which conduct impulses to the spinal cord from sensory receptors. The cell body of a sensory neuron is in the **dorsal root ganglion.** The ventral root contains the axons of motor neurons, which conduct impulses away from the cord, largely to skeletal muscles (Fig. 27.13). Each spinal nerve serves the particular region of the body in which it is located.

The Somatic System

The **somatic system** of the PNS includes the nerves that take sensory information from external sensory receptors to the CNS and motor commands away from the CNS to skeletal muscles. Voluntary control of skeletal muscles always originates in the brain. Involuntary responses to stimuli, called **reflexes,** can involve either the brain or just the spinal cord. Flying objects cause our eyes to blink, and sharp pins cause our hands to jerk away even without our having to think about it.

Figure 27.13 illustrates the path of a reflex that involves only the spinal cord. If your hand touches a sharp pin, sensory receptors in the skin generate nerve impulses that move along sensory axons toward the spinal cord. Sensory neurons that enter the spinal cord pass signals on to many interneurons. Some of these interneurons synapse with motor neurons. The short dendrites and the cell bodies of motor neurons are in the spinal cord, but their axons leave the cord. Nerve impulses travel along motor axons to an effector, which brings about a response to the stimulus. In this case, a muscle contracts so that you withdraw your hand from the pin. Various other reactions are possible—you will most likely look at the pin, wince, and cry out in pain. This whole series of responses is explained by the fact that some of the interneurons involved carry nerve impulses to the brain. Also, sense organs send messages to the brain that make us aware of our actions. The brain makes you aware of the stimulus and directs these other reactions to it.

Figure 27.13 A reflex arc showing the path of a spinal reflex.

A stimulus (e.g., a pinprick) causes sensory receptors in the skin to generate nerve impulses that travel in sensory axons to the spinal cord. Interneurons integrate data from sensory neurons and then relay signals to motor neurons. Motor axons convey nerve impulses from the spinal cord to a skeletal muscle, which contracts. Movement of the hand away from the pin is the response to the stimulus.

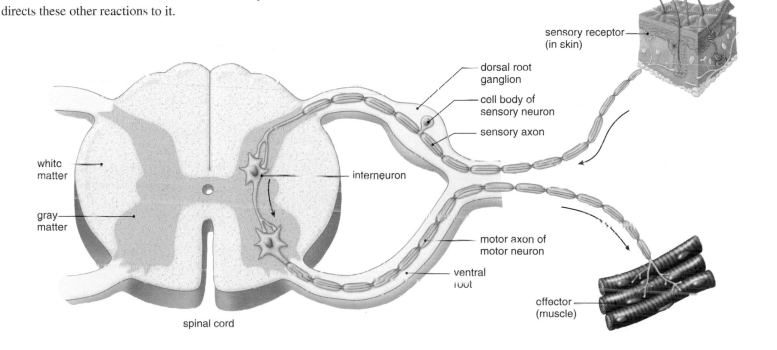

The Autonomic System

The **autonomic system** of the PNS automatically and involuntarily regulates the activity of glands and cardiac and smooth muscle. The system is divided into the parasympathetic and sympathetic divisions (Fig. 27.14). Reflex actions, such as those that regulate blood pressure and breathing rate, are especially important to the maintenance of homeostasis. These reflexes begin when the sensory neurons in contact with internal organs send information to the CNS. They are completed by motor neurons within the autonomic system.

The **parasympathetic division** includes a few cranial nerves (e.g., the vagus nerve) and also axons that arise from the last portion of the spinal cord. The parasympathetic division, sometimes called the "housekeeper division," promotes all the internal responses we associate with a relaxed state (see photograph). For example, it causes the pupil of the eye to contract, promotes digestion of food, and retards the heartbeat. The neurotransmitter utilized by the parasympathetic division is acetylcholine (ACh).

Axons of the **sympathetic division** arise from portions of the spinal cord. The sympathetic division is especially important during emergency situations and is associated with "fight or flight." If you need to fend off a foe or flee from danger, active muscles require a ready supply of glucose and oxygen. On the one hand, the sympathetic division accelerates the heartbeat and dilates the bronchi. On the other hand, the sympathetic division inhibits the digestive tract, since digestion is not an immediate necessity if you are under attack. The sympathetic nervous system utilizes the neurotransmitter norepinephrine, which has a structure like that of epinephrine (adrenaline) released by the adrenal medulla.

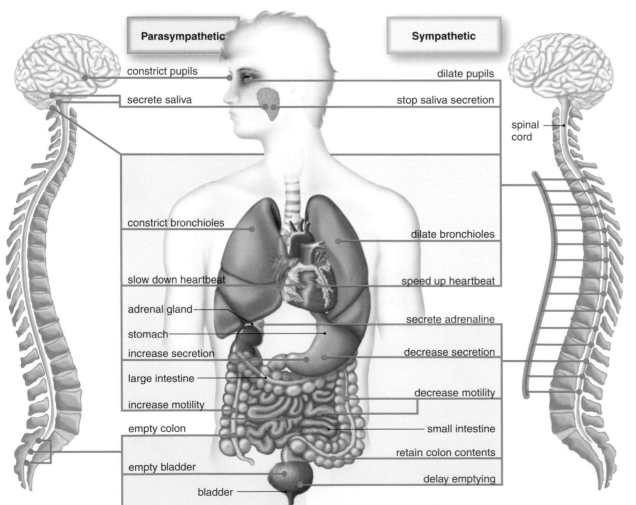

Parasympathetic

Sympathetic

constrict pupils — dilate pupils

secrete saliva — stop saliva secretion

spinal cord

constrict bronchioles — dilate bronchioles

slow down heartbeat — speed up heartbeat

adrenal gland — secrete adrenaline

stomach

increase secretion — decrease secretion

large intestine

increase motility — decrease motility

empty colon — small intestine

retain colon contents

empty bladder — delay emptying

bladder

Figure 27.14
Autonomic system.

The parasympathetic and sympathetic motor axons go to the same organs, but they have opposite effects. The parasympathetic division is active when we feel warm and cozy in the arms of someone who loves us (see photograph). The sympathetic division is active when we are stressed and feel threatened.

Check Your Progress

Explain why the parasympathetic division of the autonomic system is called the "housekeeper division."

Answer: The parasympathetic division keeps the "house" in order by maintaining the internal responses associated with a relaxed state, such as pupil contraction, food digestion, and slow heartbeat.

27.2 Endocrine System

The endocrine system consists of glands and tissues that secrete chemical signals we call hormones (Fig. 27.15). **Endocrine glands** do not have ducts; they secrete their hormones directly into the bloodstream for distribution throughout the body. They can be contrasted with exocrine glands, which have ducts and secrete their products into these ducts for transport to body cavities. For example, the salivary glands send saliva into the mouth by way of the salivary ducts.

The endocrine system and the nervous system are intimately involved in homeostasis, the relative stability of the internal environment. Several hormones directly affect the blood glucose, calcium, and sodium levels. Other hormones are involved in the maturation and function of the reproductive organs, and these are discussed in Chapter 29.

The Action of Hormones

The cells that can respond to a hormone have receptor proteins that bind to the hormone. Hormones cause these cells to undergo a metabolic change. The type of change is dependent on the chemical structure of the hormone. **Steroid hormones** (see Fig. 3.15, p. 37) are lipids, and they can pass through the plasma membrane. The hormone-receptor complex then binds to DNA, and gene expression follows—for example, a protein such as an enzyme is made by the cell. The enzyme goes on to speed a reaction in the cell (Fig. 27.16a).

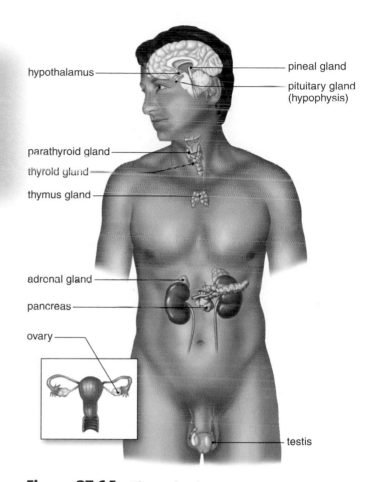

Figure 27.15 The endocrine system.

Anatomical location of major endocrine glands in the body.

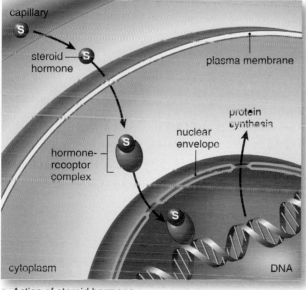

a. Action of steroid hormone

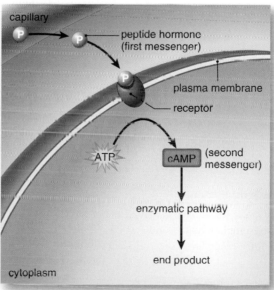

b. Action of peptide hormone

Figure 27.16 How hormones work.

a. A steroid hormone (S) is a chemical signal that is able to enter the cell. Reception of this messenger causes the cell to synthesize a product by way of the cellular machinery for protein synthesis.
b. A peptide hormone (P) is a "first messenger" that is received by a cell at the plasma membrane. Reception of the first messenger and a signal transduction pathway lead to a "second messenger" that changes the metabolism of the cell.

Peptide hormones is a category that includes peptides, proteins, glyco-proteins, and modified amino acids. Peptide hormones can't pass through the plasma membrane so they bind to a receptor protein in the plasma membrane. The peptide hormone is called the "first messenger" because a signal trans-duction pathway leads to a second molecule, i.e., the "second messenger," that changes the metabolism of the cell. The second messenger sets in motion an enzyme pathway that is sometimes called an enzyme cascade because each enzyme in turn activates another. Because enzymes work over and over, every step in an enzyme cascade leads to more reactions—the binding of a single peptide hormone molecule can result in as much as a thousandfold response.

Check Your Progress

Explain how hormones induce metabolic changes in cells.

Answer: Steroid hormones enter the cell and stimulate genes to produce enzymes that alter cell activity. Peptide hormones induce an enzyme cascade that alters cell activity.

Hypothalamus and Pituitary Gland

The **hypothalamus,** a part of the brain (see Fig. 27.9), helps regulate the inter-nal environment. For example, it is on the receiving end of information about the heartbeat and body temperature. And to correct any abnormalities, the hypo-thalamus communicates with the medulla oblongata, where the brain centers that control the autonomic system are located. The hypothalamus is also a part of the endocrine system. It controls the glandular secretions of the **pituitary gland,** a small gland connected to the brain by a stalklike structure. The pitu-itary has two portions, the anterior pituitary and the posterior pituitary, which are distinct from each other.

Anterior Pituitary

The hypothalamus controls the anterior pituitary by producing **releasing hor-mones,** most of which are stimulatory. The **anterior pituitary** in turn stimulates other glands:

1. **Thyroid-stimulating hormone (TSH)** stimulates the thyroid to produce triiodothyronine (T_3) and thyroxine (T_4).
2. **Adrenocorticotropic hormone (ACTH)** stimulates the adrenal cortex to produce the glucocorticoids.
3. **Gonadotropic hormones (FSH** and **LH)** stimulate the gonads—the testes in males and the ovaries in females—to produce gametes and sex hormones.

In these instances, a three-tiered control system develops that is described in Figure 27.17. For example, the secretion of thyroid-releasing hormone (TRH) by the hypothalamus stimulates the thyroid to produce the thyroid-stimulating hormone (TSH), and the thyroid produces its hormones (T_3 and T_4), which feed back to inhibit the release of the first two hormones mentioned.

Two other hormones produced by the anterior pituitary do not affect other endocrine glands (Fig. 27.18). **Prolactin (PRL)** is produced in quantity

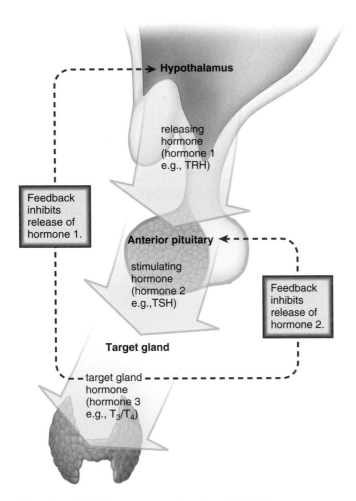

Figure 27.17   Negative feedback inhibition.

The hormones secreted by the thyroid (and also the adrenal cortex and gonads) feed back to inhibit the anterior pituitary and hypothalamic-releasing hormones so that their blood levels stay relatively constant.

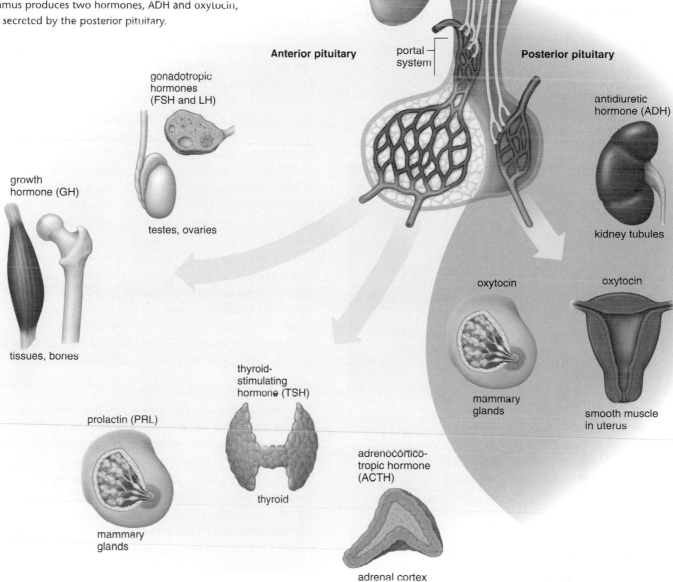

during pregnancy and after childbirth. It causes the mammary glands in the breasts to develop and produce milk. It also plays a role in carbohydrate and fat metabolism. **Growth hormone (GH)** promotes skeletal and muscular growth. It stimulates the rate at which amino acids enter cells and protein synthesis occurs. Underproduction of growth hormone leads to pituitary dwarfism, and overproduction can lead to pituitary giantism.

Posterior Pituitary

The hypothalamus produces two hormones, **antidiuretic hormone (ADH)** and **oxytocin** (Fig. 27.18). These hormones pass through axons into the **posterior pituitary,** where they are stored in axon terminals. When the hypothalamus determines that the blood is too concentrated, ADH is released from the posterior pituitary. Upon reaching the kidneys, ADH causes water to be reabsorbed. As the blood becomes dilute, ADH is no longer released. This is also an example

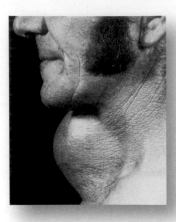

Figure 27.19   Simple goiter.

An enlarged thyroid gland can result from too little iodine in the diet. The thyroid is under constant stimulation to produce more of its hormones and so it enlarges, resulting in a simple goiter.

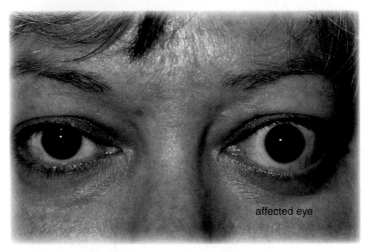

affected eye

Figure 27.20   Exopthalmic goiter.

An exopthalmic goiter is so-named because enlargement of the thyroid gland can cause the eyes to protrude. In this individual only the left eye was affected. The overactive thyroid produces too much of its hormones and the individual is hyperactive and nervous.

of control by **negative feedback** because the effect of the hormone (to dilute blood) acts to shut down the release of the hormone. Negative feedback, as discussed in Chapter 22, maintains homeostasis.

Oxytocin, the other hormone made in the hypothalamus, causes uterine contraction during childbirth and milk letdown when a baby is nursing.

Thyroid and Parathyroid Glands

The **thyroid gland** is a large gland located in the neck that produces two hormones: Triiodothyronine (T_3) contains three iodine atoms, and **thyroxine (T_4)** contains four iodine atoms. To produce these hormones, the thyroid gland needs iodine. The concentration of iodine in the thyroid gland can increase to as much as 25 times that of the blood. If iodine is lacking in the diet, the thyroid gland is unable to produce the thyroid hormones. In response to constant stimulation by the anterior pituitary, the thyroid enlarges, resulting in a **simple goiter** (Fig. 27.19). Some years ago, it was discovered that the use of iodized salt (table salt to which iodine has been added) helps prevent simple goiter.

Thyroid hormones increase the metabolic rate. They do not have a target organ; instead, they stimulate all the cells of the body to metabolize at a faster rate. More glucose is broken down, and more energy is utilized.

In the case of hyperthyroidism (oversecretion of thyroid hormone), or Graves disease, the thyroid gland is overactive, and a goiter forms. This type of goiter is called **exophthalmic goiter** (Fig. 27.20). The eyes protrude because of swelling in eye socket tissues and in the muscles that move the eyes. The patient usually becomes hyperactive, nervous, and irritable, and suffers from insomnia. Removal or destruction of a portion of the thyroid by means of radioactive iodine is sometimes effective in curing the condition.

Calcium Regulation

The thyroid gland also produces **calcitonin,** a hormone that helps regulate the blood calcium level. Calcium (Ca^{2+}) plays a significant role in both nervous conduction and muscle contraction. It is also necessary for blood clotting. Calcitonin temporarily reduces the activity and number of osteoclasts, a type of cell that breaks down bone. Therefore, more calcium is deposited in bone. When the blood calcium lowers to normal, the release of calcitonin by the thyroid is inhibited by negative feedback, but a low level stimulates the release of **parathyroid hormone (PTH)** by the parathyroid glands. The parathyroid glands are embedded in the posterior surface of the thyroid gland. Many years ago, the four **parathyroid glands** were sometimes mistakenly removed during thyroid surgery because of their size and location.

Parathyroid hormone promotes the activity of osteoclasts and the release of calcium from the bones. PTH also promotes the reabsorption of calcium by the kidneys, where it activates vitamin D. Vitamin D, in turn, stimulates the absorption of calcium from the intestine. These effects bring the blood calcium level back to the normal range so that the parathyroid glands no longer secrete PTH.

When insufficient parathyroid hormone production leads to a dramatic drop in the blood calcium level, tetany results. In **tetany,** the body shakes from continuous muscle contraction. This effect is brought about by increased excitability of the nerves, which initiate nerve impulses spontaneously and without rest.

Check Your Progress

1. Compare and contrast prolactin with growth hormone.
2. Explain why a simple goiter results from a diet low in iodine.

Answers: 1. Both are produced by the anterior pituitary and do not affect other endocrine glands. Prolactin stimulates the development of mammary glands and milk production. Growth hormone promotes skeletal and muscle growth. 2. Iodine is a component of hormones produced by the thyroid gland. These hormones cannot be made unless iodine is present. Consequently, the thyroid receives constant stimulation from the anterior pituitary, which causes it to enlarge. The result is a simple goiter.

Adrenal Glands

Two **adrenal glands** sit atop the kidneys. Each adrenal gland consists of an inner portion called the **adrenal medulla** and an outer portion called the **adrenal cortex.** These portions, like the anterior pituitary and the posterior pituitary, have no structural connection with one another.

The hypothalamus exerts control over the activity of both portions of the adrenal glands. It initiates nerve impulses that travel by way of the brain stem, spinal cord, and sympathetic nerve fibers to the adrenal medulla, which then secretes its hormones. The hypothalamus, by means of ACTH-releasing hormone, controls the anterior pituitary's secretion of ACTH, which in turn stimulates the adrenal cortex. Stress of all types, including both emotional and physical trauma, prompts the hypothalamus to stimulate both the adrenal medulla and the adrenal cortex.

Adrenal Medulla

Epinephrine (adrenaline) and **norepinephrine** (noradrenaline) produced by the adrenal medulla rapidly bring about all the body changes that occur when an individual reacts to an emergency situation. The effects of these hormones are short-term. In contrast, the hormones produced by the adrenal cortex provide a long-term response to stress.

Adrenal Cortex

The two major types of hormones produced by the adrenal cortex are the **mineralocorticoids,** such as aldosterone, and the **glucocorticoids,** such as cortisol. Aldosterone acts on the kidneys and thereby regulates salt and water balance, leading to increases in blood volume and blood pressure. Cortisol regulates carbohydrate, protein, and fat metabolism, leading to an increase in the blood glucose level. It is also an antiinflammatory agent. The adrenal cortex also secretes a small amount of both male and female sex hormones in both sexes.

When the level of adrenal cortex hormones is low due to hyposecretion, a person develops **Addison disease.** ACTH may build up as more is secreted to attempt to stimulate the adrenal cortex. The excess can cause bronzing of the skin because ACTH in excess stimulates melanin production. Without cortisol, glucose cannot be replenished when a stressful situation arises. Even a mild infection can lead to death. The lack of aldosterone results in the loss of sodium and water by the kidneys and the development of low blood pressure, and possibly severe dehydration. Left untreated, Addison disease can be fatal.

When the level of adrenal cortex hormones is high due to hypersecretion, a person develops **Cushing syndrome** (Fig. 27.21). The excess cortisol results in a tendency toward diabetes mellitus as muscle protein is metabolized and subcutaneous fat is deposited in the midsection. The trunk is obese, while the arms and legs remain a normal size. An excess of aldosterone and reabsorption of sodium and water by the kidneys lead to a basic blood pH and hypertension. The face swells and takes on a moon shape. Masculinization may occur in women because of excess adrenal male sex hormones.

Pancreas

The **pancreas** is composed of two types of tissue. Exocrine tissue produces and secretes digestive juices that pass through ducts to the small intestine. Endocrine tissue, called the **pancreatic islets** (islets of Langerhans), produces and secretes the hormones **insulin** and **glucagon** directly into the blood (Fig. 27.22).

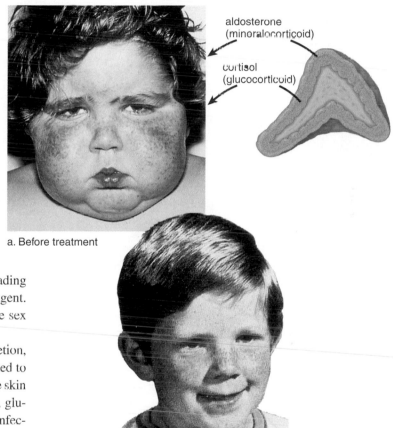

aldosterone (mineralocorticoid)

cortisol (glucocorticoid)

a. Before treatment

b. After treatment

Figure 27.21   Cushing syndrome.

Cushing syndrome results from hypersecretion of hormones by the adrenal cortex possibly due to a tumor. **a.** Patient first diagnosed with Cushing syndrome. **b.** Four months later, after treatment.

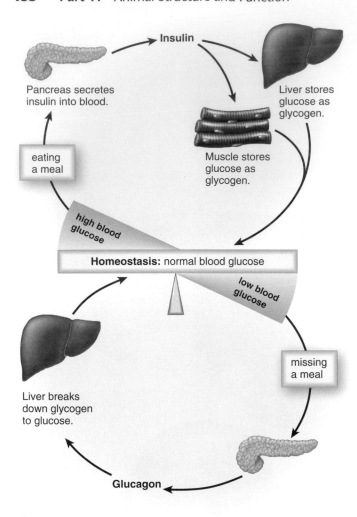

Figure 27.22   Regulation of blood glucose level.

Typical of hormones, insulin is regulated by negative feedback—once the blood glucose level is low, insulin is no longer secreted. The effect of insulin is countered by glucagon, which raises the blood glucose level. The two hormones work together to keep the blood glucose level relatively constant.

Insulin is secreted when there is a high blood glucose level, which usually occurs just after eating. Insulin stimulates the uptake of glucose by cells, especially liver cells, muscle cells, and adipose tissue cells. In liver and muscle cells, glucose is then stored as glycogen. In muscle cells, glucose supplies energy for ATP production leading to protein metabolism and muscle contraction. In fat cells the breakdown of glucose supplies glycerol and acetyl groups for the formation of fat. In these ways, insulin lowers the blood glucose level.

Glucagon is usually secreted between meals, when the blood glucose level is low. The major target tissues of glucagon are the liver and adipose tissue. Glucagon stimulates the liver to break down glycogen to glucose and to use fat and protein in preference to glucose as energy sources. The latter spares glucose and makes more available to enter the blood. In these ways, glucagon raises the blood glucose level.

Diabetes Mellitus

Diabetes mellitus is a fairly common hormonal disease in which liver cells, and indeed all body cells, do not take up and/or metabolize glucose. Therefore, the cells are in need of glucose even though there is plenty in the blood. As the blood glucose level rises, glucose, along with water, is excreted in the urine. The loss of water in this way causes the diabetic person to be extremely thirsty.

Two types of diabetes mellitus have been distinguished. In *diabetes type 1* (insulin-dependent diabetes), the pancreas is not producing insulin. The condition is believed to be brought on by exposure to an environmental agent, most likely a virus, whose presence causes cytotoxic T cells to destroy the pancreatic islets. The cells turn to the breakdown of protein and fat for energy. The metabolism of fat leads to acidosis (acid blood) that can eventually cause coma and death. As a result, the individual must have daily insulin injections. These injections control the diabetic symptoms but can still cause inconveniences, since either taking too much insulin or failing to eat regularly can bring on the symptoms of hypoglycemia (low blood sugar). These symptoms include perspiration, pale skin, shallow breathing, and anxiety. Because the brain requires a constant supply of glucose, unconsciousness can result. The treatment is quite simple: Immediate ingestion of a sugar cube or fruit juice can very quickly counteract hypoglycemia.

Of the 16 million people who now have diabetes in the United States, most have *diabetes type 2* (noninsulin-dependent diabetes). This type of diabetes mellitus usually occurs in people of any age who are obese and inactive, as discussed in Chapter 25. The pancreas produces insulin, but the liver and muscle cells do not respond to it in the usual manner. They are said to be insulin resistant. If diabetes type 2 is untreated, the results can be as serious as those of type 1. Diabetics are prone to blindness, kidney disease, and circulatory disorders. It is usually possible to prevent or at least control diabetes type 2 by adhering to a low-fat and low-sugar diet and exercising regularly.

Check Your Progress

1. Compare and contrast the adrenal medulla with the adrenal cortex.
2. Contrast the function of insulin with that of glucagon.
3. Contrast diabetes type 1 with diabetes type 2.

Answers: 1. Both are components of the adrenal glands and are controlled by the hypothalamus. The adrenal medulla provides rapid, short-term responses to stress, while the adrenal cortex provides long-term responses to stress. 2. Insulin lowers blood sugar levels by stimulating the uptake of glucose by cells. Glucagon raises blood sugar levels by stimulating the breakdown of glycogen to glucose. 3. People with diabetes type 1 do not produce insulin, so they require daily insulin injections. People with diabetes type 2 produce insulin, but the liver and muscle cells do not respond to it as they should.

Summary

27.1   Nervous System

Both the nervous and endocrine systems utilize chemical signals. The nervous system is organized into a central nervous system (CNS) and a peripheral nervous system (PNS).

Neurons
- Are composed of a cell body, an axon, and dendrites. Only axons conduct nerve impulses.
- Exist as three types: sensory (takes nerve impulses to the CNS), interneuron (takes nerve impulses between neurons of the CNS), and motor (takes nerve impulses away from the CNS).

The Nerve Impulse
- The sodium-potassium pump transports Na^+ ions out of the axon and K^+ ions into the axon. The inside of the axon has a negative charge; the outside has a positive charge.
- The nerve impulse is an action potential—there is a reversal of charge as Na^+ ions flow in, and then a return to the previous charge difference as K^+ flows out of an axon.
- The nerve impulse is much faster in myelinated axons because the impulse jumps from neurofibril node to neurofibril node.

The Synapse
- The synapse is a region of close proximity between an axon terminal and the next neuron (CNS), or between an axon terminal and a muscle cell (PNS).
- A nerve impulse causes the release of a neurotransmitter (can be excitatory or inhibitory) into the synaptic cleft.
- Neurotransmitters are ordinarily removed quickly from the synaptic cleft. AChE breaks down acetylcholine, for example.
- Drugs affect the action of neurotransmitters.

Central Nervous System
Brain
- The cerebrum functions in sensation, reasoning, learning and memory, language, and speech. The cerebral cortex has a primary sensory area in the parietal lobe that receives sensory information from each part of the body and a primary motor area in the frontal lobe that sends out motor commands to skeletal muscles. Association areas carry on integration.
- In the diencephalon, the hypothalamus helps control homeostasis; the thalamus specializes in sending sensory input on to the cerebrum.
- The cerebellum primarily coordinates skeletal muscle contractions.
- In the brain stem, the medulla oblongata has centers for vital functions, such as breathing and the heartbeat, and helps control the internal organs.

Limbic System
The limbic system blends higher functions into a united whole. The hippocampus and amygdala have roles in learning and memory and appear to be affected in Alzheimer disease.

Peripheral Nervous System
- Somatic system: Reflexes (automatic responses) involve a sensory receptor, sensory neuron, interneurons in the spinal cord, and a motor neuron.
- Autonomic system: The sympathetic division is active during times of stress, and the parasympathetic division is active during times of relaxation. Both divisions control the same internal organs.

27.2   Endocrine System

Endocrine glands secrete hormones into the bloodstream for distribution to target organs or tissues. Hormones are either steroids or peptides.

Hypothalamus and Pituitary Gland

Hypothalamus secretes

releasing and inhibiting hormones	control anterior pituitary
antidiuretic hormone (ADH)	released by posterior pituitary; causes water uptake by kidneys
oxytocin	released by posterior pituitary; causes uterine contractions

Anterior Pituitary secretes

gonadotropic hormones (FSH and LH)	stimulate gonads
thyroid stimulating (TSH)	stimulates thyroid
adrenocorticotropic hormone (ACTH)	stimulates adrenal cortex
prolactin	causes milk production
growth hormone (GH)	causes cell division, protein synthesis, bone growth; too little = *pituitary dwarfism;* too much = *pituitary giantism*

Thyroid Gland and Parathyroid Glands

Thyroid Gland secretes

thyroxine (T_4) and triiodothyronine (T_3)	increases metabolic rate; *simple goiter* when iodine is lacking, *exophthalmic goiter* when overactive
calcitonin	lowers blood calcium level

Parathyroid Glands secrete

parathyroid hormone (PTH)	raises blood calcium level

Adrenal Glands

Adrenal Medulla secretes

epinephrine and norepinephrine	response to emergency situations

Adrenal Cortex secretes

mineralocorticoids (aldosterone)	causes kidneys to reabsorb Na^+; too little as in *Addison disease* = low blood pressure; too much as in *Cushing syndrome* = high blood pressure

glucocorticoids (cortisol)	raises blood glucose level; too little as in *Addison disease* = body can't respond to stress; too much as in *Cushing syndrome* = diabetes
sex hormones	male/female differences

Pancreas secretes

insulin	causes cells to take up and liver to store glucose as glycogen; too little = *diabetes mellitus*
glucagon	causes liver to break down glycogen

Thinking Scientifically

1. Recent research indicates that Parkinson disease damages the sympathetic division of the peripheral nervous system. One test for sympathetic division function, called the Valsalva maneuver, requires the patient to blow against resistance. A functional nervous system will compensate for the decrease in blood output from the heart by constricting blood vessels. How do you suppose Parkinson patients respond to the Valsalva maneuver? How does this relate to a common condition in Parkinson patients, called orthostatic hypotension, in which blood pressure falls suddenly when the person stands up, leading to dizziness and fainting?

2. Researchers have been trying to determine the reason for a dramatic rise in diabetes type 2 in recent decades. A sedentary lifestyle and poor eating habits certainly contribute to the risk for developing the disease. In addition, however, some researchers have observed a connection between childhood vaccination and diabetes type 1. Epidemiological data from countries that have recently initiated mass immunization programs indicate that the incidence of diabetes type 1 has increased there as well. What might be the connection between vaccination and diabetes?

Testing Yourself

Choose the best answer for each question.

1. Unlike the nervous system, the endocrine system
   a. uses chemical signals as a means of communication.
   b. helps maintain equilibrium.
   c. sends messages to target organs.
   d. changes the metabolism of cells.

2. Pain receptors are at the distal ends of
   a. interneurons.
   b. intraneurons.
   c. sensory neurons.
   d. motor neurons.

3. Which of the following is not true of the nerve impulse? The nerve impulse
   a. is subject to a short refractory period before it can occur again.
   b. is slower in thick myelinated fibers.
   c. is also called an action potential.
   d. moves from neurofibril node to neurofibril node in myelinated fibers.

4. When comparing the interior of an axon at rest to the exterior, there is
   a. a charge difference.
   b. an ion concentration difference.
   c. both a charge and ion concentration difference.
   d. neither a charge nor an ion concentration difference.

5. Long-term use of cocaine causes an intense craving for the drug because the body has begun to make
   a. less dopamine.
   b. more dopamine.
   c. less anandamide.
   d. more anandamide.

6. Myelin is formed from
   a. dendrites.
   b. Schwann cells.
   c. axons.
   d. motor neurons.

7. Label the parts of the brain in the following illustration.

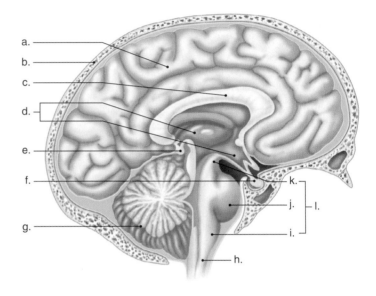

For questions 8–12, identify the part of the brain in the key that matches the description. Some answers may be used more than once. Some questions may have more than one answer.

Key:
   a. cerebrum
   b. diencephalon
   c. cerebellum
   d. brain stem

8. Regulates hunger, thirst, and sleep.

9. Receives sensory information from eyes and ears.

10. Composed of two parts; mostly white matter.

11. Regulates heartbeat, breathing, and blood pressure.

12. Responsible for sensation, voluntary movement, and higher thought processes.

13. In contrast to exocrine glands, endocrine glands

    a. secrete products.
    b. induce a response by the body.
    c. utilize ducts.
    d. produce hormones.

14. Both the adrenal medulla and the adrenal cortex are

    a. endocrine glands.
    b. in the same organ.
    c. involved in our response to stress.
    d. All of these are correct.

15. Diabetes type 1 is thought to result from a virus that

    a. interferes with gene expression in pancreatic cells.
    b. causes T cells to destroy the pancreatic islets.
    c. breaks down insulin.
    d. prevents the secretion of insulin by the pancreatic islets.

16. Label the parts of the endocrine system in the following illustration.

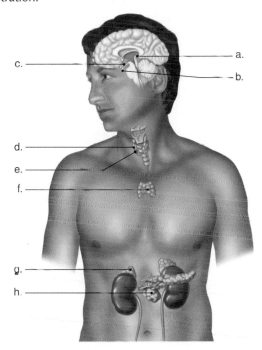

17. The cerebellum

    a. coordinates skeletal muscle movements.
    b. receives sensory input from the joints and muscles.
    c. receives motor input from the cerebral cortex.
    d. All of these are correct.

18. An interneuron can relay information

    a. from a sensory neuron to a motor neuron.
    b. only from a motor neuron to another motor neuron.
    c. only from a sensory neuron to another sensory neuron.
    d. None of these are correct.

19. The sympathetic division of the autonomic nervous system will

    a. increase heart rate and digestive activity.
    b. decrease heart rate and digestive activity.
    c. cause pupils to constrict.
    d. None of these are correct.

20. Effects of drugs can include

    a. prevention of neurotransmitter release.
    b. prevention of reuptake by the presynaptic membrane.
    c. blockages to a receptor.
    d. All of these are correct.
    e. None of these are correct.

21. Which of these correctly describes the distribution of ions on either side of an axon when it is not conducting a nerve impulse?

    a. more sodium ions (Na^+) outside and more potassium ions (K^+) inside
    b. more K^+ outside and less Na^+ inside
    c. charged protein outside; Na^+ and K^+ inside
    d. Na^+ and K^+ outside and water only inside
    e. chloride ions (Cl^-) on the outside and K^+ and Na^+ on the inside

22. When the action potential begins, sodium gates open, allowing Na^+ to cross the membrane. Now the polarity changes to

    a. negative outside and positive inside.
    b. positive outside and negative inside.
    c. neutral outside and positive inside.
    d. There is no difference in charge between outside and inside.

23. Transmission of the nerve impulse across a synapse is accomplished by the

    a. movement of Na^+ and K^+.
    b. release of a neurotransmitter by a dendrite.
    c. release of a neurotransmitter by an axon.
    d. release of a neurotransmitter by a cell body.
    e. Any one of these is correct.

24. The autonomic system has two divisions, called the

    a. CNS and PNS.
    b. somatic and skeletal systems.
    c. efferent and afferent systems.
    d. sympathetic and parasympathetic divisions.

25. The limbic system

    a. involves portions of the cerebral lobes and the diencephalon.
    b. is responsible for our deepest emotions, including pleasure, rage, and fear.
    c. is a system necessary to memory storage.
    d. is not directly involved in language and speech.
    e. All of these are correct.

26. Growth hormone is produced by the

    a. posterior adrenal gland.
    b. posterior pituitary.
    c. anterior pituitary.
    d. kidneys.
    e. None of these are correct.

27. Glucagon causes

    a. use of fat for energy.
    b. glycogen to be converted to glucose.
    c. use of amino acids to form fats.
    d. Both a and b are correct.
    e. None of these are correct.

28. Long-term complications of diabetes include
    a. blindness.
    b. kidney disease.
    c. circulatory disorders.
    d. All of these are correct.
    e. None of these are correct.

29. PTH causes the blood level of calcium to _____ , and calcitonin causes it to _____ .
    a. increase, not change
    b. increase, decrease
    c. decrease, also decrease
    d. decrease, increase
    e. not change, increase

Bioethical Issue

Recently, the Food and Drug Administration (FDA) approved the use of human growth hormone to treat "short stature." This decision implies that short stature is a medical condition with a legitimate need to be treated. In this case, medical care is not treating a disease, but enhancing a feature of an otherwise healthy person. Proponents of the FDA decision say that administering growth hormone to short people will help them avoid discrimination and live a more normal life in a world designed for taller people. Opponents say that this decision may lead to slippery-slope scenarios in which medical treatments that make us smarter or faster are assumed to make us better. Do you think human growth hormone should be used to "cure shortness"? Typically, parents will need to make decisions about growth hormone treatments for their children, since the treatments are generally administered to preschoolers. Should parents be allowed to make decisions about treatments that will determine their children's heights?

Understanding the Terms

acetylcholine (ACh)   476
acetylcholinesterase
   (AChE)   476
action potential   475
Addison disease   487
adrenal cortex   487
adrenal gland   487
adrenal medulla   487
adrenocorticotropic hormone
   (ACTH)   484
Alzheimer disease (AD)   480
anterior pituitary   484
antidiuretic hormone
   (ADH)   485
autonomic system   482
axon   474
brain stem   479

calcitonin   486
cell body   474
central nervous system
   (CNS)   473
cerebellum   479
cerebral cortex   478
cranial nerve   480
Cushing syndrome   487
dendrite   474
diabetes mellitus   488
dorsal root ganglion   481
endocrine gland   483
epinephrine   487
exophthalmic goiter   486
ganglion   480
glucagon   487
glucocorticoid   487

gonadotropic hormones
   (FSH and LH)   484
growth hormone (GH)   485
hypothalamic-releasing
   hormone   484
hypothalamus   479, 484
insulin   487
integration   476
interneuron   474
limbic system   480
medulla oblongata   479
memory   480
mineralocorticoid   487
motor neuron   474
myelin sheath   474
negative feedback   486
nerve   480
neuron   474
neurotransmitter   476
norepinephrine (NE)   476, 487
oxytocin   484
pancreas   487
pancreatic islets   487
parasympathetic division   482
parathyroid gland   486
parathyroid hormone
   (PTH)   486
peptide hormone   484

peripheral nervous system
   (PNS)   473
pituitary gland   484
posterior pituitary   485
prefrontal area   479
primary motor area   478
primary sensory area   478
prolactin (PRL)   484
reflex   481
releasing hormone   484
saltatory conduction   475
sensory neuron   474
simple goiter   486
somatic system   481
spinal cord   478
spinal nerves   481
steroid hormone   483
sympathetic division   482
synapse   476
synaptic cleft   476
tetany   486
thalamus   479
thyroid gland   486
thyroid-stimulating hormone
   (TSH)   484
thyroxine (T₄)   486
tract   478.

Match the terms to these definitions:

a. _____ Portion of a neuron that conducts nerve impulses.

b. _____ Gap between an axon terminal and the dendrite of another neuron.

c. _____ Enzyme that breaks down a neurotransmitter.

d. _____ Structure that connects the brain to the spinal nerves.

e. _____ Region of the cerebral cortex responsible for critical thinking.

f. _____ Bundle of axons.

g. _____ Stimulate the gonads to produce gametes and sex hormones.

h. _____ Inner portion of an adrenal gland.

i. _____ Disorder that results when the adrenal cortex does not secrete enough hormone.

j. _____ Endocrine tissue in the pancreas.

Sensory Input and Motor Output

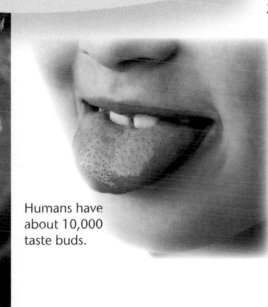

CHAPTER 28

Humans have about 10,000 taste buds.

Calcium is essential for muscle contraction.

Mature muscle cells cannot be replaced.

When you think of the importance of calcium in nutrition, you most likely think about bone. Bone serves as a depository for calcium so that there is always a reserve in the body, and as an added benefit, storing calcium makes bone tissue strong. However, calcium plays other important roles in the body as well.

One of the most essential functions of calcium involves muscle contraction. No muscle in the body (including the ▶ ▶ ▶

493

heart and the diaphragm) can contract without the presence of calcium. If you were dependent solely on obtaining calcium from your daily diet and you ran out of calcium, your heart and diaphragm wouldn't contract. Then, the inability to circulate your blood and to breathe would cause you to die. Luckily, our bones do stockpile calcium, so when we don't get enough in our diet, we get it from the reserve supply in our bones. The downside is that in order to tap into the reserve of calcium in our bones, bone tissue must be degraded. This is why obtaining enough calcium in the diet is critical. When there is an imbalance between the amount of calcium taken from bones and the amount of calcium deposited in bones, conditions such as osteoporosis often result. In osteoporosis, the bones are fragile, and a broken hip, spine, or wrist often occurs. This is why we hear so much about including plenty of calcium in the diet.

The functions of bones and muscle are closely related to each other. In this chapter, in addition to learning about these motor output systems, you will learn about sensory input systems.

28.1 The Senses

All living things respond to stimuli. Stimuli are environmental signals that tell us about the external environment or the internal environment. In a previous chapter, we learned that plants often respond to external stimuli, such as light, by changing their growth pattern. An animal's response often results in motion. Complex animals rely on sensory receptors to provide information to the central nervous system (brain and spinal cord), which integrates sensory input before directing a motor response (Fig. 28.1).

Sense organs, as a rule, are specialized to receive one kind of stimulus. The eyes ordinarily respond to light, ears to sound waves, pressure receptors to pressure, and chemoreceptors to chemical molecules. Sensory receptors transform the stimulus into nerve impulses that reach a particular section of the cerebral cortex. Those from the eye reach the visual areas, and those from the ears reach the auditory areas. The brain, not the sensory receptor, is responsible for sensation and perception, and each part of the brain interprets impulses in only one way. For example, if by accident the photoreceptors of the eye are stimulated by pressure and not light, the brain causes us to see "stars" or other visual patterns.

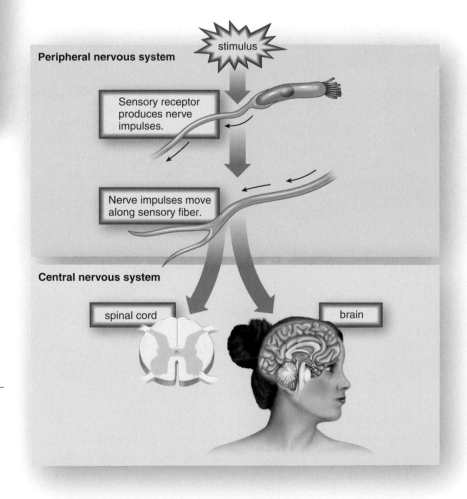

Figure 28.1   Sensory input and motor output.

After detecting a stimulus, sensory receptors initiate nerve impulses within the PNS. These impulses give the central nervous system (CNS) information about the external and internal environment. The CNS integrates all incoming information and then initiates a motor response to the stimulus.

Chemical Senses

The fundamental functions of sensory receptors include helping animals to stay safe, find food, and find mates. *Chemoreceptors* give us the ability to detect chemicals in the environment, which is believed to be our most primitive sense. Chemoreceptors occur almost universally in animals. For example, they are present all over the body of planarians (flatworms), but concentrated on the auricles at the sides of the head. Male moths have receptors for a sex attractant on their antennae. The receptors on the antennae of the male silkworm moth are so sensitive that only 40 out of 40,000 receptor proteins need to be activated in order for the male to respond to a chemical released by the female (Fig. 28.2a). Other insects, such as the housefly, have chemoreceptors largely on their feet—a fly tastes with its feet instead of its mouth. In mammals, the receptors for taste are located in the mouth, and the receptors for smell are located in the nose.

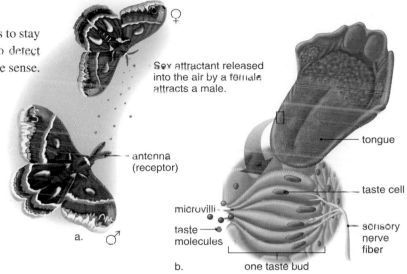

Sex attractant released into the air by a female attracts a male.

a. ♂ — antenna (receptor)

tongue
taste cell
microvilli
taste molecules
sensory nerve fiber

b.   one taste bud

Taste and Smell

In humans, **taste buds,** located primarily on the tongue, contain taste cells, and the nose contains olfactory cells (Fig. 28.2b,c). Receptors for chemicals are located on the microvilli of taste cells and on the cilia of olfactory cells. When molecules bind to these receptors, nerve impulses are generated in sensory nerve fibers that go to the brain. When they reach the appropriate cortical areas, they are interpreted as taste and smell, respectively.

There are at least four primary types of tastes (bitter, sour, salty, and sweet), and taste buds for each are located throughout the tongue, but may be concentrated in particular regions. A particular food can stimulate more than one of these types of taste buds. In this way, the response of taste buds can result in a range of sweet, sour, salty, and bitter tastes. The brain appears to survey the overall pattern of incoming sensory impulses and to take a "weighted average" of their taste messages as the perceived taste. Similarly, an odor contains many odor molecules, which activate a characteristic combination of receptors. When this complex information is communicated to the cerebral cortex, we know we have smelled a rose—or an onion!

An important role of taste and also smell is to trigger reflexes that start the digestive juices flowing. A revolting or repulsive substance in the mouth can initiate the gag reflex or even vomiting. Smell is even more important to our survival. Smells associated with danger, such as smoke, can trigger the fight-or-flight reflex. Unpleasant smells can cause us to sneeze or choke.

Have you ever noticed that a certain aroma vividly brings to mind a certain person or place? A person's perfume may remind you of someone else, or the smell of boxwood may remind you of your grandfather's farm. The olfactory bulbs have direct connections with the limbic system and its centers for emotions and memory. One investigator showed that when subjects smelled an orange while viewing a painting, they not only remembered the painting when asked about it later, but they also had many deep feelings about it

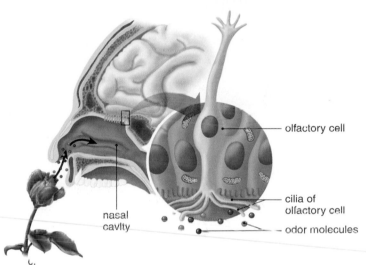

olfactory cell
cilia of olfactory cell
nasal cavity
odor molecules

c.

Figure 28.2 **Chemoreceptors.**

Chemoreceptors are common in the animal kingdom. **a.** Male moths have receptors on their antennae for a sex attractant released by the female. **b.** Humans have receptors for taste on the microvilli of taste cells and **(c)** receptors for smell on the cilia of olfactory cells.

Check Your Progress

1. Why is the term "taste cells" a misnomer?
2. Explain why a certain odor can often evoke a strong memory of a person or place.

Answers: 1. We don't taste with taste cells. After chemicals bind to the receptors of taste cells, nerve impulses go to the brain, which interprets the nerve impulses as tastes. 2. The olfactory system is directly tied to the centers for emotion and memory in the limbic system, so an odor can help us recall an emotion or memory.

Hearing and Balance

The human ear has two sensory functions: hearing and balance (equilibrium). The sensory receptors for both of these consist of *hair cells* with long microvilli called stereocilia. These microvilli, unlike those of taste cells, are sensitive to mechanical stimulation. Therefore, they are termed *mechanoreceptors*.

The similarity of the sensory receptors for balance and hearing and their presence in the same organ suggest an evolutionary relationship between them. In fact, the sense organs of the mammalian ear may have evolved from a type of sense organ in fishes.

Hearing

Most invertebrates cannot hear. Some arthropods, including insects, do have sound receptors, but they are quite simple. In insects, the ear consists of a pair of air pockets, each enclosed by a membrane, called the tympanic membrane, that passes sound vibrations to sensory receptors. The human ear has a tympanic membrane also, but it is between the outer and middle ear (Fig. 28.3*a*). The outer ear collects sound waves that cause the tympanic membrane to move back and forth (vibrate) ever so slightly. Three tiny bones in the middle ear (the ossicles) amplify the sound about 20 times as it moves from one to the other. The last of the ossicles strikes an oval membrane, causing it to vibrate, and in this way, the pressure is passed to a fluid within the hearing portion of the inner ear called the cochlea (Fig. 28.3*b*). The term cochlea means snail shell. Specifically, the sensory receptors of hearing are located in the cochlear canal of the cochlea. The sensory receptors for hearing are hair cells whose stereocilia are embedded in a gelatinous membrane. Collectively, they are called the **spiral organ** (organ of Corti) (Fig. 28.3*c,d*).

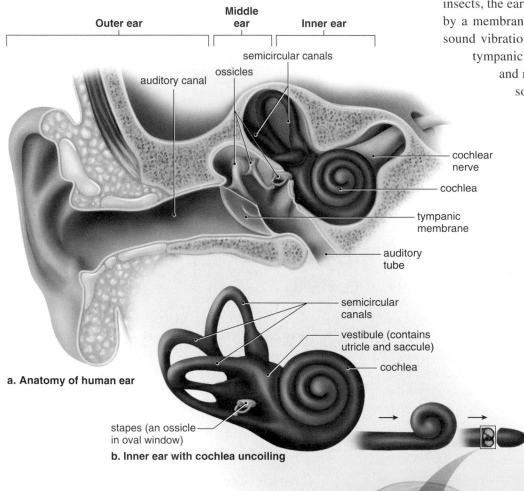

a. Anatomy of human ear

b. Inner ear with cochlea uncoiling

Figure 28.3   The human ear.

a. The outer ear collects sound waves, the middle ear amplifies sound waves, and the cochlea contains the sensory receptors for hearing. **b.** The inner ear contains the cochlea, and also the semicircular canals and the vestibule. **c.** The sensory receptors for hearing are hair cells within the spiral organ. **d.** We hear when pressure waves within the canals of the cochlea cause the hair cells to vibrate and their stereocilia to bend.

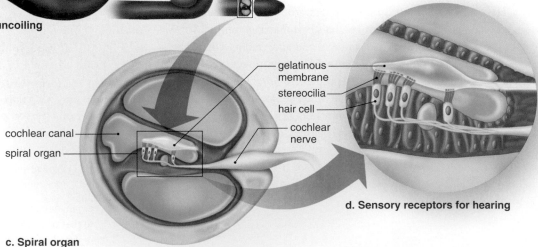

c. Spiral organ

d. Sensory receptors for hearing

The outer ear and middle ear, which collect and amplify sound waves, are filled with air. The auditory tube functions to relieve pressure in the middle ear. But the inner ear is filled with fluid; therefore, fluid pressure waves actually stimulate the spiral organ. When the last ossicle strikes an oval membrane, pressure waves cause the hair cells to move up and down, and the stereocilia of the hair cells embedded in the gelatinous membrane bend. The hair cells of the spiral organ synapse with the cochlear nerve, and when their stereocilia bend, nerve impulses begin in the cochlear nerve and travel to the brain stem. When these impulses reach the auditory areas of the cerebral cortex, they are interpreted as sound.

Each part of the spiral organ is sensitive to different wave frequencies, or *pitch*. Near the tip, the spiral organ responds to low pitches, such as a tuba, and near the base, it responds to higher pitches, such as a bell or a whistle. The nerve fibers from each region along the length of the spiral organ lead to slightly different areas in the brain. The pitch sensation we experience depends upon which region of the spiral organ vibrates and which area of the brain is stimulated.

Volume is a function of the amplitude of sound waves. Loud noises cause the spiral organ to vibrate to a greater extent. The resulting increased stimulation is interpreted by the brain as volume. It is believed that the brain interprets the *tone* of a sound based on the distribution of the hair cells stimulated.

Hearing Loss Especially when we are young, the middle ear is subject to infections that can lead to hearing impairment. Today, it is quite common for youngsters to have "tubes" put into the tympanic membrane to allow the middle ear to drain in an effort to prevent this type of hearing loss. The mobility of ossicles decreases with age, and if bone grows over the stapes, the only remedy is implantation of an artificial stapes that can move.

Deafness due to middle ear damage is called **conduction deafness.** Deafness due to spiral organ damage is called **nerve deafness.** In today's society, noise pollution is common, and even city traffic can be loud enough to damage the stereocilia of hair cells (Fig. 28.4). It stands to reason, then, that frequently attending rock concerts, constantly playing a stereo loudly, or using earphones at high volume can also damage hearing. The first hint of danger can be temporary hearing loss, a "full" feeling in the ears, muffled hearing, or tinnitus (ringing in the ears). If exposure to noise is unavoidable, specially designed noise-reduction earmuffs are available, and it is also possible to purchase earplugs made from compressible, spongelike material at the drugstore or sporting-goods store. These earplugs are not the same as those worn for swimming, and they should not be worn interchangeably. Finally, you should be aware that some medicines may damage the ability to hear. Anyone taking anticancer drugs, such as cisplatin, and certain antibiotics, such as streptomycin, should be especially careful to protect the ears from loud noises.

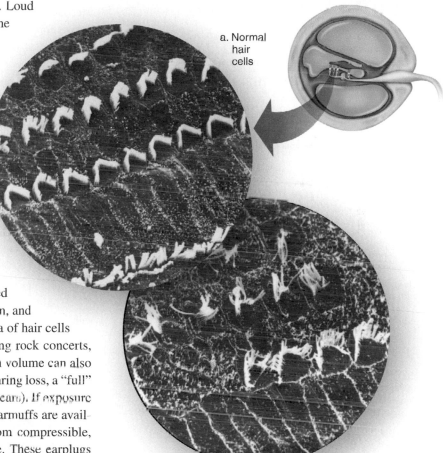

a. Normal hair cells

b. Damaged hair cells

Figure 28.4 **Effect of noise on hearing.**

a. Normal hair cells in the spiral organ of a guinea pig. **b.** Spiral organ damaged by 24-hour exposure to a noise level equivalent to that of a heavy-metal rock concert. Damaged cells cannot be replaced, so hearing is permanently impaired.

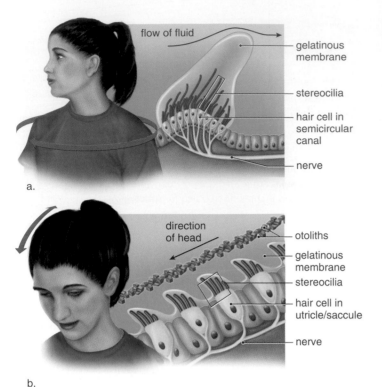

a.

b.

Figure 28.5   Sense of balance in humans.

a. Sensory receptors for rotational equilibrium. When the head rotates, the gelatinous membrane is displaced, bending the stereocilia. **b.** Sensory receptors for gravitational equilibrium. When the head bends, otoliths are displaced, causing the gelatinous membrane to sag and the stereocilia to bend.

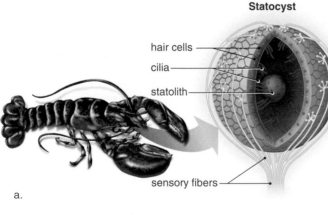

a.

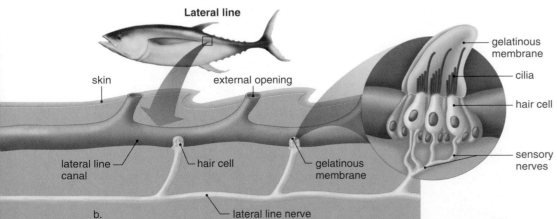

b.

Balance

Humans have two senses of balance (equilibrium): rotational and gravitational. We are able to detect the rotational and/or angular movement of the head as well as the straight-line movement of the head with respect to gravity.

Rotational equilibrium involves the semicircular canals (see Fig. 28.3b). In the base of each canal, hair cells have stereocilia embedded within a gelatinous membrane (Fig. 28.5a). Because there are three semicircular canals, each responds to head movement in a different plane of space. As fluid within a semicircular canal flows over and displaces the gelatinous membrane, the stereocilia of the hair cells bend, and the pattern of impulses carried to the central nervous system (CNS) changes. These data, usually supplemented by vision, tell the brain how the head is moving. *Vertigo* is dizziness and a sensation of rotation. It is possible to simulate a feeling of vertigo by spinning rapidly and stopping suddenly. Now the person feels like the room is spinning because of sudden stimulation of stereocilia in the semicircular canals.

Gravitational equilibrium refers to the position of the head with relation to gravity. It depends on the **utricle** and **saccule,** two membranous sacs located in the inner ear (see Fig. 28.3b). Both of these sacs also contain hair cells with stereocilia in a gelatinous membrane (Fig. 28.5b). Calcium carbonate ($CaCO_3$) granules (the otoliths) rest on this membrane. When the head moves forward or back, up or down, the otoliths are displaced, and the membrane moves, bending the stereocilia of the hair cells. This movement alters the frequency of nerve impulses to the CNS. These data, usually supplemented by vision, tell the brain the direction of the movement of the head.

Similar Receptors in Other Animals

Gravitational equilibrium organs, called **statocysts,** are found in several types of invertebrates, from cnidarians to molluscs to crustaceans. These organs give information only about the position of the head; they are not involved in the sensation of movement (Fig. 28.6a). When the head stops moving, a small particle called a statolith stimulates the cilia of the closest hair cells, and these cilia generate impulses, indicating the position of the head.

The **lateral line** system of fishes utilizes sense organs similar to those we have been studying in the human inner ear (Fig. 28.6b). In bony fishes, the system consists of sense organs located within a canal that has openings to the outside. As you might expect, the sense organ is a collection of hair cells with cilia embedded in a gelatinous membrane. Water currents and pressure waves from nearby objects cause the membrane and the cilia of the hair cells to bend. Thereafter the hair cells initiate nerve impulses that go to the brain. Fishes use these data, not for hearing or balance, but to locate other fish, including predators, prey, and mates.

Figure 28.6   Sensory receptors in other animals.

a. Invertebrates utilize statocysts to determine their position. When the statolith stops moving, cilia of the nearest hair cells are stimulated, telling the position of the head. **b.** The lateral line of fishes is not for hearing or balance. It is for knowing the location of other fishes.

Vision

Sensory receptors that are sensitive to light are called **photoreceptors.** In planarians, eyespots allow these animals to determine the direction of light only. Other photoreceptors form actual images. Image-forming eyes provide information about an object and how far away it is. Such detailed information is invaluable to an animal.

Among invertebrates, arthropods have **compound eyes** composed of many independent visual units, each of which possesses a lens to focus light rays on photoreceptors (Fig. 28.7a). The image that results from all the stimulated visual units is crude because the small size of compound eyes limits the number of visual units, which still might number as many as 28,000.

Vertebrates (including humans) and certain molluscs, such as the squid and the octopus, have a **camera-type eye** (Fig. 28.7b). A single lens focuses light on photoreceptors, which number in the millions and are closely packed together within a retina. Since molluscs and vertebrates are not closely related, the camera-type eye evolved independently in each group.

Insects have color vision, but they make use of a slightly shorter range of the electromagnetic spectrum compared to humans. However, they can see the longest of the ultraviolet rays, and this enables them to be especially sensitive to the reproductive parts of flowers, which have particular ultraviolet patterns. Some fishes, all reptiles, and most birds are believed to have color vision, but among mammals, only humans and other primates have expansive color vision. It would seem, then, that this trait was adaptive for a diurnal habit (active during the day), which accounts for its retention in only a few mammals.

The Human Eye

When looking straight ahead, each of our eyes views the same object from a slightly different angle. This slight displacement of the images permits *binocular vision,* the ability to perceive three-dimensional images and to sense depth. Like the human ear, the human eye has numerous parts, many of which are involved in preparing the stimulus for the sensory receptors. In the case of the ear, sound wave energy is magnified before it reaches the sensory receptors. In the case of the eye, light rays are brought to a focus on the photoreceptors located within the **retina.** In Figure 28.8, the **cornea** and especially the **lens** are involved in focusing light rays on the photoreceptors. The **iris,** the colored part of the eye, regulates the amount of light that enters the eye by way of the **pupil.** The retina generates nerve impulses that are sent to the visual part of the cerebral cortex, and from this information, the brain forms an image of the object.

The shape of the lens is controlled by the ciliary muscles. When we view a distant object, the lens remains relatively flat, but when we view a near object, the lens rounds up. With normal aging, the lens loses its ability to accommodate for near objects; therefore, many people need reading glasses when they reach middle age. Aging, or possibly exposure to the sun, also makes the lens subject to cataracts; the lens becomes opaque, and therefore incapable of transmitting light rays. Currently, surgery is the only viable treatment for cataracts. First, a surgeon opens the eye near the rim of the cornea. The enzyme zonulysin may be used to digest away the ligaments holding the lens in place. Most surgeons then use a cryoprobe, which freezes the lens for easy removal. An intraocular lens attached to the iris can then be implanted so that the patient does not need to wear thick glasses or contact lenses.

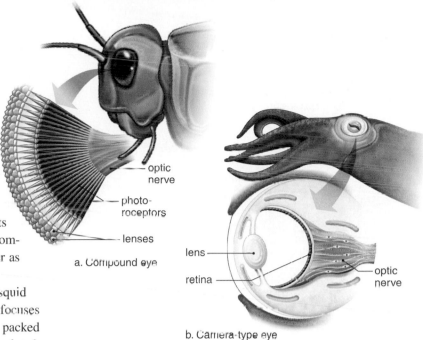

a. Compound eye

b. Camera-type eye

Figure 28.7   Eyes.

a. A compound eye has several visual units. Each visual unit has a lens that focuses light onto photoreceptor cells. **b.** A camera-type eye has one lens that focuses light onto a retina containing many photoreceptors.

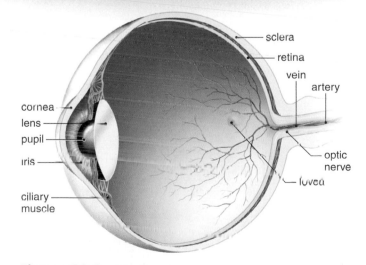

Figure 28.8   The human eye.

Light passes through the cornea and through the pupil (a hole in the iris), and is focused by the lens on the retina, which houses the photoreceptors.

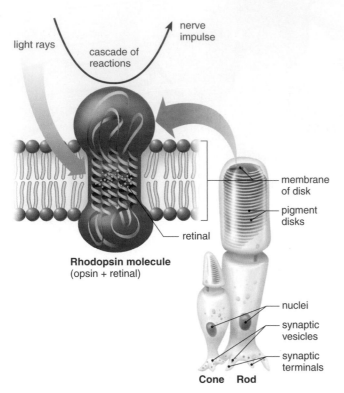

Rhodopsin molecule
(opsin + retinal)

membrane
of disk

pigment
disks

retinal

nuclei

synaptic
vesicles

synaptic
terminals

Cone   Rod

Figure 28.9 **Photoreceptors of the eye.**

In rods, the membrane of each disk contains rhodopsin, a complex molecule containing the protein opsin and the pigment retinal. When rhodopsin absorbs light energy, it splits, releasing opsin, which sets in motion a cascade of reactions that ends in nerve impulses.

Figure 28.10 **The retina.**

The retina contains a layer of rods and cones, a layer of intermediate cells, and a layer of ganglion cells. Integration of signals occurs at the synapses between the layers, and much processing occurs before nerve impulses are sent to the brain.

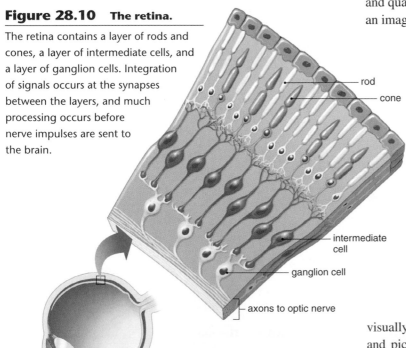

rod

cone

intermediate
cell

ganglion cell

axons to optic nerve

Photoreceptors of the Eye

Figure 28.9 illustrates the structure of the photoreceptors in the human eye, which are called **rods** and **cones.** Both types of photoreceptors contain a visual pigment similar to that found in all types of eyes throughout the animal kingdom. The visual pigment in rods is a deep-purple pigment called rhodopsin. **Rhodopsin** is a complex molecule made up of the protein opsin and a light-absorbing molecule, called *retinal,* which is a derivative of vitamin A. When a rod absorbs light, rhodopsin splits into opsin and retinal, leading to a cascade of reactions that ends in the generation of nerve impulses. Rods are very sensitive to light, and therefore are suited to night vision. (Because carrots are rich in vitamin A, it is true that eating carrots can improve your night vision.) Rod cells are plentiful throughout the entire retina; therefore, they also provide us with peripheral vision and perception of motion.

The cones, on the other hand, are located primarily in a part of the retina called the **fovea.** Cones are activated by bright light; they allow us to detect the fine detail and the color of an object. Color vision depends on three different kinds of cones, which contain pigments called the B (blue), G (green), and R (red) pigments. Each pigment is made up of retinal and opsin, but a slight difference in the opsin structure of each accounts for their specific absorption patterns. Various combinations of cones are believed to be stimulated by in-between shades of color.

Retina   The retina has three layers of cells and light has to penetrate through the first two layers to reach the photoreceptors (Fig. 28.10). The intermediate cells of the middle layer process and relay visual information from the photoreceptors to the ganglion cells that have axons forming the optic nerve. The sensitivity of cones versus rods is mirrored by how directly they connect to ganglion cells. Information from several hundred rods may converge on a single ganglion cell, while cones show very little convergence. As signals pass through the layers of the retina, integration occurs. Integration improves the overall contrast and quality of information sent to the brain which uses the information to form an image of the object.

No rods and cones occur where the optic nerve exits the retina. Therefore, no vision is possible in this area. You can prove this to yourself by putting a dot to the right of center on a piece of paper. Close your left eye, then use your right hand to move the paper slowly toward your right eye while you look straight ahead. The dot will disappear at one point—this point is your **blind spot.**

Seeing in the Dark

Some nocturnal animals rely on vision, but others use sonar (sound waves) to find their way in the dark. Bats flying in a dark room easily avoid obstacles in their path because they echolocate, like submarines. When searching for food, they emit ultrasonic sound (above the range humans can hear) in chirps that bounce back off their prey. They are able to determine the distance to dinner by timing the echo's return—a long delay means the prey is far away.

A bat-inspired sonar walking stick is being perfected to help visually impaired people sense their surroundings. It too emits ultrasonic chirps and picks up the echoes from nearby objects. Buttons on the cane's handle vibrate gently to warn a user to dodge a low ceiling or to sidestep objects blocking the path. A fast, strong signal means an obstacle is close by.

Cutaneous Receptors and Proprioceptors

Cutaneous receptors are located in the skin, and **proprioceptors** are located in the muscles and joints. These sensory receptors are connected to the primary sensory area of the cerebral cortex where each part of the body is represented. Here, sensory input is received from the skin, muscles, and joints in each part of the body (Fig. 28.11). Thus, these receptors are important to our well-being.

Cutaneous Receptors

Skin, the outermost covering of our body, contains numerous sensory receptors that help us respond to changes in our environment, be aware of dangers, and communicate with others. The sensory receptors in skin are for touch, pressure, pain, and temperature.

Skin has two regions, called the epidermis and the dermis (Fig. 28.12). The epidermis is packed with cells that become keratinized as they rise to the surface. Among these cells are free nerve endings responsive to cold or to warmth. Cold receptors are far more numerous than warmth receptors, but there are no known structural differences between the two. Also in the epidermis are pain receptors sensitive to extremes in temperature or pressure and to chemicals released by damaged tissues. Sometimes, stimulation of internal receptors is felt as pain in the skin. This is called *referred pain*. For example, pain from the heart may be felt in the left shoulder and arm. This effect most likely happens when nerve impulses from the pain receptors of internal organs travel to the spinal cord and synapse with neurons also receiving impulses from the skin.

The dermis of the skin contains sensory receptors for pressure and touch (Fig. 28.12). The Pacinian corpuscle is an onion-shaped pressure receptor that lies deep inside the dermis. Several other cutaneous receptors detect touch. A free nerve ending, called a root hair plexus, winds around the base of a hair follicle and produces nerve impulses if the hair is touched. Touch receptors are concentrated in parts of the body essential for sexual stimulation: the fingertips, the palms, the lips, the tongue, the nipples, the penis, and the clitoris.

Proprioceptors

Proprioceptors help the body maintain equilibrium and posture, despite the force of gravity always acting upon the skeleton and muscles. A **muscle spindle** consists of sensory nerve endings wrapped around a few muscle cells within a connective tissue sheath. **Golgi tendon** organs and other sensory receptors are located in the joints. The rapidity of nerve impulses by proprioceptors is proportional to the stretching of the organs they occupy. A motor response results in contraction of muscle fibers adjoining the proprioceptor.

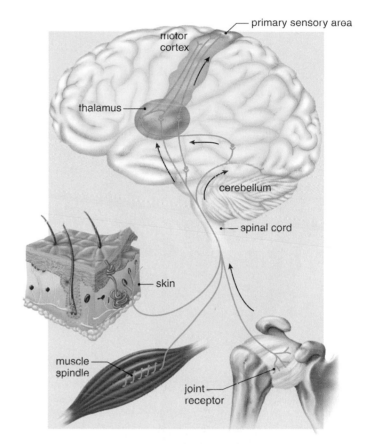

Figure 28.11   **Sensory input to primary sensory area of brain.**

Sensory receptors in the skin, muscles, and joints send nerve impulses to the CNS. Sensation and perception occur when these reach the primary sensory area of the cerebral cortex. A motor response can be initiated by the spinal cord and cerebellum without the involvement of the cerebrum.

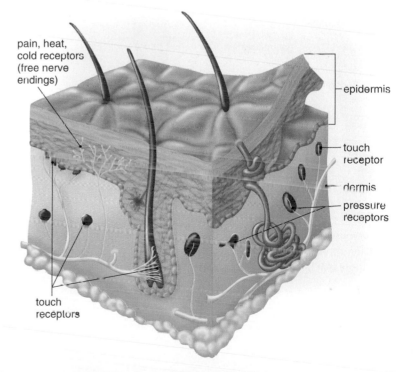

Figure 28.12   **Cutaneous receptors.**

Numerous receptors are in the skin. Free nerve endings (yellow) in the epidermis detect pain, heat, and cold. Various touch and pressure receptors (red) are in the dermis.

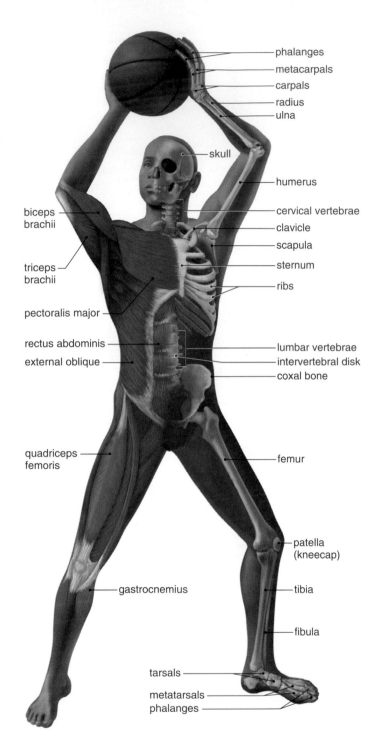

Figure 28.13   Musculoskeletal system.

Major muscles and bones of the body are labeled. The human skeleton contains bones that belong to the axial skeleton (red labels) and those that belong to the appendicular skeleton (black labels).

Labels: phalanges, metacarpals, carpals, radius, ulna, skull, humerus, biceps brachii, cervical vertebrae, clavicle, scapula, sternum, triceps brachii, ribs, pectoralis major, rectus abdominis, lumbar vertebrae, external oblique, intervertebral disk, coxal bone, quadriceps femoris, femur, patella (kneecap), gastrocnemius, tibia, fibula, tarsals, metatarsals, phalanges

28.2 The Motor Systems

In this section, we consider the motor systems of humans—the muscles and the bones—as forming a single system, namely the musculoskeletal system (Fig. 28.13). Combining the two seems appropriate because in many cases the functions of muscles and bones overlap:

Both skeletal muscles and bones support the body and make movement of body parts possible.

Both skeletal muscles and bones protect internal organs. Skeletal muscles pad the bones that protect the heart and lungs, the brain, and the spinal cord.

Both muscles and bones aid the functioning of other systems. Without the movement of the rib cage, breathing would not occur. As an aid to digestion, the jaws have sockets for teeth; skeletal muscles move the jaws so food can be chewed, and smooth muscle moves food along the digestive tract (see Fig. 24.5b). Red bone marrow supplies the red blood cells that carry oxygen, and the pumping of cardiac muscle in the wall of the heart moves the blood to the tissues, where exchanges with tissue fluid occur.

In addition to these shared functions, the muscles and bones have individual functions:

Skeletal muscle contraction assists movement of blood in the veins and lymphatic vessels (see Fig. 23.8). Without the return of lymph to the cardiovascular system and blood to the heart, circulation could not continue.

Skeletal muscles help maintain a constant body temperature (see Fig. 28.17). Skeletal muscle contraction causes ATP to break down, releasing heat that is distributed about the body.

Bones store fat and calcium. Fat is stored in yellow bone marrow (see Fig. 28.15), and the extracellular matrix of bone contains calcium. Calcium ions play a major role in muscle contraction and nerve conduction.

Check Your Progress

Describe the functions of bones and muscles when they work together.

Answer: Bones and muscles support the body and allow body parts to move, protect internal organs, and help other systems to function.

The Human Skeleton

The **endoskeleton** of vertebrates can be contrasted with the **exoskeleton** of arthropods. Both skeletons are jointed, which has helped these two groups of animals successfully live on land. The endoskeleton of humans is composed of bone, which is living material and capable of growth. Like an exoskeleton, an endoskeleton protects vital internal organs, but unlike an exoskeleton, it need not limit the space available for internal organs because it grows as the animal grows. The soft tissues that surround an endoskeleton protect it, and injuries to soft tissues are apt to be easier to repair than is the skeleton itself.

The exoskeleton of arthropods is composed of chitin, a strong, flexible, nitrogenous polysaccharide. Besides providing protection against wear and tear and against enemies, an exoskeleton also prevents drying out. Although an arthropod exoskeleton provides support for muscle contractions, it does not grow with the animal, and arthropods molt to rid themselves of an exoskeleton that has become too small (Fig. 28.14a). This process makes them vulnerable to predators.

One other type of skeleton is seen in the animal kingdom. In animals, such as worms, that lack a hard skeleton, a fluid-filled internal cavity can act as a **hydrostatic skeleton** (Fig. 28.14b). A hydrostatic skeleton offers support and resistance to the contraction of muscles so that the animal can move. As an analogy, consider that a garden hose stiffens when filled with water, and that a water-filled balloon changes shape when squeezed at one end. Similarly, an animal with a hydrostatic skeleton can change shape and perform a variety of movements.

Axial and Appendicular Skeletons

The 206 bones of the human skeleton are arranged into an axial skeleton and an appendicular skeleton.

Axial Skeleton   The labels for the bones of the **axial skeleton** are in color in Figure 28.13. The **skull** consists of the **cranium,** which protects the brain, and the facial bones. The most prominent of the **facial bones** are the lower and upper jaws, the cheekbones, and the nasal bones. The **vertebral column** (spine) extends from the skull to the **sacrum** (tailbone). It consists of a series of vertebrae separated by pads of fibrocartilage called the **intervertebral disks.** On occasion, disks can slip or even rupture. A damaged disk pressing against the spinal cord or spinal nerves causes pain. Removal of the disk may be required.

The **rib cage,** composed of the ribs and **sternum** (breastbone), demonstrates how the skeleton can be protective and flexible at the same time. The rib cage protects the heart and lungs but moves when we breathe.

Appendicular Skeleton   The **appendicular skeleton** contains the bones of two girdles and their attached limbs. The **pectoral** (shoulder) **girdle** and upper limbs are specialized for flexibility; the **pelvic** (hip) **girdle** and lower limbs are specialized for strength. The pelvic girdle also protects internal organs.

A **clavicle** (collarbone) and a **scapula** (shoulder blade) make up the shoulder girdle, but the **humerus** of the arm articulates only with the scapula. The joint is stabilized by tendons and ligaments that form a **rotator cuff.** Vigorous circular movements of the arm can lead to rotator cuff injuries. Two bones (the **radius** and **ulna**) contribute to the easy twisting motion of the forearm. The flexible hand contains bones of the wrist, palm, and five fingers.

The pelvic girdle contains two massive **coxal bones** that form a bowl, called the **pelvis,** and articulate with the longest and strongest bones of the body, the **femurs** (thighbones). The kneecap protects the knee. The **tibia** of the leg is the shinbone. The **fibula** is the more slender bone in the leg. Each foot contains bones of the ankle, instep, and five toes.

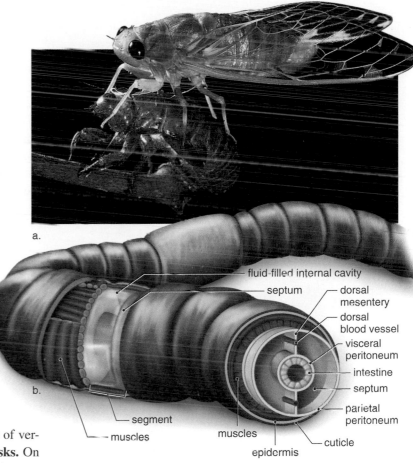

a.

fluid-filled internal cavity
septum
dorsal mesentery
dorsal blood vessel
visceral peritoneum
intestine
septum
parietal peritoneum
segment
muscles
muscles
cuticle
epidermis

b.

Figure 28.14   Types of skeletons.

a. Arthropods have an exoskeleton that must be shed as they grow. **b.** Worms have a hydrostatic skeleton, in which muscle contraction pushes against a fluid-filled internal cavity.

Check Your Progress

List the components of the appendicular skeleton.

Answer: The appendicular skeleton is composed of the shoulder girdle, the pelvic girdle, and their attached limbs.

Structure of a Bone

When a long bone is split open, as in Figure 28.15, a cavity is revealed that is bounded on the sides by compact bone. **Compact bone** contains many **osteons** (Haversian systems) where bone cells called **osteocytes** lie in tiny chambers arranged in concentric circles around central canals. The cells are separated by an extracellular matrix that contains mineral deposits, primarily calcium and phosphorous salts. Two other cell types are constantly at work in bones. **Osteoblasts** deposit bone, and **osteoclasts** secrete enzymes that digest the matrix of bone and release calcium into the bloodstream. When a person has **osteoporosis** (weak bones subject to fracture), osteoclasts are working harder than osteoblasts. Intake of high levels of dietary calcium, especially when a person is young and more active, encourages denser bones and lessens the chance of getting osteoporosis later in life.

A long bone has **spongy bone** at each end. Spongy bone has numerous bony bars and plates separated by irregular spaces. The spaces are often filled with **red bone marrow,** a specialized tissue that produces red blood cells. The cavity of a long bone is filled with yellow bone marrow and stores fat. Beyond the spongy bone are a thin shell of compact bone and a layer of hyaline cartilage, which is important to healthy joints.

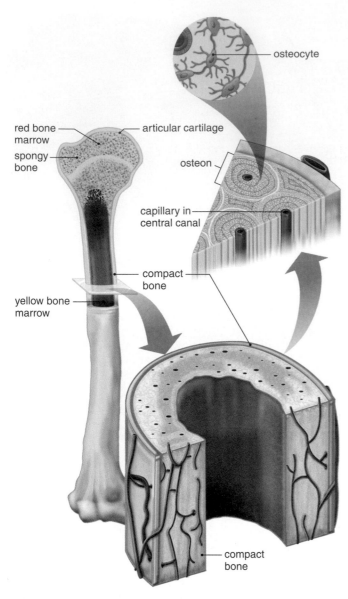

Figure 28.15   Structure of bone.

A bone has a central cavity that is usually filled with yellow bone marrow. Both spongy bone and compact bone are living tissues composed of bone cells within a matrix that contains calcium. The spaces of spongy bone contain red bone marrow.

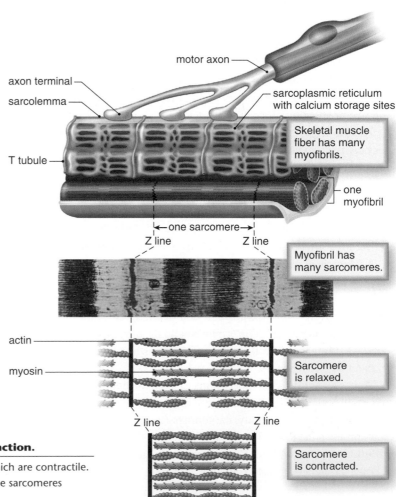

Figure 28.16   Skeletal muscle fiber structure and function.

A muscle fiber contains many myofibrils divided into sarcomeres, which are contractile. When innervated by a motor neuron, the myofibrils contract, and the sarcomeres shorten because actin filaments slide past the myosin filaments.

Skeletal Muscle Structure and Physiology

The three types of muscle—smooth, cardiac, and skeletal—have different structures (see Fig. 22.7). Smooth muscles contain sheets of long, spindle-shaped cells, each with a single nucleus. Cardiac cells are striated, each with a single nucleus. Cardiac muscle cells contain branched chains of cells that interconnect, forming a lattice network. Skeletal muscle cells, called muscle fibers, are elongated and run the length of a skeletal muscle. Skeletal muscle fibers arise during development when several cells fuse, resulting in one long, multinucleated cell. Skeletal muscle contraction has been extensively studied.

Skeletal Muscle Contraction

When skeletal muscle fibers contract, they shorten. Let's look at details of the process, beginning with the motor axon (Fig. 28.16). When nerve impulses travel down a motor axon and arrive at an axon terminal, synaptic vesicles release ACh (acetylcholine) into a synaptic cleft. ACh quickly diffuses across the cleft and binds to receptors in the plasma membrane of a muscle fiber, called the sarcolemma. The sarcolemma generates impulses that travel along its T tubules to the endoplasmic reticulum, which is called the *sarcoplasmic reticulum* in muscle fibers. The release of calcium from calcium storage sites causes muscle fibers to contract.

The contractile portions of a muscle fiber are many parallel, thread-like **myofibrils** (Fig. 28.16). An electron microscope shows that myofibrils (and therefore skeletal muscle fibers) are striated because of the placement of protein filaments within contractile units called **sarcomeres.** A sarcomere extends between two dark lines called the Z lines. Sarcomeres contain thick filaments made up of the protein **myosin** and thin filaments made up of the protein **actin.** As a muscle fiber contracts, the sarcomeres within the myofibrils shorten because actin (thin) filaments slide past the myosin (thick) filaments and approach one another. The movement of actin filaments in relation to myosin filaments is called the **sliding filament model** of muscle contraction. During the sliding process, the sarcomere shortens, even though the filaments themselves remain the same length.

Why the Filaments Slide The thick filament is a bundle of myosin molecules, each having a globular head with the capability of attaching to the actin filament when calcium (Ca^{2+}) is present. First, myosin binds to and hydrolyzes ATP. Thus, energized myosin heads attach to an actin filament. The release of ADP + P causes myosin to shift its position and pull the actin filaments to the center of the sarcomere. The action is similar to the movement of your hand when you flex your forearm (Fig. 28.17).

In the presence of another ATP, a myosin head detaches from actin. Now the heads attach at a location further along the actin filament. The cycle occurs again and again, and the actin filaments move nearer and nearer the center of the sarcomere each time the cycle is repeated. Contraction continues until nerve impulses cease and calcium ions are returned to their storage sites. The membranes of the sarcoplasmic reticulum contain active transport proteins that pump calcium ions back into the calcium storage sites, and the muscle relaxes.

Check Your Progress

List the three types of muscle.

Answer: Smooth muscle, cardiac muscle, and skeletal muscle.

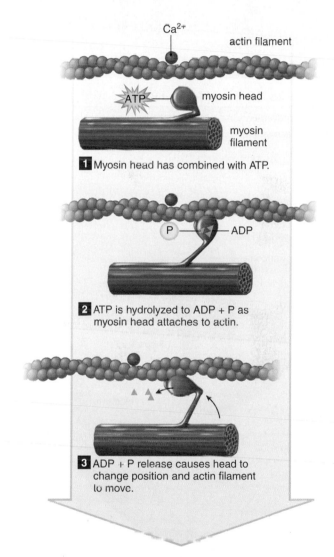

1 Myosin head has combined with ATP.

2 ATP is hydrolyzed to ADP + P as myosin head attaches to actin.

3 ADP + P release causes head to change position and actin filament to move.

Figure 28.17 Why a muscle shortens when it contracts.

The presence of calcium (Ca^{2+}) sets in motion a chain of events (1–3) that causes myosin heads to attach to and pull an actin filament toward the center of a sarcomere. After binding to other ATP molecules, myosin heads will return to their resting position. Then, the chain of events (1–3) occurs again, except that the myosin heads reattach further along the actin filament.

Figure 28.18   Muscles and bones.

Graceful movements are possible because tendons attach muscles to bones across a joint.

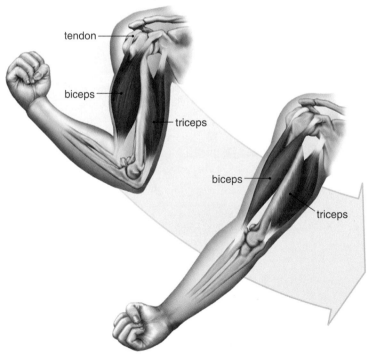

Figure 28.19   Action of muscles.

When muscles contract, they shorten. Therefore, muscles only pull—they cannot push. This causes them to work in antagonistic pairs; each member of the pair pulls on a bone in the opposite direction. For example, **(a)** when the biceps contracts, the forearm flexes (raises), and **(b)** when the triceps contracts, the forearm extends (lowers).

Muscles Move Bones at Joints

All of our movements, from those of graceful and agile ballet dancers to those of aggressive and skillful football players, occur because muscles are attached to bones by tendons that span joints (Fig. 28.18). Because muscles shorten when they contract, they have to work in antagonistic pairs. If one muscle of an antagonistic pair flexes the joint and raises the limb, the other extends the joint and straightens the limb. Figure 28.19 illustrates this principle with regard to the movement of the forearm at the elbow joint.

Figure 28.20*b* illustrates the anatomy of a freely movable **synovial joint**—that is, a joint having a cavity filled with synovial fluid, a lubricant for the joint. Ligaments connect bone to bone and support or strengthen a joint. The joint contains menisci (sing., **meniscus**), C-shaped pieces of hyaline cartilage between the bones. These give added stability and act as shock absorbers. Fluid-filled sacs called bursae (sing., **bursa**) ease friction between bare areas of bone and overlapping muscles, or between skin and tendons.

The **ball-and-socket joints** at the hips and shoulders allow movement in all planes, even rotational movement (Fig. 28.20*c*). The elbow and knee joints are synovial joints called **hinge joints** because, like a hinged door, they largely permit movement in one direction only (Fig. 28.20*d*).

Joint Disorders   *Sprains* occur when ligaments and tendons are overstretched at a joint. For example, a sprained ankle can result if you turn your ankle too much. Overuse of a joint may cause inflammation of a bursa, called *bursitis.* Tennis elbow is a form of bursitis. Cartilage injuries, often called *torn cartilage,* involve the tearing of knee menisci. Because fragments of menisci can interfere with joint movements, most physicians believe they should be removed. Today, arthroscopic surgery is possible to remove cartilage fragments or to repair ligaments or cartilage. A small instrument bearing a tiny lens and light source is inserted into a joint, as are the surgical instruments. Fluid is then added to distend the joint and allow visualization of its structure. Usually, the surgery is displayed on a monitor so that the whole operating team can see the operation. Arthroscopy is much less traumatic than surgically opening the knee with long incisions. The benefits of arthroscopy include small incisions, faster healing, a more rapid recovery, and less scarring. Because arthroscopic surgical procedures are often performed on an outpatient basis, the patient is able to return home on the same day.

Rheumatoid arthritis, discussed on page 465, is not as common as osteoarthritis (OA), which is the deterioration of an overworked joint. Constant compression and abrasion continually damage articular cartilage, and eventually it softens, cracks, and wears away entirely in some areas. As the disease progresses, the exposed bone thickens and forms spurs that cause the bone ends to enlarge and restrict joint movement. Weight loss can ease arthritis. Taking off 3 pounds can reduce the load on a hip or knee joint by 9 to 15 pounds. A sensible exercise program helps build up muscles, which stabilize joints. Low-impact activities, such as biking and swimming, are best.

Today, replacement of damaged joints with a prosthesis (artificial substitute) is often possible. Some people have found glucosamine-chondroitin supplements beneficial as an alternative to joint replacement. Glucosamine, an amino sugar, is thought to promote the formation and repair of cartilage. Chondroitin, a carbohydrate, is a cartilage component that is thought to promote water retention and elasticity and to inhibit enzymes that break down cartilage. Both compounds are naturally produced by the body.

Figure 28.20 **Synovial joints.**

a. The synovial joints of the human skeleton allow the body to be flexible and move with precision even when bearing a weight. **b.** Generalized synovial joint. Problems arise when menisci or ligaments are torn, bursae become inflamed, and articular cartilage wears away. **c.** The shoulder is a ball-and-socket joint that permits movement in three planes. **d.** The elbow is a hinge joint that permits movement in a single plane.

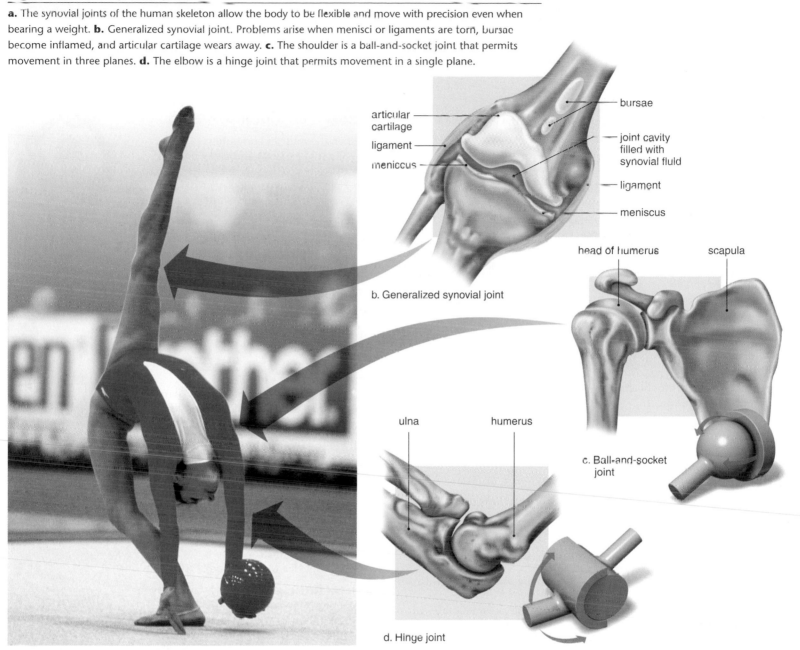

a. A gymnast depends on flexible joints.

articular cartilage

ligament

meniscus

bursae

joint cavity filled with synovial fluid

ligament

meniscus

b. Generalized synovial joint

head of humerus

scapula

c. Ball-and-socket joint

ulna

humerus

d. Hinge joint

Exercise Exercise improves muscular strength, muscular endurance, and flexibility. It improves cardiorespiratory endurance and may lower the blood cholesterol level. People who exercise are less likely to develop various types of cancer. Exercise promotes the activity of osteoblasts, and therefore helps prevent osteoporosis (see page 504). It helps prevent weight gain, not only because of increased activity, but also because as muscle mass increases, the body is less likely to accumulate fat. Exercise relieves depression and enhances the mood. A sensible exercise program provides all these benefits without the detriments of a too-strenuous program.

Check Your Progress

1. Explain why a muscle shortens when it contracts.
2. Contrast a hinge joint with a ball-and-socket joint.

Answers: 1. The globular heads of myosin molecules attach to actin filaments and pull them toward the center of the sarcomere. Then ATP allows the heads to be released, and they reattach at a new location along the actin filament, resulting in contraction. 2. A hinge joint, such as a knee or elbow, allows movement mainly in one direction. A ball-and-socket joint, such as a hip or shoulder, allows movement in all planes.

Summary

28.1   The Senses

All living things respond to stimuli. In animals, stimuli generate nerve impulses that often go to a CNS, which integrates the information before initiating motor responses:

Stimulus ⟶ Nerve impulse ⟶ CNS ⟶ Motor response

Chemical Senses

Chemoreception is found universally in animals and is therefore believed to be the most primitive sense. Human taste buds and olfactory cells are chemoreceptors.

- Taste buds have microvilli with receptors that bind to chemicals in food.
- Olfactory cells have cilia with receptors that bind to odor molecules.

The Senses of Hearing and Balance

Mammals have an ear that may have evolved from the lateral line of fishes.

The sensory receptors for hearing are hair cells with stereocilia that respond to pressure waves.

- Hair cells respond to stimuli that have been received by the outer ear and amplified by the ossicles in the middle ear.
- Hair cells are found in the spiral organ and are located in the cochlear canal of the cochlea. The spiral organ generates nerve impulses that travel to the brain.

The sensory receptors for balance (equilibrium) are also hair cells with stereocilia.
- Hair cells in the base of the semicircular canals provide rotational equilibrium.
- Hair cells in the utricle and saccule provide gravitational equilibrium.

The Sense of Vision

Arthropods have a compound eye; squids and humans have a camera-type eye. In humans, the photoreceptors:

- Respond to light that has been focused by the cornea and lens.
- Consist of two types, rods and cones. In rods, rhodopsin splits into opsin and retinal.
- Communicate with the next layer of cells in the retina. Integration occurs in the three layers of the retina before nerve impulses go to the brain.

Cutaneous Receptors and Proprioceptors

These receptors communicate with the primary sensory area of the brain. They consist of receptors for hot, cold, pain, touch, and pressure (cutaneous) and stretching (proprioceptors).

28.2   The Motor Systems

Together, the muscles and bones:
- Support the body and allow parts to move.
- Help protect internal organs.
- Assist the functioning of other systems.

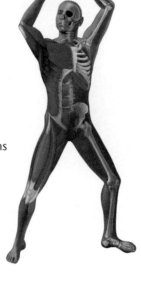

In addition:
- Skeletal muscle contraction assists movement of blood in cardiovascular veins and lymphatic vessels.
- Skeletal muscles provide heat that warms the body.
- Bones are storage areas for calcium and phosphorous salts, as well as sites for blood cell formation.

The Skeleton

Arthropods have an exoskeleton, and molting is needed to replace it as the animal grows. Humans have an endoskeleton that grows with them. Worms have a hydrostatic skeleton.

The human skeleton is divided into two parts:
- The axial skeleton is made up of the skull, vertebral column, sternum, and ribs.
- The appendicular skeleton is composed of the girdles and their appendages.

Bone contains:
- Compact bone (site of calcium storage) and spongy bone (site of red bone marrow).
- A cavity (site of fat storage).

Skeletal Muscle

Skeletal muscle is one of the three types of muscles:
- Smooth muscle is composed of spindle-shaped cells that form a sheet.
- Cardiac muscle has striated cells that form a lattice network.
- Skeletal muscle has tubular cells (fibers) that run the length of the muscle.

In a skeletal muscle cell:
- Myofibrils, myosin filaments, and actin filaments are arranged in a sarcomere.
- Myosin filaments pull actin filaments, and sarcomeres shorten.

Skeletal muscles:
- Move bones at joints.
- Work in antagonistic pairs.
- Have freely movable synovial joints that are subject to a number of disorders, such as sprains, bursitis, and torn ligaments and cartilage.

Thinking Scientifically

1. The two leading causes of blindness are age-related macular degeneration and diabetic retinopathy. Both are characterized by the development of an abnormally high number of blood vessels (angiogenesis) in the retina. Why do you suppose angiogenesis impairs vision? Progress in cancer research has led to new strategies for the treatment of these eye diseases. Do you see a connection between the causes of blindness and cancer?

2. Human genome researchers have found a family of approximately 80 genes that encode receptors for bitter-tasting compounds. The genes encode proteins made in taste receptor cells of the tongue. Typically, many types of taste receptors are expressed per cell on the tongue. However, in the olfactory system, each cell expresses only one of the 1,000 olfactory receptor genes. Different cells express different genes. How do you suppose this difference affects our ability to taste versus smell different chemicals? Why might the two systems have evolved such different patterns of gene expression?

Testing Yourself

Choose the best answer for each question.

1. The most primitive sense is probably the ability to detect
   a. light.
   b. sound.
   c. chemicals.
   d. pressure.

2. Label the parts of the ear in the following illustration.

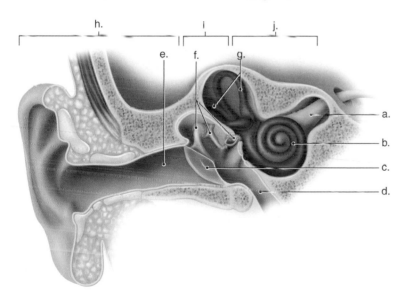

3. Loud noises generally lead to hearing loss due to damage to the
   a. outer ear.
   b. middle ear.
   c. inner ear.

4. The function of the lateral line system in fishes is to
   a. detect sound.
   b. locate other fish.
   c. maintain gravitational equilibrium.
   d. maintain rotational equilibrium.

5. The human eye focuses by
   a. changing the thickness of the lens.
   b. changing the shape of the lens.
   c. opening and closing the pupil.
   d. rotating the lens.

6. A blind spot occurs where the
   a. iris meets the pupil.
   b. retina meets the lens.
   c. optic nerve meets the retina.
   d. cornea meets the retina.

7. Label the parts of the eye in the following illustration.

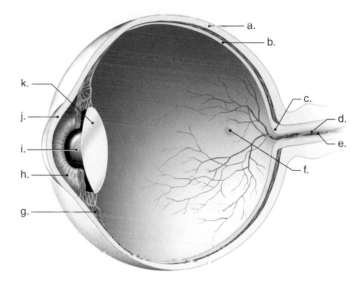

8. Cold receptors differ from warmth receptors in
   a. abundance.
   b. shape.
   c. structure.
   d. All of these are correct.

9. Unlike an exoskeleton, an endoskeleton
   a. grows with the animal.
   b. is composed of chitin.
   c. is jointed.
   d. protects internal organs.

10. A component of the appendicular skeleton is the
    a. rib cage.
    b. skull.
    c. femur.
    d. vertebral column.

For questions 11–14, identify the type of muscle in the key that matches the description. Some answers may be used more than once. Some questions may have more than one answer.

Key:

a. smooth
b. cardiac
c. skeletal

11. Composed of cells with one nucleus each.
12. Composed of fibers that result when many cells fuse.
13. Composed of a lattice network of branched cells.
14. Striated.
15. The deterioration of a joint over time can cause

a. rheumatoid arthritis.
b. a sprain.
c. bursitis.
d. tendonitis.

Bioethical Issue

A lesbian couple, both of whom are deaf, sought out a deaf sperm donor in order to have a deaf child. The women considered deafness to be a cultural identity rather than a disability. They felt that they would be better parents to a deaf child because they understood the "culture" of deafness. Should this couple have been allowed to choose a father based on his ability to contribute genes for deafness? In a situation like this, whose rights should prevail, the parents' rights to choose their mate or the child's rights to have a chance for normal hearing? Is deafness a cultural identity or a disability?

Understanding The Terms

actin   505
appendicular skeleton   503
axial skeleton   503
ball-and-socket joints   506
blind spot   500
bursa   506
camera-type eye   499
clavicle   503
compact bone   504
compound eye   499
conduction deafness   497
cone   500
cornea   499
coxal bones   503
cranium   503
cutaneous receptor   501
endoskeleton   502
exoskeleton   502
facial bones   503
femurs   503
fibula   503
fovea   500
Golgi tendon   501
gravitational equilibrium   498
hinge joint   506
hip (pelvic) girdle   503
humerus   503
hydrostatic skeleton   503
intervertebral disks   503
iris   499
lateral line   498
lens   499
ligaments   506
meniscus   506
muscle spindle   501

myofibril   505
myosin   505
nerve deafness   497
osteoblast   504
osteoclast   504
osteon   504
osteoporosis   504
photoreceptor   499
proprioceptor   501
pupil   499
radius   503
red bone marrow   504
retina   499
rhodopsin   500
rib cage   503
rod   500
rotational equilibrium   498
rotator cuff   503
saccule   498
sacrum   503
sarcomere   505
scapula   503
shoulder (pectoral) girdle   503
skull   503
sliding filament model   505
spiral organ   496
spongy bone   504
statocyst   498
sternum   503
synovial joint   506
taste buds   495
tibia   503
ulna   503
utricle   498
vertebral column   503

Match the terms to these definitions:

a. _____ Organ that contains the sensory receptors for hearing.

b. _____ Gravitational equilibrium organs found in several types of invertebrates.

c. _____ Region of the eye that regulates the amount of light that enters.

d. _____ Portion of the retina that contains cones.

e. _____ Portion of the axial skeleton that extends from the skull to the sacrum.

f. _____ Cell type that deposits bone tissue.

g. _____ Contractile portion of muscle tissue.

h. _____ Thick filaments in a sarcomere.

i. _____ Joint with a cavity filled with lubricant.

j. _____ Hyaline cartilage between bones.

Reproduction and Development

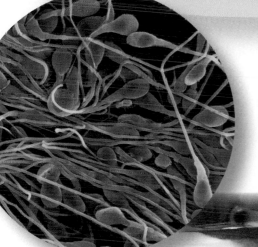

For a male to be fertile, the ejaculate usually needs to contain 40 million sperm per milliliter.

A human fetus can be viable after only 22 weeks of development (compared to the normal 40 weeks) but a premature infant has a greater risk of illness.

One out of five young women may have a sexually transmitted disease caused by herpes viruses. Babies born prematurely to these women are at greater risk of contracting the disease during birth.

Most people know how sexually transmitted diseases (STDs) are contracted, but many do not understand how they are treated or their long-term consequences. Bacterial STDs, such as chlamydia and gonorrhea, are very easy to treat, but when untreated can have lasting effects, including permanent infertility. Many people are surprised to find that they have become infertile as the result of an STD. If these diseases are easy to treat, why would anyone not be ▶ ▶ ▶

treated and take the risk of becoming infertile? There are many reasons, but one of the most important considerations is that many people with an STD do not have any symptoms of the infection. As long as everything appears normal, they do not seek treatment; meanwhile, they may pass the disease to others, and run the risk of long-term complications. This is the primary reason that testing for STDs (even when everything "seems OK") is so important. In addition to the bacterial STDs, several viral STDs are of concern—not only the well-known human immunodeficiency virus (HIV) but also human papillomavirus (HPV). HPV is the primary cause of cervical cancer (the cervix is at the entrance to the uterus). There are usually no noticeable signs of early cervical cancer, but it can be detected by having yearly checkups. Treatment for cervical cancer can result in the inability to have children if the cervix and the uterus must be surgically removed.

Reproduction is a fascinating topic with relevance not only to STDs, but also to birth control, reproductive technology, and development. In this chapter, you will learn about reproduction, including some unique issues faced by humans. You will also learn about the portion of human development that occurs before birth and about the birth process itself.

29.1 How Animals Reproduce

Animals usually reproduce sexually, but some, on occasion, can reproduce asexually. In asexual reproduction, there is only one parent, and in sexual reproduction there are two parents.

Asexual Versus Sexual Reproduction

Hydras can reproduce by budding. A new individual arises as an outgrowth (bud) of the parent (Fig. 29.1). Many flatworms can constrict into two halves; each half regenerates to become a new individual. Fragmentation, followed by **regeneration,** is also seen among sponges and echinoderms. Chopping up a sea star does not kill it; instead, each fragment grows into another animal.

Parthenogenesis is a modification of sexual reproduction in which an unfertilized egg develops into a complete individual. In honeybees, the queen bee can fertilize eggs or allow eggs to pass unfertilized as she lays them. The fertilized eggs become diploid females called workers, and the unfertilized eggs become haploid males called drones. Parthenogenesis is also observed in some amphibian species.

In sexual reproduction, animals usually produce gametes in specialized organs called **gonads.** The gonads are **testes,** which produce sperm, and **ovaries,** which produce eggs. Eggs or sperm are derived from germ cells that become specialized for this purpose during early development. During sexual reproduction, the egg of one parent is usually fertilized by the sperm of another, and a **zygote** (fertilized egg) results. Even among earthworms, which are **hermaphroditic**—each worm has both male and female sex organs—cross-fertilization occurs. In coral reef fishes called wrasses, a male has a harem of several females. If the male dies, the largest female becomes a male. Many aquatic animals practice **external fertilization**—that is, eggs and sperm join outside the body in the water (Fig. 29.2).

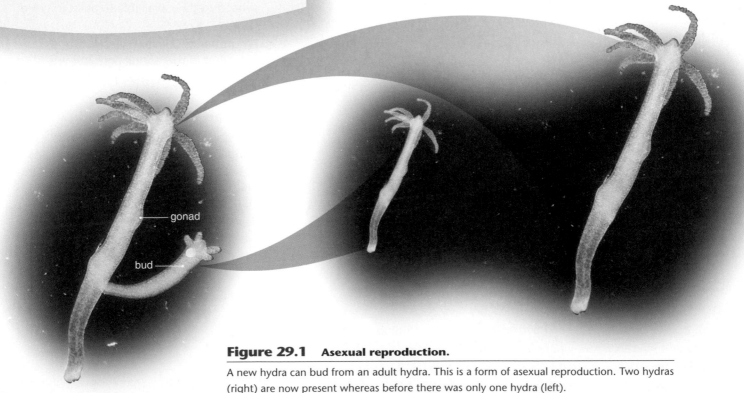

gonad

bud

Figure 29.1   Asexual reproduction.

A new hydra can bud from an adult hydra. This is a form of asexual reproduction. Two hydras (right) are now present whereas before there was only one hydra (left).

Copulation is sexual union to facilitate the reception of sperm by a female. In terrestrial animals, males typically have a penis for depositing sperm into the vagina of females (Fig. 29.3). Aquatic animals also have copulatory organs. Lobsters and crayfish have modified swimmerets. Among terrestrial animals, birds lack a penis and vagina. They have a cloaca, a chamber that receives products from the digestive, urinary, and reproductive tracts. A male transfers sperm to a female after placing his cloacal opening against hers.

Reproduction in Water Versus on Land

Many aquatic animals have a larval stage, an immature form capable of feeding. Since the larva has a different lifestyle, it is able to use a different food source than the adult. In sea stars, the bilaterally symmetrical larva undergoes metamorphosis to become a radially symmetrical juvenile. Crayfish, on the other hand, do not have a larval stage; the egg hatches into a tiny juvenile with the same form as the adult.

Reptiles, and particularly birds, provide their eggs with plentiful yolk, a rich nutrient material, and they have no larval stage. Complete development takes place within a shelled egg containing **extraembryonic membranes** to serve the needs of the embryo and prevent drying out. The shelled egg frees these animals from the need to reproduce in the water and is a significant adaptation to the terrestrial environment.

Birds, in particular, tend their eggs, and newly hatched birds usually have to be fed before they are able to fly away and seek food for themselves. Complex hormones and neural regulation are involved in the reproductive behavior of parental birds. Other animals take a different course. They do not deposit and tend their eggs; instead, they are *ovoviviparous,* meaning that their eggs are retained in the body until they hatch, releasing fully developed offspring that have a way of life like the parent. Oysters, which are molluscs, retain their eggs in the mantle cavity, and male sea horses, which are vertebrates, have a special brood pouch in which the eggs develop. Garter snakes, water snakes, and pit vipers retain their eggs in their bodies until they hatch, and thus give birth to living young.

Finally, most mammals are *viviparous,* meaning that they produce living young. After offspring are born, the mother supplies the nutrients needed for further growth. Viviparity represents the ultimate in caring for the unborn. How did viviparity among certain mammals come about? Some mammals, such as the duckbill platypus and the spiny anteater, lay eggs. In contrast, marsupial offspring are born in a very immature state; they finish their development within a pouch, where they are nourished on milk. Only among primates, including humans, does a single embryo usually develop. The **placenta** is a complex structure derived, in part, from the chorion, which first appears in a shelled egg. The evolution of the placenta allowed the developing offspring to exchange materials with the mother internally.

Figure 29.2   Reproducing in the water.

Animals that reproduce in the water have no need to protect their eggs and embryos from drying out. Here, male and female frogs are mating. They deposit their gametes in the water, where fertilization takes place. The egg contains yolk that nourishes the embryo until it is a free-swimming larva that can feed for itself.

Figure 29.3   Reproducing on land.

Animals that reproduce on land need to protect their gametes and embryos from drying out. Here, the male passes sperm to the female by way of a penis, and the developing embryo/fetus will remain in the female's body until it is capable of living independently.

29.2 Human Reproduction

In human males and females, the reproductive system consists of two components: (1) the gonads, either testes or ovaries, which produce gametes and sex hormones, and (2) accessory organs that conduct gametes, and in the case of the female, house the embryo/fetus.

Male Reproductive System

The human male reproductive system includes the testes (sing., testis), the epididymis (pl., epididymides), the vas deferens (pl., vasa deferentia), and the urethra (Fig. 29.4). (The urethra in males is a part of both the urinary system and the reproductive system.) The paired **testes,** which produce sperm, are suspended within the scrotum. The testes begin their development inside the abdominal cavity, but they descend into the scrotum as development proceeds. If the testes do not descend—and the male does not receive hormone therapy or undergo surgery to place the testes in the scrotum—sterility (the inability to produce offspring) results. This type of sterility occurs because normal sperm production is inhibited at body temperature; a slightly cooler temperature is required.

Sperm produced by the testes mature within the **epididymis,** a coiled tubule lying just outside each testis. Maturation seems to be required for the sperm to swim to the egg. Once the sperm have matured, they are propelled into the **vas deferens** by muscular contractions. Sperm are stored in both the epididymis and the vas deferens. When a male becomes sexually aroused, sperm enter first the ejaculatory duct and then the urethra, part of which is located within the penis.

The **penis** is a cylindrical organ that usually hangs in front of the scrotum. Three cylindrical columns of spongy, erectile tissue containing distensible blood spaces extend through the shaft of the penis (Fig. 29.4*b*). During sexual arousal, nervous reflexes cause an increase in arterial blood flow to the penis. This

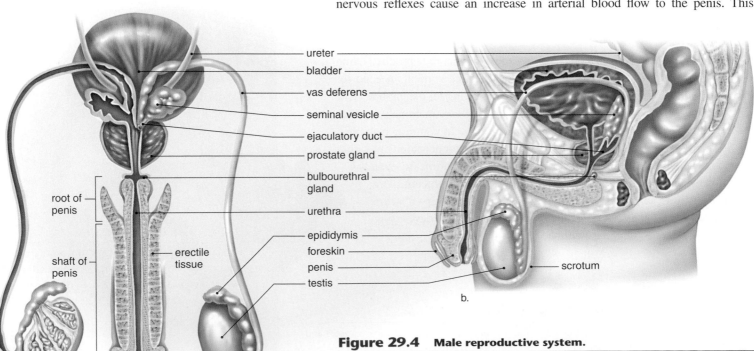

a.

b.

Figure 29.4 **Male reproductive system.**

a. Side view. **b.** Frontal view. The testes produce sperm. The seminal vesicles, the prostate gland, and the bulbourethral glands provide a fluid medium for the sperm. Circumcision is the removal of the foreskin.

increased blood flow fills the blood spaces in the erectile tissue, and the penis, which is normally limp (flaccid), stiffens and increases in size. These changes are called an **erection.** If the penis fails to become erect, the condition is called erectile disfunction. The drug Viagra works by increasing blood flow to the penis so that when a man is sexually excited, he can get and keep an erection.

Semen (seminal fluid) is a thick, whitish fluid that contains sperm and secretions from three glands: the seminal vesicles, prostate gland, and bulbourethral glands. The **seminal vesicles** lie at the base of the bladder. Each joins a vas deferens to form an ejaculatory duct that enters the urethra. As sperm pass from the vas deferens into the ejaculatory duct, these vesicles secrete a thick, viscous fluid containing nutrients for possible use by the sperm. Just below the bladder is the **prostate gland,** which secretes a milky alkaline fluid believed to activate or increase the motility of the sperm. In older men, the prostate gland may become enlarged, thereby constricting the urethra and making urination difficult. Also, prostate cancer is the most common form of cancer in men. Slightly below the prostate gland, on either side of the urethra, is a pair of small glands called **bulbourethral glands,** which have mucous secretions with a lubricating effect. Notice from Figure 29.4 that the urethra also carries urine from the bladder during urination.

If sexual arousal reaches its peak, ejaculation follows an erection. The first phase of ejaculation is called emission. During emission, the spinal cord sends nerve impulses via appropriate nerve fibers to the epididymis and vas deferens. Their subsequent motility causes sperm to enter the ejaculatory duct, whereupon the seminal vesicles, prostate gland, and bulbourethral glands release their secretions. Secretions from the bulbourethral glands occur first and may or may not contain sperm.

During the second phase of ejaculation, called expulsion, rhythmic contractions of muscles at the base of the penis and within the urethral wall expel semen in spurts from the opening of the urethra. These contractions are an example of release from muscle tension. An erection lasts for only a limited amount of time. The penis then returns to its normal flaccid state. Following ejaculation, a male may typically experience a time, called the refractory period, during which stimulation does not bring about an erection. The contractions that expel semen from the penis are a part of male **orgasm,** the physiological and psychological sensations that occur at the climax of sexual stimulation.

The Testes

A longitudinal section of a testis shows that it is composed of compartments called lobules, each of which contains one to three tightly coiled **seminiferous tubules** (Fig. 29.5a,b). A microscopic cross section of a seminiferous tubule reveals that it is packed with cells undergoing spermatogenesis, a process that involves reducing the chromosome number from diploid (2n) to haploid (n). Also present are *Sertoli cells*, which support, nourish, and regulate the production of sperm (Fig. 29.5c). A sperm has three distinct parts: a head, a middle piece, and a tail. The head contains a nucleus and is capped by a membrane-bounded acrosome. The tail is a flagellum that allows sperm to swim toward the egg, and the middle piece contains energy-producing mitochondria.

The ejaculated semen of a normal human male contains 40 million sperm per milliliter, ensuring an adequate number for fertilization to take place. Fewer than 100 sperm ever reach the vicinity of the egg, however, and only one sperm normally enters an egg.

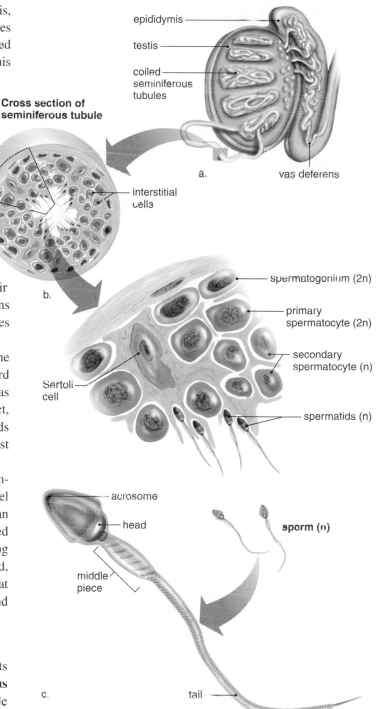

Figure 29.5 Seminiferous tubules.

a. The testes contain seminiferous tubules, where sperm are produced. **b.** Cross section of a tubule. As spermatogenesis occurs, the chromosome number is reduced to the haploid number. **c.** A sperm has a head, a middle piece, and a tail. The nucleus is in the head, which is capped by an acrosome.

Figure 29.6 **Female reproductive system.**

a. Frontal view of the female reproductive system. The ovaries produce one oocyte (egg) per month. Fertilization occurs in the oviduct, and development occurs in the uterus. The vagina is the birth canal and organ of sexual intercourse. **b.** Side view of female reproductive system plus nearby organs. **c.** Female external genitals, the vulva. At birth, the opening of the vagina is partially occluded by a membrane called the hymen. Physical activities and sexual intercourse disrupt the hymen.

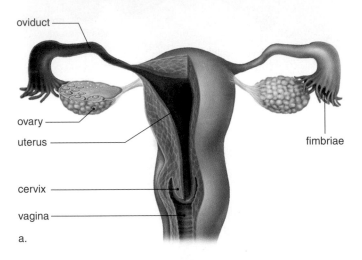

oviduct

ovary

uterus

cervix

vagina

fimbriae

a.

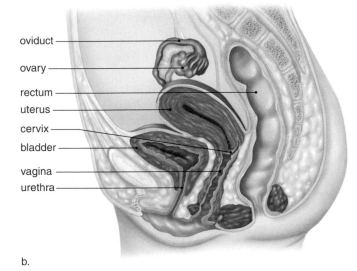

oviduct

ovary

rectum

uterus

cervix

bladder

vagina

urethra

b.

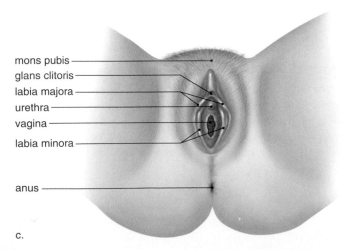

mons pubis

glans clitoris

labia majora

urethra

vagina

labia minora

anus

c.

Hormonal Regulation in Males

The hypothalamus has ultimate control of the testes' sexual function because it secretes a hormone called gonadotropin-releasing hormone, or GnRH, that stimulates the anterior pituitary to produce the gonadotropic hormones. Both males and females have two gonadotropic hormones—follicle-stimulating hormone (FSH) and luteinizing hormone (LH). In males, FSH promotes spermatogenesis in the seminiferous tubules. LH in males was formerly called interstitial cell-stimulating hormone (ICSH) because it controls the production of testosterone by the interstitial cells, which are scattered in the spaces between the seminiferous tubules.

Testosterone, the main sex hormone in males, is essential for the normal development and functioning of the sexual organs. Testosterone is also necessary for the maturation of sperm. In addition, testosterone brings about and maintains the male **secondary sex characteristics** that develop at the time of **puberty,** the time of life when sexual maturity is attained. Males are generally taller than females and have broader shoulders and longer legs relative to trunk length. The deeper voice of males compared to females is due to males having a larger larynx with longer vocal cords. Because the so-called Adam's apple is a part of the larynx, it is usually more prominent in males than in females.

Testosterone causes males to develop noticeable hair on the face, chest, and occasionally other regions of the body, such as the back. Testosterone also leads to the receding hairline and pattern baldness that occur in males. Testosterone is responsible for the greater muscular development in males. Knowing this, males and females sometimes take anabolic steroids (either the natural or synthetic form of testosterone) to build up their muscles. This practice can cause damage to the liver, kidneys, and cardiovascular system.

Check Your Progress

1. Describe the changes that occur in the penis during an erection.
2. What is the main sex hormone in males?

Answers: 1. Blood flow increases to the blood spaces in three columns of erectile tissue in the penis, causing it to stiffen and enlarge. 2. Testosterone.

Female Reproductive System

The human female reproductive system includes the ovaries, the oviducts, the uterus, and the vagina (Fig. 29.6a,b). The oviducts, also called uterine or fallopian tubes, extend from the ovaries to the uterus; however, the oviducts are not attached to the ovaries. Instead, the oviducts have fingerlike projections called **fimbriae (sing., fimbria)** that sweep over the ovaries. When an oocyte bursts from an ovary during ovulation, it usually is swept into an oviduct by the combined action of the fimbriae and the beating of cilia that line the oviducts. Fertilization, if it occurs, normally takes place in an oviduct, and the developing embryo is propelled slowly by ciliary movement and tubular muscle contraction to the uterus. The **uterus** is a thick-walled muscular organ about the size and shape of an inverted pear. The narrow end of the uterus is called the **cervix.** An embryo completes its development after embedding itself in the uterine lining, called the **endometrium.** A small opening at the cervix leads to the vaginal canal. The vagina is a tube at a 45° angle

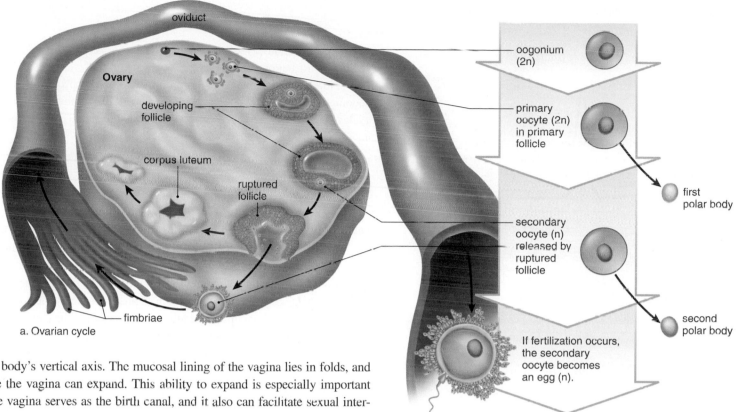

a. Ovarian cycle

b. Oogenesis

Figure 29.7   Ovarian cycle and oogenesis.

As a follicle matures in the ovary, the oocyte enlarges and is surrounded by layers of follicular cells and fluid. Eventually, ovulation occurs, the mature follicle ruptures, and the secondary oocyte is released. A single follicle actually goes through all the stages in one place within the ovary.

with the body's vertical axis. The mucosal lining of the vagina lies in folds, and therefore the vagina can expand. This ability to expand is especially important when the vagina serves as the birth canal, and it also can facilitate sexual intercourse, when the penis is inserted into the vagina.

The external genital organs of a female are known collectively as the vulva (Fig. 29.6c). The mons pubis and two folds of skin called labia minora and labia majora are on either side of the urethral and vaginal openings. At the juncture of the labia minora is the clitoris, which is homologous to the penis of males. The clitoris has a shaft of erectile tissue and is capped by a pea-shaped glans. The many sensory receptors of the clitoris allow it to function as a sexually sensitive organ.

The Ovaries

An oogonium in the ovary gives rise to an oocyte surrounded by epithelium (Fig. 29.7). This is called a primary **follicle.** An ovary contains many primary follicles, each containing an oocyte. At birth, a female has as many as 2 million primary follicles, but the number is reduced to 300,000–400,000 by the time of puberty. Only a small number of primary follicles (about 400) ever mature and produce a secondary oocyte. When mature, the follicle balloons out on the surface of the ovary and bursts, releasing the secondary oocyte surrounded by follicle cells. The release of a secondary oocyte from a mature follicle is termed **ovulation.** Oogenesis is completed when and if the secondary oocyte is fertilized by a sperm. A follicle that has lost its oocyte develops into a **corpus luteum.** If fertilization and pregnancy do not occur, the corpus luteum begins to degenerate after about ten days.

The ovarian cycle is controlled by the gonadotropic hormones FSH and LH (Fig. 29.8). During the first half, or **follicular phase** (pink), of the cycle, FSH promotes the development of follicles that primarily secrete estrogens, called *estrogen* for the sake of simplicity. As the blood level of estrogen rises, it exerts feedback control over FSH secretion, and ovulation occurs. Ovulation marks the end of the follicular phase. During the second half, or **luteal phase** (yellow), of the ovarian cycle, LH promotes the development of a corpus luteum, which primarily secretes progesterone. As the blood level of progesterone rises, it exerts feedback control over LH secretion so that the corpus luteum

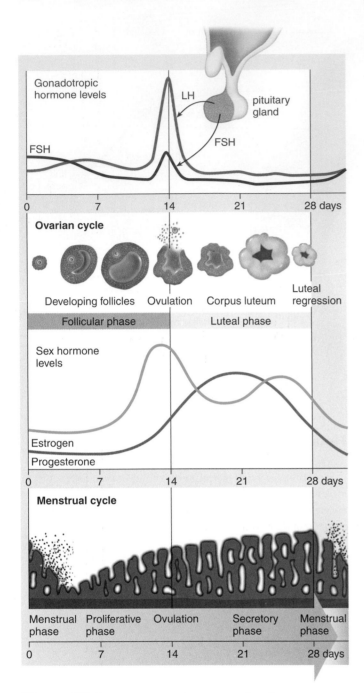

Figure 29.8   Ovarian and menstrual cycles.

During the follicular phase of the ovarian cycle (pink), FSH released by the anterior pituitary promotes the maturation of follicles in the ovary. Ovarian follicles produce increasing levels of estrogen, which causes the endometrium to thicken during the proliferative phase of the menstrual cycle (*bottom*). After ovulation and during the luteal phase of the ovarian cycle (yellow), LH promotes the development of a corpus luteum. This structure produces increasing levels of progesterone, which causes the endometrium to become secretory. Menstruation begins when progesterone production declines to a low level.

begins to degenerate if fertilization does not occur. As the luteal phase comes to an end, menstruation occurs.

Notice that the female sex hormones estrogen and progesterone affect the endometrium of the uterus, causing the series of events known as the **menstrual cycle** (Fig. 29.8, *bottom*). The twenty-eight-day menstrual cycle in the nonpregnant female is divided as follows:

During *days 1–5,* female sex hormones are at a low level in the body, causing the endometrium to disintegrate and its blood vessels to rupture. A flow of blood, known as the **menses,** passes out of the vagina during **menstruation,** also known as the menstrual period.

During *days 6–13,* increased production of estrogen by ovarian follicles causes the endometrium to thicken and to become vascular and glandular. This is called the proliferative phase of the menstrual cycle.

Ovulation usually occurs on the fourteenth day of the 28-day cycle.

During *days 15–28,* increased production of progesterone by the corpus luteum causes the endometrium to double in thickness and the uterine glands to mature, producing a thick mucoid secretion. This is called the secretory phase of the menstrual cycle. The endometrium now is prepared to receive the developing embryo. But if fertilization does not occur and no embryo embeds itself, the corpus luteum degenerates, and the low level of sex hormones in the female body causes the endometrium to break down. Menses begins, marking day one of the next cycle. Even while menstruation is occurring, the anterior pituitary begins to increase its production of FSH, and new follicles begin to mature.

Estrogen and Progesterone

Estrogen and **progesterone** are the female sex hormones. Estrogen in particular, is essential for the normal development and functioning of the female reproductive organs. Estrogen is also largely responsible for the secondary sex characteristics in females, including body hair and fat distribution. In general, females have a more rounded appearance than males because of a greater accumulation of fat beneath their skin. Also, the pelvic girdle realigns so that females have wider hips than males, and the thighs converge at a greater angle toward the knees. Both estrogen and progesterone are required for breast development as well.

Menopause, which usually occurs between ages 45 and 55, is the time in a woman's life when the ovarian and menstrual cycles cease. Menopause is not complete until menstruation has been absent for a year.

Check Your Progress

Relate the phases of the ovarian cycle to the phases of the menstrual cycle.

Answer: (1) During the follicular phase of the ovarian cycle, a follicle is producing an egg and releasing estrogen. Estrogen causes the uterine lining to thicken (proliferative phase of menstrual cycle). (2) Ovulation occurs. (3) During the luteal phase of the ovarian cycle, the corpus luteum is producing progesterone. Progesterone causes the uterine lining to become secretory (secretory phase of menstrual cycle). (4) If the egg is not fertilized by a sperm and an embryo does not implant in the uterine lining, the corpus luteum degenerates, and menstruation occurs (menstrual phase of menstrual cycle).

Aspects of Reproduction

Important topics related to human reproduction include control of reproduction, infertility, and sexually transmitted diseases.

Control of Reproduction

Table 29.1 lists various means of birth control and gives their rate of effectiveness. Figure 29.9 illustrates some birth control devices. **Abstinence**—that is, not engaging in sexual intercourse— is very reliable and has the added advantage of preventing sexually transmitted diseases. Oral contraception (the **birth control pill**) often involves taking a combination of estrogen and progesterone on a daily basis. These hormones effectively shut down the pituitary production of both FSH and LH so that no follicle in the ovary begins to develop in the ovary; since ovulation does not occur, pregnancy cannot take place. Because of possible side effects, women taking birth control pills should see a physician regularly.

Contraceptive implants utilize a synthetic progesterone to prevent ovulation by disrupting the ovarian cycle. The older version of the implant consists of six match-sized, time-release capsules that are surgically implanted under the skin of a woman's upper arm. The newest version consists of a single capsule that remains effective for about three years.

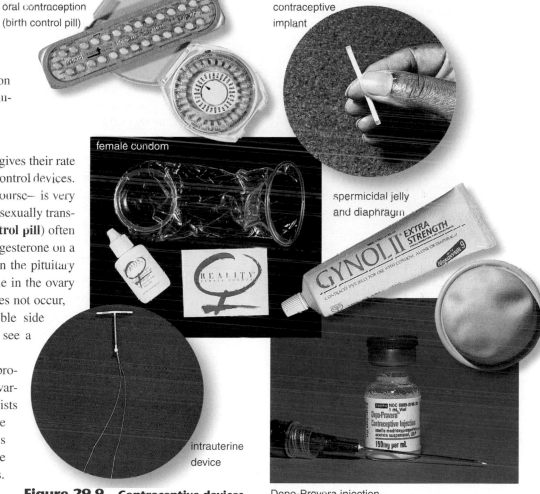

oral contraception (birth control pill)

contraceptive implant

female condom

spermicidal jelly and diaphragm

intrauterine device

Depo-Provera injection

Figure 29.9   Contraceptive devices.

Name	Procedure	Effectiveness[*]	Name	Procedure	Effectiveness[*]
Abstinence	Refrain from sexual intercourse	100%	Diaphragm	Latex cap inserted into vagina to cover cervix before intercourse	With jelly, about 90%
Sterilization					
Vasectomy	Vas deferens cut and tied	Almost 100%	Cervical cap	Latex cap held by suction over cervix	Almost 85%
Tubal ligation	Oviducts cut and tied	Almost 100%			
Combined estrogen/ progesterone available as a pill, injectable, or vaginal ring and patch	Pill is taken daily; injectable and ring last a month; patch is replaced weekly	About 100%	Male condom	Latex sheath fitted over erect penis	About 85%
			Female condom	Polyurethane liner fitted inside vagina	About 85%
Progesterone only available as a tube implant and injectable	Implant lasts three years, injectable lasts three weeks	About 95%	Coitus interruptus	Penis withdrawn before ejaculation	About 75%
			Jellies, creams, foams	Spermicidal products inserted before intercourse	About 75%
Intrauterine device (IUD)	Newest device contains progesterone and lasts up to five years	More than 90%	Natural family planning	Day of ovulation determined by record keeping; various methods of testing	About 70%
Vaginal sponge	Sponge permeated with spermicide is inserted into vagina	About 90%	Douche	Vagina cleansed after intercourse	Less than 70%

*The percentage of sexually active women per year who will not get pregnant using this method.

Figure 29.10 **Control of reproduction.**

The medical profession can offer help to control reproduction. The number of births can be decreased by the proper use of birth control, and the number of births can be increased through assisted reproductive technologies.

Contraceptive injections are available as progesterone only or a combination of estrogen and progesterone. The length of time between injections can vary from three months to a few weeks.

Interest in barrier methods of birth control has recently increased because they offer some protection against sexually transmitted diseases. A **female condom,** now available, consists of a large polyurethane tube with a flexible ring that fits onto the cervix. The open end of the tube has a ring that covers the external genitals. A **male condom** is most often a latex sheath that fits over the erect penis. The ejaculate is trapped inside the sheath, and thus does not enter the vagina. When used in conjunction with a spermicide, the protection is better than with the condom alone. The **diaphragm** is a soft latex cup with a flexible rim that lodges behind the pubic bone and fits over the cervix. Each woman must be properly fitted by a physician, and the diaphragm can be inserted into the vagina no more than two hours before sexual relations. Also, it must be used with spermicidal jelly or cream and should be left in place at least six hours after sexual relations. The **cervical cap** is a minidiaphragm.

An **intrauterine device (IUD)** is a small piece of molded plastic that is inserted into the uterus by a physician. IUDs are believed to alter the environment of the uterus and oviducts so that fertilization probably will not occur—but if it should occur, implantation cannot take place. One type of IUD has copper wire wrapped around the plastic.

Contraceptive vaccines are now being developed. For example, a vaccine intended to immunize women against human chorionic gonadotropin (HCG), a hormone necessary to maintaining the implantation of the embryo, was successful in a limited clinical trial. Since HCG is not normally present in the body, no autoimmune reaction is expected, but the immunization does wear off with time. Other researchers believe that it would also be possible to develop a safe antisperm vaccine that could be used in women.

A morning-after pill, or emergency contraception, refers to a medication that will prevent pregnancy after unprotected intercourse. One type, a kit called Preven, contains four synthetic progesterone pills; two are taken up to 72 hours after unprotected intercourse, and two more are taken 12 hours later. The medication upsets the normal menstrual cycle, making it difficult for an embryo to implant itself in the endometrium. In a recent study, it was estimated that the medication was 75% effective in preventing unintended pregnancies.

Mifepristone, better known as RU-486, is a pill that is presently used to cause the loss of an implanted embryo by blocking the progesterone receptor proteins of endometrial cells. When taken in conjunction with a prostaglandin to induce uterine contractions, RU-486 is 95% effective. It is possible that some day this medication will also be a "morning-after pill," taken when menstruation is late without evidence that pregnancy has occurred.

Infertility

Control of reproduction does not only mean preventing pregnancy. It also involves methods of treating infertility (Fig. 29.10). **Infertility** is the failure of a couple to achieve pregnancy after one year of regular, unprotected intercourse. The American Medical Association estimates that 15% of all couples are infertile. The cause of infertility can be attributed to the male (40%), the female (40%), or both (20%).

Check Your Progress

1. Name two barrier methods of birth control and their benefits.
2. How are the pill, contraceptive implants, contraceptive injections, and the morning-after pill similar?

Answers: 1. Condoms and the diaphragm can be used at the last minute, and condoms in particular offer some protection against STDs. 2. They all work by interfering with the release of and/or the normal balance of sex hormones in the body of a female.

The most frequent cause of infertility in males is low sperm count and/or a large proportion of abnormal sperm, which can be due to environmental influences. Physicians advise that a sedentary lifestyle coupled with smoking and alcohol consumption can lead to male infertility. When males spend most of the day sitting in front of a computer or television or driving, the testes temperature remains too high for adequate sperm production.

In females, extremes in body weight appear to be the most significant factor in causing infertility. When a woman is of normal weight, fat cells produce a hormone called leptin that stimulates the hypothalamus to release GnRH. But in an overweight woman, the ovaries often contain many small follicles, and the woman fails to ovulate. Other causes of infertility in females are blocked oviducts due to pelvic inflammatory disease (see page 523), and endometriosis, the presence of uterine tissue outside the uterus, particularly in the oviducts and on the abdominal organs.

Sometimes the causes of infertility can be corrected by medical intervention so that couples can have children. It is also possible to give females fertility drugs, gonadotropic hormones that stimulate the ovaries and bring about ovulation. Such hormone treatments have been known to cause multiple ovulations and multiple births.

When reproduction does not occur in the usual manner, many couples adopt a child. Others sometimes try the assisted reproductive technologies discussed in the following paragraphs.

Assisted Reproductive Technologies

Assisted reproductive technologies (ART) consist of techniques used to increase the chances of pregnancy. Often, sperm and/or eggs are retrieved from the testes and ovaries, and fertilization takes place in a clinical or laboratory setting.

Artificial Insemination by Donor (AID)   During artificial insemination, harvested sperm are placed in the vagina by a physician. Sometimes a woman is artificially inseminated by her partner's sperm. This technique is especially helpful if the partner has a low sperm count, because the sperm can be collected over a period of time and concentrated so that the sperm count is sufficient to result in fertilization. Often, however, a woman is inseminated by sperm acquired from a donor who is a complete stranger to her. At times, a combination of partner and donor sperm is used.

In Vitro Fertilization (IVF)   During IVF, conception occurs outside the body in a laboratory (Fig. 29.11). Ultrasound machines can now spot follicles in the ovaries that hold immature eggs; therefore, the latest method is to forgo the administration of fertility drugs and retrieve immature eggs by using a needle. The immature eggs are then brought to maturity in glassware before concentrated sperm are added. After about two to four days, the embryos are ready to be transferred to the uterus of the woman, who is now in the secretory phase of her menstrual cycle. If desired, the embryos can be tested for a genetic disease, and only those found to be free of disease will be used. If implantation is successful and development is normal, pregnancy continues to term.

Gamete Intrafallopian Transfer (GIFT)   The term **gamete** refers to a sex cell, either a sperm or an egg. Gamete intrafallopian transfer was devised to overcome the low success rate (15–20%) of in vitro fertilization. The method is exactly the same as in vitro fertilization, except the eggs and the sperm are placed in the oviducts immediately after they have been brought together. GIFT has the advantage of being a one-step procedure for the woman—the eggs are

Figure 29.11   In vitro fertilization.

A microscope connected to a television screen (at left) is used to carry out in vitro fertilization. On the screen, note that a pipette (at left of egg) holds the egg steady while a needle (not visible) introduces the sperm into the egg, ensuring fertilization.

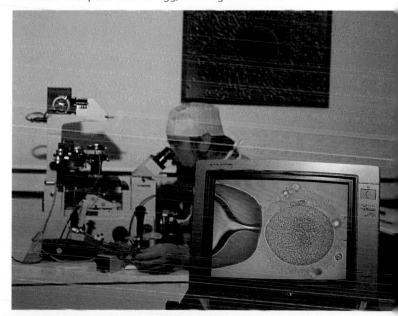

Check Your Progress

1. Contrast artificial insemination with in vitro fertilization.
2. A surrogate mother is a person who bears a child for someone else. If a surrogate mother is involved, how many possible parents could a child have?

Answers: 1. During artificial insemination, sperm are placed by a physician into a recipient's vagina, and fertilization occurs internally. During in vitro fertilization, donor eggs are fertilized by donor sperm in laboratory glassware. An embryo is then implanted in the uterus of a woman. 2. The total number of possible parents is 5: (1) the egg donor, (2) the surrogate mother, (3) the sperm donor, and (4 and 5) the woman and man who raise the child.

a. AIDS patient, Tom Moran, July 1987

b. AIDS patient, Tom Moran early January 1988

c. AIDS patient, Tom Moran, late January 1988

Figure 29.12   The course of AIDS.

This individual went through all the stages of AIDS before he died. Everyone should protect themselves from sexually transmitted diseases (STDs). If you have an STD, you are more susceptible to getting another one.

removed and reintroduced all in the same time period. A variation on this procedure is to fertilize the eggs in the laboratory and then place the zygotes in the oviducts.

Intracytoplasmic Sperm Injection (ICSI)   In this highly sophisticated procedure, a single sperm is injected into an egg. It is used effectively when a man has severe infertility problems.

Sexually Transmitted Diseases

Abstinence is the best protection against the spread of sexually transmitted diseases. For those who are sexually active, a latex condom offers some protection. Among STDs caused by viruses, treatment is available for AIDS and genital herpes, but these conditions are not curable. Only STDs caused by bacteria (e.g., chlamydia, gonorrhea, and syphilis) are curable with antibiotics.

STDs Caused by Viruses   Acquired immunodeficiency syndrome (AIDS) is caused by a virus called **human immunodeficiency virus (HIV).** HIV attacks the type of lymphocyte known as helper T cells. Helper T cells, you will recall, stimulate the activities of B lymphocytes, which produce antibodies. After an HIV infection sets in, helper T cells begin to decline in number, and the person becomes debilitated and more susceptible to other types of infections (Fig. 29.12). AIDS has three stages of infection, called categories A, B, and C.

During the category A stage, which may last about a year, the individual is an asymptomatic carrier. He or she may exhibit no symptoms, but can pass on the infection. Immediately after infection and before the blood test becomes positive, a large number of infectious viruses are present in the blood, and these could be passed on to another person. Even after the blood test becomes positive, the person remains well as long as the body produces sufficient helper T cells to keep the count higher than 500 per mm^3. With a combination therapy of several drugs, AIDS patients may remain in this stage indefinitely.

During the category B stage, which may last six to eight years, the lymph nodes swell, and the person may experience weight loss, night sweats, fatigue, fever, and diarrhea. Infections such as thrush (white sores on the tongue and in the mouth) and herpes recur.

Finally, the person may progress to category C, which is full-blown AIDS characterized by nervous disorders and the development of an opportunistic disease, such as an unusual type of pneumonia or skin cancer. Opportunistic diseases are those that occur only in individuals who have little or no capability of fighting an infection. Without intensive medical treatment, the AIDS patient dies about seven to nine years after infection.

Genital warts are caused by the human papillomaviruses (HPVs). Many times, carriers either do not have any sign of warts or merely have flat lesions. When present, the warts commonly are seen on the penis and foreskin of men and near the vaginal opening in women. Genital warts are associated with cancer of the cervix, as well as tumors of the vulva, the vagina, the anus, and the penis.

Genital herpes is characterized by painful blisters on the genitals. Once the blisters rupture, they leave painful ulcers that may take as long as three weeks or as little as five days to heal. The blisters may be accompanied by fever, pain on urination, swollen lymph nodes in the groin, and in women, a copious discharge. After the ulcers heal, the disease is latent, and blisters can recur, although usually at less frequent intervals and with milder symptoms. Fever, stress, sunlight, and menstruation are associated with recurrence of symptoms.

Hepatitis is an infection of the liver and can lead to liver failure, liver cancer, and death. Several types of hepatitis exist, some of which may be transmitted sexually. The type of hepatitis and the virus that causes it are designated by the same letter. Hepatitis A is usually acquired from sewage-contaminated drinking water, but this infection can also be sexually transmitted through oral/anal contact. Hepatitis B, which is spread in the same manner as AIDS, is even more infectious. Fortunately, a vaccine is now available for hepatitis B. Hepatitis C is spread when a person comes in contact with the blood of an infected person.

STDs Caused by Bacteria   **Chlamydia** is a bacterial infection of the lower reproductive tract that is usually mild or asymptomatic, especially in women. About 8 to 21 days after infection, men may experience a mild burning sensation on urination and a mucoid discharge. Women may have a vaginal discharge along with the symptoms of a urinary tract infection. Chlamydia also causes cervical ulcerations, which increase the risk of acquiring AIDS. If the infection is misdiagnosed or if a woman does not seek medical help, there is a particular risk of the infection spreading from the cervix to the oviducts so that **pelvic inflammatory disease (PID)** results. This very painful condition can result in blockage of the oviducts with the possibility of sterility and infertility.

Gonorrhea is easier to diagnose in males than in females because the male is more likely to experience painful urination and a thick, greenish-yellow urethral discharge. In males and females, a latent infection leads to PID, which affects the vasa deferentia or oviducts. As the inflamed tubes heal, they may become partially or completely blocked by scar tissue, resulting in sterility or infertility.

Syphilis has three stages, which are typically separated by latent periods. During the final stage, syphilis may affect the cardiovascular and/or nervous systems. An infected person may become mentally retarded, become blind, walk with a shuffle, or show signs of insanity. Gummas, which are large, destructive ulcers, may develop on the skin or within the internal organs. Syphilitic bacteria can cross the placenta, causing birth defects or stillbirth. Syphilis is easily diagnosed with a blood test.

STDs Caused by Other Organisms   Females very often have vaginitis, or infection of the vagina, caused by either the flagellated protozoan *Trichomonas vaginalis* or the yeast *Candida albicans.* Trichomoniasis is most often acquired through sexual intercourse, and the asymptomatic male is usually the reservoir of infection. *Candida albicans,* however, is a normal organism found in the vagina; its growth simply increases beyond normal under certain circumstances. For example, women taking birth control pills are sometimes prone to yeast infections. Also, the legitimate and indiscriminate use of antibiotics for infections elsewhere in the body can alter the normal balance of organisms in the vagina so that a yeast infection flares up.

Figure 29.13 lists the precautions everyone should take to avoid sexually transmitted diseases.

Figure 29.13   Preventing the spread of STDs.

Some guidelines for preventing the spread of STDs are:

1. Abstain from sexual intercourse or develop a long-term monogamous (always the same person) relationship with a person who is free of STDs.
2. Refrain from multiple sex partners or having a relationship with a person who does have multiple sex partners.
3. Be aware that having relations with an intravenous drug user is risky because the behavior of this group puts them at risk for AIDS and hepatitis B.
4. Avoid anal intercourse because HIV has easy access through the lining of the rectum.
5. Always use a latex condom if your partner has not been free of STDs for the past five years.
6. Avoid oral sex because this may be a means of transmitting AIDS and other STDs.
7. Stop, if possible, the habit of injecting drugs, and if you cannot stop, at least always use a sterile needle.

Check Your Progress

1. List the sexually transmitted diseases that are caused by viruses.
2. Why are STDs caused by viruses more problematic than those caused by bacteria?

Answers: 1. AIDS, genital warts, genital herpes, and hepatitis. 2. Bacterial infections are treatable with antibiotics, but viral infections are not.

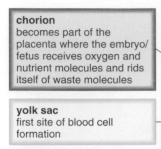

chorion
becomes part of the placenta where the embryo/fetus receives oxygen and nutrient molecules and rids itself of waste molecules

yolk sac
first site of blood cell formation

allantois
its blood vessels become the blood vessels of the umbilical cord

amnion
contains the amniotic fluid, which cushions and protects the embryo

Figure 29.14 **The extraembryonic membranes.**

Humans, like other animals that reproduce on land, are dependent on these membranes to protect and nourish the embryo (and later the fetus).

29.3 Human Development

Development encompasses all the events that occur from the time of fertilization until the animal is fully formed—in this case, a human being. Humans and other mammals have stages similar to those of all animals but with some marked differences, chiefly because of the presence of extraembryonic membranes (outside the embryo). Developing mammalian embryos (and fetuses in the case of humans), like those of reptiles and birds, depend on these membranes to protect and nourish them. Figure 29.14 shows what these membranes are and what they do in a mammal.

Fertilization

Fertilization, which results in a zygote, requires that the sperm and egg interact (Fig. 29.15). The plasma membrane of the egg is surrounded by an extracellular material termed the zona pellucida. In turn, the zona pellucida is surrounded by a few layers of adhering follicle cells.

During fertilization, a sperm moves past the leftover follicle cells, and acrosomal enzymes released by exocytosis digest a route through the zona pellucida. After a sperm binds and fuses to the egg plasma membrane, one sperm nucleus enters the egg. Only then does the secondary oocyte complete meiosis II. Finally, the haploid egg and sperm nuclei fuse. Only one sperm should enter an egg, or else the zygote will have too many chromosomes and development will not be normal. Changes in the egg's zona pellucida prevent the binding of additional sperm (called polyspermy).

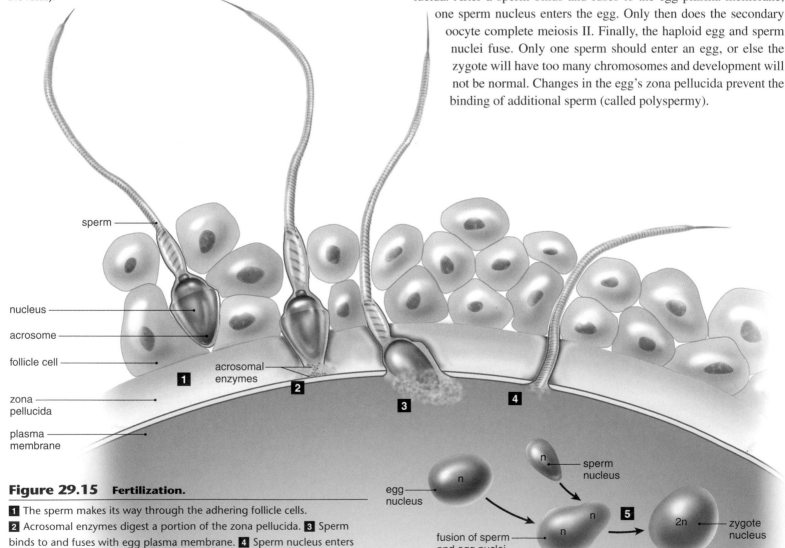

Figure 29.15 **Fertilization.**

1 The sperm makes its way through the adhering follicle cells.
2 Acrosomal enzymes digest a portion of the zona pellucida. **3** Sperm binds to and fuses with egg plasma membrane. **4** Sperm nucleus enters cytoplasm of egg. **5** Sperm and egg nuclei fuse to produce a zygote.

Early Embryonic Development

The first two months of development are considered the embryonic period. The first two weeks of development occur in the oviduct before the embryo implants itself in the uterine lining (endometrium) (Fig. 29.16). During the first stage of development, the embryo becomes multicellular. Following fertilization, the zygote undergoes **cleavage,** which is cell division without growth. DNA replication and mitotic cell division occur repeatedly, and the cells get smaller with each division. Notice that cleavage only increases the number of cells, it does not change the original volume of the egg cytoplasm. The resulting tightly packed ball of cells is called a **morula.**

The cells of the morula continue to divide, but they also secrete a fluid into the center of a ball of cells. A hollow ball of cells, called the **blastocyst,** is formed, surrounding a fluid-filled cavity called the blastocoel. Within the ball is an inner cell mass that will go on to become the embryo. The outer layer of cells is the first sign of the chorion, the extraembryonic membrane that will contribute to the development of the placenta. As the embryo implants itself in the uterine lining (endometrium), the placenta begins to form and to secrete the hormone HCG (human chorionic gonadotropin). This hormone is the basis for the pregnancy test, and it serves to maintain the corpus luteum past the time it normally disintegrates. Because of HCG, the endometrium is maintained until this function is taken over by estrogen and progesterone, produced by the placenta. Ovulation and menstruation do not occur during pregnancy.

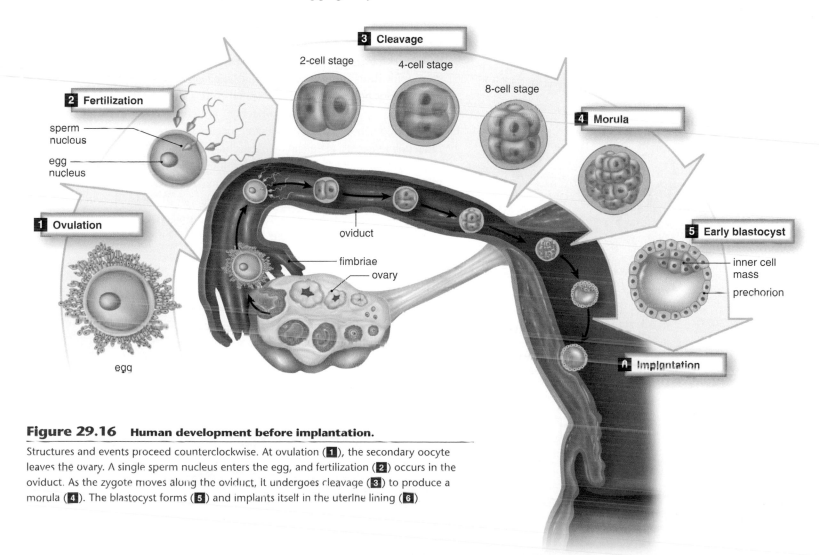

Figure 29.16   Human development before implantation.

Structures and events proceed counterclockwise. At ovulation (**1**), the secondary oocyte leaves the ovary. A single sperm nucleus enters the egg, and fertilization (**2**) occurs in the oviduct. As the zygote moves along the oviduct, it undergoes cleavage (**3**) to produce a morula (**4**). The blastocyst forms (**5**) and implants itself in the uterine lining (**6**)

a. Gastrulation is the movement of cells to establish the three germ layers: ectoderm, mesoderm, and endoderm. **b.** During neurulation, the neural tube, which becomes the central nervous system, develops right above the notochord. Later, the notochord is replaced by the vertebral column. The neural crests contain ectodermal cells that did not participate in forming the neural tube. They develop into a number of different structures, including the autonomic nervous system.

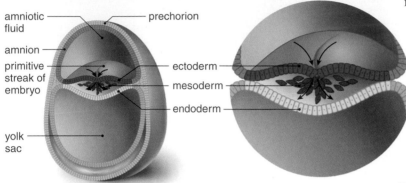

a. Gastrulation

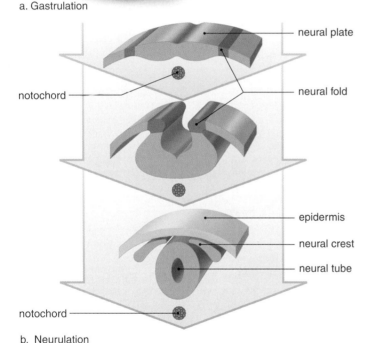

b. Neurulation

Later Embryonic Development

Gastrulation

During the third week of development, a slit called the primitive streak appears. The **gastrula** stage is evident when certain cells begin to push, or invaginate at the primitive streak, creating three layers of cells (Fig. 29.17*a*). Also evident is the **amnion,** the extraembryonic membrane that surrounds the developing embryo and later the fetus. The amnion encloses the amniotic fluid, which offers protection from sudden blows and movements, allows freedom of fetal movement, helps maintain a relatively constant temperature, and prevents drying out, which assists lung development.

Gastrulation is complete when the three layers of cells that will develop into adult organs have been produced. The outer layer is the ectoderm; the inner layer is the endoderm; and the third, or middle, layer of cells is called the **mesoderm.** Ectoderm, mesoderm, and endoderm are called the embryonic **germ layers.** As shown in Figure 29.18, the organs of an animal's body develop from these three germ layers.

Neurulation

The first organs to form are those of the central nervous system (Fig. 29.17*b*). The newly formed mesoderm cells that lie along the main longitudinal axis of the animal coalesce to form a dorsal supporting rod called the **notochord.** The central nervous system develops from midline ectoderm located just above the notochord. During **neurulation,** thickening of cells, called the **neural plate,** is seen along the dorsal surface of the embryo. Then, neural folds develop on either side of a neural groove, which becomes the **neural tube** when these folds fuse. The anterior portion of the neural tube becomes the brain, and the posterior portion becomes the spinal cord. Development of the neural tube is an example of **induction,** the process by which one tissue or organ influences the development of another. Induction occurs because the tissue initiating the induction releases a chemical that turns on genes in the tissue being induced.

Midline mesoderm cells that did not contribute to the formation of the notochord now become two longitudinal masses of tissue. These two masses become blocked off into somites, which are serially arranged on both sides along the length of the notochord. Somites give rise to the vertebrae and to muscles associated with the axial skeleton. The sequential order of the

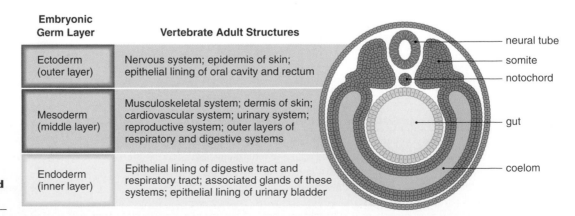

Embryonic Germ Layer	Vertebrate Adult Structures
Ectoderm (outer layer)	Nervous system; epidermis of skin; epithelial lining of oral cavity and rectum
Mesoderm (middle layer)	Musculoskeletal system; dermis of skin; cardiovascular system; urinary system; reproductive system; outer layers of respiratory and digestive systems
Endoderm (inner layer)	Epithelial lining of digestive tract and respiratory tract; associated glands of these systems; epithelial lining of urinary bladder

Figure 29.18   **The germ layers and associated organs.**

vertebrae and the muscles of the trunk testify that chordates are segmented animals. Lateral to the somites, the mesoderm splits and forms the mesodermal lining of the coelom. In addition, the neural crest consists of a band of cells that develops where the neural tube pinches off from the ectoderm. These cells migrate to various locations where they contribute to the formation of skin and muscles, in addition to the adrenal medulla and the ganglia of the peripheral nervous system.

By the end of the third week, over a dozen somites are evident, and the blood vessels and gut have begun to develop. At this point, the embryo is about 2 millimeters (mm) long.

Organ Formation Continues

A human embryo at five weeks has little flippers called limb buds (Fig. 29.19). Later, the arms and the legs develop from the limb buds, and even the hands and the feet become apparent. During the fifth week, the head enlarges, and the sense organs become more prominent. It is possible to make out the developing eyes, ears, and even the nose.

The **umbilical cord** has developed from a bridge of mesoderm called the body stalk, which connects the caudal (tail) end of the embryo with the chorion. A fourth extraembryonic membrane, the allantois, is contained within this stalk, and its blood vessels become the umbilical blood vessels. The head and the tail then lift up as the body stalk moves anteriorly by constriction. Once this process is complete, the umbilical cord, which connects the developing embryo to the placenta, is fully formed (Fig. 29.20).

A remarkable change in external appearance occurs during the sixth through eighth weeks of development—the embryo becomes easily recognized as human. Concurrent with brain development, the neck region develops, making the head distinct from the body. The nervous system is developed well enough to permit reflex actions, such as a startle response to touch. At the end of this period, the embryo is about 38 mm (1.5 inches) long and weighs no more than an aspirin tablet, even though all its organ systems are established.

Check Your Progress

1. Explain how a morula is formed.
2. What is the significance of germ layer formation during gastrulation?
3. Explain how induction occurs.

Answers: 1. The zygote divides without increasing in size, forming a ball of cells called a morula. 2. Each of the germ layers gives rise to particular structures and organs. 3. One tissue releases a chemical that turns on genes in another tissue, influencing its development.

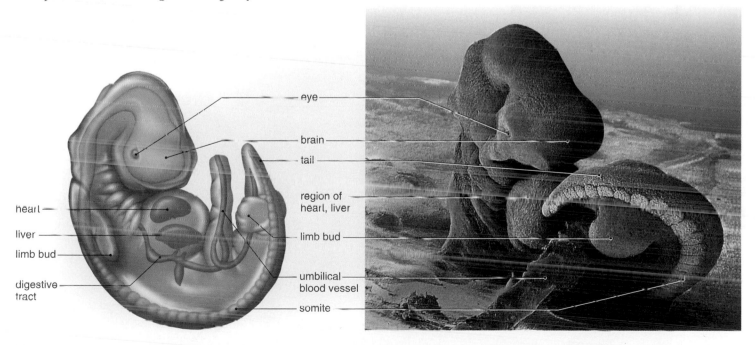

Figure 29.19 Human embryo at the beginning of the fifth week.

Labels: eye, brain, tail, region of heart, liver, limb bud, umbilical blood vessel, somite, heart, liver, limb bud, digestive tract

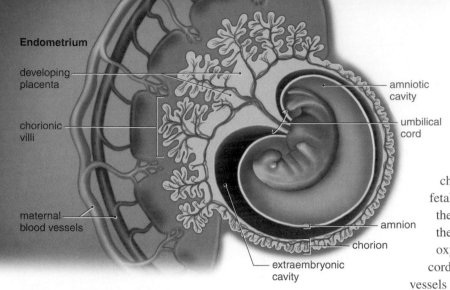

Endometrium

developing placenta

chorionic villi

maternal blood vessels

amniotic cavity

umbilical cord

amnion

chorion

extraembryonic cavity

Figure 29.20 Placenta.

Blood vessels within the umbilical cord lead to the placenta, where exchange takes place between fetal blood and maternal blood.

Figure 29.21 9-week fetus.

Placenta

The placenta has a fetal side contributed by the chorion, the outermost extraembryonic membrane, and a maternal side consisting of uterine tissues. Notice in Figure 29.20 that the chorion has treelike projections called the **chorionic villi.** The chorionic villi are surrounded by maternal blood; yet maternal and fetal blood never mix because exchange always takes place across the walls of the villi. Carbon dioxide and other wastes move from the fetal side to the maternal side of the placenta, and nutrients and oxygen move from the maternal side to the fetal side. The umbilical cord stretches between the placenta and the fetus. The umbilical blood vessels are an extension of the fetal circulatory system and simply take fetal blood to and from the placenta. Harmful chemicals can also cross the placenta. This is of particular concern during the embryonic period, when various structures are first forming. Each organ or part seems to have a sensitive period during which a substance can alter its normal development. For example, if a woman takes the drug thalidomide, a tranquilizer, between days 27 and 40 of her pregnancy, the infant is likely to be born with deformed limbs. After day 40, however, the limbs will develop normally.

Fetal Development and Birth

Fetal development encompasses the third to the ninth months (Figs. 29.21–29.24). Fetal development is marked by an extreme increase in size. Weight multiplies 600 times, going from less than 28 grams to 3 kilograms. During this time, too, the fetus grows to about 50 centimeters in length. The genitalia appear in the third month, so it is possible to tell if the fetus is male or female.

Soon after the third month, hair, eyebrows, and eyelashes add finishing touches to the face and head. In the same way, fingernails and toenails complete the hands and feet. Later, during the fifth through seventh months, a fine, downy hair (lanugo) covers the limbs and trunk, only to later disappear. The fetus looks very old because the skin is growing so fast that it wrinkles.

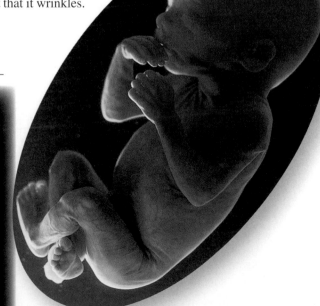

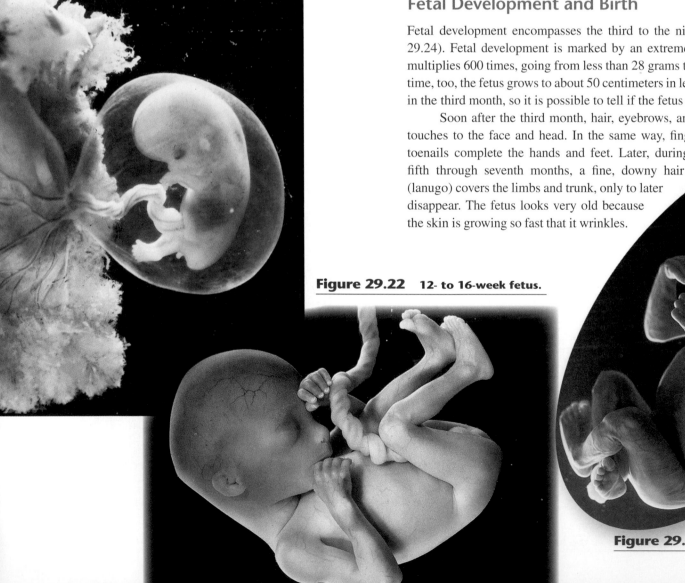

Figure 29.22 12- to 16-week fetus.

Figure 29.23 20- to 28-week fetus.

A waxy, almost cheeselike substance (called vernix caseosa) protects the wrinkly skin from the watery amniotic fluid.

The fetus at first only flexes its limbs and nods its head, but later it can move its limbs vigorously to avoid discomfort. The mother feels these movements from about the fourth month on. The other systems of the body also begin to function. As early as 10 weeks, the fetal heartbeat can be heard through a stethoscope. A fetus born at 24 weeks has a chance of surviving, although the lungs are still immature and often cannot capture oxygen adequately. Weight gain during the last couple of months increases the likelihood of survival.

The Stages of Birth

The latest findings suggest that when the fetal brain is sufficiently mature, the hypothalamus causes the pituitary to stimulate the adrenal cortex so that androgens are released into the bloodstream. The placenta utilizes androgens as a precursor for estrogen, a hormone that stimulates the production of oxytocin and prostaglandin (a molecule produced by many cells that acts as a local hormone). All three of these molecules—estrogen, oxytocin, prostaglandin—cause the uterus to contract and expel the fetus.

The process of birth (parturition) includes three stages (Fig. 29.25). During the first stage, the cervix dilates to allow passage of the baby's head and body. The amnion usually bursts about this time. During the second stage, the baby is born, and the umbilical cord is cut (Fig. 29.26). During the third stage, the placenta is delivered.

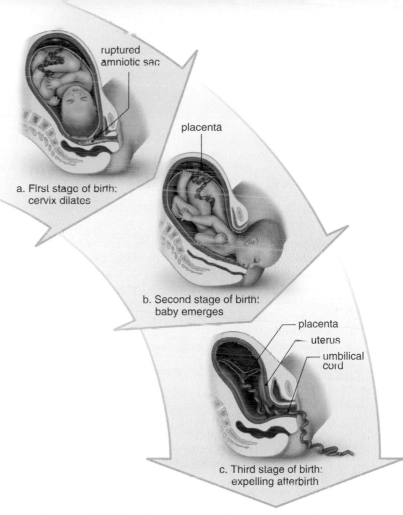

a. First stage of birth: cervix dilates

b. Second stage of birth: baby emerges

c. Third stage of birth: expelling afterbirth

Figure 29.25 **Three stages of birth (parturition).**
a. Dilation of cervix. **b.** Birth of baby. **c.** Expulsion of afterbirth (placenta).

Check Your Progress

1. Contrast embryonic development with fetal development.
2. Of what use is the afterbirth during development?

Answers: **1.** Embryonic development occurs during months 0–2 and is characterized by the establishment of organ systems. Fetal development occurs during months 3–9 and is characterized by an increase in size. **2.** The afterbirth contains the placenta, the region of exchange with the mother's bloodstream.

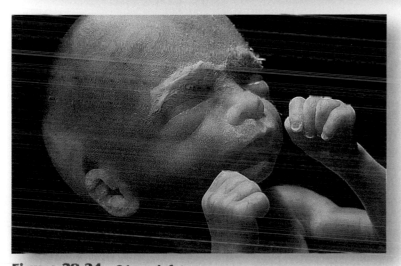

Figure 29.24 **24-week fetus.**

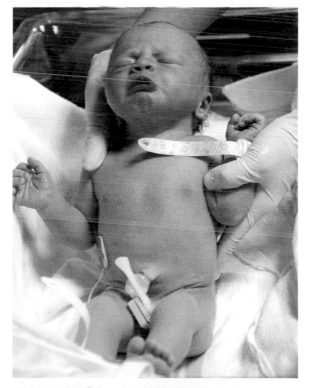

Figure 29.26 **Newborn (40 weeks).**

THE CHAPTER IN REVIEW

Summary

29.1 How Animals Reproduce

Animals usually reproduce sexually, but some can also reproduce asexually.

- In asexual reproduction, a single parent produces offspring.
- In sexual reproduction, gametes produced by gonads join to form a zygote that develops into an offspring.
- Animals adapted to reproducing in the water shed their gametes into water; fertilization and development occur in water. Animals that reproduce on land protect their gametes and embryos from drying out.
- Reptiles, birds, and mammals have extraembryonic membranes.

29.2 Human Reproduction

Male Reproductive System

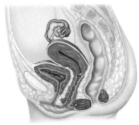

- Sperm are produced in the testis, mature in the epididymis, and are stored in the epididymis and the vas deferens.
- Sperm enter the urethra (in the penis) prior to ejaculation along with seminal fluid (produced by seminal vesicles, the prostate gland, and bulbourethral glands).
- Spermatogenesis occurs in the seminiferous tubules of the testes, which also produce testosterone in interstitial cells.

Hormone Regulation in Males

Testosterone brings about the maturation of the sex organs during puberty and promotes the secondary sex characteristics of males.

Female Reproductive System

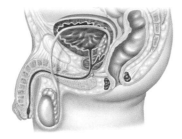

- Oocytes are produced in the ovary and move through the oviduct to the uterus. The uterus opens into the vagina. The external genital area of women includes the clitoris, the labia minora, the labia majora, and the vaginal opening.
- A follicle produces a secondary oocyte and becomes a corpus luteum. The follicle and later the corpus luteum produce estrogen and progesterone.

The ovarian and menstrual cycles last 28 days. The events of these cycles are as follows:

- Days 1–13: Menstruation occurs for five days; the anterior pituitary produces FSH, and follicles produce primarily estrogen. Estrogen causes the endometrium to increase in thickness.
- Ovulation occurs on day 14.
- Days 15–28: LH from the anterior pituitary causes the corpus luteum to produce progesterone. Progesterone causes the endometrium to become secretory.

Estrogen and Progesterone

Estrogen and progesterone bring about the maturation of the sex organs during puberty and promote the secondary sex characteristics of females.

Aspects of Reproduction

Numerous birth control methods and devices are available for those who wish to prevent pregnancy. Infertile couples are increasingly making use of assisted methods of reproduction.

Sexually transmitted diseases include the following:

- AIDS, an epidemic disease that destroys the immune system
- Genital warts, which can lead to cancer of the cervix
- Genital herpes, which repeatedly flares up
- Hepatitis, especially types A and B
- Chlamydia and gonorrhea, which cause pelvic inflammatory disease (PID)
- Syphilis, which leads to cardiovascular and neurological complications if untreated

29.3 Human Development

Development encompasses all the events that occur from fertilization to a fully-formed animal, in this case a human being.

Fertilization

The acrosome of a sperm releases enzymes that digest a pathway for the sperm through the zona pellucida. The sperm nucleus enters the egg and fuses with the egg nucleus.

Embryonic Development (months 1 and 2)

- Cleavage, which occurs in the oviduct, is cell division and formation of a morula and blastocyst. The blastocyst implants itself in the endometrium.
- Gastrulation is invagination of cells into the blastocoel, which results in formation of the germ layers. The germ layers are ectoderm, mesoderm, and endoderm.
- Organ formation can be related to the germ layers. Organ development begins with neural tube formation. Induction helps account for the steady progression of organ formation during embryonic development.
- The placenta has a fetal side and a maternal side. Gases and nutrients are exchanged at the placenta. The umbilical blood vessels are an extension of the embryo/fetal cardiovascular system and carry blood to and from the placenta.

Fetal Development and Birth (months 3–9)

- During fetal development, refinement of organ systems occurs, and the fetus adds weight.
- Birth has three stages: The cervix dilates, the baby is born, and the placenta is delivered.

Thinking Scientifically

1. In elephants, the testes never descend from the abdominal cavity. Hypothesize some ways in which elephants may differ from humans, in that they are able to produce functional sperm within the abdominal cavity. How might you determine which scenario is true?

2. Human females undergo menopause between the ages of 45 and 55. In most other animal species, both males and females maintain their reproductive capacity throughout their lives. Why do you suppose human females lose that ability in mid-life? Why don't human males go through menopause?

Testing Yourself

Choose the best answer for each question.

1. The evolutionary significance of the placenta is that it allowed
   a. animals to retain their eggs until they hatched.
   b. offspring to be born in an immature state and then develop outside the mother.
   c. offspring to exchange materials with the mother while developing inside the mother.
   d. mammals to develop the ability to produce milk.

2. In human males, sterility results when the testes are located inside the abdominal cavity instead of the scrotum because
   a. the sperm cannot pass through the vas deferens.
   b. sperm production is inhibited at body temperature.
   c. the sperm cannot travel the extra distance.
   d. digestive juices destroy the sperm.

3. Sperm are never found in the
   a. prostate gland.
   b. epididymides.
   c. urethra.
   d. vasa deferentia.

4. The function of the prostate gland is to
   a. cause sperm to reproduce.
   b. improve sperm motility.
   c. provide nutrients for sperm.
   d. extend the life of sperm.

5. Spermatogenesis produces cells that are
   a. diploid and genetically identical to each other.
   b. diploid and genetically different from each other.
   c. haploid and genetically identical to each other.
   d. haploid and genetically different from each other.

6. In the testes, the _____ produce the sperm.
   a. seminiferous tubules
   b. bulbourethral glands
   c. seminal vesicles
   d. prostate gland

7. Label the parts of the male reproductive system in the following illustration.

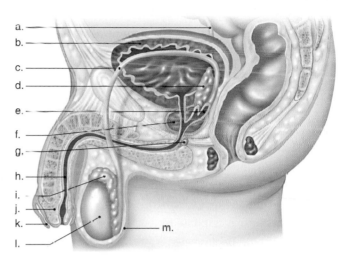

a.
b.
c.
d.
e.
f.
g.
h.
i.
j.
k.
l.
m.

8. Label the parts of the female reproductive system in the following illustration.

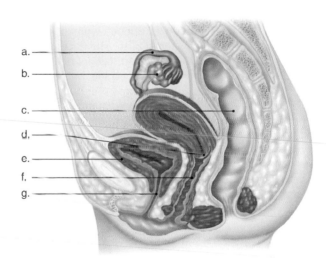

a.
b.
c.
d.
e.
f.
g.

9. Secondary sex characteristics in women are due mainly to the effects of
   a. estrogen.
   b. progesterone.
   c. follicle-stimulating hormone.
   d. luteinizing hormone.

10. Hepatitis is caused by a
    a. bacterium.
    b. fungus.
    c. protozoan.
    d. virus.

For questions 11–14, identify the stage of embryonic development in the key that matches the description.

Key:
    a. cleavage
    b. gastrulation
    c. neurulation
    d. organ formation

11. The notochord is formed.
12. The zygote divides without increasing in size.
13. Ectoderm and endoderm are formed.
14. Reflex actions, such as response to touch, develop.
15. The process of birth begins when
    a. the fetal heart is sufficiently developed.
    b. the umbilical cord can no longer support the growth of the fetus.
    c. the fetal brain is mature enough to release androgens.
    d. all organ systems have completed development.

Bioethical Issue

Recently, the news media released a story about a woman who developed complications during a vaginal birth. The doctor asked for permission to perform an emergency cesarean section, but the woman refused. The baby was delivered through the vagina, but was born dead.

There is now an effort to prosecute the woman for murder. The doctor claims that the baby would have survived the cesarean section, but the woman disagrees. Should a woman have the right to reject a medical procedure that might save her baby? If it were possible to demonstrate that this baby would have survived the cesarean section, should this woman be convicted of murder?

Understanding the Terms

abstinence   519
acquired immunodeficiency
    syndrome (AIDS)   522
amnion   526
birth control pill   519
blastocyst   525
bulbourethral glands   515
cervical cap   520
cervix   516
chlamydia   523
chorionic villi   528
cleavage   525
condom, female   520
condom, male   520
contraceptive implant   519
contraceptive injection   520
contraceptive vaccine   520
copulation   513
corpus luteum   517
diaphragm   520
ectoderm   526
endoderm   526
endometrium   517
epididymis   514
erection   515
estrogen   518
external fertilization   512
extraembryonic membranes   513
fertilization   524
fimbriae (sing., fimbria)   516
follicle   517
follicular phase   517
gamete   521
gastrula   526
gastrulation   526
genital herpes   523
genital warts   522
germ layer   526
gonads   512
gonnorhea   523
hepatitis   523

hermaphroditic   512
human immunodeficiency virus
    (HIV)   522
induction   526
infertility   520
intrauterine device (IUD)   520
luteal phase   517
menopause   518
menses   518
menstrual cycle   518
menstruation   518
mesoderm   526
morula   525
neural plate   526
neural tube   526
neurulation   526
notochord   526
orgasm   515
ovary   512
ovulation   517
parthenogenesis   512
pelvic inflammatory disease
    (PID)   523
penis   514
placenta   513
progesterone   518
prostate gland   515
puberty   516
regeneration   512
secondary sex characteristic   516
semen (seminal fluid)   515
seminal vesicle   515
seminiferous tubule   515
syphilis   523
testis (pl., testes)   512, 514
testosterone   516
umbilical cord   527
uterus   516
vas deferens   514
zygote   512

Match the terms to these definitions:

a. _____ Sexual union to deposit sperm into a female.

b. _____ Pair of glands just below the prostate gland that contribute secretions to semen.

c. _____ The main sex hormone in males.

d. _____ A thick-walled muscular organ above the cervix in the female reproductive system.

e. _____ Uterine lining.

f. _____ Release of a secondary oocyte from a follicle.

g. _____ Union of an egg and sperm.

h. _____ Product of cleavage of the zygote.

i. _____ Process by which one tissue or organ influences the development of another.

j. _____ Structure that connects a developing embryo to the placenta.

Ecology of Populations

30

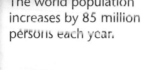

The world population increases by 85 million persons each year.

Over half the world's population lives in Asia.

To control population growth, China has imposed the one-child-per-couple policy on its citizens.

The world's human population is growing at an alarming rate. Some estimates predict it will exceed 9 billion people within 50 years. If this rate of growth continues, the planet will eventually be unable to support the human population. Different countries have found ways to encourage their citizens to limit population growth. One of the most scrutinized strategies is China's one-child policy. This policy allows each Chinese couple to have only one child.

▶ ▶ ▶

OUTLINE

30.1 The Human Population

- Humans have a clumped distribution pattern. Population densities are highest along the coasts of all the continents. 534
- The human population is undergoing rapid growth, and most of this growth takes place in the less-developed countries. 535–36
- Comparison of age structures indicates future growth trends. 536
- Most resource consumption and pollution occur in more-developed countries. 537

30.2 Characteristics of Populations

- Populations have a particular distribution and density. 538
- Population growth is dependent on the birthrate versus the death rate. 538
- Population demographics such as age and survivorship influence growth. 539
- Population growth is exponential when biotic potential expresses itself or logistic when resources are limited. 540–41
- Population growth is regulated by density-independent (abiotic) and density-dependent (biotic) factors. 542

30.3 Life History Patterns and Extinction

- Opportunistic populations have characteristics that allow them to survive unfavorable conditions and disperse easily. 544
- Equilibrium populations have characteristics that make it difficult for them to survive unfavorable conditions. 544
- Extinction is more likely in equilibrium populations with limited range. 545

30.4 The Scope of Ecology

- Ecology is the study of the interactions of organisms with each other and with the physical environment. 546
- Ecologists study these levels of biological organization: organism, population, community, ecosystem, and biosphere. 546

533

Punishments for those who do not follow the policy include imposing fines of as much as eight times the annual household income, destroying a family's home, or even forcing a pregnant female to terminate her pregnancy. While some officials believe that such a drastic policy is necessary to decrease the growth of the Chinese population, others feel the policy is too harsh and no more effective than the much less restrictive policies, such as family planning and birth control education, that are being practiced in other overpopulated countries. The situation in China poses a major ethical dilemma. Is it more important to control population growth or to respect individuals' reproductive rights?

In this chapter, you will learn about ecology and population dynamics in all sorts of populations, including humans. One important lesson to be learned is that there are no easy solutions when it comes to ecological problems.

30.1 The Human Population

The human population practically covers the face of the Earth. However, on both a large scale and a small scale, the human population has a clumped distribution. Over half of the world's people live in Asia, and most Asian populations live in China and India. Mongolia has a population density of only 1.5 persons per square kilometer, while Bangladesh has a density of nearly 1,000 persons per square kilometer. On every continent, human population densities are highest along the coasts.

Present Population Growth

For quite a while, the growth of the global human population was relatively slow, but as more reproducing individuals were added to the population, growth began to increase. At the time of the Industrial Revolution (1800s), the growth curve for the human population began to slope steeply upward, so that by now the population is undergoing rapid growth (Fig. 30.1).

Just as with populations in nature, the growth rate of the human population is determined by the difference between the number of people born and the number of people who die each year. For example, the current global birthrate is an estimated 22 per 1,000 per year, while the global death rate is approximately 9 per 1,000 per year. From these numbers we can calculate that the current annual growth rate of the human population is:

$$(22-9)/1,000 = (13)/1,000 = 1.3\%$$

Figure 30.1   **Human population growth.**

It is predicted that the world's population size may level off at 9 billion or increase to 12 billion by 2200, depending on the future growth rate. Close examination of the curve shows changes in growth that occurred in the 1300s during the black plague, in the 1800s with the Industrial Revolution, and in the 1900s with advances in science and medicine.

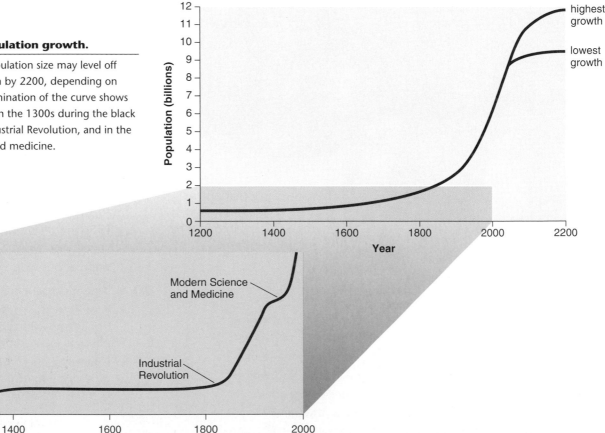

534

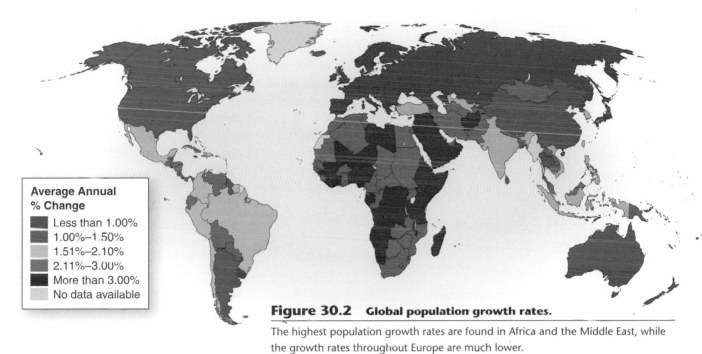

Average Annual % Change

- Less than 1.00%
- 1.00%–1.50%
- 1.51%–2.10%
- 2.11%–3.00%
- More than 3.00%
- No data available

Figure 30.2   Global population growth rates.

The highest population growth rates are found in Africa and the Middle East, while the growth rates throughout Europe are much lower.

Source: United Nations Population Division, 1993. Note: Data refer to 1990–95 (World Resource Institute).

Future Population Growth

Future growth of the human population is of great concern. It's possible that population increases will exceed the rate at which resources can be supplied. The potential for dire consequences can be appreciated by considering the *doubling time*—the length of time it takes for a population to double in size. The doubling time of the human population is now estimated to be 23 years. Will we be able to meet the extreme demands for resources by such a large population increase within such a short period of time? Already there are areas across the globe where people have inadequate access to fresh water, food, and shelter. Yet, in just 23 years, the world would need to double the amount of food, jobs, water, energy, and other resources in order to maintain the present standard of living.

Rapid growth usually begins to decline when resources, such as food and space, become scarce. Then population growth declines to zero, and the population levels off at a size appropriate to the carrying capacity of its environment. The **carrying capacity** reflects the number of individuals the environment can sustain for an indefinite period of time. The Earth's carrying capacity for humans has not been determined. Some authorities think the Earth is potentially capable of supporting far more people than currently inhabit the planet, perhaps as many as 50–100 billion people. Others believe that we may have already exceeded the number of humans the Earth can support, and that the human population may undergo a catastrophic crash to a much smaller size.

More-Developed Versus Less-Developed Countries

Complicating the issue of future human population growth is the fact that not all countries have the same growth rate (Fig. 30.2). The countries of the world today can be divided into two groups (Fig. 30.3). In the **more-developed countries (MDCs),** such as those in North America and Europe, population growth is modest, and the people enjoy a fairly good standard of living. In the **less-developed countries (LDCs),** such as those in Latin America, Asia, and Africa, population growth is dramatic, and the majority of the people live in poverty.

a.

b.

Figure 30.3   More-developed versus less-developed countries.

People in the (**a**) more-developed countries have a high standard of living and will contribute the least to world population growth, while people in the (**b**) less-developed countries have a low standard of living and will contribute the most to the world population growth.

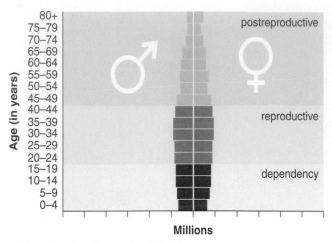

a. More-developed countries (MDCs)

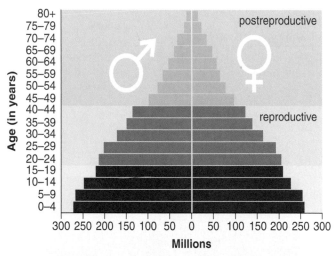

b. Less-developed countries (LDCs)

c.

Figure 30.4 **Age structure diagrams (1998).**

The diagrams predict that (**a**) the MDCs are approaching stabilization, whereas (**b**) the LDCs will expand rapidly due to their age distributions. **c.** Improved women's rights and increasing contraceptive use could change this scenario. Here a community health worker is instructing women in Bangladesh about the use of contraceptives.

(a,b) Data from the United Nations Population Division, 1998, p. 536.

The MDCs

The MDCs did not always have a low growth rate. Between 1850 and 1950, they doubled their populations, largely because of a decline in the death rate due to the development of modern medicine and improved socioeconomic conditions. The decline in the death rate was followed shortly thereafter by a decline in birthrate, so that populations in the MDCs have experienced only modest growth since 1950. This sequence of events (i.e., decreased death rate followed by decreased birthrate) is termed a demographic transition.

The growth rate for the MDCs as a whole is now about 0.1%, but populations in several countries are not growing at all, or are actually decreasing in size. The MDCs are expected to increase by 52 million between 2002 and 2050, but this modest amount will still keep their total population at just about 1.2 billion. In contrast to the other MDCs, population growth in the United States is not leveling off and continues to increase. The U.S. currently has a growth rate of 0.6%. This higher rate is due to the fact that many people immigrate to the U.S. each year, and a large number of the women are still of reproductive age. Therefore, it is unlikely that the U.S. will experience a decline in its growth rate in the near future.

The LDCs

Death rates began to decline sharply in the LDCs following World War II with the introduction of modern medicine, but the birthrate remained high. The *collective* growth rate of the LDCs peaked at 2.5% between 1960 and 1965, and since that time the rate has declined to 1.6%. Unfortunately, the growth rates in 46 of the less-developed countries have not declined. Thirty-five of these countries are in sub-Saharan Africa, where women on average are presently bearing more than five children each.

Between 2002 and 2050, the population of the LDCs is expected to jump from 5 billion to at least 8 billion. Africa will not share appreciably in this increase because of the many deaths there due to AIDS. The majority of the increase is expected to occur in Asia. Asia, with 31% of the world's arable (farmable) land, is already home to 56% of the human population. Twelve of the world's most polluted cities are located in Asia. If the human population increases as expected, Asia will experience even more urban pollution, acute water scarcity, and a significant loss of wildlife over the next 50 years.

Check Your Progress

1. Usually, why does rapid population growth begin to decline?
2. Contrast MDCs with LDCs.

Answers: 1. Historically, when resources such as food and space become scarce, population size levels off at the carrying capacity of the environment. 2. MDCs, such as those in Europe and North America, have modest population growth and a good standard of living. LDCs, such as those in Latin America, Asia, and Africa, have a fast rate of population growth, and most of the people live in poverty.

Comparing Age Structures

An age structure diagram is divided into three groups: dependency, reproductive, and postreproductive (Fig. 30.4). Many MDCs have a stable age structure, meaning that the number of individuals in each group is just about the same. Therefore, the MDC populations are expected to remain just about the same size if couples have two children, and to decline if each couple has fewer than two children. In contrast, the age structure diagram of most LDCs has a pyramid shape, with the dependency group the largest. Therefore, the LDC populations will continue to expand, even after replacement reproduction is attained. However, the more quickly replacement reproduction is achieved, the sooner zero population growth will result.

Replacement reproduction occurs when each couple in a population produces only two children. It might seem at first that replacement reproduction would cause a population to undergo zero population growth and therefore no increase in population size. However, if there are more young women entering the reproductive years than there are older women leaving them, replacement reproduction would still result in population growth.

Population Growth and Environmental Impact

Population growth is putting extreme pressure on each country's social organization, the Earth's resources, and the biosphere. Since the population of the LDCs is still growing at a significant rate, it might seem that their population increase will be the main cause of future environmental degradation. But this is not necessarily the case because the MDCs consume a much larger proportion of the Earth's resources than do the LDCs. This consumption leads to environmental degradation, which is also of the utmost concern.

Environmental Impact

The environmental impact (E.I.) of a population is measured not only in terms of population size, but also in terms of resource consumption per capita and the pollution that results because of this consumption. In other words:

E.I. = population size × resource consumption per capita
= pollution per unit of resource used

Therefore, there are two possible causes of overpopulation: population size and resource consumption. Overpopulation is more obvious in LDCs; resource consumption is more obvious in MDCs, because the per capita consumption is so much higher in the MDCs. For example, an average American family, in terms of per capita resource consumption and water requirements, is the equivalent of 30 people in India. We need to realize, therefore, that only a limited number of people can be sustained anywhere near the standard of living enjoyed in the MDCs.

The comparative environmental impacts of MDCs and LDCs are shown in Figure 30.5. The MDCs account for only about one-fourth of the world population. However, compared to the LDCs, MDCs account for 90% of the hazardous waste production due their high rate of consumption of such resources as fossil fuel, metal, and paper.

LDCs 78% MDCs 22%

a. Population

LDCs 10% MDCs 90%

b. Hazardous wastes

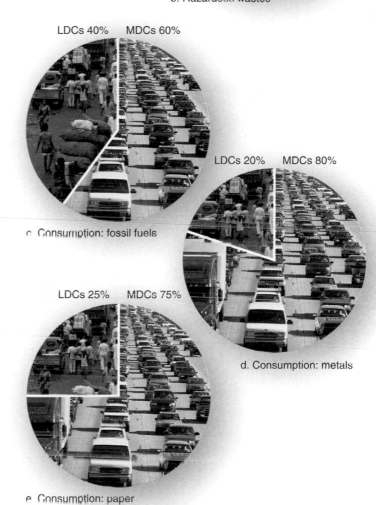
LDCs 40% MDCs 60%

c. Consumption: fossil fuels

LDCs 20% MDCs 80%

d. Consumption: metals

LDCs 25% MDCs 75%

e. Consumption: paper

Figure 30.5 Environmental impacts.

a. The combined population of the MDCs is much smaller than that of the LDCs. **b.** MDCs produce most of the world's hazardous wastes. **c–e.** MDCs consume much more fossil fuel, metals, and paper than do LDCs.

a. Mature desert shrubs

Young, small shrubs

b. Clumped

Medium shrubs

c. Random

Large shrubs

d. Uniform

Figure 30.6 **Patterns of distribution within a population.**

a. A population of mature desert shrubs. **b.** Young, small desert shrubs are clumped. **c.** Medium shrubs are randomly distributed. **d.** Large shrubs are uniformly distributed.

30.2 Characteristics of Populations

The characteristics of a population can change. For example, a population is usually subject to evolution by natural selection. During times of environmental stability, most individuals of a population are well adapted to their environment, and the population may increase in size. During times of environmental change, individuals better adapted to the environment will leave behind more offspring than those that are not as well adapted. Once the population becomes adapted to the new environment, an increase in size will occur again. So when an ecologist lists the characteristics of a population, this is a snapshot of the characteristics at a particular time and place.

Distribution and Density

Resources are the nonliving and living components of an environment that support its organisms. Light, water, space, mates, and food are some important resources for wildlife populations. The availability of resources influences the spatial distribution of individuals in a population. Three terms—**clumped, random,** and **uniform**—are used to describe the observed patterns of distribution. In a study of desert shrubs, it was found that the distribution changes from clumped to random to uniform as the plants mature (Fig. 30.6). After sufficient study, competition for belowground resources was found to cause the distribution pattern to become uniform.

Suppose we were to step back and consider the distribution of individuals, not within a single desert or forest or pond, but within a species' range. A **range** is that portion of the globe where the species can be found. On this scale, all the members of a species within the range make up the population. Members of a population are clumped within the range because they are located in areas suitable for their growth. Desert shrubs, for example, are found in deserts.

Population density is the number of individuals per unit area or volume. Population density tends to be higher in areas with plentiful resources than in areas with scarce resources. Other factors can be involved, however. In general, population density declines with increasing body size. Consider that tree seedlings live at a much higher population density than do mature trees. Therefore, more than one factor must be taken into account when explaining population densities.

Check Your Progress

1. The spatial distribution of individuals in a population can be described in what three ways?
2. List two major factors that influence population density.

Answers: 1. Clumped, random, and uniform. 2. Resource availability and body size.

Population Growth

As mentioned on page 534, the growth rate is dependent on the number of individuals born each year and the number of individuals that die each year. Populations grow when the number of births exceeds the number of deaths. To take another

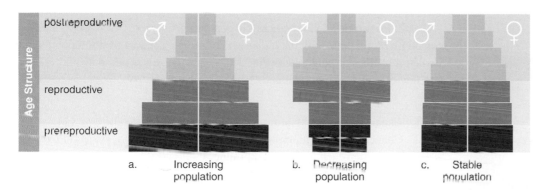

Figure 30.7   **Age structure diagrams.**

Typical age structure diagrams for hypothetical populations that are (**a**) increasing, (**b**) decreasing, or (**c**) stable. Different numbers of individuals in each age class create these distinctive shapes. In each diagram, the left half represents males while the right half represents females.

example, if the number of births is 30 per year, and the number of deaths is 10 per year per 1,000 individuals, the growth rate is 2%.

In cases where the number of deaths exceeds the number of births, the value of the growth rate is negative—the population is shrinking.

Demographics and Population Growth

Availability of resources and also certain characteristics of a population—called **demographics**—influence its growth rate.

One demographic characteristic of interest for all populations is the **age structure** of a population. As shown in Figure 30.4, it is customary to use bar diagrams to depict the age structure of a population. The population is divided into members that are *prereproductive, reproductive,* or *postreproductive* based on their age. If the prereproductive and reproductive individuals outnumber those that are postreproductive, the birthrate will exceed the death rate (Fig. 30.7*a*). In contrast, if the postreproductive individuals exceed those that are prereproductive and/or reproductive, the number of individuals in the population will decrease over time because the death rate will exceed the birthrate (Fig. 30.7*b*). If the numbers of prereproductive, reproductive, and postreproductive individuals are approximately equal, the birthrate and death rate will be equal, and the number of individuals in the population will remain stable over time (Fig. 30.7*c*).

Ecologists also study patterns of **survivorship**—how the age at death influences population size. For example, if members of a population die young, the number of reproductive individuals is lower than otherwise. First, an ecologist constructs a life table, which lists the number of individuals that die or survive at each age. The best way to arrive at a life table is to identify a large number of individuals that are born at about the same time and keep records on them from birth until death.

A life table for Dall sheep is given in Figure 30.8*a*. This table was constructed by gathering a large number of skulls and counting the growth rings on the horns to estimate each sheep's age at death. Plotting the number of survivors per 1,000 births against age produces a *survivorship curve* (Fig. 30.8*b*). While each species has its own particular pattern, three types of survivorship curves are common. In a type I curve, typical of Dall sheep and humans, survival is high until old age, when deaths increase due to illness. In a type II curve, the possibility of death is spread over all age groups. In a type III curve, death is likely among the young, and few individuals reach old age.

Figure 30.8   Life table and survivorship curves.

a. A life table for Dall sheep. **b.** Three typical survivorship curves. Among Dall sheep, with a type I curve, most individuals survive until old age, when they gradually die off. Among hydras, with a type II curve, there is an equal chance of death at all ages. Among oysters, with a type III curve, most die when they are young, and few become adults that are able to survive until old age.

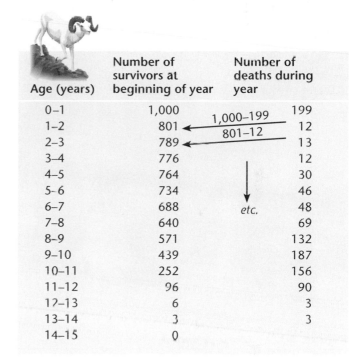

Age (years)	Number of survivors at beginning of year	Number of deaths during year
0–1	1,000	199
1–2	801	12
2–3	789	13
3–4	776	12
4–5	764	30
5–6	734	46
6–7	688	48
7–8	640	69
8–9	571	132
9–10	439	187
10–11	252	156
11–12	96	90
12–13	6	3
13–14	3	3
14–15	0	

1,000–199

801–12

etc.

a.

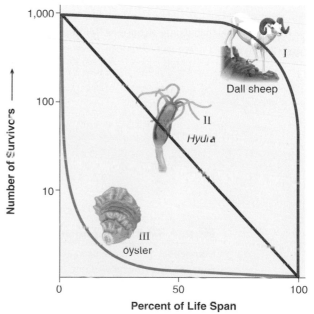

b.

Biotic Potential How quickly a population increases over time depends upon its **biotic potential,** the highest possible rate of increase for a population when resources are unlimited (Fig. 30.9). Whether the biotic potential is high or low depends on demographic characteristics of the population, such as the following:

1. Usual number of offspring per reproduction
2. Chances of survival until age of reproduction
3. How often each individual reproduces
4. Age at which reproduction begins

Patterns of Population Growth

The patterns of population growth are dependent on (1) the biotic potential of the species combined with other demographics, and (2) the availability of resources. The two fundamental patterns of population growth are exponential growth and logistic growth.

Exponential Growth

Suppose ecologists are studying the growth of a population of insects that are capable of infesting and taking over an area. Under these circumstances, **exponential growth** is expected. An exponential pattern of population growth results in a J-shaped curve (Fig. 30.10). This growth pattern can be likened to compound interest at the bank: The more your money increases, the more interest you will get. If the insect population has 2,000 individuals and the per capita rate of increase is 20% per month, there will be 2,400 insects after one month, 2,880 after two months, 3,456 after three months, and so forth.

Notice that a J-shaped curve has these phases:

Lag phase   Growth is slow because the number of individuals in the population is small.
Exponential growth phase   Growth is accelerating.

Usually, exponential growth can only continue as long as resources in the environment are unlimited. When the number of individuals in the population approaches the number that can be supported by available resources, competition among individuals for resources increases.

a.

b.

Figure 30.9   **Biotic potential.**

A population's maximum growth rate under ideal conditions—that is, its biotic potential—is greatly influenced by the number of offspring produced in each reproductive event. **a.** Pigs, which produce many offspring that quickly mature to produce more offspring, have a much higher biotic potential than (**b**) the rhinoceros, which produces only one or two offspring per infrequent reproductive events.

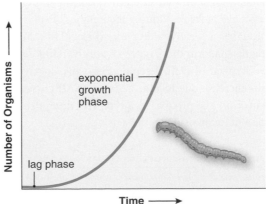

Figure 30.10   **Exponential growth.**

Exponential growth results in a J-shaped growth curve because the growth rate is positive.

Logistic Growth

As resources decrease, population growth levels off, and a pattern of population growth called **logistic growth** is expected. Logistic growth results in an S-shaped growth curve (Fig. 30.11).

Notice that an S-shaped curve has these phases:

Lag phase   Growth is slow because the number of individuals in the population is small.

Exponential growth phase   Growth is accelerating due to biotic potential (see Fig. 30.9a).

Deceleration phase   The rate of population growth slows because of increased competition among individuals for available resources.

Stable equilibrium phase   Little if any growth takes place because births and deaths are about equal.

The stable equilibrium phase is said to occur at the carrying capacity of the environment. Recall from Section 30.1 that the carrying capacity is the total number of individuals the resources can support. This number is not a constant and varies with the circumstances. For example, a large island can support a larger population of penguins than a small island, because the smaller island has a limited amount of space for nesting sites.

Applications   Our knowledge of logistic growth has practical applications. The model predicts that exponential growth will occur only when population size is much lower than the carrying capacity. So, if humans are using a fish population as a continuous food source, it would be best to maintain that population size in the exponential phase of growth when biotic potential is having its full effect and the birthrate is the highest. If people overfish, the fish population will sink into the lag phase, and it may be years before exponential growth recurs. On the other hand, if people are trying to limit the growth of a pest, it is best to reduce the carrying capacity rather than to reduce the population size. Reducing the population size only encourages exponential growth to begin once again. Farmers can reduce the carrying capacity for a pest by alternating rows of different crops instead of growing one type of crop throughout the entire field.

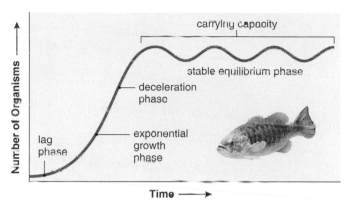

Figure 30.11   Logistic growth.

Logistic growth produces an S-shaped growth curve because competition for resources increases as the population approaches the carrying capacity.

Figure 30.12   Biotic potential.

The ability of many populations, including apple trees, to reproduce exceeds by a wide margin the number necessary to replace those that die.

Check Your Progress

1. What is the relationship between survivorship and population size?
2. List the four demographic characteristics of a population that determine its biotic potential.
3. Contrast exponential growth with logistic growth.

Answers: 1. If members of a population typically die young, the size of the population will be lower than expected. *2.* Usual number of offspring per reproduction; chances of survival until age of reproduction; how often each individual reproduces; age at which reproduction begins. *3.* Exponential growth produces a j-shaped curve and occurs when resources are unlimited. Logistic growth produces an S-shaped curve and occurs when resources are limited.

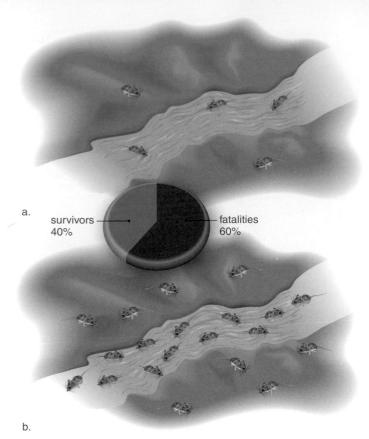

a.

b.

Figure 30.13   Density-independent effects.

The impact of a density-independent factor, such as weather or a natural disaster, is not influenced by population density. Two populations of field mice are in the path of a flash flood. (**a**) In the low-density population, 3 out of 5 mice drown, a 60% death rate. **b.** In the high-density population, 12 out of 20 mice drown—also a 60% death rate.

survivors 40%   fatalities 60%

a.

b.

Figure 30.14   Density-dependent effects—competition.

The impact of competition is directly proportional to the density of a population. **a.** When density is low, every member of the population has access to the resource. **b.** When density is high, members of the population must compete to gain access to available resources, and some fail to gain access.

Factors That Regulate Population Growth

Ecologists have long recognized that the environment contains both biotic (living) and abiotic (nonliving) components, and that these two components play an important role in regulating population size in nature.

Density-Independent Factors

Abiotic factors, such as weather or natural disasters, are typically **density-independent factors.** Abiotic factors can cause sudden and catastrophic reductions in population size. However, density-independent factors cannot in and of themselves regulate population size because the effect is not influenced by the number of individuals in the population. In other words, the intensity of the effect does not increase with increased population size. For example, the proportion of a population killed in a flash flood is independent of density—floods don't necessarily kill a larger percentage of a dense population than of a less dense population (Fig. 30.13). Of course, the larger the population, the greater the number of individuals killed, but the percentage killed does not depend on the density.

Density-Dependent Factors

Biotic factors are considered **density-dependent factors** because the percentage of the population affected does increase as the density of the population increases. Competition, predation, and parasitism are all biotic factors that increase in intensity as the density increases. We will discuss these interactions between populations again in Chapter 31 because they influence community composition and diversity.

Competition   Competition can occur when members of the same species attempt to utilize needed resources (such as light, food, or space) that are in limited supply. As a result, not all members of the population can have access to the resource to the degree necessary to ensure survival or reproduction or some other aspect of their life history.

As an example, let's consider a woodpecker population in which members have to compete for nesting sites. Each pair of birds requires a tree hole in order to raise offspring. If there are more holes than breeding pairs, each pair can have a hole in which to lay eggs and rear young birds (Fig. 30.14a). But if there are fewer holes than breeding pairs, each pair must compete to acquire a nesting site (Fig. 30.14b). Pairs that fail to gain access to holes will be unable to contribute new members to the population.

A well-known example of competition for food concerns the reindeer on St. Paul Island. Overpopulation led to an insufficiency of food to sustain all members, causing the population to crash—that is, it became drastically reduced.

Resource partitioning among different age groups is a way to reduce competition for food. During the life cycle of butterflies, the caterpillar stage requires different food than the adult stage. Caterpillars graze on leaves, while adult butterflies feed on nectar produced by flowers. Therefore, the parents are less apt to compete with their offspring for food.

Predation Predation occurs when one organism, the predator, eats another, the prey. In the broadest sense, predation can include not only animals such as lions that kill zebras, but also filter-feeding blue whales that strain krill from the ocean waters, parasitic ticks that suck blood from their hosts, and even herbivorous deer that browse on trees and bushes.

The effect of predation generally increases as the prey population grows more dense, because prey are easier to find when hiding places are limited. Consider a field inhabited by a population of mice (Fig. 30.15). Each mouse must have a hole in which to hide to avoid being eaten by a hawk. If there are 102 mice, but only 100 holes, two mice will be left out in the open. It might be hard for the hawk to find only two mice in the field. If neither mouse is caught, the predation rate is 0/2 = 0%. However, if there are 200 mice and only 100 holes, there is a greater chance that the hawk will be able to find some of these 100 mice without holes. If half of the exposed mice are caught, the predation rate is 50/100 = 50%. Therefore, increasing the density of the available prey has increased the proportion of the population preyed upon.

Predator-Prey Population Cycles Instead of remaining steady, some predator and prey populations experience an increase, and then a decrease, in population size. A famous example of such a cycle occurs between the snowshoe hare and the Canada lynx, a type of small cat (Fig. 30.16). The snowshoe hare is a common herbivore in the coniferous forests of North America, where it feeds on terminal twigs of various shrubs and small trees. Investigators first assumed that the predatory lynx brings about a decrease in the hare population, and this decrease in prey brings about a subsequent decrease in the lynx population. Once the hare population recovers, so does the lynx population, and the result is a boom–bust cycle. But some biologists noted that the decline in snowshoe hare abundance is accompanied by low growth and reproductive rates, which could be signs of food shortage. A field experiment showed that if the lynx predator was denied access to the hare population, the cycling of the hare population still occurred based on food availability. The results suggested that a hare–food cycle and a predator–hare cycle have combined to produce the pattern observed in Figure 30.16.

a.

b.

Figure 30.15 Density-dependent effects—predation.

The impact of predation on a population is directly proportional to the density of the population. **a.** In a low-density population, the chances of a predator finding the prey are low, resulting in little predation (mortality rate of 0/2, or 0%). **b.** In the higher-density population, there is a greater likelihood of the predator locating potential prey, resulting in a greater predation rate (mortality rate of 50/100, or 50%).

snowshoe hare

lynx

Figure 30.16 Predator-prey cycling of a lynx and a snowshoe hare.

The number of pelts received yearly by the Hudson Bay Company for almost 100 years shows a pattern of ten-year cycles in population densities. The snowshoe hare (prey) population reaches a peak before that of the lynx (predator) by a year or more.

Data from D. A. MacLulich, *Fluctuations in the Numbers of the Varying Hare* (Lepus americanus), University of Toronto Press, Toronto, 1937, reprinted 1974, p. 543.

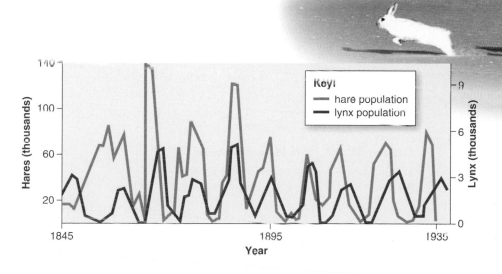

Key:
— hare population
— lynx population

Opportunistic Pattern

Small individuals
Short life span
Fast to mature
Many offspring
Little or no care of offspring

a.

30.3 Life History Patterns and Extinction

A **life history** consists of a particular mix of the characteristics we have been discussing. After studying many different types of populations, from mayflies to humans, ecologists have discovered two fundamental and contrasting patterns: the opportunistic life history pattern and the equilibrium life history pattern.

Opportunistic populations tend to exhibit exponential growth. The members of the population are small in size, mature early, have a short life span, and provide limited parental care for a great number of offspring (Fig. 30.17a). Density-independent effects tend to regulate the size of the population, which is large enough to survive an event that threatens to annihilate it. The population has a high dispersal capacity. Various types of insects and weeds provide the best examples of opportunistic species.

Equilibrium populations exhibit logistic population growth, and the size of the population remains close to or at the carrying capacity (Fig. 30.17b). Resources are relatively scarce, and members best able to compete—those with phenotypes best suited to the environment—have the largest number of offspring. Members allocate energy to their own growth and survival, and to the growth and survival of their few offspring. Therefore, they are fairly large, slow to mature, and have a fairly long life span. The growth of equilibrium populations tends to be regulated by density-dependent effects. Various birds and mammals provide the best examples of equilibrium species.

Equilibrium Pattern

Large individuals
Long life span
Slow to mature
Few offspring
Much care of offspring

Figure 30.17   Life history patterns.

a. Dandelions are an opportunistic species, whereas **(b)** mountain gorillas are an equilibrium species.

b.

Extinction

Extinction is the total disappearance of a species or higher group. Which species shown in Figure 30.17, the dandelion or the mountain gorilla, is apt to become extinct? Because the dandelion matures quickly, produces many offspring at one time, and has seeds that are dispersed widely by wind, it can more easily withstand a local decimation than can the mountain gorilla.

A study of equilibrium species shows that three other factors—namely, size of geographic range, degree of habitat tolerance, and size of local populations—can help determine whether an equilibrium species is in danger of extinction. Figure 30.18 compares several equilibrium species on the basis of these three factors. The mountain gorilla has a restricted geographic range, narrow habitat tolerance (few preferred places to live), and small local population. This combination of characteristics makes the mountain gorilla very vulnerable to extinction. The possibility of extinction increases depending on whether a species is similar to the gorilla in one, two, or three ways. Such population studies can assist conservationists and others who are trying to preserve the biodiversity of the biosphere.

Check Your Progress

1. Contrast opportunistic populations with equilibrium populations.
2. List five factors that can determine whether an equilibrium population is in danger of extinction.

Answers: 1. Opportunistic populations exhibit exponential growth and are regulated by density-independent effects. Equilibrium populations exhibit logistic growth and remain near carrying capacity. 2. Time to maturity; number of offspring produced; size of geographic range; degree of habitat tolerance; and size of local populations.

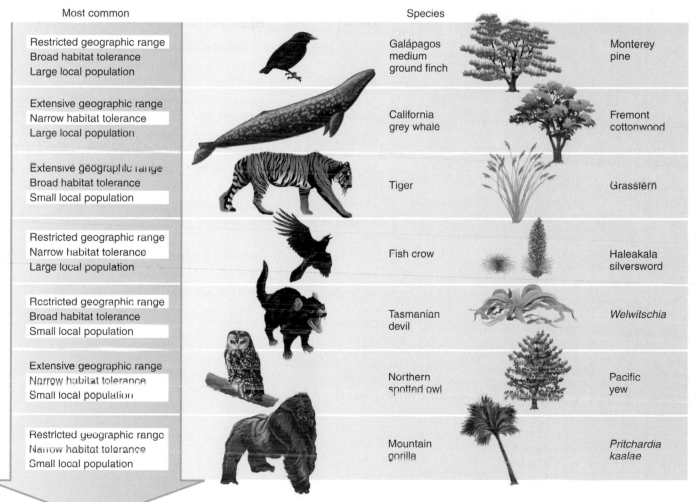

Figure 30.18 Vulnerability to extinction.

Vulnerability is particularly tied to range, habitat, and size of population. The amount of white-highlighted text in the boxes indicates the chances and causes of extinction.

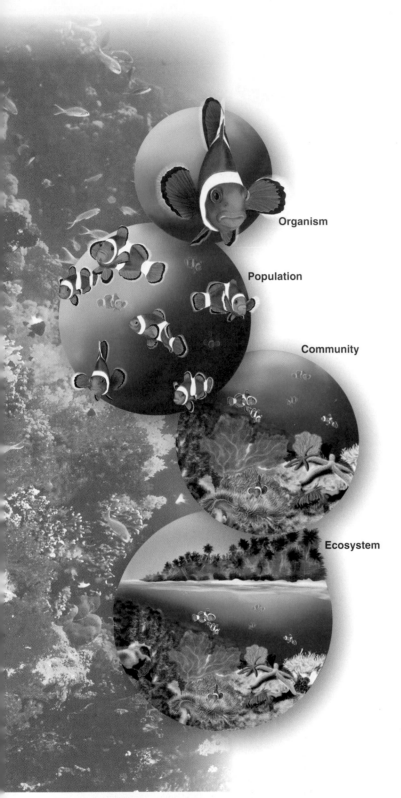

Organism

Population

Community

Ecosystem

Figure 30.19 **Levels of organization.**

The study of ecology encompasses the organism, population, community, and ecosystem levels of biological organization.

30.4 The Scope of Ecology

Ecology is the scientific study of the interactions of organisms with each other and with the physical environment. Ecology is one of the two biological sciences of most interest to the public today, the other being genetics. Genetics captures our attention because of its amazing advances, but ecology is of urgent concern because of a newly gained recognition: Unless we learn to live with the environment and not against it, our society cannot endure. Understanding ecology will help you make informed decisions, ranging from what kind of car to drive and whether to use pesticides on your lawn, to how to support the preservation of a forested area in your town. Your ecological decisions will affect not only your life, but also the lives of generations to come.

Ecology involves the study of several levels of biological organization (Fig. 30.19). Some ecologists study **organisms,** focusing on their adaptations to a particular environment. For example, organismal ecologists might investigate what attributes of a parrotfish in a coral reef allow it to live in warm tropical waters.

In this chapter, we learned that organisms are usually part of a **population,** a group of individuals of the same species in a given location. At the population level of study, ecologists describe changes in population size over time. For example, a population ecologist might compare the number of parrotfishes living in a given location today to data she obtained from that same location 20 years ago.

A **community** consists of all the various populations at a particular locale. A coral reef contains numerous populations of fishes, crustaceans, corals, algae, and so forth. At the community level, ecologists study how the interactions between populations affect the populations' well-being. For example, a community ecologist might study how a decrease in algal populations affects the population sizes of crustaceans and fishes living on the coral reef.

An **ecosystem** consists of a community of living organisms as well as their physical environment. As an example of how the physical environment affects a community, consider that the presence of suspended particles in the water decreases the amount of sunlight reaching algae living on the coral reef. Without solar energy, algae cannot produce the organic nutrients they and the other populations require.

Finally, the **biosphere** is the portion of the Earth's surface—air, water, land—where living things exist. Having information about the many levels of organization within a coral reef allows ecologists to understand how a coral reef contributes to the biodiversity and dynamics of the biosphere.

Ecology: A Biological Science

Ecology began as a part of **natural history,** the discipline of observing and describing organisms in their environment, but today ecology is also an experimental science. A central goal of ecology is to develop models that explain and predict the distribution and abundance of populations based on their interactions within an ecosystem. Achieving such a goal involves testing hypotheses. For example, ecologists might formulate and test hypotheses about the role fire plays in maintaining a lodgepole pine forest (Fig. 30.20). To test these hypotheses, ecologists could compare the characteristics of a community before and after a prescribed burn of the area by professionals under controlled conditions. Ultimately, ecologists wish to develop models about the distribution and abundance of ecosystems within the biosphere.

Ecology and Environmental Science

The field of environmental science applies ecological principles to practical human concerns. Comprehension of ecological principles helps us understand why a functioning biosphere is critical to our survival and why our population of 6 billion people and our overconsumption of resources pose threats to the biosphere.

Conservation biology is a new discipline that studies all aspects of biodiversity with the goal of conserving natural resources, including wildlife, for the benefit of this generation and future generations. Conservation biology recognizes that wildlife species are an integral part of a well-functioning biosphere on which human life depends.

> ### Check Your Progress
>
> 1. Which levels of biological organization are of interest to ecologists?
> 2. How is conservation biology like environmental science?
>
> *Answers: 1. Ecologists study organisms, populations, communities, ecosystems, and the biosphere. 2. Both of these fields apply the principles of ecology to matters of practical concern to humans.*

a. After a burn.

b. Before a burn.

Figure 30.20   Studying the effects of fire.

Fire is a natural occurrence in lodgepole pine forests, such as these in Yellowstone National Park. Over time, the recently burned forest (**a**) will mature and look like the unburned forest (**b**).

THE CHAPTER IN REVIEW

Summary

30.1 The Human Population

The human population exhibits clumped distribution (both on a large scale and a small scale) and is undergoing rapid growth.

Future Population Growth

At the present growth rate, the doubling time for the human population is estimated to be 23 years. This rate of growth will put extreme demands on resources, and growth will decline due to resource scarcity. Eventually the population will most likely level off at its carrying capacity.

More-Developed Countries (MDCs)   The growth rate for MDCs is about 0.1%. The total MDC population between 2002 and 2050 will remain at around 1.2 billion.

Less-Developed Countries (LDCs)   The growth rate for LDCs is about 1.6%. The LDC population between 2002 and 2050 is expected to increase from 5 billion to 8 billion.

Comparing Age Structures

The age structure of populations is divided into three age groups: dependency, reproductive, and postreproductive. The MDCs are approaching stabilization with just about equal numbers in each group. The LDCs will continue to expand because their prereproductive group is the largest.

Population Growth and Environmental Impact

Two types of environmental impact can occur: The LDCs put stress on the biosphere due to population growth, and the MDCs put stress on the biosphere due to resource consumption and waste production.

30.2 Characteristics of Populations

Distribution patterns and population density are both dependent on resource availability.

Distribution Patterns   Clumped, random, uniform.

Population Density   The number of individuals per unit area is higher in areas with abundant resources, and lower when resources are limited.

Demographics and Population Growth

Population growth can be calculated based on the annual birthrate and death rate. Population growth is determined by:

- Resource availability
- Demographics (age structures, survivorship, and biotic potential—the highest rate of increase possible)

Patterns of Population Growth

The two patterns of population growth are exponential growth and logistic growth.

Exponential Growth   Exponential growth results in a J-shaped curve. The two phases of exponential growth are lag phase (slow growth) and exponential growth phase (accelerating growth).

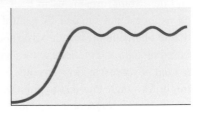

Logistic Growth   Logistic growth results in an S-shaped curve. The four phases of logistic growth are: lag phase (slow growth), exponential growth phase (accelerating growth), deceleration phase (slowing growth), and stable equilibrium phase (little growth).

Factors That Regulate Population Growth

The two factors that regulate population growth are density-independent and density-dependent factors.

Density-independent factors include abiotic factors, such as weather and natural disasters. The effect of the factor is not dependent on density.

Density-dependent factors include biotic factors, such as competition and predation. The effect of the factor is dependent on density.

30.3 Life History Patterns and Extinction

The two fundamental life history patterns are exhibited by opportunistic populations and equilibrium populations.

Opportunistic populations are characterized by small individuals with short life spans, who mature quickly, produce many offspring, have strong dispersal ability, provide little or no care of offspring, and exhibit exponential growth.

Equilibrium populations are characterized by large individuals with long life spans, who mature slowly, produce few offspring, provide much care of offspring, and exhibit logistic growth.

Extinction

Extinction is the total disappearance of a species or higher group. Opportunistic populations are less likely than equilibrium populations to become extinct. Three factors in particular influence vulnerability of equilibrium populations to extinction: size of geographic range, degree of habitat tolerance, and size of local populations.

30.4 The Scope of Ecology

Ecology is an experimental science that studies the interactions of organisms with each other and with the physical environment. The levels of biological organization studied by ecologists are:

- Organisms
- Population
- Community
- Ecosystem
- Biosphere

Thinking Scientifically

1. In Sri Lanka, the death rate is 6 per 1,000, while the birthrate is 19 per 1,000. Calculate the current population growth rate.

   The formula for the doubling time of a population is:

   $t = 0.69/r$, where t is the doubling time and r is the growth rate.

   Determine the doubling rate (the number of years it will take to double the population size) in Sri Lanka. Be sure to convert your growth rate from a percentage (e.g., 2.1%) to a fraction (0.021). If the birthrate drops to 10 per 1,000, what will be the doubling rate?

2. Assume that a population of dandelions grows exponentially so that it doubles in size every week. The population (beginning with one plant) expands to fill a field in 20 weeks. How many weeks will it take to fill one-quarter of the field? How long will it take to fill half the field?

Testing Yourself

Choose the best answer for each question.

1. Decreased death rate followed by decreased birthrate has occurred in
   a. MDCs.
   b. LDCs.
   c. MDCs and LDCs.
   d. neither MDCs nor LDCs.

2. Human societies at present are characterized by
   a. ever-increasing population growth.
   b. unsustainable practices.
   c. overreliance on fossil fuels.
   d. All of these are correct.

3. A population's maximum growth rate is also called its
   a. carrying capacity.
   b. biotic potential.
   c. growth curve.
   d. replacement rate.

4. Which of these levels of ecological study involves both abiotic and biotic components?
   a. organisms
   b. populations
   c. communities
   d. ecosystem
   e. All of these are correct.

5. The level of biological organization subject to evolution by natural selection is called
   a. organism.
   b. population.
   c. community.
   d. ecosystem.
   e. biosphere.

6. Within a range, members of a population exhibit this type of spatial distribution.
   a. variable   c. random
   b. clumped   d. uniform

7. Which are likely to have the highest population density?
   a. zebras on the African savanna during the dry season
   b. mice in a city park
   c. moose in a Canadian forest
   d. earthworms in an organically rich soil

8. Calculate the growth rate of a population of 500 individuals in which the birthrate is 10 per year and the death rate is 5 per year.
   a. 5%
   b. −5%
   c. 1%
   d. −1%

9. If the human birthrate were reduced to 15 per 1,000 per year and the death rate remained the same (9 per thousand), what would be the growth rate?
   a. 9%
   b. 6%
   c. 10%
   d. 0.6%
   e. 15%

10. Replacement reproduction in a population with a pyramid-shaped age structure diagram results in
    a. no population growth.
    b. population growth.
    c. a decline in the population.

11. Label the following age structure diagrams to indicate whether the population is stable, increasing, or decreasing.

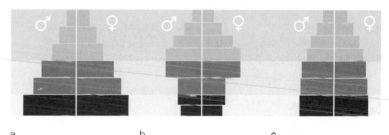

   a. _____   b. _____   c. _____

12. If the number of prereproductive and reproductive members of a population exceeds the number of postreproductive members, the population will
    a. grow.
    b. remain stable.
    c. decline.

13. A J-shaped growth curve indicates
    a. logistic growth.
    b. logarithmic growth.
    c. exponential growth.
    d. additive growth.

14. Exponential growth occurs
    a. when population size is increasing to an ever higher amount.
    b. at the carrying capacity of the environment.
    c. when people are poor and don't have enough to eat.

For statements 15–18, indicate the type of factor in the key exemplified by the scenario.

Key:

    a. density-independent factor
    b. competition
    c. predation
    d. predator-prey cycle

15. A severe drought destroys the entire food supply of a herd of gazelle.

16. A population of feral cats increases in size as the mouse population increases and then crashes regularly after the mouse population enters periods of decline.

17. Only the swiftest coyotes are able to catch the limited supply of rabbits available as a food source. The remaining animals are not strong enough to reproduce.

18. Deer in a forest damage a dense thicket of oak saplings more severely than a few young oak trees.

19. Which of the following is not a feature of an opportunistic life history pattern?

    a. many offspring
    b. little or no care of offspring
    c. long life span
    d. small individuals
    e. fast to mature

20. The distribution of the human population is

    a. variable.
    b. clumped.
    c. random.
    d. uniform.

21. When the carrying capacity of the environment is exceeded, the population typically

    a. increases, but at a slower rate.
    b. stabilizes at the highest level reached.
    c. decreases.
    d. dies off entirely.

22. A pyramid-shaped age distribution means that

    a. the prereproductive group is the largest group.
    b. the population will grow for some time in the future.
    c. the country is more likely an LDC than an MDC.
    d. fewer women are leaving the reproductive years than entering them.
    e. All of these are correct.

23. Which of these is a density-independent factor?

    a. competition
    b. predation
    c. weather
    d. resource availability

Go to www.mhhe.com/maderessentials for more quiz questions.

Bioethical Issue

Species that are more prone to certain risk factors are more likely to become extinct than others. For example, species with a unique lineage, such as the giant panda, are at severe risk for extinction. Should our limited resources for species protection be focused on species that are at the highest risk for extinction? Some argue that high-risk species are less successful products of evolution and should not receive extraordinary protection. Consequently, all species at risk for extinction should be equally protected. Which camp are you in? Support your position.

Understanding the Terms

age structure 539	life history 544
biosphere 546	logistic growth 541
biotic potential 540	more-developed country
carrying capacity 535	(MDC) 535
clumped 538	natural history 547
community 546	opportunistic population 544
competition 542	organism 546
demographics 539	population 546
density-dependent factor 542	population density 538
density-independent factor 542	predation 543
ecology 546	random 538
ecosystem 546	range 538
equilibrium population 544	resource 538
exponential growth 540	survivorship 539
extinction 545	uniform 538
less-developed country	
(LDC) 535	

Match the terms to these definitions:

a. _____ All the populations at a particular locale.

b. _____ Nonliving and living components of the environment that support organisms.

c. _____ Portion of the globe in which a species is found.

d. _____ Number of individuals per unit area or volume.

e. _____ The highest possible rate of increase for a population when resources are unlimited.

f. _____ Total number of individuals that available resources can support.

g. _____ Members of the same species attempting to use the same limited resource.

h. _____ Life history pattern in which population growth is logistic.

i. _____ The total disappearance of a group.

j. _____ Country where population growth is modest and people enjoy a good standard of living.

Communities and Ecosystems

Approximately 70% of the Earth is covered by water, but less than 1% of that water is drinkable.

Photosynthesizers produce almost all of the organic nutrients for the biosphere.

Too much of a nutrient such as nitrogen can harm an ecosystem.

Nitrogen is an essential element for all living things. Unfortunately, the nitrogen gas in the atmosphere is not biologically available to plants until it has been "fixed." This fixing process is part of the nitrogen cycle, and much of the fixing is performed by soil bacteria. Once this process is complete, plants can absorb the nitrogen in the form of ammonium and nitrate ions, and other organisms receive this nitrogen when they eat plants. Humans fix nitrogen ▶ ▶ ▶

OUTLINE

when they make fertilizer, and this provides more than double the amount of nitrogen produced by natural nitrogen fixation. While commercial fertilizers benefit agriculture by causing increased growth of crops, they can also have negative consequences. When excess water runs off agricultural fields, it carries fertilizer into nearby bodies of water. The nitrogen in the fertilizer causes aquatic algae and plants to undergo a huge population increase and then die off. Their subsequent decomposition by organisms of decay leads to a lack of oxygen that can cause a major fish kill.

Excessive use of fertilizer contributes to other ecological problems as well. The nitrates in fertilizer become nitric acid (the primary component of acid rain); cause increasing levels of nitrous oxide gas, which produces smog; and lead to the depletion of certain nutrients in the soil. The question that arises is: Do the benefits of artificial nitrogen fixation outweigh its risks to the environment?

In this chapter, you will learn about communities, aquatic and terrestrial ecosystems, and chemical cycling, including the nitrogen cycle.

31.1 Ecology of Communities

A **community** is an assemblage of populations, each a different species, interacting with one another within a single environment. For example, the various species living on and within a fallen log, such as plants, fungi, worms, and insects, interact with one another and form a community. The fungi break down the log and provide food for the earthworms and insects living in and on the log. Those insects may feed on one another, too. If birds flying throughout the forest feed on the insects and worms living in and on the log, they are also part of the larger forest community.

Communities come in different sizes, and it is sometimes difficult to decide where one community ends and another one begins. The relationships and interactions between species in a community form over time. Some of the relationships between species are products of **coevolution,** by which an evolutionary change in one species results in an evolutionary change in another. As discussed on page 558, flowering plants and their pollinators are co-adapted (Fig. 31.1). A striking example is the flower of the Australian orchid, *Chiloglottis trapeziformis.* This flower resembles the body of a wasp, and its odor mimics the pheromones of a female wasp. Therefore, male wasps are attracted to the flower, and when they attempt to mate with it, they become covered with pollen, which they transfer to the next orchid flower (Fig. 31.1*b*). The orchid is dependent upon wasps for pollination because neither wind nor other insects pollinate these flowers.

All species in a community possess adaptations suitable to the conditions of the particular physical environment. An **ecosystem** consists of these species interacting with each other *and* with the physical environment. If

Figure 31.1 Coevolution.

Flowers and pollinators have evolved to be suited to one another. **a.** Hummingbird pollinated flowers are usually red, a color that these birds can see, and the petals are recurved to allow the stamens to dust the birds' heads. **b.** The reward offered by the flower is not always food. This orchid looks and smells like the female of this wasp's species. The male tries to copulate with flower after flower and in the process transfers pollen. **c.** Bats are nocturnal, and the flowers they pollinate are white or light-colored making them visible in moonlight. The flowers smell like bats and are large and sturdy enabling them to withstand insertion of the bat's head as it uses its long, bristly tongue to lap up nectar and pollen.

a.

b.

c.

the physical environment of an ecosystem changes, corresponding changes will most likely occur in the species comprising the community and in the relationships between these species. Extinction of species can occur when environmental change is too rapid for suitable adaptations to evolve.

Rapid environmental changes can be detrimental to humans too, even though technology increases our ability to adapt. Sometimes the economy of an area is dependent upon the climate and, in some cases, the species composition, of an aquatic or terrestrial ecosystem. Therefore, human activities that negatively impact the community of the area can also negatively affect the economy of that area. Knowledge of community and ecosystem ecology will help you better understand how human activities resulting in, say, climate change can in the end be detrimental to ourselves.

Community Composition and Diversity

Two characteristics of communities—species composition and diversity—allow us to compare communities. The species composition of a community, also known as **species richness,** is simply a listing of the various species found in that community. The **diversity** includes both species richness and species evenness, or the relative abundance of individuals of different species.

Species Composition

It is apparent by comparing the photographs in Figure 31.2 that a coniferous forest has a different species composition than a tropical rain forest. Narrow-leaved evergreen tree species are prominent in the coniferous forest, whereas broad-leaved evergreen tree species are numerous in the tropical rain forest. As the list of mammals demonstrates, a coniferous community and a tropical rain forest community contain different types of mammals. Ecologists comparing these two communities would go on to find differences in other plants and animals too. In the end, ecologists would conclude that not only are the species compositions of these two communities different, but the tropical rain forest has more species and therefore higher species richness.

Diversity

The diversity of a community goes beyond species richness to include abundance, the number of individuals of each species per unit area. For example, suppose a deciduous forest in West Virginia has, among individuals of other species, 76 yellow poplar trees but only one American elm. If you were simply walking through this forest, you might miss the lone American elm. If, instead, the forest had 36 yellow poplar trees and 41 American elm trees, the forest would seem more diverse to you, and indeed would have a higher diversity value. The greater the species richness and the more even the distribution of the species in the community, the greater the diversity.

moose
lynx
snowshoe hare
bear
red fox
wolf

a.

monkey
sloth
anteater
kinkajou
jaguar
tapir
bat

b.

Figure 31.2 Community species composition.

Communities differ in their species composition, as exemplified by the predominant plants and animals in (**a**) a coniferous forest and (**b**) a tropical rain forest. Some mammals found in coniferous forests and in tropical rain forests are listed.

a.

b.

Figure 31.3 **Climax communities.**

Does succession in a particular area always lead to the same climax community? For example, temperate forests (**a**) only occur where there is adequate rainfall, and deserts (**b**) occur where rainfall is minimal. Even so, the exact same mix of plants and animals may not always arise because the assemblage of organisms depends on which organisms, by chance, migrate to the area.

Ecological Succession

Community species composition and diversity do change over time, although they may seem to remain static because it can take decades—even longer than the human life span—for noticeable changes to occur. Natural forces, such as glaciers, volcanic eruptions, lightning-ignited forest fires, hurricanes, tornadoes, and floods, bring about community changes. Communities also change because of human activities, such as logging, road building, sedimentation, and farming. A more or less orderly process of community change is known as **ecological succession.**

Ecologists have developed models to explain why succession occurs and to predict patterns of succession following a natural or human-made disturbance. The climax-pattern model says that the climate of an area always leads to the same stable **climax community,** a specific assemblage of bacterial, fungal, plant, and animal species (Fig. 31.3). For example, a coniferous forest community is expected in northern latitudes, a deciduous forest in temperate zones, and a tropical forest in areas with a tropical climate. Now that we know that disturbances influence community composition and diversity, and that despite the climate, the composition of a community is not always the same, the climax model of succession is being modified.

Two Types of Succession

Ecologists define two types of ecological succession: primary and secondary (Fig. 31.4). **Primary succession** occurs where soil has not yet formed. For example, hardened lava flows or the scraped bedrock that remains following a glacial retreat are subject to primary succession. **Secondary succession** begins, for example, in a cultivated field that is no longer farmed, where soil is already present. With both primary and secondary succession, a progression of species occurs over time as first spores of fungi and then seeds

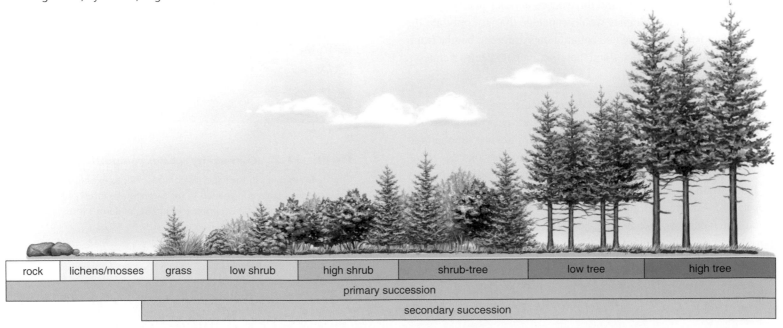

rock	lichens/mosses	grass	low shrub	high shrub	shrub-tree	low tree	high tree

primary succession

secondary succession

Figure 31.4 **Primary and secondary succession.**

Primary succession begins on areas of bare rock. Secondary succession begins in areas where soil remains following natural or human-caused disturbance.

of nonvascular plants, followed by seeds of gymnosperms and/or angiosperms, are carried into the area by wind, water, or animals from the surrounding regions (Figs. 31.4 and 31.5).

The first species to appear in an area undergoing either primary or secondary succession are called opportunistic pioneer species. These species are small in stature, short-lived, quick to mature, and they produce numerous offspring per reproductive event. The first pioneer species to arrive are photosynthetic organisms such as lichens and mosses. The fungal partners of lichens play a critical role by breaking down rock or lava into usable mineral nutrition, not only for their algal partners but also for pioneer plants. The mycorrhizal fungal partners of plants (see Fig. 18.26) pass minerals directly to plants so that they can grow successfully in poor soil. Pioneer plant species that become established in an area are accompanied by pioneer herbivore species (e.g., insects) and then carnivore species (e.g., small mammals). As the community continues to change, equilibrium species become established in the area. Equilibrium species, such as deer, wolves, and bears, are larger in size, long-lived, and slow to mature, and they produce few offspring per reproductive event.

Check Your Progress

1. Describe what is meant by coevolution.
2. Contrast a community with an ecosystem.
3. Contrast species richness with diversity.
4. Contrast primary succession with secondary succession.

Answers: 1. An evolutionary change in one species results in an evolutionary change in another species. 2. A community is a group of populations, while an ecosystem is all the species in a community interacting with each other and with the environment. 3. Species richness is the list of species in a community (species composition), while diversity encompasses both composition and the relative abundance of each species. 4. Primary succession occurs where soil has not yet been formed, while secondary succession occurs where soil is present.

a. During the first year, only the remains of corn plants are seen.

b. During the second year, wild grasses have invaded the area.

c. By the fifth year, the grasses look more mature, and sedges have joined them.

d. After twenty years, the juniper trees are mature, and there are also birch and maple trees in addition to the blackberry shrubs.

Figure 31.5   Secondary succession.

Secondary successional changes in a western Pennsylvania field from (**a**) first year, (**b**) second year, (**c**) fifth year, and (**d**) after twenty years.

Table 31.1	Species Interactions
Interaction	**Expected Outcome**
Competition (− −)	Abundance of both species decreases.
Predation (+ −)	Abundance of predator increases, and abundance of prey decreases.
Parasitism (+ −)	Abundance of parasite increases, and abundance of host decreases.
Commensalism (+ 0)	Abundance of one species increases, and the other is not affected.
Mutualism (+ +)	Abundance of both species increases.

Figure 31.6 **Niche of a backswimmer.**

Backswimmers require warm, clear pond water containing insects that they can eat and vegetation where they can hide from predators.

Interactions in Communities

Species interactions, especially competition for resources, fashion a community into a dynamic system of interspecies relationships. In Table 31.1, the plus and minus signs tell how the relationship affects the abundance of the two interacting species. **Competition** between two species for limited resources has a negative effect on the abundance of both species. In **predation,** one animal, the predator, feeds on another, the prey; in **parasitism,** one species obtains nutrients from another species, called the host, but does not kill the host. **Commensalism** is a relationship in which one species benefits while the second species is not harmed. Commensalism often occurs when one species provides a home or transportation for another. In **mutualism,** two species interact in a way that benefits both of them.

Ecological Niche

Each species occupies a particular position in the community, both in a spatial and a functional sense. Spatially, species live in a particular area of the community, or **habitat,** such as underground, in the trees, or in shallow water. Functionally, each species plays a role, such as whether it is a photosynthesizer, predator, prey, or parasite. The **ecological niche** of a species incorporates the role the species plays in its community, its habitat, and its interactions with other species. The niche includes the living and nonliving resources that individuals in the population need to meet their energy, nutrient, and survival demands. The habitat of an insect, called a backswimmer, is a pond or lake, where it eats other insects (Fig. 31.6). The pond or lake must contain vegetation where the backswimmer can hide from its predators, including fish and birds. The pond water must be clear enough for the backswimmer to see its prey and warm enough for it to maintain a good metabolic rate. Since it is difficult to describe and measure the total niche of a species in a community, ecologists often focus on a certain aspect of a species' niche, as with the birds featured in Figure 31.7.

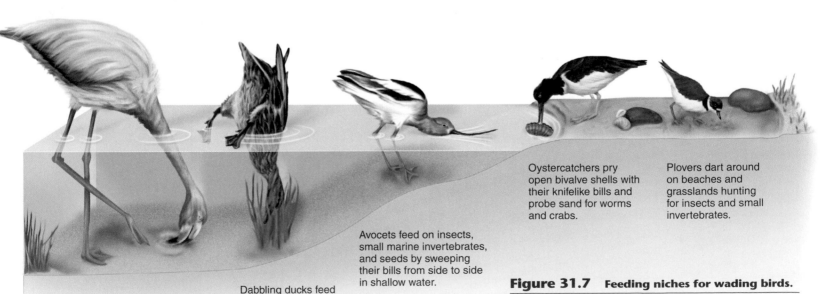

Oystercatchers pry open bivalve shells with their knifelike bills and probe sand for worms and crabs.

Plovers dart around on beaches and grasslands hunting for insects and small invertebrates.

Avocets feed on insects, small marine invertebrates, and seeds by sweeping their bills from side to side in shallow water.

Flamingos feed on small molluscs, crustaceans, and vegetable matter strained from mud pumped through their bills by their powerful tongues.

Dabbling ducks feed by tipping, tail up, to reach aquatic plants, seeds, snails, and insects.

Figure 31.7 **Feeding niches for wading birds.**

Flamingos feed in deeper water by filter feeding; dabbling ducks feed in shallower areas by upending; avocets feed by sifting. Oystercatchers and plovers have adaptations for feeding in shallows, such as shorter legs.

Competition

Competition for resources such as light, space, or nutrients contributes to the niche of each species and helps structure the community. Laboratory experiments helped ecologists formulate the **competitive exclusion principle,** which states that no two species can occupy the same niche at the same time. In the 1930s, G. F. Gause grew two species of *Paramecium* in one test tube containing a fixed amount of bacterial food. Although populations of each species survived when grown in separate test tubes, only one species, *Paramecium aurelia*, survived when the two species were grown together (Fig. 31.8). *P. aurelia* acquired more of the food resource and had a higher population growth rate than did *P. caudatum*. Eventually, as the *P. aurelia* population grew and obtained an increasingly greater proportion of the food resource, the number of *P. caudatum* individuals decreased, and the population died out.

Niche Specialization   Competition for resources does not always lead to localized extinction of a species. Multiple species coexist in communities by partitioning, or sharing, resources. In another laboratory experiment using other species of *Paramecium,* Gause found that the two species could survive in the same test tube if one species consumed bacteria at the bottom of the tube and the other ate bacteria suspended in solution. This **resource partitioning** decreased competition between the two species, leading to increased niche specialization. What could have been one niche became two more specialized niches due to species differences in feeding behavior.

When three species of ground finches of the Galápagos Islands live on the same islands, their beak sizes differ, and each feeds on a different-sized seed (Fig. 31.9). When the finches live on separate islands, their beaks tend to be the same intermediate size, enabling each to feed on a wider range of seeds. Such so-called **character displacement** often is viewed as evidence that competition and resource partitioning have taken place.

The niche specialization that permits coexistence of multiple species can be very subtle. Species of warblers that occur in North American forests are all about the same size, and all feed on budworms, a type of caterpillar found on spruce trees. Robert MacArthur recorded the length of time each warbler species spent in different regions of spruce canopies to determine where each species did most of its feeding. He discovered that each species primarily used different parts of the tree canopy and, in that way, had a more specialized niche. As another example, consider that three types of birds—swallows, swifts, and martins—all eat flying insects and parachuting spiders. These birds even frequently fly in mixed flocks. But each type of bird has a different nesting site and migrates at a slightly different time of year.

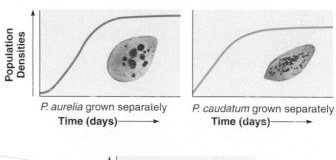

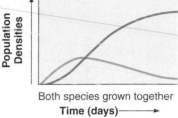

Both species grown together
Time (days) ⟶

Figure 31.8   Competitive exclusion principle demonstrated by *Paramecium*.

The competitive exclusion principle states that no two species occupy the exact same niche. When grown separately, *Paramecium caudatum* and *Paramecium aurelia* exhibit logistic growth. When grown together, *P. aurelia* excludes *P. caudatum*.

Data from G.F. Gause, *The Struggle for Existance*, 1934, Williams & Wilkins Company, Baltimore, MD. p. 557.

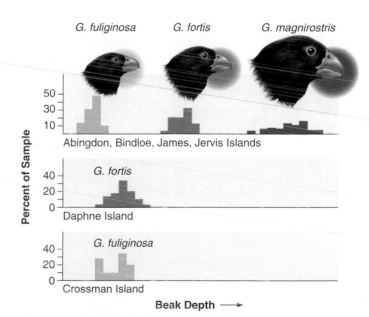

Figure 31.9   Character displacement in finches on the Galápagos Islands.

When *G. fuliginosa, G. fortis,* and *G. magnirostris* coexist on the same island, their beak sizes are appropriate for eating small-, medium-, and large-sized seeds, respectively. When *G. fortis* and *G. fuliginosa* are on separate islands, their beaks have the same intermediate size, which allows them to eat seeds of various sizes. Character displacement is evidence that resource partitioning has occurred.

Mutualism

Mutualism, a symbiotic relationship in which both members benefit, is now recognized to be at least as important as competition in shaping community structure. The relationship between plants and their pollinators mentioned previously is a good example of mutualism. Perhaps the relationship began when herbivores feasted on pollen. The provision of nectar by the plant may have spared the pollen, and at the same time allowed the animal to become an agent of pollination. By now, pollinator mouthparts are adapted to gathering the nectar of a particular plant species, and this species is dependent on the pollinator for dispersing pollen. As also mentioned previously, lichens can grow on rocks because the fungal member leaches minerals that are provided to the algal partner. The algal partner, in turn, photosynthesizes and provides organic food for both members of the relationship.

In tropical America, ants form mutualistic relationships with certain plants. The bullhorn acacia tree is adapted to provide a home for ants of the species *Pseudomyrmex ferruginea* (Fig. 31.10). Unlike other acacias, this species has swollen thorns with a hollow interior where ant larvae can grow and develop. In addition to housing the ants, acacias provide them with food. The ants feed from nectaries at the base of the leaves and eat fat- and protein-containing nodules called Beltian bodies, which are found at the tips of the leaves. For 24 hours a day, ants constantly protect the tree from herbivores that would like to feed on it. The ants are so critical to the trees' survival that when the ants on experimental trees were poisoned, the trees died.

The outcome of mutualism is an intricate web of species interdependencies critical to the community. For example, in areas of the western United States, the branches and cones of whitebark pine are turned upward, meaning that the seeds do not fall to the ground when the cones open. Birds called Clark's nutcrackers eat the seeds of whitebark pine trees and store them in the ground (Fig. 31.11). Therefore, Clark's nutcrackers are critical seed dispersers for the trees. Also, grizzly bears find the stored seeds and consume them. Whitebark pine seeds do not germinate unless their seed coats are exposed to fire. When natural forest fires in the area are suppressed, whitebark pine trees decline in number, and so do Clark's nutcrackers and grizzly bears. When lightning-ignited fires are allowed to burn, or prescribed burning is used in the area, the whitebark pine populations increase, as do the populations of Clark's nutcrackers and grizzly bears.

Community Stability

As witnessed by our discussion of succession, community stability is fragile. However, some communities have one species that stabilizes the community, helps maintain its characteristics, and essentially helps hold its web of interactions together. Such a species is known as a **keystone species,** or species on which the existence of a large number of other species in the ecosystem depends. The term "keystone" comes from the name for the center stone of an arch that holds the other stones in place so that the arch can keep its shape.

Figure 31.10   Mutualism.

The bullhorn acacia tree is adapted to provide nourishment for a mutualistic ant species. **a.** The thorns are hollow, and the ants live inside. **b.** The bases of the leaves have nectaries (openings) where ants can feed. **c.** The tips of the leaves of the bullhorn acacia have Beltian bodies, which ants harvest for larval food.

Figure 31.11   Interdependence of species.

Clark's nutcrackers feed on the seeds of whitebark pines. But the storing of the seeds by the nutcrackers is the primary means of seed dispersal for whitebark pine.

Keystone species are not always the most numerous in the community. However, the extinction of a keystone species can lead to other species extinctions and a loss of diversity. For example, bats are designated as keystone species in tropical forests. Bats are pollinators and also disperse the seeds of certain tropical trees. When bats are killed off or their roosts destroyed, these trees fail to reproduce. The grizzly bear is a keystone species in the northwestern United States and Canada. Grizzly bears disperse as many as 7,000 berry seeds in one dung pile. Grizzlies also kill the young of many herbivorous mammals, such as deer, and thereby keep their populations under control.

The sea otter is a keystone species of a kelp forest ecosystem. Kelp forests, created by large brown seaweeds, provide a home for a vast assortment of organisms. The kelp forests occur just off the coast and protect coastline ecosystems from damaging wave action. Among other species, sea otters eat sea urchins, keeping their population size in check. Otherwise, sea urchins feed on the kelp, causing the kelp forest and its associated species to severely decline. Fishermen don't like sea otters because they also prey on abalone, a mollusc prized for its commercial value. They do not realize that without the otters, abalone and many other species would not be around because their natural habitat, a kelp forest, would no longer exist.

Native Versus Exotic Species

Native species are species indigenous to an area. They colonize an area without intentional or accidental human assistance. For example, you naturally find maple trees in Vermont and many other states of the eastern United States. The introduction of **exotic species,** or nonnative species, into a community greatly disrupts normal interactions, and therefore a community's web of species. Populations of exotic species tend to grow exponentially because they are better competitors, or because their numbers are not controlled by predators or disease. The unique assemblage of native species on an island often cannot compete well against an exotic species. For example, myrtle trees, introduced into the Hawaiian Islands from the Canary Islands, are mutualistic with a type of bacterium capable of fixing atmospheric nitrogen. The bacterium provides the plant with a form of nitrogen it can utilize. This feature allows myrtle trees to become established on nitrogen-poor volcanic soil, a distinct advantage in Hawaii. Once established, myrtle trees halt the normal succession of native plants on volcanic soil.

Exotic species disrupt communities in continental areas as well. The red fox was deliberately imported into Australia to prey on the previously introduced European rabbit, but instead, the red fox has now reduced the populations of native small mammals (Fig. 31.12a). The brown tree snake was introduced onto a number of islands in the Pacific Ocean (Fig. 31.12b). The brown tree snake eats eggs, nestlings, and adult birds. On Guam, it has reduced ten native bird species to the point of extinction. On the Galápagos Islands, black rats accidentally carried to the islands by ships have reduced populations of the giant tortoise. Goats and feral pigs have changed the vegetation on the islands from highland forest to pampas-like grasslands and destroyed stands of cactus. In the United States, gypsy moths, zebra mussels, the Chestnut blight fungus, fire ants, and African bees are well-known exotic species that have killed native species. At least two species, fire ants and African bees, have attacked humans with serious consequences.

a.

b.

Figure 31.12   Exotic species.

Human introduction of exotic species, such as (**a**) the red fox and (**b**) the brown tree snake, have disrupted communities in Australia and Guam, respectively.

Check Your Progress

1. List the five major types of species interactions in a community.
2. Distinguish between habitat and ecological niche.
3. Why is character displacement a form of resource partitioning?
4. Describe the role of a keystone species.

Answers: 1. Competition, predation, parasitism, commensalism, and mutualism. 2. A habitat is the special location of a species. It is one component of an ecological niche, which also includes the role the species plays in the community and its interactions with other species. 3. Due to character displacement, species are able to feed on different types of food. 4. A keystone species stabilizes the community, helps maintain its characteristics, and helps hold the web of interactions together.

Figure 31.13 **Producers.**

Green plants and algae are photoautotrophs.

a. Herbivores

b. Carnivores

Figure 31.14 **Consumers.**

a. Caterpillars and giraffes are herbivores. **b.** A praying mantis and an osprey are carnivores.

31.2 Ecology of Ecosystems

An ecosystem is more inclusive than a community because when studying a community, we only consider how species interact with one another. When studying an ecosystem, we also consider interactions with the physical environment. For example, one important aspect of ecological niche is how an organism acquires food. It is obvious that autotrophs interact with the physical environment but so do heterotrophs.

Autotrophs

Autotrophs take in only inorganic nutrients (e.g., CO_2 and minerals) and an outside energy source to produce organic nutrients for their own use and for all the other members of a community. They are called **producers** because they produce food. *Photoautotrophs,* often called photosynthetic organisms, produce most of the organic nutrients for the biosphere (Fig. 31.13). Algae of all types possess chlorophyll and carry on photosynthesis in freshwater and marine habitats. Algae make up the phytoplankton, which are photosynthesizing organisms suspended in water. Green plants are the dominant photosynthesizers on land. All photosynthesizing organisms release O_2 to the atmosphere.

Some bacteria are *chemoautotrophs.* They obtain energy by oxidizing inorganic compounds such as ammonia, nitrites, and sulfides, and they use this energy to synthesize organic compounds. Chemoautotrophs have been found to support communities in some caves and also at hydrothermal vents along deep-sea oceanic ridges.

Heterotrophs

Heterotrophs need a source of preformed organic nutrients, and they release CO_2 to the atmosphere. They are called **consumers** because they consume food. Herbivores are animals that graze directly on algae or plants (Fig. 31.14*a*). In aquatic habitats, zooplankton, such as protozoans, are herbivores; in terrestrial habitats, insects, including caterpillars, play that role. Among larger herbivores, giraffes browse on trees. Carnivores eat other animals; for example, a praying mantis catches and eats caterpillars, and an osprey preys on and eats fish (Fig. 31.14*b*). These examples illustrate that there are primary consumers (e.g., giraffes), secondary consumers (e.g., praying mantis), and tertiary consumers (e.g., osprey). Sometimes tertiary consumers are called top predators. Omnivores are animals that eat both plants and animals. As you likely know, most humans are omnivores.

The **decomposers** are heterotrophic bacteria and fungi, such as molds and mushrooms, that break down dead organic matter, including animal wastes (Fig. 31.15). They perform a very valuable service because they release inorganic nutrients (CO_2 and minerals) that are then taken up by plants once more. Otherwise, plants would

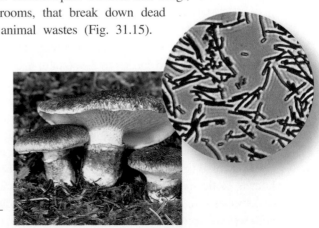

Figure 31.15 **Decomposers.**

Mushrooms and bacteria are decomposers.

of dead organisms plus the bacteria and fungi of decay. Fanworms feed on detritus floating in marine waters, while clams take detritus from the sea bottom. Earthworms and some beetles, termites, and maggots are soil detritus feeders.

Energy Flow and Chemical Cycling

The living components of ecosystems process energy and chemicals. Energy flow through an ecosystem begins when producers absorb solar energy, and chemical cycling begins when producers take in inorganic nutrients from the physical environment (Fig. 31.16). Thereafter, via photosynthesis, producers convert the solar energy and inorganic nutrients into chemical energy in the form of organic nutrients, such as carbohydrates. Producers synthesize organic nutrients directly for themselves and indirectly for the heterotrophic components of the ecosystem. Energy flows through an ecosystem because as organic nutrients pass from one component of the ecosystem to another, as when a herbivore eats a plant or a carnivore eats a herbivore, a portion is used as an energy source. Eventually, the energy dissipates into the environment as heat. Therefore, the vast majority of ecosystems cannot exist without a continual supply of solar energy.

Only a portion of the organic nutrients made by producers is passed on to consumers because plants use organic molecules to fuel their own cellular respiration. Similarly, only a small percentage of nutrients consumed by lower-level consumers, such as herbivores, is available to higher-level consumers, or carnivores. As Figure 31.17 demonstrates, a certain amount of the food eaten by a herbivore is never digested and is eliminated as feces. Metabolic wastes are excreted as urine. Of the assimilated energy, a large portion is utilized during cellular respiration for the production of ATP and thereafter becomes heat. Only the remaining energy, which is converted into increased body weight or additional offspring, becomes available to carnivores.

The elimination of feces and urine by a heterotroph, and indeed the death of all organisms, does not mean that organic nutrients are lost to an ecosystem. Instead, they represent the organic nutrients made available to decomposers. Decomposers convert the organic nutrients back into inorganic chemicals and release them to the soil or atmosphere. Chemicals complete their cycle within

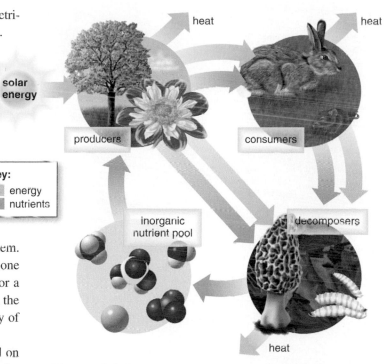

Figure 31.16   Chemical cycling and energy flow.

Chemicals cycle within, but energy flows through, an ecosystem. As energy is repeatedly passed from one component to another, all the chemical energy derived from solar energy dissipates as heat.

Key:
□ energy
■ nutrients

Figure 31.17   Energy balances.

Only about 10% of the nutrients and energy taken in by a herbivore is passed on to carnivores. A large portion goes to detritus feeders. Another large portion is used for cellular respiration

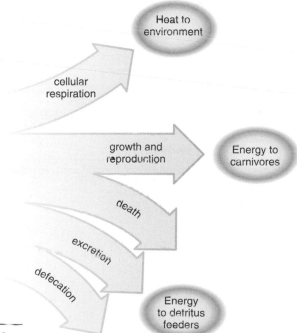

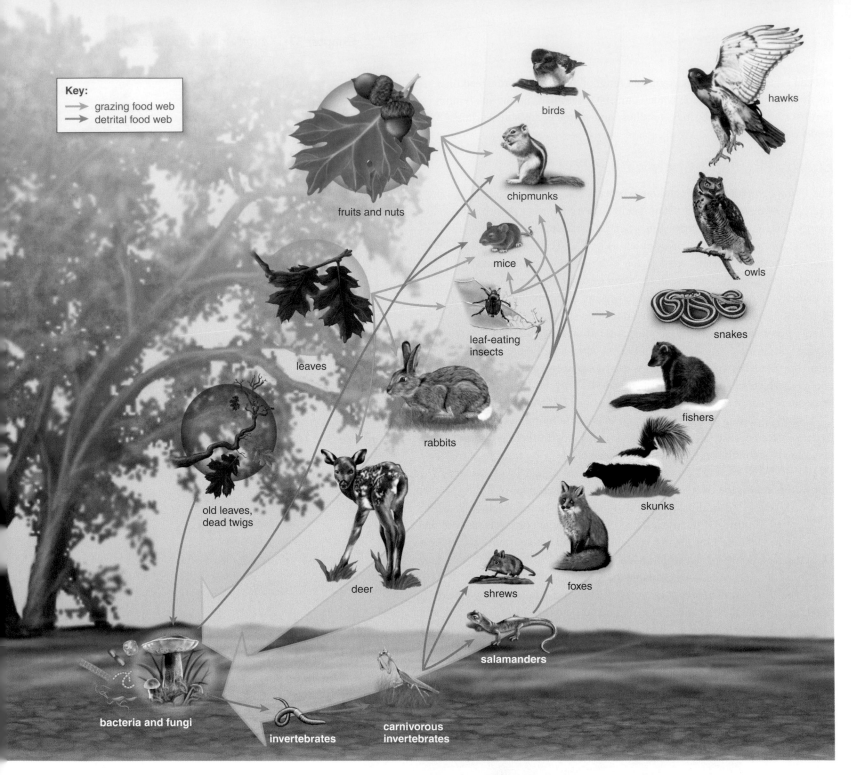

Key:
→ grazing food web
→ detrital food web

hawks

birds

fruits and nuts

chipmunks

owls

mice

leaf-eating insects

snakes

leaves

rabbits

fishers

old leaves, dead twigs

deer

skunks

foxes

shrews

salamanders

bacteria and fungi

invertebrates

carnivorous invertebrates

Figure 31.18   Food webs.

The grazing and detrital food webs of ecosystems are linked.

an ecosystem when inorganic chemicals are absorbed by the producers from the atmosphere or from soil.

Energy Flow

Applying the principles discussed so far to a temperate deciduous forest, ecologists can draw a **food web** to represent the interconnecting paths of energy flow between the component species of the ecosystem. In Figure 31.18, the green arrows are part of a **grazing food web** because the web begins with plants, such as the oak trees depicted. A **detrital food web** (brown arrows) begins with bacteria and fungi. In the grazing food web, caterpillars and other herbivorous insects feed on the leaves of the trees, while other herbivores, including mice, rabbits, and deer, feed on leaves at or near the ground. Birds, chipmunks, and

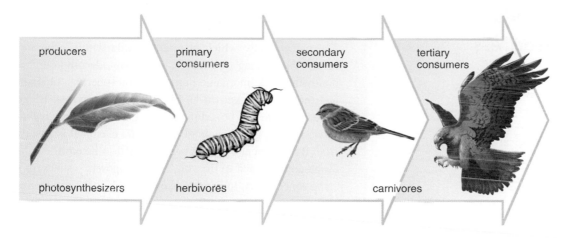

producers	primary consumers	secondary consumers	tertiary consumers
photosynthesizers	herbivores		carnivores

Figure 31.19   Food chain.

A food chain diagrams a single path of energy flow in an ecosystem.

mice feed on fruits and nuts of the trees, but they are in fact omnivores because they also feed on caterpillars and other insects. These herbivores and omnivores all provide food for a number of different carnivores. In the detrital food web, detritus, which includes smaller decomposers such as bacteria and fungi, is food for larger organisms. Because some of these organisms, such as shrews and salamanders, become food for aboveground animals, the detrital and the grazing food webs are joined.

We tend to think that the aboveground parts of trees are the largest storage form of organic matter and energy, but this is not necessarily the case. In temperate deciduous forests, the organic matter lying on the forest floor and mixed into the soil, along with the belowground roots of the trees, contains over twice the energy of the leaves, branches, and trunks of living trees combined. Therefore, more energy and matter in a forest may be stored in or funneled through the detrital food web than the grazing food web.

Trophic Levels and Ecological Pyramids   The arrangement of component species in Figure 31.19 suggests that organisms are linked to one another in a straight line according to feeding relationships, or who eats whom. Diagrams that show a single path of energy flow in an ecosystem are called **food chains** (Fig. 31.19). A **trophic level** is a level of nourishment within a food web or chain. In the grazing food web (see Fig. 31.18), going from left to right: The trees are producers (the first trophic level), the first series of animals are herbivores (the second trophic level), and many of the animals in the next series are carnivores (the third and possibly fourth trophic levels). Food chains are short because energy is lost between trophic levels. In general, only about 10% of the energy of one trophic level is available to the next trophic level. Therefore, if an herbivore population consumes 1,000 kg of plant material, only about 100 kg is converted to herbivore tissue, 10 kg to first-level carnivores, and 1 kg to second-level carnivores. The so-called 10% rule of thumb explains why few carnivores can be supported in a food web. The flow of energy with large losses between successive trophic levels is sometimes depicted as an **ecological pyramid** (Fig. 31.20).

A pyramid based on the number of organisms can run into problems because, for example, one tree can support many herbivores. Pyramids of **biomass,** which is the number of organisms multiplied by their weight, eliminate size as a factor. Even then, apparent inconsistencies can arise. In aquatic ecosystems, such as lakes and open seas where algae are the only producers, the herbivores at some point in time may have a greater biomass than the producers. Why? Because even though the algae reproduce rapidly, they are also consumed at a high rate.

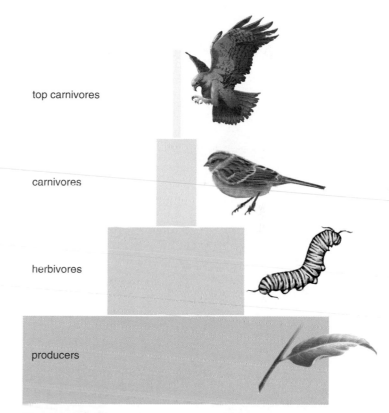

top carnivores

carnivores

herbivores

producers

Figure 31.20   Ecological pyramid.

An ecological pyramid depicts the loss of nutrients and energy from one trophic level to the next.

Check Your Progress

1. Contrast the two types of autotrophs.
2. Contrast herbivores with carnivores.
3. Contrast the first organisms in the grazing food web with those in the detrital food web.
4. Contrast a food chain with an ecological pyramid.
5. Contrast a sedimentary biogeochemical cycle with a gaseous one.

Answers: 1. Photoautotrophs use photosynthesis to produce organic nutrients, while chemoautotrophs use energy from inorganic compounds to produce organic nutrients. 2. Herbivores are animals that feed on autotrophs, while carnivores feed on other animals. 3. The first organisms in the grazing food web are herbivorous insects, while those in the detrital food web are bacteria and fungi. 4. A food chain depicts a single path of energy flow in an ecosystem, while an ecological pyramid depicts the entire flow of energy between trophic levels. 5. A sedimentary cycle involves the exchange of chemicals in the soil, while a gaseous cycle involves chemicals in the atmosphere.

Chemical Cycling

The pathways by which chemicals cycle within ecosystems involve both living (producers, consumers, decomposers) and nonliving (rock, inorganic nutrients, atmosphere) components, and therefore are known as **biogeochemical cycles.** Biogeochemical cycles can be sedimentary or gaseous. In a *sedimentary cycle,* such as the phosphorus cycle, the chemical is absorbed from the sediment by plant roots, passed to heterotrophs, and eventually returned to the soil by decomposers, usually in the same general area. In a *gaseous cycle,* such as the nitrogen and carbon cycles, the element returns to and is withdrawn from the atmosphere as a gas.

Chemical cycling of an element may involve reservoirs and exchange pools as well as the biotic community (Fig. 31.21). A *reservoir* is a source normally unavailable to organisms. For example, much carbon is found in calcium carbonate shells in ocean bottom sediments. An *exchange pool* is a source from which organisms generally take elements. For example, photosynthesizers can utilize carbon dioxide in the atmosphere for their carbon needs. The *biotic community* consists of the autotrophic and heterotrophic species of an ecosystem that feed on one another. Human activities, such as mining or burning fossil fuels, increase the amounts of chemical elements removed from reservoirs and cycling within ecosystems. As a result, the physical environment of the ecosystem contains excess chemicals that, in turn, can alter the species composition and diversity of the biotic community.

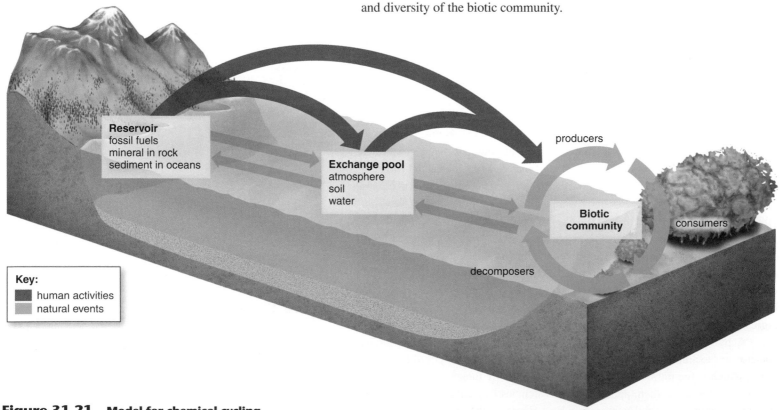

Figure 31.21   Model for chemical cycling.

Chemical nutrients cycle between these components of ecosystems. Reservoirs, such as fossil fuels, minerals in rocks, and sediments in oceans, are normally relatively unavailable sources, but exchange pools, such as those in the atmosphere, soil, and water, are available sources of chemicals for the biotic community. When human activities (purple arrows) remove chemicals from reservoirs and make them available to the biotic community, pollution can result.

Phosphorus Cycle

On land, the very slow weathering of rocks fostered by fungi adds phosphates (PO_4^{3-} and HPO_4^{2-}) to the soil, some of which become available to terrestrial plants for uptake (Fig. 31.22). Phosphates made available by weathering also run off into aquatic ecosystems, where algae absorb phosphates from the water before they become trapped in sediments. Phosphates in sediments become available again only when a geological upheaval exposes sedimentary rocks to weathering once more.

Producers use phosphates in a variety of molecules, including phospholipids, ATP, and the nucleotides that become a part of DNA and RNA. Animals consume producers and incorporate some of the phosphates into teeth, bones, and shells that take many years to decompose. Decomposition of dead plant and animal material and animal wastes does, however, make phosphates available to producers at a faster rate than weathering. Because much of the available phosphates are utilized within food chains, phosphate is usually a limiting inorganic nutrient for ecosystems. In other words, the limited supply of phosphates limit plant growth, and therefore primary productivity.

The importance of phosphates and calcium to population growth is demonstrated by considering the fate of lemmings every four years (Fig. 31.23). You may have heard that lemmings dash mindlessly over cliffs into the sea; ecologists tell us that actually these lemmings are migrating to find food. What happened? Every four years or so, grasses and sedges of the tundra (see Fig. 31.28) become rich in minerals, and the lemming population starts to explode. Once the lemmings number in the millions, the grasses and sedges of the tundra suffer a decline caused by a lack of minerals. Now the lemming population suffers a crash, but it takes about four years before the animals decompose in this cold region and minerals return to the producers. Then the cycle begins again.

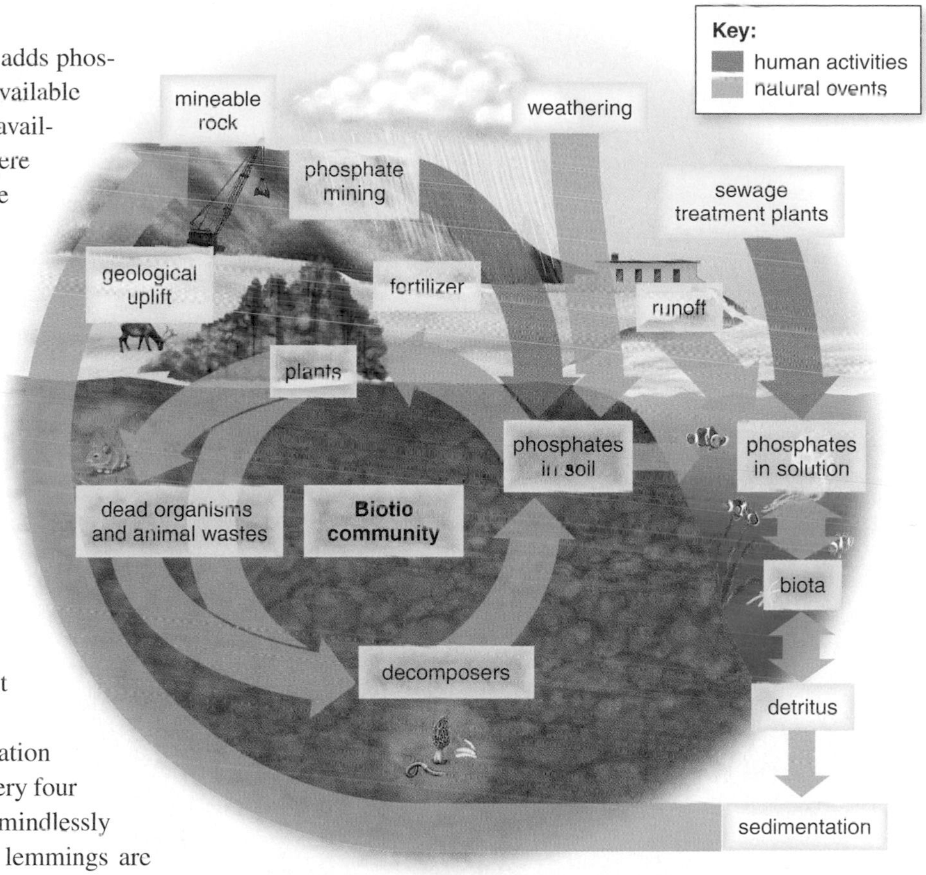

Human Activities A **transfer rate** is defined as the amount of a nutrient that moves from one component of the environment to another within a specified period of time. Human activities alter the dynamics of a community by changing transfer rates. For example, humans mine phosphate ores and use them to make fertilizers, animal feed supplements, and detergents. Phosphate ores are slightly radioactive, and therefore mining phosphate poses a health threat to all organisms, including the miners. Animal wastes from livestock feedlots, fertilizers from lawns and cropland, and untreated and treated sewage discharged from cities all add excess phosphates to nearby waters. The result is **eutrophication,** or overenrichment, of a body of water, which causes an algal overpopulation called an algal bloom. When the algae die and decay, oxygen is consumed, causing fish kills. In the mid-1970s, Lake Erie was dying because of eutrophication. Control of nutrient phosphates, particularly in sewage effluent and household detergents, reversed the situation.

Figure 31.22 The phosphorus cycle.

The phosphorus cycle is a sedimentary biogeochemical cycle. Globally, phosphates flow into large bodies of water and become a part of sedimentary rocks. Thousands or millions of years later, the seafloor can rise; the phosphates are then exposed to weathering and become available. Locally, phosphates cycle within a community when plants on land and algae in the water take them up. Animals gain phosphates when they feed on plants or algae. Decomposers return phosphates to plants or algae, and the cycle within the community begins again.

Figure 31.23 Lemmings.

Lemmings, small furry rodents, feed on plant species in an Arctic community called the tundra.

Nitrogen Cycle

Nitrogen, in the form of nitrogen gas (N_2), comprises about 78% of the atmosphere by volume. But plants cannot make use of nitrogen gas. Instead, plants rely on various types of bacteria to make nitrogen available to them. Therefore, nitrogen, like phosphorus, is also a limiting inorganic nutrient of producers in ecosystems.

Plants can take up both ammonium (NH_4^+) and nitrate (NO_3^-) from the soil and incorporate the nitrogen into amino acids and nucleic acids. Two processes, nitrogen fixation and nitrification, convert nitrogen gas, N_2, into NH_4^+ and NO_3^-, respectively (Fig. 31.24). *Nitrogen fixation* occurs when nitrogen gas is converted to ammonium. Some cyanobacteria in aquatic ecosystems and some free-living, nitrogen-fixing bacteria in soil are able to fix nitrogen in this way. Other nitrogen-fixing bacteria live in nodules on the roots of legumes, plants such as peas, beans, and alfalfa. They make organic compounds containing nitrogen available to the host plant.

Nitrification is the production of nitrates. Ammonium in the soil is converted to nitrate by certain nitrifying soil bacteria in a two-step process. (First, nitrite-producing bacteria convert ammonium to nitrite (NO_2^-), and then nitrate-producing bacteria convert nitrite to nitrate.) In Figure 31.24, notice that the biotic community subcycle in the nitrogen cycle does not depend on the presence of nitrogen gas.

Denitrification is the conversion of nitrate to nitrogen gas, which enters the atmosphere. Denitrifying bacteria are chemoautotrophs living in the anaerobic mud of lakes, bogs, and estuaries that carry out this process as a part of their own metabolism. In the nitrogen cycle, denitrification counterbalanced nitrogen fixation until humans started making fertilizer.

Human Activities Human activities significantly alter the transfer rates in the nitrogen cycle by producing fertilizers from N_2—in fact, humans nearly double the fixation rate. Unfortunately, industrial nitrogen fixation requires tremendous heat and pressure, usually produced by burning great quantities of fossil fuels, with accompanying air pollution. The nitrate in fertilizers, just like phosphate, can leach out of agricultural soils into surface waters, leading to eutrophication. Deforestation by humans also causes a loss of nitrogen to groundwater and makes regrowth of the forest difficult. The underground water supplies in farming areas today are apt to contain excess nitrate. Infants below the age of six months who drink water containing excessive amounts of nitrate can become seriously ill and, if untreated, may die.

To cut back on fertilizer use, it might be possible to genetically engineer soil bacteria with increased nitrogen fixation rates. Also, farmers could grow legumes that increase the nitrogen content of soil (Fig. 31.25). In one study, the rotation of legumes and winter wheat produced a better yield than fertilizers after several years.

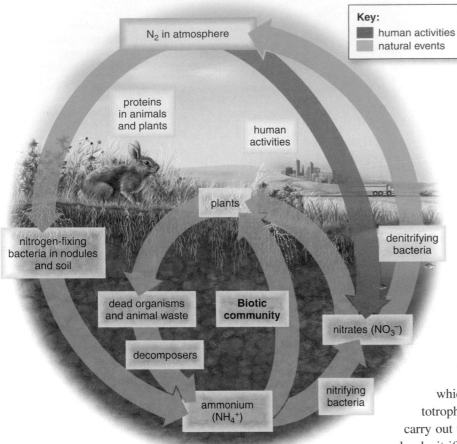

Figure 31.24 **The nitrogen cycle.**

The nitrogen cycle is a gaseous biogeochemical cycle normally maintained by the work of several populations of soil bacteria.

Figure 31.25 **Root nodules.**

Bacteria that live in nodules on the roots of plants in the legume family, such as pea plants, convert nitrogen in the air to a form that plants can use to make proteins.

Carbon Cycle

In the carbon cycle, organisms in both terrestrial and aquatic ecosystems exchange carbon dioxide with the atmosphere (Fig. 31.26). On land, plants take up carbon dioxide from the air, and through photosynthesis, they incorporate carbon into organic nutrients that are used by autotrophs and heterotrophs alike. When aerobic organisms respire, a portion of this carbon is returned to the atmosphere as carbon dioxide, a waste product of cellular respiration.

In aquatic ecosystems, the exchange of carbon dioxide with the atmosphere is indirect. Carbon dioxide from the air combines with water to produce bicarbonate ion (HCO_3^-), a source of carbon for algae that produce food for themselves and for heterotrophs. Similarly, when aquatic organisms respire, the carbon dioxide they give off becomes bicarbonate ion. The amount of bicarbonate in the water is in equilibrium with the amount of carbon dioxide in the air.

Living and dead organisms contain organic carbon and serve as one of the reservoirs for the carbon cycle. The world's biotic components, particularly trees, contain billions of tons of organic carbon, and additional tons are estimated to be held in the remains of plants and animals in the soil. If dead plant and animal remains fail to decompose, they are subjected to extremely slow physical processes that transform them into coal, oil, and natural gas, the **fossil fuels.** Most of the fossil fuels were formed during the Carboniferous period, 286–360 MYA, when an exceptionally large amount of organic matter was buried before decomposing. Another reservoir is the calcium carbonate ($CaCO_3$) that accumulates in limestone and in shells. Many marine organisms have calcium carbonate shells that remain in bottom sediments long after the organisms have died. Geological forces change these sediments into limestone.

Human Activities The transfer rates of carbon dioxide due to photosynthesis and cellular respiration are just about even. However, more carbon dioxide is being deposited in the atmosphere than is being removed. This increase is largely due to the burning of fossil fuels and the destruction of forests to make way for farmland and pasture. When we do away with forests, excess carbon dioxide enters the atmosphere. But only about half of this excess CO_2 remains in the atmosphere; it is believed that the rest is dissolved in the ocean.

The increased amount of carbon dioxide (and other gases) in the atmosphere is causing a rise in temperature called **global warming.** These gases allow the sun's rays to pass through, but they absorb and reradiate heat back to the Earth, a phenomenon called the **greenhouse effect.** Scientists predict that if the ice at the poles melts and sea levels rise as a result, many of the world's most populous cities will be flooded. Furthermore, weather pattern changes might cause the American Midwest to become a dust bowl.

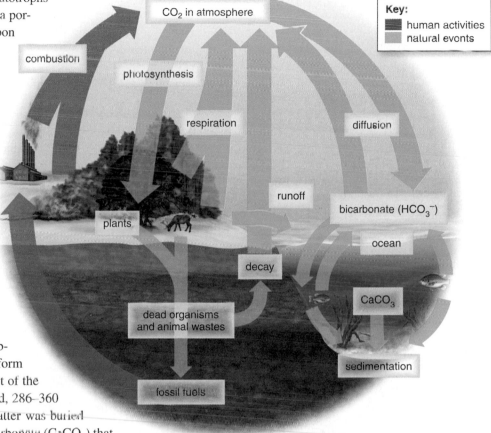

Figure 31.26   The carbon cycle.

The carbon cycle is a gaseous biogeochemical cycle. Producers take in carbon dioxide from the atmosphere and convert it to organic molecules that feed all organisms. Fossil fuels arise when organisms die but do not decompose. The burning of fossil fuels releases carbon dioxide and causes environmental pollution.

Check Your Progress

1. Describe how phosphate enters the phosphorus cycle.
2. Describe the significance of nitrogen fixation and nitrification to the nitrogen cycle.
3. Describe the relationship between the greenhouse effect and global warming.

Answers: 1. Most phosphate comes from decomposition of organisms, but some comes from weathering. *2.* Both processes allow nitrogen to enter the nitrogen cycle. *3.* The greenhouse effect (absorption of the sun's rays) is believed to contribute to global warming (a rise in the temperature of the atmosphere).

31.3 Ecology of Major Ecosystems

The **biosphere,** which encompasses all the ecosystems on planet Earth, is the final level of biological organization. **Aquatic ecosystems** are divided into those composed of fresh water and those composed of salt water (marine ecosystems) (Fig. 31.27). The ocean is a marine ecosystem that covers 70% of the Earth's surface. Two types of freshwater ecosystems are those with standing water, such as lakes and ponds, and those with running water, such as rivers and streams. The richest marine ecosystems lie near the coasts. Coral reefs are located offshore, while marshes occur where rivers meet the sea.

Scientists recognize several distinctive major types of **terrestrial ecosystems,** also called *biomes* (Fig. 31.28). Temperature and rainfall define the biomes, which contain communities adapted to the regional climate. The tropical rain forests, which occur at the equator, have a high average temperature and the greatest amount of rainfall of all the biomes. They are dominated by large evergreen, broad-leaved trees. The savanna is a tropical grassland with alternating wet and dry seasons. Temperate grasslands receive less rainfall than temperate forests (in which trees lose their leaves during the winter) and more water than deserts, which lack trees. The taiga is a very cold northern coniferous forest, and the tundra, which borders the North Pole, is also very cold, with long winters and a short growing season. A permafrost persists even during the summer in the tundra and prevents large plants from becoming established.

Figure 31.27 **The major aquatic ecosystems.**

Aquatic ecosystems are divided into those that have salt water, such as the ocean (**a**) and those that have fresh water, such as a river (**b**). Saltwater, or marine, ecosystems also include coral reefs (**c**) and marshes (**d**).

Figure 31.28   The major terrestrial ecosystems.

The tundra is the northernmost terrestrial ecosystem and has the lowest average temperature of all the terrestrial ecosystems, with minimal to moderate rainfall. The taiga, a coniferous forest that encircles the globe, also has a low average temperature, but moderate rainfall. Temperate forests have moderate temperatures and occur where rainfall is moderate, yet sufficient to support trees. A tropical grassland (savanna) has high temperatures and moderate/seasonal rainfall. A temperate grassland (prairie) has low to high temperatures, with low annual rainfall. Deserts have changeable temperatures with minimal rainfall. Tropical rain forests, which generally occur near the equator, have a high average temperature and the greatest amount of rainfall of all the terrestrial ecosystems.

Primary Productivity

One way to compare ecosystems is based on **primary productivity,** the rate at which producers capture and store energy as organic nutrients over a certain length of time. Temperature and moisture, and secondarily the nature of the soil, influence the primary productivity and, as already discussed, the assemblage of species in an ecosystem. In terrestrial ecosystems, primary productivity is generally lowest in high-latitude tundras and deserts, and highest at the equator where tropical forests occur (Fig. 31.29). The high productivity of tropical rain forests provides varied niches and much food for consumers. The number and diversity of species in tropical rain forests are the highest of all the terrestrial ecosystems. Therefore, conservation biologists are interested in preserving as much of this biome as possible.

The primary productivity of aquatic communities is largely dependent on the availability of inorganic nutrients. Estuaries, swamps, and marshes are rich in organic nutrients and in decomposers that convert those organic nutrients into their inorganic chemical components. Estuaries, swamps, and marshes also contain a large number of varied species, particularly in the early stages of their development before they venture forth into the sea. Therefore, all of these coastal regions are in great need of preservation. The open ocean has a productivity somewhere between that of a desert and the tundra because it lacks a concentrated supply of inorganic nutrients. Coral reefs exist near the coasts in warm tropical waters where currents and waves bring nutrients and where sunlight penetrates to the ocean floor. Coral reefs are areas of remarkable biological abundance, equivalent to that of tropical rain forests.

Check Your Progress

1. Describe the two major types of ecosystems of the biosphere.
2. List the terrestrial ecosystems of the world.
3. Explain why swamps have higher levels of primary productivity than open oceans.

Answers: 1. The biosphere is divided into aquatic ecosystems and terrestrial ecosystems. 2. Taiga, savanna, prairie, temperate forest, desert, tropical rain forest, and tundra. 3. Swamps have a more concentrated supply of nutrients.

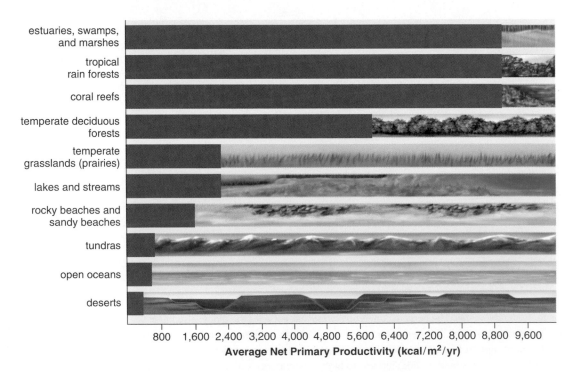

Figure 31.29   Primary productivity.

Ecologists can compare ecosystems based on primary productivity, the rate at which producers convert and store solar energy as chemical energy.

Summary

31.1   Ecology of Communities

Knowledge of community and ecosystem ecology is important for understanding the impacts of human alterations to the environment.

Community   A community is an assemblage of the populations of different species interacting with each other in a given area.

Ecosystem   An ecosystem consists of species interacting with one another and with the physical environment.

Ecological Succession

bare rock → lichens/mosses → grasses → shrubs → trees

The two types of ecological succession are primary succession (begins on bare rock) and secondary succession (following a disturbance; begins where soil is present). Ecological succession leads to a stable climax community.

Interactions in Communities

Species in communities interact with one another in the following ways:

- **Competition**   Species vie with one another for resources, such as light, space, and nutrients. Aspects of competition are the competitive exclusion principle, resource partitioning, and character displacement.
- **Predation**   One species (predator) eats another species (prey).
- **Parasitism**   One species (parasite) obtains nutrients from another species (host) but does not kill the host species.
- **Commensalism**   One species benefits from the relationship, while the other species is not harmed.
- **Mutualism**   Two species interact in a way that benefits both.

Ecological Niche   The ecological niche of a species is defined by the role it plays in its community, the habitat, and its interactions with other species.

Keystone Species   The interactions of a keystone species in the community hold the community and its species together. Removal of a keystone species can lead to species extinctions and loss of diversity. An example of a keystone species is the grizzly bear.

Native Versus Exotic Species   Native species are indigenous to a given area and thrive without assistance. Exotic species are introduced into an area, and greatly disrupt the balance and interactions between native species in that area's community

31.2   Ecology of Ecosystems

In the food chain of an ecosystem, some populations are autotrophs and some are heterotrophs.

Autotrophs are the producers. They require only inorganic nutrients (e.g., CO_2 and minerals) and an outside energy source to produce organic nutrients for their own use and for the use of other members of the community. Examples of autotrophs are algae, cyanobacteria, and plants.

Heterotrophs are the consumers. They require a preformed source of organic nutrients and give off CO_2. Examples of consumers are herbivores (feed on plants), carnivores (feed on other animals), and omnivores (feed on both plants and animals). Other heterotrophs are the decomposers (the bacteria and fungi of decay).

Energy Flow and Chemical Cycling

Energy flows through an ecosystem, while chemicals cycle within an ecosystem.

Food Webs and Food Chains   Energy flows in an ecosystem through food chains and detrital and grazing food webs.

Trophic Level   A trophic level is a level of nourishment in a food web or chain.

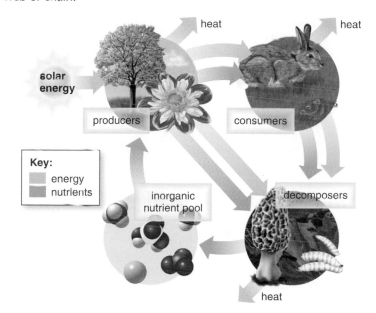

Ecological Pyramid   An ecological pyramid illustrates the energy losses that occur between trophic levels.

- Only about 10% of the energy of one trophic level is available to the next trophic level.
- Top carnivores occupy the last and smallest trophic level.

Biogeochemical Cycle   Chemicals cycle within an ecosystem through various biogeochemical cycles, such as the phosphorus cycle, the nitrogen cycle, and the carbon cycle. Human activities significantly alter the transfer rates in these cycles.

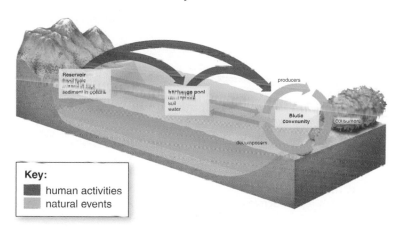

31.3   Ecology of Major Ecosystems

The biosphere encompasses all the major ecosystems of the Earth.

Aquatic Ecosystems
Aquatic ecosystems are classified as freshwater ecosystems (rivers, streams, lakes, ponds) and saltwater, or marine, ecosystems (oceans, coral reefs, saltwater marshes).

Terrestrial Ecosystems
The terrestrial ecosystems are called biomes. The major biomes are:

- Tundra
- Taiga
- Temperate forests
- Tropical grasslands (savanna)
- Temperate grasslands (prairie)
- Deserts
- Tropical rain forests

Primary Productivity
Primary productivity is the rate at which producers capture solar energy and convert it to chemical energy over a specified length of time. The number of species in an ecosystem is positively related to its primary productivity.

Thinking Scientifically

1. One of the most striking examples of coevolution is between insects and flowers. The earliest angiosperms produced wind-pollinated flowers, which released large quantities of pollen. The ovules exuded tiny droplets of sugary sap to catch passing pollen. Outline the course of events that probably took place between insects and flowers to result in the highly specialized interactions we see today.

2. Over 200 wildlife species have been observed around prairie dog colonies in the Great Plains of the United States. The prairie dog, which burrows, forages, and feeds in the area, acts as a keystone species. If the prairie dogs are destroyed, many other species will die as well. How do you think the activities of the prairie dogs influence the survival of so many other species?

Testing Yourself

Choose the best answer for each question.

1. As diversity increases,
   a. species richness increases, and distribution of species becomes more even.
   b. species richness decreases, and distribution of species becomes more even.
   c. species richness increases, and distribution of species becomes less even.
   d. species richness decreases, and distribution of species becomes less even.

2. Which is not a feature of an opportunistic pioneer species?
   a. long life span
   b. short time to maturity
   c. small size
   d. high reproductive output

For statements 3–7, indicate the type of interaction described in each scenario.

Key:
   a. competition
   b. predation
   c. parasitism
   d. commensalism
   e. mutualism

3. An alfalfa plant gains fixed nitrogen from the bacterial species *Rhizobium* in its root system, while *Rhizobium* gains carbohydrates from the plant.

4. Both foxes and coyotes in an area feed primarily on a limited supply of rabbits.

5. Roundworms establish a colony inside a cat's digestive tract.

6. A fungus captures nematodes as a food source.

7. An orchid plant lives in the treetops, gaining access to sun and pollinators, but not harming the trees.

8. The abundance of both species is expected to increase as a result of which type of interaction?

   a. predation   d. competition
   b. commensalism   e. parasitism
   c. mutualism

9. According to the competitive exclusion principle,
   a. one species is always more competitive than another for a particular food source.
   b. competition excludes multiple species from using the same food source.
   c. no two species can occupy the same niche at the same time.
   d. competition limits the reproductive capacity of species.

10. Fungi are examples of
    a. autotrophs.   d. omnivores.
    b. herbivores.   e. decomposers.
    c. carnivores.

11. In the following diagram, fill in the components of chemical cycling and nutrient flow.

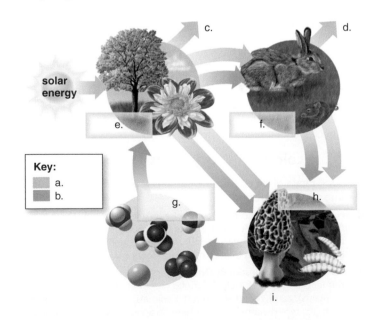

12. An ecological pyramid depicts the amount of _____ in various trophic levels.

    a. food
    b. organisms
    c. energy
    d. nutrients

13. Which of the following would be a primary consumer in a vegetable garden?

    a. aphid sucking sap from cucumber leaves
    b. lady beetle eating aphids
    c. songbird eating lady beetles
    d. fox eating songbirds
    e. All of these are correct.

14. Detritus always contains

    a. bacteria and fungi.
    b. leaf litter.
    c. decaying logs.
    d. animal carcasses.

15. Producers are

    a. autotrophs.
    b. herbivores.
    c. omnivores.
    d. carnivores.
    e. Both a and b are correct.

16. The first trophic level in a food web is occupied by the

    a. producers.
    b. primary consumers.
    c. secondary consumers.
    d. tertiary consumers.

17. Which of the following represents a grazing food chain?

    a. leaves → detritus feeders → deer → owls
    b. birds → mice → snakes
    c. nuts → leaf-eating insects → chipmunks → hawks
    d. leaves → leaf-eating insects → mice → snakes

18. In a grazing food web, carnivores that eat herbivores are

    a. producers.
    b. primary consumers.
    c. secondary consumers.
    d. tertiary consumers.

19. Identify the components of the ecological pyramid in the following diagram.

    a. _____

    b. _____

    c. _____

    d. _____

20. Minerals in rocks are considered members of this component of an ecosystem.

    a. exchange pool
    b. community
    c. reservoir
    d. More than one of these are correct.

21. Which of the following is a sedimentary biogeochemical cycle?

    a. carbon
    b. nitrogen
    c. phosphorus

22. Underground oil is an example of a carbon

    a. cycle.
    b. pathway.
    c. reservoir.
    d. exchange pool.

For questions 23–25, match the description to the process in the key.

Key:

    a. nitrogen fixation
    b. nitrification
    c. denitrification

23. Nitrate to nitrogen gas.

24. Nitrogen gas to nitrate.

25. Nitrogen gas to ammonium.

26. Which of the following is not a component of the nitrogen cycle?

    a. proteins
    b. ammonium
    c. decomposers
    d. photosynthesis
    e. bacteria in root nodules

27. Which biome is characterized by a coniferous forest with low average temperature and moderate rainfall?

    a. taiga
    b. savanna
    c. tundra
    d. tropical rain forest
    e. temperate forest

28. Which biome has the lowest primary productivity?

    a. tundra
    b. lake
    c. sandy beach
    d. prairie
    e. temperate forest

For questions 29–34, match the description to the biome in the key.

Key:

    a. tundra
    b. taiga
    c. tropical rain forest
    d. temperate grassland
    e. tropical grassland
    f. desert

29. Very cold northern coniferous forest.

30. Prairie.

31. Northernmost terrestrial ecosystem; persisting permafrost.

32. Changeable temperatures with minimal rainfall; low primary productivity.

33. Occurs near the equator; high temperatures, large amount of rainfall; high primary productivity.

34. Savanna, alternating wet and dry seasons.

For questions 35–38, match the description to the type of aquatic ecosystem in the key.

Key:
   a. ocean
   b. marshes
   c. rivers, lakes
   d. coral reefs

35. Lie near the coast; high primary productivity.

36. Lowest primary productivity.

37. Freshwater ecosystems.

38. Lie offshore; high primary productivity.

Go to www.mhhe.com/maderessentials for more quiz questions.

Bioethical Issue

Many exotic species, such as zebra mussels and sea lampreys, are so obviously troublesome that most people do not object to programs aimed at controlling their populations. However, some exotic species eradication programs meet with more resistance. For example, the mute swan, one of the world's largest flying birds, is beautiful, graceful, and makes an impressive presence. However, it is very aggressive and territorial. The mute swan was introduced to the United States from Asia and Europe in the nineteenth century as an ornamental bird. The birds consume large amounts of aquatic vegetation and displace native birds from feeding and nesting areas. The U.S. Fish and Wildlife Service believes it will be necessary to kill 3,000 mute swans in Maryland in the next two years in order to protect native bird populations. Attempts to limit the size of the mute swan populations in Maryland and other states have been met with opposition by citizens who find the birds beautiful.

Do you feel that native populations need not be protected as long as the exotic species serves a suitable human purpose? Or, do you feel that native species should be protected regardless?

Understanding the Terms

aquatic ecosystem 568	exotic species 559
biogeochemical cycle 564	food chain 563
biomass 563	food web 562
biosphere 568	fossil fuel 567
character displacement 557	global warming 567
climax community 554	grazing food web 562
coevolution 552	greenhouse effect 567
commensalism 556	habitat 556
community 552	keystone species 558
competition 556	mutualism 556
competitive exclusion principle 557	native species 559
consumer 560	parasitism 556
decomposer 560	predation 556
detrital food web 562	primary productivity 570
detritus 560	primary succession 554
diversity 553	producer 560
ecological niche 556	resource partitioning 557
ecological pyramid 563	secondary succession 554
ecological succession 554	species richness 553
ecosystem 552	terrestrial ecosystem 568
eutrophication 565	transfer rate 565
	trophic level 563

Match the terms to these definitions:

a. _____ Assemblage of populations of different species.

b. _____ Relationship in which one species obtains nutrients from another species but does not kill it.

c. _____ Combination of the role a species plays in its community, its habitat, and its interactions with other species.

d. _____ Tendency for characteristics to be more divergent when populations of different species belong to the same community than when they are isolated.

e. _____ One species in a community that stabilizes the community, helps maintain its characteristics, and helps hold the web of interactions together.

f. _____ Remains of dead organisms plus the bacteria and fungi of decay.

g. _____ All the organisms that feed at a particular link in a food chain.

h. _____ Amount of a nutrient that moves from one component of the environment to another within a certain period of time.

i. _____ All the ecosystems on planet Earth.

Human Impact on the Biosphere

The Earth loses 4,000–6,000 species due to rain forest deforestation each year.

The Earth loses 12–17 million acres of farmland due to erosion each year.

The creation of a successful, self-sustaining ecosystem has been attempted.

Between 1987 and 1989, scientists attempted to develop a sustainable ecosystem from scratch on a piece of land in Arizona. This incredibly expensive project, called Biosphere 2, consisted of a series of airtight domes designed to house thousands of plant and animal species, including several humans, for two years. Some of the early supporters were interested in the project's potential for the future creation of closed systems capable of ▶ ▶ ▶

supporting colonies of people on the moon or on Mars.

After careful planning and construction, Biosphere 2 was sealed, and the experiment began. What happened was unexpected. Oxygen levels got so low that the humans inside were in danger of asphyxiation, and pure oxygen gas had to be pumped in. In addition, invasive plant and animal species destroyed other species, carbon dioxide levels were elevated, and soil problems developed. Although people did manage to live in Biosphere 2 for two years, most of the experimenters concluded that humans are not able to build a sustainable ecosystem.

Still, the project yielded some benefits. Columbia University transformed Biosphere 2 into a teaching and learning center for public education on sustainable ecosystems. It's now possible to visit Biosphere 2 and tour rain forest, grassland, desert, marsh, and marine ecosystems. Visitors gain both an appreciation for the Earth's ecosystems, with the aim of preserving what we have, and an understanding of the choices people must make in order to manage the planet well.

In this chapter, you will learn about how humans negatively impact ecosystems, and some of the potential solutions that will help us move toward a sustainable society.

32.1 Resources and Pollution

Human beings have certain basic needs, and they use resources to meet these needs. Land, water, food, energy, and minerals are the maximally used resources (Fig. 32.1).

Some resources are nonrenewable, and some are renewable. **Nonrenewable resources** are limited in supply. For example, the amount of land, fossil fuels, and minerals is finite and can be exhausted. Better extraction methods and efficient use can make the supply last longer, but eventually these resources will run out. **Renewable resources** are not limited in supply. We can use water and certain forms of energy (e.g., solar energy) or harvest plants and animals for food, and more supply will always be forthcoming. However, even though these resources are renewable, we must be careful not to squander them. Consider, for example, that most species have population thresholds below which they cannot recover, as when the huge herds of buffalo that once roamed the west disappeared after being overexploited.

Unfortunately, a side effect of resource consumption can be pollution. **Pollution** is any alteration of the environment in an undesirable way. Pollution is often caused by human activities. The impact of humans on the environment is proportional to the size of the population. As the population grows, so does the need for resources and the amount of pollution caused by using these resources. Consider that six people adding waste to the ocean might not be alarming, but six billion people doing so would certainly affect its cleanliness. In modern times, the consumption of mineral and energy resources has grown faster than population size, most likely because people in the less-developed countries (LDCs) have increased their use of them.

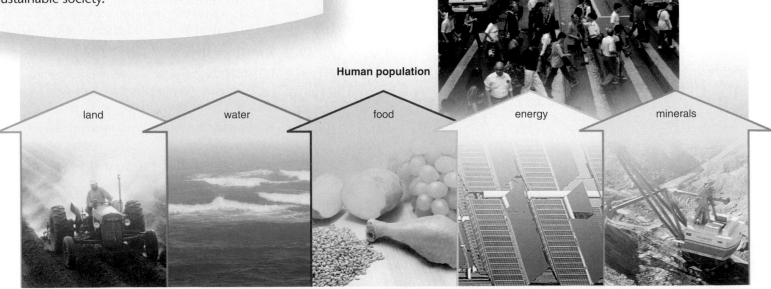

Human population

land   water   food   energy   minerals

Figure 32.1   **Resources.**

Human beings use land, water, food, energy, and minerals to meet their basic needs, including a place to live, food to eat, and products that make their lives easier.

Land

People need a place to live. Worldwide, there are currently more than 32 persons for each square kilometer (km) (83 persons per square mile) of all available land, including Antarctica, mountain ranges, jungles, and deserts. Naturally, land is also needed for a variety of uses aside from homes, such as agriculture, electric power plants, manufacturing plants, highways, hospitals, schools, and so on.

Beaches and Human Habitation

At least 40% of the world population lives within 100 km (60 mi) of a coastline, and this number is expected to increase. In the United States today, over one-half of the population lives within 80 km (50 mi) of the coasts (including the Great Lakes). Living right on the coast is an unfortunate choice because it leads to beach erosion and loss of habitat for marine organisms.

Beach Erosion   An estimated 70% of the world's beaches are eroding; Figure 32.2 shows how extensive the problem is in the United States. The seas have been rising for the past 12,000 years, ever since the climate turned warmer after the last Ice Age. Authorities are concerned that global warming, discussed later in this chapter (see page 583), is also contributing to the melting of ice caps and glaciers and, therefore, to an increase in sea level.

Humans carry on other activities that divert more water to the oceans, contributing to rising seas and beach erosion. For example, humans have filled in coastal wetlands, such as mangrove swamps in the southern United States and saltwater marshes in the northern United States. With growing recognition of the services provided by wetlands, the tide of wetland loss has been stemmed in the United States during the past 40 years, but it is just starting in South America, where a project to straighten the Parana River will drain the world's largest wetland. One reason to protect coastal wetlands is that they are spawning areas for fish and other forms of marine life. They are also habitats for certain terrestrial species, including many types of birds.

Humans often try to stabilize beaches by building groins (structures that extend from the beach into the water) and seawalls. Groins trap sand on one side, but erosion is worse on the other side. Seawalls, in the end, also increase erosion because ocean waves remove sand from in front of and to the side of the seawalls. Importing sand is a better solution, but it is very costly and can disturb plant and animal populations It's estimated that today, the U.S. shoreline loses 40% more sediment than it receives, especially because the building of dams prevents sediment from reaching the coast.

Coastal Pollution   The coasts are particularly subject to pollution because toxic substances placed in freshwater lakes, rivers, and streams may eventually find their way to the coast. Oil spills at sea cause localized harmful effects also.

a.

b.

Figure 32.2   Beach erosion.

a. Most of the U.S. coastline is subject to beach erosion.
b. Therefore, people who choose to live near the coast may eventually lose their homes.

Check Your Progress

1. **Distinguish between renewable and nonrenewable resources.**
2. **Describe two problems related to human habitation near oceans.**

Answers: 1. Renewable resources are not limited in supply, while nonrenewable resources are. *2.* Beach erosion occurs due to land development on the shore, and coastal pollution occurs due to human activities in the vicinity of the water.

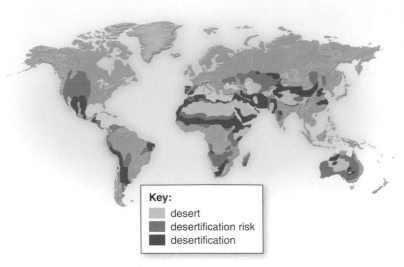

Key:
- desert
- desertification risk
- desertification

Figure 32.3 Desertification.

Desertification is a worldwide occurrence that reduces the amount of land suitable for human habitation.

Source: Data from A. Goudie and J. Wilkinson, *The Warm Desert Environment.* Copyright 1977 by Cambridge University Press, New York.

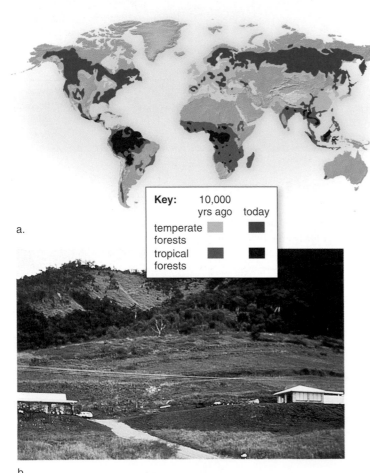

Key: 10,000 yrs ago / today

- temperate forests
- tropical forests

a.

b.

Figure 32.4 Deforestation.

a. Nearly half of the world's forest lands have been cleared for farming, logging, and urbanization. **b.** The soil of tropical rain forests is not suitable for long-term farming.

Sources: United Nations Environment Program, World Resources Institute.

Semiarid Lands and Human Habitation

Forty percent of the Earth's lands are already deserts, and any land adjacent to a desert is in danger of becoming unable to support human life if it is improperly managed by humans. **Desertification** is the conversion of semiarid land to desertlike conditions (Fig. 32.3).

Quite often, desertification begins when humans allow animals to overgraze the land. The soil can no longer hold rainwater, which runs off instead of keeping the remaining plants alive or recharging wells. Humans then remove whatever vegetation they can find to use as fuel or fodder for their animals. The end result is a lifeless desert, which is then abandoned as people move on to continue the process someplace else. Some estimate that nearly three-quarters of all rangelands worldwide are in danger of desertification. The recent famine in Ethiopia was due, at least in part, to degradation of the land to the point that it could no longer support human beings and their livestock.

Tropical Rain Forest and Human Habitation

Deforestation, the removal of trees, has long allowed humans to live in areas where forests once covered the land. The concern recently has been that people are settling in tropical rain forests, such as the Amazon, following the building of roads (Fig. 32.4). This land, too, is subject to desertification. Soil in the tropics is often thin and nutrient-poor because all the nutrients are tied up in the trees and other vegetation. When the trees are felled and the land is used for agriculture or grazing, it quickly loses its fertility and becomes subject to desertification.

Loss of biodiversity also sometimes results from the destruction of rain forests. For example, logging in the Congo Republic has contributed to the growth of a city called Pokola, whose residents routinely eat wild bush animals.

Water

In the water-poor areas of the world, people may not have ready access to drinking water, and if they do, the water may be unclean. Access to clean drinking water is considered a human right, but actually most fresh water is utilized by agriculture and industry (Fig. 32.5). Worldwide, 70% of all fresh water is used to irrigate crops! Much of a recent surge in demand for water stems from increased industrial activity and irrigation-intensive agriculture, the type of agriculture that now supplies about 40% of the world's food crops. Domestically, in the more-developed countries (MDCs), more water is usually used for bathing, flushing toilets, and watering lawns than for drinking and cooking.

Increasing Water Supplies

Although the needs of the human population overall do not exceed the renewable supply, this is not the case in certain regions of the United States and the world. As illustrated in Figure 32.3, about 40% of the world's land is desert, and deserts are bordered by semiarid land. When necessary, humans increase the supply of fresh water by damming rivers and withdrawing water from aquifers.

Dams   The world's 45,000 large dams catch 14% of all precipitation runoff, provide water for up to 40% of irrigated land, and give some 65 countries more than half their electricity. Damming of certain rivers has been so extensive that they no longer flow as they once did. The Yellow River in China fails to reach the sea most years; the Colorado River barely makes it to the Gulf of California; and even the Rio Grande dries up before it can merge with the Gulf of Mexico. The Nile in Egypt and the Ganges in India are also so overexploited that at some times of the year, they hardly make it to the ocean

Dams have other drawbacks: (1) They lose water due to evaporation and seepage into underlying rock beds. The amount of water lost sometimes equals the amount made available! (2) The salt left behind by evaporation and agricultural runoff can make a river's water unusable further downstream. (3) Sediment buildup causes dams to hold back less water; with time, dams may become useless for storing water. (4) The impact on the wildlife in estuaries is devastating.

Aquifers   To meet their freshwater needs, people are pumping vast amounts of water from **aquifers,** which are natural reservoirs found just below or as much as 1 km below the surface. Aquifers hold about 1,000 times the amount of water that falls on land as precipitation each year. This water accumulates from rain that fell in far-off regions as many as hundreds of thousands of years ago. In the past 50 years, groundwater depletion has become a problem in many areas of the world. In substantial portions of the High Plains Aquifer, which stretches from South Dakota to the Texas Panhandle, more than half the water has been pumped out. In the 1950s, India had 100,000 motorized pumps in operation; today, India has 20 million pumps, a huge increase in groundwater pumping.

Environmental Consequences   Removal of water is causing land **subsidence,** settling of the soil as it dries out. In California's San Joaquin Valley, an area of more than 13,000 square km has subsided, and in the worst spot, the surface has dropped more than 9 meters (m)! In some parts of Gujarat, India, the water table has dropped as much as 7 m. Subsidence damages canals, buildings, and underground pipes. Withdrawal of groundwater can cause sinkholes, as when an underground cavern collapses because water no longer holds up its roof.

Saltwater intrusion is another consequence of groundwater depletion. The flow of water from streams and aquifers usually keeps them fairly free of seawater. But as water is withdrawn, the water table can lower to the point that seawater backs up into streams and aquifers. Saltwater intrusion reduces the supply of fresh water along the coast.

Conservation of Water

By 2025, two-thirds of the world's population may be living in countries that are facing serious water shortages. Some solutions for expanding water supplies have been suggested. Planting drought- and salt-tolerant crops would help a lot. Using drip irrigation delivers more water to crops and saves about 50% over traditional methods while increasing crop yields as well (Fig. 32.6). Although the first drip systems were developed in 1960, they are used on less than 1% of irrigated land. Most governments subsidize irrigation so heavily that farmers have little incentive to invest in drip systems or other water-saving methods. Reusing water and adopting conservation measures could help the world's industries cut their water demands by more than half.

a. Agriculture uses most of the fresh water consumed.

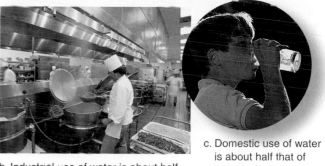

b. Industrial use of water is about half that of agricultural use.

c. Domestic use of water is about half that of industrial use.

Figure 32.5   Global water use.

a. Agriculture primarily uses water for irrigation. **b.** Industry uses water variously. **c.** People use water for drinking, showering, flushing toilets, and watering lawns.

a. Drought-resistant plants

tubing

drop of water

b. Drip irrigation

Figure 32.6   Conservation measures to save water.

a. Planting drought-resistant plants in parks and gardens and drought-resistant crops in the field cuts down on the need to irrigate. **b.** Drip irrigation delivers water directly to the roots.

Figure 32.7   Conservation methods.

a. Polyculture reduces the ability of one parasite to wipe out an entire crop and reduces the need to use a herbicide to kill weeds. This farmer has planted alfalfa in between strips of corn, which also replenishes the nitrogen content of the soil (instead of adding fertilizers). **b.** Contour farming with no-till conserves topsoil because water has less tendency to run off. **c.** Instead of pesticides, it is sometimes possible to use a natural predator. Here ladybugs are eating cottony-cushion scale insects on citrus trees.

a. Polyculture

b. Contour with no-till farming

c. Biological pest control

Food

In 1950, the human population numbered 2.5 billion, and there was only enough food to provide less than 2,000 calories per person per day; now, with over 6 billion people on Earth, the world food supply provides more than 2,500 calories per person per day. Generally speaking, food comes from three activities: growing crops, raising animals, and fishing the seas. Unfortunately modern farming methods, which have increased the food supply, include some harmful practices:

1. *Planting of a few genetic varieties.* The majority of farmers practice monoculture. Wheat farmers plant the same type of wheat, and corn farmers plant the same type of corn. Unfortunately, monoculture means that a single type of parasite can cause much devastation.
2. *Heavy use of fertilizers, pesticides, and herbicides.* Fertilizer production is energy-intensive, and fertilizer runoff contributes to water pollution. Pesticides reduce soil fertility because they kill beneficial soil organisms as well as pests, and some pesticides and herbicides are linked to the development of cancer.
3. *Generous irrigation.* As already discussed, water is sometimes taken from aquifers whose water content may in the future become so reduced that it could be too expensive to pump out any more.
4. *Excessive fuel consumption.* Irrigation pumps remove water from aquifers, and large farming machines are used to spread fertilizers, pesticides, and herbicides, as well as to sow and harvest the crops. In effect, modern farming methods transform fossil fuel energy into food energy.

Figure 32.7 shows ways to minimize the harmful effects of modern farming practices.

Soil Loss and Degradation

Land suitable for farming and for grazing animals is being degraded worldwide. Topsoil, the topmost portion of the soil, is the richest in organic matter and the most capable of supporting grass and crops. When bare soil is acted on by water and wind, soil erosion occurs and topsoil is lost. As a result, marginal rangeland becomes desertized, and farmland loses its productivity.

The custom of planting the same crop in straight rows that facilitate the use of large farming machines has caused the United States and Canada to have one of the highest rates of soil erosion in the world. Conserving the nutrients now being lost could save farmers $20 billion annually in fertilizer costs. Much of the eroded sediment ends up in lakes and streams, where it reduces the ability of aquatic species to survive.

Between 25% and 35% of the irrigated western croplands are thought to have undergone **salinization,** an accumulation of mineral salts due to the evaporation of excess irrigation water. Salinization makes the land unsuitable for growing crops.

Green Revolutions

About 50 years ago, research scientists began to breed tropical wheat and rice varieties specifically for farmers in the LDCs. The dramatic increase in yield due to the introduction of these new varieties around the world was called "the green revolution." These plants helped the world food supply keep pace with the rapid increase in world population. Most green revolution plants are called "high responders" because they need high levels of fertilizer, water, and pesticides in order to produce a high yield. In other words, they require the same subsidies and create the same ecological problems as do modern farming methods.

Genetic Engineering   Genetic engineering can produce transgenic plants with new and different traits, among them, resistance to both insects and herbicides. When herbicide-resistant crops are planted, weeds are easily controlled, less tillage is needed, and soil erosion is minimized. Researchers also want to produce crops that tolerate salt, drought, and cold. In addition, some progress has been made in increasing the food quality of crops so that they will supply more of the proteins, vitamins, and minerals people need. Genetically engineered crops are resulting in another green revolution.

Nevertheless, some citizens are opposed to the use of genetically engineered crops, fearing that they will damage the environment and lead to health problems in humans.

Domestic Livestock

A low-protein, high-carbohydrate diet consisting only of grains such as wheat, rice, or corn can lead to malnutrition. In the LDCs, kwashiorkor, a condition caused by a severe protein deficiency, is seen in infants and children ages 1–3, usually after a new baby arrives in the family and the older children are no longer breast-fed. Such children are lethargic, irritable, and have bloated abdomens. Mental retardation is expected.

In the MDCs, many people tend to have more than enough protein in their diet. Almost two-thirds of U.S. cropland is devoted to producing livestock feed. This means that a large percentage of the fossil fuel, fertilizer, water, herbicides, and pesticides we use are actually for the purpose of raising livestock. Typically, cattle are range-fed for about four months, and then they are brought to crowded feedlots where they receive growth hormone and antibiotics while feeding on grain or corn. Most pigs and chickens spend their entire lives cooped up in crowded pens and cages (Fig. 32.8).

Check Your Progress

1. List the harmful effects of modern agricultural practices.
2. Describe the two green revolutions.

Answers: 1. Limited genetic diversity, heavy use of chemicals, frequent irrigation, and excessive fuel consumption. 2. The first occurred in the 1950s and resulted from the breeding of high-yielding rice and wheat varieties. The second is under way now and is the result of the creation of transgenic crops.

Figure 32.8   Crowding of livestock.

Hogs milling in a feedlot pen.

Just as livestock eat a large proportion of the crops in the United States, raising livestock accounts for much of the pollution associated with farming. Consider also that presently, fossil fuel energy is needed not just to produce herbicides and pesticides and to grow food, but also to make the food available to the livestock. Raising livestock is extremely energy-intensive in the MDCs. In addition, water is used to wash livestock wastes into nearby bodies of water, where they add significantly to water pollution. Whereas human wastes are sent to sewage treatment plants, raw animal wastes are not.

For these reasons, it is prudent to recall the ecological energy pyramid (see Fig. 31.20), which shows that as you move up the food chain, energy is lost. As a rule of thumb, for every 10 calories of energy from a plant, only 1 calorie is available for the production of animal tissue in a herbivore. In other words, it is extremely wasteful for the human diet to contain more protein than is needed to maintain good health. It is possible to feed ten times as many people on grain as on meat.

Fishing

Worldwide, between 1970 and 1990, the number of large boats devoted to fishing doubled to 1.2 million. The U.S. fishing fleet participated in this growth due to the availability of federal loans for building fishing boats. The new boats have sonar and depth recorders, and their computers remember the sites of previous catches so that the boats can go there again. Helicopters, planes, and even satellite data are used to help find fish. The result of the increased number and efficiency of fishing boats is a severe reduction in fish catch (Fig. 32.9). For example, the number of North Atlantic swordfish caught in the United States declined 70% from 1980 to 1990, and the average weight fell from 115 to 60 pounds. Many believe that the Atlantic bluefin tuna is so overfished that it will never recover and will instead become extinct.

Modern fishing practices negatively impact biodiversity because a large number of marine animals are caught by chance in the huge nets some fishing boats use. These animals are discarded. The world's shrimp fishery has an annual catch of 1.8 million tons, but the other animals caught and discarded in the process amount to 9.5 million tons.

a.

Figure 32.9 **Fishing.**

a. The world fish catch has declined in recent years (insert) because (**b**) modern fishing methods are overexploiting fisheries.

Data from U.S. Marine Fisheries Service.

b.

Energy

Presently, about 6% of the world's energy supply comes from nuclear power, and 75% comes from fossil fuels; both of these are finite, nonrenewable sources. Although it was once predicted that the nuclear power industry would fulfill a significant portion of the world's energy needs, this has not happened for two reasons: (1) People are very concerned about nuclear power dangers, such as the meltdown that occurred in 1986 at the Chernobyl nuclear power plant in Russia. (2) Radioactive wastes from nuclear power plants remain a threat to the environment for thousands of years, and we still have not determined the best way to safely store them.

As you learned in Chapter 31, oil, natural gas, and coal are **fossil fuels**—the compressed remains of organisms that died many thousands of years ago. The MDCs presently consume more than twice as much fossil fuel as the LDCs, yet there are many more people in the LDCs than in the MDCs. It has been estimated that each person in the MDCs uses approximately as much energy in one day as a person in an LDC does in one year.

Among the fossil fuels, oil burns more cleanly than coal, which may contain a considerable amount of sulfur. So despite the fact that the United States has a goodly supply of coal, imported oil is our preferred fossil fuel today. Even so, the burning of any fossil fuel causes environmental problems because as it burns, pollutants are emitted into the air.

Fossil Fuels and Global Climate Change

In 1850, the level of carbon dioxide in the atmosphere was about 280 parts per million (ppm), and today it is about 350 ppm. This increase is largely due to the burning of fossil fuels and the burning and clearing of forests to make way for farmland and pasture. Human activities are causing the emission of other gases as well. For example, the amount of methane given off by oil and gas wells, rice paddies, and all sorts of organisms, including domesticated cows, is increasing by about 1% a year. These gases are known as **greenhouse gases** because, just like the panes of a greenhouse, they allow solar radiation to pass through but hinder the escape of infrared heat back into space.

Today, data collected around the world show a steady rise in the concentration of the various greenhouse gases. These data are used to generate computer models that predict the Earth may warm to temperatures never before experienced by living things. The global climate has already warmed about 0.6°C since the Industrial Revolution. Computer models are unable to consider all possible variables, but the Earth's temperature may rise 1.5°–4.5°C by 2060 if greenhouse emissions continue at the current rates (Fig. 32.10).

It is predicted that as the oceans warm due to global warming, temperatures in the polar regions will rise to a greater degree than in other regions. If so, glaciers will melt, and sea levels will rise, not only due to this melting but also because water expands as it warms. Water evaporation will increase, and most likely precipitation will increase along the coasts, while conditions inland become drier. The occurrence of droughts will reduce agricultural yields, and will also cause trees to die off. Coastal agricultural lands, such as the deltas of China and Bangladesh, India, will be inundated, and billions will have to be spent to keep U.S. coastal cities, such as New York, Boston, Miami, and Galveston, from disappearing into the sea.

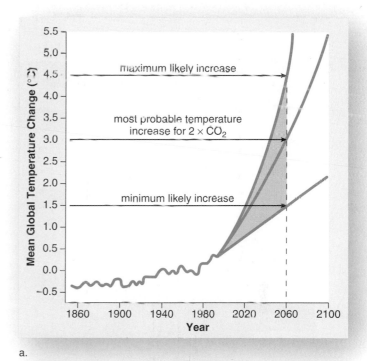

a.

b.

Figure 32.10   Global warming.

a. Mean global temperature is expected to rise due to the introduction of greenhouse gases into the atmosphere. Global warming has the potential to significantly affect the world's biodiversity. **b.** A temperature rise of only a few degrees causes coral reefs to "bleach" and become lifeless.

Renewable Energy Sources

Renewable types of energy include hydropower, geothermal, wind, and solar (Fig. 32.11).

Wind Power Wind power is expected to account for a significant percentage of our energy needs in the future. Despite the common belief that a huge amount of land is required for the "wind farms" that produce commercial electricity, the actual amount of space for a wind farm compares favorably to the amount of land required by a coal-fired power plant or a solar thermal energy system.

A community that generates its own electricity by using wind power can solve the problem of uneven energy production by selling electricity to a local public utility when an excess is available and buying electricity from the same facility when wind power is in short supply.

Hydropower Hydroelectric plants convert the energy of falling water into electricity. Hydropower accounts for about 10% of the electric power generated in the United States and almost 98% of the total renewable energy used. Brazil, New Zealand, and Switzerland produce at least 75% of their electricity with water power, but Canada is the world's leading hydropower producer.

a.

b.

c.

d.

Figure 32.11 **Renewable energy sources.**

a. Hydropower dams provide a clean form of energy but can be ecologically disastrous in other ways. **b.** Wind power requires land on which to place enough windmills to generate energy. **c.** Photovoltaic cells on rooftops and (**d**) sun-tracking mirrors on land can now collect diffuse solar energy more cheaply than in the past.

Worldwide, hydropower presently generates 19% of all electricity utilized, but this percentage is expected to rise due to increased use in certain countries. For example, Iceland has an ambitious hydropower project under way because presently it uses only 10% of its potential capacity.

Much of the hydropower development in recent years has been due to the construction of enormous dams, which are known to have detrimental environmental effects. The better choice is believed to be small-scale dams that generate less power per dam but do not have the same environmental impact.

Geothermal Energy Elements such as uranium, thorium, radium, and plutonium undergo radioactive decay below the Earth's surface and then heat the surrounding rocks to hundreds of degrees Centigrade. When the rocks are in contact with underground streams or lakes, huge amounts of steam and hot water are produced. This steam can be piped up to the surface to supply hot water for home heating or to run steam-driven turbogenerators. The California's Geysers project is the world's largest geothermal electricity-generating complex.

Energy and the Solar-Hydrogen Revolution Solar energy is diffuse energy that must be (1) collected, (2) converted to another form, and (3) stored if it is to compete with other available forms of energy. Passive solar heating of a house is successful when the windows of the house face the sun, the building is well insulated, and heat can be stored in water tanks, rocks, bricks, or some other suitable material.

In a **photovoltaic (solar) cell,** a wafer of an electron-emitting metal is in contact with another metal that collects the electrons and passes them along into wires in a steady stream. Spurred by the oil shocks of the 1970s, the U.S. government has been supporting the development of photovoltaics ever since. As a result, the price of buying a system has dropped from about $100 per watt to around $4. The photovoltaic cells placed on roofs, for example, generate electricity that can be used inside a building and/or sold back to a power company.

Several types of solar power plants are now operational in California. In one type, huge reflectors focus sunlight on a pipe containing oil. The heated pipes boil water, generating steam that drives a conventional turbogenerator. In another type, 1,800 sun-tracking mirrors focus sunlight onto a molten salt receiver mounted on a tower. The hot salt generates steam that drives a turbogenerator.

Scientists are working on the possibility of using solar energy to extract hydrogen from water via electrolysis. The hydrogen can then be used as a clean-burning fuel; when it burns, water is produced. Presently, cars have internal combustion engines that run on gasoline. In the future, vehicles are expected to be powered by fuel cells, which use hydrogen to produce electricity (Fig. 32.12). The electricity runs a motor that propels the vehicle. Fuel cells are now powering buses in Vancouver and Chicago, and more buses are planned.

Hydrogen fuel can be produced locally or in central locations, using energy from photovoltaic cells. If produced in central locations, hydrogen can be piped to filling stations using the natural gas pipes already plentiful in the United States. The advantages of a solar-hydrogen revolution would be at least twofold: (1) The world would no longer be dependent on oil, and (2) environmental problems, such as global warming, acid rain, and smog, would begin to lessen.

a.

b.

c.

Figure 32.12 Solar-hydrogen revolution.

a. Hydrogen fuel cells. **b.** This bus is powered by hydrogen fuel. **c.** The use of fuel-cell hybrid vehicles, such as this prototype, will reduce air pollution and dependence on fossil fuels.

Figure 32.13   Modern mining capabilities.

Giant mining machines—some as tall as a 20-story building—can remove an enormous amount of the Earth's crust in one scoop in order to mine for coal or a metal.

Minerals

Minerals are nonrenewable raw materials in the Earth's crust that can be mined (extracted) and used by humans. Nonrenewable minerals include fossil fuels; nonmetallic raw materials, such as sand, gravel, and phosphate; and metals, such as aluminum, copper, iron, lead, and gold.

Nonrenewable resources are subject to depletion—that is, the supply that is mineable will eventually run out. A depletion curve is dependent on how fast the resource is used, whether new reserves can be found, and whether recycling and reuse are possible. We can extend our supply of fossil fuels if we conserve our use and if we find new reserves. In addition to these possibilities, metals can be recycled.

Most of the metals mined each year are consumed by the United States, Japan, and Europe, despite the fact that they are primarily mined in South America, South Africa, and countries of the former Soviet Union. One of the greatest threats to the maintenance of ecosystems and biodiversity is surface mining, called strip mining. In the United States, huge machines can go as far as removing mountaintops in order to reach a mineral (Fig. 32.13). The land, devoid of vegetation, takes on a surreal appearance, and rain washes toxic waste deposits into nearby streams and rivers. Legislation now requires that strip miners restore the land to its original condition, a process that can take years to complete.

The most dangerous metals to human health are the heavy metals: lead, mercury, arsenic, cadmium, tin, chromium, zinc, and copper. They are used to produce batteries, electronics, pesticides, medicines, paints, inks, and dyes. In the ionic form, they enter the body and inhibit vital enzymes. That's why these items should be discarded carefully and taken to hazardous waste sites.

Other Sources of Pollution

Synthetic organic compounds and wastes are also pollutants of concern.

Synthetic Organic Compounds

In addition to metals, synthetic organic compounds are another area of considerable ecological concern due to their detrimental effects on the health of living things, including humans. Synthetic organic compounds play a role in the production of plastics, pesticides, herbicides, cosmetics, coatings, solvents, wood preservatives, and hundreds of other products.

Synthetic organic compounds include halogenated hydrocarbons, in which halogens (chlorine, bromine, fluorine) have replaced certain hydrogens. **Chlorofluorocarbons (CFCs)** are a type of halogenated hydrocarbon in which both chlorine and fluorine atoms replace some of the hydrogen atoms. CFCs have brought about a thinning of the Earth's ozone shield, which protects terrestrial life from the dangerous effects of ultraviolet radiation. In most MDCs, legislation has been passed to prevent the production of any more CFCs. Hydrofluorocarbons, which contain no chlorine, are expected to take their place in coolants and other products. The ozone shield is predicted to recover by 2050; in the meantime, many more cases of skin cancer are expected to occur.

Other synthetic organic chemicals pose a direct and serious threat to the health of living things, including humans. Rachel Carson's book *Silent Spring,* published in 1962, made the public aware of the deleterious effects of pesticides.

Wastes

Every year, the countries of the world discard billions of tons of solid wastes, some on land and some in fresh and marine waters. Solid wastes are visible wastes, some of which are hazardous to our health.

Industrial Wastes   Industrial wastes are generated during the mining and production of a product. Clean-water and clean-air legislation in the early 1970s prevented venting industrial wastes into the atmosphere and flushing them into waterways. Industry turned to land disposal, which was unregulated at the time. Utilization of deep-well injection, pits with plastic liners, and landfills led to much water pollution and human illness, including cancer. An estimated 5 billion metric tons of highly toxic chemicals were improperly discarded in the United States between 1950 and 1975. The public's concern was so great that the Environmental Protection Agency (EPA) came into existence. Using an allocation of monies called the superfund, the EPA oversees the cleanup of hazardous waste disposal sites in the United States.

Among the most commonly found contaminants are heavy metals (lead, arsenic, cadmium, chromium) and organic compounds (trichloroethylene, toluene, benzene, polychlorinated biphenyls [PCBs], and chloroform). Some of the chemicals used in pesticides, herbicides, plastics, food additives, and personal hygiene products are classified as endocrine disrupters. These products can affect the endocrine system and interfere with reproduction. In the environment, they occur at a level 1,000 times greater than the hormone levels in human blood.

Decomposers are unable to break down these wastes. They enter and remain in the bodies of organisms because they are not excreted. Therefore, they become more concentrated as they pass along a food chain, a process termed **biological magnification** (Fig. 32.14). This effect is most apt to occur in aquatic food chains, which have more links than terrestrial food chains. Humans are the final consumers in both types of food chains, and in some areas, human milk contains detectable amounts of the polychlorinated hydrocarbons DDT and PCBs.

Sometimes industrial wastes accumulate in the mud of deltas and estuaries of highly polluted rivers and cause environmental problems if disturbed. Industrial pollution is being addressed in many MDCs, but usually has low priority in LDCs.

Sewage   Raw sewage causes oxygen depletion in lakes and rivers. As the oxygen level decreases, the diversity of life is greatly reduced. Also, human feces can contain pathogenic microorganisms that cause cholera, typhoid fever, and dysentery. In regions of the LDCs where sewage treatment is practically nonexistent, many children die each year from these diseases.

Typically, sewage treatment plants use bacteria to break down organic matter to inorganic nutrients, such as nitrates and phosphates, which then enter surface waters. The result can be cultural eutrophication; first there is an algal bloom, and then when the algae die off, decomposition robs the water of oxygen, which can result in a massive fish kill.

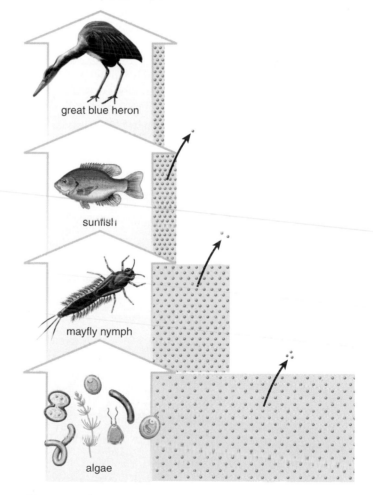

great blue heron

sunfish

mayfly nymph

algae

Figure 32.14   Biological magnification.

A poison (*dots*), such as DDT, that is excreted in relatively small amounts (*arrows*) becomes maximally concentrated as it passes along a food chain due to the reduced size of the trophic levels.

Figure 32.15 **Habitat loss.**

a. In a study examining records of imperiled U.S. plants and animals, habitat loss emerged as the greatest threat to wildlife.
b. Macaws that reside in South American tropical rain forests are endangered for the reasons listed in the graph in (**a**).

Wild species, like the rosy periwinkle, are sources of many medicines.

Wild species, like the nine-banded armadillo, play a role in medical research.

Figure 32.16 **Medicinal value of biodiversity.**

Many wildlife species are sources of medicines for today's ills.

32.2 Biodiversity

Biodiversity can be defined as the variety of life on Earth, described in terms of the number of different species. We are presently in a biodiversity crisis—the number of extinctions (loss of species) expected to occur in the near future is unparalleled in the history of the Earth. In the United States, researchers have found that habitat loss was involved in 85% of the extinction cases (Fig. 32.15a). Exotic species had a hand in nearly 50%, pollution was a factor in 24%, over-exploitation in 17%, and disease in 3%. The percentages add up to more than 100% because most of these species are imperiled for more than one reason. For example, not only have macaws had their habitat reduced by encroaching timber and mining companies, but they are also hunted for food and collected for the pet trade (Fig. 32.15b).

Conservation biology strives to reverse the trend toward the possible extinction of thousands of plants and animals. To bring this about, it is necessary to make all people aware that biodiversity is a resource of immense value—both direct and indirect.

Direct Values of Biodiversity

Figure 32.16 depicts some of the direct values of biodiversity, which include medicines, foods, and other products that benefit human beings.

Medicinal Value

Most of the prescription drugs used in the United States were originally derived from organisms. The rosy periwinkle from Madagascar is an excellent example of a tropical plant that has provided us with useful medicines (Fig. 32.16a). Potent chemicals from this plant are now used to treat two forms of cancer: leukemia and Hodgkin disease. Due to these drugs, the survival rate for childhood leukemia has gone from 10% to 90%, and Hodgkin disease is now usually curable. Although the value of saving a life cannot be calculated, it is still sometimes easier for us to appreciate the worth of a resource if it is explained in monetary terms. Thus, researchers tell us that, judging from the success rate in the past, an additional 328 types of drugs are yet to be found in tropical rain forests, and the value of this resource to society is probably $147 billion.

You may already know that the antibiotic penicillin is derived from a fungus and that certain species of bacteria produce the antibiotics tetracycline and streptomycin. These drugs have proven to be indispensable in the treatment of diseases, including certain sexually transmitted diseases.

Leprosy is among the diseases for which there is, as yet, no cure. The bacterium that causes leprosy will not grow in the laboratory, but scientists discovered that it grows naturally in the nine-banded armadillo (Fig. 32.16b). Having a source for the bacterium may make it possible to find a cure for leprosy. The blood of horseshoe crabs contains a substance called limulus amoebocyte lysate, which is used to ensure that medical devices, such as pacemakers, surgical implants, and prostheses, are free of bacteria. Blood is taken from 250,000 horseshoe crabs a year, and then they are returned to the sea unharmed.

Agricultural Value

Crops such as wheat, corn, and rice are derived from wild plants that have been modified to be high producers. The same high-yield, genetically similar strains tend to be grown worldwide. When rice crops in Africa were being devastated by a virus, researchers grew wild rice plants from thousands of seed samples until they found one that contained a gene for resistance to the virus. These wild plants were then used in a breeding program to transfer the gene into high-yield rice plants. If this variety of wild rice had become extinct before its resistance could be discovered, rice cultivation in Africa might have collapsed.

Biological pest controls—specifically, natural predators and parasites—are often preferable to chemical pesticides (Fig. 32.17a). When a rice pest called the brown planthopper became resistant to pesticides, farmers began to use natural enemies of the brown planthopper instead. The economic savings were calculated at well over $1 billion. Similarly, cotton growers in Cañete Valley, Peru, found that pesticides were no longer working against the cotton aphid due to resistance. Research identified natural predators that are now being used to an ever greater degree by cotton farmers. Again, savings have been enormous.

Most flowering plants are pollinated by animals, such as bees, wasps, butterflies, beetles, birds, and bats (Fig. 32.17b). The honeybee has been domesticated, and it pollinates almost $10 billion worth of food crops annually in the United States. The danger of this dependency on a single species is exemplified by mites that have now wiped out more than 20% of the commercial honeybee population in the United States. Where can we get resistant bees? From the wild, of course. The value of wild pollinators to the U.S. agricultural economy has been calculated at $4.1 to $6.7 billion a year.

Consumptive Use Value

We have had much success in cultivating crops, keeping domesticated animals, growing trees on plantations, etc. But so far, aquaculture, the growing of fish and shellfish for human consumption, has contributed only minimally to human welfare. Instead, most freshwater and marine harvests depend on the catching of wild animals, such as fishes (e.g., trout, cod, tuna, and flounder), crustaceans (e.g., lobsters, shrimps, and crabs), and mammals (e.g., whales) (Fig. 32.17c). These aquatic organisms are an invaluable biodiversity resource.

The environment provides all sorts of other products that are sold in the marketplace worldwide, including wild fruits and vegetables, skins, fibers, beeswax, and seaweed. Also, some people obtain their meat directly from the environment. In one study, researchers calculated that the economic value of wild pig in the diet of native hunters in Sarawak, East Malaysia, was about $40 million per year.

Similarly, many trees are still felled in the natural environment for their wood. Researchers have calculated that a species-rich forest in the Peruvian Amazon is worth far more if the forest is used for fruit and rubber production than for timber production (Fig. 32.17d). Fruit and the latex needed to produce rubber can be brought to market for an unlimited number of years, whereas once the trees are gone, no more timber can be harvested.

Wild species, like ladybugs, play a role in biological control of agricultural pests.

a.

Wild species, like the long-nosed bat, are pollinators of agricultural and other plants.

b.

c.

Wild species, like many marine species, provide us with food.

Wild species, like rubber trees, can provide a product indefinitely if the forest is not destroyed.

Figure 32.17 Agriculture and consumptive value of wildlife.

Many wildlife species help account for our bountiful harvests and are sources of food for today's world.

d.

Figure 32.18   Indirect value of ecosystems.

Forests and the oceans perform many of the functions listed as the indirect value of ecosystems.

Indirect Values of Biodiversity

The wild species we have been discussing live in ecosystems. If we want to preserve them, it is more economical to save the ecosystems than the individual species. Ecosystems perform many services for modern humans, who increasingly live in cities. These services are said to be indirect because they are pervasive and not easily discernible (Fig. 32.18). Even so, our very survival depends on the functions that ecosystems perform for us.

Biogeochemical Cycles

You'll recall from Chapter 31 that ecosystems are characterized by energy flow and chemical cycling. The biodiversity within ecosystems contributes to the workings of the water, carbon, nitrogen, phosphorus, and other biogeochemical cycles. We are dependent on these cycles for fresh water, removal of carbon dioxide from the atmosphere, uptake of excess soil nitrogen, and provision of phosphate. When human activities upset the usual workings of biogeochemical cycles, the dire environmental consequences include the release of excess pollutants that are harmful to us. Technology is unable to artificially contribute to or create any of the biogeochemical cycles.

Waste Disposal

Decomposers break down dead organic matter and other types of wastes to inorganic nutrients that are used by the producers within ecosystems. This function aids humans immensely because we dump millions of tons of waste material into natural ecosystems each year. If not for decomposition, waste would soon cover the entire surface of our planet. We can build sewage treatment plants, but they are expensive, and few of them break down solid wastes completely to inorganic nutrients. It is less expensive and more efficient to water plants and trees with partially treated wastewater and let soil bacteria cleanse it completely.

Biological communities are also capable of breaking down and immobilizing pollutants, such as heavy metals and pesticides, that humans release into the environment. A review of wetland functions in Canada assigned a value of $50,000 per hectare (2.471 acres or 10,000 square meters) per year to the ability of natural areas to purify water and take up pollutants.

Provision of Fresh Water

Few terrestrial organisms are adapted to living in a salty environment—they need fresh water. The water cycle continually supplies fresh water to terrestrial ecosystems. Humans use fresh water in innumerable ways, including drinking it and irrigating their crops. Freshwater ecosystems, such as rivers and lakes, also provide us with fish and other types of organisms for food.

Unlike other commodities, there is no substitute for fresh water. We can remove salt from seawater to obtain fresh water, but the cost of desalination is about four to eight times the average cost of fresh water acquired via the water cycle.

Forests and other natural ecosystems exert a "sponge effect." They soak up water and then release it at a regular rate. When rain falls in a natural area, plant foliage and dead leaves lessen its impact, and the soil slowly absorbs it, especially if the soil has been aerated by organisms. The water-holding capacity of forests reduces the possibility of flooding. The value of a marshland outside Boston, Massachusetts, has been estimated at $72,000 per hectare per year solely on its ability to reduce floods. Forests release water slowly for days or weeks after the rains have ceased. Compared to rivers from West African coffee plantations, rivers flowing through forests release twice as much water halfway through the dry season and between three and five times as much at the end of the dry season.

Prevention of Soil Erosion

Intact ecosystems naturally retain soil and prevent soil erosion. The importance of this ecosystem attribute is especially noticeable following deforestation. In Pakistan, the world's largest dam, the Tarbela Dam, is losing its storage capacity of 12 billion cubic meters many years sooner than expected because silt is building up behind the dam due to deforestation. At one time, the Philippines were exporting $100 million worth of oysters, mussels, clams, and cockles each year. Now, silt carried down rivers following deforestation is smothering the mangrove ecosystem that serves as a nursery for the sea. Most coastal ecosystems are not as bountiful as they once were because of deforestation and a myriad of other assaults.

Regulation of Climate

At the local level, trees provide shade and reduce the need for fans and air conditioners during the summer.

Globally, forests regulate the climate because they take up carbon dioxide. The leaves of trees use carbon dioxide when they photosynthesize, and the bodies of the trees store carbon. When trees are cut and burned, carbon dioxide is released into the atmosphere. Carbon dioxide makes a significant contribution to global warming, which is expected to be stressful for many plants and animals. Only a small percentage of wildlife will be able to move northward where the weather will be suitable for them.

Ecotourism

Almost everyone prefers to vacation in the natural beauty of an ecosystem (Fig. 32.19). In the United States, nearly 100 million people enjoy vacationing in a natural setting. To do so, they spend $4 billion each year on fees, travel, lodging, and food. Many tourists want to go sport fishing, whale watching, boat riding, hiking, birdwatching, and the like. Some merely want to immerse themselves in the beauty of a natural environment.

Figure 32.19   Ecotourism.

Whale watchers experience the thrill of seeing an orca surfacing off the coast of Washington state.

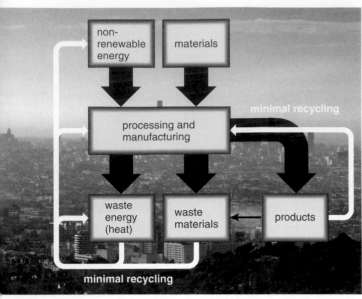

a. Human society at present

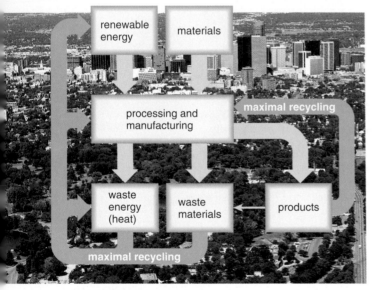

b. Sustainable society

Figure 32.20   Current human society versus a sustainable society.

a. At present, our "throw-away" society is characterized by high input of energy and raw materials, large output of waste materials and energy in the form of heat, and minimal recycling (*white arrows*). **b.** A sustainable society would be characterized by the use of only renewable energy sources, reuse of heat and waste materials, and maximal recycling of products (*blue arrows*).

32.3 Working Toward a Sustainable Society

A **sustainable society** would be able to provide the same goods and services for future generations of human beings as it does now. At the same time, biodiversity would be preserved.

Today's Society

The following evidence indicates that at present, human society is most likely not sustainable (Fig. 32.20*a*):

- A considerable proportion of land, and therefore natural ecosystems, is presently being used for human purposes (homes, agriculture, factories, etc.).
- Agriculture requires large inputs of nonrenewable fossil fuel energy, fertilizer, and pesticides, which create much pollution. More fresh water is used for agriculture than in homes.
- At least half of the agricultural yield in the United States goes toward feeding animals. According to the ten-to-one rule of thumb, it takes 10 pounds of grain to grow one pound of meat. Therefore, it is wasteful for citizens in MDCs to eat as much meat as they do. Also, animal sewage pollutes water.
- Even though fresh water is a renewable resource, we are running out of the available supply within a given time frame.
- Our society primarily utilizes nonrenewable fossil fuel energy, which leads to global warming, acid precipitation, and smog.
- Minerals are nonrenewable, and the mining, processing, and use of minerals are responsible for much environmental pollution.

Characteristics of a Sustainable Society

A natural ecosystem can offer clues as to what a sustainable human society would be like. A natural ecosystem is characterized by use of renewable solar energy, and its materials cycle through the various populations back to the producers once again. It is clear that if we want to develop a sustainable society, we too should use renewable energy sources and recycle materials (Fig. 32.20*b*).

While we are sometimes quick to perceive that the growing populations of the LDCs are straining the environment, we should realize that the excessive resource consumption of the MDCs also stresses the environment. More than likely, sustainability is incompatible with the kinds of consumption and waste patterns currently practiced in the MDCs.

Check Your Progress

1. **List the characteristics of today's society that make it unsustainable.**
2. **List the characteristics of a sustainable society.**

Answers: 1. Large proportions of natural ecosystems are being used for human purposes; agriculture requires large inputs; much agriculture is used for animal production; fresh water is being used up; most energy comes from fossil fuels; and minerals are causing pollution. 2. Use of renewable energy resources, reuse of heat and waste materials, and maximal recycling.

Summary

32.1 Resources and Pollution

Five resources are maximally used by humans:

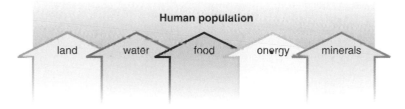

Resources are either nonrenewable or renewable.

Nonrenewable resources are limited in supply:
- Land
- Fossil fuel
- Minerals

Renewable resources are not limited in supply:
- Water
- Solar energy
- Food

Land
Human activities, such as habitation, farming, and mining, contribute to erosion, pollution, desertification, deforestation, and loss of biodiversity.

Water
Industry and agriculture use most of the freshwater supply. Water supplies are increased by damming rivers and drawing from aquifers. As aquifers are depleted, subsidence, sinkhole formation, and saltwater intrusion can occur. If used by industries, water conservation methods could cut world water consumption by half.

Food
Food comes from growing crops, raising animals, and fishing.

- Modern farming methods increase the food supply, but some methods harm the land, pollute water, and consume fossil fuels excessively.
- Transgenic plants can increase the food supply and reduce the need for chemicals.
- Raising livestock contributes to water pollution and uses fossil fuel energy.
- The increased number and high efficiency of fishing boats have caused the world fish catch to decline.

Energy
Most of the world's energy is supplied by the burning of fossil fuels, a nonrenewable resource. The burning of fossil fuels causes pollutants and gases to enter the air.

- Greenhouse gases include CO_2 and other gases emitted into the atmosphere. The increase in CO_2 is the result of burning fossil fuels and burning to clear land for farming. Other gases

are emitted by various means. Greenhouse gases cause global warming because solar radiation can pass through, but infrared heat cannot escape back into space.
- Renewable energy sources include hydropower, geothermal, wind, and solar power.

Minerals
Minerals are a nonrenewable resource that can be mined. These raw materials include sand, gravel, phosphate, and metals. The act of mining causes destruction of the land by erosion, loss of vegetation, and toxic runoff into bodies of water. Some metals are dangerous to human health. Land ruined by mining can take many years to recover.

Synthetic Organic Compounds
Compounds such as chlorofluorocarbons (CFCs) are detrimental to the ozone layer and to the health of living things, including humans.

Wastes
Raw sewage and industrial wastes can pollute land and bodies of water. Some industrial wastes cause biological magnification, by which toxins become more concentrated as they pass along a food chain.

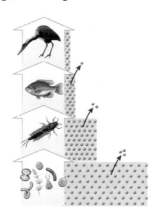

32.2 Biodiversity

Biodiversity is the variety of life on Earth. It has both direct value and indirect value.

Direct Value
Direct values of biodiversity are:

- Medicinal value (medicines derived from living organisms)
- Agricultural value (crops derived from wild plants)
- Biological pest controls and animal pollinators
- Consumptive use values (food production)

Indirect Value
Biodiversity in ecosystems contributes to:

- The function of biogeochemical cycles (water, carbon, nitrogen, phosphorus, and others)
- Waste disposal (through the action of decomposers and the ability of natural communities to purify water and take up pollutants)
- Fresh water provision through the water biogeochemical cycle
- Prevention of soil erosion, which occurs naturally in intact ecosystems
- Climate regulation (plants take up carbon dioxide)
- Ecotourism (human enjoyment of a beautiful ecosystem)

32.3 Working Toward a Sustainable Society

A sustainable society would use only renewable energy sources, would reuse heat and waste materials, and would recycle almost everything. It would also provide the same goods and services presently provided and would preserve biodiversity.

Thinking Scientifically

1. In the early 1960s, Soviet officials built dams on two large Siberian rivers to support a developing agricultural economy based on cotton, a crop with a high demand for water. Since that time, the Aral Sea (which is fed by those two rivers) in central Asia has shrunk to 25% of its original size. Now plans are being discussed to build a canal to divert water from two other Siberian rivers into the dammed ones in order to keep up with agricultural water demands. What might be the positive and negative environmental impacts of connecting the two river systems for the purpose of increasing the supply of water for cotton production? Another option is to remove the dams from the first two rivers and let them flow freely again. What might be the positive and negative impacts of this strategy?

2. Every time a new species becomes extinct, we lose a potential source of human medicine. Since we do not know which species are likely to provide valuable medicines, it is difficult to know where to focus conservation efforts. If you were in charge of determining high priorities for species conservation in a country, with the goal of identifying plants with medicinal value, what factors would you consider in making your decisions?

Testing Yourself

Choose the best answer for each question.

1. Beach erosion is caused by
   a. rising sea levels.
   b. global warming.
   c. human development.
   d. filling of coastal wetlands.
   e. All of these are correct.

2. Desertification typically happens because
   a. deserts naturally expand in size.
   b. humans allow overgrazing.
   c. desert animals wander into adjacent areas for food.
   d. humans tap into limited water supplies for water needs in the nearby desert.

3. Soils in tropical rain forests are typically nutrient-poor because
   a. they are over bedrock.
   b. they are sandy.
   c. nitrogen-fixing bacteria are absent.
   d. nutrients are tied up in vegetation.

4. Most fresh water in the world is used for
   a. drinking.
   b. supporting industry.
   c. irrigating crops.
   d. cooking.
   e. bathing.

5. Which of the following will help conserve fresh water?
   a. saltwater intrusion
   b. salt-tolerant crops
   c. sprinkler irrigation
   d. More than one of these are correct.

6. The first green revolution resulted from the development of
   a. high-responder wheat and rice varieties.
   b. effective irrigation systems.
   c. transgenic crop plants.
   d. crops designed for animal feed.

7. The preferred fossil fuel in the United States is
   a. coal because it produces less pollution than oil.
   b. oil because it produces less pollution than coal.
   c. coal because it is more abundant than oil.
   d. oil because it is more abundant than coal.

For questions 8–12, indicate the type of energy associated with each statement. Some questions may have more than one answer.

Key:
   a. fossil fuels
   b. hydropower
   c. geothermal energy
   d. wind power
   e. solar energy

8. May require the building of dams.

9. Environmental impact is minimal.

10. Use will probably lead to inland droughts.

11. Must be stored at times.

12. May be used to extract hydrogen from water.

13. A major negative effect of the dumping of raw sewage into lakes and rivers is
   a. oxygen depletion.
   b. the buildup of carbon.
   c. a reduction of light penetration into the water.
   d. an increase in populations of small fish.

14. Show how the following diagram must change in order to develop a sustainable society.

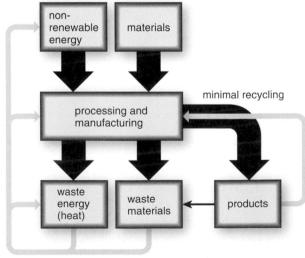

minimal recycling

15. The preservation of ecosystems indirectly provides fresh water because

     a. trees produce water as a result of photosynthesis.
     b. animals excrete water-based products.
     c. forests soak up water and release it slowly.
     d. ecosystems promote the growth of bacteria that release water into the environment.

16. A renewable energy source is

     a. hydropower.
     b. natural gas.
     c. coal.
     d. solar energy.
     e. Both a and d are correct.

17. A nonrenewable energy source is

     a. wind.
     b. geothermal energy.
     c. solar energy.
     d. fossil fuels.

18. A transition to hydrogen fuel technology will

     a. be long in coming and not likely be of major significance.
     b. lessen many current environmental problems.
     c. not be likely since it will always be expensive and as polluting as fossil fuels.
     d. be of major consequence, but resource limitations for obtaining hydrogen will hinder its progress.

19. Most fresh water is held in

     a. groundwater.
     b. lakes.
     c. rivers.
     d. ice and snow.
     e. oceans.

20. Subsidence involves

     a. the overexploitation of land.
     b. the removal of water from aquifers.
     c. the removal of plant life.
     d. desertification or deforestation.

21. Evaporation of excess water from irrigation of farmland causes

     a. salinization.
     b. accumulation of mineral salts.
     c. subsidence.
     d. loss of topsoil.
     e. Both a and b are correct.

22. Which of these are indirect values of species?

     a. participation in biogeochemical cycles
     b. participation in waste disposal
     c. provision of fresh water
     d. prevention of soil erosion
     e. All of these are indirect values.

Go to www.mhhe.com/maderessentials for more quiz questions.

Bioethical Issue

Beach erosion is a serious problem that is difficult to address. One control strategy has been to build hard structures such as seawalls and groins to reduce the impact of waves. However, these structures are expensive to construct and actually accelerate beach erosion by concentrating the power of waves. Construction of new barriers has recently been banned in several states. The opposite approach to saving beaches is "managed retreat." This method allows coastal erosion to occur naturally by removing man-made beach structures. Several parks have adopted this strategy by removing all hard structures. Should private property owners on beaches be encouraged or forced to remove physical barriers as well? If so, should they be compensated for the loss of beach that will occur as a result of natural erosion? Should incentives be put in place to encourage the planting of native vegetation to reduce erosion?

Understanding the Terms

aquifer   579
biodiversity   588
biological magnification   587
chlorofluorocarbons (CFCs)   586
deforestation   578
desertification   578
fossil fuel   583
greenhouse gases   583
mineral   586
nonrenewable resources   576
photovoltaic (solar) cell   585
pollution   576
renewable resources   576
salinization   581
saltwater intrusion   579
subsidence   579
sustainable society   592

Match the terms to these definitions:

a. _____ Alteration of the environment in an undesirable way.

b. _____ Removal of trees.

c. _____ Reservoir of water just below the surface of the Earth.

d. _____ Settling of soil due to the removal of water from aquifers.

e. _____ Flow of seawater into streams and aquifers as a result of lowering the water table.

f. _____ Accumulation of mineral salts in soil due to evaporation of irrigation water.

g. _____ Emissions that allow solar radiation to reach the Earth but not escape back into space.

h. _____ Collects electrons and passes them into wires.

i. _____ Nonrenewable raw material in the Earth's crust.

j. _____ Type of halogenated hydrocarbon.

Appendix A

Answer Key

Chapter 1

Thinking Scientifically

1. Plant a section of your garden in the same variety of tomato and plant seedlings that are the same age and size. One possibility would be to plant the seedlings in square groups of four, with space between each group. It would be best to plant the groups in a block. If you have room, you could plant 16 groups in a 4 by 4 square. Then, randomly assign 8 groups to receive the name-brand fertilizer and 8 groups to receive the generic brand. Treat all the groups the same otherwise, by watering and weeding them uniformly. Fertilize them at the same time and with the same amount of fertilizer. At the end of the growing season, weigh the fruit that you harvest from each plot and compare the total weight of the name-brand treatment with that of the generic treatment. 2. Biscoe showed treatment effects that increased consistently from 21 to 7 days before harvest. The growth regulator did not work equally effectively in both cultivars. It worked better for Biscoe because the difference between the untreated and treatment was greater. The most effective treatment was 7 days before harvest.

Testing Yourself

1. c; 2. a. small molecules; b. large molecules; c. cells; d. tissues; e. organs; f. organ systems; g. complex organisms; see Fig. 1.1, p. 2, in text.; 3. b; 4. c; 5. d; 6. b; 7. b; 8. d; 9. c; 10. a; 11. c; 12. a; 13. g

Understanding the Terms

a. hypothesis; b. model; c. adaptation; d. species; e. taxonomy; f. biosphere; g. data

Chapter 2

Thinking Scientifically

When the bird enters the water, surface tension produced by hydrogen bonding keeps the surface smooth and continuous. When the bird flies out of the water, the cohesiveness of water molecules causes drops of water to be pulled out with the bird.

Testing Yourself

1. b; 2. c; 3. c; 4. d; 5. b; 6. a; 7. b

Understanding the Terms

a. electrons; b. matter; c. polar; d. hydrophilic; e. isotopes; f. tracer; g. ion

Chapter 3

Thinking Scientifically

1. a. plasma membrane; b. increases, because the plants are improving their ability to function at cold (but not freezing) temperatures by increasing the proportion of linolenic acid in their plasma

membranes. Chilling injury results mainly from a loss of membrane function. The plasma membrane is composed of proteins floating in a "sea" of phospholipids. If those proteins can not flow freely, then the membrane can not function properly. c. Canola oil remains liquid; butter is solid. d. no; e. fatty acids; f. The fatty acids in the plasma membrane of cold tolerant plants, such as temperate plants, are mainly unsaturated. They remain fluid at cold temperatures, allowing the membrane to function properly. The fatty acids in the plasma membrane of plants that are not cold tolerant, such as many tropical plants, have high levels of saturated fatty acids. At cold temperatures, the membrane lipids solidify, inactivating membrane-bound proteins. This results in a loss of membrane function and the cells are damaged. 2. A protein's ability to function correctly is a factor of its shape. A substituted amino acid with a polar *R* group may be more likely to bind to other amino acids in the protein, and hence change its shape, than substituted amino acids with nonpolar *R* groups.

Testing Yourself

1. c; 2. c; 3. d; 4. d; 5. b; 6. b; 7. d; 8. a; 9. c; 10. b; 11. d; 12. b; 13. d; 14. c; 15. a; 16. a; 17. c; 18. b; 19. d; 20. e; 21. b; 22. a; 23. d; 24. c; 25. d; 26. c; 27. d; 28. d; 29. c; 30. d; 31. a; 32. c; 33. a; 34. d; 35. a; 36. d; 37. b

Understanding the Terms

a. isomer; b. polymer; c. steroid; d. enzyme; e. phospholipid; f. denatured

Chapter 4

Thinking Scientifically

1. a. A dietician may be able to suggest a diet that would not include the offending long-chain fatty acids. b. Lysosomes are similar to peroxisomes in that they are single membrane organelles containing enzymes that break down cellular compounds. 2. The plastid must be necessary to the life of the parasite because the parasite dies if the plastid no longer functions.

Testing Yourself

1. e; 2. a; 3. c; 4. d; 5. c; 6. f; 7. b; 8. d; 9. e; 10. h; 11. c; 12. b; 13. a; 14. b; 15. c; 16. d; 17. c; 18. a. vesicle; b. centrioles; c. mitochondrion; d. rough endoplasmic reticulum; e. smooth endoplasmic reticulum; f. lysosome; g. Golgi apparatus; h. nucleus; see Fig. 4.7, p. 54, in text.

Understanding the Terms

a. cell theory; b. endomembrane system; c. apoptosis; d. granum; e. Golgi apparatus; f. cytoskeleton; g. capsule

Chapter 5

Thinking Scientifically

1. On a calm, dark night on the ocean, collect a water sample and count the number of dinoflagellate-eaters and predators of the dinoflagellate-eaters in the sample. Shine a light into the water for a period of time and then take another sample. Determine whether the number of dinoflagellate-eaters has decreased and whether the number of predators of the dinoflagellate-eaters has increased. 2. A normal channel protein allows chloride to leave the cells. When chloride leaves, water will follow, keeping the outer cell surfaces hydrated. If chloride does not leave cells, then water will not exit and the cell surface will be covered with a thick mucus. The cholera bacterium produces a toxin that opens chloride channels in cells of the small intestine. As salt leaves the intestinal cells, water follows, resulting in diarrhea. The defective cystic fibrosis protein closes chloride channels, so the toxin can not open the channels.

Testing Yourself

1. d; 2. e; 3. b; 4. d; 5. c; 6. a,d; 7. b,c; 8. b,c; 9. a,d; 10. a. without enzyme; b. with enzyme; see Fig. 5.8, p. 75, in text; 11. a; 12. a,b,c; 13. b,c,d; 14. d; 15. d; 16. a; 17. b; 18. c; 19. b; 20. a; 21. c; 22. d; 23. a; 24. c; 25. b; 26. a; 27. c; 28. b; 29. b; 30. c; 31. e

Understanding the Terms

a. entropy; b. plasmolysis; c. metabolic pathway; d. enzyme; e. substrate

Chapter 6

Thinking Scientifically

1. a. No. b. At first it did, but above a certain level—approximately 1/3 of the final light intensity—additional light did not increase the photosynthetic rate; c. Yes. CO_2 enrichment had to be combined with increased temperature to have its maximum effect on growth. d. Yes, especially at 30°C. Again, boosting all factors that contribute to photosynthetic rate produced the maximum effect. e. Yes, an increase in CO_2 levels resulted in a dramatic increase in photosynthetic rate. 2. a.–d. below are suggested answers to this question. a. There may be other limiting factors for growth, such as mineral nutrients and water. In a greenhouse study, plants are typically supplied with adequate levels of water and nutrients.
b. High levels of CO_2 may allow soil microorganisms to grow faster and outcompete plants for limiting nutrients, such as phosphorus or potassium. In greenhouse studies, plants are usually grown in pasteurized soil, so microorganisms are not a variable. c. Mature trees may not take up CO_2 as effectively as young, herbaceous greenhouse

plants. Trees store large amounts carbon that can be tapped for growth, so not as much needs to be taken in via photosynthesis. Greenhouse studies focus on non-woody plants that don't have much capacity for carbon storage. d. Changes in plant community composition may result from competition among species that differ in their response to elevated CO_2 and temperature. This may influence total CO_2 utilization in the ecosystem. Greenhouse studies typically focus on individual species. 3. The alga was producing oxygen as a byproduct wherever its chlorophyll was able to absorb the light energy needed for photosynthesis. The bacteria were using the oxygen for cellular respiration.

Testing Yourself

1. c; 2. a; 3. d; 4. a; 5. a; 6. a; 7. b; 8. d; 9. a. stroma; b. electron transport chain; c. ATP; d. NADPH; e. ATP synthetase complex; f. thylakoid membrane; g. thylakoid space; see Fig. 6.7, p. 89, in text; 10. d; 11. c; 12. a, b, c; 13. b; 14. b; 15. c

Understanding the Terms

a. stroma; b. stomata; c. ATP synthase; d. thylakoid; e. coenzymes; f. carbon dioxide fixation; g. photosystem; h. carotenoid; i. CAM

Chapter 7

Thinking Scientifically

1. Physiologically active kernels and microbes carry out respiration, producing heat and water as by-products. The warmer temperatures and additional moisture increase physiological activity (including respiration) even more, producing ever-increasing levels of heat and moisture. Flammable gases produced by physiological activity will eventually ignite as the temperature rises above their ignition point. 2. The mitochondria of older people exhibit lower levels of metabolic activity than those of younger people. Since the mitochondria are less active, they are less able to break down fatty acids in muscle cells. This leads to the accumulation of fatty acids in cells, resulting in insulin resistance. To reduce the risk of diabetes, the mitochondrial activity in muscle cells must be increased. This could be accomplished by increasing either the number or activity of mitochondria. Recent research indicates that exercise increases the number of mitochondria in cells. These cells, therefore, are better equipped to metabolize fatty acids.

Testing Yourself

1. a; 2. c; 3. a; 4. d; 5. b; 6. a; 7. b; 8. a. Pyruvate is broken down to an acetyl group.; b. Acetyl group is taken up and a C_6 molecule results.; c. Oxidation results in NADH and CO_2.; d. ATP is produced by substrate-level and ATP synthesis.; e. Oxidation produces more NADH and $FADH_2$; see Fig. 7.6, p. 103, in text.; 9. c; 10. a, c, d; 11. a; 12. a, b, c; 13. d; 14. c; 15. a; 16. b, c, d; 17. c; 18. d; 19. c; 20. b; 21. b; 22. b; 23. b; 24. c; 25. c; 26. a; 27. c; 28. a. citric acid cycle; b. anaerobic; c. pyruvate; d. fermentation

Understanding the Terms

a. glycolysis; b. preparatory reaction; c. fermentation; d. intermembrane space; e. citric acid cycle

Chapter 8

Thinking Scientifically

1. Cancer results from a series of muations ("hits") that accumulate in somatic cells over a person's lifetime. People exposed to high levels of radiation are likely to have a higher number of somatic mutations than the general population. They would then require fewer additional random somatic mutations to occur before they develop cancer. A reasonable explanation for the data above is that some cancers require a larger number of hits in order to be expressed. That is, more genes need to be mutated in a pathway before cancer develops. A type of cancer that requires more hits (random somatic mutations) would generally develop later in life. 2. Sister chromatids of chromosomes (inside the nucleus) must link to microtubules produced by centrosomes (outside the nucleus) in order to be pulled apart during mitosis. Therefore, microtubules cannot attach to and separate chromatids if the nuclear membrane is present. The current hypothesis is that cytoplasmic dynein, a molecular motor protein, tears the nuclear envelope and transports pieces of the membrane away from the area along microtubules. Once pieces are torn away, the nuclear envelope loses shape and cytoplasmic proteins flow into the nuclear region. The rush of cytoplasmic proteins is believed to stimulate chromosome condensation and spindle formation. So, dismantling of the nuclear membrane is coordinated with chromosome behavior at mitosis.

Testing Yourself

1. b; 2. c; 3. a; 4. c; 5. a; 6. b; 7. a. G_1; b. S; c. G_2; d. prophase; e.metaphase; f. anaphase; g. telophase; h. cytokinesis; see Fig. 8.3, p. 114, in text; 8. a; 9. d; 10. c; 11. c; 12. b; 13. a, b; 14. b; 15. b; 16. c; 17. a; 18. a; 19. c; 20. c; 21. d; 22. e; 23. e; 24. d

Understanding the Terms

a. histones; b. centromere; c. spindle; d. centrosome; e. signal

Chapter 9

Thinking Scientifically

1. a. Normal $\times$ mutant = 10 + 20 = 30 (That is, a triploid plant, with 3 copies of every chromosome.) b. Mutant $\times$ mutant = 20 + 20 = 40 (That is, a tetraploid plant, with 4 copies of every chromosome.) 2. a. To disprove the hypothesis, look at the incidence of Down syndrome in families of various sizes. Down syndrome is correlated with maternal age, but not family size. b. Hypotheses for the relationship between maternal age and incidence of Down syndrome include: (1) The older a woman is, the longer her oocytes have been arrested in meiosis. Exposure to mutagens during this time may result in nondisjunction; (2) A woman may have a pool of oocytes resulting from nondisjunction, but is better able to prevent them maturing at a young age; (3) Estrogen levels (which control the rate of meiosis in developing oocytes) drop with maternal age and may slow down the rate of meiosis. This might allow nondisjunction to happen more frequently in older women.

Testing Yourself

1. a; 2. d; 3. d; 4. a; 5. b; 6. a; 7. c; 8. d; 9. a; 10. f; 11. h; 12. c; 13. g; 14. b; 15. e; 16. a. homologues condense and undergo synapsis, chromosomes condense and do not pair; b. tetrads at spindle equator, dyads at spindle equator; c. homologues separate, daughter chromosomes separate; d. two daughter cells from meiosis I; (bottom) four daughter cells following meiosis II, two daughter cells following mitosis; see Fig. 9.7, p.135, in text.; 17. a; 18. c

Understanding the Terms

a. oogenesis; b. dyad; c. synapsis; d. crossing-over; e. nondisjunction; f. Turner syndrome; g. Barr body

Chapter 10

Thinking Scientifically

1. a. $CcPp \times CcPp$ would produce 9:16 purple ($C\_\_P\_\_$) and 7:16 white offspring; b. All white offspring. 2. That chromosomal segment contained the *SRY* gene. XX men have the *SRY* gene on one of the X chromosomes, so they have the male phenotype. XY women contain a Y chromosome in which *SRY* has been lost, so they have the female phenotype. The *SRY* gene probably moved from the Y chromosome to the X chromosome following a rare recombination event.

Testing Yourself

1. b; 2. c; 3. d; 4. d; 5. b; 6. b; 7. e; 8. a. *TG*; b. *tg*; c. *TtGg*; d. *TTGG*; e. *TtGg*; f. *TTgg*; g. *Ttgg*; h. *TtGG*; i. *ttGG*; j. *TtGg*; k. *ttGg*; l. *ttgg*; see Fig. 10.6, p. 146, in text.; 9. b; 10. c; 11. c; 12. b; 13. a; 14. c, 15. d; 16. e; 17. d; 18. a; 19. a, b; 20. a, b, c, d; 21. c, d; 22. a, b, d; 23. c

Understanding the Terms

a. alleles; b. homozygous; c. phenotype; d. wild-type; e. rule of multiplication; f. incomplete dominance; g. linkage group

Chapter 11

Thinking Scientifically

1. DNA replication occurs as a part of the cell cycle and replication errors sometimes lead to mutations. Skin cells undergo the cell cycle more frequently than nerve cells in the brain; therefore, skin cells may be mutating faster than nerve cells. Some of these mutations may be cancer-causing. This hypothesis would be supported if there are more differences in DNA base sequences in skin cells compared to brain cells. However, you need a standard to compare your base sequence data to. You should choose another tissue, like muscle tissue, whose cells also don't divide often. 2. A vertebrate has tens of thousands of genes. Many of the genes are for basic cellular functions, and would have no role in the specific developmental program of the animal. So even with 30 complete genes it is highly unlikely that the genes would have much to do with development. Supposing that you did have developmental genes, they could only function properly if they were active at specific points in devleopment. This would require other specific genes that would turn the developmental

genes on and off at appropriate times. Lastly, even if one did have the luck to possess developmental genes and regulatory genes, these developmental genes would require specific target molecules which would probably not exist in a modern reptile. Therefore, even in the highly improbable circumstance that one had 100 intact genes it is extremely unlikely that these genes could be used to recreate a dinosaur.

Testing Yourself

1. a. sugar-phosphate backbone; b. complementary base pairing or rungs; c. hydrogen bonds; d. sugar; see Fig. 11.5, p. 163, in text.; 2. a; 3. d; 4. d; 5. a, b, c; 6. b; 7. a; 8. c; 9. c; 10. b; 11. a. DNA template strand; b. RNA polymerase; c. mRNA; see Fig. 11.11, p. 167, in text.; 12. d; 13. b; 14. c; 15. a; 16. b

Understanding the Terms

a. adenine, guanine; b. complementary base pairing; c. template; d. DNA polymerase; e. promoter

Chapter 12

Thinking Scientifically

1. Not all genes on the second X chromosome are inactivated. Expression of genes on the second X chromosome is apparently necessary for normal growth and sexual development. 2. A tortoiseshell male must have two X chromosomes (one carrying the orange allele and one carrying the black allele) and a Y chromosome (so it is male). This happens only if nondisjunction of the sex chromosomes happened in one of its parents. The cat will be sterile because the three sex chromosomes do not separate normally at meiosis, producing nonviable gametes.

Testing Yourself

1. b; 2. c; 3. a; 4. a. regulatory gene; b. promoter; c. operator; d. lactose metabolizing genes; e. mRNA; f. repressor; g. RNA polymerase; see Fig. 12.5, p. 182, in text.; 5. b; 6. c; 7. d; 8. c; 9. a; 10. b; 11. c; 12. a; 13. d; 14. d; 15. a; 16. e; 17. c; 18. c; 19. e; 20. e; 21. b; 22. b; 23. a

Understanding the Terms

a. operon; b. repressor; c. heterochromatin; d. transcription factor; e. enhancer; f. growth factor; g. telomere

Chapter 13

Figure Legends

Fig. 13.7. The individual is normal, but she has a child that is affected. Fig. 13.8. The individual is affected, but he has a child that is normal.

Thinking Scientifically

1. The gene for CFTR is large, and many different mutations can produce an altered protein. Some mutations result in a protein that is somewhat functional. People carrying these mutations will have mild forms of cystic fibrosis. Other mutations prevent the CFTR gene from functioning at all, resulting in severe expression of cystic fibrosis. A genetic test will identify a specific mutation. People with that mutation will test positive, however, other people will have mutations in different parts of the CFTR gene. These mutations will not be detected by the same genetic test. 2. The adenovirus probably inserted itself into a gene that regulates the cell cycle. The gene may have been a tumor-suppressor gene or a proto-oncogene. The insertion of the virus into the cell cycle gene resulted in a mutation, preventing that gene from producing a normal product. In fact, the virus was found to have inserted itself in the proximity of a proto-oncogene called LMO-2.

Testing Yourself

1. d, 2. c; 3. d; 4. a. aa; b. A?; c. Aa; d. Aa; e. Aa; f. Aa; g. aa; h. aa; i. A?; see Fig. 13.7, p. 200, in text.; 5. a. Aa; b. Aa; c. aa; d. Aa; e. aa; f. Aa; g. Aa; h. aa; see Fig. 13.8, p. 200, in text.; 6. a. X^{h}Y; b. X^{B}X^b; c. X^{B}Y; d. X^{B}Y; e. X^{b}Y; see Fig. 13.9, p. 201, in text.; 7. b; 8. a; 9. b, d; 10. a, c; 11. b; 12 e, f; 13. g; 14. f; 15. i; 16. b; 17. h, i, j; 18. a; 19. c; 20. d; 21. b; 22. c; 23. b; 24. 50%, 0%; 25. parents Aa, child aa; 26. Yes, the son may inherit the father's recessive allele.; 27. X^{a}X^a, father X^{a}Y, mother X^{A}X^a or X^{a}X^a.

Understanding the Terms

a. karyotype; b. chorionic villi sampling; c. translocation; d. pedigree; e. proteomics; f. gene therapy

Chapter 14

Thinking Scientifically

1. The genes involved in male reproduction are directly and immediately affected by natural selection. These genes are under continuous pressure to outcompete other males and father offspring. Natural selection acts on other traits, such as the ability to compete for food and other resources. However, these traits are likely to be influenced by the environment and not consistently selected in every generation. 2. Some genes simply can not tolerate much change. These genes, called housekeeping genes, produce products necessary for basic metabolic processes. Mutations would be so catastrophic to the organism that they would be lethal. Therefore, they would not be perpetuated. Other genes may code for products in which some variation is tolerated. These genes would accumulate mutations faster than housekeeping genes.

Testing Yourself

1. b; 2. b; 3. d; 4. e; 5. c; 6. b; 7. b; 8. a. Originally, giraffes had short necks.; b. Giraffes stretched their necks in order to reach food.; c. Today most giraffes have long necks.; d. Originally, giraffe neck length varied.; e. Struggle to exist causes long-necked giraffes to have the most offspring.; f. Today most giraffes have long necks.; see Figure 14.8, p. 222, in text.; 9. e; 10. d; 11. c; 12. c; 13. b; 14. a; 15. c; 16. b; 17. c; 18. c; 19. e; 20. e; 21. c; 22. e

Understanding the Terms

a. homologous structure; b. uniformitarianism; c. natural selection; d. vestigial structure; e. adaptation

Chapter 15

Thinking Scientifically

There are several reasonable answers. First of all, although the disease is lethal, individuals with CF may reproduce (and pass on the defective allele) before they die. Second, the recessive (defective) allele may be hidden and perpetuated by carriers. Finally, the defective allele may confer a selective advantage in certain environments. Heterozygotes actually survive cholera epidemics better than homozygous dominant individuals because they resist the opening of chloride channels by the cholera bacterial toxin.

Testing Yourself

1. a; 2. d; 3. c; 4. a; 5. a; 6. c; 7. d; 8. b; 9. a; 10. a; 11. b; 12. b; 13. a; 14. a; 15. a; 16. c; 17. c; 18. b, 19. d; 20. a; 21. a. Wave will be in the center with shaded area in center of rise; b. Wave will shift to the right with small shaded area to the left of rise.; see Fig. 15.10, p. 239, in text.; 22. a. Wave has narrowed at base.; b. Wave has narrowed still with light shading; see Fig 15.11, p. 240, in text.; 23. a. Wave has dipped in the middle separating the shaded area from the nonshaded area.; b. Waves are two distinct waves, one shaded and not; see Fig. 15.12, p. 240, in text.

Understanding the Terms

a. gene pool; b. assortative mating; c. sexual selection; d. bottleneck effect

Chapter 16

Thinking Scientifically

1. a. Ferns; b. seed plants; c. naked seeds; d. needle-like leaves, Conifers; e. fan-shaped leaves, Gingkos; f. enclosed seeds; g. one embryonic leaf, Monocots; h. two embryonic leaves, Eudicots. 2. The Keys are too close to the mainland for populations to become isolated and undergo speciation. Immigrants from the mainland are constantly mating with individuals on the islands, allowing continuous gene exchange. In addition, the Keys do not exhibit the environmental diversity seen on the Hawaiian Islands. Dramatic habitat differences do not exist on each island, so specialization to unique environments does not exist.

Testing Yourself

1. c; 2. c; 3. b; 4. f; 5. a; 6. g; 7. d; 8. e; 9. h; 10. a. species 1; b. geographic barrier; c. genetic changes; d. species 2; e. genetic changes; f. species 3; see Fig. 16.7, p. 250, in text.; 11. b; 12. a; 13. c; 14. a, b, c; 15. d; 16. b; 17. e; 18. e; 19. a. three; designated by color; b. all animals. vertebra; only snakes and lizards: amniotic egg and internal fertilization; c. the animals in a clade; they share the same derived characters; 20. h; 21. a

Understanding the Terms

a. adaptive radiation; b. paleontology; c. mass extinction; d. binomial name; e. evolutionary tree; f. domain; g. homologous structure

Chapter 17

Thinking Scientifically

1. The genetic engineers remove the genes that cause disease and replace them with the beneficial genes that they would like to introduce into the plant. Then, they infect the plant with the modified bacterium and allow the bacterium to introduce the

foreign genes into the plant genome. 2. It appears that the plant-produced viral capsid blocks an early stage of the viral infection process. This capsid may prevent the virus from uncoating when it first enters the plant cell. If it cannot uncoat, then it cannot express the viral genome and cause disease. Since infection with the RNA alone can cause disease, the expression of the capsid does not appear to inhibit the expression of the viral genome.

Testing Yourself

1. c; 2. a. capsid; b. protein unit; c. DNA; d. spike; see Fig. 17.1, p. 266, in text.; 3. d; 4. b; 5. a; 6. d; 7. d; 8. c; 9. a; 10. c; 11. a; 12. b; 13. a; 14. d; 15. b; 16. d; 17. a. contractile vacuole (partially full); b. pellicle; c. cilia; d. food vacuoles; e. oral groove; f. macronucleus; g. micronucleus; h. gullet; i. anal pore; j. contractile vacuole (full); see Fig. 17.21, p. 278, in text.

Understanding the Terms

a. lysogenic cycle; b. retrovirus; c. prion; d. peptidoglycan; e. binary fission

Chapter 18
Thinking Scientifically

1. An open field is not likely to contain the mycorrhizal fungi that colonize pine tree roots. However, in areas where pine trees are already growing, those fungi are likely to be prevalent in the soil. Therefore, if you transplant a native seedling along with some of its soil, you introduce valuable mycorrhizal fungi to the open field. The seedling now has the opportunity to establish a mycorrhizal association that will help it to obtain nutrients and water. 2. The males are fooled into believing that the orchid flowers are female wasps, so they fly to the flowers and try to copulate with them. This is a good strategy because it attracts only one type of pollinator. Each male wasp will fly from flower to flower within a single plant species, carrying only pollen from that species.

Testing Yourself

1. b; 2. a; 3. a. sporophyte; b. meiosis; c. spore; d. gametophyte; e. gametes; f. zygote; g. 2n; h. n; see Fig. 18.3, p. 286, in text.; 4. b, c, d; 5. a, b, c, d; 6. c, d; 7. a; 8. c; 9. a, b; 10. d; 11. a. anther; b. filament; c. stigma; d. style; e. ovary; f. ovule; g. sepals (calyx); h. petals (corolla); see Fig. 18.15, p. 292, in text.; 12. b; 13. b; 14. a; 15. b; 16. e; 17. e; 18. c; 19. c; 20. e; 21. b; 22. b; 23. e

Understanding the Terms

a. seed; b. anther; c. xylem; d. fronds; e. pollen grains; f. mycelium; g. saprotrophs; h. sporophyte; i. fruit

Chapter 19
Thinking Scientifically

1. The recent findings support the idea that a constant lineage exists in Africa from primitive to modern humans. However, it does not prove that this lineage exists only in Africa. The discovery of similar fossils in Asia and Europe would provide support for the multiregional continuity hypothesis. 2. Animals that are sessile tend to be radially

symmetrical because their food comes to them from all directions. There is no need to have anterior and posterior body regions. Animals that move through their environment are bilaterally symmetrical, with the anterior portion containing sensory organs. This allows the animal to sense and respond to the environment as it travels through it.

Testing Yourself

1. b; 2. a. ectoderm; b. mesoderm; c. endoderm; d. ectoderm; e. mesoderm; f. endoderm; g. pseudocoelom; h. ectoderm; i. mesentery; j. endoderm; k. coelom; l. mesoderm; see Fig. 19.5, p. 309, in text.; 3. d; 4. g; 5. a, b, c, d, e, g; 6. a, b, c; 7. a; 8. e; 9. c; 10. f; 11. b; 12. e; 13. c; 14. b; 15. a; 16. a; 17. c; 18. a; 19. d; 20. d; 21. a; 22. b; 23. a; 24. e; 25. a; 26. e; 27. a; 28. b; 29. d; 30. d; 31. b; 32. d

Understanding the Terms

a. cephalization; b. coelom; c. hermaphrodite; d. radial symmetry; e. endothermic

Chapter 20
Thinking Scientifically

1. Only the nitrogen-fixing bacteria produce polysaccharide derivatives that the plant recognizes and latches onto. Then, root hairs adjacent to the bacteria fold around them and the plant engulfs the bacteria. 2. The plants get most of their water from fog that rolls off the nearby ocean at night. Therefore at night, the plants open all their stomata and take in both moisture and carbon dioxide for photosynthesis. The stomata are closed during the day. The large number of stomata, in this unusual case, actually helps the plant to survive in a very dry environment.

Testing Yourself

1. c; 2. e; 3. a; 4. b; 5. a. terminal bud; b. blade (leaf); c. axillary bud; d. node; e. shoot system; f. vascular tissues; g. root system; see Fig. 20.1, p. 338, in text.; 6. d; 7. a; 8. a, b, c, d; 9. b; 10. a, d; 11. b; 12. c; 13. a; 14. a. cuticle; b. upper epidermis; c. palisade mesophyll; d. leaf vein; e. spongy mesophyll; f. lower epidermis; g. guard cell; see Fig. 20.9, p. 343, in text.; 15. a. epidermis; b. cortex; c. vascular bundle; d. xylem; e. phloem; f. pith; see Fig. 20.11*a*, p. 344, in text.; 16. a. endodermis; b. pericycle; c. xylem; d. phloem; e. cortex; f. epidermis; g. root hair; h. apical meristem protected by root cap; i. root cap; j. zone of cell division; k. zone of elongation; l. zone of maturation; see Fig. 20.13, p. 346, in text.

Understanding the Terms

a. perennial; b. phloem; c. stomata; d. wood; e. meristem

Chapter 21
Thinking Scientifically

1. When the plants are shipped long distances, they produce ethylene as a result of the stress. The ethylene concentration builds up in the truck and leaf abscission is stimulated. The plastic sleeves help to retain high concentrations of

ethylene near the plant, increasing the abscission response. 2. You could determine whether flowers or stems are responsible by cutting off the flowers and looking to see whether the stems alone exhibit heliotropism. If they do, then the stems rather than the flowers sense the sun. You could determine which portion of the stem is responsible for heliotropism by shading different portions of the stem and determining which shading treatment blocks the heliotropic response. You could determine whether auxin is responsible by looking at a stem under a microscope to see whether the differential cell elongation has occurred.

Testing Yourself

1. d; 2. a; 3. e; 4. b; 5. a; 6. e; 7. b; 8. c; 9. b; 10. c; 11. d; 12. d; 13. a. sporophyte; b. meiosis; c. microspore; d. megaspore; e. male gametophyte; f. female gametophyte; g. egg; h. sperm; i. zygote; see Fig. 21.10, p. 362, in text.; 14. a; 15. c; 16. d; 17. b; 18. c; 19. d; 20. d; 21. b; 22. d; 23. a; 24. c; 25. e; 26. c; 27. e; 28. b; 29. a. stigma; b. style; c. carpel; d. ovary; e. ovule; f. receptacle; g. peduncle; h. sepal; i. petal; j. stamen; k. filament; l. anther; see Fig. 21.11, p. 362, in text.; 30. b; 31. b; 32. e; 33. d; 34. a. seed coat; b. embryo; c. cotyledon; d. seed coat; e. first true leaves; f. cotyledons (two); g. seed coat; see Fig. 21.19, p. 368, in text.; 35. d; 36. a; 37. b

Understanding the Terms

a. auxin; b. abscisic acid; c. photoperiod; d. phytochrome; e. microspore

Chapter 22
Thinking Scientifically

1. Both are diseases of muscle tissue. Researchers have found that all cardiomyopathy gene mutations also lead to skeletal myopathy. They are involved in fundamental pathways that lead to muscle development. 2. One possibility is that socioeconomic factors such as compliance with post-operative procedures and availability of health insurance may differentially influence survival. To test this hypothesis, you could look at patients' records to determine whether there is a relationship between, for example, availability of health insurance and survival rate. Another explanation is that there is a genetically-based physiological difference between African Americans and white Americans. To test this hypothesis, you would need to perform inheritance studies to determine whether there is a genetic basis for liver transplant success.

Testing Yourself

1. a. cell; b. tissue; c. organ; d. organ system; e. organism; see Fig. 22.1, p. 378, in text.; 2. a; 3. b; 4. a, b; 5. a; 6. a, b, c, d; 7. a, b; 8. d; 9. c; 10. a. plasma; b. blood cells; c. white blood cells; d. platelets; e. red blood cells; see Fig. 22.6, p. 383, in text.; 11. d; 12. b; 13. a. skeletal, b. cardiac, c. smooth; 14. a; 15. a. dendrite; b. cell body; c. nucleus; d. nucleus of Schwann cells; e. axon; see Fig. 22.8, p. 385, in text.; 16. e; 17. d; 18. c; 19. c; 20. d; 21. d; 22. b; 23. a; 24. d; 25. a; 26. d; 27. b; 28. e; 29. e; 30. c; 31. a; 32. e; 33. c; 34. e; 35. b; 36. e

Understanding the Terms

a. ligaments; b. bone; c. platelets; d. skeletal muscle; e. striated; f. neuroglia; g. integumentary system; h. endocrine system; i. homeostasis; j. negative feedback

Chapter 23

Thinking Scientifically

1. Endothermic animals use ten times more energy than ectothermic animals of similar size, so their cardiovascular systems need to deliver large amounts of oxygen and fuel and remove large amounts of carbon dioxide and waste products. The separation of the systemic and pulmonary systems, along with a large, powerful heart provides the capacity to carry the large volume of blood necessary to maintain an endothermic lifestyle. 2. At high altitudes, there is less oxygen in the air, so the body must find a way to deliver more oxygen to tissues. In response to low oxygen levels, the kidneys secrete erythropoietin. This stimulates red blood cell production in the red bone marrow, resulting in an increase in the number of oxygen-carrying red blood cells. When the athlete returns to a lower altitude, he or she has a higher than normal red blood cell count, providing a high oxygen-carrying capacity at normal atmospheric oxygen levels.

Testing Yourself

1. c; 2. a. pulmonary vein; b. aorta; c. renal artery; d. lymphatic vessel; e. pulmonary artery; f. superior vena cava; g. inferior vena cava; h. hepatic vein; i. hepatic portal vein; j. renal vein; see Fig. 23.9, p. 402, in text.; 3. a; 4. a. superior vena cava; b. aortic semilunar valve; c. pulmonary semilunar valve; d. right atrium; e. tricuspid valve; f. right ventricle; g. inferior vena cava; h. aorta; i. pulmonary artery; j. pulmonary arteries; k. pulmonary veins; l. left atrium, m. bicuspid mitral valve; n. septum; o. left ventricle; see Fig. 23.4, p. 399, in text.; 5. d; 6. b; 7. c; 8. e; 9. d; 10. b; 11. a; 12. a, b; 13. a; 14. b; 15. c, d; 16. b; 17. d; 18. a; 19. b; 20. b; 21. c; 22. b; 23. c; 24. e; 25. a; 26. c; 27. a; 28. d; 29. c; 30. e; 31. c; 32. d; 33. a. blood pressure; b. osmotic pressure; c. blood pressure; d. osmotic pressure; see Fig. 23.17, p. 408, in text.

Understanding the Terms

a. systemic circuit; b. atria; c. heart murmur; d. SA (sinoatrial) node; e. capillary; f. aorta, venae cavae; g. lymph; h. plasma; i. macrophages; j. platelets

Chapter 24

Thinking Scientifically

1. A camel maintains its plasma volume at the expense of tissue fluid. Therefore, circulation is not impeded by a lack of water. The red blood cells of camels are oval-shaped, allowing them to circulate when the blood becomes more viscous. Camels can drink up to 20 gallons of water at a time and slowly absorb it from their stomach and intestines. In addition, red blood cells can "store" water by swelling up to more than double their normal size without bursting. Camels' kidneys can concentrate urine to reduce water loss. Finally, water can be extracted from their fecal pellets. 2. According to Dr. Atkins, overweight people eat too many

carbohydrates. If we severely restrict carbohydrates in our diet, our bodies will begin to burn stored body fat more efficiently, leading to weight loss. However, a recent study published in the *New England Journal of Medicine* indicated that, while initial weight loss was rapid in Atkins dieters, there was no difference in weight loss between Atkins dieters and conventional dieters after one year. One explanation for this observation is that the initial weight loss is due to water loss. The body responds to a shortage of glucose by storing it as glycogen. The synthesis of glycogen results in the production of water. Negative side effects of the diet include increased risk of colorectal cancer (due to the high intake of red meat), increased incidence of heart disease (due to high levels of saturated fat), reduced kidney function (due to the processing of large amounts of protein), and osteoporosis (due to loss of urinary calcium).

Testing Yourself

1. a. salivary glands; b. mouth; c. liver; d. gallbladder; e. duodenum; f. cecum; g. appendix; h. anus; i. pharynx; j. esophagus; k. diaphragm; l. stomach; m. pancreas; n. small intestine; o. large intestine; p. anal canal; see Fig. 24.2*b*, p. 415, in text.; 2. b; 3. a; 4. d; 5. e; 6. b; 7. a. nostril; b. nasal cavity; c. pharynx; d. epiglottis; e. glottis; f. larynx; g. trachea; h. bronchus; i. bronchiole; j. capillary network; k. lung; l. diaphragm; see Fig. 24.14, p. 423, in text.; 8. d; 9. d; 10. a. kidney; b. ureter; c. urinary bladder; d. urethra; see Fig. 24.23*a*, p. 427, in text.; 11. c; 12. d; 13. b; 14. a; 15. a; 16. d; 17. b; 18. a; 19. c; 20. b; 21. d; 22. c; 23. c; 24. d

Understanding the Terms

a. pharynx; b. pepsin; c. chyme; d. gallbladder; e. larynx; f. trachea; g. alveoli; h. heme; i. nephrons

Chapter 25

Thinking Scientifically

Soda Then: Total calories 84; Soda Now: Fat 0; Carbohydrate 93; Protein 0; Total calories 372; Calorie increase 288; Fries Then: Total calories 223; Fries Now: Fat 25; Carbohydrate 64; Protein 7; Total calories 509; Calorie increase 286; Bagels Then: Total calories 188; Bagels Now: Fat 1; Carbohydrate 100; Protein 15; Total calories 469; Calorie increase 281

Testing Yourself

1. b; 2. a; 3. f; 4. a; 5. b; 6. e; 7. c; 8. b; 9. d; 10. e; 11. c; 12. a; 13. a. grains; b. 6 oz.; c. vegetables; d. 2.5 cups; e. fruit; f. 2 cups; g. oils; h. 1–2 Tbs.; i. milk; j. 3 cups (2 for kids aged 2–8); k. meat and beans; l. 5.5 oz.; see Fig. 25.14, p. 449, in text.; 14. d; 15. c; 16. a; 17. d; 18. a

Understanding the Terms

a. nutrient; b. fiber; c. triglycerides; d. cholesterol; e. vitamins; f. antioxidants; g. body mass index; h. anorexia nervosa; i. bulimia nervosa; j. muscle dysmorphia

Chapter 26

Thinking Scientifically

1. Each antibody gene is composed of several discrete segments. The antibody sites that

recognize antigens are pieced together from various combinations of these segments. The ability to combine multiple segments into many different combinations allows the genome to produce antibodies that interact with a multitude of antigens. Mutations in antibody genes produce additional variability. 2. Drugs such as cyclosporine inhibit IL-2, therefore suppressing the production of natural killer cells and cytotoxic T cells. However, it does not affect other components of the immune or healing systems, including the production of other types of white blood cells.

Testing Yourself

1. c; 2. a. red bone marrow; b. thymus gland; c. lymph nodes; d. spleen; e. lymphatic vessels; see Fig. 26.1, p. 456, in text.; 3. e; 4. a; 5. c; 6. a; 7. b; 8. d; 9. b; 10. c; 11. d; 12. c; 13. d; 14. e; 15. a; 16. b; 17. e; 18. c; 19. d; 20. c; 21. b; 22. c; 23. a. hair shaft; b. oil (sebaceous) gland; c. sweat glands; d. dermis; e. epidermis; see Fig. 26.2, p. 458, in text.; 24. e; 25. a; 26. a; 27. a; 28. e; 29. b; 30. c; 31. b; 32. d; 33. a; 34. b; 35. a; 36. c; 37. a

Understanding the Terms

a. appendix; b. red bone marrow; c. macrophage; d. histamine; e. B cells; f. agglutination; g. cytotoxic T cells; h. vaccine; i. allergens; j. Autoimmune disease

Chapter 27

Thinking Scientifically

1. Parkinson's patients do not exhibit a constriction of blood vessels, so their blood pressure slowly decreases. The loss of function of the sympathetic division; prevents the body from regulating blood pressure, so rapid changes in blood pressure can occur, leading to orthostatic hypotension. 2. Diabetes type 1 is believed to result when an environmental agent such as a virus causes T cells to destroy the pancreatic islets. Vaccines that use a live virus may introduce that agent. For example, the mumps and rubella viruses may remain in the body for years after vaccination and can infect pancreatic islet cells. Infection lowers the levels of insulin secreted by these cells. In addition, exposure to the viruses may lead to an autoimmune disorder that causes the destruction of pancreatic islet cells.

Testing Yourself

1. d; 2. c; 3. b; 4. c; 5. a; 6. b; 7. a. cerebrum; b. skull; c. corpus callosum; d. diencephalon; e. pineal gland; f. pituitary gland; g. cerebellum; h. spinal cord; i. medulla oblongata; j. pons; k. midbrain; l. brain stem; see Fig. 27.9*b*, in text.; 8. b; 9. a; 10. a; 11. d; 12. a; 13. d; 14. d; 15. b; 16. a. pineal gland; b. pituitary gland (hypophysis); c. hypothalamus; d. parathyroid gland; e. thyroid gland; f. thymus gland; g. adrenal gland; h. pancreas; see Fig. 27.15, p. 483, in text.; 17. d; 18. a; 19. d; 20. d; 21. a; 22. a; 23. c; 24. d; 25. e; 26. c; 27. d; 28. d; 29. b

Understanding the Terms

a. axon; b. synaptic cleft; c. acetylcholinesterase; d. spinal cord; e. prefrontal area; f. nerve; g. gonadotropic hormones; h. adrenal medulla; i. Addison disease; j. pancreatic islets

Chapter 28

Thinking Scientifically

1. The blood vessels interfere with the transmission of light to the back of the eye, consequently impairing the signal sent to the brain. Both cancer and eye diseases involve the growth of new blood vessels. Anti-angiogenesis drugs that have been developed for cancer treatment are currently being studied for the treatment of eye diseases. 2. Our olfactory system allows us to recognize and discriminate between many different odors. Therefore, we can detect the difference between a dangerous odor, such as smoke, and a harmless one, such as flower fragrance. On the other hand, every natural toxin tastes bitter, so it is more important to be able to sense bitterness than to discriminate between bitter tastes. Therefore, humans can identify bitter compounds by taste, but cannot distinguish between different compounds.

Testing Yourself

1. c; 2. a. cranial nerve; b. cochlea; c. tympanic membrane; d. auditory tube; e. auditory canal; f. ossicles; g. semicircular canals; h. outer ear; i. middle ear; j. inner ear; see Fig. 28.3a, p. 496; 3. c; 4. b; 5. b; 6. c; 7. a. sclera; b. retina; c. optic nerve; d. vein; e. artery; f. fovea; g. ciliary muscle; h. iris; i. pupil; j. cornea; k. lens; see Fig. 28.8, p. 499, in text.; 8. a; 9. a; 10. c; 11. a, b; 12. c; 13. b; 14. b, c; 15. a

Understanding the Terms

a. spiral organ; b. statocysts; c. iris; d. fovea; e. vertebral column; f. osteoblast; g. myofibrils; h. myosin; i. synovial joint; j. meniscus

Chapter 29

Thinking Scientifically

1. The two most obvious potential differences are that the sperm can tolerate higher temperatures or that the body temperature of elephants is lower than that of humans. To test the high temperature hypothesis, compare the temperatures at which sperm are viable in elephants with that in humans. To test the low body temperature hypothesis, compare the body temperatures of elephants with that of humans. (The latter is true.) 2. The most logical (although controversial) explanation is that menopause in women developed early in human evolution. It allowed women to channel their efforts into caring for their existing children, increasing the survival rate of these children. This would give the woman the best chance to pass her genes through generations, so it increased her fitness. Since men were not the main caregivers for their children, their fitness would not be increased if they evolved the ability to undergo menopause.

Testing Yourself

1. c; 2. b; 3. a; 4. b; 5. d; 6. a; 7. a. ureter; b. bladder; c. vas deferens; d. seminal vesicle; e. ejaculatory duct; f. prostate gland; g. bulbourethral gland; h. urethra; i. epididymis; j. penis; k. foreskin; l. testis; m. scrotum; see Fig. 29.4b, p. 514, in text.; 8. a. oviduct; b. ovary; c. rectum; d. cervix; e. bladder; f. vagina; g. urethra; see Fig. 29.6b, p. 516, in text.; 9. a; 10. d; 11. c; 12. a; 13. b; 14. d; 15. c

Understanding the Terms

a. copulation; b. bulbourethral glands; c. testosterone; d. uterus; e. endometrium; f. ovulation; g. fertilization; h. morula; i. induction; j. umbilical cord

Chapter 30

Thinking Scientifically

1. Current growth rate = 1.3%; doubling rate = 53 years; growth rate at 10 births per 1,000 = 0.4%; new doubling rate = 172 years. 2. At the 19th week, the field will be half full. It will double during the 19th week to fill the field. At the 18th week, it will be one quarter full and will double to fill half of the field by the 19th week. Therefore, a significant population of dandelions will not be observed until the last week.

Testing Yourself

1. a; 2. d; 3. b; 4. d; 5. b; 6. b; 7. d; 8. c; 9. d; 10. b; 11. a. increasing, b. decreasing, c. stable; see Fig. 30.7, p. 539, in text. 12. a; 13. c; 14. a; 15. a; 16. d; 17. b; 18. c; 19. c; 20. b; 21. c; 22. e; 23. c

Understanding the Terms

a. community; b. resources; c. range; d. population density; e. biotic potential; f. carrying capacity; g. competition; h. equilibrium population; i. extinction; j. more-developed countries

Chapter 31

Thinking Scientifically

1. First, insects probably found the sap on the wind-pollinated flowers and used it as a food source. Because it was nutritious and easy to find, the insects gradually became dependent on it. The insects accidentally picked up and carried pollen as they moved from flower to flower in search of food. Plants that were pollinated by these insects were more successful (produced more seeds) than those that were wind-pollinated. Therefore, plants that evolved mechanisms to encourage visits by insects were more fit. Nectaries evolved to bring insects to flowers, and then showy petals and fragrance evolved to advertise the presence of flowers to insects. Insects continued to evolve to utilize flowers of certain species, and those plant species evolved flower features to attract and reward those insects. 2. Their burrows act as homes to many other animals, while their burrowing loosens soil, increasing the ability of plants to grow. Their foraging and feeding habits encourage a diversity of plants to grow. The plants that grow as a result of prairie dog activity are food for other animals. Finally, prairie dogs themselves are major food sources for several predatory species.

Testing Yourself

1. a; 2. a; 3. e; 4. a; 5. c; 6. b; 7. d; 8. c; 9. c; 10. e; 11. a. energy; b. nutrients; c. heat; d. heat; e. producers; f. consumers; g. inorganic nutrient pool; h. decomposers; i. heat; see Fig. 31.16, p. 561, in text.; 12. c; 13. a; 14. a; 15. a; 16. a; 17. d; 18. c; 19. a. top carnivores; b. carnivores; c. herbivores; d. producers; see Fig. 31.20, p. 563, in text.; 20. c; 21. c; 22. c; 23. c; 24. b; 25. a; 26. d; 27. a; 28. a; 29. b; 30. e; 31. a; 32. g; 33. d; 34. f; 35. b; 36. a; 37. c; 38. d

Understanding the Terms

a. community; b. parasitism; c. ecological niche; d. character displacement; e. keystone species; f. detritus; g. trophic level; h. transfer rate; i. biosphere

Chapter 32

Thinking Scientifically

1. **Canal option.** Positive environmental impacts are: refilling the Aral Sea; combating desertification that has been occurring in recent years; providing ample water for irrigation, and therefore maintaining the livelihood of farmers. Negative environmental impacts are: Water flow downstream of the diversion canal on the second two rivers will be lessened. **Dam removal option.** Positive environmental impacts are: refilling the Aral Sea; combating desertification that has been occurring in recent years. Negative environmental impacts are: Farmers who rely on irrigation will no longer be able to grow their crops. 2. These would most likely be high priorities: regions with the highest levels of biodiversity; regions containing species used for traditional medicine; regions containing species used for medicine by indigenous people; regions containing species that are related to species with known medicinal value; regions containing species that produce chemical compounds related to those of known medicinal value.

Testing Yourself

1. e; 2. b; 3. d; 4. c; 5. b; 6. a; 7. b; 8. b; 9. c, d, e; 10. a; 11. a, d, e; 12. e; 13. a; 14. Use of only renewable energy sources, reuse of heat and waste materials, and maximal recycling of products. See Fig. 32.20b, p. 592, in text.; 15. c; 16. e; 17. d; 18. b; 19. d; 20. b; 21. e; 22. e

Understanding the Terms

a. pollution; b. deforestation; c. aquifer; d. subsidence; e. saltwater intrusion; f. salinization; g. greenhouse gases; h. photovoltaic cell; i. mineral; j. chlorofluorocarbon

Appendix B

Metric System

Unit and Abbreviation	Metric Equivalent	Approximate English-to-Metric Equivalents	Units of Temperature
Length			
nanometer (nm)	$= 10^{-9}\,m\ (10^{-3}\,\mu m)$		
micrometer (μm)	$= 10^{-6}\,m\ (10^{-3}\,mm)$		
millimeter (mm)	$= 0.001\ (10^{-3})\,m$		
centimeter (cm)	$= 0.01\ (10^{-2})\,m$	1 inch = 2.54 cm 1 foot = 30.5 cm	
meter (m)	$= 100\ (10^{2})\,cm$ $= 1,000\,mm$	1 foot = 0.30 m 1 yard = 0.91 m	
kilometer (km)	$= 1,000\ (10^{3})\,m$	1 mi = 1.6 km	
Weight (mass)			
nanogram (ng)	$= 10^{-9}\,g$		
microgram (μg)	$= 10^{-6}\,g$		
milligram (mg)	$= 10^{-3}\,g$		
gram (g)	$= 1,000\,mg$	1 ounce = 28.3 g 1 pound = 454 g	
kilogram (kg)	$= 1,000\ (10^{3})\,g$	= 0.45 kg	
metric ton (t)	$= 1,000\,kg$	1 ton = 0.91 t	
Volume			
microliter (μl)	$= 10^{-6}\,l\ (10^{-3}\,ml)$		
milliliter (ml)	$= 10^{-0}\,liter$ $= 1\,cm^{3}\ (cc)$ $= 1,000\,mm^{3}$	1 tsp = 5 ml 1 fl oz = 30 ml	
liter (l)	$= 1,000\,ml$	1 pint = 0.47 liter 1 quart = 0.95 liter 1 gallon = 3.79 liter	
kiloliter (kl)	$= 1,000\,liter$		

°C	°F	
100	212	Water boils at standard temperature and pressure.
71	160	Flash pasteurization of milk
57	134	Highest recorded temperature in the United States, Death Valley, July 10, 1913
41	105.0	Average body temperature of a marathon runner in hot weather
37	98.6	Human body temperature
13.7	56.66	Human survival is still possible at this temperature.
0	32.0	Water freezes at standard temperature and pressure.

To convert temperature scales:

$$°C = \frac{(°F - 32)}{1.8}$$

$$°F = 1.8\,(°C) + 32$$

Glossary

A

abdominal cavity Located in the ventral cavity below the thoracic cavity and is separated from the thoracic cavity by the diaphragm; contains most other internal organs except the heart and the lungs. 416

abscisic acid (ABA) Plant hormone that causes stomata to close and initiates and maintains dormancy. 358

abscission Dropping of leaves, fruits, or flowers from a plant. 358

abstinence Method of birth control; the practice of not engaging in sexual intercourse. 519

acetylcholine (ACh) Neurotransmitter active in both the peripheral and central nervous systems. 476

acetylcholinesterase (AChE) Enzyme that breaks down acetylcholine within a synapse. 476

acetyl CoA Molecule made up of a 2-carbon acetyl group attached to coenzyme A. During cellular respiration, the acetyl group enters the citric acid cycle for further breakdown. 102

acid Molecules tending to raise the hydrogen ion concentration in a solution and to lower its pH numerically. 25

acoelomate Animals without a coelom, as in flatworms. 308

acquired immunodeficiency syndrome (see AIDS)

actin Muscle protein making up the thin filaments in a sarcomere; its movement shortens the sarcomere, yielding muscle contraction. Actin filaments play a role in the movement of the cell and its organelles. 505

actin filament Cytoskeletal filament of eukaryotic cells composed of the protein actin; also refers to the thin filaments of muscle cells. 62

action potential Electrochemical changes that take place across the axomembrane; the nerve impulse. 475

active immunity Resistance to disease due to the immune system's response to a microorganism or a vaccine. 164

active site Region on the surface of an enzyme where the substrate binds and where the reaction occurs. 76

active transport Use of a plasma membrane carrier protein to move a molecule or ion from a region of lower concentration to one of higher concentration; it opposes equilibrium and requires energy. 78

adaptation Organism's modification in structure, function, or behavior suitable to the environment. 4, 221

adaptive radiation Evolution of several species from a common ancestor into new ecological or geographical zones. 251

Addison disease Condition resulting from a deficiency of adrenal cortex hormones; characterized by low blood glucose, weight loss, and weakness. 487

adenine (A) One of four nitrogen-containing bases in nucleotides composing the structure of DNA and RNA. 161

adenosine triphosphate (ATP) Nucleotide with three phosphate groups. The breakdown of ATP into ADP + $\textcircled{P}$ makes energy available for energy-requiring processes in cells. 60

adhesion junction Junction between cells in which the adjacent plasma membranes do not touch but are held together by intercellular filaments attached to buttonlike thickenings. 65

adrenal cortex Located in the adrenal gland; produces the glucocorticoid and mineralocorticoid hormones. 487

adrenal gland Gland that lies atop a kidney; the *adrenal medulla* produces the hormones epinephrine and norepinephrine, and the *adrenal cortex* produces the glucocorticoid and mineralocorticoid hormones. 487

adrenal medulla Inner portion of the adrenal gland; secretes the hormones epinephrine and norepinephrine. 487

adrenocorticotropic hormone (ACTH) Hormone secreted by the anterior lobe of the pituitary gland that stimulates activity in the adrenal cortex. 484

age structure In demographics, a display of the age groups of a population; a growing population has a pyramid shaped diagram. 539

agglutination Clumping of red blood cells due to a reaction between antigens on red blood cell plasma membranes and antibodies in the plasma. 461

AIDS (acquired immunodeficiency syndrome) Disease caused by a retrovirus and transmitted via body fluids; characterized by failure of the immune system. 466, 522

algae (sing., alga) Type of protist that carries on photosynthesis; unicellular forms are a part of phytoplankton, and multicellular forms are called seaweed. 276

allele Alternative form of a gene — alleles occur at the same locus on homologous chromosomes. 128, 145

allergen Foreign substance capable of stimulating an allergic response. 465

allergy Immune response to substances that usually are not recognized as foreign. 465

allopatric speciation Origin of new species between populations that are separated geographically. 250

alternation of generations Life cycle, typical of plants, in which a diploid sporophyte alternates with a haploid gametophyte. 286, 362

alveolus (pl., alveoli) In humans, terminal, microscopic, grapelike air sac found in lungs. 423

Alzheimer disease (AD) Brain disorder characterized by a general loss of mental abilities. 480

amino acid Organic molecule composed of an amino group and an acid group; covalently bonds to produce peptide molecules. 39

amniocentesis Procedure for removing amniotic fluid surrounding the developing fetus for testing of the fluid or cells within the fluid. 197

amnion Extraembryonic membrane of birds, reptiles, and mammals that forms an enclosing, fluid-filled sac. 526

amoeboid Cell that moves and engulfs debris with pseudopods. 278

amphibian Member of a class of vertebrates that includes frogs, toads, and salamanders; they are still tied to a watery environment for reproduction. 323

analogous structure Structure that has a similar function in separate lineages but differs in anatomy and ancestry. 226, 257

analogy Similarity of characters due to convergent evolution. 257

anaphase Mitotic phase during which daughter chromosomes move toward the poles of the spindle. 115

angiogenesis Formation of new blood vessels; one mechanism by which cancer spreads. 121, 190

angiosperm Flowering plant; the seeds are borne within a fruit. 286, 292

animal Multicellular, heterotrophic organism belonging to the animal kingdom. 6

annelid Member of a phylum of invertebrates that contains segmented worms, such as the earthworm and the clam worm. 315

annual ring Layer of wood (secondary xylem) usually produced during one growing season. 345

anorexia nervosa Eating disorder characterized by a morbid fear of gaining weight. 448

anterior pituitary Portion of the pituitary gland that is controlled by the hypothalamus and produces six types of hormones, some of which control other endocrine glands. 484

anther In flowering plants, pollen-bearing portion of stamen. 292, 363

anthropoid Group of primates that includes monkeys, apes, and humans. 328

antibody Protein produced in response to the presence of an antigen; each antibody combines with a specific antigen. 406, 456

anticodon Three-base sequence in a transfer RNA molecule base that pairs with a complementary codon in mRNA. 168

antidiuretic hormone (ADH) Hormone secreted by the posterior pituitary that increases the permeability of the collecting ducts in a kidney. 485

antigen Foreign substance, usually a protein or a polysaccharide, that stimulates the immune system to react, such as to produce antibodies. 406, 456

antigen-presenting cell (APC) Cell that displays the antigen to certain cells of the immune system so they can defend the body against that particular antigen. 462

antioxidant Substances, such as vitamins C, E, and A, which defend the body against free radicals. 442

anus Outlet of the digestive tube. 420

aorta In humans, the major systemic artery that takes blood from the heart to the tissues. 399, 402

aortic body Structure located in the walls of the aorta; contains chemoreceptors sensitive to hydrogen ion and carbon dioxide concentrations in the blood. 424

apical dominance Influence of a terminal bud in suppressing the growth of lateral buds. 357

apical meristem In vascular plants, masses of cells in the root and shoot that reproduce and elongate as primary growth occurs. 341

apoptosis Programmed cell death involving a cascade of specific cellular events leading to death and destruction of the cell. 58, 120

appendicular skeleton Part of the vertebrate skeleton forming the appendages, shoulder girdle, and hip girdle. 503

appendix In humans, small, tubular appendage that extends outward from the cecum of the large intestine. 419, 456

aquatic ecosystem Ecosystems that are freshwater ecosystems (rivers, streams, lakes, ponds) and saltwater (marine) ecosystems (oceans, coral reefs, saltwater marshes). 568

aquifer Rock layers that contain water and will release it in appreciable quantities to wells or springs. 579

arachnid Group of arthropods that contains spiders, scorpions, and ticks. 317

Archaea One of the three domains of life; contains prokaryotic cells that often live in extreme habitats and have unique genetic, biochemical, and physiological characteristics; its members are sometimes referred to as *archaea*. 6, 261, 275

arteriole Vessel that takes blood from an artery to capillaries. 401

artery Blood vessel that transports blood away from the heart. 399, 401

arthropod Member of a phylum of invertebrates that contains, among other groups, crustaceans and insects that have an exoskeleton and jointed appendages. 316

assortative mating Individuals tend to mate with those that have the same phenotype with respect to certain characteristics. 236

aster Short, radiating fibers produced by the centrosomes in animal cells. 115

atom Smallest particle of an element that displays the properties of the element. 17

atomic number Number of protons within the nucleus of an atom. 17

atomic symbol One or two letters that represent the name of an element—e.g., H stands for a hydrogen atom, and Na stands for a sodium atom. 17

ATP synthase Enzyme that is part of an *ATP synthase complex* and functions in the production of ATP in chloroplasts and mitochondria. 89

atrioventricular valve Heart valve located between an atrium and a ventricle. 399

atrium (pl., atria) Chamber; particularly an upper chamber of the heart lying above a ventricle. 399

australopithecine One of several species of *Australopithecus,* a genus that contains the first generally recognized hominids. 330

autoimmune disease Disease that results when the immune system mistakenly attacks the body's own tissues. 465

autonomic system Portion of the peripheral nervous system that regulates internal organs. 482

autosome Any chromosome other than the sex-determining pair. 129

autotroph Organism that can capture energy and synthesize organic molecules from inorganic nutrients. 560

auxin Plant hormone regulating growth, particularly cell elongation; also called indoleacetic acid (IAA). 356

AV (atrioventricular) node Small region of neuromuscular tissue that transmits impulses received from the SA node to the ventricular walls. 400

axial skeleton Part of the vertebrate skeleton forming the vertical support or axis, including the skull, the rib cage, and the vertebral column. 503

axon Elongated portion of a neuron that conducts nerve impulses, typically from the cell body to the synapse. 474

B

bacteria One of the three domains of life; contains prokaryotic cells that differ from archaea because they have their own unique genetic, biochemical, and physiological characteristics. 6, 261, 270

bacteriophage Virus that infects bacteria. 267

bark External part of a tree, containing cork, cork cambium, and phloem. 345

Barr body Dark-staining body (discovered by M. Barr) in the nuclei of female mammals that contains a condensed, inactive X chromosome. 137

base Molecules tending to lower the hydrogen ion concentration in a solution and raise the pH numerically. 25

B-cell receptor (BCR) Complex on the surface of a B cell that binds an antigen and stimulates the B cell. 460

bicarbonate ion Ion that participates in buffering the blood, and the form in which carbon dioxide is transported in the bloodstream. 426

bilateral symmetry Body plan having two corresponding or complementary halves. 308

bile Secretion of the liver that is temporarily stored and concentrated in the gallbladder before being released into the small intestine, where it emulsifies fat. 418

binary fission Splitting of a parent cell into two daughter cells; serves as an asexual form of reproduction in bacteria. 271

binge-eating disorder Condition characterized by overeating episodes that are not followed by purging. 448

binomial name Scientific name of an organism, the first part of which designates the genus and the second part of which designates the specific epithet. 7, 256

biodiversity Total number of species, the variability of their genes, and the communities in which they live. 7, 588

biogeochemical cycle Circulating pathway of elements such as carbon and nitrogen involving exchange pools, storage areas, and biotic communities. 564

biogeography Study of the geographical distribution of organisms. 219

bioinformatics Computer technologies used to study the genome. 209

biological magnification Process by which substances become more concentrated in organisms in the higher trophic levels of a food web. 587

biology Scientific study of life. 8

biomass Weight of one or more organisms. 563

bioremediation Cleanup of the environment using bacteria to break down pollutants such as oil spills and Agent Orange. 273

biosphere Zone of air, land, and water at the surface of the Earth in which organisms are found. 5, 546, 568

biotechnology Term that encompasses genetic engineering and other techniques that make use of natural biological systems to create a product or achieve a particular result desired by humans. 175

biotic potential Maximum population growth rate under ideal conditions. 540

bipedalism Walking erect on two feet. 329

bird Endothermic vertebrate that has feathers and wings, is often adapted for flight, and lays hard-shelled eggs. 324

birth control pill Oral contraception containing estrogen and progesterone. 519

bivalve Type of mollusc with a shell composed of two valves; includes clams, oysters, and scallops. 314

blade Broad, expanded portion of a plant leaf that may be single or compound leaflets. 338

blastocyst Early stage of human embryonic development that consists of a hollow, fluid-filled ball of cells. 525

blind spot Region of the retina, lacking rods or cones, where the optic nerve exits the retina. 500

blood Fluid circulated by the heart through a closed system of vessels. 383, 405

blood pressure Force of blood pushing against the inside wall of blood vessels. 398

B lymphocyte (B cell) Lymphocyte that matures in the bone marrow and, when stimulated by the presence of a specific antigen, gives rise to antibody-producing plasma cells. 457

body mass index (BMI) Calculation used to determine whether or not a person is overweight or obese. 444

bone Connective tissue having protein fibers and a hard matrix of inorganic salts, notably calcium salts. 383

bony fishes Member of a class of vertebrates (class Osteichthyes) containing numerous diverse fishes, with a bony rather than cartilaginous skeleton. 322

bottleneck effect Cause of genetic drift; occurs when a majority of genotypes are prevented from participating in the production of the next generation as a result of a natural disaster or human interference. 237

brain stem In mammals; portion of the brain consisting of the medulla oblongata, pons, and midbrain. 479

bronchiole In terrestrial vertebrates, small tube that conducts air from a bronchus to the alveoli. 423

bronchus (pl., bronchi) In terrestrial vertebrates, branch of the trachea that leads to the lungs. 423

brown algae Marine photosynthetic protists with a notable abundance of xanthophyll pigments; this group includes well-known seaweeds of northern rocky shores. 277

bryophyte Member of one of three phyla of nonvascular plants—the mosses, liverworts, and hornworts; these plants have no vascular tissue and occur in moist locations. 304

buffer Substance or group of substances that tend to resist pH changes of a solution, thus stabilizing its relative acidity and basicity. 26

bulbourethral glands Two small structures located below the prostate gland in males; add secretions to semen. 515

bulimia nervosa Eating disorder characterized by binge eating followed by purging via self-induced vomiting or use of a laxative. 448

bursa Saclike, fluid-filled structure, lined with synovial membrane, that occurs near a synovial joint. 506

C

C₃ plant Plant that fixes carbon dioxide via the Calvin cycle; the first stable product of C₃ photosynthesis is a 3-carbon compound. 92

C₄ plant Plant that fixes carbon dioxide to produce a C₄ molecule that releases carbon dioxide to the Calvin cycle. 92

calcitonin Hormone secreted by the thyroid gland that increases the blood calcium level. 486

calorie Amount of heat energy required to raise the temperature of water 1°C. 70

Calvin cycle reaction Portion of photosynthesis that takes place in the stroma of chloroplasts and can occur in the dark; it uses the products of the light reactions to reduce CO₂ to a carbohydrate. 86

calyx The sepals collectively; the outermost flower whorl. 292

CAM Crassulacean-acid metabolism; plant fixes carbon dioxide at night to produce a C₄ molecule that releases carbon dioxide to the Calvin cycle during the day. 93

Cambrian explosion The dramatic appearance of animal fossils 543 MYA; possibly related to each of the animal phyla during some 10 million years. 307

camera-type eye Type of eye found in vertebrates and certain molluscs; a single lens focuses an image on closely packed photoreceptors. 499

cancer Malignant tumor whose nondifferentiated cells exhibit loss of contact inhibition, uncontrolled growth, and the ability to invade tissue and metastasize. 121

capillary Microscopic blood vessel; gases and other substances are exchanged across the walls of a capillary between blood and tissue fluid. 401

capsid Protective protein container of the genetic material of a virus. 266

capsule Gelatinous layer surrounding the cells of blue-green algae and certain bacteria. 50

carbohydrate Class of organic compounds that includes monosaccharides, disaccharides, and polysaccharides. 33, 438

carbon dioxide (CO₂) fixation Photosynthetic reaction in which carbon dioxide is attached to an organic compound. 90

carcinogenesis Development of cancer. 121

cardiac cycle One complete cycle of systole and diastole for all heart chambers. 400

cardiac muscle Striated, involuntary muscle tissue found only in the heart. 384

cardiovascular system Organ system in which blood vessels distribute blood under the pumping action of the heart. 386, 397

carnivore Consumer in a food chain that eats other animals. 415, 560

carotenoid Yellow or orange pigment that serves as an accessory to chlorophyll in photosynthesis. 87

carotid body Structure located at the branching of the carotid arteries; contain chemoreceptors sensitive to hydrogen ion and carbon dioxide concentrations in blood. 424

carpel Ovule-bearing unit that is a part of a pistil. 292, 363

carrier Heterozygous individual who has no apparent abnormality but can pass on an allele for a recessively inherited genetic disorder. 153

carrying capacity Largest number of organisms of a particular species that can be maintained indefinitely by a given environment. 535

cartilage Connective tissue in which the cells lie within lacunae embedded in a flexible, proteinaceous matrix. 383

cartilaginous fishes Member of a class of vertebrates (class Chondrichthyes) with a cartilaginous rather than bony skeleton; includes sharks, rays, and skates. 322

cell Smallest unit that displays the properties of life; composed of cytoplasm surrounded by a plasma membrane. 2, 48

cell body Portion of a neuron that contains a nucleus and from which dendrites and an axon extend. 474

cell cycle Repeating sequence of events in eukaryotes that involves cell growth and nuclear division; consists of the stages G₁, S, G₂, and M. 114

cell plate Structure across a dividing plant cell that signals the location of new plasma membranes and cell walls. 118

cell-signaling pathway Mechanism used by a signal molecule leading to a signal transduction pathway, where a series of molecular changes converts a signal into a specific response in the cell. 187

cell theory One of the major theories of biology, which states that all organisms are made up of cells; cells are capable of self-reproduction and come only from preexisting cells. 50

cellular respiration Metabolic reactions that use the energy from carbohydrates, fatty acid, or amino acid breakdown to produce ATP molecules. 61

cellulose Polysaccharide that is the major complex carbohydrate in plant cell walls. 34

cell wall Structure that surrounds a plant, protistan, fungal, or bacterial cell and maintains the cell's shape and rigidity. 50, 64

central nervous system (CNS) Portion of the nervous system consisting of the brain and spinal cord. 473

centriole Cell organelle, existing in pairs, that occurs in the centrosome and may help organize a mitotic spindle for chromosome movement during animal cell division. 62

centrosome Central microtubule organizing center of cells. In animal cells, it contains two centrioles. 62, 115

cephalization Having a well-recognized anterior head with a brain and sensory receptors. 308

cephalopod Type of mollusc in which a modified foot develops into the head region; includes squids, cuttlefish, octopuses, and nautiluses. 314

cerebellum In terrestrial vertebrates, portion of the brain that coordinates skeletal muscles to produce smooth, graceful motions. 479

cerebral cortex Outer layer of cerebral hemispheres; receives sensory information and controls motor activities. 478

cervical cap Birth control device made of latex or rubber in the shape of a cup which covers the cervix, considered a minidiaphragm. 520

cervix Narrow end of the uterus leading into the vagina. 516

character displacement Tendency for characteristics to be more divergent when similar species belong to the same community than when they are isolated from one another. 557

checkpoint In the cell cycle, one of several points where the cell cycle can stop or continue on, depending on the internal signal it receives; ensures that each step of the cell cycle is completed before the next one begins. 119

chemoautotroph Organism able to synthesize organic molecules by using carbon dioxide as the carbon source and the oxidation of an inorganic substance (such as hydrogen sulfide) as the energy source. 272

chemoheterotroph Organism that is unable to reproduce its own organic molecules, and therefore requires organic nutrients in its diet. 272

chitin Strong but flexible nitrogenous polysaccharide found in the exoskeleton of arthropods. 34, 316

chlorofluorocarbons (CFCs) Organic compounds containing carbon, chlorine, and fluorine atoms. CFCs such as Freon can deplete the ozone shield by releasing chlorine atoms in the upper atmosphere. 386

chlorophyll Green pigment that absorbs solar energy and is important in algal and plant photosynthesis; occurs as chlorophyll a and chlorophyll b. 85, 87

chloroplast Membrane-bounded organelle in algae and plants with chlorophyll-containing membranous thylakoids; where photosynthesis takes place. 60, 85

cholesterol One of the major lipids found in animal plasma membranes; makes the membrane impermeable to many molecules. 37, 440

chordate Animal in the phylum Chordata that has a dorsal tubular nerve cord, a notochord, pharyngeal pouches, and postanal tail at some point in its life cycle. 320

chorionic villi Treelike extensions of the chorion of the embryo; project into the maternal tissues at the placenta. 528

chorionic villi sampling Prenatal test in which a sample of chorionic villi cells is removed for diagnostic purposes. 197

chromatin Network of fibrils consisting of DNA and associated proteins observed within a nucleus that is not dividing. 56, 113

chromosomal mutation Alteration in the chromosome structure or number typical of the species. 198

chromosome Structure consisting of DNA complexed with proteins that transmits genetic information from the previous generation of cells and organisms to the next generation. 56, 113

chromosome map Sequence that shows the relative distance between gene loci on a chromosome. 154

chyme Thick, semiliquid food material that passes from the stomach to the small intestine. 418

ciliate Complex unicellular protist that moves by means of cilia and digests food in food vacuoles. 278

cilium (pl., cilia) Short, hairlike projection from the plasma membrane, occurring usually in larger numbers. 63

citric acid cycle Cycle of reactions in mitochondria that begins with citric acid. It breaks down an acetyl group and produces CO_2, ATP, NADH, and $FADH_2$; also called the Krebs cycle. 99, 102

cladistics School of systematics that uses derived characters to determine monophyletic groups and construct cladograms. 259

class One of the categories, or taxa, used by taxonomists to group species; class is the taxon above the order level. 6, 256

cleavage Cell division without cytoplasmic addition or enlargement; occurs during the first stage of animal development. 525

climax community In ecology, community that results when succession has come to an end. 554

closed circulatory system Cardiovascular system in all vertebrates and some invertebrates, which is composed of a muscular heart and blood vessels. 397

clumped Spatial distribution of individuals in a population in which individuals are more dense in one area than in another. 538

cnidarian Invertebrate in the phylum Cnidaria existing as either a polyp or medusa with two tissue layers and radial symmetry. 310

codominance Inheritance pattern in which both alleles of a gene are equally expressed. 149

codon Three-base sequence in messenger RNA that causes the insertion of a particular amino acid into a protein, or termination of translation. 166

coelom Body cavity lying between the digestive tract and body wall that is completely lined by mesoderm. 308, 414

coenzyme Nonprotein organic molecule that aids the action of the enzyme to which it is loosely bound. 89

coenzyme A (CoA) Molecule that helps oxidate pyruvate in the preparatory (prep) reaction during cellular respiration. 99

coevolution Joint evolution in which one species exerts selective pressure on the other species. 552

cohesion-tension model Explanation for upward transport of water in xylem based upon transpiration-created tension and the cohesive properties of water molecules. 349

collecting duct Duct within the kidney that receives fluid from several nephrons; the reabsorption of water occurs here. 428

collenchyma Plant tissue composed of cells with unevenly thickened walls; supports growth of stems and petioles. 341

color blindness Deficiency in one or more of the three kinds of cone cells responsible for color vision. 204

commensalism Symbiotic relationship in which one species is benefited, and the other is neither harmed nor benefited. 556

common ancestor Ancestor held in common by at least two lines of descent. 247, 256

community Assemblage of populations interacting with one another within the same environment. 5, 546, 552

compact bone Type of bone that contains osteons consisting of concentric layers of matrix and osteocytes in lacunae. 383, 504

competition Interaction between two organisms in which both require the same limited resource, which results in harm to both. 542, 556

competitive exclusion principle Theory that no two species can occupy the same niche. 557

complementary base pairing Hydrogen bonding between particular purines and pyrimidines in DNA. 162

complement system Series of proteins in plasma that form a nonspecific defense mechanism against a pathogen invasion; it complements the antigen-antibody reaction. 459

compound Substance having two or more different elements united chemically in a fixed ratio. 19

compound eye Type of eye found in arthropods; it is composed of many independent visual units. 499

conclusion Statement made following an experiment as to whether or not the results support the hypothesis. 9

condom, female Birth control method that blocks the entrance of sperm to the uterus; also prevents STDs. 520

condom, male Sheath used to cover the penis during sexual intercourse; used as a contraceptive and, if latex, to minimize the risk of transmitting infection. 520

conduction deafness Deafness due to middle ear damage. 497

cone Photoreceptor cell in vertebrate eyes that responds to bright light and makes color vision possible. 500

cone In conifers, structure that bears either pollen (male gametophyte) or seeds (female gametophyte). 291

conifer Member of a group of cone-bearing gymnosperm plants that includes pine, cedar, and spruce trees. 291

conjugation Transfer of genetic material from one cell to another. 272

connective tissue Type of animal tissue that binds structures together, provides support and protection, fills spaces, stores fat, and forms blood cells; adipose tissue, cartilage, bone, and blood are types of connective tissue. 382

consumer Organism that feeds on another organism in a food chain; primary consumers eat plants, and secondary consumers eat animals. 560

contact inhibition In cell culture, the point where cells stop dividing when they become a one-cell-thick sheet. 120

continuous ambulatory peritoneal dialysis (CAPD) Dialysis which takes place inside the body using the peritoneum, the natural lining of the abdomen, as the dialysis membrane. 430

contraceptive implant Birth control method utilizing synthetic progesterone; prevents ovulation by disrupting the ovarian cycle. 519

contraceptive injection Birth control method utilizing progesterone or estrogen and progesterone; prevents ovulation by disrupting the ovarian cycle. 520

contraceptive vaccine Under development, this birth control method immunizes against the hormone HCG, crucial to maintaining implantation of the embryo. 520

control group Sample that goes through all the steps of an experiment but does not contain the experimental variable being tested; a standard against which the results of an experiment are checked. 8

convergent evolution Similarity in structure in distantly related groups due to adaptation to the environment. 257

copulation Sexual union between a male and a female. 513

cork cambium Lateral meristem that produces cork. 345

cork cell Dead cells found in the outer covering of the bark of trees; may be sloughed off. 345

cornea Transparent, anterior portion of the outer layer of the eyeball. 499

corolla Petals, collectively; usually the conspicuously colored flower whorl. 292

corpus luteum Follicle that has released an egg and increases its secretion of progesterone. 517

cortex In plants, ground tissue bounded by the epidermis and vascular tissue in stems and roots; in animals, outer layer of an organ, such as the cortex of the kidney or adrenal gland. 347

cotyledon Seed leaf for embryo of a flowering plant; provides nutrient molecules for the developing plant before photosynthesis begins. 293, 340, 366

coupled reaction Reaction that occurs simultaneously; one is an exergonic reaction that releases energy, and the other is an endergonic reaction that requires an input of energy in order to occur. 73

covalent bond Chemical bond in which atoms share one pair of electrons. 20

cranial nerve Nerve that arises from the brain. 180

cristae Short, fingerlike projections formed by the folding of the inner membrane of mitochondria. 61

Cro-Magnon Common name for the first fossils to be designated *Homo sapiens*. 332

crossing-over Exchange of segments between nonsister chromatids of a tetrad during meiosis. 131

crustacean Member of a group of marine arthropods that contains, among others, shrimps, crabs, crayfish, and lobsters. 316

culture Total pattern of human behavior; includes technology and the arts, and is dependent upon the capacity to speak and transmit knowledge. 330

Cushing syndrome Condition resulting from hypersecretion of glucocorticoids; characterized by thin arms and legs and a "moon face," and accompanied by high blood glucose and sodium levels. 487

cutaneous receptor Sensory receptors for pressure and touch found in the dermis of the skin. 501

cuticle Waxy layer covering the epidermis of plants that protects the plant against water loss and disease-causing organisms. 341

cyanobacteria Photosynthetic bacteria that contain chlorophyll and release O_2; formerly called blue-green algae. 272

cyclin Protein that cycles in quantity as the cell cycle progresses; combines with and activates the kinases that function to promote the events of the cycle. 119, 189

cystic fibrosis Generalized, autosomal recessive disorder of infants and children, in which there is widespread dysfunction of the exocrine glands. 202

cytokine Type of protein secreted by a T cell that attacks viruses, virally infected cells, and cancer cells. 463

cytokinesis Division of the cytoplasm following mitosis and meiosis. 114

cytokinin Plant hormone that promotes cell division; often works in combination with auxin during organ development in plant embryos. 358

cytoplasm Contents of a cell between the nucleus (nucleoid region of bacteria) and the plasma membrane. 50

cytosine (C) One of four nitrogen-containing bases in the nucleotides composing the structure of DNA and RNA; pairs with guanine. 161

cytoskeleton Internal framework of the cell, consisting of microtubules, actin filaments, and intermediate filaments. 55, 62

cytotoxic T cell T cell that attacks and kills antigen-bearing cells. 462

D

data (sing., datum) Facts or information collected through observation and/or experimentation. 9

day-neutral plant Plant whose flowering is not dependent on day length—e.g., tomato and cucumber. 360

deciduous Shedding leaves annually. 339

decomposer Organism, usually a bacterium or fungus, that breaks down organic matter into inorganic nutrients that can be recycled in the environment. 560

deficiency disorder Disorders caused by the lack of certain vitamins or minerals in the diet. Examples are bleeding gums, rickets (vitamin deficiencies); osteoporosis, anemia (mineral deficiencies). 436

deforestation Removal of trees from a forest in a way that ever reduces the size of the forest. 578

dehydration reaction Chemical reaction resulting in a covalent bond with the accompanying loss of a water molecule. 33

delayed allergic response Allergic response initiated at the site of the allergen by sensitized T cells, involving macrophages and regulated by cytokines. 465

deletion Change in chromosome structure in which the end of a chromosome breaks off or two simultaneous breaks lead to the loss of an internal segment; often causes abnormalities—e.g., cri du chat syndrome. 198

demographics Study of human populations, their characteristics, and their changes. 539

denatured Loss of an enzyme's normal shape so that it no longer functions; caused by a less than optimal pH and temperature. 40

dendrite Part of a neuron that sends signals toward the cell body. 474

dense fibrous connective tissue Type of connective tissue containing many collagen fibers packed together; found in tendons and ligaments, for example. 382

density-dependent factor Biotic factor, such as disease or competition, that affects population size according to the population's density. 542

density-independent factor Abiotic factor, such as fire or flood, that affects population size independent of the population's density. 542

deoxyribose Pentose sugar found in DNA that has one less hydroxyl group than ribose. 33

desertification Denuding and degrading a once-fertile land, initiating a desert-producing cycle that feeds on itself and causes long-term changes in the soil, climate, and biota of an area. 578

detrital food web Complex pattern of interlocking and crisscrossing food chains that begins with a population of detritivores. 562

detritus Organic matter produced by decomposition of substances such as tissues and animal wastes. 560

deuterostome Group of coelomate animals in which the second embryonic opening is associated with the mouth; the first embryonic opening, the blastopore, is associated with the anus. 313

diabetes mellitus Condition characterized by a high blood glucose level and the appearance of glucose in the urine, due to a deficiency of insulin production and failure of cells to take up glucose. 488

dialysate Material that passes through the membrane in dialysis. 430

diaphragm In mammals, dome-shaped muscularized sheet separating the thoracic cavity from the abdominal cavity. Also, a birth control device consisting of a soft rubber or latex cup that fits over the cervix. 424, 520

diarrhea Excessively frequent and watery bowel movements. 420

diastole Relaxation period of a heart chamber during the cardiac cycle. 400

diet A person's typical food choices; a balanced diet contains all the nutrients in the right proportions to maintain a healthy body. 437

dietary supplement Nutrients and plant products (i.e., herbal teas, protein supplements) that are used to enhance health; do not need to undergo the same safety and effectiveness testing required for prescription drugs. 450

digestive system Organ system that includes the mouth, esophagus, stomach, small intestine, and large intestine (colon) that receives food and digests it into nutrient molecules. Also has associated organs: teeth, tongue, salivary glands, liver, gallbladder, and pancreas. 386

dihybrid Individual that is heterozygous for two traits; shows the phenotype governed by the dominant alleles but carries the recessive alleles. 147

diploid (2n) number Cell condition in which two of each type of chromosome are present. 129

directional selection Outcome of natural selection in which an extreme phenotype is favored, usually in a changing environment. 239

disaccharide Sugar that contains two units of a monosaccharide; e.g., maltose. 33

disruptive selection Outcome of natural selection in which the two extreme phenotypes are favored over the average phenotype, leading to more than one distinct form. 240

distal tubule Final portion of a nephron that joins with a collecting duct; associated with tubular secretion. 428

diversity Amount of each specific species in a community. 553

DNA (deoxyribonucleic acid) Nucleic acid polymer produced from covalent bonding of nucleotide monomers that contain the sugar deoxyribose; the genetic material of nearly all organisms. 41

DNA chip Very small glass square containing several rows of DNA probes. 205

DNA fingerprinting The use of DNA fragment lengths resulting from restriction enzyme cleavage to identify particular individuals. 174

DNA ligase Enzyme that links DNA fragments; used during production of recombinant DNA to join foreign DNA to vector DNA. 173

DNA polymerase During replication, an enzyme that joins the nucleotides complementary to a DNA template. 164

DNA probe Piece of single stranded DNA that will bind to a complementary piece of DNA. 203

DNA replication Synthesis of a new DNA double helix prior to mitosis and meiosis in eukaryotic cells and during prokaryotic fission in prokaryotic cells. 113, 164

domain Largest of the categories, or taxa, used by taxonomists to group species; the three domains are Archaea, Bacteria, and Eukarya. 6, 256

dominant allele Allele that exerts its phenotypic effect in the heterozygote; it masks the expression of the recessive allele. 145

dormancy In plants, a cessation of growth under conditions that seem appropriate for growth. 357

dorsal root ganglion Mass of sensory neuron cell bodies located in the dorsal root of a spinal nerve. 481

double fertilization In flowering plants, one sperm nucleus unites with the egg nucleus, and a second sperm nucleus unites with the polar nuclei of an embryo sac. 293, 365

Duchenne muscular dystrophy Chronic progressive disease affecting the shoulder and hip (pelvic) girdles, commencing in early childhood; characterized by increasing weakness of the muscles followed by atrophy and a peculiar swaying gait with the legs kept wide apart. Transmitted as an X-linked trait; affected individuals, predominantly males, rarely survive to maturity. Death is usually due to respiratory weakness or heart failure. 204

duodenum First part of the small intestine where chyme enters from the stomach. 420

duplication Change in chromosome structure in which a particular segment is present more than once in the same chromosome. 198

dyad Chromosome composed of two sister chromatids. 130

E

echinoderm Marine animals in phylum Echinodermata that include sea stars, sea urchins, and sand dollars; characterized by radial symmetry and a water vascular system. 319

ecological niche Role an organism plays in its community, including its habitat and its interactions with other organisms. 556

ecological pyramid Pictorial graph based on the biomass, number of organisms, or energy content of various trophic levels in a food web—from the producer to the final consumer populations. 563

ecological succession Gradual replacement of communities in an area following a disturbance (secondary succession) or the creation of new soil (primary succession). 554

ecology Study of the interactions of organisms with other organisms and with the physical and chemical environment. 546

ecosystem Biological community together with the associated abiotic environment; characterized by a flow of energy and a cycling of inorganic nutrients. 5, 546, 552

ectoderm Outermost primary tissue layer of an animal embryo; gives rise to the nervous system and the outer layer of the integument. 526

ectothermic Having a body temperature that varies according to the environmental temperature. 324

edema Swelling due to tissue fluid accumulation in the intercellular spaces. 430

electrocardiogram (ECG) Recording of the electrical activity associated with the heartbeat. 400

electron Negative subatomic particle, moving about in an energy level around the nucleus of an atom. 17

electronegativity Ability of an atom to attract electrons toward itself in a chemical bond. 22

electron shell Concentric energy levels in which electrons orbit. 18

electron transport chain Passage of electrons along a series of electron carriers from a higher to lower energy level; the energy released is used for the synthesis of ATP. 88, 99

element Substance that cannot be broken down into substances with different properties; composed of only one type atom. 16

embryo sac Female gametophyte of flowering plants. 365

emerging virus Causative agent of a disease that is new or is demonstrating increased prevalence, such as the viruses that cause AIDS, SARS, and avian influenza. 269

emulsification Breaking up of fat globules into smaller droplets by the action of bile salts or any other emulsifier. 418

endocrine gland Ductless organ that secretes hormone(s) into the bloodstream. 420, 483

endocrine system Organ system involved in the coordination of body activities; uses hormones as chemical signals secreted into the bloodstream. 387

endocytosis Process by which substances are moved into the cell from the environment by phagocytosis (cellular eating) or pinocytosis (cellular drinking); includes receptor-mediated endocytosis. 79

endoderm Innermost primary tissue layer of an animal embryo that gives rise to the linings of the digestive tract and associated structures. 526

endodermis Internal plant root tissue forming a boundary between the cortex and the vascular cylinder. 347

endomembrane system Collection of membranous structures involved in transport within the cell. 58

endometrium Mucous membrane lining the interior surface of the uterus. 517

endoplasmic reticulum (ER) System of membranous saccules and channels in the cytoplasm, often with attached ribosomes. 57

endoskeleton Protective internal skeleton, as in vertebrates. 307, 502

endosperm In flowering plants, nutritive storage tissue that is derived from the union of a sperm nucleus and polar nuclei in the embryo sac. 357

endospore Spore formed within a cell; certain bacteria form endospores. 272

endothermic Maintenance of a constant body temperature independent of the environmental temperature. 325

energy Capacity to do work and bring about change; occurs in a variety of forms. 3, 70

energy laws Two laws explaining energy, and its relationships and exchanges. The first, also called the "law of conservation," says that energy cannot be created or destroyed but can only be changed from one form to another; the second says that energy cannot be changed from one form to another without a loss of usable energy. 70

energy of activation Energy that must be added in order for molecules to react with one another. 75

enhancer Elements that regulate transcription from nearby genes. Function by acting as binding sites for transcription factors. 186

entropy Measure of disorder or randomness. 71

enzyme Organic catalyst, usually a protein, that speeds a reaction in cells due to its particular shape. 38, 75

enzyme inhibition Means by which cells regulate enzyme activity; may be competitive or noncompetitive inhibition. 76

epidermal tissue (epidermis) In plants, tissue that covers roots, leaves, and stems of nonwoody organisms. 341

epididymis Coiled tubule next to the testes where sperm mature and may be stored for a short time. 514

epiglottis Structure that covers the glottis and closes off the air tract during the process of swallowing. 416

epinephrine Hormone secreted by the adrenal medulla in times of stress; adrenaline. 487

epithelial tissue Tissue that lines hollow organs and covers surfaces. 380

equilibrium population Population demonstrating a life history pattern in which members exhibit logistic population growth and the population size remains at or near the carrying capacity. Its members are large in size, slow to mature, have a long life span, have few offspring, and provide much care to offspring (e.g., bears, lions). 544

erection Increase in blood flow to the penis during sexual arousal, causing the penis to stiffen and become erect. 515

esophagus Muscular tube for moving swallowed food from the pharynx to the stomach. 416

essential nutrient Substance in the diet that contributes to good health; must be supplied by the diet because the body either cannot synthesize it or makes insufficient amounts to meet the body's needs. 437

estrogen Female sex hormone that helps maintain sexual organs and secondary sex characteristics. 518

ethylene Plant hormone that causes ripening of fruit and is also involved in abscission. 359

euchromatin Chromatin that is extended and accessible for transcription. 184

eudicot Abbreviation of eudicotyledon. Flowering plant group; members have two embryonic leaves (cotyledons), net-veined leaves, vascular bundles in a ring, flower parts in fours or fives and their multiples, and other characteristics. 340

Eukarya One of the three domains of life, consisting of organisms with eukaryotic cells and further classified into the kingdoms Protista, Fungi, Plantae, and Animalia. 6, 261

eukaryotic cell Type of cell that has a membrane-bounded nucleus and membranous organelles; found in organisms within the domain Eukarya. 50

eutrophication Enrichment of water by inorganic nutrients used by phytoplankton. Often, overenrichment caused by human activities leads to excessive bacterial growth and oxygen depletion. 565

evolution Descent of organisms from common ancestors with the development of genetic and phenotypic changes over time that make them more suited to the environment. 4, 215

exocrine gland Gland which discharges its secretion into ducts; the pancreas is an exocrine gland when it secretes pancreatic juice into the duodenum. 420

exocytosis Process in which an intracellular vesicle fuses with the plasma membrane so that the vesicle's contents are released outside the cell. 79

exophthalmic goiter Enlargement of the thyroid gland accompanied by an abnormal protrusion of the eyes. 486

exoskeleton Protective external skeleton, as in arthropods. 307, 502

exotic species Species that is new to a community, nonnative. 559

experimental design Artificial situation devised to test a hypothesis. 8

experimental variable In a scientific experiment, a condition of the experiment that is deliberately changed. 8

exponential growth Growth, particularly of a population, in which the increase occurs in the same manner as compound interest. 540

external fertilization Fertilization of an egg by sperm that occurs outside of the body, as in aquatic animals. 512

extinction Total disappearance of a species or higher group. 7, 545

extracellular matrix Meshwork of polysaccharides and proteins that provides support for an animal cell and affects its behavior. 64

extraembryonic membranes Membranes that are not a part of the embryo but are necessary to the continued existence and health of the embryo. 513

F

facilitated diffusion Passive transfer of a substance into or out of a cell along a concentration gradient by a process that requires a carrier. 77

familial hypercholesterolemia (FH) Inability to remove cholesterol from the bloodstream; predisposes individual to heart attack. 204

family One of the categories, or taxa, used by taxonomists to group species; the taxon above the genus level. 6, 256

fat Organic molecule that contains glycerol and fatty acids and is found in adipose tissue of vertebrates. 35, 439

fatty acid Molecule that contains a hydrocarbon chain and ends with an acid group. 35

feedback inhibition Mechanism for regulating metabolic pathways in which the concentration of the product is kept within a certain range until binding shuts down the pathway, and no more product is produced. 76

female gametophyte In seed plants, the gametophyte that produces an egg; in flowering plants, an embryo sac. Sometimes called a megagametophyte. 365

fermentation Anaerobic breakdown of glucose that results in a gain of two ATP and end products such as alcohol and lactate. 107

fern Member of a group of plants that have large fronds; in the sexual life cycle, the independent gametophyte produces flagellated sperm, and the vascular sporophyte produces windblown spores. 289

fertilization Fusion of sperm and egg nuclei, producing a zygote that develops into a new individual. 524

fiber Structure resembling a thread; also plant material that is nondigestible. 438

fibroblast Connective tissue cell that synthesizes fibers and ground substance. 382

fibrous protein Structural protein with only a secondary structure (i.e., keratin, silk, collagen). 40

filament End-to-end chains of cells that form as cell division occurs in only one plane; in plants, the elongated stalk of a stamen. 292, 363

filter feeder Method of obtaining nourishment by certain animals which strain minute organic particles from the water in a way that deposits them in the digestive tract. 310

filtration Movement of small molecules from a blood capillary into the nephron capsule due to the action of blood pressure. 428

fimbriae (sing., fimbria) In bacteria, small, bristle-like fibers on bacterial cell surface that enable bacteria to adhere to surfaces. 51, 516

five-kingdom system System of classification that contains the kingdoms Monera, Protista, Plantae, Animalia, and Fungi. 261

flagellum (pl., flagella) Long, slender extension used for locomotion by some bacteria, protozoans, and sperm. 51, 63, 271

flatworm Unsegmented worm lacking a body cavity; phylum Platyhelminthes. 311

flower Reproductive organ of a flowering plant, consisting of several kinds of modified leaves arranged in concentric rings and attached to a modified stem called the receptacle. 285, 362

follicle Structure in the ovary of animals that contains an oocyte; site of oocyte production. 317

follicular phase First half of the ovarian cycle, during which the follicle matures and much estrogen (and some progesterone) is produced. 517

food chain Order in which one population feeds on another in an ecosystem, thereby showing the flow of energy from a detritivore (detrital food chain) or a producer (grazing food chain) to the final consumer. 563

food web In ecosystems, a complex pattern of linked and crisscrossing food chains. 562

foraminiferan Protozoan; marine amoeba having a calcium carbonate skeleton, many of which make limestone formations such as the White Cliffs of Dover. 278

formed elements Constituents of blood that are either cellular (red blood cells and white blood cells) or at least cellular in origin (platelets). 405

fossil Any past evidence of an organism that has been preserved in the Earth's crust. 253

fossil fuel Fuel such as oil, coal, and natural gas that is the result of partial decomposition of plants and animals coupled with exposure to heat and pressure for millions of years. 567, 583

fossil record History of life recorded from remains from the past. 224

founder effect Cause of genetic drift due to colonization by a limited number of individuals who, by chance, have different gene frequencies than the parent population. 238

fovea Region of the retina consisting of densely packed cones; responsible for the greatest visual acuity. 500

frameshift mutation Alteration in a gene due to deletion of a base, so that the reading "frame" is shifted; can result in a nonfunctional protein. 172

frond Fern leaf. 289

fruit In flowering plants, the structure that forms from an ovary and associated tissues and encloses seeds. 293, 366

functional group Specific cluster of atoms attached to the carbon skeleton of organic molecules that enters into reactions and behaves in a predictable way. 31

fungus (pl., fungi) Saprotrophic decomposer; the body is made up of filaments called hyphae that form a mass called a mycelium. 6, 296

G

gallbladder Organ attached to the liver that serves to store and concentrate bile. 418

gamete Haploid sex cell; e.g., egg and sperm. 521

gametophyte Haploid generation of the alternation of generations life cycle of a plant; produces gametes that unite to form a diploid zygote. 286, 362

ganglion Collection or bundle of neuron cell bodies usually outside the central nervous system. 480

gap junction Junction between cells formed by the joining of two adjacent plasma membranes; it lends strength and allows ions, sugars, and small molecules to pass between cells. 65

gastropod Mollusc with a broad flat foot for crawling (e.g., snails and slugs). 314

gastrula Stage of animal development during which the germ layers form, at least in part, by invagination. 526

gastrulation Formation of a gastrula from a blastula; characterized by an invagination of the cell layers to form a caplike structure. 526

gene Unit of heredity existing as alleles on the chromosomes; in diploid organisms, typically two alleles are inherited—one from each parent. 4

gene flow Sharing of genes between two populations through interbreeding. 236

gene linkage Existence of several alleles on the same chromosome. 154

gene locus Specific location of a particular gene on homologous chromosomes. 145

gene mutation Change in the sequence of bases in a gene. 172

gene pool Total of all the genes of all the individuals in a population. 233

gene therapy Correction of a detrimental mutation by the addition of new DNA and its insertion in a genome. 210

genetically modified plant (GMP) Plant that carries the genes of another organism as a result of DNA technology; also transgenic plant. 371

genetic counseling Prospective parents consult a counselor who determines the genotype of each and whether an unborn child will have a genetic disorder. 196

genetic drift Mechanism of evolution due to random changes in the allelic frequencies of a population; more likely to occur in small populations or when only a few individuals of a large population reproduce. 237

genetic engineering Alteration of genomes for medical or industrial purposes. 173

genetic marker Abnormality in the sequence of a base at a particular location on a chromosome signifying a disorder. 205

genetic profile An individual's complete genotype, including any possible mutations. 209

genotype Genes of an organism for a particular trait or traits; often designated by letters—for example, *BB* or *Aa*. 145

genus One of the categories, or taxa, used by taxonomists to group species; contains those species that are most closely related through evolution. 6, 256

germinate Beginning of growth of a seed, spore, or zygote, especially after a period of dormancy. 368

germ layer Primary tissue layer of a vertebrate embryo—namely, ectoderm, mesoderm, or endoderm. 526

gibberellin Plant hormone promoting increased stem growth; also involved in flowering and seed germination. 357

global warming Predicted increase in the Earth's temperature due to human activities that promote the greenhouse effect. 567

globular protein Polypeptides in this protein have a tertiary structure (i.e., enzymes). 40

glottis Opening for airflow in the larynx. 423

glucagon Hormone secreted by the pancreas which causes the liver to break down glycogen and raises the blood glucose level. 420, 487

glucocorticoid Type of hormone secreted by the adrenal cortex that influences carbohydrate, fat, and protein metabolism; see cortisol. 487

glucose Six-carbon sugar that organisms degrade as a source of energy during cellular respiration. 33

glycerol Three-carbon carbohydrate with three hydroxyl groups attached; a component of fats and oils. 35

glycogen Storage polysaccharide found in animals; composed of glucose molecules joined in a linear fashion but having numerous branches. 34

glycolysis Anaerobic breakdown of glucose that results in a gain of two ATP and the end product pyruvate. 99

golden-brown algae Marine photosynthetic protist with elaborate shells of silica; represented by diatoms. 277

Golgi apparatus Organelle consisting of saccules and vesicles that processes, packages, and distributes molecules about or from the cell. 58

Golgi tendon organ Proprioceptor located in the joints; helps maintain equilibrium and posture. 50

gonadotropic hormones (FSH and LH) Substances secreted by the anterior pituitary that regulate the activity of the ovaries and testes; principally, follicle-stimulating hormone (FSH) and luteinizing hormone (LH). 484

gonads Organs that produce gametes; the ovary produces eggs, and the testis produces sperm. 512

granum (pl., grana) Stack of chlorophyll-containing thylakoids in a chloroplast. 60, 85

gravitational equilibrium Maintenance of balance when the head and body are motionless. 498

gravitropism Growth response of roots and stems of plants to the Earth's gravity; roots demonstrate positive gravitropism, and stems demonstrate negative gravitropism. 360

grazing food web Complex pattern of linked and crisscrossing food chains that begins with a population of photosynthesizers serving as producers. 562

green algae Member of a diverse group of photosynthetic protists that contains chlorophylls *a* and *b* and has other biochemical characteristics like those of plants. 277

greenhouse effect Reradiation of solar heat toward the Earth, caused by gases such as carbon dioxide, methane, nitrous oxide, water vapor, ozone, and nitrous oxide in the atmosphere. 567

greenhouse gases Gases in the atmosphere such as carbon dioxide, methane, nitrous oxide, water vapor, ozone, and nitrous oxide that are involved in the greenhouse effect. 583

ground tissue Tissue that constitutes most of the body of a plant; consists of parenchyma, collenchyma, and sclerenchyma cells that function in storage, basic metabolism, and support. 341

growth factor Chemical signal that regulates mitosis and differentiation of cells that have receptors for it; important in such processes as fetal development, tissue maintenance and repair, and hematopoiesis; sometimes a contributing factor in cancer. 189

growth hormone (GH) Substance secreted by the anterior pituitary; controls size of an individual by promoting cell division, protein synthesis, and bone growth. 485

guanine (G) One of four nitrogen-containing bases in nucleotides composing the structure of DNA and RNA; pairs with cytosine. 161

gymnosperm Type of woody seed plant in which the seeds are not enclosed by fruit and are usually borne in cones, such as those of the conifers. 286, 291

H

habitat Place where an organism lives and is able to survive and reproduce. 556

halophile Type of archaea that lives in extremely salty habitats. 275

haploid (n) number Cell condition in which only one of each type of chromosome is present. 129

heart Muscular organ whose contraction causes blood to circulate in the body of an animal. 396

heart attack Damage to the myocardium due to blocked circulation in the coronary arteries; myocardial infarction. 404

heart murmur Clicking or swishy sounds, often due to leaky valves. 399

heat Type of kinetic energy; captured solar energy eventually dissipates as heat in the environment. 71

helper T cell Secretes lymphokines, which stimulate all kinds of immune cells. 463

heme Iron-containing group found in hemoglobin. 426

hemodialysis Cleansing of blood by using an artificial membrane that causes substances to diffuse from blood into a dialysis fluid. 430

hemoglobin Iron-containing respiratory pigment occurring in vertebrate red blood cells and in the blood plasma of some invertebrates. 38, 405

hemophilia Most common of the severe clotting disorders caused by the absence of a blood clotting factor. 204

herbaceous stem Nonwoody stem. 344

herbivore Primary consumer in a grazing food chain; a plant eater. 415, 560

hermaphrodite Type of animal that has both male and female sex organs. 312, 512

heterochromatin Highly compacted chromatin that is not accessible for transcription. 184

heterotroph Organism that cannot synthesize organic compounds from inorganic substances, and therefore must take in organic food. 560

heterozygous Possessing unlike alleles for a particular trait. 145

hip (pelvic) girdle Portion of the vertebrate skeleton to which the lower (hind) limbs are attached; consists of the coxal bones. 503

histamine Substance, produced by basophils in blood and mast cells in connective tissue, that causes capillaries to dilate. 458

histone Protein molecule responsible for packing chromatin. 113

HIV (human immunodeficiency virus) Virus responsible for AIDS. 466, 522

homeostasis Maintenance of normal internal conditions in a cell or an organism by means of self-regulating mechanisms. 3, 388

hominid Member of the family Hominidae, which contains australopithecines and humans. 329

Homo erectus Hominid who used fire and migrated out of Africa to Europe and Asia. 330

Homo habilis Hominid of two million years ago who is believed to have been the first tool user. 330

homologous chromosome Member of a pair of chromosomes that are alike and come together in synapsis during prophase of the first meiotic division; a *homologue*. 128

homologous structure In evolution, a structure that is similar in different types of organisms because these organisms are derived from a common ancestor. 226, 257

homologue Member of a pair of chromosomes that are alike and come together in synapsis during prophase of the first meiotic division; a *homologous chromosome*. 128

Homo sapiens Modern humans. 331

homozygous Possessing two identical alleles for a particular trait. 145

Huntington disease Genetic disease marked by progressive deterioration of the nervous system due to deficiency of a neurotransmitter. 203

hyaline cartilage Cartilage whose cells lie in lacunae separated by a white translucent matrix containing very fine collagen fibers. 383

hybridization Crossing of different species. 370

hydrogen (H) bond Weak bond that arises between a slightly positive hydrogen atom of one molecule and a slightly negative atom of another molecule or between parts of the same molecule. 22

hydrogen ion (H⁺) Hydrogen atom that has lost its electron and therefore bears a positive charge (H⁺). 25

hydrolysis reaction Splitting of a compound by the addition of water, with the H⁺ being incorporated in one fragment and the OH⁻ in the other. 33

hydrophilic Type of molecule that interacts with water by dissolving in water and/or by forming hydrogen bonds with water molecules. 23, 31

hydrophobic Type of molecule that does not interact with water because it is nonpolar. 23, 31

hydrostatic skeleton Fluid-filled body compartment that provides support for muscle contraction resulting in movement; seen in cnidarians, flatworms, roundworms, and segmented worms. 503

hydroxide ion (OH⁻) One of two ions that results when a water molecule dissociates; it has gained an electron, and therefore bears a negative charge (OH⁻). 25

hypertension Elevated blood pressure, particularly the diastolic pressure. 403

hypertonic solution Higher solute concentration (less water) than the cytoplasm of a cell; causes cell to lose water by osmosis. 78

hypothalamic-releasing hormone One of many hormones produced by the hypothalamus that stimulates the secretion of an anterior pituitary hormone. 484

hypothalamus In vertebrates, part of the brain that helps regulate the internal environment of the body—for example, heart rate, body temperature, and water balance. 479, 484

hypothesis Supposition established by reasoning after consideration of available evidence; it can be tested by obtaining more data, often by experimentation. 8

hypotonic solution Lower solute (more water) concentration than the cytoplasm of a cell; causes cell to gain water by osmosis. 78

I

immediate allergic response Allergic response that occurs within seconds of contact with an allergen; caused by the attachment of the allergen to IgE antibodies. 465

Immune system All the cells in the body that protect the body against foreign organisms and substances and also against cancerous cells. 387, 456

Immunity Ability of the body to protect itself from foreign substances and cells, including disease-causing agents. 456

incomplete dominance Inheritance pattern in which the offspring has an intermediate phenotype, as when a red-flowered plant and a white-flowered plant produce pink-flowered offspring. 149

induced fit model Change in the shape of an enzyme's active site that enhances the fit between the active site and its substrate(s). 76

induction Ability of a chemical or a tissue to influence the development of another tissue. 526

inductive reasoning Using specific observations and the process of logic and reasoning to arrive at a hypothesis. 8

industrial melanism Increased frequency of darkly pigmented (melanic) forms in a population when soot and pollution make lightly pigmented forms easier for predators to see against a pigmented background. 234

infertility Inability to have as many children as desired. 520

inflammatory response Tissue response to injury that is characterized by redness, swelling, pain, and heat. 458

inorganic chemistry The study of compounds not having a carbon basis; chemistry of the nonliving world. 30

insect Type of arthropod. The head has antennae, compound eyes, and simple eyes; the thorax has three pairs of legs and often wings; and the abdomen has internal organs. 318

insulin Hormone secreted by the pancreas that lowers the blood glucose level by promoting the uptake of glucose by cells and the conversion of glucose to glycogen by the liver and skeletal muscles. 420, 487

integration Summing up of excitatory and inhibitory signals by a neuron or by some part of the brain. 476

integumentary system Organ system consisting of skin various organs, such as hair, which are found in skin. 387

intercalated disk Region that holds adjacent cardiac muscle cells together; disks appear as dense bands at right angles to the muscle striations. 384

interkinesis Period of time between meiosis I and meiosis II during which no DNA replication takes place. 133

intermembrane space Space that occurs between the outer and inner membrane of a mitochondrion. 105

interneuron Neuron located within the central nervous system that conveys messages between parts of the central nervous system. 474

internode In vascular plants, the region of a stem between two successive nodes. 339

interphase Stages of the cell cycle (G₁, S, G₂) during which growth and DNA synthesis occur when the nucleus is not actively dividing. 114

intestinal enzyme Enzyme, produced by the epithelial cells on the surface of villi, which functions in the digestion of small organic molecules. 418

intrauterine device (IUD) Birth-control device consisting of a small piece of molded plastic inserted into the uterus, and believed to alter the uterine environment so that fertilization does not occur. 520

inversion Change in chromosome structure in which a segment of a chromosome is turned around 180°; this reversed sequence of genes can lead to altered gene activity and abnormalities. 198

ion Charged particle that carries a negative or positive charge. 20

ionic bond Chemical bond in which ions are attracted to one another by opposite charges. 20

Iris Muscular ring that surrounds the pupil and regulates the passage of light through this opening. 499

isomer Molecules with the same molecular formula but a different structure, and therefore a different shape. 31

isotonic solution Solution that is equal in solute concentration to that of the cytoplasm of a cell; causes cell to neither lose nor gain water by osmosis. 78

isotope Atom of the same element having the same atomic number but a different mass number due to the number of neutrons. 18

J

jawless fishes Type of fish(es) that have no jaws, includes today's hagfishes and lampreys. 322

K

karyotype Chromosomes arranged by pairs according to their size, shape, and general appearance in mitotic metaphase. 196

keystone species Species whose activities significantly affect community structure. 558

kidneys Paired organs of the vertebrate urinary system that regulate the chemical composition of the blood and produce a waste product called urine. 427

kilocalorie Caloric value of food; 1,000 calories. 70

kinase Enzyme that activates another enzyme by adding a phosphate group. 119

kinetic energy Energy associated with motion. 70

kingdom One of the categories, or taxa, used by taxonomists to group species; the taxon above phylum. 6, 256

Klinefelter syndrome Condition caused by the inheritance of XXY chromosomes. 137

L

lacteal Lymphatic vessel in an intestinal villus; aids in the absorption of fats. 418

lacuna Small pit or hollow cavity, as in bone or cartilage, where a cell or cells are located. 383

lancelet Invertebrate chordate with a body that resembles a lancet and has the four chordate characteristics as an adult. 320

large intestine In vertebrates, portion of the digestive tract that follows the small intestine; in humans, consists of the cecum, colon, rectum, and anal canal. 419

larynx Cartilaginous organ located between the pharynx and the trachea; in humans, contains the vocal cords; sometimes called the voice box. 423

lateral bud Site on a stem where lateral branches grow. 339

lateral line Canal system containing sensory receptors that allow fishes and amphibians to detect water currents and pressure waves from nearby objects. 498

law Theory that is generally accepted by an overwhelming number of scientists. 9

law of independent assortment Alleles of unlinked genes assort independently of each other during meiosis so that the gametes contain all possible combinations of alleles. 146

law of segregation Separation of alleles from each other during meiosis so that the gametes contain one from each pair. Each resulting gamete has an equal chance of receiving either allele. 145

leaf Lateral appendage of a stem, highly variable in structure, often containing cells that carry out photosynthesis. 338

leaf vein Vascular tissue within a leaf. 340

leech Blood-sucking annelid, usually found in fresh water, with a sucker at each end of a segmented body. 315

lens Clear, membranelike structure found in the vertebrate eye behind the iris; brings objects into focus. 499

less-developed country (LDC) Country that is becoming industrialized; typically, population growth is expanding rapidly, and the majority of people live in poverty. 535

lichen Symbiotic relationship between certain fungi and algae, in which the fungi possibly provide inorganic food or water and the algae provide organic food. 299

life cycle Recurring pattern of genetically programmed events by which individuals grow, develop, maintain themselves, and reproduce. 129

life history Adaptations in characteristics that influence an organism's biology, such as how many offspring it produces, its survival, and factors such as age and size that determine its reproductive maturity. 544

ligament Tough cord or band of dense fibrous tissue that binds bone to bone at a joint. 382, 506

light reaction Portion of photosynthesis that captures solar energy and takes place in thylakoid membranes of chloroplasts; it produces ATP and NADPH. 86

lignin Chemical that hardens the cell walls of plants. 288

limbic system In humans, functional association of various brain centers, including the amygdala and hippocampus; governs learning and memory and various emotions such as pleasure, fear, and happiness. 480

lineage Evolutionary line of descent. 328

linkage group Alleles of different genes that are located on the same chromosome and tend to be inherited together. 154

lipase Fat-digesting enzyme secreted by the pancreas. 418

lipid Class of organic compounds that tends to be soluble in nonpolar solvents; includes fats and oils. 35

liver Large, dark red internal organ that produces urea and bile, detoxifies the blood, stores glycogen, and produces the plasma proteins, among other functions. 418

lobe-finned fishes Type of fish(es) with limblike fins. 323

logistic growth Population increase that results in an S-shaped curve; growth is slow at first, steepens, and then levels off due to environmental resistance. 541

long-day plant Plant that flowers when day length is longer than a critical length; e.g., wheat, barley, clover, and spinach. 360

loose fibrous connective tissue Tissue composed mainly of fibroblasts widely separated by a matrix containing collagen and elastic fibers. 382

lungs Internal respiratory organs containing moist surfaces for gas exchange. 323, 422

luteal phase Second half of the ovarian cycle, during which the corpus luteum develops and much progesterone (and some estrogen) is produced. 517

lymph Fluid, derived from tissue fluid, that is carried in lymphatic vessels. 403

lymphatic organs Organs other than a lymphatic vessel that are part of the lymphatic system; the lymphatic organs are the lymph nodes, tonsils, spleen, thymus gland, and bone marrow. 456

lymphatic system Organ system consisting of lymphatic vessels and lymphatic organs; transports lymph and lipids, and aids the immune system. 386, 403

lymphatic vessel Vessel that carries lymph. 403

lymph node Mass of lymphatic tissue located along the course of a lymphatic vessel. 457

lymphocyte Specialized white blood cell that functions in specific defense; occurs in two forms—T lymphocytes and B lymphocytes. 406

lysogenic cycle Bacteriophage life cycle in which the virus incorporates its DNA into that of a bacterium; occurs preliminary to the lytic cycle. 267

lysosome Membrane-bounded vesicle that contains hydrolytic enzymes for digesting macromolecules. 58

lytic cycle Bacteriophage life cycle in which the virus takes over the operation of the bacterium immediately upon entering it and subsequently destroys the bacterium. 267

M

macroevolution Large-scale evolutionary change, such as the formation of new species. 246

macronutrient Essential element needed in large amounts by plants and humans. In plants, nitrogen, calcium, and sulfur are needed for plant growth; in humans, carbohydrates, lipids, and proteins supply the body's energy needs. 437

macrophage In vertebrates, large phagocytic cell derived from a monocyte that ingests microbes and debris. 406, 457

male gametophyte In seed plants, the gametophyte that produces sperm; a pollen grain. Sometimes called a microgametophyte. 364

Malpighian tubule Blind, threadlike excretory tubule near the anterior end of an insect's hindgut. 429

mammal Endothermic vertebrate characterized especially by the presence of hair and mammary glands. 326

Marfan syndrome Congenital disorder of connective tissue characterized by abnormal length of the extremities. 203

marsupial Member of a group of mammals bearing immature young nursed in a marsupium, or pouch—for example, kangaroo and opossum. 326

mass extinction Episode of large-scale extinction in which large numbers of species disappear in a few million years or less. 254

mass number Sum of the number of protons and neutrons in an atom's nucleus. 17

mast cell Connective tissue cell that releases histamine in allergic reactions. 458

matrix Unstructured semifluid substance that fills the space between cells in connective tissues or inside organelles. 61, 382

matter Anything that takes up space and has mass. 16

medulla oblongata In vertebrates, part of the brain stem that is continuous with the spinal cord; controls heartbeat, blood pressure, breathing, and other vital functions. 479

megaspore One of the two types of spores produced by seed plants; develops into a female gametophyte (embryo sac). 362, 364

meiosis, meiosis I, meiosis II Type of nuclear division that occurs as part of sexual reproduction in which the daughter cells receive the haploid number of chromosomes in varied combinations. 129, 130

memory Capacity of the brain to store and retrieve information about past sensations and perceptions; essential to learning. 480

meniscus Cartilaginous wedges that separate the surfaces of bones in synovial joints. 506

menopause Termination of the ovarian and menstrual cycles in older women. 518

menses Flow of blood during menstruation. 518

menstrual cycle Cycle that runs concurrently with the ovarian cycle; it prepares the uterus to receive a developing zygote. 518

menstruation Periodic shedding of tissue and blood from the inner lining of the uterus in primates. 518

meristem Undifferentiated embryonic tissue in the active growth regions of plants. 341

mesoderm Middle primary tissue layer of an animal embryo that gives rise to muscle, several internal organs, and connective tissue layers. 526

mesophyll Inner, thickest layer of a leaf consisting of palisade and spongy mesophyll; the site of most of photosynthesis. 342

messenger RNA (mRNA) Type of RNA formed from a DNA template and bearing coded information for the amino acid sequence of a polypeptide. 165

mRNA transcript Faithful copy of the sequence of bases in DNA. 167

metabolic pathway Series of linked reactions, beginning with a particular reactant and terminating with an end product. 75

metabolism All of the chemical reactions that occur in a cell during growth and repair. 3

metaphase Mitotic phase during which chromosomes are aligned at the spindle equator. 115

metastasis Spread of cancer from the place of origin throughout the body; caused by the ability of cancer cells to migrate and invade tissues. 121, 190

methanogen Type of archaea that lives in oxygen-free habitats, such as swamps, and releases methane gas. 275

microevolution Change in gene frequencies between populations of a species over time. 232

micronutrient Essential element needed in small amounts by plants and humans. In plants, boron, copper, and zinc are needed for plant growth; in humans, vitamins and minerals help regulate metabolism and physiological development. 437

microspore One of the two types of spores produced by seed plants; develops into a male gametophyte (pollen grain). 362, 364

microtubule Small, cylindrical organelle composed of tubulin protein around an empty central core; present in the cytoplasm, centrioles, cilia, and flagella. 62

microvillus (pl., microvilli) Cylindrical process that extends from an epithelial cell of a villus and serves to increase the surface area of the cell. 418

mineral Naturally occurring inorganic substance containing one or more elements; certain minerals are needed in the diet. As raw materials in the Earth, they are nonrenewable resources. 441, 586

mineralocorticoids Hormones secreted by the adrenal cortex that regulate salt and water balance, leading to increases in blood volume and blood pressure. 487

mitochondrion Membrane-bounded organelle in which ATP molecules are produced during the process of cellular respiration. 60

mitosis Process in which a parent nucleus produces two daughter nuclei, each having the same number and kinds of chromosomes as the parent nucleus. 114

model Simulation of a process that aids conceptual understanding until the process can be studied firsthand; a hypothesis that describes how a particular process could possibly be carried out. 8

molecule Union of two or more atoms of the same element; also, the smallest part of a compound that retains the properties of the compound. 19

molting Periodic shedding of the exoskeleton in arthropods. 316

monocot Abbreviation of monocotyledon. Flowering plant group; members have one embryonic leaf (cotyledon), parallel-veined leaves, scattered vascular bundles, flower parts in threes or multiples of three, and other characteristics. 340

monocyte Type of a granular leukocyte that functions as a phagocyte, particularly after it becomes a macrophage. 406

monomer Small molecule that is a subunit of a polymer—e.g., glucose is a monomer of starch. 32

monosaccharide Simple sugar; a carbohydrate that cannot be decomposed by hydrolysis—e.g., glucose. 33

monotreme Egg-laying mammal—e.g., duckbill platypus and spiny anteater. 326

more-developed country (MDC) Country that is industrialized; typically, population growth is low, and the people enjoy a good standard of living. 535

morula Spherical mass of cells resulting from cleavage during animal development prior to the blastula stage. 525

mosaic evolution Concept that human characteristics did not evolve at the same rate; for example, some body parts are more humanlike than others in early hominids. 330

motor neuron Nerve cell that conducts nerve impulses away from the central nervous system and innervates effectors (muscle and glands). 474

mouth In humans, organ of the digestive tract where food is chewed and mixed with saliva. 415

multicellular Organism composed of many cells; usually has organized tissues, organs, and organ systems. 2

multifactorial trait Trait controlled by polygenes subject to environmental influences; each dominant allele contributes to the phenotype in an additive and like manner. 150

multiregional continuity hypothesis Proposal that modern humans evolved separately in at least three different places: Asia, Africa, and Europe. 331

muscle dysmorphia Mental state where a person thinks his or her body is underdeveloped, and becomes preoccupied with body-building and diet; affects more men than women. 448

muscle spindle Proprioceptor wrapped around a few muscle cells within a connective tissue sheath; can respond to changes in muscle length. 501

muscular system System of muscles that produces movement, both within the body and of its limbs; principal components are skeletal muscle, smooth muscle, and cardiac muscle. 387

muscle tissue Type of animal tissue composed of fibers that shorten and lengthen to produce movements. 384

mutagen Agent, such as radiation or a chemical, that brings about a mutation in DNA. 172

mutation Alternation in chromosome structure or number and also an alteration in a gene due to a change in DNA composition. 235

mutualism Symbiotic relationship in which both species benefit in terms of growth and reproduction. 556

mycelium Tangled mass of hyphal filaments composing the vegetative body of a fungus. 296

mycorrhizal fungi, mycorrhiza Mutualistic relationship between fungal hyphae and roots of vascular plants. 299, 348

myelin sheath White, fatty material—derived from the membranes of tightly spiraled cells— that forms a covering for nerve fibers. 474

myofibril Specific muscle cell organelle containing a linear arrangement of sarcomeres, which shorten to produce muscle contraction. 505

myosin Muscle protein making up the thick filaments in a sarcomere; it pulls actin to shorten the sarcomere, yielding muscle contraction. 505

N

nasal cavity One of two canals in the nose, separated by a septum. 422

native species Indigenous species that colonize an area without human assistance. 559

natural history The study of how organisms are influenced by factors such as climate, predation, competition, and evolution; uses field observations instead of experimentation. 547

natural killer (NK) cell Lymphocyte that causes an infected or cancerous cell to burst. 459

natural selection Mechanism of evolution caused by environmental selection of organisms most fit to reproduce; results in adaptation to the environment. 4, 221, 238

Neandertal Hominid with a sturdy build that lived during the last Ice Age in Europe and the Middle East; hunted large game and left evidence of being culturally advanced. 332

negative feedback Mechanism of homeostatic response by which the output of a system suppresses or inhibits activity of the system. 389, 486

nematocyst In cnidarians, a capsule that contains a threadlike fiber, the release of which aids in the capture of prey. 311

nephridium Segmentally arranged, paired excretory tubules of many invertebrates, as in the earthworm. 315

nephron Microscopic kidney unit that regulates blood composition by filtration, reabsorption, and secretion. 427

nephron capsule Cuplike structure that is the initial portion of a nephron. 428

nephron loop Portion of a nephron between the proximal and distal tubules; functions in water reabsorption. 428

nerve Bundle of long axons outside the central nervous system. 385, 480

nerve deafness Deafness due to spiral organ damage. 497

nervous system Organ system consisting of the brain, spinal cord, and associated nerves that coordinates the other organ systems of the body. 387

nervous tissue Tissue that contains nerve cells (neurons), which conduct impulses, and neuroglia, which support, protect, and provide nutrients to neurons. 385

neural plate Region of the dorsal surface of the chordate embryo that marks the future location of the neural tube. 526

neural tube Tube formed by closure of the neural groove during development. In vertebrates, the neural tube develops into the spinal cord and brain. 526

neuroglia Nonconducting nerve cells that are intimately associated with neurons and function in a supportive capacity. 385

neuron Nerve cell that characteristically has three parts: dendrites, cell body, and an axon. 385, 474

neurotransmitter Chemical stored at the ends of axons that is responsible for transmission across a synapse. 476

neurulation Development of the central nervous system organs in an embryo. 526

neutron Neutral subatomic particle, located in the nucleus and assigned one atomic mass unit. 17

neutrophil Granular leukocyte that is the most abundant of the white blood cells; first to respond to infection. 406, 458

node In plants, the place where one or more leaves attach to a stem. 339

nondisjunction Failure of homologous chromosomes or daughter chromosomes to separate during meiosis I and meiosis II, respectively. 136

nonrandom mating Mating among individuals on the basis of their phenotypic similarities or differences, rather than mating on a random basis. 236

nonrenewable resources Minerals, fossil fuels, and other materials present in essentially fixed amounts (within human time scale) in our environment. 576

nonvascular plant Bryophytes such as mosses that have no vascular tissue and either occur in moist locations or have special adaptations for living in dry locations. 287

norepinephrine (NE) Neurotransmitter of the postganglionic fibers in the sympathetic division of the autonomic system; also, a hormone produced by the adrenal medulla. 476, 487

notochord Cartilaginous-like supportive dorsal rod in all chordates sometime in their life cycle; replaced by vertebrae in vertebrates. 320, 526

nuclear envelope Double membrane that surrounds the nucleus in eukaryotic cells and is connected to the endoplasmic reticulum; has pores that allow substances to pass between the nucleus and the cytoplasm. 57

nuclear pore Opening in the nuclear envelope that permits the passage of proteins into the nucleus and ribosomal subunits out of the nucleus. 57

nuclease Enzyme that catalyzes decomposition of nucleic acids. 418

nucleic acid Polymer of nucleotides; both DNA and RNA are nucleic acids. 41

nucleoid Region of prokaryotic cells where DNA is located; it is not bounded by a nuclear envelope. 51, 271

nucleolus Dark-staining, spherical body in the nucleus that produces ribosomal subunits. 56

nucleosome In the nucleus of a eukaryotic cell, a unit composed of DNA wound around a core of eight histone proteins, giving the appearance of a string of beads. 113

nucleotide Monomer of DNA and RNA consisting of a 5-carbon sugar bonded to a nitrogenous base and a phosphate group. 41

nucleus Membrane-bounded organelle within a eukaryotic cell that contains chromosomes and controls the structure and function of the cell. 17, 54

nutrient Chemical substances in foods that are essential to the diet and contribute to good health. 436

nutrition The Council on Food and Nutrition of the American Medical Association defines nutrition as "the science of food; the nutrients and the substances therein; their action, interaction, and balance in relation to health and disease; and the process by which the organism (i.e., body) ingests, digests, absorbs, transports, utilizes, and excretes food substances." 436

O

obesity Characterized by excess adipose tissue; exceeding desirable weight by more than 20%. 439

observation Step in the scientific method by which data are collected before a conclusion is drawn. 8

octet rule States that an atom other than hydrogen tends to form bonds until it has eight electrons in its outer shell; an atom that already has eight electrons in its outer shell does not react and is inert. 19

oil Triglyceride, usually of plant origin, that is composed of glycerol and three fatty acids and is liquid in consistency due to many unsaturated bonds in the hydrocarbon chains of the fatty acids. 35

omnivore Organism in a food chain that feeds on both plants and animals. 415, 560

oncogene Cancer-causing gene. 189

oogenesis Production of eggs in females by the process of meiosis and maturation. 129

open circulatory system Circulatory system, such as that found in a grasshopper, where a tubular heart pumps hemolymph through channels and body cavities. 396

operon Group of structural and regulating genes that function as a single unit. 182

opportunistic population Population demonstrating a life history pattern in which members exhibit exponential population growth. Its members are small in size, mature early, have a short life span, produce many offspring, and provide little or no care to offspring (e.g., dandelions). 544

order One of the categories, or taxa, used by taxonomists to group species; the taxon above the family level. 6, 256

organ Combination of two or more different tissues performing a common function. 2

organelle Small, often membranous structure in the cytoplasm having a specific structure and function. 54

organic Molecule that always contains carbon and hydrogen, and often contains oxygen as well; organic molecules are associated with living things. 30

organic chemistry The study of carbon compounds; chemistry of the living world. 30

organism Individual living thing. 2, 30, 546

organ system Group of related organs working together. 2

orgasm Physiological and psychological sensations that occur at the climax of sexual stimulation. 515

osmosis Diffusion of water through a differentially permeable membrane. 77

osteoblast Bone-forming cell. 504

osteoclast Cell that causes erosion of bone. 504

osteon Cylindrical-shaped unit containing bone cells that surround an osteonic canal; also called Haversian system. 504

osteoporosis Condition in which bones break easily because calcium is removed from them faster than it is replaced. 504

out-of-Africa hypothesis Proposal that modern humans originated only in Africa; then they migrated and supplanted populations of *Homo* in Asia and Europe about 100,000 years ago. 331

ovary Female gonad in animals that produces an egg and female sex hormones; in flowering plants, the enlarged, ovule-bearing portion of the carpel that develops into a fruit. 292, 363, 512

ovulation Bursting of a follicle when a secondary oocyte is released from the ovary; if fertilization occurs, the secondary oocyte becomes an egg. 517

ovule In seed plants, a structure that contains the female gametophyte and has the potential to develop into a seed. 290, 363

oxygen deficit Amount of oxygen needed to metabolize lactate, a compound that accumulates during vigorous exercise. 107

oxytocin Hormone released by the posterior pituitary that causes contraction of uterus and milk letdown. 484

P

paleontology Study of fossils that results in knowledge about the history of life. 253

palisade mesophyll Layer of tissue in a plant leaf containing elongated cells with many chloroplasts. 342

pancreas Internal organ that produces digestive enzymes and the hormones insulin and glucagon. 418, 487

pancreatic amylase Enzyme that digests starch to maltose. 418

pancreatic islets Masses of cells that constitute the endocrine portion of the pancreas. 487

parallel evolution Similarity in structure in related groups that cannot be traced to a common ancestor. 257

parasitism Symbiotic relationship in which one species (the *parasite*) benefits in terms of growth and reproduction to the detriment of the other species (the *host*). 556

parasympathetic division Division of the autonomic system that is active under normal conditions; uses acetylcholine as a neurotransmitter. 482

parathyroid gland Gland embedded in the posterior surface of the thyroid gland; it produces parathyroid hormone. 486

parathyroid hormone (PTH) Hormone secreted by the four parathyroid glands that increases the blood calcium level and decreases the phosphate level. 486

parenchyma Plant tissue composed of the least-specialized of all plant cells; found in all organs of a plant. 341

parthenogenesis Development of an egg cell into a whole organism without fertilization. 512

passive immunity Protection against infection acquired by transfer of antibodies to a susceptible individual. 464

pathogen Disease-causing agent such as viruses, parasitic bacteria, fungi, and animals. 274

pedigree Chart showing the relationships of relatives and which ones have a particular trait. 200

pelvic inflammatory disease (PID) Latent infection of gonorrhea or chlamydia in the vasa deferentia or uterine tubes. 523

penis Male copulatory organ; in humans, the male organ of sexual intercourse. 514

pepsin Enzyme secreted by gastric glands that digests proteins to peptides. 417

peptide Two or more amino acids joined together by covalent bonding. 40

peptide bond Type of covalent bond that joins two amino acids. 39

peptide hormone Type of hormone that is a protein, a peptide, or derived from an amino acid. 484

peptidoglycan Unique molecule found in bacterial cell walls. 271

perennial plant Flowering plant that lives more than one growing season because the underground parts regrow each season. 339

pericycle Layer of cells surrounding the vascular tissue of roots; produces branch roots. 347

peripheral nervous system (PNS) Nerves and ganglia that lie outside the central nervous system. 473

peristalsis Wavelike contractions that propel substances along a tubular structure such as the esophagus. 416

petal Flower part that occurs just inside the sepals; often conspicuously colored to attract pollinators. 292, 363

petiole Part of a plant leaf that connects the blade to the stem. 338

pH Hydrogen ion concentration. 26

phagocytosis Process by which amoeboid-type cells engulf large substances, forming an intracellular vacuole. 79

pharynx In vertebrates, common passageway for both food intake and air movement; located between the mouth and the esophagus. 416, 423

phenotype Visible expression of a genotype—e.g., brown eyes or attached earlobes. 145

phenylketonuria (PKU) Result of accumulation of phenylalanine characterized by mental retardation, light pigmentation, eczema, and neurological manifestations unless treated by a diet low in phenylalanine. 202

phloem Vascular tissue that conducts organic solutes in plants; contains sieve-tube members and companion cells. 288, 340

phospholipid Molecule that forms the *phospholipid bilayer* of plasma membranes has a polar, hydrophilic head bonded to two nonpolar, hydrophobic tails. 36

photoautotroph Organism able to synthesize organic molecules by using carbon dioxide as the carbon source and sunlight as the energy source. 272

photoperiod Relative lengths of daylight and darkness that affect the physiology and behavior of an organism. 360

photoreceptor Sensory receptor that responds to light stimuli. 199

photosynthesis Process occurring usually within chloroplasts whereby chlorophyll-containing organelles trap solar energy to reduce carbon dioxide to carbohydrate. 3, 84

photosystem Photosynthetic unit where solar energy is absorbed and high-energy electrons are generated; contains a pigment complex and an electron acceptor; occurs as PS (photosystem) I and PS II. 88

phototropism Growth response of plant stems to light; stems demonstrate positive phototropism. 360

photovoltaic (solar) cell Energy-conversion device that captures solar energy and directly converts it to electrical current. 585

pH scale Measurement scale for hydrogen ion concentration. 26

phylogenetic tree Diagram that indicates common ancestors and lines of descent among a group of organisms. 256

phylogeny Evolutionary history of a group of organisms. 256

phylum One of the categories, or taxa, used by taxonomists to group species, the taxon above the class level. 6, 256

phytochrome Photoreversible plant pigment that is involved in photoperiodism and other responses of plants, such as etiolation. 361

pinocytosis Process by which vesicle formation brings macromolecules into the cell. 79

pith Parenchyma tissue in the center of some stems and roots. 347

pituitary gland Small gland that lies just inferior to the hypothalamus; consists of the anterior and posterior pituitary, both of which produce hormones. 484

placenta Organ formed during the development of placental mammals from the chorion and the uterine wall; allows the embryo, and then the fetus, to acquire nutrients and rid itself of wastes; produces hormones that regulate pregnancy. 513

placental mammal Mammal characterized by the presence of a placenta during the development of an offspring. 327

planarian Free-living flatworm with a ladder-like nervous system. 311

plant Multicellular, usually photosynthetic, organism belonging to the plant kingdom. 6

plaque Accumulation of soft masses of fatty material, particularly cholesterol, beneath the inner linings of the arteries. 404

plasma In vertebrates, the liquid portion of blood; contains nutrients, wastes, salts, and proteins. 405

plasma membrane Membrane surrounding the cytoplasm that consists of a phospholipid bilayer with embedded proteins; functions to regulate the entrance and exit of molecules from cell. 52

plasmid Self-duplicating ring of accessory DNA in the cytoplasm of bacteria. 271

plasmodesmata In plants, cytoplasmic strands that extend through pores in the cell wall and connect the cytoplasm of two adjacent cells. 64

plasmolysis Contraction of the cell contents due to the loss of water. 78

platelet Component of blood that is necessary to blood clotting. 383, 406

pleiotropy Inheritance pattern in which one gene affects many phenotypic characteristics of the individual. 151

point mutation Alteration in a gene due to a change in a single nucleotide; results of this mutation vary. 172

polar In chemistry, bond in which the sharing of electrons between atoms is unequal. 22

pollen grain In seed plants, structure that is derived from a microspore and develops into a male gametophyte. 290, 364

pollen tube In seed plants, a tube that forms when a pollen grain lands on the stigma and germinates. The tube grows, passing between the cells of the stigma and the style to reach the egg inside an ovule, where fertilization occurs. 293

pollination In gymnosperms, the transfer of pollen from pollen cone to seed cone; in angiosperms, the transfer of pollen from anther to stigma. 290, 365

pollution Any environmental change that adversely affects the lives and health of living things. 576

polygenic inheritance Pattern of inheritance in which a trait is controlled by several allelic pairs; each dominant allele contributes to the phenotype in an additive and like manner. 150

polymer Macromolecule consisting of covalently bonded monomers; for example, a polypeptide is a polymer of monomers called amino acids. 32

polymerase chain reaction (PCR) Technique that uses the enzyme DNA polymerase to produce millions of copies of a particular piece of DNA. 174

polymorphism Genes that have more than one wild-type allele. 241

polyp Small, abnormal growth that arises from the epithelial lining. 420

polypeptide Polymer of many amino acids linked by peptide bonds. 40

polyribosome String of ribosomes simultaneously translating regions of the same mRNA strand during protein synthesis. 57, 169

polysaccharide Polymer made from sugar monomers; the polysaccharides starch and glycogen are polymers of glucose monomers. 34

population Group of organisms of the same-species occupying a certain area and sharing a common gene pool. 5, 232, 546

population density The number of individuals per unit area or volume living in a particular habitat. 538

portal system Pathway of blood flow that begins and ends in capillaries, such as the portal system located between the small intestine and liver. 403

posterior pituitary Portion of the pituitary gland that stores and secretes oxytocin and antidiuretic hormone produced by the hypothalamus. 485

postzygotic isolating mechanism Anatomical or physiological difference between two species that prevents successful reproduction after mating has taken place. 249

potential energy Stored energy as a result of location or spatial arrangement. 70

predation Interaction in which one organism (the *predator*) uses another (the *prey*) as a food source. 543, 556

prefrontal area Association area in the frontal lobe that receives information from other association areas and uses it to reason and plan actions. 479

preparatory (prep) reaction Reaction that oxidizes pyruvate with the release of carbon dioxide; results in acetyl CoA and connects glycolysis to the citric acid cycle. 99, 102

pressure-flow model Explanation for phloem transport; osmotic pressure following active transport of sugar into phloem brings a flow of sap from a source to a sink. 350

prezygotic isolating mechanism Anatomical or behavioral difference between two species that prevents the possibility of mating. 248

primary motor area Area in the frontal lobe where voluntary commands begin; each section controls a part of the body. 478

primary productivity Amount of biomass produced primarily by photosynthesizers. 570

primary sensory area Area located in the parietal lobe where sensory information arrives from the skin and skeletal muscles. 478

primary succession Stage in ecological succession, which involves the creation of new soil. 554

principle Theory that is generally accepted by an overwhelming number of scientists; also called a law. 9

prion Infectious particle consisting of protein only and no nucleic acid. 270

producer Photosynthetic organism at the start of a grazing food chain that makes its own food—e.g., green plants on land and algae in water. 560

product Substance that forms as a result of a reaction. 21

progesterone Female sex hormone that helps maintain sexual organs and secondary sex characteristics. 518

prokaryote Organism that lacks the membrane-bounded nucleus and membranous organelles typical of eukaryotes. 270

prokaryotic cell Lacking a membrane-bounded nucleus and organelles; the cell type within the domains Bacteria and Archaea. 50

prolactin (PRL) Hormone secreted by the anterior pituitary that stimulates the production of milk from the mammary glands. 484

promoter In an operon, a sequence of DNA where RNA polymerase binds prior to transcription. 167

prophase Mitotic phase during which chromatin condenses so that chromosomes appear; chromosomes are scattered. 115

proprioceptor Sensory receptor that assists the brain in knowing the position of the limbs. 501

prosimian Group of primates that includes lemurs and tarsiers, and may resemble the first primates to have evolved. 328

prostate gland Gland located around the male urethra below the urinary bladder; adds secretions to semen. 515

protein Molecule consisting of one or more polypeptides; a macronutrient in the diet that is digested to amino acids used by cells to synthesize cellular proteins. 38, 440

proteome Collection of proteins resulting from the translation of genes into proteins. 209

proteomics Study of the structure, function, and interaction of proteins. 209

protist Member of the kingdom Protista. 6, 276

proton Positive subatomic particle located in the nucleus and assigned one atomic mass unit. 17

proto-oncogene Normal gene that can become an oncogene through mutation. 189

protostome Group of coelomate animals in which the first embryonic opening (the blastopore) is associated with the mouth. 313

protozoan Heterotrophic, unicellular protist that moves by flagella, cilia, or pseudopodia, or is immobile. 276

proximal tubule Portion of a nephron following the nephron capsule where reabsorption of filtrate occurs. 428

pseudocoelom Body cavity lying between the digestive tract and body wall that is incompletely lined by mesoderm. 309

pseudopod Cytoplasmic extension of amoeboid protists; used for locomotion and engulfing food. 278

puberty Period of life when secondary sex changes occur in humans; marked by the onset of menses in females and sperm production in males. 516

pulmonary artery Blood vessel that takes blood away from the heart to the lungs. 399

pulmonary circuit Circulatory pathway between the lungs and the heart. 398

pulmonary trunk Large blood vessel that divides into the pulmonary arteries; takes blood away from the heart to the lungs. 399

pulmonary vein Blood vessel that takes blood to the heart from the lungs. 399

pulse Vibration felt in arterial walls due to expansion of the aorta following ventricle contraction. 400

Punnett square Grid used to calculate the expected results of simple genetic crosses. 144

pupil Opening in the center of the iris of the vertebrate eye. 499

R

radial symmetry Body plan in which similar parts are arranged around a central axis, like spokes of a wheel. 308

radiolarian Protozoan; marine amoeba having a calcium carbonate skeleton, many of which make limestone formations such as the White Cliffs of Dover. 278

random Spatial distribution of individuals in a population in which individuals have an equal chance of living anywhere within an area. 538

range That portion of the globe where a certain species can be found. 538

ray-finned fishes Group of bony fishes with fins supported by parallel bony rays connected by webs of thin tissue. 322

reabsorption Movement of primarily nutrient molecules and water from the contents of the nephron into blood at the proximal tubule. 428

reactant Substance that participates in a reaction. 21

receptacle Area where a flower attaches to a floral stalk. 362

receptor-mediated endocytosis Selective uptake of molecules into a cell by vacuole formation after they bind to specific receptor proteins in the plasma membrane. 79

recessive allele Allele that exerts its phenotypic effect only in the homozygote; its expression is masked by a dominant allele. 145

recombinant DNA (rDNA) DNA that contains genes from more than one source. 173

recombinant gamete New combination of alleles incorporated into a gamete during crossing-over. 154

red algae Marine photosynthetic protist with a notable abundance of phycobilin pigments; includes coralline algae of coral reefs. 277

red blood cell Erythrocyte; contains hemoglobin and carries oxygen from the lungs or gills to the tissues in vertebrates. 383, 405

red bone marrow Vascularized, modified connective tissue that is sometimes found in the cavities of spongy bone; site of blood cell formation. 457, 504

reflex Automatic, involuntary response of an organism to a stimulus. 481

regeneration Ability of sponges and echinoderms to "generate" into another organism if separated into pieces. 512

regulatory gene In an operon, a gene that codes for a protein that regulates the expression of other genes. 183

renal cortex Outer portion of the kidney that appears granular. 427

renal medulla Inner portion of the kidney that consists of renal pyramids. 427

renal pelvis Hollow chamber in the kidney that lies inside the renal medulla and receives freshly prepared urine from the collecting ducts. 427

renewable resources Resources normally replaced or replenished by natural processes; resources not depleted by moderate use. Examples include solar energy, biological resources, such as forests and fisheries, biological organisms, and some biogeo-chemical cycles. 576

repressor In an operon, protein molecule that binds to an operator, preventing transcription of structural genes. 183

reproduce To produce a new individual of the same kind. 3

reproductive cloning Genetically identical to the original individual. 180

reproductive system Organ system that contains male or female organs and specializes in the production of offspring. 387

reptile Member of a class of terrestrial invertebrates with internal fertilization, scaly skin, and an egg with a leathery shell; includes snakes, lizards, turtles, and crocodiles. 324

resource In economic terms, anything with potential use in creating wealth or giving satisfaction. 538

resource partitioning Mechanism that increases the number of niches by apportioning the supply of a resource such as food or living space between species. 557

respiration Sequence of events that results in gas exchange between the cells of the body and the environment. 422

respiratory system Organ system consisting of the lungs and tubes that bring oxygen into the lungs and take carbon dioxide out. 386

restriction enzyme Bacterial enzyme that stops viral reproduction by cleaving viral DNA; used to cut DNA at specific points during production of recombinant DNA. 173

retina Innermost layer of the vertebrate eyeball containing the photoreceptors—rods and cones. 499

retrovirus RNA virus containing the enzyme reverse transcriptase that carries out RNA/DNA transcription. 268

rhodopsin Light-absorbing molecule in rods and cones that contains a pigment and the protein opsin. 500

ribose Pentose sugar found in RNA. 33

ribosomal RNA (rRNA) Type of RNA found in ribosomes that translate messenger RNAs to produce proteins. 165

ribosome RNA and protein in two subunits; site of protein synthesis in the cytoplasm. 51, 56, 165

RNA (ribonucleic acid) Nucleic acid produced from covalent bonding of nucleotide monomers that contain the sugar ribose; occurs in three forms: messenger RNA, ribosomal RNA, and transfer RNA. 41, 164

RNA polymerase During transcription, an enzyme that joins nucleotides complementary to a DNA template. 167

rod Photoreceptor in vertebrate eyes that responds to dim light. 500

root cap Protective cover of the root tip, whose cells are constantly replaced as they are ground off when the root pushes through rough soil particles. 347

root hair Extension of a root epidermal cell that collectively increases the surface area for the absorption of water and minerals. 339

root nodule Structure on plant root that contains nitrogen-fixing bacteria. 348

root system Includes the main root and any and all of its lateral (side) branches. 338

rotational equilibrium Maintenance of balance when the head and body are suddenly moved or rotated. 498

rough ER Membranous system of tubules, vesicles, and sacs in cells; has attached ribosomes. 58

roundworm Member of the phylum Nematoda with a cylindrical body that has a complete digestive tract and a pseudocoelom; some forms are free-living in water and soil, and many are parasitic. 312

RuBP carboxylase (rubisco) Enzyme that is required for carbon dioxide fixation (atmospheric CO_2 attaches to RuBP) in the Calvin cycle. 90

rule of multiplication The chance of two (or more) independent events occurring together is the product of their chance occurring separately. 147

S

saccule Saclike cavity in the vestibule of the vertebrate inner ear; contains sensory receptors for gravitational equilibrium. 498

salinization Process in which mineral salts accumulate in the soil, killing plants; occurs when soils in dry climates are irrigated profusely. 581

salivary amylase In humans, enzyme in saliva that digests starch to maltose. 416

salivary gland In humans, gland associated with the mouth that secretes saliva. 416

salt Compound produced by a reaction between an acid and a base. 20

saltatory conduction Movement of nerve impulses from one neurofibral node to another along a myelinated axon. 475

saltwater intrusion Movement of saltwater into freshwater aquifers in coastal areas where groundwater is withdrawn faster than it is replenished. 579

SA (sinoatrial) node Small region of neuromuscular tissue that initiates the heartbeat; also called the pacemaker. 400

saprotroph Organism that secretes digestive enzymes and absorbs the resulting nutrients back across the plasma membrane. 272, 298

sarcomere One of many units, arranged linearly within a myofibril, whose contraction produces muscle contraction. 505

saturated-fatty acid Fatty acid molecule that lacks double bonds between the carbons of its hydrocarbon chain. The chain bears the maximum number of hydrogens possible. 36

scientific theory Concept supported by a broad range of observations, experiments, and data. 9

sclerenchyma Plant tissue composed of cells with heavily lignified cell walls; functions in support. 341

secondary sex characteristic Trait that is sometimes helpful but not absolutely necessary for reproduction and is maintained by the sex hormones in males and females. 516

secondary succession In ecological succession, the stage which is the gradual replacement of communities in an area following a disturbance. 554

secretion In the cell, release of a substance by exocytosis from a cell that may be a gland or part of a gland; in the urinary system, movement of certain molecules from blood into the distal tubule of a nephron so that they are added to urine. 58, 428

seed Mature ovule that contains an embryo, with stored food enclosed in a protective coat. 285, 366

segmentation Repetition of body units as seen in the earthworm. 309

semen (seminal fluid) Thick, whitish fluid consisting of sperm and secretions from several glands of the male reproductive tract. 515

semiconservative Duplication of DNA resulting in two double helix molecules, each having one parental and one new strand. 164

semilunar valve Valve resembling a half moon located between the ventricles and their attached vessels. 399

seminal vesicle Convoluted, saclike structure attached to the vas deferens near the base of the urinary bladder in males; adds secretions to semen. 515

seminiferous tubule Long, coiled structure contained within chambers of the testis where sperm are produced. 515

senescence Sum of the processes involving aging, decline, and eventual death of a plant or plant part. 358

sensory neuron Nerve cell that transmits nerve impulses to the central nervous system after a sensory receptor has been stimulated. 474

sepal Outermost leaflike covering of the flower; usually green in color. 292, 363

septum Partition or wall that divides two areas; the septum in the heart separates the right half from the left half. 399

sex chromosome Chromosome that determines the sex of an individual; in humans, females have two X chromosomes, and males have an X and a Y chromosome. 129

sex pili In a bacterium, elongated, hollow appendage used to transfer DNA to other cells. 51

sexual selection Changes in males and females, often due to male competition and female selectivity, leading to increased fitness. 236

shoot system Aboveground portion of a plant consisting of the stem, leaves, and flowers. 338

short-day plant Plant that flowers when day length is shorter than a critical length—e.g., cocklebur, poinsettia, and chrysanthemum. 360

shoulder (pectoral) girdle Portion of the vertebrate skeleton that provides support and attachment for the upper (fore) limbs; consists of the scapula and clavicle on each side of the body. 503

sickle cell disease Hereditary disease in which red blood cells are narrow and curved so that they are unable to pass through capillaries and are destroyed; causes chronic anemia. 203

sieve-tube member Member that joins with others in the phloem tissue of plants as a means of transport for nutrient sap. 342

signal Molecule that stimulates or inhibits an event in the cell cycle. 119

signal transduction pathway Activation and inhibition of intracellular targets after binding of growth factors. 187

simple diffusion Movement of molecules or ions from a region of higher to lower concentration; it requires no energy and tends to lead to an equal distribution. 77

simple goiter Condition in which an enlarged thyroid produces low levels of thyroxine. 486

sinus Cavity or hollow space in an organ such as the skull. 422

sister chromatid One of two genetically identical chromosomal units that are the result of DNA replication and are attached to each other at the centromere. 113

skeletal muscle Striated, voluntary muscle tissue that comprises skeletal muscles; also called striated muscle. 384

skeletal system System of bones, cartilage, and ligaments that works with the muscular system to protect the body and provide support for locomotion and movement. 387

sliding filament model An explanation for muscle contraction based on the movement of actin filaments in relation to myosin filaments. 505

slime mold Protists that decompose dead material and feed on bacteria by phagocytosis; vegetative state is mobile and amoeboid. 279

small intestine In vertebrates, the portion of the digestive tract that precedes the large intestine; in humans, consists of the duodenum, jejunum, and ileum. 418

smooth ER Membranous system of tubules, vesicles, and sacs in eukaryotic cells; lacks attached ribosomes. 58

smooth (visceral) muscle Nonstriated, involuntary muscles found in the walls of internal organs. 384

sodium-potassium pump Carrier protein in the plasma membrane that moves sodium ions out of and potassium ions into animal cells; important in nerve and muscle cells. 79

solute Substance that is dissolved in a solvent, forming a solution. 77

solution Fluid (the solvent) that contains a dissolved solid (the solute). 77

solvent Liquid portion of a solution that serves to dissolve a solute. 77

somatic cell Body cell; excludes cells that undergo meiosis and become sperm or egg. 120

somatic system Portion of the peripheral nervous system containing motor neurons that control skeletal muscles. 481

speciation Origin of new species due to the evolutionary process of descent with modification. 220, 246

species Group of similarly constructed organisms capable of interbreeding and producing fertile offspring; organisms that share a common gene pool; the taxon at the lowest level of classification. 4, 6, 256

species richness List of difference species in a community. 553

specific epithet In the binomial system of taxonomy, the second part of an organism's name; it may be descriptive. 256

spermatogenesis Production of sperm in males by the process of meiosis and maturation. 129

spinal cord In vertebrates, the nerve cord that is continuous with the base of the brain and housed within the vertebral column. 478

spinal nerves Nerves that arises from the spinal cord. 481

spindle Microtubule structure that brings about chromosomal movement during nuclear division. 115

spindle equator Disk formed during metaphase in which all of a cell's chromosomes lie in a single plane at right angles to the spindle fibers. 115

spiral organ Structure in the vertebrate inner ear that contains auditory receptors (also called organ of Corti). 496

spleen Large, glandular organ located in the upper left region of the abdomen; stores and purifies blood. 457

sponge Invertebrate animal of the phylum Porifera; pore-bearing filter feeder whose inner body wall is lined by collar cells. 310

spongy bone Type of bone that has an irregular, meshlike arrangement of thin plates of bone. 504

spongy mesophyll Layer of tissue in a plant leaf containing loosely packed cells, increasing the amount of surface area for gas exchange. 342

spore Asexual reproductive or resting cell capable of developing into a new organism without fusion with another cell, in contrast to a gamete. 286

sporophyte Diploid generation of the alternation of generations life cycle of a plant; produces haploid spores that develop into the haploid generation. 286, 362

sporozoan Spore-forming protist that has no means of locomotion and is typically a parasite with a complex life cycle having both sexual and asexual phases. 278

stabilizing selection Outcome of natural selection in which extreme phenotypes are eliminated and the average phenotype is conserved. 240

stamen In flowering plants, the portion of the flower that consists of a filament and an anther containing pollen sacs where pollen is produced. 292, 363

starch Storage polysaccharide found in plants that is composed of glucose molecules joined in a linear fashion with few side chains. 34

statocyst Gravitational equilibrium organ found in some invertebrates; give information about head position. 498

stem Usually the upright, vertical portion of a plant that transports substances to and from the leaves. 339

steroid Type of lipid molecule having a complex of four carbon rings—e.g., cholesterol, estrogen, progesterone, and testosterone. 37

steroid hormone Type of hormone that has the same complex of four carbon rings, but each one has different side chains. 483

stigma In flowering plants, portion of the carpel where pollen grains adhere and germinate before fertilization can occur. 292, 363

stoma (pl., stomata) Small opening between two guard cells on the underside of leaf epidermis through which gases pass. 85, 341

stomach In vertebrates, muscular sac that mixes food with gastric juices to form chyme, which enters the small intestine. 417

striated Having bands; in cardiac and skeletal muscle, alternating light and dark bands produced by the distribution of contractile proteins. 384

stroke Condition resulting when an arteriole in the brain bursts or becomes blocked by an embolism; cerebrovascular accident. 404

stroma Fluid within a chloroplast that contains enzymes involved in the synthesis of carbohydrates during photosynthesis. 60, 85

style Elongated, central portion of the carpel between the ovary and stigma. 292, 363

subsidence Occurs when a portion of the Earth's surface gradually settles downward. 579

substrate Reactant in a reaction controlled by an enzyme. 75

surface-area-to-volume ratio Ratio of a cell's outside area to its internal volume. 49

survivorship Probability of newborn individuals of a cohort surviving to particular ages. 539

sustainable society Ability of a society or ecosystem to maintain itself while also providing services to human beings. 592

symbiotic Relationship that occurs when two different species live together in a unique way; it may be beneficial, neutral, or detrimental to one and/or the other species. 273

sympathetic division Division of the autonomic system that is active when an organism is under stress; uses norepinephrine as a neurotransmitter. 482

sympatric speciation Origin of new species in populations that overlap geographically. 250

synapse Junction between neurons consisting of the axon membrane, the synaptic cleft, and a dendrite. 476

synapsis Pairing of homologous chromosomes during meiosis I. 130

synaptic cleft Small gap between membranes of a synapse. 476

synovial joint Freely moving joint in which two bones are separated by a cavity. 506

systematics Study of the diversity of organisms to classify them and determine their evolutionary relationships. 256

systemic circuit Circulatory pathway of blood flow between the tissues and the heart. 398, 402

systole Contraction period of the heart during the cardiac cycle. 400

T

taste buds Structures in the vertebrate mouth containing sensory receptors for taste; in humans, most taste buds are on the tongue. 495

taxon (pl., taxa) Group of organisms that fills a particular classification category. 256

taxonomy Branch of biology concerned with identifying, describing, and naming organisms. 6, 256

Tay-Sachs disease Lethal genetic disease in which the newborn has a faulty lysosomal digestive enzyme. 202

T-cell receptor (TCR) Receptor on the T cell surface consisting of two antigen-binding peptide chains; it is associated with a large number of other glycoproteins. Binding of antigen to the TCR, usually in association with MHC, activates the T cell. 462

telomere Long, repeating DNA sequence at the ends of chromosomes; function like a cap and chromosomes from fusing with each other. 120, 190

telophase Mitotic phase during which daughter cells are located at each pole. 115

template Parental strand of DNA that serves as a guide for the complementary daughter strand produced during DNA replication. 164

tendon Strap of fibrous connective tissue that connects skeletal muscle to bone. 382

terminal bud Bud that develops at the apex of a shoot. 338

terrestrial ecosystem Also called biomes, these are the tundra, the taiga, temperate forests, tropical grasslands (savanna), temperate grasslands (prairie), deserts, and tropical rain forests. 568

testcross Cross between an individual with the dominant phenotype and an individual with the recessive phenotype. The resulting phenotypic ratio indicates whether the dominant phenotype is homozygous or heterozygous. 144

testis (pl., testes) Male gonad that produces sperm and the male sex hormones. 512, 514

testosterone Male sex hormone that helps maintain sexual organs and secondary sex characteristics. 516

tetany Severe twitching caused by involuntary contraction of the skeletal muscles due to a calcium imbalance. 486

tetrad Homologous chromosomes, each having sister chromatids that are joined during meiosis; also called bivalent. 130

thalamus In vertebrates, the portion of the diencephalon that passes on selected sensory information to the cerebrum. 479

therapeutic cloning Used to create mature cells of various cell types. Also, used to learn about specialization of cells and provide cells and tissue to treat human illnesses. 181

thermoacidophile Type of archaea that lives in hot, acidic, aquatic habitats, such as hot springs or near hydrothermal vents. 275

thoracic cavity Located in the ventral cavity above the abdominal cavity and is separated from the abdominal cavity by the diaphragm; contains the heart and the lungs. 416

three-domain system System of classification that recognizes three domains: Bacteria, Archaea, and Eukarya. 261

thrombin Enzyme that converts fibrinogen to fibrin threads during blood clotting. 407

thylakoid Flattened sac within a granum whose membrane contains chlorophyll and where the light reactions of photosynthesis occur. 60, 85

thymine (T) One of four nitrogen-containing bases in nucleotides composing the structure of DNA; pairs with adenine. 161

thymus gland Lymphatic organ involved in the development and functioning of the immune system; T cells mature in the thymus gland. 457

thyroid gland Large gland in the neck that produces several important hormones, including thyroxine, triiodothyronine, and calcitonin. 486

thyroid-stimulating hormone (TSH) Substance produced by the anterior pituitary that causes the thyroid to secrete thyroxine and triiodothyronine. 484

thyroxine (T_4) Hormone secreted from the thyroid gland that promotes growth and development; in general, it increases the metabolic rate in cells. 486

tight junction Junction between cells when adjacent plasma membrane proteins join to form an impermeable barrier. 65

tissue Group of similar cells combined to perform a common function. 2, 378

tissue culture Process of growing tissue artificially, usually in a liquid medium in laboratory glassware. 369

tissue fluid Fluid that surrounds the body's cells; consists of dissolved substances that leave the blood capillaries by filtration and diffusion. 386

T lymphocyte (T cell) Lymphocyte that matures in the thymus and exists in four varieties, one of which kills antigen-bearing cells outright. 457

tonsils Partially encapsulated lymph nodules located in the pharynx. 456

totipotent Cell that has the full genetic potential of the organism, including the potential to develop into a complete organism. 369

tracer Substance having an attached radioactive isotope that allows a researcher to track its whereabouts in a biological system. 18

trachea (pl., tracheae) In tetrapod vertebrates, air tube (windpipe) that runs between the larynx and the bronchi. 423

tracheid In vascular plants, type of cell in xylem that has tapered ends and pits through which water and minerals flow. 342

tract Bundles of myelinated axons in the central nervous system. 478

traditional systematics School of systematics that takes into consideration the degree of difference between derived characters to construct phylogenetic trees. 256

transcription Process whereby a DNA strand serves as a template for the formation of mRNA. 166

transcription activator Protein that speeds transcription. 186

transcription factor In eukaryotes, protein required for the initiation of transcription by RNA polymerase. 186

transduction Exchange of DNA between bacteria by means of a bacteriophage. 272

transfer rate Amount of a substance that moves from one component of the environment to another within a specified period of time. 565

transfer RNA (tRNA) Type of RNA that transfers a particular amino acid to a ribosome during protein synthesis; at one end, it binds to the amino acid, and at the other end it has an anticodon that binds to an mRNA codon. 165

transformation Taking up of extraneous genetic material from the environment by bacteria. 272

transgenic organism Free-living organism in the environment that has had a foreign gene inserted into it. 173

transgenic plant Plant that carries the genes of another organism as a result of DNA technology; also genetically modified plant. 371

translation Process whereby ribosomes use the sequence of codons in mRNA to produce a polypeptide with a particular sequence of amino acids. 166

translocation Movement of a chromosomal segment from one chromosome to another nonhomologous chromosome, leading to abnormalities—e.g., Down syndrome. 170, 199

transpiration Plant's loss of water to the atmosphere, mainly through evaporation at leaf stomata. 349

transposon DNA sequence capable of randomly moving from one site to another in the genome. 172

trichomoniasis Sexually transmitted disease caused by the parasitic protozoan *Trichomonas vaginalis*. 523

triglyceride Neutral fat composed of glycerol and three fatty acids. 439

triplet code During gene expression, each sequence of three nucleotide bases stands for a particular amino acid. 167

trophic level Feeding level of one or more populations in a food web. 563

tropism In plants, a growth response toward or away from a directional stimulus. 360

trypanosome Parasitic zooflagellate that causes severe disease in human beings and domestic animals, including a condition called sleeping sickness. 278

trypsin Protein-digesting enzyme secreted by the pancreas. 418

tumor Cells derived from a single mutated cell that has repeatedly undergone cell division; benign tumors remain at the site of origin, while malignant tumors metastasize. 121

tumor suppressor gene Gene that codes for a protein that ordinarily suppresses cell division; inactivity can lead to a tumor. 189

tunicate Type of primitive invertebrate chordate. 320

Turner syndrome Condition caused by the inheritance of a single X chromosome. 137

U

umbilical cord Cord connecting the fetus to the placenta through which blood vessels pass. 527

unicellular Made up of but a single cell, as in the bacteria. 2

uniform Spatial distribution of individuals in a population in which individuals are dispersed uniformly through the area. 538

uniformitarianism Belief espoused by James Hutton that geological forces act at a continuous, uniform rate. 219

unsaturated-fatty acid Fatty acid molecule that has one or more double bonds between the carbons of its hydrocarbon chain. The chain bears fewer hydrogens than the maximum number possible. 36

uracil (U) Pyrimidine base that replaces thymine in RNA; pairs with adenine. 164

ureter Tubular structure conducting urine from the kidney to the urinary bladder. 427

urethra Tubular structure that receives urine from the bladder and carries it to the outside of the body. 427

urinary bladder Organ where urine is stored. 427

urinary system Organ system consisting of the kidneys and urinary bladder; rids the body of nitrogenous wastes and helps regulate the water-salt balance of the blood. 386

urine Liquid waste product made by the nephrons of the vertebrate kidney through the processes of filtration, reabsorption, and secretion. 427

uterus In mammals, expanded portion of the female reproductive tract through which eggs pass to the environment or in which an embryo develops and is nourished before birth. 516

utricle Saclike cavity in the vestibule of the vertebrate inner ear; contains sensory receptors for gravitational equilibrium. 498

V

vaccine Antigens prepared in such a way that they can promote active immunity without causing disease. 464

vacuole Membrane-bounded sac, larger than a vesicle; usually functions in storage and can contain a variety of substances. In plants, the central vacuole fills much of the interior of the cell. 59

valence shell Outer shell of an atom. 19

vascular bundle In plants, primary phloem and primary xylem enclosed by a bundle sheath. 340

vascular cambium In plants, lateral meristem that produces secondary phloem and secondary xylem. 341

vascular plant Plant that has vascular tissue (xylem and phloem); includes seedless vascular plants (e.g., ferns) and seed plants (gymnosperms and angiosperms). 288

vascular tissue Transport tissue in plants, consisting of xylem and phloem. 285, 341

vas deferens Tube that leads from the epididymis to the urethra in males. 514

vector Piece of DNA that can have a foreign DNA attached to it; a common vector is a plasmid. 173

vein Blood vessel that arises from venules and transports blood toward the heart. 399, 401

vena cava (pl., venae cavae) Large systemic vein that returns blood to the right atrium of the heart in tetrapods; either the superior or inferior vena cava. 399, 402

ventricle Cavity in an organ, such as a lower chamber of the heart or the ventricles of the brain. 399

venule Vessel that takes blood from capillaries to a vein. 399

vertebral column Portion of the vertebrate endoskeleton that houses the spinal cord; consists of many vertebrae separated by intervertebral disks. 503

vertebrate Chordate in which the notochord is replaced by a vertebral column. 320

vesicle Small, membrane-bounded sac that stores substances within a cell. 58

vessel element Cell that joins with others to form a major conducting tube found in xylem. 342

vestigial structure Remains of a structure that was functional in some ancestor but is no longer functional in the organism in question. 226

villus (pl., villi) Small, fingerlike projection of the inner small intestinal wall. 418

viroid Infectious strand of RNA devoid of a capsid and much smaller than a virus. 270

virus Noncellular parasitic agent consisting of an outer capsid and an inner core of nucleic acid. 266

vitamin Essential requirement in the diet, needed in small amounts. Vitamins are often part of coenzymes. 442

vocal cord In humans, folds of tissue within the larynx; create vocal sounds when they vibrate. 423

W

water mold Filamentous organisms having cell walls made of cellulose; typically decomposers of dead freshwater organisms, but some are parasites of aquatic or terrestrial organisms. 279

white blood cell Leukocyte, of which there are several types, each having a specific function in protecting the body from invasion by foreign substances and organisms. 383, 406

wild-type Allele or phenotype that is the most common for a certain gene in a population. 146

wood Secondary xylem that builds up year after year in woody plants and becomes the annual rings. 345

X

X-linked Allele that is located on an X chromosome but may control a trait that has nothing to do with the sex characteristics of an animal. 152

xylem Vascular tissue that transports water and mineral solutes upward through the plant body; it contains vessel elements and tracheids. 288, 340

Z

zooflagellate Nonphotosynthetic protist that moves by flagella; typically zooflagellates enter into symbiotic relationships, and some are parasitic. 278

zygote Diploid cell formed by the union of two gametes; the product of fertilization. 129, 512

Credits

Photographs

Chapter 1
Opener (Soil): © Barry Runk/Stan/Grant Heilman; (Trees): © Barbara von Hoffman/Tom Stack & Associates; (Butterfly): © Kjell Sandved/Butterfly Alphabet; (Toucan): © Ed Reschke/Peter Arnold; (Frog): © Kevin Schaefer and Martha Hill/Tom Stack & Associates; (Orchid): © Max and Bea Hunn/Visuals Unlimited; (Jaguar): © BIOS (Seitre)/Peter Arnold; (Bacteria): © David Scharf/SPL/Photo Researchers, Inc.; 1.2a: © agefotostock/SuperStock; 1.2b: © Ed Young/Corbis; 1.3 (Top): © Vol. 113/PhotoDisc/Getty Images; 1.3 (Bottom): © David Mack/SPL/Photo Researchers, Inc.; 1.5 (background): © Ralph Robinson/Visuals Unlimited; 1.5 (inset): © M. Rhode/GBF/SPL/Photo Researchers; 1.6 (background): © A.B. Dowsett/SPL/Photo Researchers; 1.6 (inset): © Biophoto Assoc/Photo Researchers; 1.7: © John D. Cunningham/Visuals Unlimited; 1.8: © Rob Planck/Tom Stack & Associates; 1.9: © Farell Grehan/Photo Researchers; 1.10: © Leonard L. Rue; 1.11a: © Kathy Sloane/Photo Researchers, Inc.; 1.11b: © Edgar Bernstein/Peter Arnold, Inc.; 1.11c: © Luiz C. Marigo/Peter Arnold, Inc.; 1.13a–c: From Bidlack, J.E., S.C. Rao, and D.H. Demezas 2001. Nodulation, nitrogenase activity, and dry weight of chickpea and pigeon pea cultivars using different Bradyrhizobium strains. *Journal of Plant Nutrition:* 24:549–560.

Chapter 2
Opener (Girders): © Royalty-Free/Corbis; (Blood): © PhotoLink/Getty Images; (Ocean): © Evelyn Jo Hebert/Pinkletink Photo; (Sunbathing): © Adam smith/CORBIS; (Melanoma): © James Stevenson/SPL/Photo Researchers; (Radiographers): © SPL/Photo Researchers, Inc.; 2.1: © Corbis Royalty-Free; 2.4a: © Biomed Commun./Custom Medical Stock Photo; 2.4b (Top): © Hank Morgan/Rainbow; 2.4b (Bottom): © Mazzlota etal./Photo Researchers; 2.5a: © Tony Freeman/Photo Edit; 2.5b: © Geoff Tompkinson/SPL/Photo Researchers; 2.7b: © Charles M. Falco/Photo Researchers; 2.9: © Fritz Polking/Peter Arnold, Inc.; 2.10: © Corbis Royalty-Free; 2.11: © Claude Nuridsany & Marie Perennou/Photo Researchers, Inc.; 2.12a: © Martin Dohrn/SPL/Photo Researchers; 2.12b: © Amanda Langford/SuperStock RF; 2.18a: © Frederica Georgia/Photo Researchers; 2.18b: © Ray Pfortner/Peter Arnold.

Chapter 3
Opener (Omlet): © Dennis Gray/Getty Images; (FiberCon): © The McGraw Hill Companies, Inc./John Thoeming, photographer; (Woman): © Dennis Gray/Cole Group/Getty Images; (Eggs): © Dennis Gray/Cole Group/Getty Images; 3.1a: © Lary Lefever/Grant Heilman Photo; 3.1b: © Leonard Lee Rue/Photo Researchers; 3.1c: © H. Pol/CNRI SPL/Photo Researchers; 3.4, 3.5, 3.6: © Dwight Kuhn; 3.8: © Renee Lynn/Photo Researchers; 3.10a: © Jeremy Burgess/SPL/Photo Researchers; 3.10b: © Don W. Fawcett/Photo Researchers; 3.10c: © J.D. Litvay/Visuals Unlimited; 3.11: ©LIGHTWAVE PHOTOGRAPHY, INC./Animals Animals - Earth Scenes; 3.15b,c: © CORBIS RF; 3.16a (Top): © Reuters/Corbis; 3.16a (Web): © OSF/Animals Animals/Earth Scenes; 3.16b: © P. Motta & S. Correr/Photo Researchers, Inc.; 3.16c: © Duomo/Corbis; 3.21: © Eye of Science/Photo Researchers, Inc.

Chapter 4
Opener (Canoeing): © Warren Morgan/Corbis; (Osteocyte): © Biophoto Associates/Photo Researchers, Inc.; (Apoptosis): © David McCarthy/SPL/Photo Researchers, Inc.; (Neuron): © Ed Reschke; (Cone cell): © Lennart Nilsson, from "The Incredible Machine"; (Oocyte): © Ed Reschke/Peter Arnold, Inc.; (RBC): © Ed Reschke; 4.1 (LM): © Matt Meadows/SPL/Photo Researchers, Inc.; 4.1 (Simple squamous): 4.1 (TEM): © Inga Spence/Visuals Unlimited; 4.1 (Leaf cells): © Ray F. Evert, University of Wisconsin, Madison; 4.1 (Lymphocyte): Courtesy Dr. Joseph R. Goodman; 4.1b (Euglena): © T.E. Adams/Visuals Unlimited; 4.4: © Ralph A. Slepecky/Visuals Unlimited; 4.7a: © Alfred Pasieka/Photo Researchers, Inc.; 4.8a: © Newcomb/Wergin/BPS; 4.9: © Ron Milligan, Scripps Research Institute; 4.11: © R. Bolender and D. Fawcett/Visuals Unlimited; 4.13a: © Roland/Birke/Peter Arnold, Inc.; 4.13b: © Newcomb/Wergin/BPS; 4.14: Courtesy Herbert W. Israel, Cornell University; 4.15b: Courtesy Keith Porter; 4.18: Courtesy Kent McDonald; 4.19a (left): © Manfred Kage/Peter Arnold; 4.19a (right): © David M. Phillips/Photo Researchers; 4.19b (both): © William L. Dentler/BPS; 4.20: © E.H. Newcomb/BPS; 4.22a: From Douglas E. Kelly, Journal of Cell Biology, 28 (1966:51). Reproduced by permission of The Rockefeller University Press.

Chapter 5
Opener (Antibiotic): © Tom Pantages/ Phototake; (Staph): © Oliver Meckes/SPL/Photo Researchers, Inc.; (Athlete): © agefotostock/SuperStock; (Temperature): © Vol. 41/PhotoDisc/Getty Images; 5.1: © GoodShoot/SuperStock RF; 5.2 (both): © Spencer Grant/Photo Edit; 5.6 (top): © Comstock/Punch Stock; 5.6 (bottom): © Karl Weatherly/Getty Images; 5.7 (Potato, grain): © McGraw Hill Companies, Inc./John Thoeming, photographer; 5.7 (Apple): © Getty Images; 5.7 (Meat): © Foodcollection/Getty Images RF; 5.16 (both): Courtesy Mark Bretscher.

Chapter 6
Opener (Light): Courtesy Hydrofarm Horticultural Products; (Leaf): © Comstock/Punch Stock; (Diatom): © Carolina Biological Supply/Phototake; (Dinoflagellate):© SPL/Photo Researchers, Inc.; 6.1 (Tree): © Connie Coleman/Stone/Getty Images; 6.1 (Sunflower): © Grant Heilman Photography; 6.1 (Euglena): © T.E. Adams/Visuals Unlimited; 6.1 (Moss): © Bruce Iverson; 6.1 (Kelp): © Chuck Davis/Stone/Getty Images; 6.1 (Diatom): © Ed Reschke/Peter Arnold; 6.1 (Cyanobacteria): © Sherman Thomas/Visuals Unlimited; 6.2 (top): © Manfred Kage/Peter Arnold; 6.2 (bottom): Courtesy Herbert W. Israel, Cornell University; 6.5 (both): © Digital Vision RF/Getty; 6.10: © Jim Steinberg/Photo Researchers; 6.11: © Charlie Waite; 6.12: © Beverly Factor Photography; p. 95: © Eric V. Grave/Photo Researchers, Inc.

Chapter 7
Opener (Girls): © Alan Carey/The Image Works; (Exhaust): © Bernard Wittich/Visuals Unlimited; (Athlete): © Bill Robbins/Index Stock Imagery; 7.1: © Royalty-Free Corbis; 7.5: Courtesy Dr. Keith Porter; 7.10: © C Squared Studios/Getty Images; 7.11a: © Rubber Ball Productions/RF Index Stock Imagery, Inc.; 7.11b,c: © C Squared Studios/Getty Images.

Chapter 8
Opener (Patient): © Antonia Reeve/SPL/Photo Researchers, Inc.; (Cells): © VVG/SPL/Photo Researchers, Inc.; (MRI): © Simon Fraser/Royal Victoria Infirmary, Newcastle Upon Tyne/SPL/Photo Researchers, Inc.; 8.1a: © CORBIS RF; 8.1b (top): © McDonald Wildlife Photography/Animals Animals Earth Scenes; 8.1b (bottom): © Kevin Hanley/Animals Animals Earth Scenes; 8.1c: © Biophoto Associates/Photo Researchers, Inc.; 8.1d (both): Lennart Nilsson, "A Child is Born," 1990 Delacorte Press, pg. 111 (top frame); 8.2 (bottom): Courtesy O.L. Miller, Jr. and Steve L. McKnight; 8.2 (top): © Biophoto Associates/Photo Researchers; 8.5a–e: Photo Courtesy Dr. Andrew S. Bajer, University of Oregon; 8.6 (all): © Ed Reschke; 8.7 (both): Copyright by R.G. Kessel and C.Y. Shih, *Scanning Electron Microscopy in Biology: A Students' Atlas on Biological Organization,* Springer-Verlag, 1974.; 8.8: © B.A. Palevitz and E.H. Newcomb/BPS/Tom Stack & Associates; 8.11 (top): Courtesy Kalus Hahn, Cecelia Subauste, and Gary Bikoch from *Science,* Vol. 280, April 3, 1998 p. 32; 8.11 (bottom): Courtesy Douglas R. Green, LaJolla Institute for Allergy and Immunology, San Diego; 8.12b: © Breast Scanning Unit, Kings College Hospital, London/SPL/Photo Researchers, Inc.; 8.13: © Nancy Kedersha/Immunogen/SPL/Photo Researchers, Inc.; 8.14 (Chard): © Roy Morsch/Corbis; 8.14 (Cabbage): © C. McIntyre/PhotoLink/Getty Images; 8.14 (Berries): © M. Lamotte/Cole Group/Getty Images; 8.14 (Broccoli): © Royalty Free/Corbis; 8.14 (Oranges): © Steve Lupton/CORBIS.

Chapter 9

Opener (Child): © Gary Parker/SPL/Photo Researchers, Inc.; (Sperm, Egg): © Pascal Goetgheluck/SPL/Photo Researchers, Inc.; (Couple): © CORBIS/RF; 9.1: © CNRI/SPL/Photo Researchers, Inc.; 9.9 (top): © Jill Cannefax Photography; 9.9 (bottom): © CNRI/SPL/Photo Researchers, Inc.

Chapter 10

Opener (Hand): © Science Photo Library/Photo Researchers, Inc.; (Chromosomes): © Photo Researchers, Inc; (Couple): © Colin Paterson/SuperStock RF; (Smoking): © Royalty-Free/Corbis; 10.1: © Bettmann/CORBIS; 10.8 (Ears): © McGraw-Hill Higher Education, John Thoeming, photographer; 10.8 (Widow's peak): © SuperStock; 10.8 (Straight): © Michal Grecco/Stock Boston; 10.12: © Jane Burton/Bruce Coleman; 10.13 (left): © AP/Wide World Photos; 10.13 (right): © Ed Reschke; 10.14 (all): © Ryan McVay/Getty Images.

Chapter 11

Opener: (The Eppendorf®): Courtesy of Eppendorf North America. The Eppendorf® Thermal Cycler is an Authorized Thermal Cycler and may be used with PCR licenses available from Applied Biosystems. Its use with Authorized Reagents also provides a limited PCR license in accordance with the label rights accompanying such reagents; (Neurons): © Sercomi/Photo Researchers, Inc.; (Hyaline): © Innerspace Imaging/Science Photo Library/Photo Researchers, Inc.; (Pigs): Courtesy Norrie Russell, The Roslin Institute; 11.1: © Lee Simon/Photo Researchers, Inc.; 11.3a: © Photo Researchers, Inc.; 11.3b: © Science Source/Photo Researchers,Inc.; 11.4: © A. Barrington Brown/Photo Researchers, Inc.; 11.5a: Courtesy Earling Corporation; 11.14b: Courtesy Alexander Rich; 11.18: Courtesy of Cold Spring Harbor Laboratory; 11.22a: © Nita Winter; 11.22b: Courtesy Robert H. Devlin, Fisheries and Oceans Canada; 11.22c: © Richard Shade; 11.22d: Bio-Rad; 11.22e: © Jerry Mason/Photo Researchers, Inc.

Chapter 12

Opener: (Blood, Lymphoma, Cervical, Colon): © Steve Gschmeissner/SPL/Photo Researchers, Inc.; (Smear): © Science Photo Library/Photo Researchers, Inc.; (HPV): © James Cavallini/Photo Researchers, Inc.; 12.1.1: © Barry Runk/STAN/Grant Heilman; 12.1.2: © Grant Heilman; 12.1.3: © E. Webber/Visuals Unlimited; 12.3: © AP/Wide World Photos; 12.7: © Chanan Photo 2004; 12.12: Courtesy Prof. Walter Gehring; 12.18: © Pam Francis/Time Pix/Getty; 12.19: © Dr. M.A. Ansary/Science Photo Library/Photo Researchers, Inc.

Chapter 13

Opener (Scientist): Courtesy of the U.S. Department of Energy Joint Genome Institute. ©2005 The Regents of the University of California; (DNA): © Peter Menzel Photography; (IVF): © Pascal Goetgheluck/SPL/Photo Researchers, Inc.; 13.1: © Martha Cooper/Peter Arnold, Inc.; 13.2c: © CNRI/SPL/Photo Researchers, Inc; 13.3b: Courtesy The Williams Syndrome Association; 13.4b: Courtesy Kathy Wise; 13.5b: From N.B. Spinner, et al., *American Journal of Human Genetics,*

55 (1994): p. 239. The University of Chicago Press; 13.10 (top): Courtesy Daniel S. Friend; 13.10 (bottom): Courtesy Dr. Robert Terry, USCD-School of Medicine; 13.11: © Pat Pendarvis; 13.12: © Dr. Gopal Murti/SPL/Photo Researchers, Inc.; 13.13 (both): Courtesy Dr. Hemachandra Reddy; 13.14: © Mediscan/Medical-On-Line; 13.15 (both): Courtesy Dr. Rabi Tawil; 13.15 (middle): Courtesy Muscular Dystrophy; 13.17: Courtesy of Mergen, Ltd; 13.18: © Bernard Bennot/SPL/Photo Researchers, Inc.; 13.19, 13.20, 13.21a (both): © Royalty Free/Getty Images; 13.21b-2: © The McGraw-Hill Companies/Bob Coyle, photographer; 13.21c (both): © Royalty Free/Getty Images; 13.23: © Cindy Charles/Photo Edit; 13.25a: © AP Photo/Elizabeth Fulford, 1994; 13.25b: © AP Photo/Muscular Dystrophy Association/Jay LaPrete.

Chapter 14

Opener (Dogs): © Yoav Levy/Phototake; (Surgeons): © CORBIS Royalty Free; (Akiapolaau): © Jack Jeffrey Photography; (Finch): Courtesy Dr. Sheila Conant, Department of Zoology, University of Hawaii, Manoa, 14.1 (Ship): © Mary Evans Picture Library/The Image Works; 14.1 (Iguana): © Norbert Wu/Peter Arnold; 14.1 (Rhea): © Tom Stack & Associates; 14.1 (Darwin): © American Museum of Natural History, photo courtesy Geoffrey West. Neg. no 326697.; 14.2a: © Henry W. Robinson/Visuals Unlimited; 14.2b: © Shiela Terry/Science Photo Library/Photo Researchers, Inc.; 14.3b: © J & L Weber/Peter Arnold, Inc.; 14.4b: © PhotoDisc/Getty Images; 14.5: © Juan & Carmecita Munoz/Photo Researchers, Inc.; 14.6a: © Adrienne T. Gibson/Animals Animals/Earth Scenes; 14.6b: © Joe McDonald/Animals Animals/Earth Scenes; 14.6c: © Leonard L. Rue/Animals Animals/Earth Scenes; 14.7: Courtesy David M. Hillis, University of Texas, Austin; 14.10: © James Watt/Animals Animals/Earth Scenes; 14.9a,b: © Edward S. Ross; 14.10 (bottom): © Haroldo Palo, Jr./NHPA; 14.10 (top): © Michael & Patricia Fogden/Minden Pictures; 14.11: © Joe Tucciarone/Interstellar Illustrations; 14.13 (Sugar glider): © A.N.T. Photo Library/NHPA; 14.13 (Wombat): © Adrienne Gibson/Animals Animals/Earth Scenes; 14.13 (Koala): © Ann B. Kelser/SuperStock; 14.13 (Kangaroo): © George Holton/Photo Researchers, Inc.; 14.15: © CBS/Phototake.

Chapter 15

Opener (Zebras): © Peter Chadwick/SPL/Photo Researchers, Inc.; (Peacocks): © Ernest A. Janes/Bruce Coleman Inc.; (Douglas): © Reuters/Corbis; 15.1: © Francois Gohier/Photo Researchers, Inc.; 15.2: © Vol. 25/CORBIS; 15.3a: Hardy: National Library; 15.3b: Weinberg: National Library; 15.5a,b: © Michael Tweedie/Photo Researchers, Inc.; 15.6 (Bairdi): © Visuals Unlimited; 15.6 (Obsoleta): © William Weber/Visuals Unlimited; 15.6 (Quadrivittate): © Zig Leszczynski/Animals Animals Earth Scenes; 15.6 (Rossaleni): © Dale Jackson/Visuals Unlimited; 15.6 (Spiloides): © Joseph Collins/Photo Researchers, Inc.; 15.6 (Lindheimeri): © Zig Leszczynski/Animals Animals/Earth Scenes; 15.9: Courtesy Victor McKusick; 15.12: © Bob Evans/Peter Arnold; 15.13b: © Stanley Flegler.

Chapter 16

Opener (Green): © Zig Leszczynsky/Animals Animals Earth Scenes; (Rhinocerous): © Ken Lucas/Visuals Unlimited; (Marine): © Kelvin Aitken/Peter Arnold, Inc.; 16.1: The Field Museum of Natural History. Neg. CK9T, Chicago; 16.2 (Acadian): © Karl Maslowski/Visuals Unlimited; 16.2 (Least): © Stanley Maslowski/Visuals Unlimited; 16.2 (Traaill's): © Ralph Reinhold/Animals Animals/Earth Scenes; 16.3a: © Sylvia Mader; 16.3b: © B & C Alexander/Photo Researchers, Inc.; 16.5: © Barbara Gerlach/Visuals Unlimited; 16.6 (top left): © Superstock, Inc./SuperStock; 16.6 (top right): © Robert J. Erwin/Photo Researchers, Inc.; 16.6c (left): © Jorg & Petra Wegner/Animals Animals Earth Scenes; 16.10a: © Carolina Biological Supply/Phototake; 16.10b: © Alfred Pasieka/SPL/Photo Researchers; 16.10c: © Sinclair Stammers/SPL/Photo Researchers, Inc.; 16.10d: © Gary Retherford/Photo Researchers, Inc.; 16.10e: © Phil Degginger/Animals Animals/Earth Scenes; 16.10f: © Scott Berner/Visuals Unlimited.

Chapter 17

Opener (Orphan): © Steve Raymer/Corbis; (AIDS): © A. Ramey/Photo Edit; (X-ray): © AP Photo/Lori Waselchuk; (Leprosy): © Kelly Fajack/Peter Arnold, Inc.; (Masks): © Wolfgang Schmidt/Peter Arnold, Inc.; (Hands): © Aaron Haupt/Photo Researchers, Inc.; 17.2b: © Lee Simon/Photo Researchers; 17.4: © Jack Bostrack/Visuals Unlimited; 17.7 (inset): © Phanie/Photo Researchers, Inc.; 17.7 (background): © AP Photo/ho; 17.20a: © L.L. Sims/Visuals Unlimited; 17.20b: © Daniel V. Gotschall/Visuals Unlimited; 17.8a: © Alfred Pasieka/SPL/Photo Researchers, Inc.; 17.20c: © D.P. Wilson & Eric David Hosking/Photo Researchers, Inc.; 17.20d: © Biophoto Associates/Photo Researchers, Inc.; 17.8b: © David Scharf/SPL/Photo Researchers, Inc.;17.20e: © Pascal Goethgheluck/SPL/Photo Researchers, Inc.; 17.20f: © John D. Cunningham/Visuals Unlimited; 17.8c: © Science Source/Photo Researchers, Inc.; 17.20g: © M.I. Walker/Science Source/Photo Researchers, Inc.; 17.20h: © D.P. Wilson/Photo Researchers, Inc.; 17.10c: © Alfred Pasieka/SPL/Photo Researchers, Inc.; 17.20i: © Randy Morse/Animals Animals/Earth Scenes; 17.20j: © Andrew J. Martinez/Photo Researchers, Inc.; 17.10b: © CNRI/SPL/Photo Researchers, Inc.; 17.11a: © Eric Graves/Photo Researchers, Inc.; 17.11b: © Science VU/Visuals Unlimited; 17.12: Courtesy Nitragin Company, Inc.; 17.13a: © Vol. 29/PhotoDisc/Getty Images; 17.13b: Ken Graham/Accent Alaska; 17.13 (left): © Science VU/Visuals Unlimited; 17.14: © Paul Webster/Getty Images; 17.15a: © Susan Rosenthal/CORBIS; 17.15 (inset): © Ralph Robinson/Visuals Unlimited; 17.16a: © John Sohlden/Visuals Unlimited; 17.16b: From J.T. Staley, et al., "Bergey's Manual of Systematic Bacteriology," Vol. 13, © 1989 Williams and Wilkins Co., Baltimore. Prepared by A.L. Usted Photography by Dept. of Biophysics, Norwegian Institute of Technology; 17.17a: © Jeff Lepore/Photo Researchers, Inc.; 17.17b (inset): Courtesy Dennis W. Grogan, University of Cincinnati; 17.22a: © Eric Grave/SPL/Photo Researchers, Inc.; 17.22b: © Manfred Kage/Peter Arnold; 17.21: © Dr. David Patterson/SPL/Photo Researchers, Inc.; 17.22c: © Omikron/Photo Researchers, Inc.; 17.22d: © Ed Reschke/Peter

Arnold; 17.22e: © Dr. Fred Hossler/Visuals Unlimited; 17.23 (top): © Cabisco/Visuals Unlimited; 17.23 (bottom): © V. Duran/Visuals Unlimited.

Chapter 18

Opener (Background): © Photodisc Red/Getty Images; (Goshawk): © Leonard Rue Enterprises/Animals Animals/Earth Scenes; (Wheatfield): © Pixtal/age foto; (Cotton plant): © S. Solum/PhotoLink/Getty Images; 18.1a (Moss): © John Shaw/Tom Stack & Associates; 18.1a (inset): Courtesy Graham Kent; 18.1b (Fern): © William E. Ferguson; 18.1b (inset): Courtesy Graham Kent; 18.1c (Conifer): © Kent Dannen/Photo Researchers; 18.1c (inset): Courtesy Graham Kent; 18.1d (Cherry blossom): © E. Webber/Visuals Unlimited; 18.1d (inset): © Richard Shiell/Animals Animals/Earth Scenes; 18.5 (right): © Heather Angel; 18.5 (left): © Bruce Iverson; 18.6 (top): © Kingsley Stern; 18.6 (middle): © Runk/Schoenberger/Grant Heilman; 18.6 (bottom): © Ed Reschke; 18.7: The Field Museum of Natural History. Neg. no. GEO75400C, Chicago; 18.8 (Royal): © Forest W. Buchanan/Visuals Unlimited; 18.8 (Maidenhair): © John Gerlach/Visuals Unlimited; 18.8 (Hart's tongue): © Walter H. Hodge/Peter Arnold, Inc.; 18.10: © Dwight Kuhn; 18.12a: © Kingsley Stern; 18.12b: © Alan G. Nelson/Animals Animals/Earth Scenes; 18.13a (Pine tree): © Michael P. Gadomski/Photo Researchers, Inc.; 18.13 (Seed cones): © Geoff Kidd/Science Photo Library/Photo Researchers, Inc.; 18.13a (Pollen cone): © Barry Runk/STAN/Grant Heilman; 18.13b (Spruce): © Ed Reschke, Peter Arnold, Inc.; 18.13c (Juniper tree): © T. Daniel/Bruce Coleman; 18.13c (Berries): © Edward R. Degginger/Bruce Coleman; 18.14: Courtesy Stephen McCabe/Arboretum at University of California Santa Cruz; 18.17a: © Comstock; 18.17b: © H. Eisenbelss/Photo Researchers, Inc.; 18.17c: © Anthony Mercieca/Photo Researchers, Inc.; 18.17d: © Merlin D. Tuttle/Bat Conservation International; 18.18b–d: © Walther H. Hodge/Peter Arnold; 18.19 (all): © Peter Harholdt/SuperStock; 18.20a: © Biophoto Assoc./Photo Researchers; 18.20b: © Bill Keogh/Visuals Unlimited; 18.20c: © Gary R. Robinson/Visuals Unlimited; 18.20d: © Michael Vivard/Peter Arnold; 18.21a: © Gary T. Cole/Biological Photo Service; 18.22 (Bread): © Precision Graphics/Kim Brucker, Photographer; 18.22 (Micrograph): © James W. Richardson/Visuals Unlimited; 18.23b: © Glenn Oliver/Visuals Unlimited; 18.24: © L. West/Photo Researchers, Inc.; 18.25 (Fruitcose): © John Shaw/Tom Stack & Associates; 18.25 (Foliose): © Kerry T. Givens/Tom Stack & Associates; 18.26: © R. Roncandore/Visuals Unlimited; 18.27 (Wine): © Barry Runk/STAN/Grant Heilman; 18.28a: © Kingsley Stern; 18.29a: © Everett S. Beneke/Visuals Unlimited; 18.29b: © John Hadfield/SPL/Photo Researchers, Inc.; 18.29c: © P. Marazzi/SPL/Photo Researchers, Inc.

Chapter 19

Opener (Platypus): © D. Parer & E. Parer-Cook; (Kitten): © Royalty-Free/Getty Images; (Penguin): © Johnny Johnson/Animals Animals Earth Scenes; (Gull): © Tony Camacho/SPL/Photo Researchers, Inc.; 19.1a (a–d lower, e): © Dwight Kuhn; 19.1 (a–d top): © Cabisco/Phototake; 19.7a: © Andrew J. Martinez/Photo Researchers, Inc.; 19.8a: © CABISCO/Visuals Unlimited; 19.8b: © CABISCO/Phototake; 19.9a: © Runk/Schoenberger/Grant Heilman; 19.9b: © Ron Taylor/Bruce Coleman; 19.9c: © Under Watercolours; 19.10e: © Tom E. Adams/Peter Arnold; 19.11a: © James Webb/Phototake; 19.11b: © NIBSC/SPL/Photo Researchers, Inc.; 19.12a: © Arthur Siegelman/Visuals Unlimited; 19.12 (Trichinella): © John Burbidge/SPL/Photo Researchers, Inc.; 19.15 (Snail): © Bob Coyle; 19.15 (Nudibranch): © Kenneth W. Fink/Bruce Coleman; 19.15 (Octopus): © Alex Kerstitch/Bruce Coleman; 19.15 (Nautilus): © Douglas Faulkner/Photo Researchers; 19.15 (Scallop): Courtesy of Larry S. Roberts; 19.15 (Mussels): © Rick Harbo; 19.17a: © Heather Angel; 19.17b: © James Carmichael/Nature Photographics; 19.17c: © St. Bartholomews Hospital/Photo Researchers, Inc.; 19.19a–e: © John Shaw/Tom Stack & Associates; 19.20a: © Kim Taylor/Bruce Coleman; 19.20b: © James Carmichael/Nature Photographics; 19.21a: © C. Allan Morgan/Peter Arnold, Inc.; 19.20c: © Natural History Photographic Agency; 19.21b: © John Bell/Bruce Coleman Inc.; 19.20d: © Kjell Sandved/Butterfly Alphabet; 19.21c (lower): © Daniel Lyons/Bruce Coleman Inc.; 19.21c (top): © Ken Lucas/Ardea; 19.21d: © David M. Dennis/Animals Animals/Earth Scenes; 19.21e (Millipede): © Geof DeFeu/ImageState; 19.22 (Scale): © Science VU/Visuals Unlimited; 19.22 (Beetle): © Kjell Sandved/Bruce Coleman; 19.22 (Lacewing): © Glenn Oliver/Visuals Unlimited; 19.22 (Fly): © L. West/Bruce Coleman; 19.22 (Walking stick): © Art Wolfe; 19.22 (Honeybee): © John Shaw/Tom Stack & Associates; 19.22 (Flea, Bug): © Edward S. Ross; 19.22 (Dragonfly): © John Gerlach/Visuals Unlimited; 19.22 (Moth): © Clevland P. Hickman; 19.22 (Grasshopper): © Gary Meszaros/Visuals Unlimited; 19.22 (Leaf hopper): © Farley Bridges; 19.23b (left): © P. Danna/Peter Arnold, Inc.; 19.23b (right): © Carl Roessler/Tom Stack & Associates; 19.23c: © Alex Kerstitch/Visuals Unlimited; 19.23d: © Jim Greenfield/Imagequestmarine.com; 19.23e (left): © Randy Morse/Animals Animals/Earth Scenes; 19.23e (right): © Roger Steene/ Imagequestmarine. com; 19.25a: © Rick Harbo; 19.25b, 19.28a: © Heather Angel; 19.28b: © Norbert Wu; 19.28c: © Ron & Valerie Taylor/Bruce Coleman; 19.29b: © John Dudak/Phototake; 19.30: © Zig Leszczynski/Animals Animals/Earth Scenes; 19.31a: © Martin Harvey; Gallo Images/Corbis; 19.32 (both): © Daniel J. Cox; 19.33a: © Kirtley Perkins/Visuals Unlimited; 19.33b: © Thomas Kitchin/Tom Stack & Associates; 19.33c: © Brian Parker/Tom Stack & Associates; 19.34a: © B.G. Thompson/Photo Researchers, Inc.; 19.34b: © Leonard Lee Rue/Photo Researchers; 19.34c: © Fritz Prenzel/Animals Animals/Earth Scenes; 19.35a: © Leonard Lee Rue; 19.35b: © Stephen J. Krasemann/DRK Photo; 19.35c: © Denise Tackett/Tom Stack & Associates; 19.35d: © Mike Bacon/Tom Stack & Associates; 19.38a: © Dan Dreyfus and Associates; 19.38b: © John Reader/Photo Researchers; 19.40: Courtesy American Museum of Natural History.

Chapter 20

Opener (Fertilizer): © Evelyn Jo Herbert; (Redwood): © Tim Davis/Photo Researchers; (Pitcher plant): © Vol. 6/PhotoDisc/Getty Images; 20.2a: © Patti Murray/Animals Animals/Earth Scenes; 20.2b: © Michael Gadomski/Photo Researchers; 20.2c: © CABISCO/Phototake; 20.3a: © G.R. Roberts/Natural Sciences Image Library; 20.3b: © Ed Degginger/Color Pic; 20.5a: © Runk Schoenberger/Grant Heilman; 20.3c: © David Newman/Visuals Unlimited; 20.5b: © J.R. Waaland/BPS; 20.5c, 20.6a–c: © Biophoto Associates/Photo Researchers, Inc.; 20.7a: Courtesy Wilfred A. Cote, SUNY College of Environmental Forestry; 20.7b: © Randy Moore/Visuals Unlimited; 20.9: © Jeremy Burgess/SPL/Photo Researchers, Inc.; 20.10: © J. Robert Waaland/BPS; 20.11a: Courtesy Ray F. Evert, University of Wisconsin, Madison; 20.11b: © CABISCO/Phototake; 20.12: © Ed Reschke/Peter Arnold, Inc.; 20.13 (top): © CABISCO/Phototake; 20.13 (bottom): © Terry Ashley/Tom Stack & Associates; 20.14a: © Ken Wagner/Phototake; 20.14b: © Ken Wagner/Phototake; 20.15: Courtesy Mary E. Doohan; 20.16a: © Dwight Kuhn; 20.16b: © Runk Schoenberger/Grant Heilman.

Chapter 21

Opener (Light): © Runk/Schoenberger/Grant Heilman; (Chrysanthemum): © Vol. DT09/PhotoDisc/Getty Images; (Soybean): © Photodisc/Getty Images; 21.2a–b: Courtesy Prof. Malcolm B. Wilkins; 21.3: © Robert E. Lyons/Visuals Unlimited; 21.4a: Courtesy John Selier Virginia Tech Forestry Department; 21.4b: © Jeremy Burgess/SPL/Photo Researchers, Inc.; 21.5a,b: © Kingsley Stern; 21.5c: © Anton Vengo/SuperStock RF; 21.6: © Runk/Schoenberger/Grant Heilman, Inc.; 21.7: © Kingsley Stern; 21.9: © David Young-Wolff/Photo Edit; 21.12a: Courtesy Timothy J. Fehr; 21.12b: © Ed Reschke; 21.13a: © Arthur C. Smith, III/Grant Heilman; 21.13b: © Grant Heilman/Grant Heilman; 21.14 (top): Courtesy Graham Kent; 21.14 (bottom): © Ed Reschke; 21.15a: © Vol. 7/PhotoDisc/Getty Images; 21.15b: © Simko/Visuals Unlimited; 21.15c: © Dwight Kuhn; 21.16f: © Jack Bostrack/Visuals Unlimited; 21.17 (Walnuts): © Harry Haralambou/Peter Arnold, Inc.; 21.17 (Pea pod): © Andrew Thomas/Getty Images; 21.17 (Peach): © Age fotostock/SuperStock; 21.17 (Apple): © Age fotostock/SuperStock; 21.18: © Vol. 39/PhotoDisc/Getty Images; 21.22a–f: Courtesy Prof. Dr. Hans-Ulrich Koop, from *Plant Cell Reports*, 17:601-604; 21.23: © Kingsley Stern; 21.24a–c: Courtesy Monsanto; 21.25: © William Freese; 21.26: Courtesy Dr. Eduardo Blumwald, From Zhang, H.X. & Blumwald, E. (2001) "Transgenic salt tolerant tomato plants accumulate salt in the foliage but not in the fruits" in *Nature Biotechnology*, 19: 765-768 (http://www.nature.com/).

Chapter 22

Opener (Finger): © Vol. SS36/PhotoDisc/Getty Images; (Insulin): © SIU/Visuals Unlimited; 22.3 (Pseudostratified, Simple squamous): © Ed Reschke; 22.3 (Simple columnar): © Ed Reschke/Peter Arnold; 22.3 (Simple cuboidal): © Ed Reschke; 22.4 (left): © Jim Merli/Visuals Unlimited; 22.4 (right), 22.5a–d, 22.7a–c, 22.8: © Ed Reschke.

Chapter 23

Opener (Blood): © Susan Van Etten/Photo Edit; (Treadmill): © Fotopic/Miles Simons/Phototake; (Pacemaker): © Vol. 59/PhotoDisc/Getty Images; 23.6b-c: © Ed Reschke; 23.11: © Biophoto Associates/Photo Researchers; 23.12: © Pascal Goethgheluck/SPL/Photo Researchers, Inc.; 23.14: © Prof P.M. Motta, and S. Correr/SPL/Photo Researchers, Inc.; 23.16d: © Manfred Kage/Peter Arnold.

Chapter 24

Opener (Urine): © Vol. 54/PhotoDisc/Getty Images; (Preemie): © Ansel Horn/Phototake; 24.6: © Ardea London Ltd; 24.8c (bottom): © Fawcett, Hirokawa, Heuser/SPL/Photo Researchers, Inc.; 24.13: © Photo Insolite Realite/SPL/Photo Researchers, Inc.; 24.15b: © Ed Reschke.

Chapter 25

Opener (Steak): © Royalty-Free CORBIS; (Pills): © Michael Keller/CORBIS; (Oreo): © Reuters/CORBIS; 25.1: © Ryan McVay/Getty Images; 25.2a: "A Colour Atlas and Text of Nutritional Disorders," by Dr. Donald D. McLaren/Mosby-Wolfe Europe Lt.; 25.2b: © ISM/Phototake; 25.3: © Precision Graphics/Kim Brucker, Photographer; 25.4: © McGraw-Hill Higher Education, John Thoeming photographer; 25.5: Courtesy ® Quaker Oates; 25.6: © Vol. 20/Photo Disc/Getty Images; 25.7: © Biophoto Assoicates/Photo Researchers, Inc.; 25.9a: © Royalty-Free/CORBIS; 25.9b: © Lester Lefkowitz/The Image Bank/Getty Images; 25.9c: © Medical Graphics, Inc.; 25.11: Ryan McVay/Getty Images; 25.12: © Corbis Royalty-Free; 25.13a: © Tony Freeman/Photo Edit; 25.13b: © Donna Day/Stone/Getty Images.

Chapter 26

Opener (HIV): © Lennart Nilsson/Boehringer Ingelheim International GmbH; (Cyclosporine): © Brad Nelson/Phototake; (Nurse): © Vol. 41/Getty Images; 26.2: © David Young-Wolff/Photo Edit; 26.4a: © Boehringer Ingelheim Intl, GmbH, photo by Lennart Nilsson; 26.6: Courtesy Dr. Arthur J. Olson, Scripps Institute; 26.8b: © Boehringer Ingelheim Intl, GmbH, photo by Lennart Nilsson; 26.9: © Michael Newman/Photo Edit; 26.10 (Pollen): © David Scharf/SPL/Photo Researchers, Inc.; 26.10 (Girl): © Damien Lovegrove/SPL/Photo Researchers, Inc.; 26.11: © Richard Anderson; 26.12: © A. Ramey/Photo Edit.

Chapter 27

Opener (Woman, Couple): © Royalty-Free CORBIS; (Brain): © Neil Bordon/Photo Researchers, Inc.; 27.7 (both): © Science VU/Visuals Unlimited; 27.8: © Reuters/CORBIS; 27.11: © Ryan McVay/Getty Images; 27.14: © Tom McHugh/Photo Researchers, Inc.; 27.19: © Biophoto Associates/Photo Researchers, Inc.; 27.20: © Dr. P. Marazzi/Science Photo Library; 27.21a–b: "Atlas of Pediatric Physical Diagnosis," 2nd edition by Zitelli and Davis, 1992. Mosby-Wolfe Europe LTd., London, U.K..

Chapter 28

Opener (Weights): © Royalty-Free/CORBIS; (Tongue): © William Radcliffe/Science Faction/Getty Images; (Calcium): © Vol. 67/PhotoDisc/Getty Images; Fig. 28.4a,b: Courtesy Robert S. Preston and Joseph E. Hawkins, Kresge Research Institute, University of Michigan; 28.14a: © Michael Fogden/OSF/Animals Animals/Earth Scenes; 28.16: Courtesy Hugh Huxley; 28.18: © Julie Lemberger/Corbis; 28.19a: © Gerard Vandystadt/Photo Researchers, Inc.

Chapter 29

Opener (Spermatazoa): © Dennis Kunkel/Phototake; (Baby): © Royalty-Free/CORBIS; (Gonorrhhea): © CNRI/SPL/Photo Researchers,Inc.; 29.1: © Dennis Kunkel/Visuals Unlimited; 29.2: © Hans Pfletschinger/Peter Arnold, Inc.; 29.3: © Leonard Lee Rue, Jr./Photo Researchers, Inc.; 29.9 (BCP): © McGraw-Hill Higher Education, Bob Coyle photographer; 29.9 (IUD): © McGraw-Hill Higher Education, Vincent Ho photographer; 29.9 (Depo-Provera, Spermicidal jelly, Condom): © McGraw-Hill Higher Education, Bob Coyle photographer; 29.9 (Implant): © Population Council/Karen Tweedy-Holmes; 29.10: © Myrleen Ferguson Cate/Photo Edit; 29.11: © CC Studio/SPL/Photo Researchers, Inc.; 29.12: © Nicholas Nixon; 29.13: © Vol. 62/CORBIS; 29.19a: © Lennart Nilsson "A Child is Born" Dell Publishing Company; 29.21: © Lennart Nilsson/A Child is Born, Dell Publishing; 29.22: © John Watney/Photo Researchers, Inc.; 29.23f: © James Stevenson/SPL/Photo Researchers, Inc.; 29.24: Lennart Nilsson, "A Child is Born" 1990 Delacorte Press, Pg. 111 top frame; 29.26: © Dennis MacDonald/Photo Edit.

Chapter 30

Opener (Newborn): © ER Productions/Corbis; (Hong Kong): © D. Normark/Photo Link/Getty Images RF; (Billboard): © Bohemian Nomad Picturemakers/Corbis; 30.3a: © Jeff Greenberg/Peter Arnold, Inc.; 30.3b: © Ben Osborne/OSB/Animals Animals/Earth Scenes; 30.4c: © Still Pictures/Peter Arnold, Inc.; 30.5 (Cars): © Comstock Images/Getty RF; 30.5 (Crowd): © Earl & Nazima Kowall/CORBIS; 30.6a: © Richard Weymouth Brooks/Photo Researchers, Inc.; 30.9a: © age fotostock/SuperStock; 30.9b: © Royalty-Free/CORBIS; 30.12: © The McGraw Hill Companies, Inc./Bob Coyle, photographer; 30.16: © Alan Carey/Photo Researchers, Inc.; 30.17 (top): © Ted Levin/Animals Animals/Earth Scenes; 30.17 (lower): © Martin Harvey;Gallo Images/Corbis; 30.19: © Mike Bacon/Tom Stack & Associates; 30.20a: © Jeff Henry/National Park Service; 30.20b: © William P. Cunningham.

Chapter 31

Opener (Earth): © Stock Trek/Getty Images RF; (Kelp): © Chuck Davis/Stone/Getty Images; (Lake): © Michael Gadomski/Animals Animals/Earth Scenes; 31.1a: © Anthony Mercieca/Photo Researchers, Inc.; 31.1b: Courtesy Dr. Rod Peakall, School of Botany and Zoology, Australian National University; 31.1c: © Merlin D. Tuttle/Bat Conservation International; 31.2a: © Charlie Ott/Photo Researchers, Inc.; 31.2b: © Michael Graybill and Jan Hodder/BPS; 31.3 (Temperate, Desert): © Vol. 63/CORBIS; 31.5a–c,e: © Breck P. Kent/Animals Animals/Earth Scenes; 31.6: © Steve Austin Papilio/Corbis; 31.10a–c: Courtesy Daniel Janzen; 31.11: © George Armistead/VIREO; 31.12a: © Howard Buffett/Grant Heilman; 31.12b: © John Mithcell/Photo Researchers; 31.13 (both): © Hermann Eisenbeiss/Photo Researchers; 31.13 (Diatom): © Ed Reschke/Peter Arnold; 31.14 (Giraffe): © George W. Cox; 31.14 (Caterpillar): © Muridsany et Perennou/Photo Researchers; 31.14 (Osprey): © Joe McDonald/Visuals Unlimied; 31.14 (Mantis): © Scott Camazine; 31.15 (Mushroom): © Michael Beug; 31.15 (Acetobacter): © David M. Phillips/Visuals Unlimited; 31.17: © Sylvia Mader; 31.23: © Tom McHugh/Photo Researchers, Inc.; 31.25: © Biophoto Associates/Photo Researchers, Inc.; 31.27a: © Vol. 121/CORBIS; 31.27b: © McGraw-Hill Higher Education, Carlyn Iverson, photographer; 31.27c: © David Hall/Photo Researchers; 31.27d: © Vol. 121/CORBIS; 31.28 (Tundra): © Vol. 121/CORBIS; 31.28 (Taiga): © Vol. 63/CORBIS; 31.28 (Temperate): © Vol. 63/CORBIS; 31.28 (Prairie): © Vol. 86/CORBIS; 31.28 (Savannah): © Gregory Dimijan/Photo Researchers, Inc.; 31.28 (Desert): © Vol. 63/CORBIS; 31.28 (Rain forest): © Vol. 121/CORBIS.

Chapter 32

Opener (Deforestation): © L. Hobbs/PhotoLink/Getty Images RF; (Bulldozer): © PhotoLink/Getty Images RF; (Biosphere): © James Marshall/Corbis; 32.1 (Hong Kong): © D. Normark/Photolink/Getty Images; (Ocean): Evelyn Jo Hebert; (Food): © The McGraw-Hill Companies, Inc./John Thoeming, Photographer; (Solar): © Gerald and Buff Corsi/Visuals Unlimited; (Mining): © James P. Blair, National Geographic Image Collection; (Farming): © Vol. 39/PhotoDisc/Getty Images; 32.2: © Melvin Zucker/Visuals Unlimited; 32.4b: Courtesy of Carla Montgomery; 32.5a: © Gregg Vaughn/Tom Stack & Associates; 32.5b: © Jeremy Samuelson/FoodPix/Getty Images; 32.5c: © Gregg Vaughn/Tom Stack & Associates; 32.6a: © Jodi Jacobson/Peter Arnold; 32.6b: © Peter Essick/Aurora; 32.7a: © Laish Bristo/Visuals Unlimited; 32.7b: © Inga Spence/Visuals Unlimited; 32.7c: Courtesy V. Jane Windsor, Division of Plant Industry, Florida Department of Agriculture and Consumer Services; 32.8: © CORBIS RF; 32.9a: © Shane Moore/Animals Animals/Earth Scenes; 32.9b: © PhotoLink/Getty Images RF; 32.10b: Courtesy Walther C. Japp, Florida Fish and Wildlife Conservation Commission; 32.11a: © David L. Pearson/Visuals Unlimited; 32.11b: © S.K. Patrick/Visuals Unlimited; 32.11c: © Argus Foto Archiv/Peter Arnold; 32.11d: © Gerald and Buff Corsi/Visuals Unlimited; 32.12a: © Matt Stiveson/NREL/DOE; 32.12b: Courtesy of DaimlerChrysler AG; 32.12c: © Warren Gretz/NREL/DOE; 32.13: © James P. Blair, National Geographic Image Collection; 32.15b: © Gunter Ziesler/Peter Arnold; 32.16 (Periwinkle): © Kevin Schaefer/Peter Arnold; 32.16 (Armadillo): © John Cancalosi/Peter Arnold; 32.17 (Lady bug): © Anthony Mercieca/Photo Researchers, Inc.; 32.17 (Bat): © Merlin D. Tuttle/Bat Conservation International; 32.17 (Fishermen): © Herve Donnezan/Photo Researchers, Inc.; 32.17d: © Bryn Campbell/Stone/Getty; 32.18 (top): © William M. Smithey, Jr.; 32.18 (bottom): © Corbis/Vol. 121; 32.19: © Stuart Westmorland/Corbis; 32.20a: © Kent Knudson/PhotoLink/Getty Images RF; 32.20b: © Scenics of America/PhotoLink/Getty Images.

Page numbers followed by f refer to figures, t refer to tables, respectively.

pace of, 253–254, 254f
principle of, 9
*punctuated equilibrium model of,
    254, 254f*
theory of, 9, 216–223, 228
exchange pool, in chemical cycling,
    564, 564f
excretory system
    of annelids, 315, 315f
    of planarians, 311f
exercise, 447, 447f
    benefits of, 507
exocrine gland(s), 420
exocytosis, 79, 79f, 80
exons, 168, 168f, 185, 185f
exophthalmic goiter, 486
exoskeleton, 307, 502, 503f
    of arthropods, 316
    of insects, 318
    of molluscs, 314, 314f
exotic species, 559, 559f, 571
experiment(s), 8–9, 9f
experimental design, 8
experimental variable, 8
expiration, 424, 424f, 431
exponential growth, 540, 540f, 548
exponential growth phase
    of exponential growth, 540, 540f
    of logistic growth, 541, 541f
expulsion, in ejaculation, 515
external fertilization, 512, 513f
extinction(s), 7, 221, 246, 246f,
    252t, 545, 548, 553
    causes of, 588, 588f
    mass, 7
    potential, number of, 588
    rate of, 7
    vulnerability to, 545, 545f, 550
extracellular matrix, 64, 67
    of animal cells, 64, 64f
extraembryonic membranes, 513,
    524, 524f, 525, 526
    of mammals, 327
    of reptiles, 324, 324f
eye(s)
    of arthropods, 316, 316f
    of cephalopods, 314, 314f
    human, anatomy of, 499, 499f
    types of, 499, 499f
eyespots, of planarians, 311, 311f
Ey gene, 186, 186f

F

facial bones, 503
facilitated diffusion, 77, 80
FAD
    as coenzyme, 99
    in electron transport chain, 104, 104f
FADH₂, 99, 99f
    in citric acid cycle, 102, 103f
    *and electron transport chain,
        104, 104f*
fairy ring, 298, 298f
fallopian tubes. *See* oviduct(s)
familial hypercholesterolemia (FH), 204
    characteristics of, 204, 204f
    gene therapy for, 210

family, 6, 256
family (taxonomic), 256f
family pedigree. *See* pedigree(s)
farming
    conservation methods for, 580, 580f
    contour, 580f
    fossil fuel use in, 580
    harmful practices used in, 580
    no-till, 580f
fat(s), 35, 43
    breakdown of, 35–36, 35f
    dietary, 439–440
        reduction of, 440, 440t
    metabolism of, 106, 106f
    synthesis of, 35–36, 35f, 106
fatty acid(s), 35, 36f, 43
    formation of, in plants, 91, 91f
    metabolism of, 106, 106f
    omega-3 unsaturated, 447
feathers, 321f, 324
feedback inhibition, 76, 76f
female gametophyte, 362, 362f, 364,
    364f, 365
    of angiosperms, 293f
    of seed plants, 290, 290f
femur, 503
fermentation, 33, 107, 107f, 108
    fungal, 300
    yeast, 300
fern(s), 284f, 288, 289
    adaptations of, 289, 289f
    diversity of, 289, 289f
    evolution of, 285f
    as food, 289
    gametophyte of, 286, 286f
    life cycle of, 289, 289f
    medicinal value of, 289
    uses of, 289
fertilization
    of angiosperms, 293, 293f
    in ferns, 289f
    *in human reproduction, 515, 516,
        524, 524f, 525f, 530*
    of seed plants, 290, 290f
fertilizer(s), 566
    controlled study of, 10–11, 10f–11f
    in farming, 580
fetal cells, testing, for genetic
        disorders, 206
fetal development, 528–529, 528f,
    529f, 530
fetus
    age of viability, 511
    testing, for genetic disorders, 206
    ultrasound of, 206, 206f
fever, 455–456
F₂ fitness, species and, 248f, 249
F₁ generation, 144, 144f
F₂ generation, 144, 144f
fiber, dietary, 29, 34, 91,
    438–439, 439f
fibrillin, 151
fibrin, 407, 407f
fibrinogen, 407, 407f
fibroblast(s), 382, 382f
fibrous protein, 40, 40f
fibula, 503

filaments, 363
    in cytoskeleton, 55f
    of stamen, 292, 292f
filarial worm, 313
filter feeder, 310
filtration, 428
fimbriae, 51, 51f, 66, 516, 517f
fin(s), of fish, 322, 322f
finches, of Galápagos Islands, 220,
    220f, 251, 251f
    *niche specialization among,
        557, 557f*
fishes, 334
    circulatory pathways in, 398, 398f
    classes of, 322, 322f
    evolution of, 321, 321f, 322, 322f
fishing, 582, 582f
fitness, 222
five-kingdom system, 6, 13, 261, 262
flagellum (pl., flagella), 51, 51f, 63,
    63f, 66, 67
    amoebic, 278
    bacterial, 271, 271f
    of protozoans, 276
    of sponges, 310, 310f
flame cells, of planarians, 311f
flatworm(s), 309, 309f, 311–312, 333
    parasitic, 312, 312f
Fleming, Alexander, 8
flood(s), prevention of, 295, 591
flower(s), 285, 362
    anatomy of, 362, 362f
    complete, 363
    evolution of, 285, 285f
    incomplete, 363
    parts of, 292, 292f
flowering plant(s)
    asexual reproduction in, 369–372
    eudicot, 340, 340f
        flowers of, 362, 363, 363f
        germination in, 368, 368f
        roots of, 346f, 347, 347f, 352
        *seed development in,
            366, 366f*
        stems of, 344, 344f
    life cycle of, 362, 364, 364f, 373
    monocot, 340, 340f
        flowers of, 362, 363, 363f
        germination in, 368, 368f
        roots of, 347
        seed development in, 366
        stems of, 344, 344f
    as photosynthesizers, 85
    responses in, 356–361
    *sexual reproduction in,
        362–368, 373*
flowering plants, 284f.
        See also angiosperm(s)
    evolution of, 285, 285f
    gametophyte of, 286, 286f
fluid-mosaic model, 52, 66
flukes, 312, 312f
fluoroquinolone, 274
flycatchers, species of, 246, 247f
follicle(s), 517
    number of, in human female, 517
    primary, 517, 517f

follicle-stimulating hormone (FSH), 516
    in ovarian cycle, 517, 518f
follicular phase, of ovarian cycle,
    517–518, 518f
food, 3, 3f
    empty-calorie, 437
    energy from, 74, 74f
    production of, 96
    *as resource, 576, 576f,
        580–582, 593*
food chain, 563, 563f, 571, 582
food guide pyramid, 449–450, 449f,
    449t, 451
food poisoning, 274
food science, bacteria in, 274, 274f
food web, 562, 562f, 571
foot, of molluscs, 314, 314f
foraminiferan, 278, 278f
foreskin, 514f
forest(s)
    in climate regulation, 591
    ecological benefits of, 295
    effects of fire on, 547, 547f
    old-growth, logging of, 354
    temperate, 568, 569f
        *primary productivity in,
            570, 570f*
    water-holding by, 591
fossil(s), 217f, 253, 253f
    Darwin's study of, 219, 219f
fossil fuel, 85, 567, 567f, 583
    consumption of, 583
    as energy source, 583
    *and global climate change,
        583, 583f*
    use in farming, 580
fossil fuel energy, in agriculture, 96
fossil record, 224
founder effect, 238
    and microevolution, 238, 238f
fovea, 499f, 500
frameshift mutation, 172
Franklin, Rosalind, 160, 162, 162f
freckles, 233f
free radicals, 16
freshwater ecosystems, 568, 568f,
    590–591
frog(s), 323, 323f
    characteristics of, 306–307, 306f
frond(s), 289, 289f
fructose, 33
fruit(s), 284f, 293, 293f, 366
    as food, 294f, 295
    and seed dispersal, 294
    types of, 366–367, 367f, 373
fruiting body, of mushroom,
    298, 298f
functional group, 31, 31f
fungus (pl. fungi), 6, 262, 296–301
    *and animals, differences between,
        296, 296t, 306*
    antibiotic from, 588
    biology of, 296–298, 302
    carnivorous, 298, 298f
    disease-causing, 300–301, 302
    diversity of, 296, 296f
    ecological benefits of, 298–299, 302